Digital Circuits

by

William J. Streib

Publisher
The Goodheart-Willcox Company, Inc.
Tinley Park, Illinois

Library of Congress Catalog Card Number 96-22364
International Standard Book Number 1-56637-338-7

3 4 5 6 7 8 9 10 11 97 07 06 05 04 03 02

Library of Congress Cataloging in Publication Data

Streib, William J.
Digital Circuits / by William J. Streib

p. cm.
Includes index.
ISBN 1-56637-338-7
1. Digital electronics. I. Title
TK7868.D5S78 1997
621.39'5--dc20 96-22364
CIP

Introduction

Digital Circuits is an introductory text for courses in computer design, circuit theory, troubleshooting, and service. It serves as an introduction to the devices and circuits used to build computers and other digital equipment. Whether you are studying to be an electrician, engineer, technologist, or technician, or if your field is computer science, the concepts presented in this text serve as a foundation for your work with digital hardware. The things you will learn have direct application in industry and serve as a basis for additional course work.

The study of digital circuits is serial in nature. That is, each new concept builds upon the previous. For example, early in this book, you will be introduced to three circuits—AND, OR, and NOT elements. The rest of your study of digital circuits is based on the very limited number of devices. They will be modified and interconnected to form the devices and circuits that make up computers and other digital equipment.

Learning objectives at the beginning of each chapter provide focus for your studies. Each chapter also ends with a *Summary,* a list of *Important Terms, Test Your Knowledge questions,* and *Study Problems.* Although your instructor may assign any or all of the preceding as homework, it would serve you well to use them on your own as a review.

The **Digital Circuits Study Guide** is also available for use with the textbook. It includes numerous questions and hands-on problems that will help you in preparation for the laboratory activities.

The **Digital Circuits** text takes advantage of a unique characteristic of digital electronics. Unlike most areas of electronics, a great deal can be learned about computer and other digital circuits without first studying electricity and mathematics. Although a background in electricity/electronics would be an asset, only the most elementary knowledge of electronics and mathematics is required to use this book.

William J. Streib

Contents

Digital Technology

Chapter 1

LEARNING OBJECTIVES

After studying this chapter and completing lab assignments and study problems, you will be able to:

- Describe the characteristics of the industrial and electronic revolutions.
- List the members of the technical team in electronics and describe their typical educational backgrounds.
- Define such terms as analog signal, digital signal, bus, MSB, LSB, DIP, IC, TTL, and CMOS.
- Count to at least 20 in the binary system and convert binary numbers to decimal.
- List names given to the two logic levels.

The world is moving from the industrial revolution to an information and communication revolution based on electronics. This change is likely to affect almost every aspect of modern life. As a worker in digital electronics, you will contribute to the electronics revolution.

This chapter is an introduction to digital technology. The place of electronics in industry is described. Basic technical terms are defined, and the numbering system used in digital electronics is outlined. Also, integrated circuits (the basic component of digital circuits) are described.

ELECTRONICS AND SOCIETY

Productivity measures the goods and services people produce per hour of work. When productivity is high, there is more food, clothing, shelter, and entertainment available. The quality of life is based on high productivity.

During the past two hundred years, much of the increase in productivity has resulted from the methods of the industrial revolution. Machines have replaced hand operations. Animals and humans are no longer used as sources of power. Oil, coal, and atomic power have replaced horses and sails in transportation.

However, increased productivity required large amounts of energy and natural resources. Pollution increased, and worker attitude was hurt by repetitive tasks. As a result, many of the methods of the industrial revolution can no longer be used to increase productivity.

ELECTRONIC REVOLUTION

Important questions need to be asked: "Can electronics support continued growth in productivity? Can it do this without the disadvantages of the industrial revolution?"

INCREASED PRODUCTIVITY

Electronics has already demonstrated its ability to increase productivity. Many automatic and almost all automated machines are controlled by electronics. See Fig. 1-1 and Fig. 1-2. Such machines increase productivity by doing repetitive tasks once accomplished by humans.

However, the effects of electronics are not limited to manufacturing. Computers have increased the

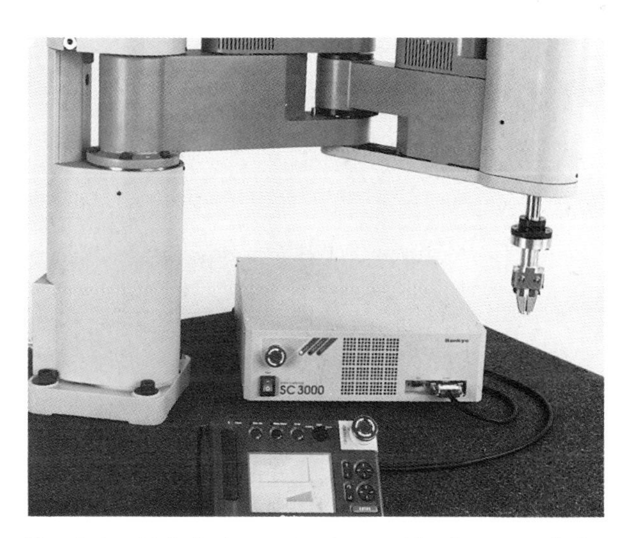

Fig. 1-1. Digital electronics is used in the control circuit of this robot. (Sankyo Robotics)

Fig. 1-2. Computers are very much a part of modern machine and process controls. (Ingersoll Milling Machine Company)

efficiency of record keeping, and the availability and speed of modern communications are due to electronics.

In homes, electronics improves the standard of living. Radios, television sets, stereos, recorders, computers, and electronic games are examples. Other electronic devices are less obvious. Electronic clocks, microwave ovens, and electronically controlled hand tools and appliances are in wide use.

THE ELECTRONIC REVOLUTION HAS FEW DISADVANTAGES

Unlike the methods of the industrial revolution, electronics has few disadvantages. Production of electronic circuits consumes few natural resources, and such circuits use little power. Except for limited radiation, electronic equipment produces no pollution.

ELECTRONICS AND THE WORLD OF WORK

Most electronic equipment can be operated by people with no knowledge of electronics. Since the time of early radio, equipment has been designed for use by people with no training in electronics. Computers are a high-technology example. Many people in business and industry use computers, yet few of them have knowledge of electronics. See Fig. 1-3.

Those who design, test, construct, sell, install, and repair electronic equipment are called hardware personnel. Fig. 1-4 shows the work of a person involved with hardware. Their jobs usually require a knowledge of electrical/electronic circuits and devices. This knowledge can usually be divided into theory and practice.

Electronic theory represents the general knowledge of the field. That is, theory explains how classes of circuits and devices function. Without theory, each circuit must be treated as a special case. Using theory, the operations of seemingly different circuits can be explained on the basis of common characteristics. Mathematics is often used to describe electronic

Fig. 1-3. Most computer-controlled equipment is user-friendly and operators only need basic knowledge of electronics. (Wyoming Technical Institute)

Fig. 1-4. Those working with electronic hardware must have a firm knowledge of electronic theory and practice. (Hewlett-Packard Company)

theory. Except for design situations, however, advanced mathematics is seldom required. Usually, basic algebra is all that is needed.

Practice involves standard solutions to problems. Practices are often used to speed the construction and repair of equipment. They permit previous experience to be passed on to future workers. When standard fabrication and installation methods are used, detailed instructions are not required for each task. This is particularly true of the work performed by electricians. Sometimes standard practices cannot cover all situations. When faced with a new problem, hardware personnel depend on theory to suggest new solutions. Effective workers in electronics understand both electronic theory and practice.

EDUCATION FOR ELECTRONICS

In electronics, the technical team is composed of engineers, technologists, technicians, and electricians. See Fig. 1-5. Their common bond is a knowledge of electronics. They use the same technical vocabulary (list of words), and they understand the symbols that appear on electrical/electronic drawings.

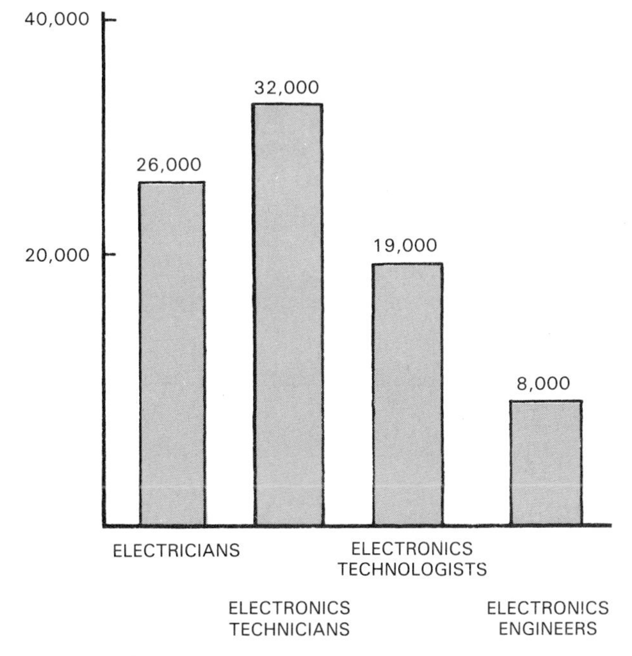

Fig. 1-5. The numbers of technical people in the United States varies.

The amount of theory needed compared to the amount of skill gained in hands-on practice is important in describing these positions. In turn, this ratio of theory to practice determines the form and length of education required for entry-level jobs.

ELECTRICIANS

To install and repair electronic equipment quickly, safely, and at a low cost, electricians must be highly skilled in applying electronic practices. These practices are best learned on the job. Apprenticeship is the usual form of preparation for an electrician.

However, electronic theory is not easily learned on the job. It is better taught in classrooms and student laboratories. Therefore, courses in electrical/electronic theory (including digital circuits) are usually part of an apprentice's related training.

Training may be taken in community colleges and technical schools. Increasing numbers of electrical apprentices work toward two-year associate degrees.

TECHNICIANS

Tasks assigned to electronic technicians tend to have a nearly equal mix of theory and practice. The work of laboratory technicians is an example. Typically, they build and test prototype (unit which is the first of its kind) circuits. As a result, technicians must be skilled in electronic practices. Yet, they must have a firm understanding of theory. The circuits they work with often have an advanced design. To aid in the design process, laboratory technicians must understand the theory behind such circuits.

Two-year college programs and training provided by the military are primary sources of technician preparation. Community colleges and technical school programs usually lead to associate degrees.

TECHNOLOGISTS

Technologists are new to the technical team. They need an amount of theoretical training midway between that for engineers and that for technicians.

The word engineer often appears in the job titles given to technologists. For example, a technologist employed in a company manufacturing electronic parts might have the title Quality Control Engineer. However, by training, interests, and job assignments, technologists are a unique part of the technical team. They should not be confused with engineers and technicians.

College programs for technologists are four years long and lead to bachelor's degrees. Credit from two-year technician programs is usually accepted toward these degrees.

GRADUATE ENGINEERS

The term graduate engineer means an engineer possessing at least a bachelor's degree from an engineering college. Graduate engineers are prepared to do design work and applied research. Theory is very important in their education. If a task does not require deep knowledge of theory, it might better be left to technologists, technicians, or electricians.

Engineering programs are at least four years long. Many students work toward master's degrees. Mathematics and science are emphasized, for these are the tools of design and research work.

CLASSIFICATION OF ELECTRONIC CIRCUITS

The basic laws of electricity apply to all electronic circuits. However, such circuits are usually divided into two groups for study and application. These groups are called ANALOG CIRCUITS and DIGITAL CIRCUITS.

ANALOG CIRCUITS

If two physical quantities vary in the same manner, one is said to be the ANALOG of the other. For example, the length of the mercury in a thermometer is the analog of temperature.

Electrical signals that follow the smooth changes of physical quantities are called *ANALOG SIGNALS.* The voltage that usually makes up such signals is likely to have a one-to-one relation to the original variable. For example, you may have spoken into a microphone and had your voice displayed on an oscilloscope. The electrical signal was analog in nature. See Fig. 1-6.

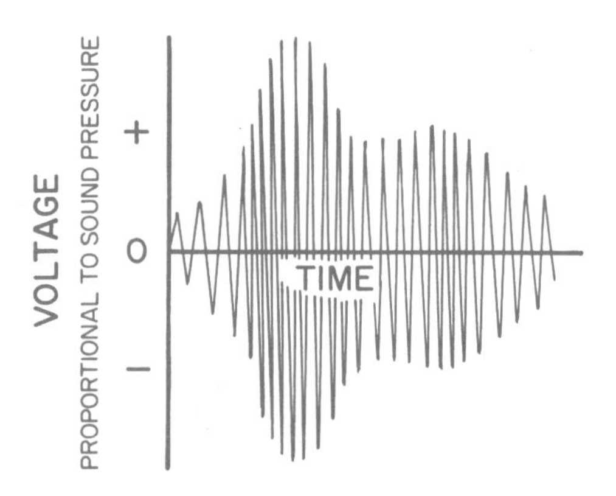

Fig. 1-6. The voltage from a microphone is proportional to the sound intensity. The voltage is an analog signal.

DIGITAL CIRCUITS

In digital circuits, voltages do not directly represent physical quantities. Rather, they represent numbers. Fig. 1-7 shows a signal which is either off or on. The number 0 represents off, while the number 1 represents on.

In computers, all information must be in number form. For example, the letter A cannot be stored directly. It must be converted to a number (usually 65) and then stored.

This book deals with digital circuits. That is, it describes circuits that handle numbers.

In the past, most electronic circuits were analog. If a worker understood an analog circuit, he or she understood electronics.

The analog circuit is still widely used in radio, television, and certain control applications. However, digital circuits are now an important part of electronics. Workers must know both analog and digital circuits.

NUMBERS IN COMPUTERS

When you count or do computations, you use the decimal system. The prefix DECI means ten. The decimal system has ten symbols—0, 1, 2, 3, 4, 5, 6, 7, 8, and 9.

Computers and other digital circuits could be built using the decimal system. In such circuits, signals would have ten voltage levels. For example, 4 volts might represent the number 4.

There is, however, a problem. It is difficult and expensive to build circuits that produce an output having ten distinct voltage levels. In fact, it is difficult to build circuits with five or even three levels.

The solution is to use only two voltage levels. These levels or *STATES* are usually called OFF and ON. A digital circuit might have an output of 0 volts for OFF and some voltage (say +5 volts) for ON. Such circuits are inexpensive and easy to build.

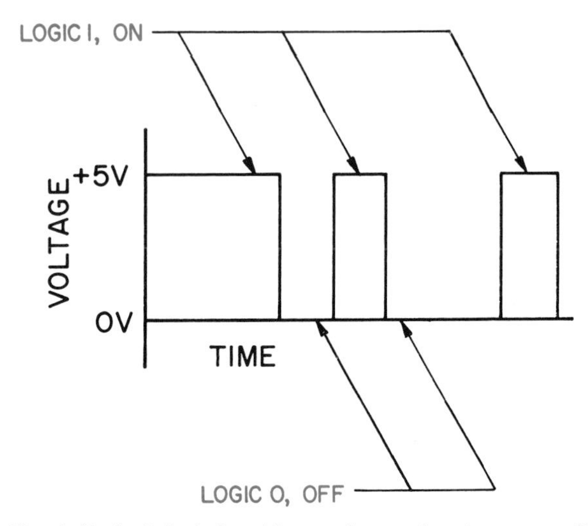

Fig. 1-7. A digital signal has only two levels, on or off.

Instead of calling the two states OFF and ON, they can be named 0 and 1. That is, they can represent numbers.

But, can a numbering system with only two symbols, 0 and 1, have any practical value? It turns out that it can. Almost all digital computers use such a system. It is called the *BINARY SYSTEM.* The prefix BI means two.

Counting in binary is nearly the same as counting in decimal. Therefore, a review of decimal counting will be used to introduce the binary system.

DECIMAL COUNTING

Everyone can count. Yet, numbering systems are a bit more complex than they appear.

When the decimal system is used for counting, you start at 0 and count to 9. At that point, the symbol set has been depleted. To count beyond 9, the concept of *CARRY* is used. On the tenth count, the symbol 0 is again used. To indicate that this is the second time through the symbol set, a carry is generated. That is, a 1 is placed in the column at the left of the 0. The number 10 results.

The number 10 might be thought of as a compound symbol. It is not part of the symbol set. Rather it is made up of two symbols from the set.

To continue counting beyond 10, you proceed through the symbol set. When 19 is reached, the set is again depleted. On the next count, a 0 results, and another carry is generated. The resulting number is 20.

Consider what happens at 99. On the next count, 1 is added to the 9 at the right. A 0 results, and a carry is generated. That carry is added to the 9 at the left. Again a 0 results, and a carry is generated. The number 100 is produced.

BINARY COUNTING

Counting in binary is much like counting in decimal. You count until all symbols in the symbol set have been used. Then, a 0 results, and a carry is generated. In the binary system, however, there are only two symbols in the set.

Binary counting starts at 0. The first count is 1, but at this point, the symbol set has been depleted.

As with the decimal system, the next count after the depletion of the symbol set results in a 0 and the generation of a carry. The result is the symbol 10. That is, in binary 1 + 1 = 10. Refer to Fig. 1-8. Binary 10 equals decimal 2.

The third count results in 10 + 1 = 11. That is binary 11 equals decimal 3.

The fourth count is more complex. Two carries are generated. That is, 11 + 1 = 100. (3 + 1 = 4 in decimal).

On a piece of paper, attempt to count to ten in binary. Check your results against Fig. 1-8. Then continue your count to decimal 20. Again check your results.

Remember, binary counting is used because digital circuits normally have only two states – OFF and ON.

Decimal	Binary	Decimal	Binary
0	0	18	10010
1	1	19	10011
2	10	20	10100
3	11	21	10101
4	100	22	10110
5	101	23	10111
6	110	24	11000
7	111	25	11001
8	1000	26	11010
9	1001	27	11011
10	1010	28	11100
11	1011	29	11101
12	1100	30	11110
13	1101	31	11111
14	1110	32	100000
15	1111		
16	10000		
17	10001		

Fig. 1-8. Counting in decimal and binary.

BITS AND BIT NOTATION

The language of bits and that of bit notation is needed for simple representation of information. Most technicians break data into bits or bytes.

BIT

Each position in a decimal number is called a digit. That is, 4,932 is a four-digit number. And, 576 is a three-digit number.

In binary, the term BIT is used. It stands for BINARY DIGIT. For example, 10110 is a five-bit, binary number. When the term bit is used, a binary number must be involved.

BIT NOTATION

In digital circuits, it is often necessary to refer to specific bits within a binary number. To make this easier, bits are numbered from the right as follows.

D4 D3 D2 D1 D0

In the number 10100, bit D2 is a 1. Bit D0 is a 0.

For the number 100,101, determine the values of bits D3 and D2. If you found them to be D3 = 0 and D2 = 1, you are correct. The reason for starting at D0 rather than D1 will be explained later.

EVALUATING BINARY NUMBERS

Binary numbers are just as valid as decimal numbers. They can be added, subtracted, multiplied, or divided. Anything that can be accomplished in the decimal system can be done in the binary system.

However, people are so used to using the decimal system that binary numbers have little meaning. For example, it may be difficult to look at the number 10,110 (binary) and recognize it as 22 (decimal). The following method for quickly converting from binary to decimal is widely used.

Again refer to Fig. 1-8. Note the following:

Binary	Decimal
1	1
10	2
100	4
1000	8
10000	16

That is, a 1 in position D0 contributes 1 to the decimal value of a binary number. A 1 in position D1 contributes 2. A 1 in position D2 contributes 4, etc. The amount contributed by a given bit is called its *PLACE VALUE.* The place values for the first six bits are:

32	16	8	4	2	1
D5	D4	D3	D2	D1	D0

To find the decimal value of a binary number, merely add the place values of the bits where 1s appear. For example, the number 100,101 is converted to a decimal value below:

1	0	0	1	0	1	= 37 (decimal)
32	16	8	4	2	1	

It is usually best to start the conversion at the left (it is easier to add smaller numbers to larger numbers). In this case, 32 + 4 + 1 = 37. The bits where 0s appear are ignored.

Here is another example. Determine the decimal value of the binary number 1,100.

1	1	0	0	= 12 (decimal)
8	4	2	1	

On a sheet of paper, evaluate the following binary numbers.

a. 101
b. 1,011
c. 10,111

If your decimal numbers are a. 5, b. 11, and c. 23, you are correct.

Place values for high-order bits can be determined by noting the relationships between place values already known.

32	16	8	4	2	1
D5	D4	D3	D2	D1	D0

Starting with D0, place values double as you go to the left. This means that the place values for the next two bits are:

128	64
D7	D6

What is the place value for D8? If you said 256, you are correct.

LSB AND MSB

In an ordinary binary number, D0 is called the *LEAST SIGNIFICANT BIT (LSB).* The most this bit can contribute to the value of a number is 1. (Of course, it contributes nothing if D0 = 0.)

The bit at the extreme left is called the *MOST SIGNIFICANT BIT (MSB).* In the number 10,011, the MSB contributes 16 (decimal) to the value of the number. That is, the place value of D4 is 16.

If an error were made in copying a number, in which location, MSB or LSB, would it have the larger affect on the value of the number? An error in the MSB would result in the greater change.

NUMBERS IN DIGITAL CIRCUITS

To move numbers within a computer or other digital circuit, wires or *LEADS* (pronounced like "beads") are used. Usually, each bit has its own lead. An inspection of a printed circuit board (PCB) reveals parallel leads running from each electronic component to another electronic component. See Fig. 1-9. The transfer of binary numbers depends on the patterns of 1s and 0s on these leads. Fig. 1-10 shows the results of making voltage measurements on a circuit. What binary number is being transferred, and what is its decimal value? If you found the binary number to be 00,110,001, you are correct. The decimal value of the binary number is 49.

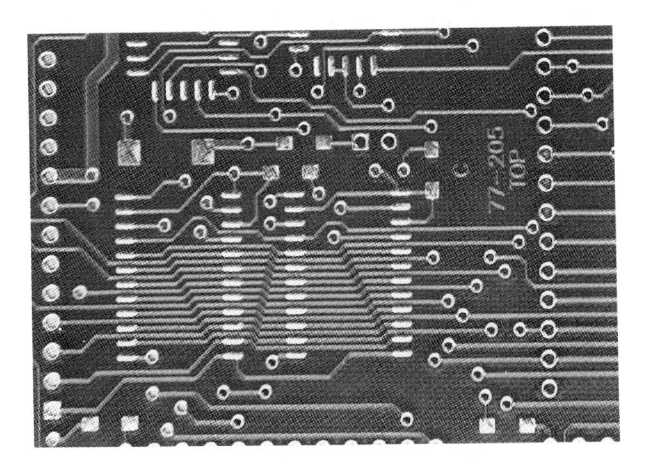

Fig. 1-9. The parallel leads on this printed circuit board are buses for carrying signals in a digital circuit. (General Technology Corporation)

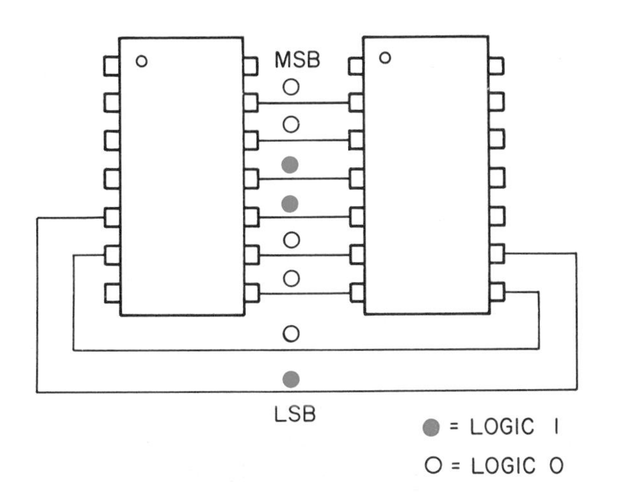

Fig. 1-10. Eight leads are used to transfer a binary number from the device at the left to the one at the right.

BUS

A group of leads used to transport numbers is often described as a *BUS.* Note that vehicles that pick up and discharge passengers along a fixed route are called buses. In circuits, buses perform similar electronic functions.

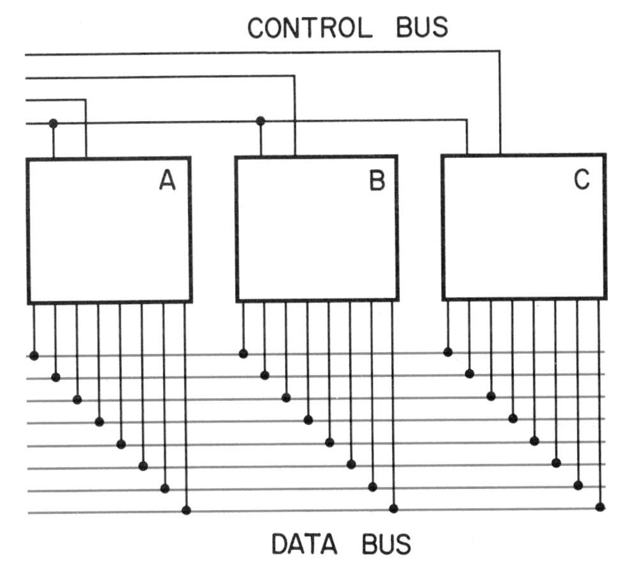

Fig. 1-11. Buses are used to send control and data signals to and from these digital devices.

DATA BUS

In Fig. 1-11, an 8-bit bus runs along the bottom of the drawing. It can be used to transfer data between the three attached devices. For example, device B might place a number on the bus, and device C might accept it. Data would be transferred from B to C by such a process.

WORD SIZE

The data bus in Fig. 1-11 can transport an 8-bit number at one time. In computer terminology, this circuit has an 8-bit *WORD.*

CONTROL LEADS

Each device in this circuit must be told whether it is to place a number on the bus, accept a number from the bus, or remain passive. The upper leads perform this function. They are called *CONTROL* leads.

These leads often connect only two devices, so they do not truly form a bus. Yet, as a group, control leads are often referred to as a control bus.

Signals on control leads do not represent numbers. Only patterns of 1s and 0s within a device represent numbers. As a result, state names other than 1 and 0 are often used. Some of these are:

1	0
ON	OFF
HIGH (H)	LOW (L)
TRUE (T)	FALSE (F)
ACTIVE	NOT ACTIVE

A third type of bus, the ADDRESS BUS, will be described later.

LOGIC ELEMENTS

The basic building blocks of computer and other digital circuits are logic elements. The word logic suggests reasoning, but of course, logic elements do not think. They can, however, make simple decisions.

One of the logic elements studied in the next chapter is called an AND. In its simplest form, an AND has two inputs and a single output. It will output a 1 only when BOTH inputs are 1. If a 0 appears at either or both inputs, it outputs a 0. This is low-level decision making, but it is decision making.

Other basic logic elements are the OR and NOT. They make equally low-level decisions. Although elements are simple, complex circuits result when large numbers of logic elements are interconnected.

INTEGRATED CIRCUITS

The electronic components that make up logic elements are usually in integrated circuit (IC) form. Here, integrated means formed as a whole. Transistors and other parts are created directly within a thin DIE or CHIP. They cannot be separated.

Fig. 1-12 shows an early IC viewed through a microscope. This chip is about 1/16 in. on a side and contains a single logic element. The white squares around the outside are *PADS*. They are used to make electrical connections between the IC and the outside world. Connections to pads are shown in Fig. 1-13. In Fig. 1-12, the U-shaped forms containing two parallel bars are transistors. The areas that appear to be mazes are resistors.

Fig. 1-14 shows a modern IC chip. It is about 3/16″ on a side and contains thousands of elements. A whole computer can be placed on such a chip, yet each chip costs only a few dollars.

Photographic methods are used to manufacture IC chips. The process begins with computer-generated circuit drawings. These drawings are reduced to the size of the chip, and several hundred chips may be manufactured at a time on thin disks. See Fig. 1-15. Lines are scribed between individual circuits, and the chips are separated.

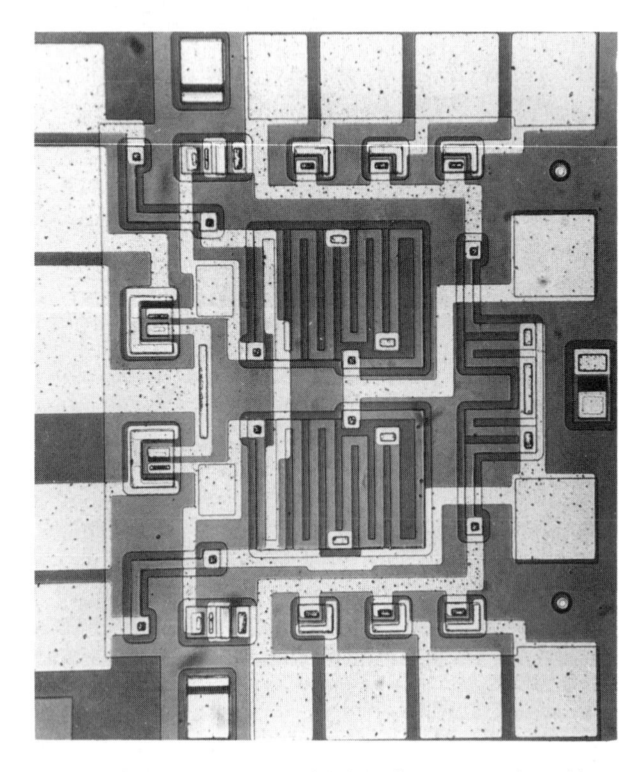

Fig. 1-12. This is a very old chip. On present-day chips, circuit details are difficult to see even through a microscope.

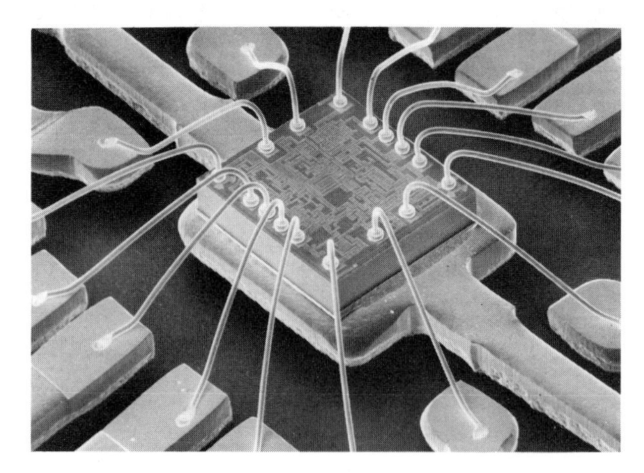

Fig. 1-13. A scanning electron microscope was used to obtain the picture of a chip. It shows the electrical connections between the chip and the outside world. (Delco Electronics Div., GMC)

Fig. 1-14. This chip is a processor and is the heart of many present-day computers. Note its size in comparison to the paper clips. (Intel Corporation)

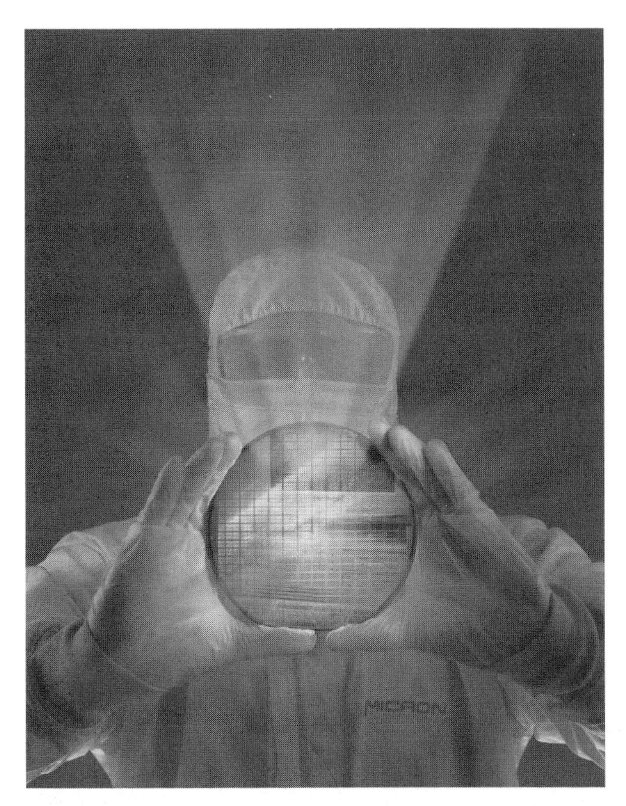

Fig. 1-15. This wafer contains hundreds of chips. They were formed at the same time and will be separated later. (Micron Electronics, Inc.)

IC PACKAGES

IC chips are too small and fragile to be used directly. They must be mounted in holders or packages. These are available in a number of standard forms.

The DUAL-IN-LINE PACKAGE (DIP) is the most common mounting. Its name comes from the two rows of pins used on such packages. The outline of a package is shown in Fig. 1-16. Signals and power enter and leave through the pins on the package.

The number of pins a DIP has depends on circuit complexity. There may be as few as four, or more than 80 pins on a DIP.

Digital circuits are usually assembled by mounting DIPs on printed circuit (PC) boards. These consist of plastic boards with metal leads formed on their surfaces. DIP pins are inserted into holes in these boards and soldered to the leads. In laboratories, solderless sockets, Fig. 1-17, are often used to temporarily construct circuits.

LOGIC FAMILIES

Although two AND elements made by two different processes may perform exactly the same logic function, the transistor circuits on their chips may differ greatly. When this is the case, they are said to belong to different logic families.

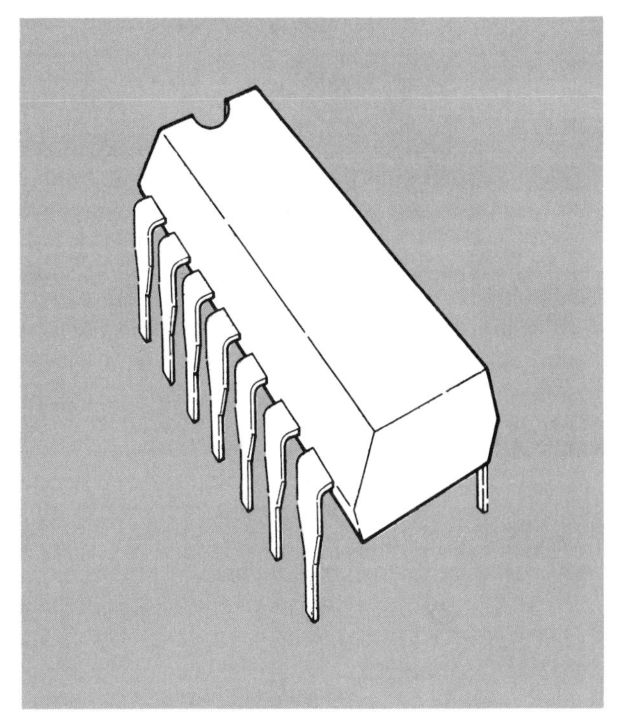

Fig. 1-16. Power and signals enter this DIP (dual-in-line package) through the pins on each side.

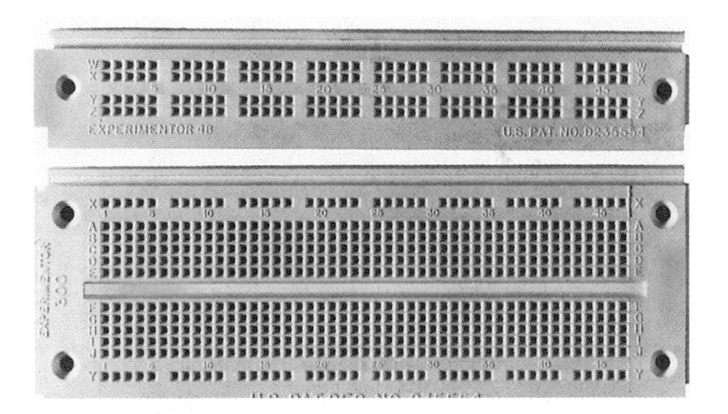

Fig. 1-17. Sockets such as these are widely used in industry and student laboratories. (Global Specialties)

Chips from a given family operate well together. They have the same signal levels, and they can use a common power supply.

This is not true of chips from different families. They may use different voltages to represent 1s and 0s, and they may need separate power supplies. When signals flow from one family to another, special circuits are often required at the *INTERFACE* (the point where a signal passes from one family to the other).

Two of the more common families are TTL and CMOS. You will probably use one of these in your student laboratory, and the TTL type of circuit is emphasized in this book.

The term TTL stands for transistor-transistor logic. This is a general-purpose family. When an application has few special requirements, TTL is usually used. TTL chips are inexpensive, and they are not easily damaged. They are found throughout industry, so repair personnel are familiar with them.

Power to operate TTL chips must come from a +5 volt supply. TTL logic levels (the voltages used to represent 0s and 1s) are 0 volts and +5 volts. However, the 5 volt level can drop as low as +2 volts and still be accepted as a logic 1.

The term CMOS stands for complementary metal oxide semiconductor. The main advantage of a CMOS circuit is low power. A CMOS circuit may use 100 times less power than a similar TTL circuit.

CMOS components are slow. Care must be taken when CMOS chips are used in high-speed computers and digital circuits.

For CMOS, the supply voltage may be anywhere between +3 and +18 volts. Logic levels are a percentage of the supply voltage, so they vary from circuit to circuit.

SUMMARY

Electronics has an important place in modern life. It may be the basis of the revolution that will replace the industrial revolution.

Engineers, technologists, technicians, and electricians are the industrial workers that design, test, construct, sell, install, and repair electronic equipment. Formal education for each member of the technical team reflects the mix of theory and practice needed on the job.

Electronic circuits are divided into two groups—analog and digital. At one time, most circuits were analog in nature. However, more and more tasks are being accomplished by digital circuits.

In digital circuits, information is in the form of numbers. The binary numbering system is used. TTL and CMOS are the logic families used most.

TEST YOUR KNOWLEDGE

1. List at least three ways that the methods of the industrial revolution increased productivity.
2. Describe at least three problems resulting from the use of the methods of the industrial revolution to increase productivity.
3. Give at least two ways in which an electronics revolution might increase productivity.
4. Almost everyone that works with electronic equipment should understand the theory and practice of electronics. (True or False?)
5. Members of the electronics technical team are listed below. Rearrange this list to show the relative amounts of electronic theory usually in each worker's education. Start the list with the worker that usually has the largest knowledge of electronic theory.
   a. Technologist
   b. Technician
   c. Engineer
   d. Electrician
6. Repeat Question 5. This time, indicate the relative knowledge of electronic practice.
7. Set up a table with two columns. In the first, number from 0 to 20 using the decimal system. In the second, number from 0 to 10,100 in binary.
8. Indicate the number of bits in each of the following numbers.
   a. 1,011
   b. 101,100
9. Based on the number 101,001, indicate the values of the following bits.
   a. D1 = ________
   b. D3 = ________
   c. D4 = ________
10. Copy the following on a piece of paper. Indicate the place values of the bits of a 10-bit binary number in such a pattern.
    D9 D8 D7 D6 D5 D4 D3 D2 D1 D0
    __ __ __ __ __ __ __ __ __ __
11. Evaluate the following binary numbers. That is, determine their values in the decimal system.
    a. 100
    b. 1,100
    c. 100,101
    d. 1,100,010
12. Copy the binary number 101,110 onto a sheet of paper. Circle its LSB. Draw a square around the MSB of the number.
13. Copy the following table onto a sheet of paper. Place each of the following state names in the proper columns: L, H, T, F, ACTIVE, INACTIVE.

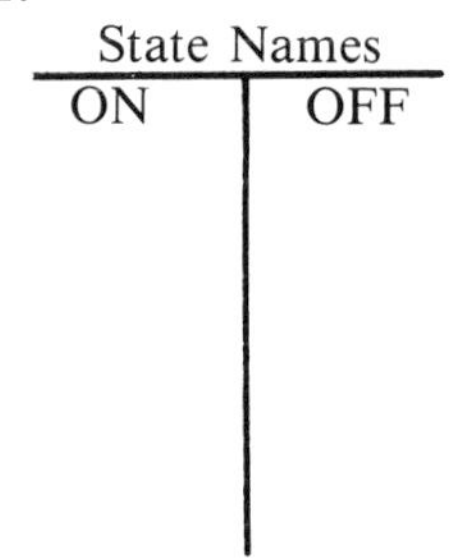

14. What does the abbreviation DIP stand for?

15. Which two of the following logic families are most used in industry and student laboratories?
    a. RTL
    b. DTL
    c. TTL
    d. CMOS
    e. ECL

STUDY PROBLEMS

1. An 8-bit data bus is shown in the figure marked Fig. P1-1. The leads are in the proper order between LSB and MSB. A solid dot above a lead indicates the presence of +5 volts (logic 1). An open dot indicates 0 volts (logic 0). For the measurements shown, what is the binary number on this bus?

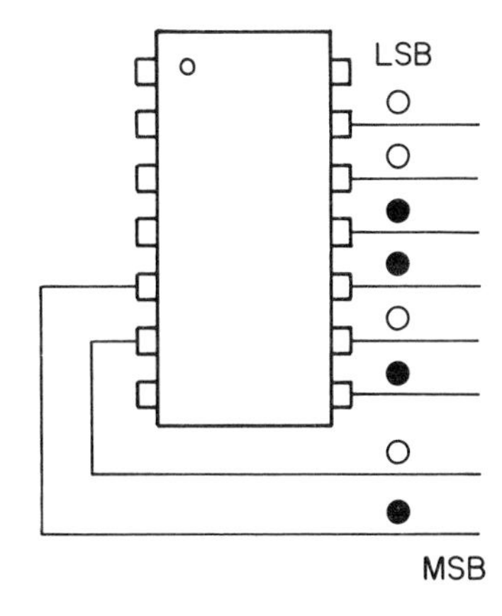

Fig. P1-1. Determine the binary number represented in this circuit.

High productivity and quality are obtained by combining high-tech controls and state-of-the-art machines. Note that the operators are at the computer-based controls rather than at the machines themselves. (Russell T. Gilman, Inc.)

Logic Elements

Chapter

2

LEARNING OBJECTIVES

After studying this chapter and completing lab assignments and study problems, you will be able to:

- Recognize switch-based AND, OR, and NOT circuits and explain their actions in terms of truth tables and Boolean algebra expressions.
- Recognize symbols for integrated circuit AND, OR, and NOT logic elements and explain their actions in terms of truth tables, Boolean expressions, and timing diagrams.
- Describe the actions of multi-input logic elements.
- Predict logic levels at all points in circuits containing AND, OR, and NOT elements.

Three logic elements are basic to the study of digital circuits. They are called AND, OR, and NOT. These elements are found in all digital circuits. They are also used to explain the actions of advanced logic elements. Therefore, to study and work in digital electronics, you must be completely familiar with AND, OR, and NOT logic elements.

SWITCH-BASED AND

An AND is a logic element. As such, it is one of the building blocks of computers and other digital circuits. In its simplest form, an AND has two inputs and one output.

Fig. 2-1 shows an AND that uses pushbutton switches as inputs and a lamp as an output. This circuit is described as a switch-based AND.

CIRCUIT SYMBOLS

Electrical symbols describe the parts they represent. For example, the actions of push-button switches A and B are suggested by their symbols. Pressing A closes switch A.

Lines on the lamp symbol suggest light rays. These are part of the symbol and are present whether or not the lamp is lit.

The battery symbol represents the internal construction of this device. The battery supplies power to light the lamp. The longer of the two bars is its positive terminal.

Lines between symbols are called *LEADS.* These represent wires used to connect the electrical parts when the circuit is assembled.

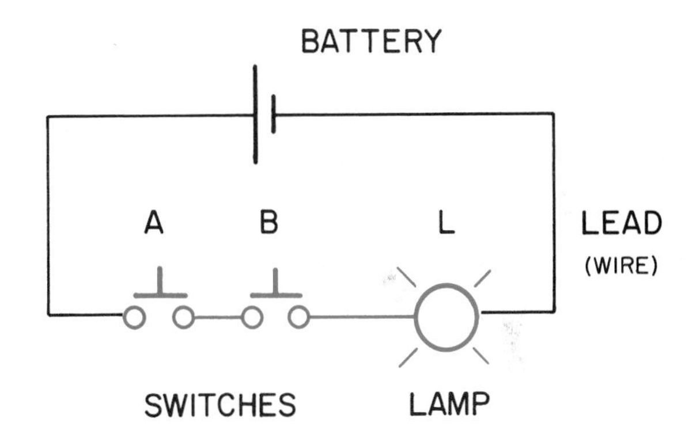

Fig. 2-1. Switches in switch-based AND circuits are connected in series.

CIRCUIT ACTION FOR A SWITCH-BASED AND

The circuit for a switch-based AND is a series circuit, since there is only one path through which current can flow. Current leaves one terminal of the battery, flows through the electrical parts one after the other, then returns to the battery.

For current to flow, the circuit must be complete. That is, there must be a metal path from one terminal of the battery to the other. If there is an opening, the circuit is said to be OPEN, and current will not flow. For the positions of the switches shown in Fig. 2-1, the circuit is open, and the lamp is out.

To light the lamp, pushbutton switches A and B must be pressed at the same time. This completes the circuit. If A or B or both are open, the circuit is not complete. Only when A AND B are pressed will lamp L light. This is the source of the name of this logic element.

APPLICATION OF SWITCH-BASED AND

Switch-based ANDs are widely used to control machines. The punch press in Fig. 2-2 is an example. Two switches control the downward motion of this press. These switches are mounted at the ends of the control panel, and they are connected as an AND. That is, they are in series. To operate this machine, these switches must be pressed at the same time.

Note the safety aspect of this switch placement. To bring the press down, the operator's hands must be on the switches—safely out of the machine.

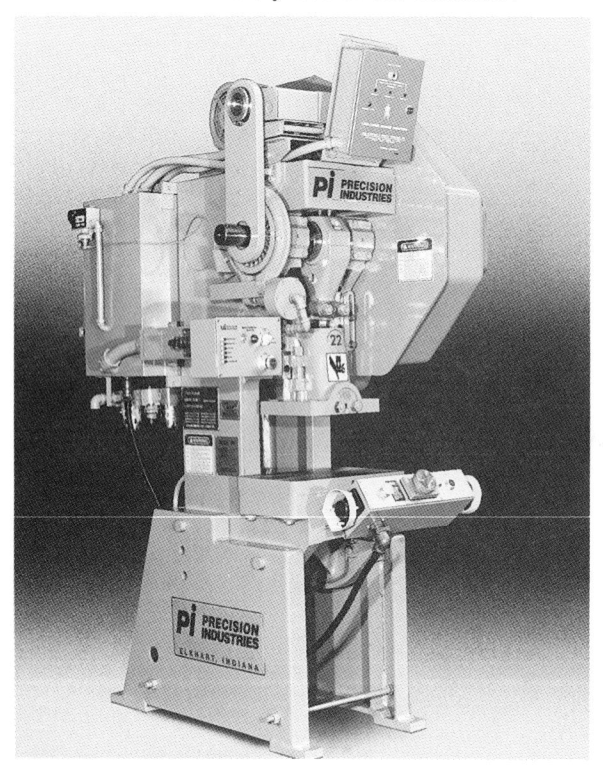

Fig. 2-2. To ensure the operator's hands are safely out of this press, switches at each end of the control panel must be pressed at the same time. (Precision Industries Corp.)

TRUTH TABLE

A truth table is a tabular way of describing the action of a digital circuit. Fig. 2-3 is the truth table for the 2-input, switch-based AND in Fig. 2-1. Inputs A and B are at the left; output L is at the right.

A 1 (one) in an input column indicates an activated switch. That is, a 1 in column A means that this switch

INPUTS		OUTPUT
A	B	L
0	0	0
0	1	0
1	0	0
1	1	1

Fig. 2-3. This is the truth table for the switch-based AND circuit in Fig. 2-1.

is being pressed. A 0 (zero) indicates an unactivated switch.

A 1 in the output column means that lamp L is lit. A 0 means that it is out.

Truth tables show all possible input combinations. In this case there are two inputs, so there are four such combinations. See the first two columns of Fig. 2-3.

This truth table restates what is already known about this circuit. Lamp L will be lit only when A and B are pressed at the same time.

In such simple circuits, circuit action is easily determined from the drawing. When complex circuits are involved, circuit action is better described by one or more truth tables.

BOOLEAN ALGEBRA

Boolean algebra is the third method used to describe the action of digital circuits. It involves an interesting form of mathematics. Boolean algebra is easy to learn when simple steps are studied.

This algebra is named after George Boole (1815-1864). He was an English mathematician and philosopher. Of course, digital circuits did not exist in the middle of the 19th century. Boole developed his algebra to better understand true/false arguments in philosophy.

In 1938, a scientist at Bell Laboratories saw that the true/false statements of Boolean algebra were much like the on/off nature of switches. That scientist was Claude Shannon. Since that time, Boolean algebra has been used in the development of digital circuits and computers.

The Boolean statement for the AND in Fig. 2-1 is:

$$A \times B = L$$

This is read, "A AND B equals L." The $\times$ sign does not mean multiplication. It should not be read as "times." Often, the $\times$ sign is omitted. That is: $AB = A \times B$.

This Boolean expression restates the AND truth table. That is:

$$0 \times 0 = 0$$
$$0 \times 1 = 0$$
$$1 \times 0 = 0$$
$$1 \times 1 = 1$$

The last statement would be read, "1 AND 1 equals 1."

These statements are identical to those from mathematics. However, statements for other logic elements are rather different and interesting.

Boolean algebra is often used in the design of digital circuits. All workers in digital electronics should be familiar with this method of describing circuits.

To review, three ways of describing an AND have been suggested. They are: the circuit's diagram, its truth table, and its Boolean expression. They contain the same information, but each description has its own uses.

SWITCH-BASED OR

An OR is another logic element. Like the AND, it is used in computers and other digital circuits. Fig. 2-4 shows a 2-input, switch-based OR.

Switches in this circuit are in PARALLEL. They are connected across each other.

The closing of either switch completes this circuit. If A is closed, current flows from the battery, through switch A, through the lamp, then back to the battery. If B is closed, the lamp will again light. In this case, current flows through switch B.

If A and B are pressed at the same time, the lamp will, of course, light. Current will divide at the switches. Half of the current will flow through each switch.

Again, the name of this circuit is descriptive. Closing either A or B will light L. Only when both A and B are open will lamp L be out.

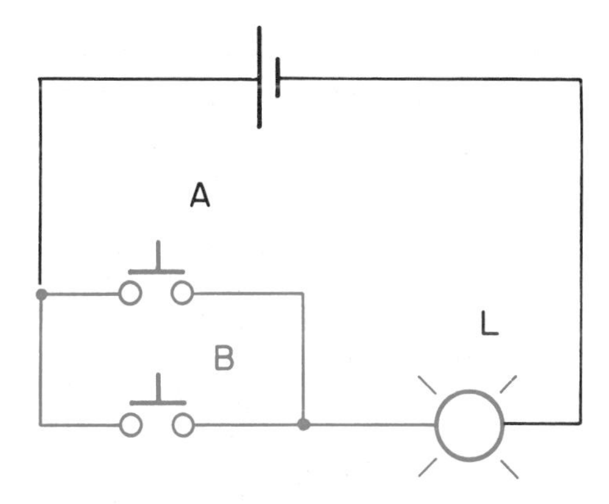

Fig. 2-4. Switches in a switch-based OR are connected in parallel.

APPLICATION OF SWITCH-BASED OR

An example of a use for a switch-based OR circuit is for doors separating two parts of a factory. Workers must be able to operate the motor-driven openers from both sides. It is likely that the switches that start the motor on a given door are connected as an OR. That is, they are in parallel.

TRUTH TABLE FOR A SWITCH-BASED OR CIRCUIT

Fig. 2-5 is the truth table for a 2-input OR. Again, a 1 in an input column implies an activated switch. The presence of a 1 in the output column means that the output is active.

BOOLEAN ALGEBRA STATEMENT

The Boolean algebra statement for a 2-input OR is:

$$A + B = L$$

This is read, "A OR B equals L." The + sign does not imply addition or mean a positive number. It should not be read as "plus."

This expression restates the truth table. That is:

$$0 + 0 = 0$$
$$0 + 1 = 1$$
$$1 + 0 = 1$$
$$1 + 1 = 1$$

Note the last line. It does not follow the rules of arithmetic. The last line accurately describes the action of switches in parallel. When both switches are closed (A = 1 and B = 1), the lamp is lit.

BOOLEAN LANGUAGE VERSUS BINARY LANGUAGE

Boolean algebra and binary numbers both use 1s and 0s. They are both closely related to digital circuits. Yet, they are completely different concepts.

INPUTS		OUTPUT
A	B	L
0	0	0
0	1	1
1	0	1
1	1	1

Fig. 2-5. This is the truth table for the switch-based OR circuit in Fig. 2-4.

BINARY ARITHMETIC

As indicated in Chapter 1, binary is a numbering system. The 1 and 0 used in binary are numbers. They can be added, subtracted, multiplied, and divided. They can be used for counting. Counting to any number, say six, proceeds as: 0, 1, 10, 11, 100, 101, and 110.

In binary, the + sign stands for addition. The statement $A + B = Y$ is read, "A plus B equals Y." The truth table for binary addition is at the left in Fig. 2-6. Note that $1 + 1 = 10$ in binary.

BOOLEAN ALGEBRA

Boolean algebra uses 1s and 0s, but these are not numbers. They are symbols used in place of ON and OFF, or HIGH and LOW, etc. The 1s and 0s used in Boolean algebra cannot be added, subtracted, multiplied, or divided.

In Boolean algebra, the + sign stands for OR. To avoid confusion, a U-shaped symbol is sometimes used rather than the + sign.

The truth table for the Boolean expression $A + B = Y$ is at the right in Fig. 2-6. Note that $1 + 1 = 1$.

Boolean algebra is used to design and explain digital circuits. Binary arithmetic is used to count and do mathematical operations.

When the two come together in a computer, the tie is the truth table. A truth table is set up to depict the binary operation that is to be accomplished. Then Boolean algebra is used to design digital circuits to do those operations. Although they both use 1s and 0s, keep in mind that binary arithmetic and Boolean algebra are different concepts.

BINARY ARITHMETIC

A + B = Y

+ IMPLIES PLUS

INPUTS A	INPUTS B	OUTPUT Y
0	0	0
0	1	1
1	0	1
1	1	10

BOOLEAN ALGEBRA

A + B = Y

+ IMPLIES OR

INPUTS A	INPUTS B	OUTPUT Y
0	0	0
0	1	1
1	0	1
1	1	1

Fig. 2-6. These truth tables compare binary arithmetic and Boolean algebra.

SWITCH-BASED NOT

The NOT is the third basic logic element. Unlike ANDs and ORs, a NOT has only one input. When that input is 1, its output is 0. When that input is 0, a NOT will output a 1. A NOT's output is always the opposite of its input.

Fig. 2-7 introduces a second type of pushbutton switch. The switches used up to this point are called *NORMALLY OPEN* (NO). See the first column in the figure. When unactivated (in its normal position) a NO switch is open. When activated, it is closed.

The switch at the right in the figure is *NORMALLY CLOSED* (NC). When unactivated ($A = 0$), NC switches are closed. They open when activated.

A normally-closed switch can be used to produce a NOT circuit. See Fig. 2-8.

When $A = 0$ (switch is not activated), the circuit is complete, and the lamp is lit ($L = 1$). When $A = 1$ (switch is activated), the lamp goes out ($L = 0$).

	NORMALLY OPEN (NO)	NORMALLY CLOSED (NC)
UNACTIVATED A = 0		
ACTIVATED A = 1		
ELECTRICAL ACTION	A	$\bar{A}$

Fig. 2-7. Pushbutton switches can be normally open or normally closed.

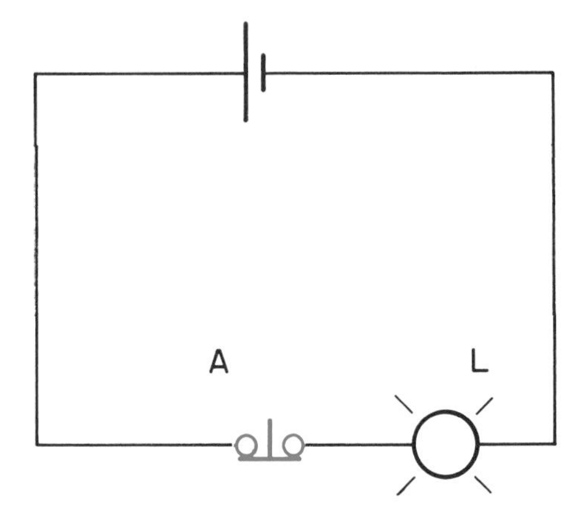

Fig. 2-8. Normally-closed switches are used to obtain NOT logic circuits.

TRUTH TABLE AND BOOLEAN EXPRESSION FOR A NOT CIRCUIT

The NOT has a simple truth table, Fig. 2-9. With only one input, there are only two possible input conditions. L is always the opposite of A.

The Boolean statement for this NOT is:

$$\overline{A} = L$$

This is read, "A NOT equals L." Note the line over A. This Boolean expression restates the truth table.

$$\overline{0} = 1$$
$$\overline{1} = 0$$

INPUT A	OUTPUT L
0	1
1	0

Fig. 2-9. This is the truth table for the switch-based NOT circuit in Fig. 2-8.

APPLICATION OF SWITCH-BASED NOT

Fig. 2-10 shows NO and NC switches connected in series. To light the lamp, only B need be pressed, since the NC switch is already closed.

If this circuit were part of a machine control, switch A might be a safety switch. When pressed, it prevents the machine from starting. A is said to override B. While A is active, B cannot start the machine.

Switches A and B are in series, so they are an AND. Taking into account the NOTing of A, the Boolean expression for this circuit is:

$$\overline{A}B = L$$

This is read, "A NOT AND B equals L." L is 1 only when A = 0, and B = 1.

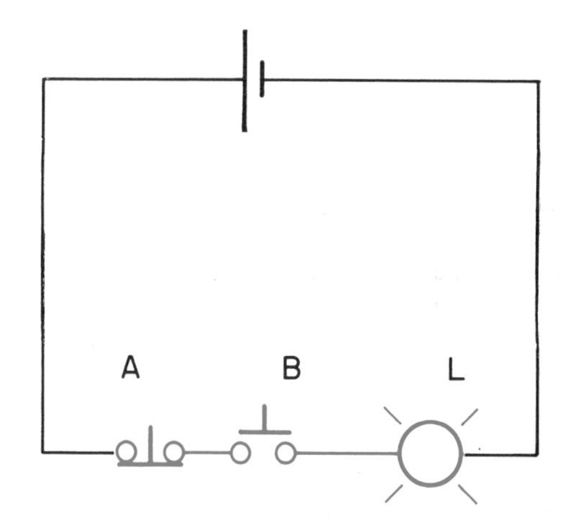

Fig. 2-10. To light the lamp, only switch B is pressed. Pressing A overrides the action of B.

	AND	OR	NOT
CIRCUIT	A B L	A B L	A L
BOOLEAN EXPRESSION	AB = L	A + B = L	$\overline{A}$ = L

TRUTH TABLE — AND

INPUTS A	INPUTS B	OUTPUT L
0	0	0
0	1	0
1	0	0
1	1	1

TRUTH TABLE — OR

INPUTS A	INPUTS B	OUTPUT L
0	0	0
0	1	1
1	0	1
1	1	1

TRUTH TABLE — NOT

INPUT A	OUTPUT L
0	1
1	0

Fig. 2-11. Circuit diagrams, Boolean expressions, and truth tables are used to describe the actions of AND, OR, and NOT circuits.

REVIEW OF DESCRIPTIONS

Fig. 2-11 shows descriptions of AND, OR, and NOT circuits side by side. The first row shows the circuits. Switches in series form ANDs; switches in parallel form ORs; and normally closed contacts result in NOTs.

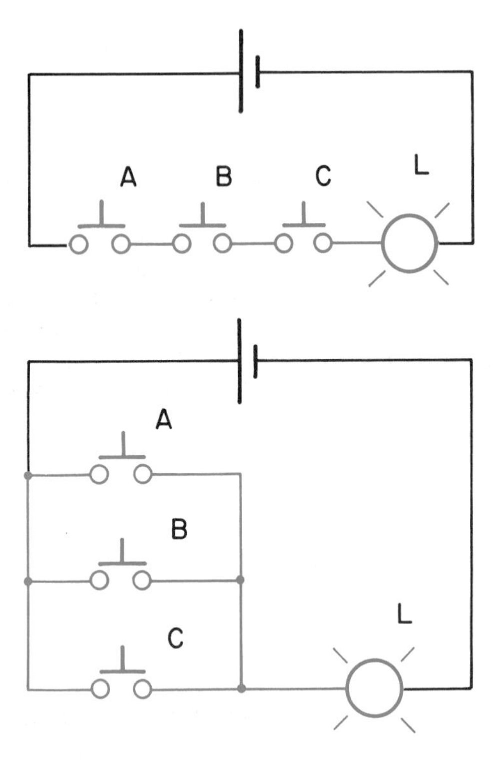

Fig. 2-12. Logic circuits can have multiple inputs.

MULTI-INPUT CIRCUITS

AND and OR circuits can have many inputs. Fig. 2-12 shows circuits with three inputs. The Boolean expression for the AND is:

$$ABC = L$$

This is read, "A AND B AND C equals L."

The Boolean expression for the 3-input OR is:

$$A + B + C = L$$

This is read, "A OR B OR C equals L."

Truth tables for these circuits are found in Fig. 2-13. No matter how many inputs there are, the output of an AND is 1 only when all inputs are 1. For an OR, its output is 1 when any input or combination of inputs is 1.

Truth tables lengthen quickly as inputs are added. With n as the number of inputs, the formula for the number of rows in a truth table is:

AND

INPUTS A	B	C	OUTPUT Y
0	0	0	0
0	0	1	0
0	1	0	0
0	1	1	0
1	0	0	0
1	0	1	0
1	1	0	0
1	1	1	1

OR

INPUTS A	B	C	OUTPUT Y
0	0	0	0
0	0	1	1
0	1	0	1
0	1	1	1
1	0	0	1
1	0	1	1
1	1	0	1
1	1	1	1

Fig. 2-13. Truth tables for 3-input logic circuits.

$$\text{Rows in truth table} = 2^n$$

That is, the number of possible input combinations equals 2 raised to the n power. This is easily computed on a calculator. The above gives the following results:

Inputs	Rows in Truth Table
1	2
2	4
3	8
4	16

INTEGRATED CIRCUIT LOGIC

Most of what you have learned about switch-based logic applies to IC (integrated circuit) logic. Truth tables and Boolean expressions are the same for switch-based logic and for IC logic. However, there are a few differences.

SYMBOLS

Switch-based symbols are descriptive. See part "a" of Fig. 2-14. The action of this circuit is obvious. This is not true for IC logic symbols.

Part "b" of Fig. 2-14 shows the circuit for an IC AND. It is complex. It is difficult to follow the action of this circuit. Imagine twenty or so of these circuits connected together to form a logic circuit.

Part "c" of Fig. 2-14 shows the solution to the resulting problem. IC-logic elements are represented by symbols. Unlike switch-based symbols, however, IC-logic symbols are abstract. They tell little of how a given element works. You must memorize their meanings.

SIGNALS

Switch-based and IC logic elements differ in the ways 1s and 0s are represented. Switches control current flow. See part "a" of Fig. 2-15. Shown is the flow of current through the output (in this case the lamp) that represents a logic 1.

In IC logic, voltage is used to represent signals. See part "b" of Fig. 2-15. The voltage between A and G represents a signal that could be applied to a logic element. Switch position I places +5 volts between terminal A and ground. For TTL, this is a logic 1. Position II grounds terminal A. A logic 0 results. That is, there is 0 volts between terminal A and ground. A TTL element interprets a voltage as low as 2.8 V as a logic 1.

For TTL, terminal A must be grounded to produce a logic 0. Merely removing the +5 volts is not enough. For example, moving the switch to position III results in a FLOATING lead (lead which sometimes has noise on it). Terminal A is connected to neither +5 volts nor ground. TTL logic elements will treat floating leads as logic 1s.

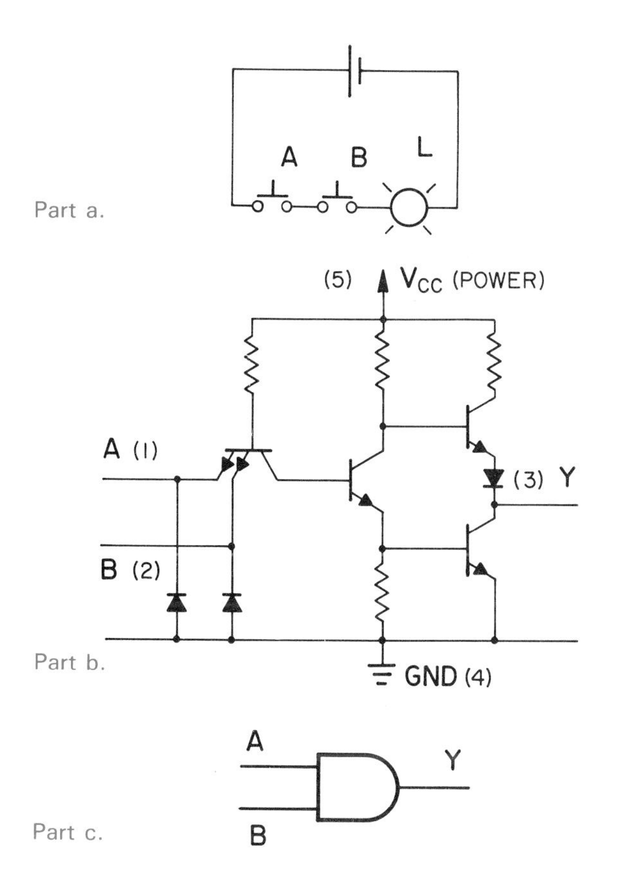

Fig. 2-14. Circuits within integrated circuits are so complex that symbols must be used to represent a pattern of logic elements.

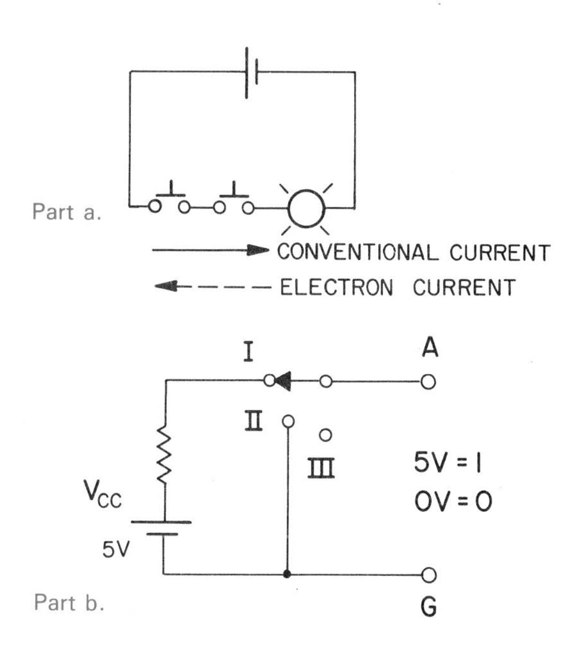

Fig. 2-15. Current flow is used to carry signals in switch-based logic circuits. In IC logic circuits, voltage is used.

INTEGRATED CIRCUIT AND

Three descriptions of an IC AND are shown in Fig. 2-16. Note that the truth table and Boolean expression are identical to those of the switch-based AND.

SYMBOL	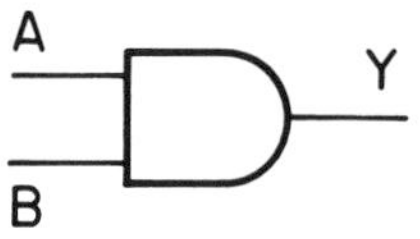
BOOLEAN EXPRESSION	AB = Y

TRUTH TABLE

INPUTS A	INPUTS B	OUTPUT Y
0	0	0
0	1	0
1	0	0
1	1	1

Fig. 2-16. These are three methods of describing IC ANDs.

AND SYMBOL

Templates assist in drawing logic symbols. A typical template is shown in Fig. 2-17. If you must draw AND symbols freehand, the three steps shown in Fig. 2-18 will help you draw a readable symbol. Note that this symbol contains portions of a rectangle and a circle.

The logic diagram in Fig. 2-19 contains several logic elements. How many ANDs are shown? If you found three, you found them all.

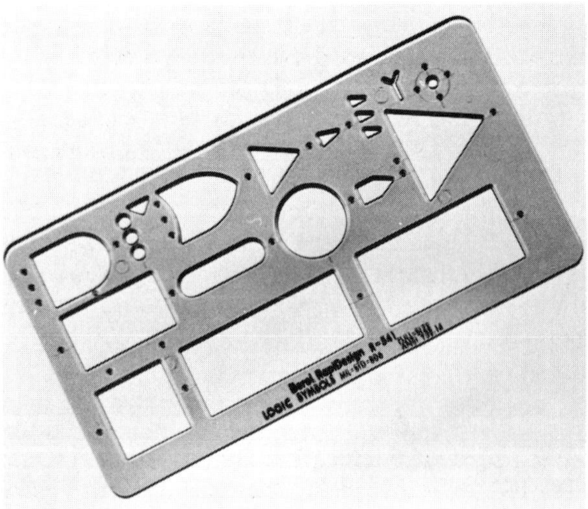

Fig. 2-17. Templates aid in drawing logic symbols. (Berol RapiDesign)

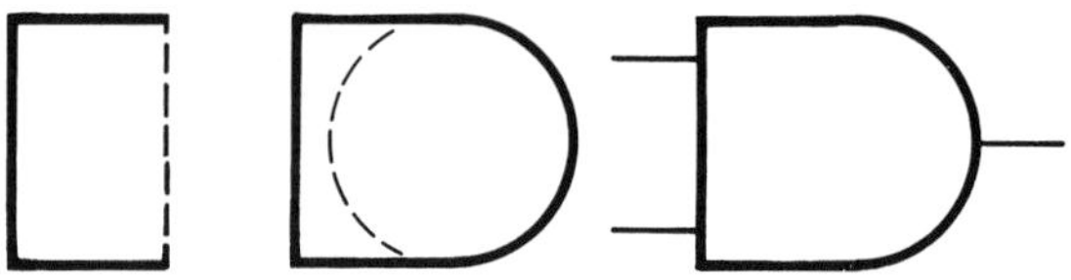

Fig. 2-18. AND symbols are made up of portions of a rectangle and a circle.

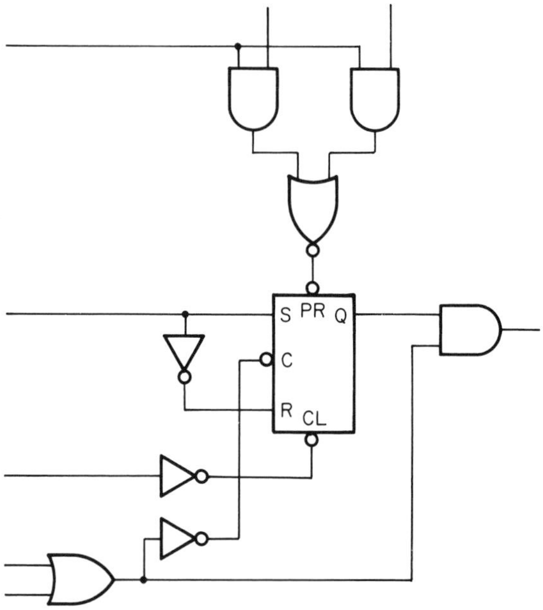

Fig. 2-19. This logic diagram contains three AND symbols.

GROUND AND POWER

Return to Fig. 2-14, part "b." Five leads enter this circuit (refer to the numbers in parentheses). Two are inputs (A and B); a third is the output (Y); another is marked GND; and the last one is power (Vcc).

GROUND

GND stands for ground. However, this is misleading, for this point in the circuit may or may not be connected to an earth ground.

Although the term ground is widely used, it would be better to call this point "common." Note the number of electrical/electronic parts connected to GND. Also, all external voltages are measured with respect to GND.

Note the symbol for ground. It too usually means a common point in a circuit. In any circuit, all ground symbols are connected together even though they may appear in widely different parts of a drawing.

On most chips, ground is brought out on only one lead. On drawings like the one in part "b" of Fig. 2-14, however, ground is often extended to emphasize that all input and output signals are always measured with respect to GND.

POWER SUPPLY

The arrow at the top of Fig. 2-14 part "b" points to a power supply. This is an electronic circuit that converts ac power into dc power at the proper voltage. This power operates the transistors, resistors, and diodes in the logic elements.

LOGIC SYMBOLS

Power and ground leads are not shown on logic symbols. See part "a" of Fig. 2-20. However, workers in digital electronics are aware of this omission and make the connections shown at "b."

Part "c" of Fig. 2-20 emphasizes that the power supply is connected between the Vcc lead and ground. Often, battery symbols are used to depict power supplies, for diode-based supplies and batteries are both sources of dc power.

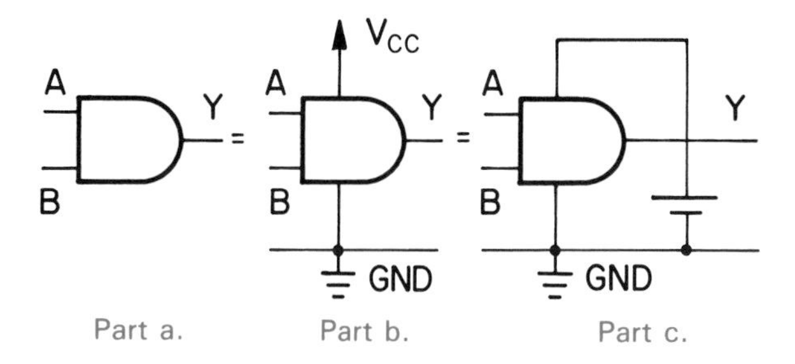

Fig. 2-20. Although power and ground connections are not shown on logic symbols, these connections must be made in logic circuits.

SWITCH EQUIVALENT

A switch equivalent can be used to suggest the action of an AND. Part "a" of Fig. 2-21 shows the signal source described in part "b" of Fig. 2-15. With the switch in the position shown, the voltage between A and GND is +5 volts (logic 1). With the switch in the other position, this voltage is 0 volts (logic 0).

The other drawing shows the same circuit using the arrow notation for the power supply. The actions of the two circuits are identical.

Fig. 2-22 shows a switch equivalent of an AND. Of course, there is no mechanical switch in an IC AND.

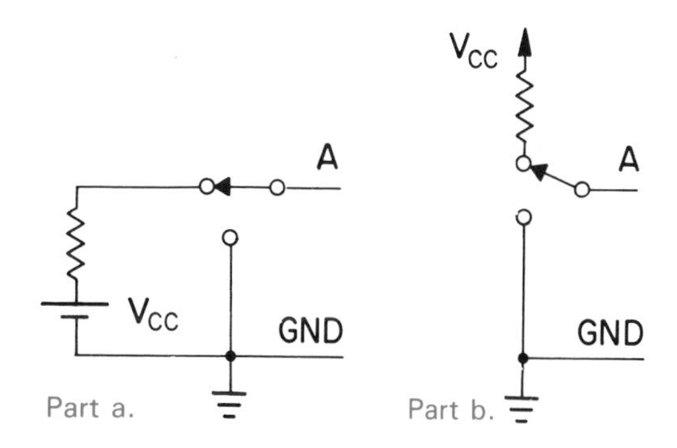

Fig. 2-21. These signal sources are electrically identical.

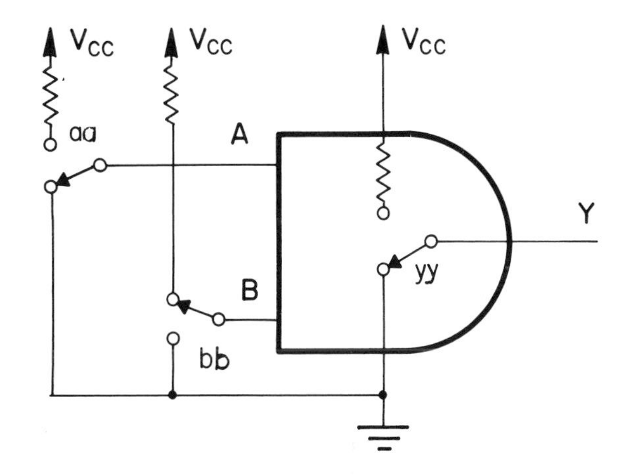

Fig. 2-22. The action of the output transistors of an IC logic element can be represented by a mechanical switch.

This is merely a way of representing the action of the transistors in such a circuit.

For the input signals indicated (A = 0, B = 1), output Y should equal 0. Note the position of switch yy. It grounds output Y and produces the required 0.

If A and B were both 1, output Y would equal 1. Switch yy would move to the upper position and connect the output to +5 volts.

Note that the output circuit of the IC looks just like the input signal sources. That is, output Y can act as an input to another logic element, as shown in Fig. 2-23. This concept is used to explain and describe complex circuit action.

TIMING DIAGRAMS

So far, three methods have been used to describe an AND—symbols, truth tables, and Boolean expressions. These methods do not take time into account, so they are said to be static. They tell nothing of the order of arrival of signals.

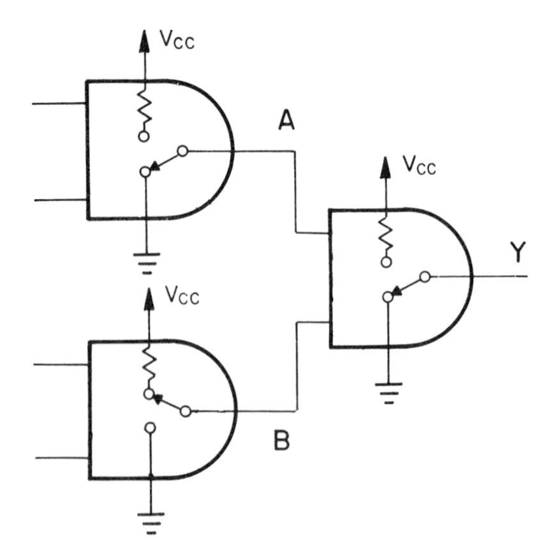

Fig. 2-23. The outputs of the two elements at the left serve as signal sources for the element at the right.

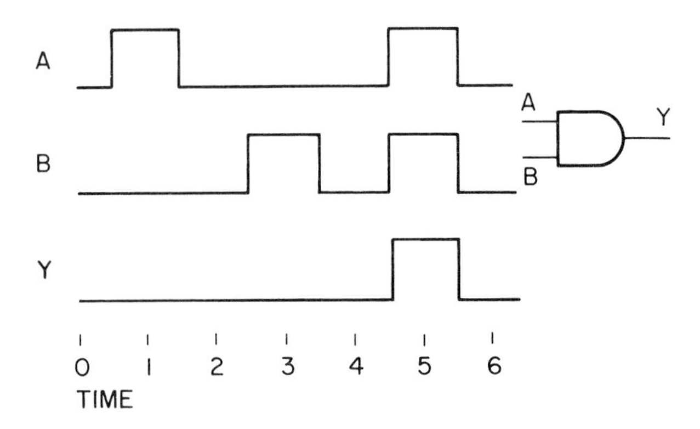

Fig. 2-25. In most timing diagrams, output graphs are drawn below those of the inputs.

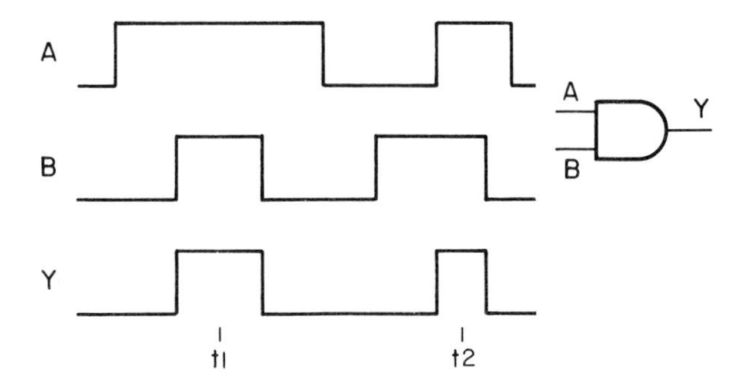

Fig. 2-26. The output of an AND is 1 only when 1s appear at all inputs.

In high speed digital circuits, the time of arrival is often important. To show this, descriptions called dynamic descriptions are needed. Timing diagrams give enough information, Fig. 2-24. Time is plotted from left to right on the three horizontal axes. Signal voltages are plotted vertically.

At any given time, values of A, B, and Y can be read from these graphs. For example, at time 1, A = 1, B = 0, and Y = 0. Timing diagrams restate the information in truth tables and Boolean expressions, but they express this information as a function of time.

In Fig. 2-24, the graph for Y is at the output of the AND symbol. As a result, it may be difficult to relate Y to A and B. To correct this, output graphs are often drawn below input graphs, Fig. 2-25. Care must be taken to distinguish input and output signals.

Fig. 2-26 is the timing diagram for another set of input signals. It emphasizes that Y = 1 only when A and B are both 1. Refer to time t1. Although A is 1 for a relatively long period, Y is 1 only during the time that B also equals 1.

Also refer to time t2. Here, Y is 1 only during the overlap of A and B, when both A and B are 1.

AND GATE

AND logic elements are often referred to as AND gates. Fig. 2-27 suggests the source of this term. Input A is a series of pulses. Before time t1, B = 0, so

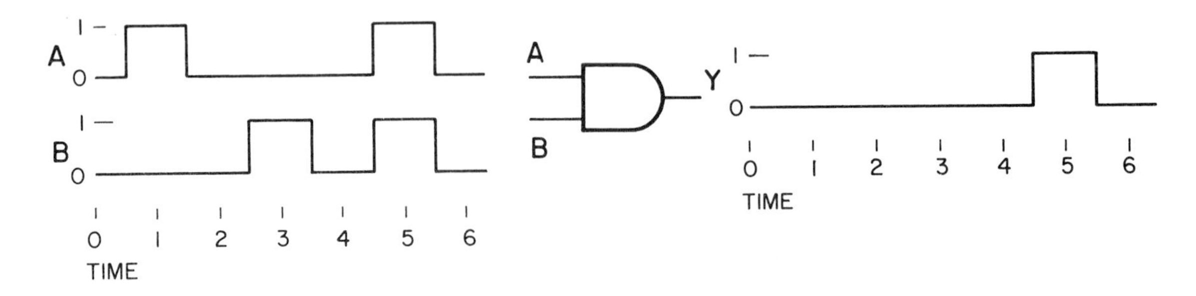

Fig. 2-24. Timing diagrams provide dynamic descriptions of the actions of logic elements.

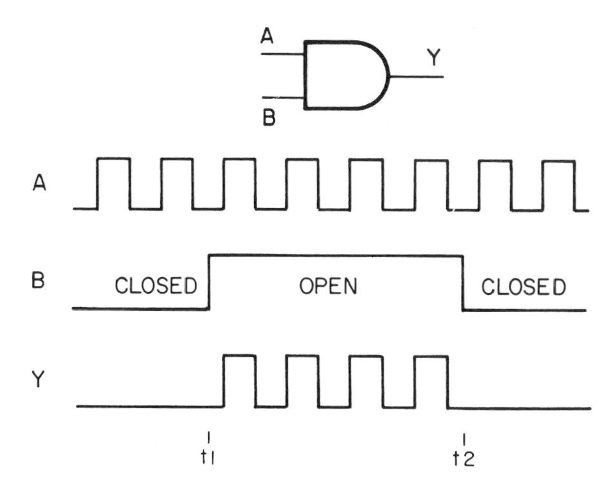

Fig. 2-27. This AND is being used as a gate. The signal at A appears at Y only while B is 1.

the output of the AND is held at 0. Expressed in Boolean form:

$$AB = Y$$
$$A0 = 0$$

No matter what the value of A (1 or 0), output Y is 0. That is, the gate is closed. The signal at input A cannot get through to Y.

Between t1 and t2, B = 1. The Boolean expression for this condition is:

$$AB = Y$$
$$A1 = A$$

That is, during this period, Y takes on the value of A. The gate is open.

Input B might be called an *ENABLE*. B "enables" the gate. When it is 1 (active), the signal at A reaches the output. When B = 0 (inactive), A cannot reach Y. Y remains 0.

APPLICATION OF IC AND

IC ANDs are used in computers and other digital circuits. However, only one application will be described here.

In computers, there is usually a circuit called an *ADDER*. It adds binary numbers. ANDs are used to construct a portion of an adder.

RULES OF BINARY ADDITION

Rules for binary addition can be developed from a knowledge of binary counting. As in any numbering system, 0 + 0 = 0, 0 + 1 = 1, and 1 + 0 = 1. The special nature of the binary system comes into play when 1 is added to 1. From the first chapter, 1 + 1 = 10. That is, when 1 is added to 1 in binary, a 0 results and a carry is generated. These rules of binary addition are summarized in Fig. 2-28.

Based on this table, an adder has two outputs: one for the sum (S) and one for the carry (C). See part "a" of Fig. 2-29.

Examine the truth table for the carry operation. Note that C = 1 only when A and B are both 1. This is the action of an AND. In a computer, the carry portion of this adder could be an AND, as in part "b" of Fig. 2-29.

INPUTS		OUTPUTS	
A	B	C (CARRY)	S (SUM)
0	0	0	0
0	1	0	1
1	0	0	1
1	1	1	0

Fig. 2-28. This truth table describes the binary addition of two numbers.

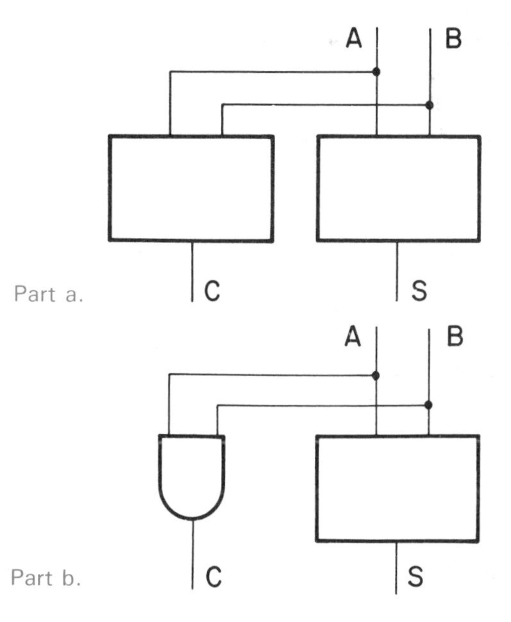

Fig. 2-29. In this binary adder, an AND is used in the carry circuit.

USING IC ANDS

IC logic elements are described in data books, handbooks, and technical literature. These are prepared by IC manufacturers and are sold or given to technical personnel.

Data books describe the 7408 as a TTL, quadruple, 2-input, positive-logic AND. That is, it is from the TTL family. Quadruple means that it contains four elements. Each element has two inputs. And, positive logic means that logic 1 is a positive voltage compared to logic 0.

This chip is mounted in a 14-pin DIP, but other packages are available. Functions of individual pins are described by a *PINOUT* (from the term "pin outline diagram"). See Fig. 2-30. In addition to inputs and outputs, power supply (Vcc) and ground (GND) pins are shown.

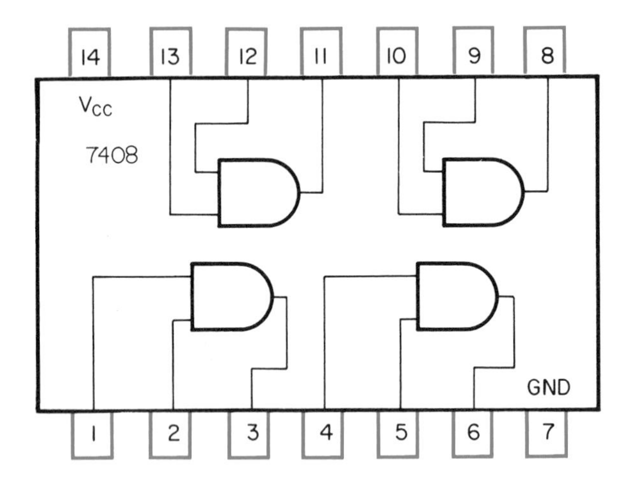

Fig. 2-30. Pin outline diagrams (sometimes called "pinouts") show the functions of individual pins. A 7408, TTL, quadruple, 2-input, positive-logic AND package is shown.

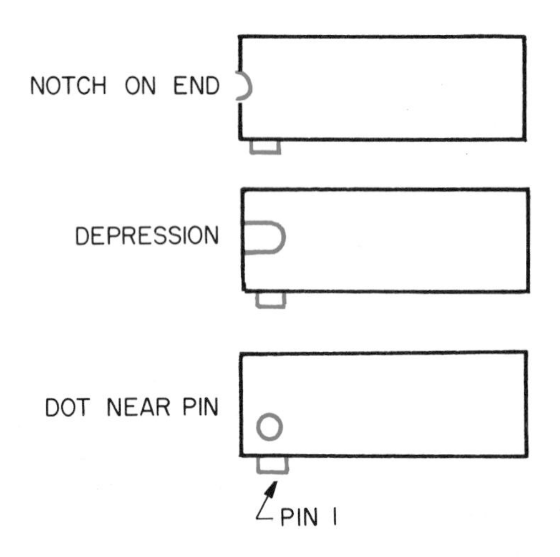

Fig. 2-31. Three methods are used to identify pin 1 on DIPs.

Viewed from above, pins are numbered counterclockwise (CCW). Three methods of indicating pin 1 are shown in Fig. 2-31.

Except for the common power supply and ground, the ANDs in a 7408 are independent. They can be used in different circuits without interacting. If less than four ANDs are needed, unused elements are ignored (unless power usage must be limited).

INTEGRATED CIRCUIT OR

Three descriptions of the IC OR are shown in Fig. 2-32. Again, the truth table and Boolean expression are the same as those for switch-based ORs.

SYMBOL	A, B → Y
TRUTH TABLE	see below
BOOLEAN EXPRESSION	A + B = Y

INPUTS A	INPUTS B	OUTPUT Y
0	0	0
0	1	1
1	0	1
1	1	1

Fig. 2-32. The action of an IC OR is described using the three representation methods.

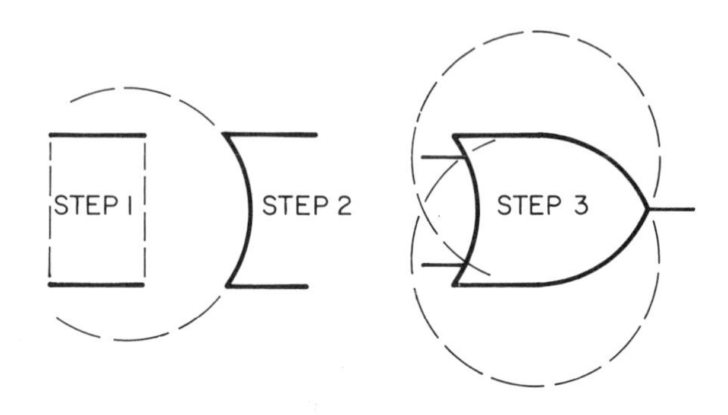

Fig. 2-33. OR symbols contain portions of a rectangle and three circles.

OR SYMBOL

OR symbols are more difficult to draw than AND symbols. See Fig. 2-33. If poorly drawn, OR symbols are easily confused with ANDs.

How many ORs are there in Fig. 2-34? You should find two.

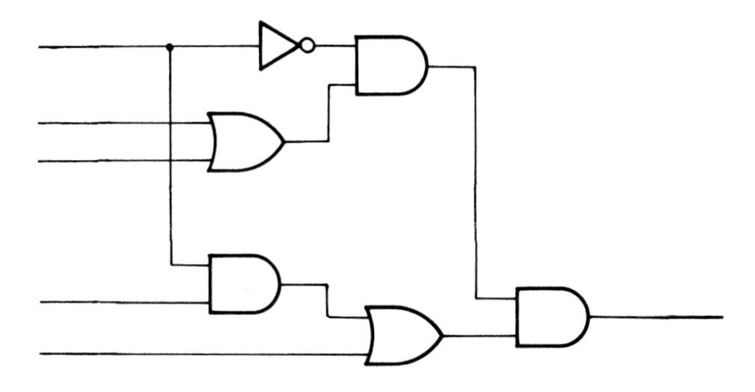

Fig. 2-34. This logic diagram contains two OR symbols.

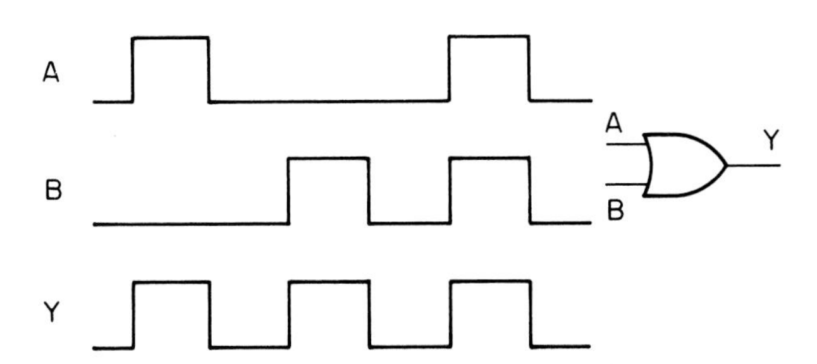

Fig. 2-35. This timing diagram is a dynamic description of the action of an OR.

TIMING DIAGRAM FOR AN IC OR

The timing diagram in Fig. 2-35 is a dynamic description of the action of an OR. It restates the OR truth table and Boolean expression but adds time as another dimension. The OR truth table and Boolean expression are called static representations. The diagrams with time added are called dynamic representations.

Fig. 2-36 emphasizes that the output of an OR is 1 whenever an input is 1. Only when all inputs are 0 will Y = 0.

OR elements are often referred to as OR gates. See Fig. 2-37. Input A is a series of pulses. Until time t1, B = 1. This forces output Y to a constant 1, and the pulses at A cannot get through the closed gate. Expressed in Boolean form, this becomes:

$$A + B = Y \qquad A + 1 = 1$$

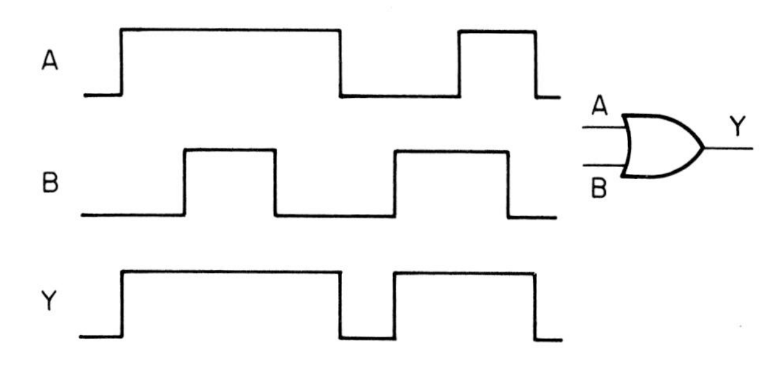

Fig. 2-36. This OR logic element timing diagram stresses that Y = 0 only when A and B are both 0.

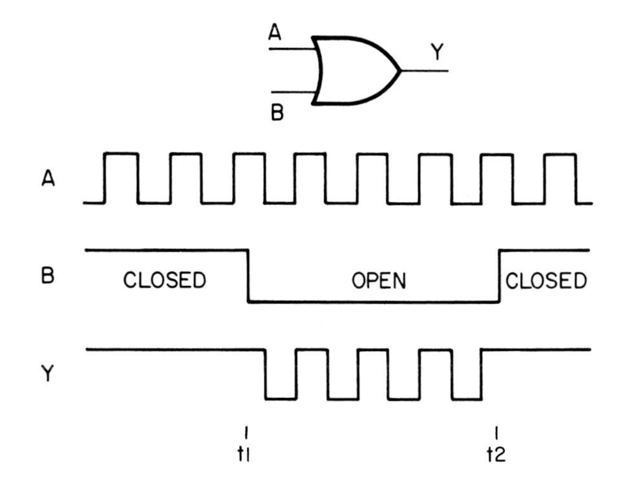

Fig. 2-37. An OR can be used as a gate. Y = A only while B = 0.

Between t1 and t2, B = 0, so the gate is open. Output Y follows the variations in A. In Boolean form:

$$A + B = Y$$
$$A + 0 = A$$

IC OR ACTION

A switch equivalent of a TTL OR is shown in Fig. 2-38. As before, there is no mechanical switch in an IC OR. This equivalent merely represents the action of the device's output transistors.

APPLICATION OF AN IC OR

Fig. 2-39 shows a display that might be used on a digital meter. When properly lighted, the crossed bars can be a + or − sign. Each bar glows when a positive voltage between +3 and +5 volts is applied. Current-limiting resistors and ground leads have been omitted for simplicity.

To produce a negative sign, a logic 1 is applied to

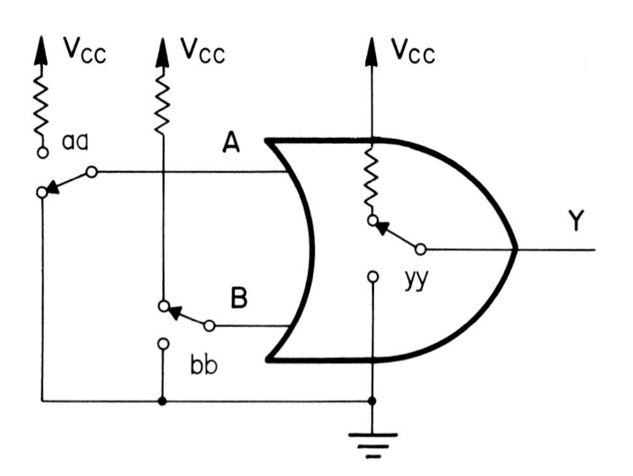

Fig. 2-38. IC OR action can be represented by a mechanical switch.

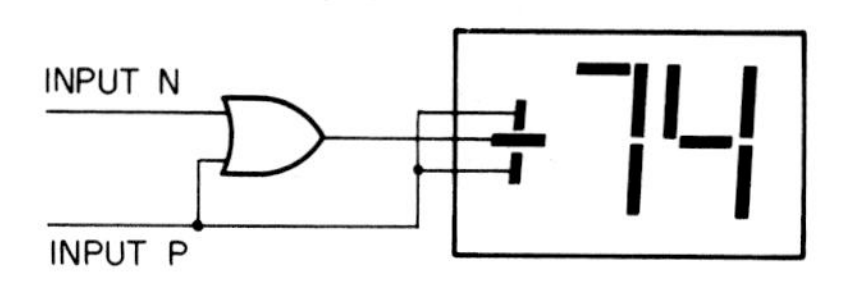

Fig. 2-39. The OR lights the horizontal bar of the +/− sign for both positive and negative numbers. N stands for negative and P stands for positive.

input N (N stands for negative), while zero volts are applied to P (positive). With a 1 at an input, the OR will output the proper positive voltage to light the horizontal bar.

If a 1 is applied to input P, a plus sign will light. The voltage at P directly lights the vertical bar, and the 1 at the lower input of the OR causes its output to go to 1. This lights the horizontal bar. If 0s are applied to both N and P, the sign does not light.

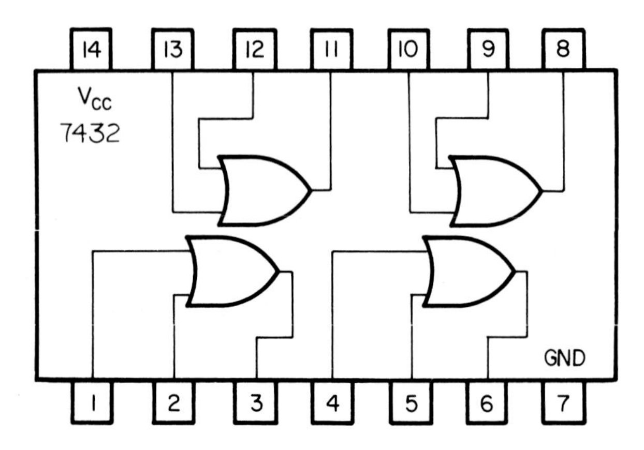

Fig. 2-40. The 7432 is a TTL, quad, 2-input, positive-logic OR package of logic elements.

USING IC ORS

Except for identifying numbers, AND and OR chips look alike. Fig. 2-40 is the pin outline (pinout) of a 7432. This package is a TTL, quad (short for quadruple), 2-input, positive-logic OR unit.

INTEGRATED CIRCUIT NOT

The function of a NOT is to invert signals. When a 1 is applied to a NOT, it outputs a 0. When a 0 is applied, it outputs a 1. See Fig. 2-41 for descriptions of an IC NOT.

SYMBOL	A —▷o— Y
TRUTH TABLE	INPUT A: 0, 1 — OUTPUT Y: 1, 0
BOOLEAN EXPRESSION	$\bar{A} = Y$

INPUT A	OUTPUT Y
0	1
1	0

Fig. 2-41. The action of an IC NOT can be described by the three static methods (the static methods do not show time and are not dynamic).

NOT SYMBOL

The NOT symbol is smaller than symbols for ANDs and ORs, Fig. 2-42. The NOT symbol has two parts—a triangle and a circle or *BUBBLE* (a small circle at the right end).

A bare triangle represents an amplifier, part "a" of Fig. 2-43. The output of such a device follows its input signal. When A = 1, Y = 1, etc. However, the power available at the output of an amplifier is higher than at its input. As a result, amplifiers are often called *DRIVERS* or *BUFFERS*. Drivers can drive the inputs of 10 to 25 elements. Buffers cut down interference between 2 or 3 high-speed circuits.

The addition of a bubble changes an amplifier symbol into that of a NOT. The bubble, not the triangle, indicates the inversion. The bubble may be placed on

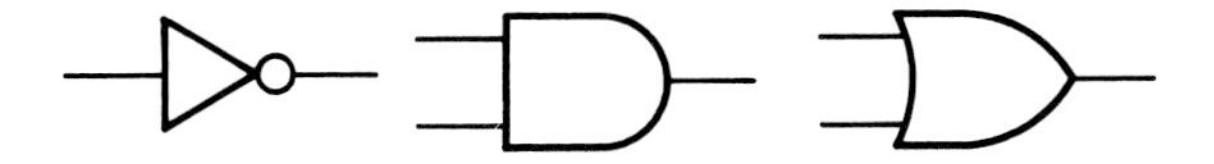

Fig. 2-42. NOT symbols are smaller than AND and OR symbols. The symbol has a bubble (small circle) at the right end.

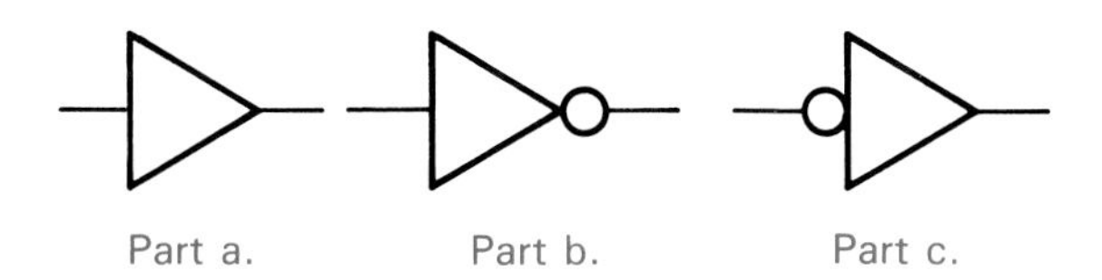

Fig. 2-43. An amplifier symbol (at part "a") is changed into a NOT symbol by adding a bubble. Note that the components shown in parts "b" and "c" operate the same.

either the input or output, but it is usually placed at the symbol's output. See parts "b" and "c" in Fig. 2-43.

TIMING DIAGRAM FOR AN IC NOT CIRCUIT

As expected, NOT timing diagrams are simple, as shown in Fig. 2-44. Output Y is always the inverse or complement of the lone input.

APPLICATION OF AN IC NOT CIRCUIT

Part "a" of Fig. 2-45 shows a *DECODER* (one type of decoder is a binary-to-decimal converter). The input of the decoder is any 2-bit binary number B,A. Its output is the corresponding decimal number. For example, if B,A = 1,0, a 1 will appear on lead OUT2. That is, the binary number 10 equals decimal 2. The other outputs will be 0.

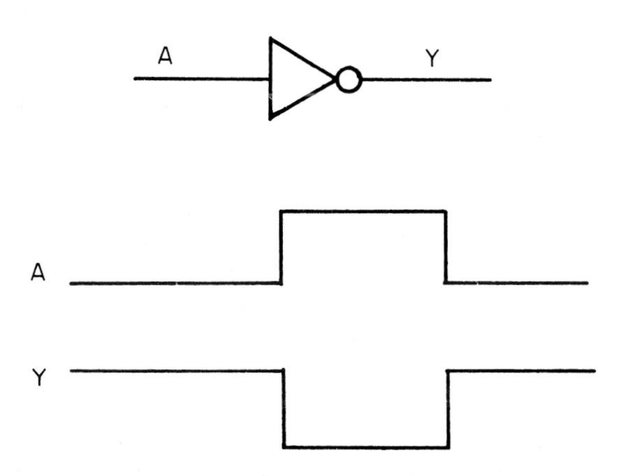

Fig. 2-44. The NOT timing diagram is very simple.

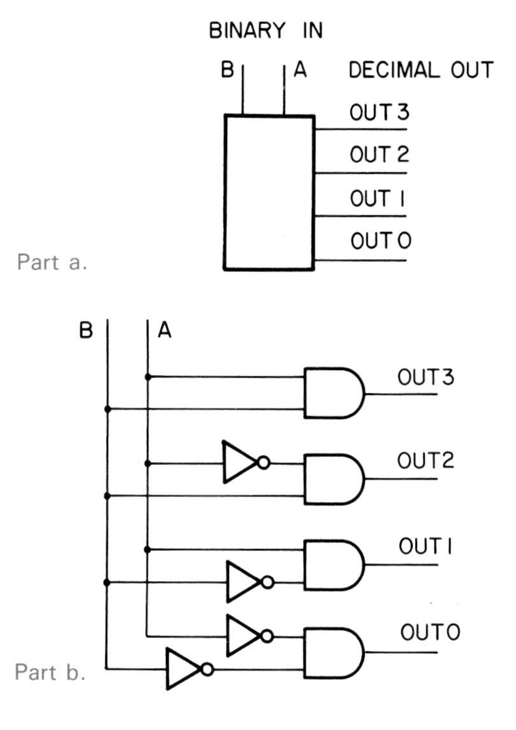

Fig. 2-45. This circuit uses NOTs to decode 2-bit binary numbers (convert them to decimal form).

The circuit of the decoder is shown in part "b" of Fig. 2-45. Note the use of NOTs. If B,A = 1,0, a 1 appears at the lower input of the second AND. The 0 at A is inverted by the NOT, so a 1 is applied to the upper input of this AND. As a result, a 1 appears at OUT2. Test the other three outputs.

Fig. 2-46 shows the truth table for the decoder circuit in Fig. 2-45. The Boolean expressions for this circuit are:

$$BA = OUT3$$
$$B\bar{A} = OUT2$$
$$\bar{B}A = OUT1$$
$$\bar{B}\bar{A} = OUT0$$

MULTI-INPUT ELEMENTS

Logic elements with more than two inputs are available or can be assembled. Their truth tables and Boolean expressions are identical to those of switch-based elements. Static descriptions of a 3-input AND are shown in Fig. 2-47.

SYMBOLS FOR MULTI-INPUT ELEMENTS

Symbols for multi-input ANDs and ORs are shown in Fig. 2-48. When there are more than three inputs,

INPUTS		OUTPUTS			
B	A	OUT0	OUT1	OUT2	OUT3
0	0	1	0	0	0
0	1	0	1	0	0
1	0	0	0	1	0
1	1	0	0	0	1

Fig. 2-46. The truth table for a 2-bit binary-to-decimal decoder identifies the decimal numbers 0, 1, 2, and 3.

SYMBOL	A, B, C → Y
TRUTH TABLE	(see below)
BOOLEAN EXPRESSION	ABC = Y

INPUTS			OUTPUT
A	B	C	Y
0	0	0	0
0	0	1	0
0	1	0	0
0	1	1	0
1	0	0	0
1	0	1	0
1	1	0	0
1	1	1	1

Fig. 2-47. These three static methods can be used to describe the actions of multi-input elements.

extenders are used. These prevent crowding of the leads without increasing the symbol size. All symbols on a given drawing should be the same size.

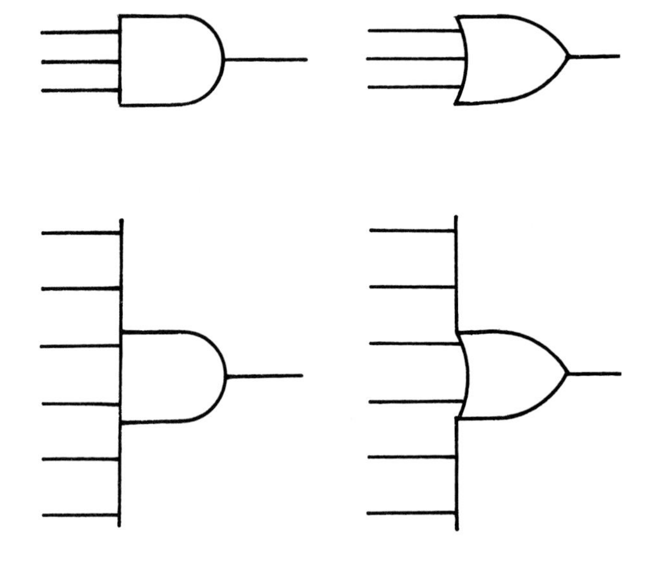

Fig. 2-48. Extenders are often used on symbols with four or more leads.

TIMING DIAGRAMS FOR MULTI-INPUT ELEMENTS

Timing diagrams for multi-input elements are created by adding input graphs, Fig. 2-49. Output Y of this OR is 1 whenever any input is 1.

Fig. 2-50 is a timing diagram for a 3-input AND. Its output is 1 only when all inputs are 1.

DECREASING THE NUMBER OF INPUTS

If all inputs of a logic element are not used, care must be taken to properly connect the unused leads. Several methods will be described in later chapters, but for now, the method shown in Fig. 2-51 is recommended. Tie unused leads to an active lead.

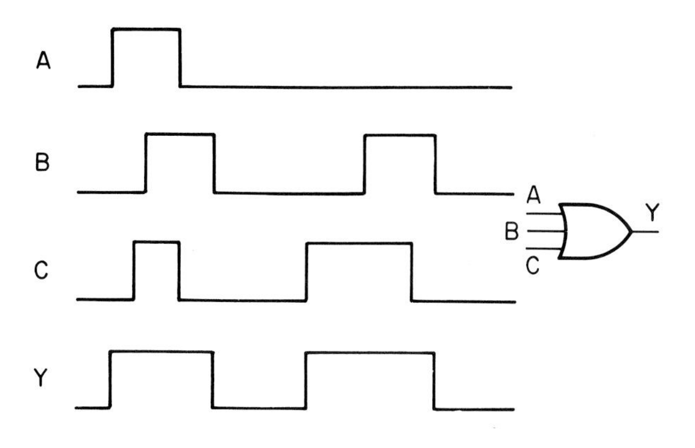

Fig. 2-49. The output of the 3-input OR shown is 0 only when all of the inputs are 0.

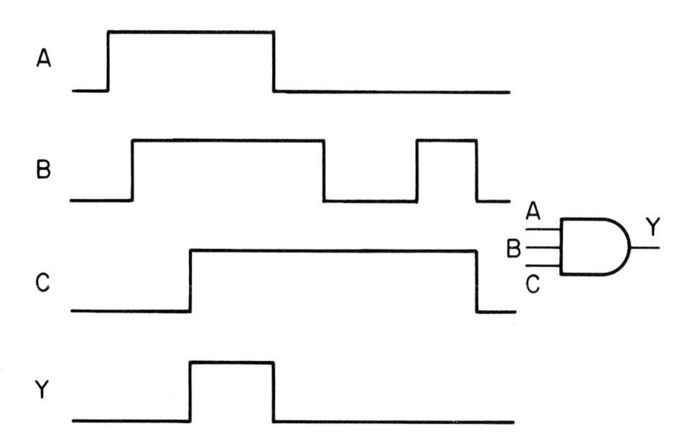

Fig. 2-50. The output of the 3-input AND shown is 1 only when all of the inputs are 1.

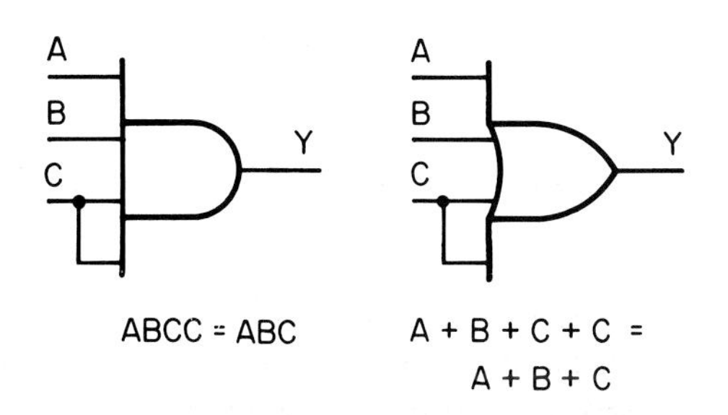

Fig. 2-51. Unused inputs should not float (have a fixed logic 1). Connecting such leads to active leads solves this problem.

Refer to the 4-input AND. With two inputs tied together, its Boolean expression is:

$$ABCC = Y$$

Because the Cs will always have the same value, this statement can be simplified to:

$$ABC = Y$$

A similar simplification is valid for multi-input ORs.

TRUTH TABLE CIRCUIT ANALYSIS

It is often necessary to predict the signals that should be present within a circuit. If the measured and the predicted values disagree, the circuit is probably faulty.

A direct method of predicting signals through a circuit might be described as truth table circuit analysis.

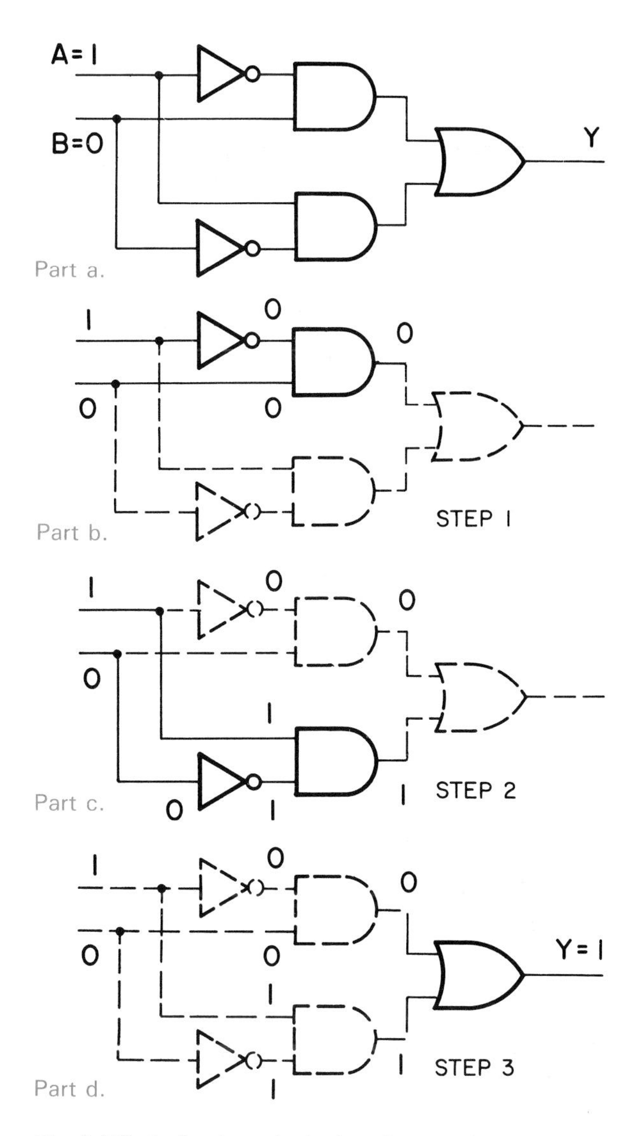

Fig. 2-52. A circuit analysis done by a truth table determines signals at all points within a logic circuit composed of AND, OR, and NOT elements. The circuit shown is the "sum" part of a 2-bit binary adder.

To use this method three things are needed:
1. Truth tables of all elements.
2. A logic diagram of the circuit.
3. Values of input signals.

The circuit in part "a" of Fig. 2-52 will be used to describe this method. The circuit contains only AND, OR, and NOT elements, so you are familiar with the required truth tables. The input signals to be used (A = 1, B = 0) are shown. The problem is one of determining the signals present at each point in this circuit.

This analysis is best approached in small pieces. The drawing in part "b" of Fig. 2-52 shows the first step. Based on NOT and AND truth tables, the output of the upper AND is found. For the given values of A and B, this AND outputs a 0.

At part "c," the process is repeated for the lower AND. With 1s present at both inputs, it outputs a 1.

Finally, the truth table for the OR is used to determine the circuit's output. It is found to be a 1. This completes the task, because signals at all points have been predicted.

The circuit in Fig. 2-53 is a 2-bit adder. That is, it adds binary digits input at A and B and outputs their sum and carry at S and Co. The circuit is special in that it uses only ANDs and NOTs. Most adders require ORs as well as ANDs and NOTs.

In part "a" of Fig. 2-53, truth table circuit analysis has been used to determine the circuit's outputs for A = 0 and B = 0. Trace part "b" of that figure onto a piece of paper and use truth table circuit analysis to determine the circuit's output for A = 1 and B = 0.

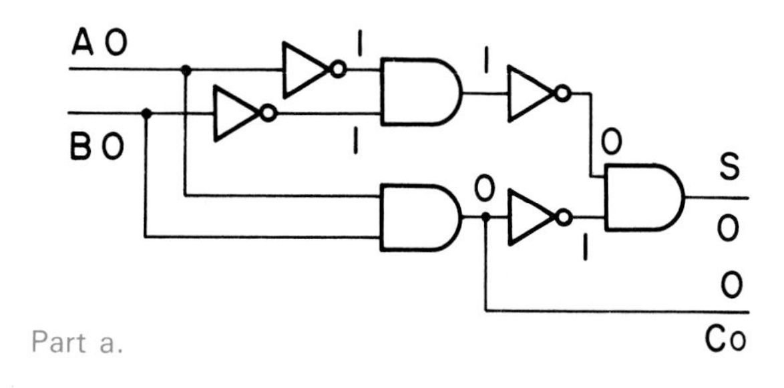

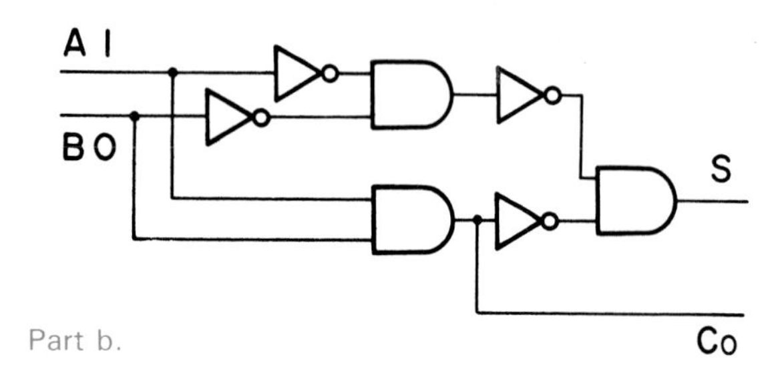

Fig. 2-53. This 2-bit binary adder has two outputs, SUM and CARRY. Truth table circuit analysis is not altered by the presence of additional outputs.

Fig. 2-54 shows a circuit for a more complex adder. Truth table analysis has again been applied, but there is a problem. The predicted outputs are incorrect. S should be 1, and C should be 0. A mistake has been made somewhere in the analysis. An ability to find errors is an important skill for technical personnel, so attempt to find the source of this error. If you found it to be at the output of element F, you are correct.

SUMMARY

Knowledge of AND, OR, and NOT logic elements is basic to the study of and work in digital electronics.

Switches can be used to construct logic elements.

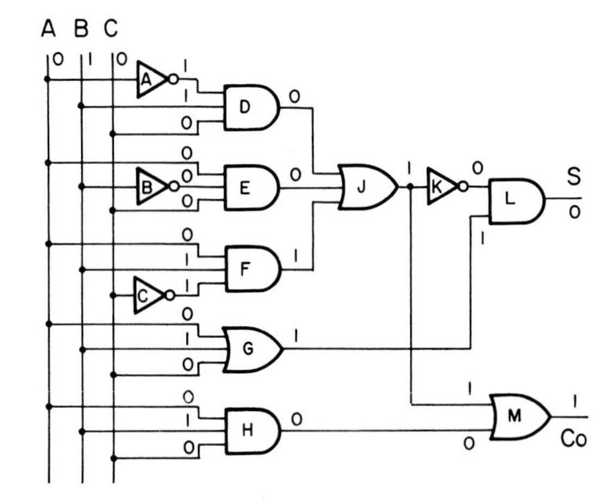

Fig. 2-54. This adder has three inputs and two outputs. When truth table circuit analysis was applied, an error was made. Although the error might be considered minor, it altered the predicted outputs at both S and Co.

Contacts placed in series produce ANDs; when placed in parallel, ORs result. Normally-closed contacts are used to produce NOTs.

Integrated circuit logic elements are used in modern computers and other digital circuits. Logic symbols, truth tables, and Boolean expressions are static methods of describing IC elements. They relate input and output signals at fixed points in time. Timing diagrams are dynamic descriptions. They show how signals vary with time. See Fig. 2-55.

Signal values at each point in a circuit can be determined by applying truth table circuit analysis. When compared with measured signals, the predicted values can be used to determine whether or not a circuit is working properly.

IMPORTANT TERMS

Activated, Adder, AND, AND gate, Binary arithmetic, Boolean algebra, Dynamic, Integrated circuit logic, Leads, Normally closed, Normally open, NOT, Open, OR, Override, Parallel, Pinout, Series, Static, Switch-based elements, Switch equivalent, Template, Timing diagram, Truth table circuit analysis, Unactivated.

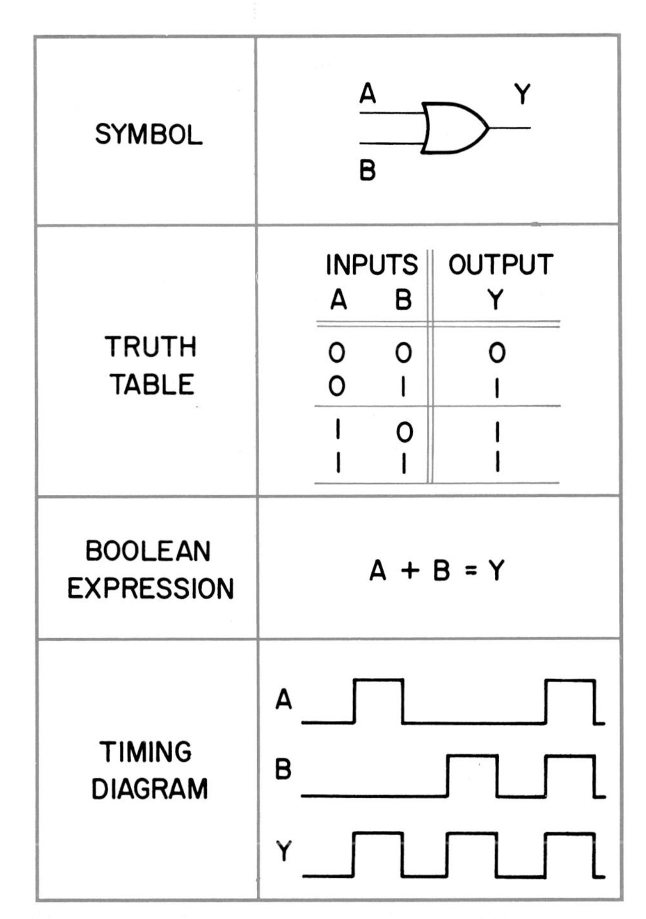

Fig. 2-55. These four methods for describing the actions of logic elements are used in industry.

TEST YOUR KNOWLEDGE

1. List the three logic elements that are basic to the study of digital circuits.
2. An AND element outputs a 1 when ________ (all, any) inputs are 1.
3. An OR circuit outputs a 0 when ________ (all, any) inputs are 0.
4. A NOT circuit outputs a 1 when its input is a ________.
5. A floating ground is a proper type of input to an IC logic element. (True or False?)
6. Boolean algebra follows the rules of binary arithmetic. (True or False?)
7. Does a static circuit representation show the effects of time?
8. The Boolean statement "A × B = L" is read "A ________ (AND, OR, NOT) B equals L."
9. The Boolean statement "A + B = L" is read "A ________ (AND, OR, NOT) B equals L."
10. What two drawing components make up the symbol for an AND circuit?

STUDY PROBLEMS

1. Draw the circuit for a switch-based AND. Use two push-button switches (A and B) for inputs, and a lamp (L) for its output. Use a battery as a power source.
2. Based on the circuit in Problem 1, construct a truth table for a switch-based AND, and write its Boolean expression.
3. Repeat Problems 1 and 2 for a switch-based OR.
4. Repeat Problems 1 and 2 for a switch-based NOT.
5. Copy the circuit in Fig. P2-1 onto a piece of paper. Use a series of arrows to show the complete current path through the battery, switches, and lamp. Do this for A = 1 and B = 0.

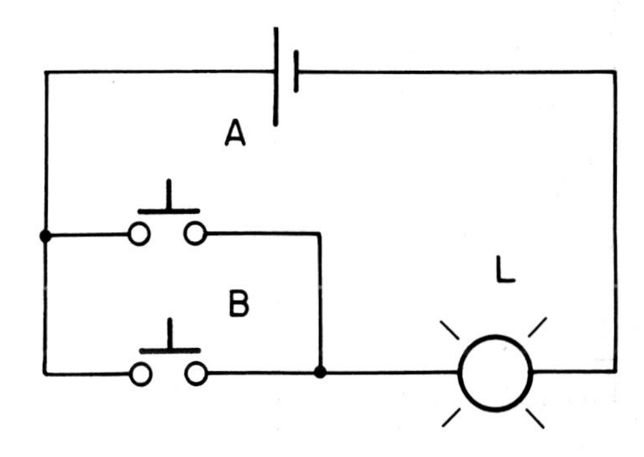

Fig. P2-1. Use arrows to show where the current flows when the switches are set for A = 1 and B = 0.

6. Write the Boolean expression for each of the circuits in Fig. P2-2.
7. Copy the truth table in Fig. P2-3 onto a piece of paper and complete it for the AND shown. Note

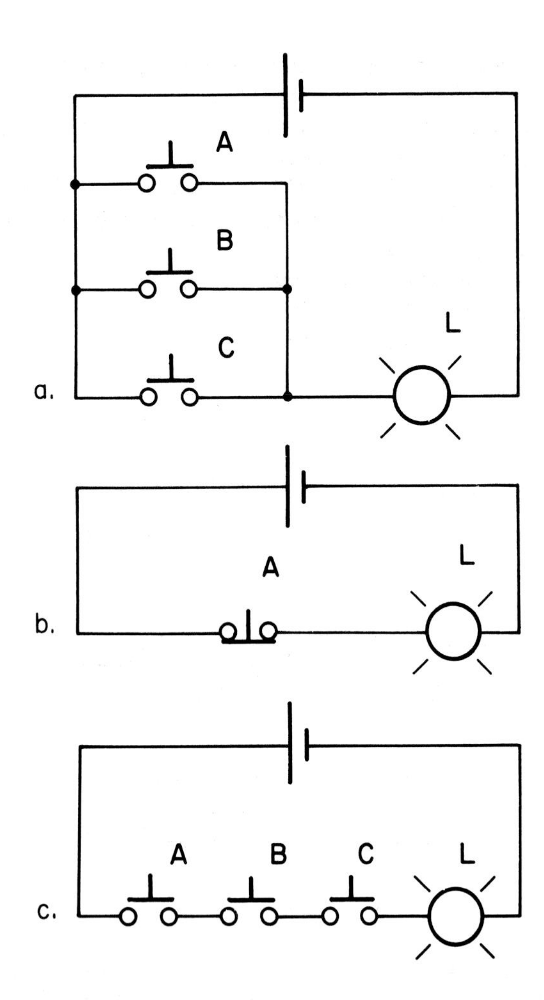

Fig. P2-2. Write the Boolean expression for each of the circuits shown.

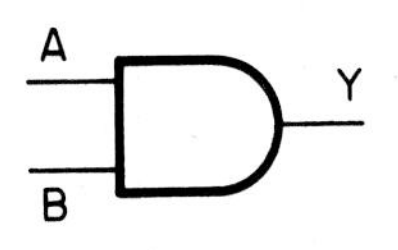

INPUTS		OUTPUT
A	B	Y
1	0	
		1
0	0	
	1	0

Fig. P2-3. For the AND circuit shown, complete the set of entries begun in the table.

that the entries are not in standard order.

8. Repeat Problem 7 for the OR shown in Fig. P2-4.
9. Determine Y for each of these Boolean statements.

a. $1 \times 0 = Y$ g. $1 \times 0 \times 1 = Y$
b. $1 + 0 = Y$ h. $0 + 0 + 1 = Y$
c. $0 + 0 = Y$ i. $1 \times 1 \times 1 \times 0 = Y$
d. $1 \times 1 = Y$ j. $0 + 0 + 0 + 1 = Y$
e. $\overline{1} = Y$ k. $\overline{0} \times 1 \times 1 = Y$
f. $\overline{0} = Y$ l. $\overline{0} + 0 + 0 = Y$

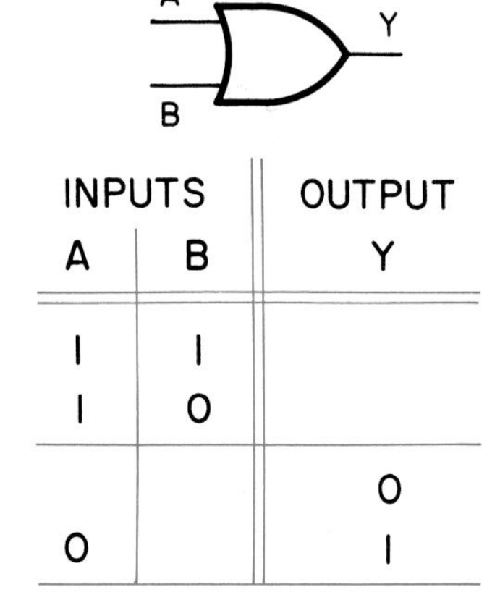

INPUTS		OUTPUT
A	B	Y
1	1	
1	0	
		0
0		1

Fig. P2-4. For the OR circuit shown, complete the set of entries begun in the table.

10. Trace the timing diagram in Fig. P2-5 onto a sheet of paper and carefully construct the expected output signal.

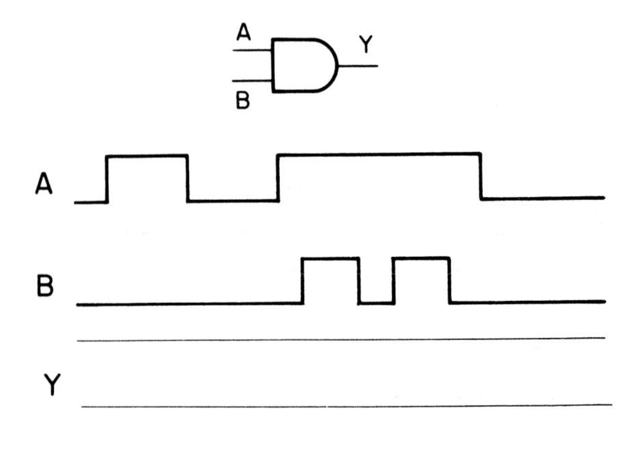

Fig. P2-5. Trace the timing diagram shown for this 2-input AND circuit and add the output graph between the two fine lines following the "Y." The upper fine line means 1 and the lower fine line means 0.

11. Repeat Problem 10 for the logic element and set of input signals shown in Fig. P2-6.

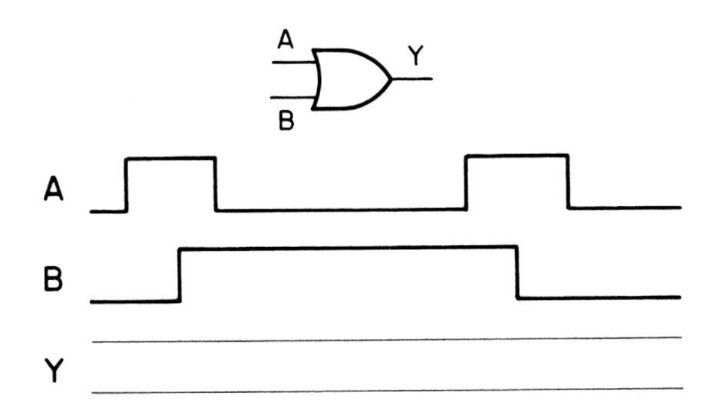

Fig. P2-6. Trace the timing diagram for this 2-input OR circuit and add the output graph for Y.

12. Repeat Problem 10 for the logic element and set of input signals in Fig. P2-7.

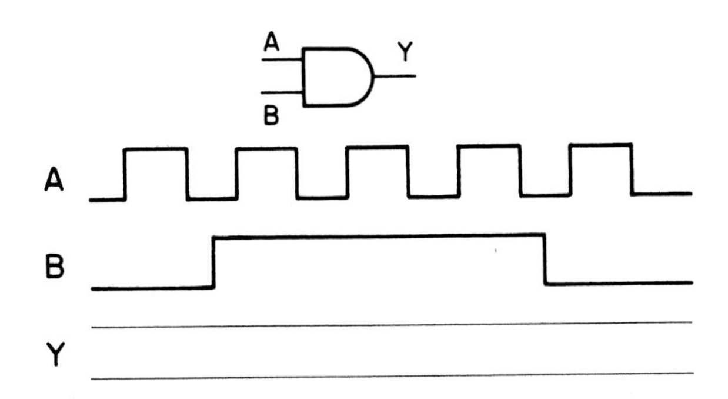

Fig. P2-7. Trace the timing diagram for this 2-input AND circuit used as a gate and add the output graph.

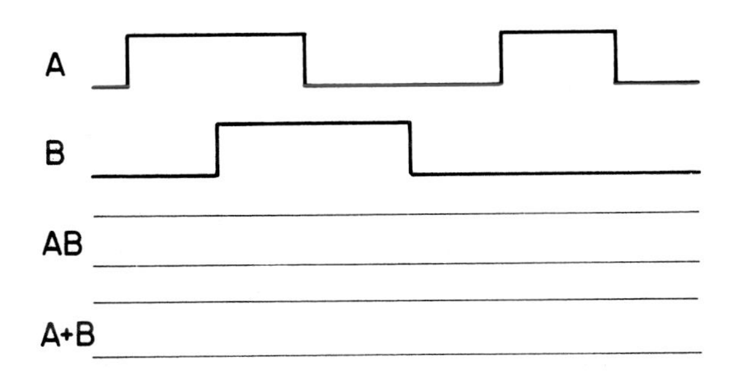

Fig. P2-8. Construct the output graph for the AND circuit represented by "AB." Then construct the output graph for the OR circuit represented by "A + B."

13. Assume the set of signals in Fig. P2-8 have been applied to an AND. Trace the timing diagram onto a sheet of paper and carefully construct the expected output signal.
14. Repeat Problem 13. This time, assume the signals shown in the top portion of Fig. P2-8 have been applied to an OR.
15. TTL logic elements treat floating inputs as logic 1s. For each of the elements shown in Fig. P2-9, determine the resulting output signals. Unmarked leads are floating.

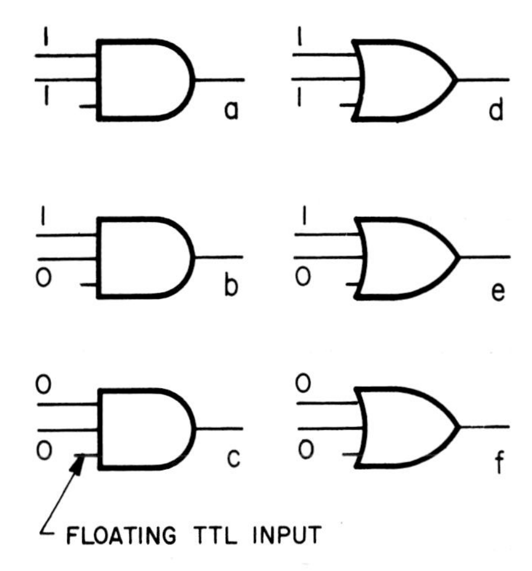

Fig. P2-9. The TTL logic elements shown have 1 floating lead per element. Show whether the outputs are 0s or 1s.

16. Copy the circuit in Fig. P2-10 onto a sheet of paper. For the inputs shown, predict the signals at all points in the circuit.
17. Repeat Problem 16 for the circuit and input signals shown in Fig. P2-11.

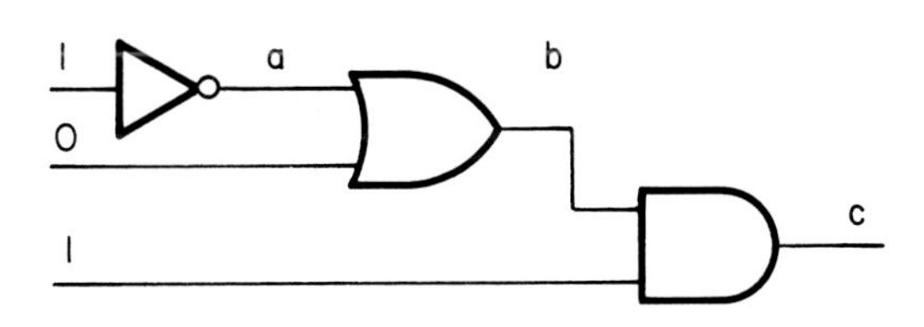

Fig. P2-10. Copy the TTL circuit and show the logic signals (1 or 0) at points a, b, and c. The inputs are given at the left of the diagram.

18. Repeat Problem 16 for the circuit and input signals shown in Fig. P2-12.

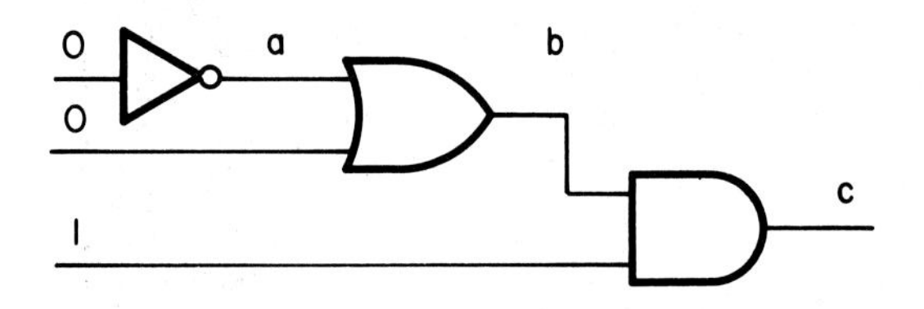

Fig. P2-11. Copy the TTL circuit and show the logic signals (1 or 0) at points a, b, and c.

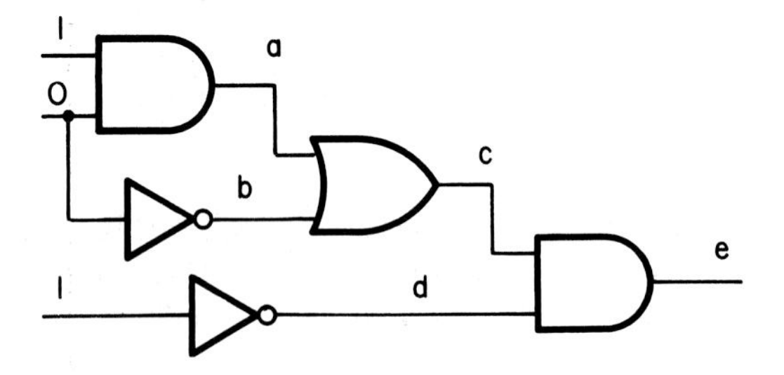

Fig. P2-12. Copy the circuit and show the signals (1 or 0) at points a, b, c, d, and e.

19. Repeat Problem 16 for the circuit and input signals shown in Fig. P2-13.

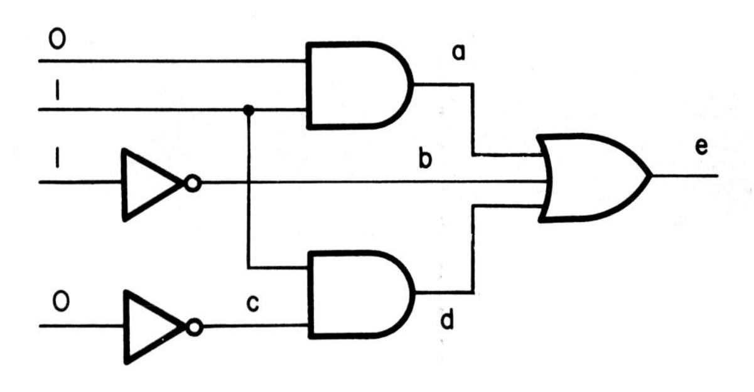

Fig. P2-13. Copy the circuit and show the signals (1 or 0) at points a, b, c, d, and e.

Introduction to Troubleshooting

Chapter 3

LEARNING OBJECTIVES

After studying this chapter and completing lab assignments and study problems, you will be able to:

- Classify digital circuit faults according to:
  - Design versus circuit faults.
  - Intermittent versus permanent faults.
  - External versus internal faults.
  - Parametric versus logic faults.
  - Static versus dynamic faults.
- Describe basic digital instruments and classify them as static or dynamic.
- Do basic troubleshooting in circuits containing AND, OR, and NOT elements.

Electronic equipment is reliable. That is, it operates properly for long periods of time. When it does fail, there is a search for the cause of the problem. The search is called troubleshooting.

An ability to troubleshoot is an asset to all digital electronics workers. Industrial electricians repair digital controls. Technicians may have to locate faults in newly constructed circuits. Technologists may work in a quality control department in an electronic firm. Engineers may have to search for design errors in circuits being developed.

Circuits assembled by students in laboratories often contain wiring errors and other errors. Without some troubleshooting ability, finding such errors is difficult. As a result, you should begin developing troubleshooting skills early in your study of digital circuits.

FAULTS

A defect that prevents a circuit from operating properly is called a fault. Other defects are described as cosmetic. For example, a scratch on a printed circuit board that only hurts its appearance is a cosmetic defect. If that scratch opened a conductor, it would be termed a fault.

CLASSIFICATION OF FAULTS

Knowledge of the ways circuits can fail is an aid to troubleshooting. Digital circuits can fail in many ways. Only the common modes of failure will be described.

DESIGN FAULTS VERSUS CIRCUIT FAULTS

When power is first applied to a new circuit, it may not operate. Is the fault in the components and wiring of the circuit, or is it in the design?

DESIGN FAULTS

A design fault is a mistake in the logic diagram of a circuit. Even if a circuit is built exactly as shown with fault-free parts, it will not operate. The error was made by the engineer, designer, or drafter. Circuit changes are needed to correct design faults.

Design faults tend to be found in new circuits. Those that have been operating for some time are assumed to be free of design errors.

Do not assume that circuit diagrams in books and technical publications are free of errors. They often contain mistakes. When using such diagrams, refer to the backs of books and later issues of magazines to see if corrections have been listed. These are often found under the heading Errata.

CIRCUIT FAULTS

Wiring errors and faulty components are classified as circuit faults. When circuits that have been operating for some time fail, it is almost always a circuit fault. Such faults can be corrected by replacing parts or repairing open or shorted leads.

Faults in newly constructed circuits are more likely to be circuit faults than design faults. A search for wiring errors, broken leads, and faulty parts should come first. If this effort does not suggest the source of the problem, the circuit's design should be checked.

INTERMITTENT FAULTS VERSUS PERMANENT FAULTS

Almost everyone has experienced equipment that stops and starts working for no apparent reason. Such equipment contains intermittent faults. Equipment that fails and will not operate again without corrective action is said to contain permanent faults.

INTERMITTENT FAULTS

A basic rule of troubleshooting is: symptoms must be observable if a fault is to be isolated. In circuits that are working properly, but only at the present time, there is generally no way to tell what parts or connections are faulty.

There are two approaches to working on intermittent equipment. Repair personnel can wait for equipment to fail and then attempt to find the fault before it starts operating again. The other approach involves forcing failure. Tapping such equipment can cause failure. Also, chemical sprays (available in pressurized cans) are used. These chill suspected parts and force failure by thermal shock. See Fig. 3-1.

PERMANENT FAULTS

Permanent faults are unaffected by temperature or mechanical vibration. Circuits containing such faults will not function properly until the fault has been corrected by repair personnel.

The observability rule also applies to permanent faults. Only when symptoms are observable can a fault be isolated.

Fig. 3-2 demonstrates this rule. Assume that the circuit in Fig. 3-2 has been built in a laboratory. When placed in use, it appears to operate properly most of the time. Yet, every once in a while an error is made. Does the circuit contain a permanent fault or does it have an intermittent fault?

INPUTS A	INPUTS B	PREDICTED Y	MEAS. Y
0	0	0	0
0	1	1	0*
1	0	1	1
1	1	0	0

*ERROR

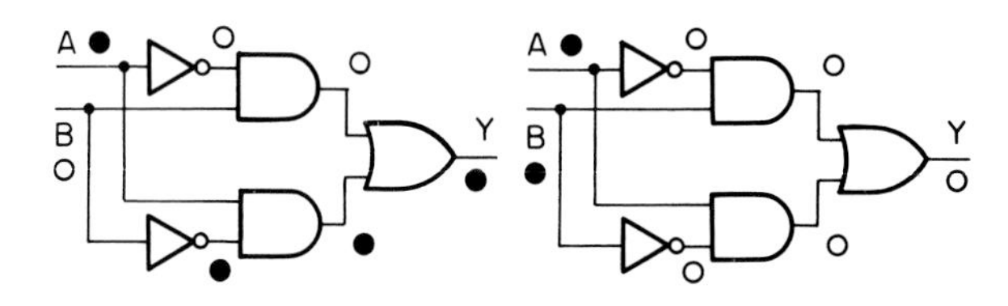

Fig. 3-2. To locate a fault, the circuit must be in its fault condition. In this case, the set of inputs A = 0, B = 1 must be applied.

Fig. 3-1. Sprays are often used to chill intermittent faults so they become permanent faults. (Mill Stephenson Chemical Co., Inc.)

Tests show that only when the set of input signals A = 0, B = 1 is applied will the incorrect output appear. See the truth table. This is a permanent fault. The error will appear every time this input set is applied.

To find this fault, the proper input set must be applied. The circuit must be in its fault condition if the bad part is to be found.

EXTERNAL FAULTS VERSUS INTERNAL FAULTS

Faults may be external (in wiring or sockets) or internal (inside ICs and other parts). Internal faults are common in older equipment. In newly constructed circuits, faults are often external. Wiring errors, ICs improperly inserted in sockets, and open and shorted leads are mistakes made during circuit construction. Such faults are common in student laboratories.

EXTERNAL FAULTS

Many external faults can be located through visual inspection. In student laboratories, it may be helpful to have one person wire a circuit and another check it for errors.

Taking a circuit apart and reassemblying it is an ineffective troubleshooting method. Faulty components and wires from the first circuit are likely to appear in the second attempt. In addition, new faults are likely to appear. Two errors are more than twice as difficult to find.

The likelihood of error increases with circuit size. Therefore, complex circuits should be assembled and tested section by section.

External faults are often repairable, Fig. 3-3. Experimental circuits are often assembled on solderless circuit boards (called strip sockets in this text). With strip sockets, wiring errors are easily corrected.

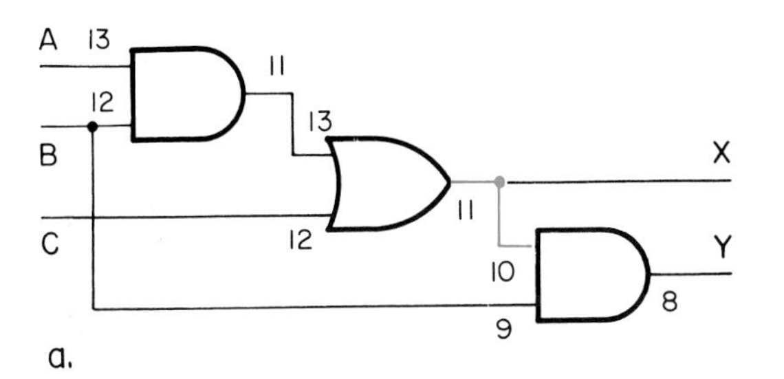

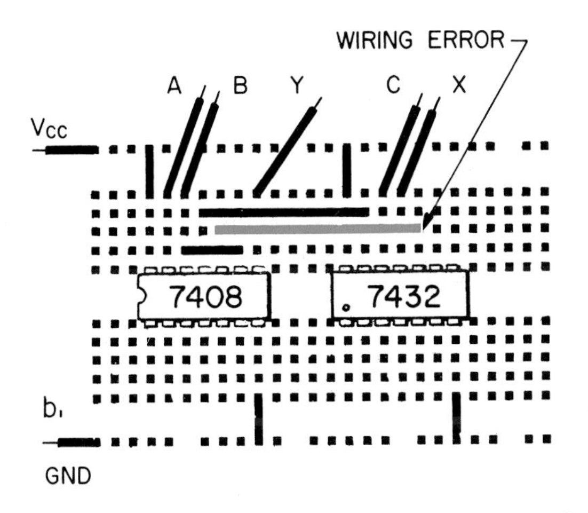

Fig. 3-3. Experimental circuits are often constructed on strip sockets, so external faults can be repaired. The wire from pin 10 on the 7408 (the AND) should go to pin 11 on the 7432 (the output of the OR).

Fig. 3-4 shows a common printed circuit problem, a *SOLDER BRIDGE.* A solder bridge is an unwanted extension of solder. During assembly, solder can short or bridge adjacent leads. Solder bridges can often be removed by reheating the connection. A knife can also be used to remove such faults.

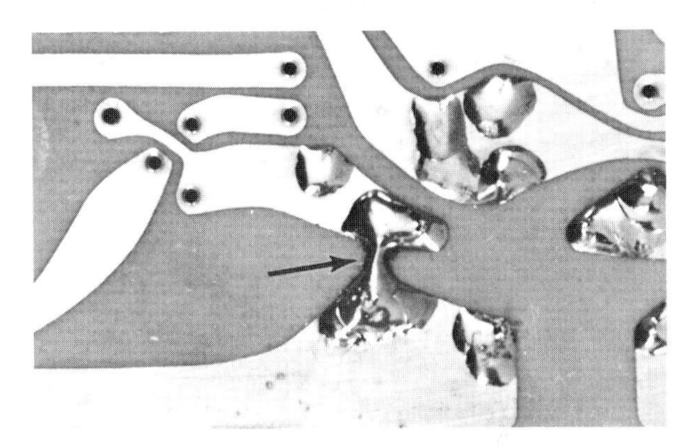

Fig. 3-4. Solder bridges are external faults.

INTERNAL FAULTS

Internal faults are usually not visible. Electrical measurements are usually needed to isolate parts with internal faults.

When internal faults cause overheating, smoke and surface texture changes may result. In such cases, visual inspections aid in detecting internal faults.

Touching components to detect overheating is poor practice, for they may be hot enough to cause severe burns. Pay close attention to the safety instructions given by your instructor.

Internal faults are seldom repairable. Faulty ICs and other parts are usually replaced.

PARAMETRIC FAULTS VERSUS LOGIC FAULTS

A faulty thermostat will be used to describe differences between logic and parametric faults. A broken thermostat that cannot signal for heat, no matter how cold a room becomes, contains a logic fault. A thermostat that turns the heat on and off but permits uncomfortable temperature swings contains a parametric fault. That is, devices with parametric faults work but do not work well.

PARAMETRIC FAULTS

A *PARAMETER* (pa-ram-e-ter) describes the action of a device or system. It is often a number. For example, the gain of an amplifier (ratio of input to output signals) is a parameter.

Just as amplifier gain can decrease with age, temperature, or line voltage, the outputs of IC logic elements can vary. When an element's output has decreased enough to hurt circuit actions, a parametric fault exists.

Symptoms of parametric faults usually appear in the output signals of logic elements. Part "a" of Fig. 3-5 shows signal voltages at the output of a TTL element. Voltages between 0.4 and 2.4 usually suggest parametric faults. Logic 1s at the output of a properly operating TTL element will be between 2.4 and 5 volts. Logic 0s will be between 0.4 and 0 volts.

Part "b" of Fig. 3-5 shows signal levels accepted by TTL inputs. Note the wider range. Voltages as high as 0.8 are accepted as 0s; those as low as 2 are accepted as 1s. This means that any properly operating TTL output can drive any properly operating TTL input. Signals between 0.8 and 2 volts are considered invalid.

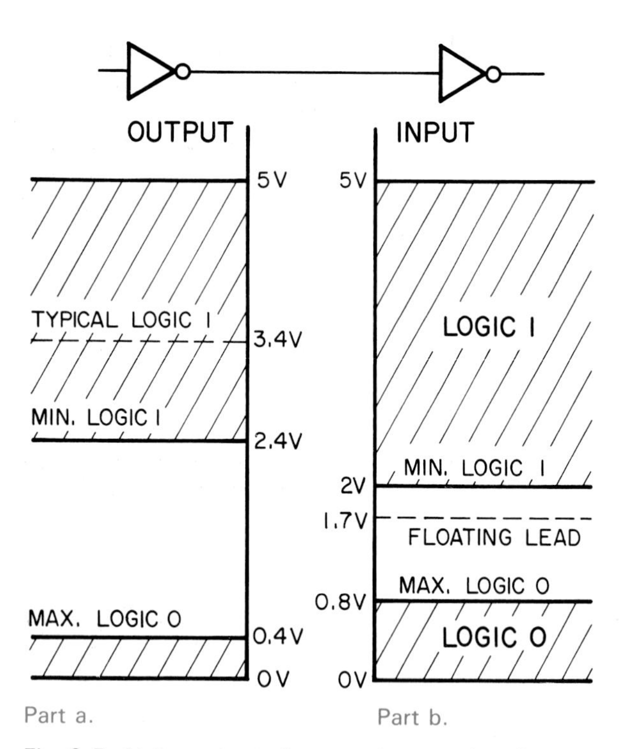

Fig. 3-5. At the output of a properly operating element, the signal voltage for a logic 1 will be between 2.4 volts and 5 volts. A logic 0 will be between 0.4 volts and 0 volts.

One voltage in the invalid region has special meaning. It represents a logic fault rather than a parametric fault. If about 1.7 volts appears between a TTL input and ground, that input is probably floating. See Fig. 3-6. Although this voltage is in the invalid region, TTL inputs treat it as a logic 1.

LOGIC FAULTS

When the output of a circuit or device is 0 when it should be 1, or 1 when it should be 0, a logic fault exists.

Logic faults usually result in logic signals that stay at one level no matter what input signals are applied. These are described as *STUCK-AT-1* (s-a-1) and *STUCK-AT-0* (s-a-0) faults. Most logic faults are either s-a-1 or s-a-0.

Five types of logic faults are common. These are:

- A short to ground.
- A short to the power supply.
- An open lead.
- Deep internal faults.
- Shorts between leads.

Short circuits to ground produce s-a-0 logic faults. See Fig. 3-7. Such short circuits may be internal or external.

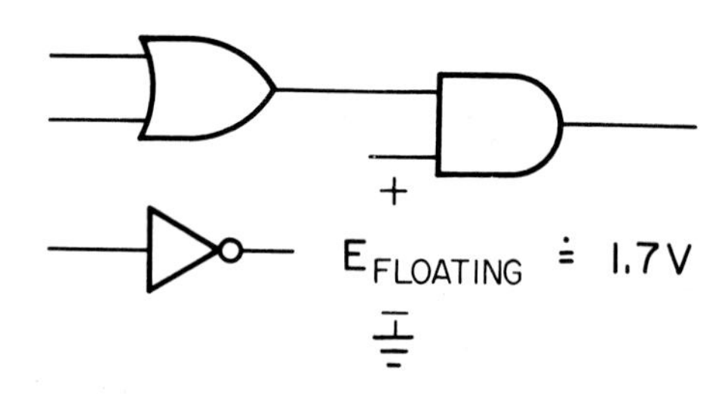

Fig. 3-6. The voltage between a floating TTL lead and ground is about 1.7 Volts.

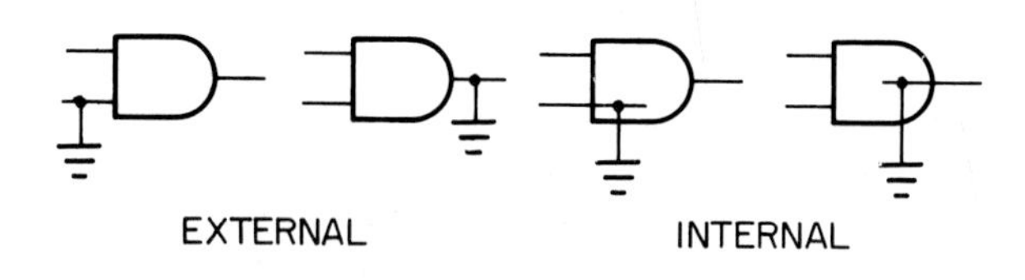

Fig. 3-7. Stuck-at-0 faults may result from internal or external shorts to ground.

A short to ground at an element's output results in an output that is fixed at 0. If the short to ground is at an element's input, the effect at the output depends on the type of element involved. If an input to an AND is s-a-0, its output will also be s-a-0. See part "a" of 3-8. If an input to an OR is grounded, its output responds to signals on its other input leads, part "b" of Fig. 3-8. However, signals on the shorted lead are lost.

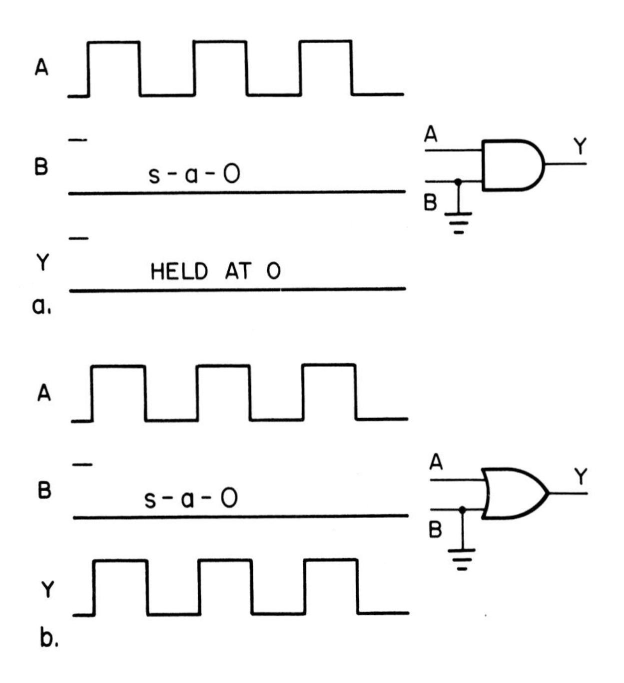

Fig. 3-8. The output of a logic element with an s-a-0 input depends on the type of element.

If a short to ground is suspected, voltage measurements may be helpful. In TTL circuits, valid 0s are usually between 0.07 and 0.4 volts. A voltage near or below 0.07 suggests a short to ground.

A short circuit to the power supply (to Vcc) results in an s-a-1 logic fault, Fig. 3-9. In TTL circuits, valid logic 1s are usually near 3.4 volts, but they can be anywhere between 2.4 and 5 volts. Voltages near +5 volts suggest a short to the supply.

Open leads result in floating inputs, Fig. 3-10. In TTL circuits, such inputs are treated as logic 1s, so s-a-1 logic faults result.

If a floating lead is suspected, voltage measurements can be helpful. A reading of about 1.7 volts between point A and ground suggests an open input. Note that this approach cannot be used to detect internal open inputs. Leads inside chips are not available for testing.

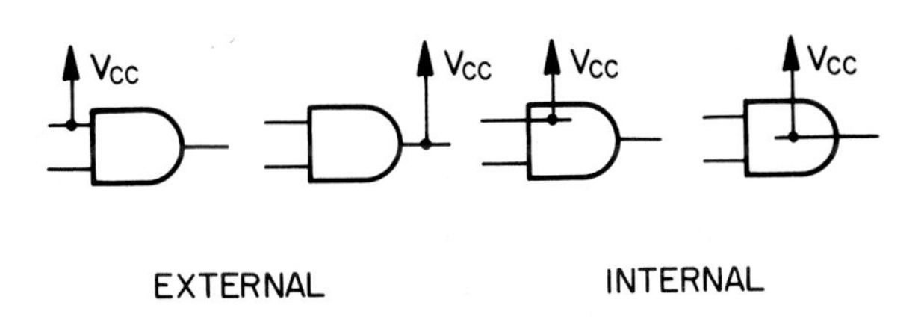

Fig. 3-9. Short circuits to the supply voltage result in the s-a-1 type of fault.

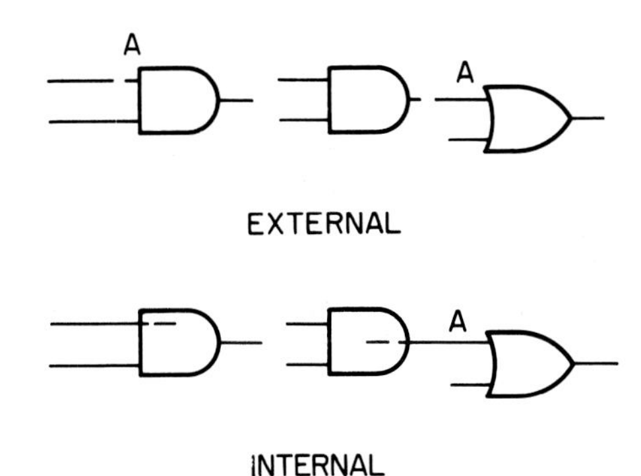

Fig. 3-10. In TTL circuits, open leads act like s-a-1 faults.

Deep internal faults like open and short circuits can occur deep within an IC. These usually result in outputs that are s-a-0 or s-a-1. See Fig. 3-11. Logic levels for such faults are usually within accepted voltage limits, but merely stay at one level no matter what inputs are applied.

Shorts between leads can occur. The lead spacing on IC logic elements is very small. As a result, shorts between leads are possible. See Fig. 3-12.

Unlike other logic faults, shorts between leads normally do not produce "stuck-at" faults. Rather, they appear to add logic elements to a circuit. These are called *WIRED-LOGIC* elements.

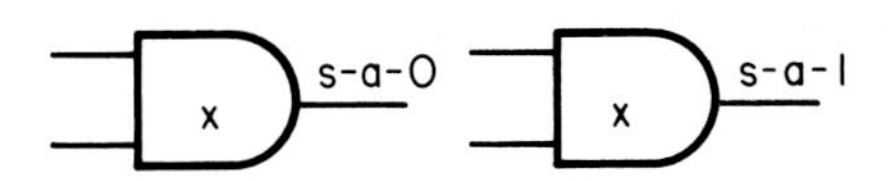

Fig. 3-11. Deep internal faults usually act like s-a-0 and s-a-1 faults at the element's output.

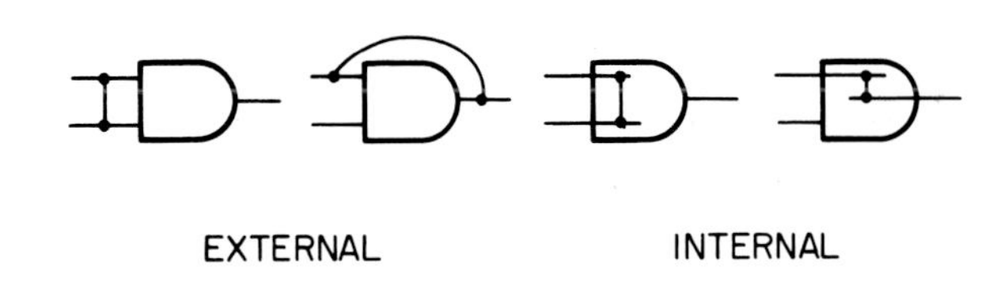

Fig. 3-12. Short circuits between signal leads are possible in logic circuits.

To review, most logic faults are "stuck-at" faults. If a detailed analysis of a fault is needed, voltmeter readings are helpful. Fig. 3-13 reviews voltages associated with certain faults. Voltage readings are

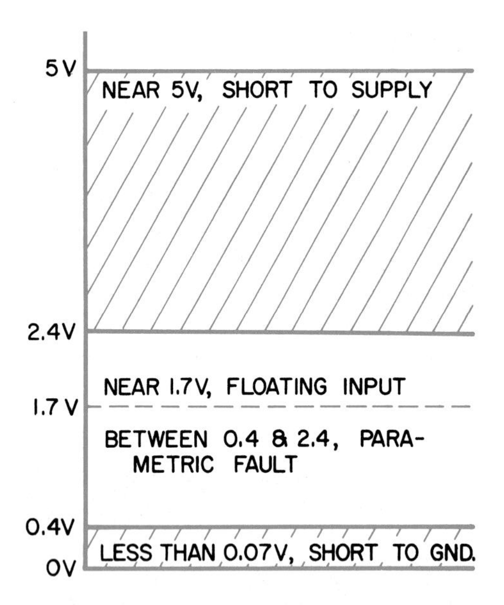

Fig. 3-13. Voltage measurements can be helpful when analyzing TTL faults.

normally not helpful in analyzing deep-internal faults and pin-to-pin faults.

STATIC FAULTS VERSUS DYNAMIC FAULTS

In general, static faults do not change with time. That is, such faults are present whether signal levels are changing or fixed. Dynamic faults depend on time.

Fig. 3-14 shows a dynamic fault. For inputs A and B, output Y should remain 0. However, a short *SPIKE* (a thin pulse) can be seen on Y. Such a spike is called a *GLITCH.* For some reason, the leading edge of the signal on A arrived too soon, or the trailing edge of the signal on B lasted a bit too long. This is called a *RACE CONDITION.* The result is a brief overlap of A and B.

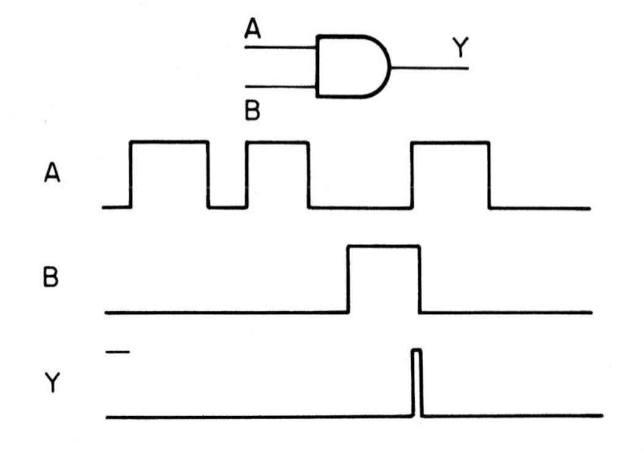

Fig. 3-14. Overlap of signals A and B causes glitch (thin pulse) at Y. A race condition can cause a glitch.

In time, glitches are short. Special equipment is needed to detect them. These unwanted signals cause errors in computers and digital circuits.

Race conditions are not circuit faults. They are design faults. Replacing the logic element in Fig. 3-14 would not correct this problem. The problem is in the design of the circuit.

TEST EQUIPMENT

Digital test instruments may be roughly divided into static and dynamic. Static instruments provide truth-table related information—information that is not time dependent. They measure signal voltages at one instant of time. Dynamic instruments are used when timing diagram information is needed. They graph signals as a function of time.

STATIC MEASUREMENTS

Instruments for making static measurements need not be complex. They are often handheld units. Three types of static instruments are common. These are:

1. A logic probe built by the user.
2. Commercial logic probes.
3. Clip-on indicators.

LOGIC PROBE BUILT BY THE USER

A logic probe is a device for detecting and indicating logic levels within a circuit. In its simplest form, it consists of an indicator lamp and a length of wire. The lamps available on digital trainers in student laboratories can be used as logic probes.

COMMERCIAL LOGIC PROBES

Fig. 3-15 shows a purchased logic probe. When the tip of this probe is placed on a circuit lead, it indicates the presence of a logic 1 or 0. Because indicator lamps are mounted on the probe, the user need not look away from the circuit under test.

Such probes often detect floating leads and invalid logic levels. Built-in memories permit the detection of short pulses and glitches.

CLIP-ON INDICATORS

Fig. 3-16 shows an indicator that clips directly to a DIP. It is a group of logic probes in one package. Signals at all pins are displayed at one time.

Circuits that automatically locate Vcc and ground are often available in clip-on indicators. When this is the case, external power is not needed, since the instrument takes power from the chip under test.

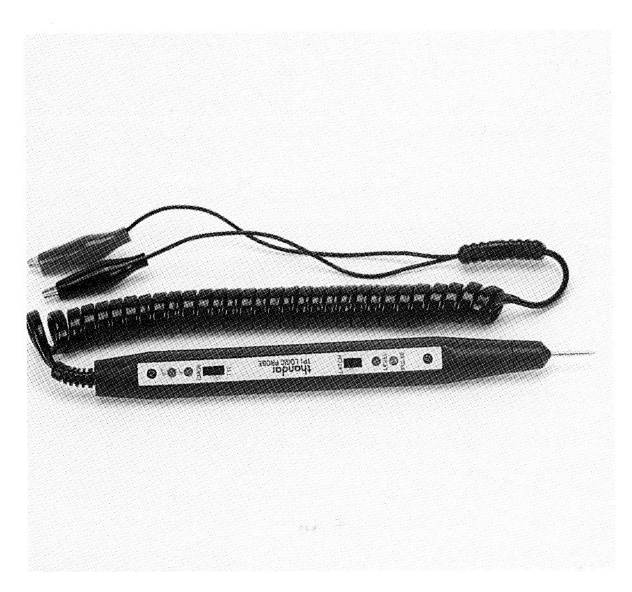

Fig. 3-15. Logic probes detect the presence of 1s and 0s in digital circuits. This probe does not need its own power supply. The clips are used to obtain power from the circuit under test. (Jameco Electronic Components)

DYNAMIC MEASUREMENTS

Dynamic measurements involve time. Changes in time are often displayed on a screen. As a result, most dynamic instruments have screens like television screens. Dynamic instruments display measurements in timing-diagram form. Two types of dynamic instruments are the oscilloscope and the logic analyzer.

Fig. 3-16. Current probes are used under certain conditions for troubleshooting. The device on the left indicates currents greater than 1 mA. The device on the right is a pulser. It injects signals into circuits to aid in troubleshooting. (Hewlett-Packard Company)

OSCILLOSCOPE

An oscilloscope, Fig. 3-17, is an instrument that uses a glowing spot on a screen to draw lines. The vertical position of the spot represents the voltage at the instrument's input. The spot moves horizontally with time. When attached to a digital circuit, the resulting pattern is a timing diagram.

Oscilloscopes are designed for viewing signals that repeat. This situation may not exist in digital circuits. As a result, oscilloscopes have limited use in digital troubleshooting.

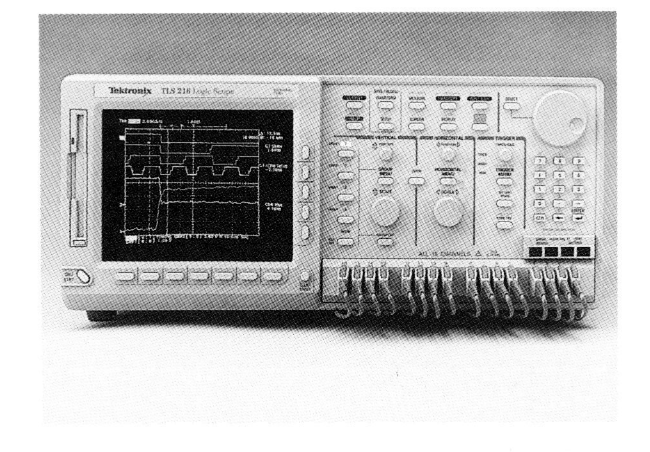

Fig. 3-17. Oscilloscopes display voltage as a function of time. (Tektronix, Inc. 1996©)

LOGIC ANALYZER

Logic analyzers are instruments like oscilloscopes designed for use on digital circuits. See Fig. 3-18.

Oscilloscopes are analog devices, so they display sizes of signal voltages. In digital circuits, however, the size of a signal is usually unimportant. It is the presence or absence of voltage that determines whether a signal is 1 or 0. Logic analyzers have circuits that test whether a signal is within the limits for a logic 0 or whether a signal is within the limits for a logic 1. An oscilloscope cannot do this.

Logic analyzers accept signals from many points in a circuit, convert these to patterns of 1s and 0s, and store them within the instrument. These patterns can then be displayed in timing-diagram form, part "a" of Fig. 3-19. Data can also be displayed in tabular form. See part "b" of Fig. 3-19.

Logic analyzers do not replace logic probes and oscilloscopes. Rather, they are useful tools for studying actions of complex digital circuits. Logic analyzers save time when doing repetitive jobs.

Fig. 3-18. This logic analyzer displays timing information in both timing diagram and tabular forms. (Tektronix, Inc. 1996©)

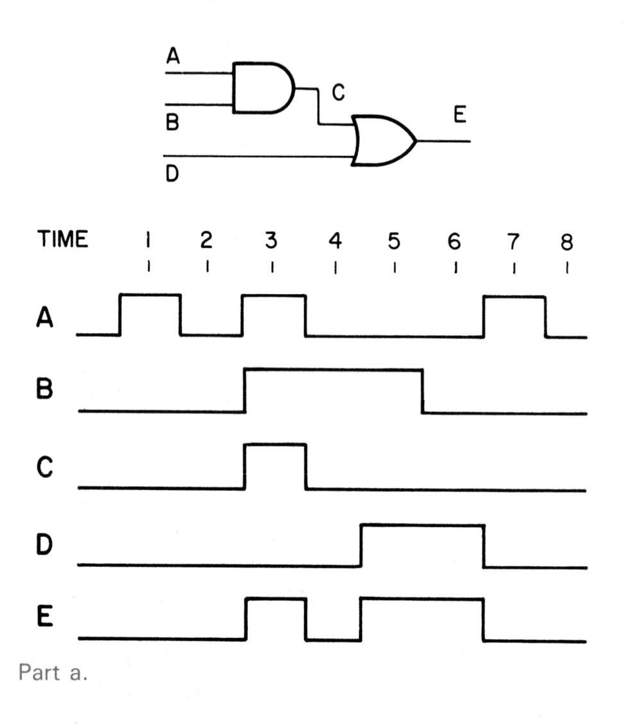

TIME	1	2	3	4	5	6	7	8
A	1	0	1	0	0	0	1	0
B	0	0	1	1	1	0	0	0
C	0	0	1	0	0	0	0	0
D	0	0	0	0	1	1	0	0
E	0	0	1	0	1	1	0	0

Part b.

Fig. 3-19. Timing diagrams and tabular display are available on most logic analyzers.

FAULT LOCALIZATION AND FAULT ISOLATION

Troubleshooting is best accomplished in two steps. First, a troubleshooter determines the approximate location of a fault. This is called fault localization. The person then searches the selected area to determine the faulty part or connection. This is fault isolation.

There are two basic rules associated with troubleshooting. They are:

1. Information on how the circuit is supposed to work must be available.
2. The circuit must be in the fault condition.

The first rule states that you must be able to predict the correct signals at critical points in a faulty circuit. Without this information, you do not know if measured signals are correct or incorrect. In circuits containing AND, OR, and NOT elements, truth-table analysis (introduced in Chapter 2) can be used to predict signals within a circuit.

The second rule restates the observability concept. If a circuit is not in a fault condition, signals will be correct. The fault cannot be detected.

Each failure tends to be unique, so it is not possible to describe all faults. However, the following examples suggest troubleshooting methods.

Example: Return to the circuit in Fig. 3-2. You may remember that this circuit produced outputs at the correct values for Y for all input signal sets except A = 0, B = 1. That fault condition has been reproduced in part "a" of Fig. 3-20. Solid dots represent measured 1s; open dots represent measured 0s.

A fault localization process is done first. Based on AND, OR, and NOT truth tables, part "b" of Fig. 3-20 shows predicted signals at all points in this circuit. Comparing predicted and measured signals, the output of the upper AND is incorrect. It appears that the lead between the AND and the OR is s-a-0. The fault has been localized.

Fault isolation is carried out next. The problem is one of pinpointing the fault. Is the lead between the two elements shorted to ground? Is the output of the AND s-a-0? Is the input to the OR s-a-0?

Assume this circuit is experimental and has been assembled on a strip socket. Fault isolation begins with a visual inspection. The lead between the AND and OR should receive the most attention. Let us assume that no wiring error or short to ground is found. The fault is probably in the AND or OR.

To determine which element contains the fault, the lead between the AND and OR is opened. See part "c" of Fig. 3-20. The dots indicate measurements made

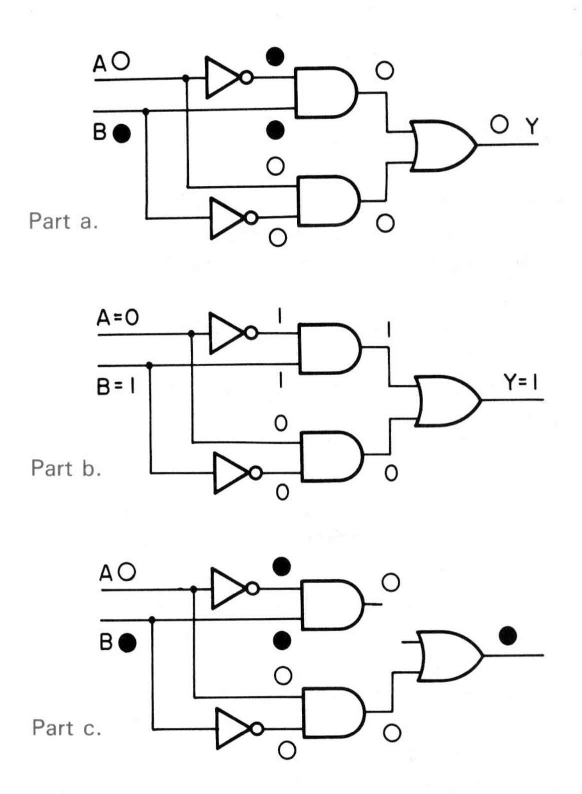

Fig. 3-20. Troubleshooting is based on fault localization and isolation.

with this lead open. Note that the output of the AND is still 0. Its output may be shorted to ground or it may contain a deep internal fault.

As an aside, the output of the OR changed to a 1 when its input lead was opened. This is as it should be. TTL logic treats open inputs as 1s, so this OR outputs a 1.

Replacing the AND chip and reconnecting the lead should correct the situation. To be sure the fault has indeed been found, the circuit would be retested.

Example: Fig. 3-21 shows measurements made on a circuit for two input signal sets. In the first drawing, measured and predicted signals agree. For the second input signal set, they do not agree. The circuit contains a fault.

Fault localization is the first step. The measured and predicted signals in part "b" of Fig. 3-21 suggest that the fault is near the output of the AND.

Fault isolation is next. The output of the AND and input of the last OR are neither s-a-0 nor s-a-1, since the measured signal on this lead changed when the input signal set changed. The fault must be at the input of the AND. But note that both inputs to this AND changed when the input set changed. The fault must be inside the AND.

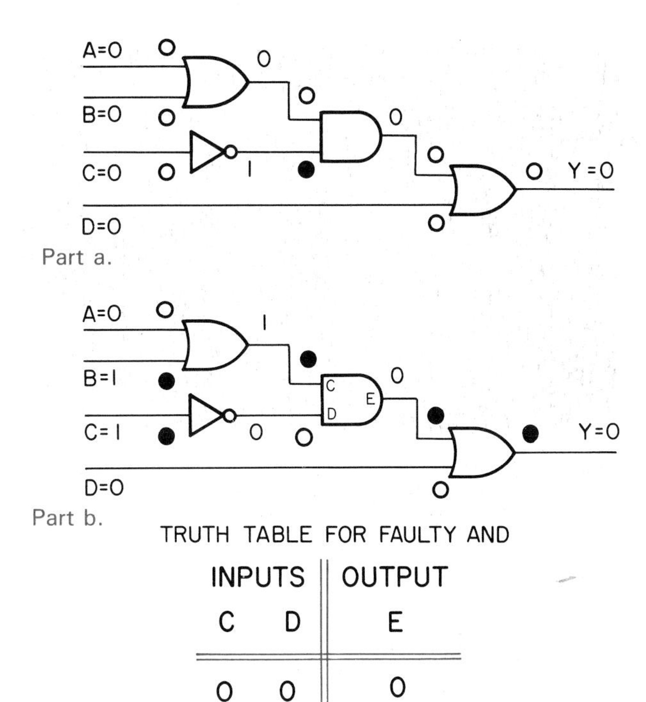

TRUTH TABLE FOR FAULTY AND

INPUTS C	INPUTS D	OUTPUT E
0	0	0
0	1	0
1	0	1*
1	1	1

*ERROR

Part c.

Fig. 3-21. This fault produces no external "stuck-at" signals. Analysis suggests that lead D of the AND is internally s-a-1.

What sort of fault would result in the observed error? The first set of inputs resulted in the AND outputting the proper number. That is, 0 AND 1 = 0. When the second input set was applied, the AND output an incorrect number. That is, 1 AND 0 = 1. Note that this 1 would be correct if the lower input to this AND was s-a-1. Although a 0 was applied, the AND saw a 1. Input D could be open on the inside of the DIP.

Replacing the AND should correct the problem. After it is replaced, the circuit should be retested to be sure the problem has been corrected.

Each fault has its own characteristics, so troubleshooting is a skill. With experience, patterns become apparent, and the process becomes easier.

Example: Part "a" of Fig. 3-22 shows measured and predicted signals at all points in a circuit. Note that it has two outputs and that they are both outputting incorrect numbers.

Fault localization is simple. The output of the OR

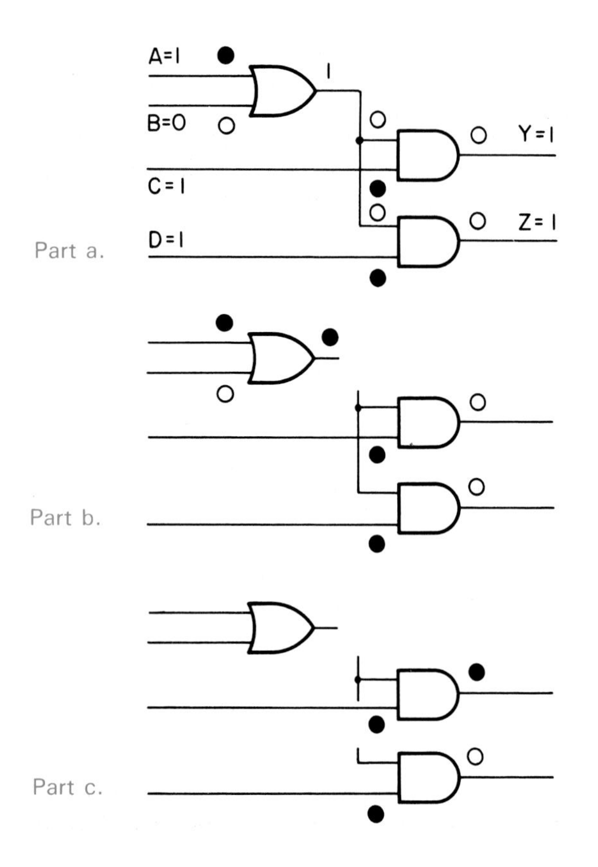

Fig. 3-22. In the circuit shown, an input lead of the lower AND element is s-a-0.

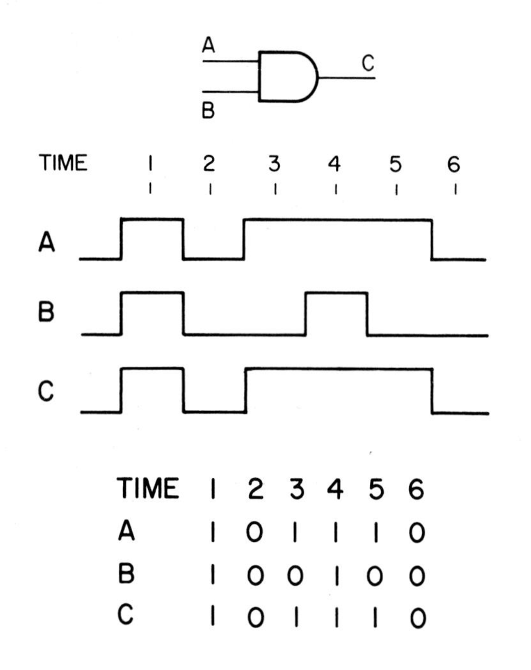

TIME	1	2	3	4	5	6
A	1	0	1	1	1	0
B	1	0	0	1	0	0
C	1	0	1	1	1	0

Fig. 3-23. Logic analyzers can be used for fault detection.

is incorrect, so the fault is easily localized.

Fault isolation begins. The fault appears to involve the lead between the OR and the two ANDs. A visual inspection of this lead is the first step. Assume no problem is found during this inspection.

Part "b" of Fig. 3-22 shows measurements made with the output lead from the OR open. The output of the OR has gone to 1, so this element appears to be operating properly. The problem must be in one of the AND inputs.

In part "c" of Fig. 3-22, one of the AND inputs has been opened. Note the results. The upper AND is outputting a 1, while the lower is still outputting a 0. The input of the lower AND is probably s-a-0 (shorted to ground).

Replacing the AND chip should correct the problem. Again, the leads that were disconnected must be carefully replaced and the circuit must be retested to be sure the problem has been corrected.

TESTING FOR DYNAMIC FAULTS

Of the fault conditions requiring dynamic measurements, one has already been mentioned. Race conditions (glitches and spikes) may require dynamic instruments.

Dynamic methods may also be required when the flow of input signals cannot be stopped. In some computers and digital circuits, input signals are not directly controllable by users. Troubleshooting must be based on signals that are generated by the equipment.

Fig. 3-23 shows an AND that is deep within a computer. A logic analyzer supplied the timing diagram and tabular representation of the data. Based on the AND truth table, this element produced an output with the proper numbers at times t1, t2, t4, and t6. Errors appear at times t3 and t5.

What fault would produce the observed error? If input B were s-a-1 internally, the output would be as shown. Replacing this chip would probably correct the fault.

SUMMARY

An ability to troubleshoot is an asset to almost every worker in digital electronics. Skills develop with experience.

Digital faults can be classified according to:

- Design versus circuit faults.
- Intermittent versus permanent faults.
- External versus internal faults.
- Parametric versus logic faults.
- Static versus dynamic faults.

A logic probe is a static test instrument. A logic analyzer is dynamic in nature.

Truth-table circuit analysis can be used as a basis for troubleshooting. The process is divided into fault localization and fault isolation.

TEST YOUR KNOWLEDGE

1. A (an) __________ fault is a temporary fault.
2. A "glitch" is a __________ (static, dynamic) fault.
3. A circuit must be in the __________ (fault, no-fault) condition for testing to proceed.
4. A __________(cold, hot) spray can force a circuit fault to appear.
5. A __________ __________ is an unwanted solder connection.
6. What do "s-a-1" and "s-a-0" mean?
7. List three of the five common types for pairs of logic faults.
8. A floating input to a TTL logic element acts as a logic __________ (1 or 0).
9. A __________ __________ is an instrument like an oscilloscope designed for use on digital circuits.
10. Troubleshooting is best accomplished in two steps. These are fault __________ and fault __________.

STUDY PROBLEMS

1. The wire at A in Fig. P3-1 was nicked when it was stripped, and it broke when inserted into the strip socket. From time to time, the broken ends of this wire make contact, so the circuit jumps back and forth between working properly and a fault condition.
   Classify this fault by selecting one item from each of the following pairs:
   1a. Design. 1b. Circuit.
   2a. Intermittent. 2b. Permanent.
   3a. External. 3b. Internal.
   4a. Parametric. 4b. Logic.
   5a. Static. 5b. Dynamic.

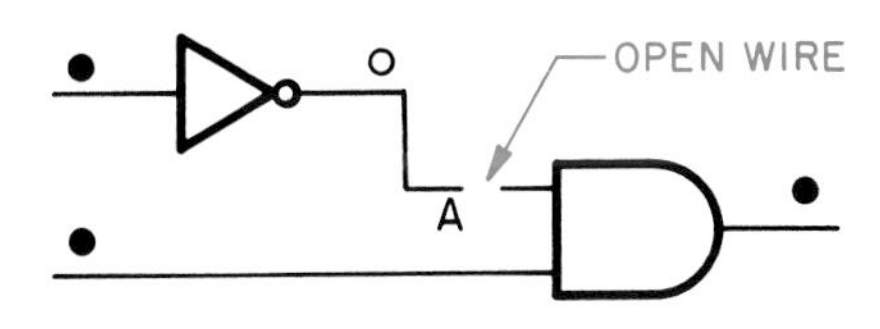

Fig. P3-1. An open wire can be classified by one term from each of the five pairs of terms describing a fault.

2. During receiving tests (the testing of purchased devices), the output voltages indicated in Fig. P3-2 were measured. Note the output of 0.9 volts. This was a fixed voltage that did not change with temperature, etc. Classify this fault by selecting one item from each of the pairs of faults in Problem 1.

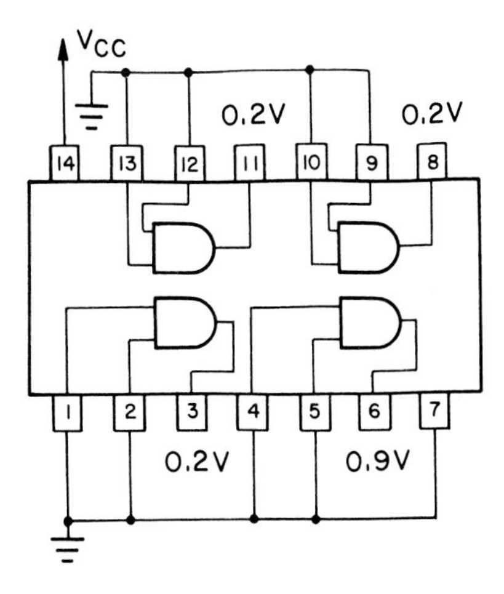

Fig. P3-2. Notice the 0.9 V output of the lower right AND. This fault can be classified by five pairs of terms.

3. The device at the right of the circuit in Fig. P3-3 is called a toggle flip-flop. In this circuit, its output should go to 1 when the output of the OR (at T) goes from 1 to 0 at t2. However, it goes to 1 at time t1. It appears that a race problem exists at t1. Using the list from Problem 1, classify this fault. Omit item 3 (external/internal).

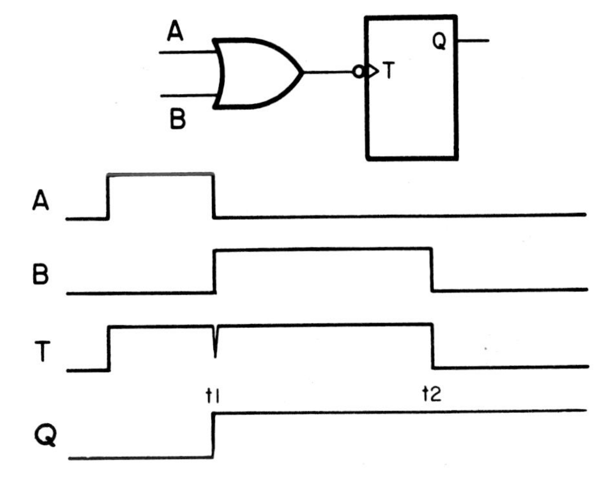

Fig. P3-3. A race condition exists at t1. Classify the fault. Omit the third pair of terms (external/internal).

4. The output of the AND in Fig. P3-4 should be 0 at all times. However, a glitch appears at its output. At what time does the race condition that produces this glitch most likely occur?

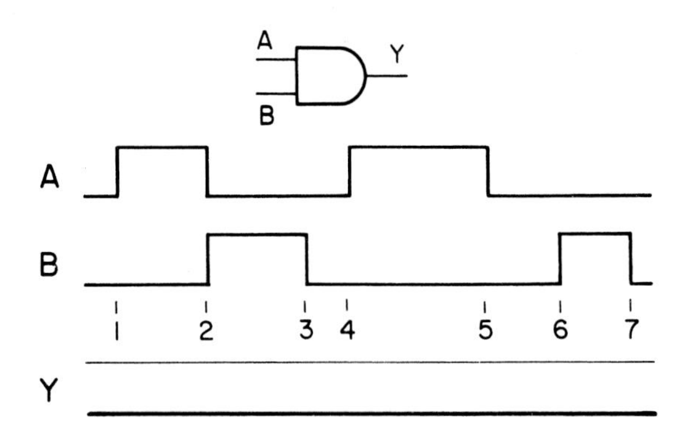

Fig. P3-4. The output of the AND should be 0 at all times. Determine the likely time when a race condition occurs.

5. The output of the OR in Fig. P3-5 should be 1 at all times. However, a negative-going glitch occurs. At what time does the race condition that produces this glitch most likely occur?
6. Using voltages from the graph in Fig. P3-6 or from the text, indicate the voltages associated with each of the following circuit conditions:
   1. Most likely output voltage of a properly operating logic element that is outputting a logic 1.
   2. Floating input lead.
   3. Short to Vcc.
   4. Short to ground.
   5. Parametric fault.

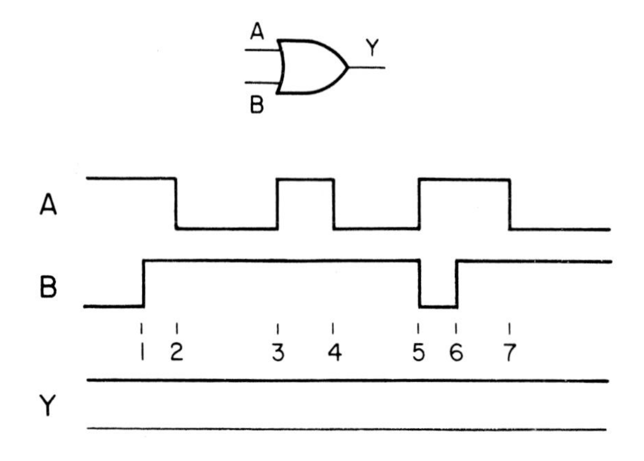

Fig. P3-5. The output of the OR should be 1 at all times. When will a race condition most likely occur?

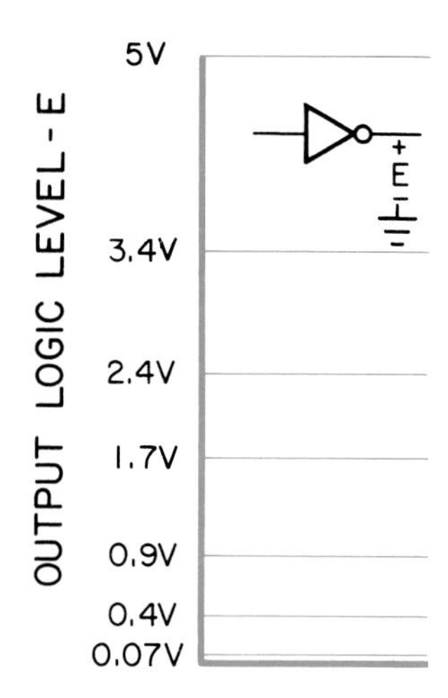

Fig. P3-6. Note the common voltages which occur in TTL circuits. Check with the text material to clarify their meanings.

7. For the inputs shown in Fig. P3-7, the output of the circuits should be 1. Which circuit contains a design fault?

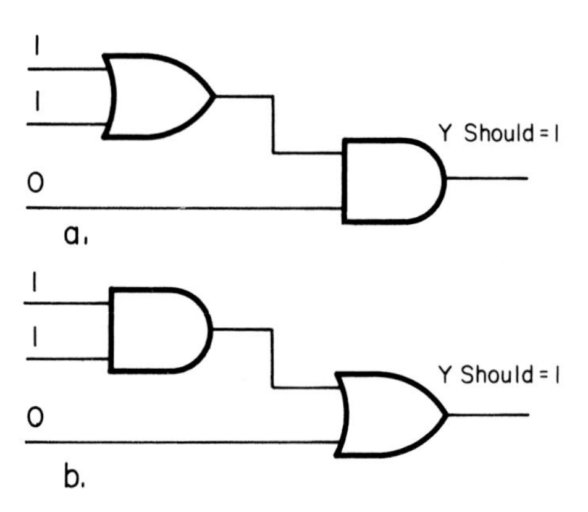

Fig. P3-7. For the inputs shown, the outputs should be 1. Which circuit has a design fault?

8. The output of the circuit in Fig. P3-8 should be 1. Localize the fault by determining which element is outputting an incorrect logic level.
9. Repeat Problem 8 for the circuit shown in Fig. P3-9.
10. For the inputs shown in Fig. P3-10, is the circuit working properly? If not, which element has an incorrect output?
11. The circuit in Fig. P3-11 is faulty. Which input signal set should be applied during the test to determine the source of the problem?

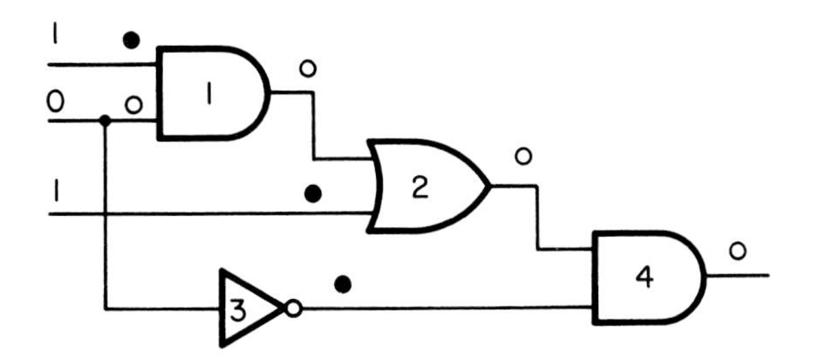

Fig. P3-8. The output should be 1. The output of which element is wrong?

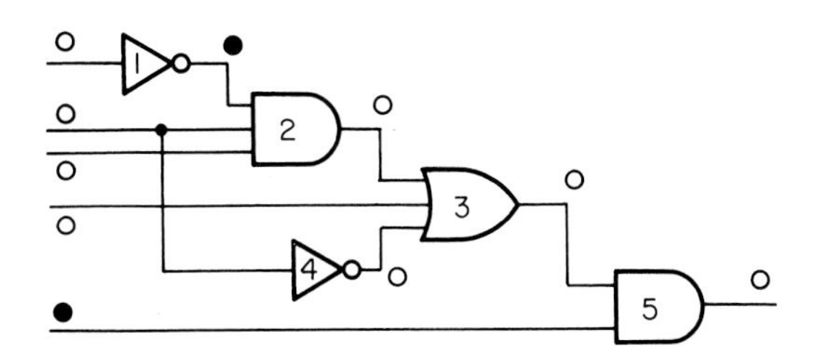

Fig. P3-9. The output should be 1. The output of which element is wrong?

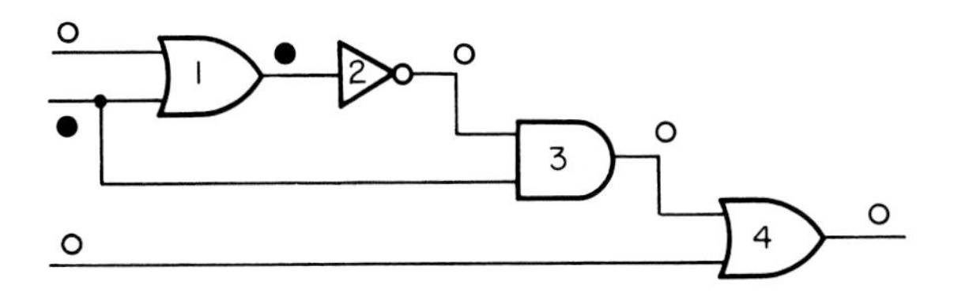

Fig. P3-10. For the inputs shown, is the circuit working properly? If not, which element has an incorrect output?

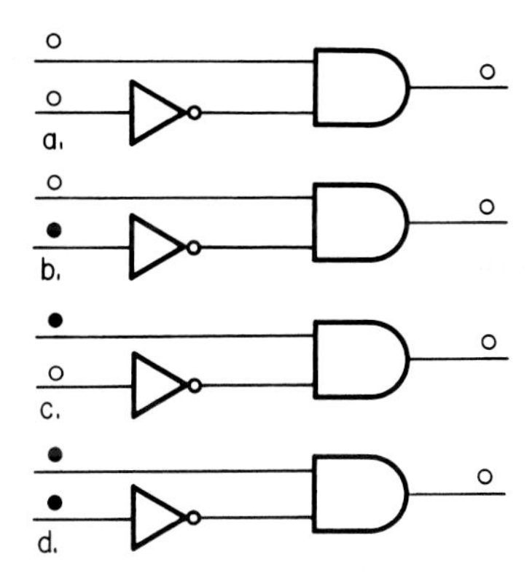

Fig. P3-11. A faulty circuit is shown. Which input signal set (a, b, c, or d) should be applied in a test to find the fault?

12. The upper circuit in Fig. P3-12 shows measurements made to localize the fault in the circuit. The output of the AND is incorrect. To isolate the fault, the lead between the AND and OR was opened. Based on the measurements in the lower circuit, which of the elements is at fault, the AND element or the OR element?

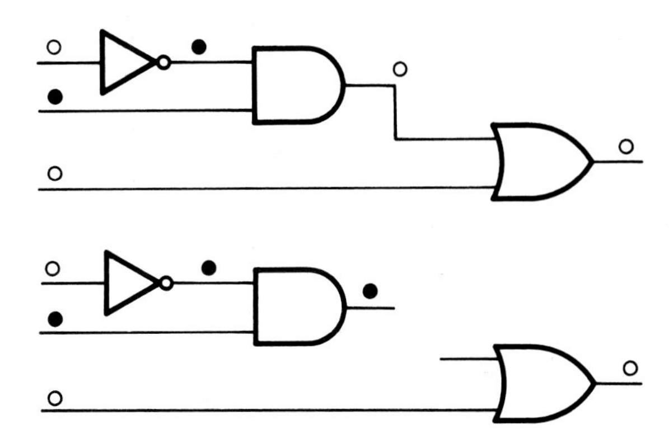

Fig. P3-12. The output of the AND is incorrect. Which element is at fault, the AND or the OR?

13. Repeat Problem 12 for the circuit in Fig. P3-13. Should the NOT or the AND be replaced?
14. Based on the inputs shown in the timing diagram, Fig. P3-14, draw the expected output (Y "expected") at Ye. Based on the measured output (Y "measured") shown at Ym, which of the following most likely describes the fault in the OR?
    1. Input A is s-a-1.
    2. Input A is s-a-0.
    3. Input B is s-a-1.
    4. Input B is s-a-0.
    5. The element is not faulty.

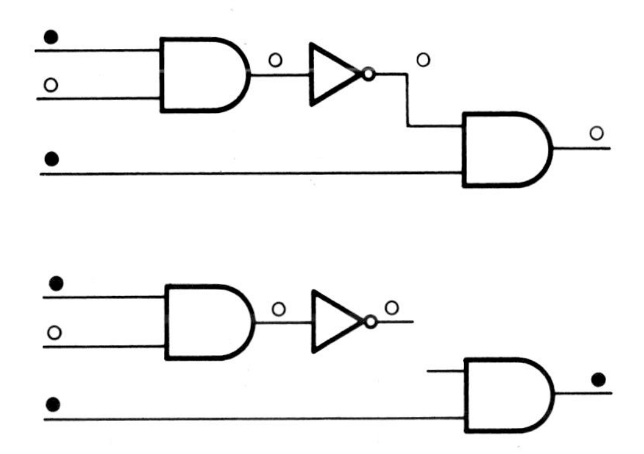

Fig. P3-13. The output of the NOT is incorrect. Which element should be replaced, the NOT or the AND?

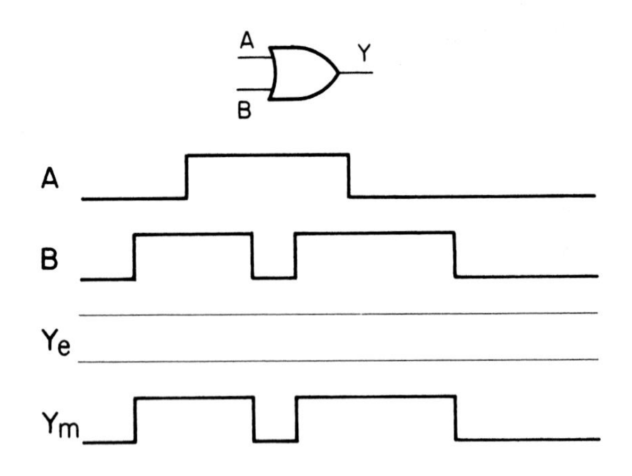

Fig. P3-14. Draw the expected output at Ye (Y "expected"). Based on the measured output at Ym (Y "measured"), find the fault at one of the inputs. Is it s-a-0 or s-a-1?

15. Repeat Problem 14 for the AND in Fig. P3-15.
16. Trace the timing diagram in Fig. P3-16 onto a sheet of paper. Based on the input signals shown, predict the signals at points B, D, and F.
17. Copy the table in Fig. P3-17 onto a sheet of paper. Using your solution to Problem 16 as a data source, complete the table.

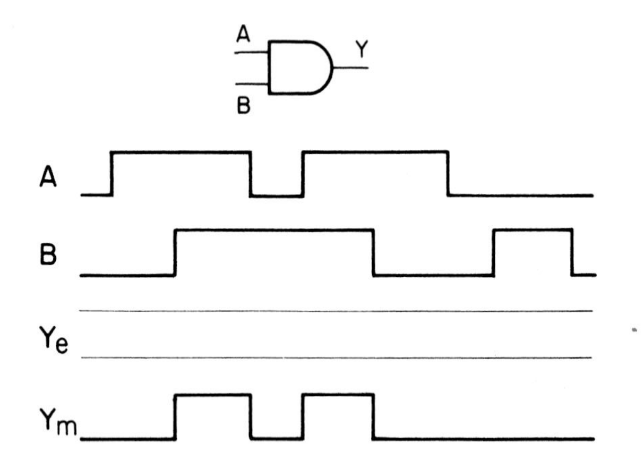

Fig. P3-15. Draw the expected output at Ye. Based on Ym, find the fault at one of the inputs. Is it s-a-0 or s-a-1?

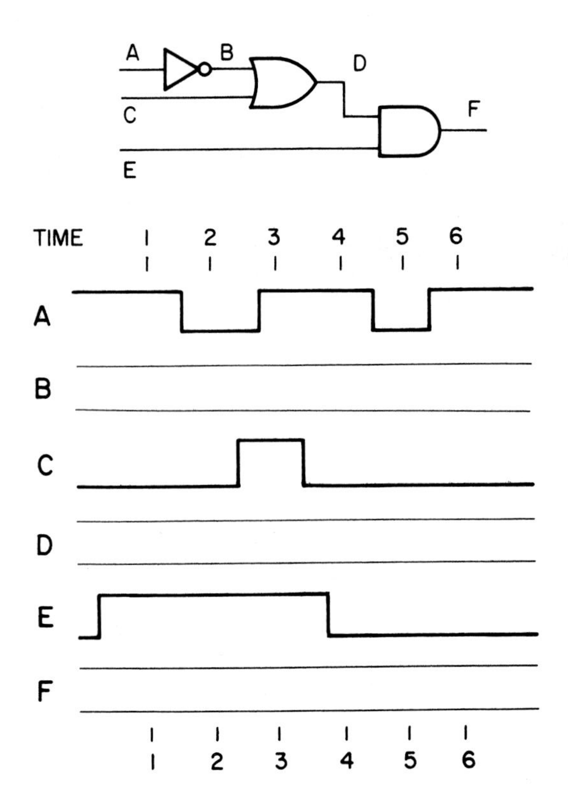

Fig. P3-16. Trace the timing diagram onto a sheet of paper. Predict the signals at points B, D, and F.

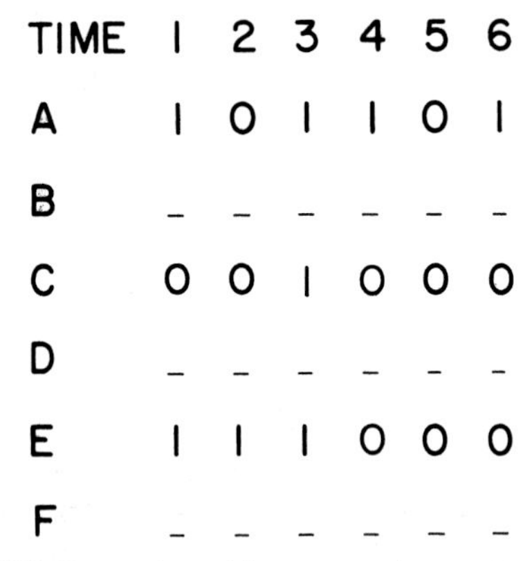

TIME	1	2	3	4	5	6
A	1	0	1	1	0	1
B	_	_	_	_	_	_
C	0	0	1	0	0	0
D	_	_	_	_	_	_
E	1	1	1	0	0	0
F	_	_	_	_	_	_

Fig. P3-17. Copy the table onto a sheet of paper. Complete the table, using your solution to Problem 16 as a data source.

Chapter 4

Combinational Logic—AND, OR, NOT

LEARNING OBJECTIVES

After studying this chapter and completing lab assignments and study problems, you will be able to:

- Construct truth tables for combinational logic circuits containing AND, OR, and NOT elements based on their logic diagrams.
- Construct truth tables for such circuits based on their Boolean expressions.
- Write Boolean expressions for combinational logic circuits composed of AND, OR, and NOT elements based on their logic diagrams.
- Draw logic diagrams for such circuits based on their Boolean expressions.
- Write Boolean expressions for combinational logic circuits composed of AND, OR, and NOT elements based on their truth tables using sum-of-products and product-of-sums methods.

The circuit in Fig. 4-1 is typical of those studied up to this point. It is called a combinational logic circuit. Its output is a combination of its input signals. When its input signals change, its output changes in a predictable way.

Logic diagrams, truth tables, and Boolean expressions are used to describe the actions of combinational logic circuits. Given one description, the others can be obtained. Methods of transforming a given description into the other two forms will be covered in this chapter. See Fig. 4-2. There are five types of transformations. These are as follows:

- Logic diagrams to truth tables.
- Boolean expressions to truth tables.
- Logic diagrams to Boolean expressions.
- Boolean expressions to logic diagrams.
- Truth tables to Boolean expressions.

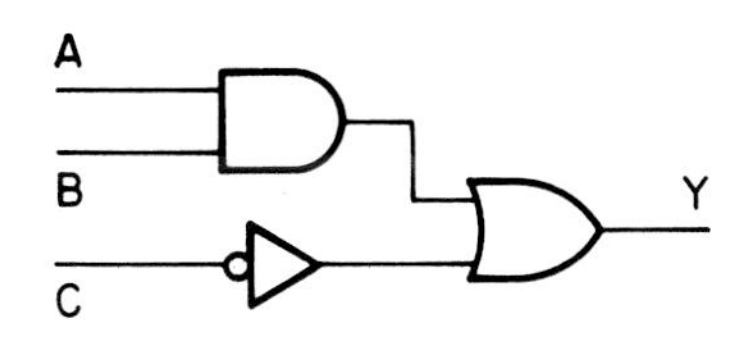

Fig. 4-1. Circuits such as this are called combinational logic circuits.

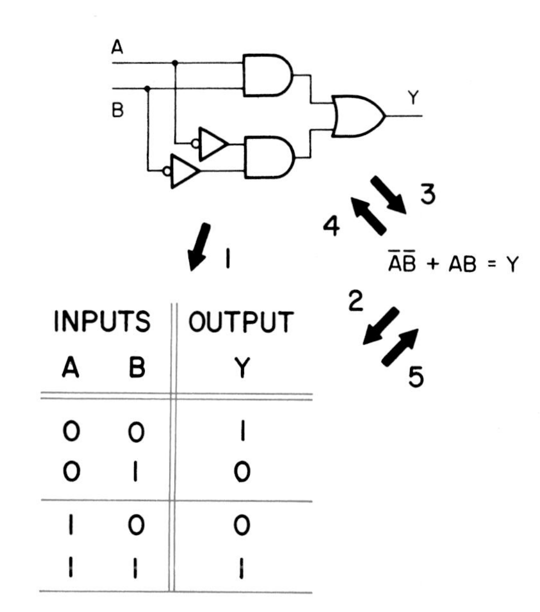

INPUTS		OUTPUT
A	B	Y
0	0	1
0	1	0
1	0	0
1	1	1

Fig. 4-2. Given one description of a combinational logic circuit, the other two can be obtained.

LOGIC DIAGRAMS TO TRUTH TABLES

The first transformation of the five types is the transformation from logic diagrams to truth tables. A logic diagram or truth table can be used to describe the action of a combinational logic circuit. That is, the two descriptions in Fig. 4-3 contain the same information. This suggests that, given a circuit's logic diagram, a person should be able to construct the truth table for the circuit.

A process of accomplishing this first transformation is shown in Fig. 4-4. For each input signal set, a truth table circuit analysis was used to determine the output Y. Copying the input signal sets and corresponding Ys into the table completes the circuit's truth table.

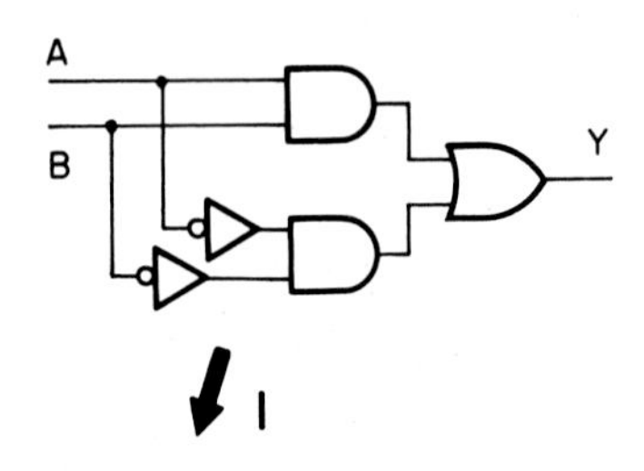

1

INPUTS A	INPUTS B	OUTPUT Y
0	0	
0	1	
1	0	
1	1	

Fig. 4-3. A circuit's truth table can be constructed from its logic diagram.

BOOLEAN EXPRESSIONS TO TRUTH TABLES

Boolean algebra expressions can be used to describe the actions of combinational logic circuits. As a result, it should be possible to construct truth tables based on Boolean expressions. See Fig. 4-5.

The expression (A + B) C = Y will be used to demonstrate this transformation. See Fig. 4-6. To evaluate expressions, three rules must be introduced. These are as follows:

1. Quantities within brackets are evaluated first.
2. Quantities under NOT signs are evaluated after those within brackets.
3. After brackets and NOTs, ANDs are evaluated.

Based on the first rule, A must be ORed with B before C is considered. BC does not exist by itself, but only with AC.

To determine the value of Y for the first input signal set, A = 1, B = 0, and C = 1 are substituted. See

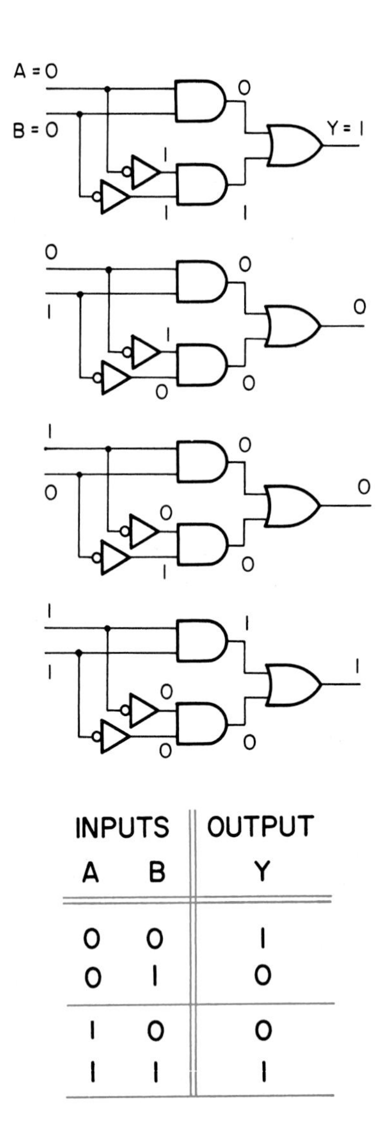

INPUTS A	INPUTS B	OUTPUT Y
0	0	1
0	1	0
1	0	0
1	1	1

Fig. 4-4. The transformation from logic diagrams to truth tables can be accomplished by analyzing the circuit's action for each input signal set.

$\overline{A}\overline{B} + AB = Y$

2

INPUTS A	INPUTS B	OUTPUT Y
0	0	
0	1	
1	0	
1	1	

Fig. 4-5. A circuit's Boolean expression can be used to construct a truth table.

$(A + B)C = Y$

	INPUTS A B C	EVALUATION $(A + B)C = Y$	OUTPUT Y
Part a.	1 0 1	$(1 + 0)1 = (1)1 =$	1
Part b.	1 1 0	$(1 + 1)0 = (1)0 =$	0
Part c.	0 0 1	$(0 + 0)1 = (0)1 =$	0
		etc.	

Fig. 4-6. The portion of the expression within the parentheses or brackets must be evaluated first.

part "a" of Fig. 4-6. Based on the first rule, the 1 and 0 within the brackets are ORed first. The result of that operation is then ANDed with the 1 representing C. For this input set, Y = 1.

At b, another input set is considered. Evaluate Y for A = 1, B = 1, and C = 0. If you find Y to equal 0, you are correct.

Another input signal set is shown at c. Again evaluate the expression. If you find Y to equal 0, you are correct. The process would be continued until all possible input sets had been evaluated.

To evaluate the expression in Fig. 4-7, the second rule is needed (quantities under NOT signs are evaluated after those within brackets). In the expression in Fig. 4-7, A and B must be ORed then NOTed before C is ORed with the results. For the input signal set at a, A = 1, B = 0, and C = 0, the result of ORing A and B is 1. When NOTed, this 1 becomes a 0. And when ORed with the value of C, output Y is found to be 0.

The input set at b results in Y = 1. Look through the mathematics of this evaluation to convince yourself that this is the correct result.

$\overline{A + B} + C = Y$

	INPUTS A B C	EVALUATION $\overline{A + B} + C = Y$	OUTPUT Y
Part a.	1 0 0	$\overline{1 + 0} + 0 = \overline{1} + 0 = 0 + 0 =$	0
Part b.	0 0 0	$\overline{0 + 0} + 0 = \overline{0} + 0 = 1 + 0 =$	1
Part c.	0 1 1	$\overline{0 + 1} + 1 = \overline{1} + 1 = 0 + 1 =$	1
		etc.	

Fig. 4-7. The quantity under the NOT must be evaluated first.

For the third input set shown, A = 0, B = 1, and C = 1, determine the value of Y. If you find it to be 1, you are correct.

Fig. 4-8 shows a NOT under a NOT. The deepest NOT, the one over A + B, must be evaluated first. That is, A must be ORed with B and the result NOTed before the ORing with C can be considered.

Fig. 4-9 suggests the third rule (after brackets and NOTs, ANDs are evaluated). The quantity within the bracket must be evaluated first. To do this, however, $\overline{A}$ must be evaluated. See part "a" of Fig. 4-9. When a NOT has been determined, it can be ORed with B to determine the value of the quantity within the brackets.

$\overline{\overline{A + B} + C} = Y$

EVALUATE FOR: A = 0, B = 1, C = 0

$\overline{\overline{0 + 1} + 0} = Y$

$\overline{\overline{1} + 0} = Y$

$\overline{0 + 0} = \overline{0} = 1$

Fig. 4-8. The deepest NOT must be evaluated first.

$(\overline{A} + B)C + D = Y$

Part a. EVALUATE FOR: A = 0, B = 0, C = 0, D = 1

$(\overline{0} + 0)0 + 1 = Y$

$(1 + 0)0 + 1 = Y$

$1 \cdot 0 + 1 = Y$

$0 + 1 = 1$

Part b. EVALUATE FOR: A = 0, B = 0, C = 1, D = 0

Fig. 4-9. ANDs are evaluated after brackets and NOTs.

Should C be ORed with D or ANDed with the bracketed quantity? Rule 3 indicates that the ANDing must be accomplished first.

Determine the value of Y for the input signal set at b. If you find that Y = 1, you are correct.

ORs are evaluated last. This does not imply that they are unimportant. For example, D plays an important role in the following expression:

$$\bar{A}BC + D = Y$$

If D = 1, the value of Y will be 1 no matter what the values of A, B, and C are. A 1 ORed with anything results in a 1.

COMPLEX EVALUATION

Fig. 4-10 is an example of the evaluation of a complex expression. The indicated signal set will be used to demonstrate the process. The steps are:

$$\bar{A} + B(\overline{AB} + C) = Y$$

EVALUATE FOR: A = 0
B = 0
C = 0

Part a. $\bar{0} + 0(\overline{0 \cdot 0} + 0) = Y$

Part b. $\bar{0} + 0(1 + 0) = Y$

Part c. $\bar{0} + 0(1) = Y$

Part d. $1 + 0(1) = Y$

Part e. $1 + 0 = Y$

Part f. $1 = Y$

Fig. 4-10. The evaluation of this expression begins with the NOT within the brackets.

1. In part "a" of Fig. 4-10, the input set is substituted.
2. In part "b" of Fig. 4-10, the quantity in brackets is evaluated first. Within the brackets, the NOTed quantity has the highest priority. Here, 0 AND 0 = 0. Then, $\bar{0}$ = 1.
3. In part "c," ORing 1 and 0 completes the evaluation of the bracketed quantity.
4. In "d," the NOT is removed next. The 0 at the left becomes a 1.
5. In "e," the AND must be evaluated before the final OR is evaluated.
6. In "f," the last OR operation results in Y = 1.

LOGIC DIAGRAMS TO BOOLEAN EXPRESSIONS

The third transformation is from logic diagrams to Boolean expressions. Logic diagrams and Boolean expressions contain the same information about circuit action. Therefore, one can be used to derive the other. Fig. 4-11 suggests that Boolean expressions for combinational logic circuits can be obtained from logic diagrams. Examples will be used to describe this process.

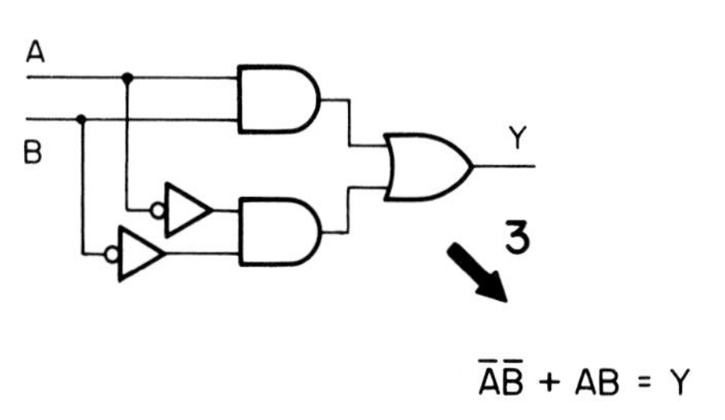

Fig. 4-11. Given the logic diagram for a circuit, its Boolean expression can be written.

Example: Part "a" of Fig. 4-12 shows a simple combinational logic circuit. The process of writing its Boolean expression begins at its inputs.

Based on the Boolean description of ORs, the quantities at the outputs of the two input elements can be written. See the circuit at b.

The output quantities of the ORs serve as inputs to the AND. Its output (and the output of the circuit) is therefore:

$$(A + B)(C + D) = Y$$

Note the brackets. They are necessary, since A will be ORed with B before it is applied to the AND. This is also true of C and D. Without the brackets, the following incorrect expression would result:

$$A + BC + D = Y \text{ (incorrect)}$$

Example: Part "a" of Fig. 4-13 shows an interesting situation. In this circuit, the lead to output Y is NOTed. NOTs are also present in the portion of the circuit that produced output Z. However, these NOTs are before the OR. Will the output at Z equal the output at Y?

The circuit at b shows the first step in finding the expressions for Y and Z. At c, the two expressions are complete. Are the two expressions equal?

To test for equality, the expressions have been evaluated at d. The input set A = 1 and B = 1 was used, and it appears that the two expressions are indeed equal.

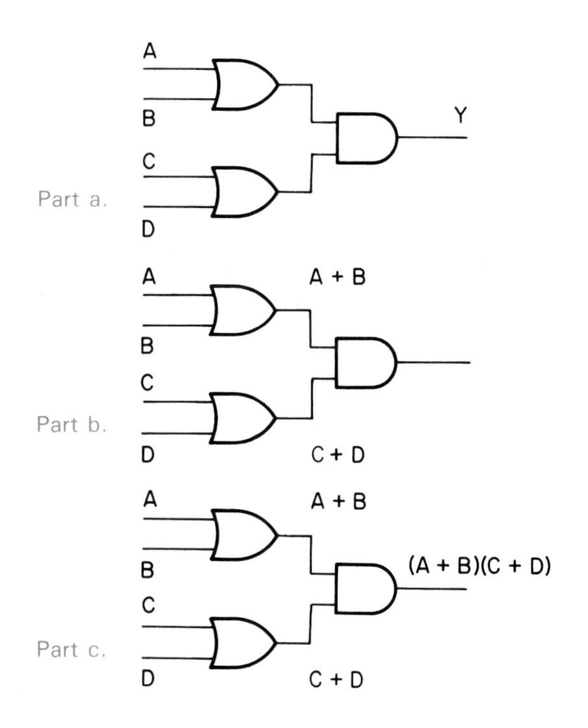

Fig. 4-12. The expression for a circuit can be obtained by writing expressions for every point in the circuit.

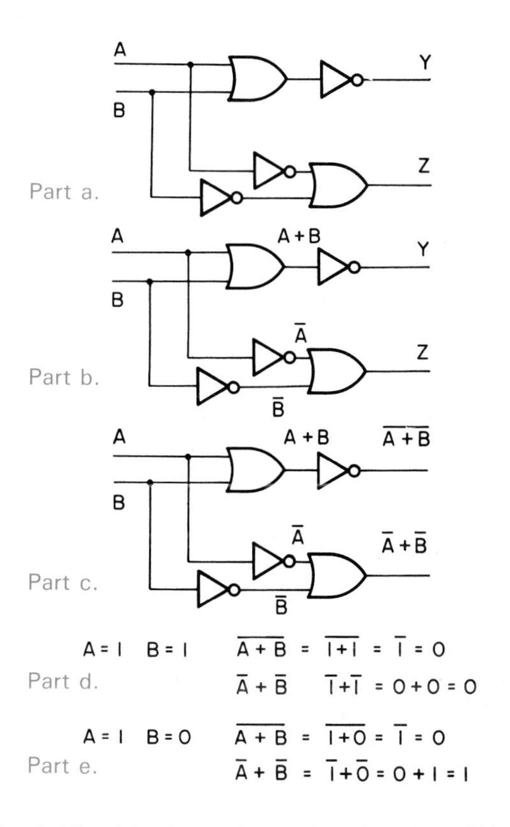

Fig. 4-13. This circuit shows that there is a difference between NOTing before and after an element.

At e, however, the input set A = 1 and B = 0 has been used. Note the results. The two expressions result in different values for Y and Z. It makes a difference whether NOTs are placed before or after a logic element.

Example: Fig. 4-14 is a complex circuit. Follow the process for obtaining the expression for Y in terms of A, B, C, and D.

Note that the bar at the output of the second NOT (at the right in Fig. 4-14) extends across the entire quantity. In the final expression, brackets around the NOTed quantity emphasize that this is a single number, a 1 or a 0. However, they are not needed, since the NOT must be evaluated before the OR.

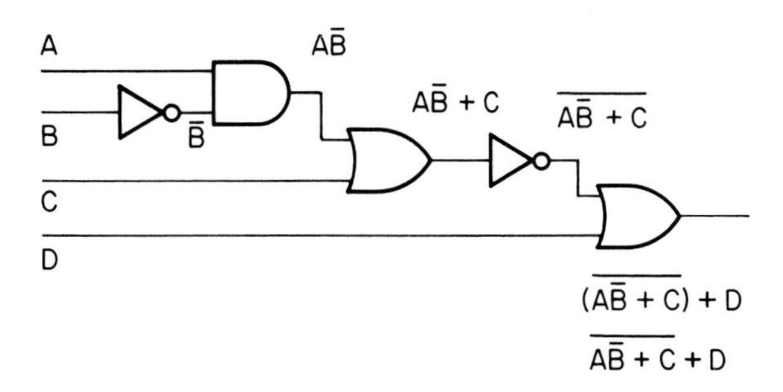

Fig. 4-14. A person writes the Boolean expression for this circuit in a direct way.

Fig. 4-15 shows a simple test of the validity of the expression. If it is correct, it should result in the same

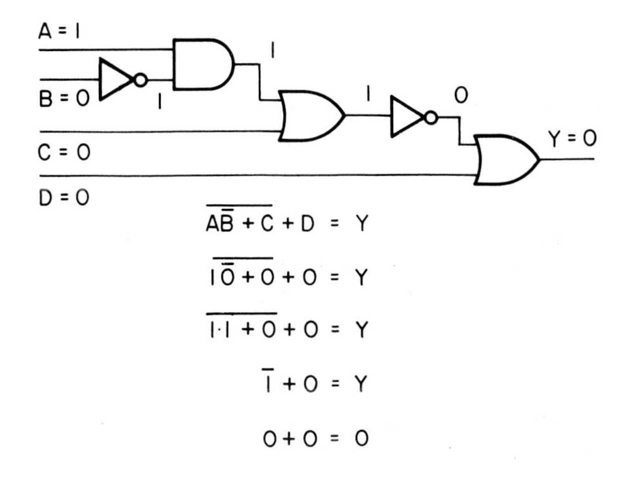

Fig. 4-15. The validity of an expression can be tested by using a truth table circuit analysis on the logic diagram.

predicted output as an analysis of the logic diagram. For the given input signal set, a truth table circuit analysis indicates that Y = 0. Evaluation of the Boolean expression also gives Y = 0. While this does not prove that the two are equivalent, it is a quick test of its validity.

Example: The circuit in Fig. 4-16 has some interesting features. Some of its inputs are used more than once, and logic elements with more than two inputs are used.

The logic diagram has been transformed into its Boolean expression. To convince yourself that the Boolean expression for the logic diagram is correct, follow the process used to obtain the circuit's expression. Note the brackets around AB + C. They are necessary because this quantity represents one of the three inputs to the last AND.

BOOLEAN EXPRESSIONS TO LOGIC DIAGRAMS

In the last section, Boolean expressions were obtained from logic diagrams. Fig. 4-17 suggests that the reverse process is possible.

Example: The expression at the top of Fig. 4-18 has enough information to design a circuit that will produce the required output. The expression can be *IMPLEMENTED* (a circuit can be drawn based on the expression) in three steps.

The order of implementation for Fig. 4-18 is identical to that used when evaluating expressions:

1. Bracketed quantities first.
2. NOTed quantities next.
3. ANDed quantities next.
4. ORs last.

The steps are:

1. In part "a" of Fig. 4-18, A and B are in brackets, so they are implemented first.

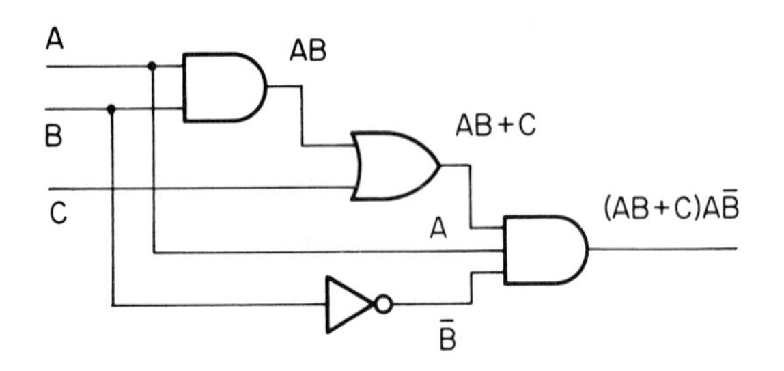

Fig. 4-16. The approach to writing expressions from logic diagrams does not change when input variables (letters standing for input signals or input numbers) appear more than once.

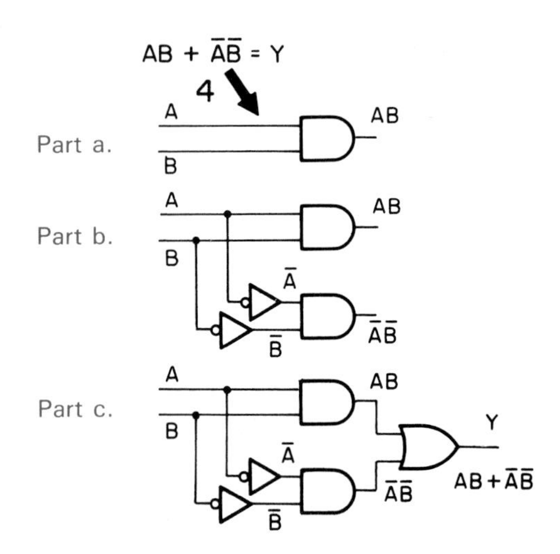

Fig. 4-17. Given a circuit's Boolean expression, its logic diagram can be drawn.

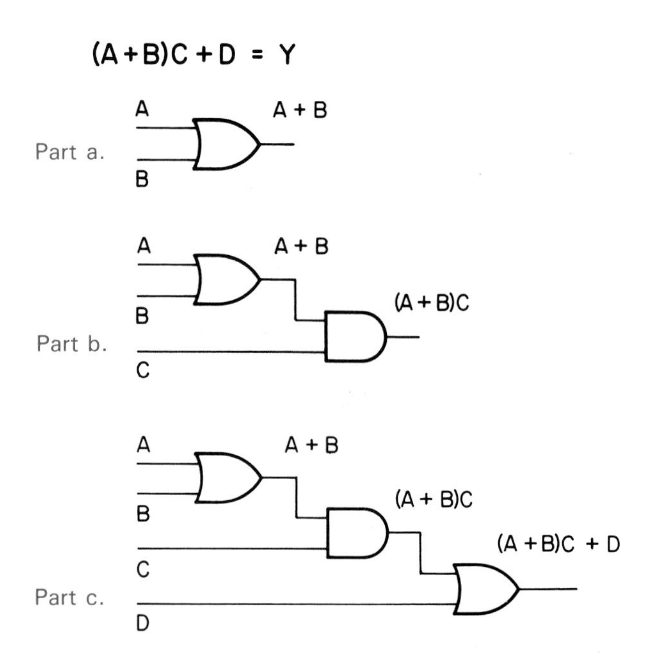

Fig. 4-18. This diagram was drawn using the Boolean expression for the circuit as a starting point.

2. In part "b," C is ANDed with the output of the OR.
3. In "c," D is ORed to produce the required output.

Fig. 4-19 shows a quick test of the validity of this circuit. For the given input signals, the Boolean expression indicates that the output will be 1. For the same signal set, a truth table circuit analysis also indicates that Y = 1.

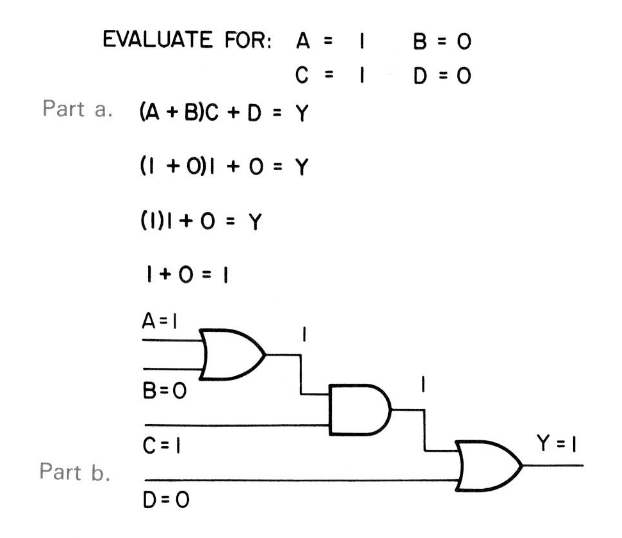

Fig. 4-19. This quick check of the logic diagram against the circuit's expression does not prove the diagram is correct. It does, however, aid in finding errors.

Example: The expression in Fig. 4-20 contains several interesting features. Its inputs are not nicely ordered, and it contains two sets of brackets. These quantities must be implemented separately and then combined.

Either bracketed quantity can be worked on first. Here, BC + A will be the starting point. The following steps are used:

1. In part "a" of Fig. 4-20, a bracket must be done first. Within the first bracket, the AND has priority.
2. In part "b," the first bracket needs to be completed. When A is ORed with BC, the quantity within this set of brackets is complete.
3. In "c," the quantity in the second set of brackets must be implemented next. The NOT takes precedence over the OR, so this is accomplished first.
4. In "d," ORing C with $\overline{A}$ completes the second bracketed quantity.
5. In "e," ANDing the bracketed quantities completes the circuit.

Example: Notice the NOTs in the expression in Fig. 4-21. The bar over the A is separate from the one over the rest of the expression. The two quantities must be created and NOTed separately. In effect the long NOT places brackets around the first portion of the expression.

The quantity under the long bar will be developed first. The steps are as follows:

1. In part "a" of Fig. 4-21, the ANDing of C and D must come first.

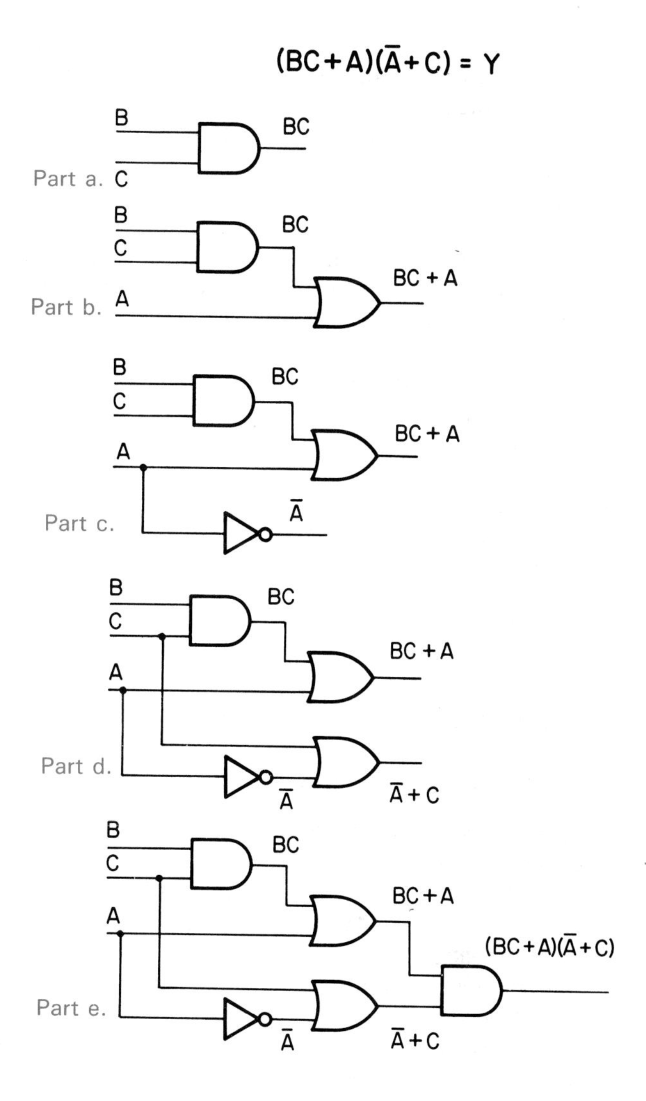

Fig. 4-20. The drawing of a logic diagram does not have to start with input A.

2. In part "b," the OR is added. Note the use of a 3-input element.
3. In "c," the NOT completes the first part of the expression.
4. In "d," the NOTing of A takes precedence over the final ANDing, so it is accomplished next.
5. In "e," the ANDing of the two NOTed quantities completes the circuit.

Example: NOTs within NOTs are found in the expression in Fig. 4-22.

The deepest NOT must be implemented first, so $\overline{C}$ was created in part "a" of Fig. 4-22. The steps are as follows:

1. In part "a" of Fig. 4-22, C is NOTed first.
2. In "b," $\overline{C}$ is ANDed with B.

$$\overline{A + B + CD}\,\overline{A} = Y$$

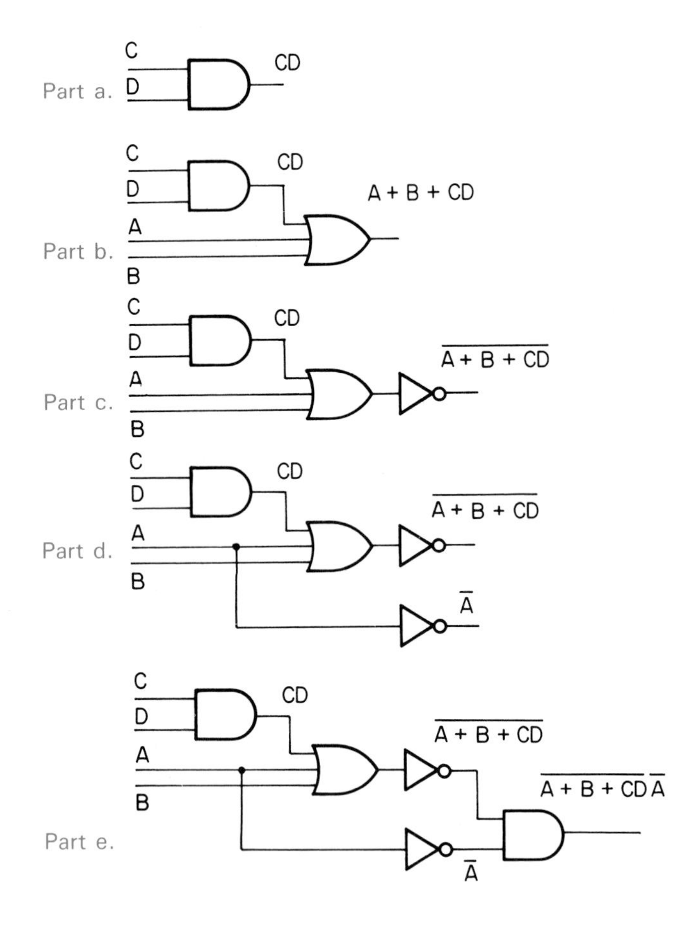

Fig. 4-21. The NOTed quantities in this expression must be implemented separately and then ANDed.

3. In "c," $B\overline{C}$ is ORed with A.
4. In "d," the NOTing of $A + B\overline{C}$ completes the second NOT.
5. In "e," B is ORed with the previous output.
6. In "f," the final NOT completes the circuit.

Test this circuit against the original expression to see if the circuit is valid. Use $A = 0$, $B = 1$, and $C = 0$.

TRUTH TABLES TO BOOLEAN EXPRESSIONS

Fig. 4-23 suggests that Boolean expressions can be written to describe the content of truth tables. This is an important transformation, because it is often used to design digital circuits. The design process often begins with a truth table that describes the action of the proposed circuit. This table is then transformed into a Boolean expression. Then, the expression is implemented to obtain a logic diagram. Based on that diagram, the circuit is constructed and tested.

A number of methods are used to make the transformation from truth tables to Boolean expressions. Two methods will be described here: the SUM-OF-PRODUCTS method and the PRODUCT-OF-SUMS method.

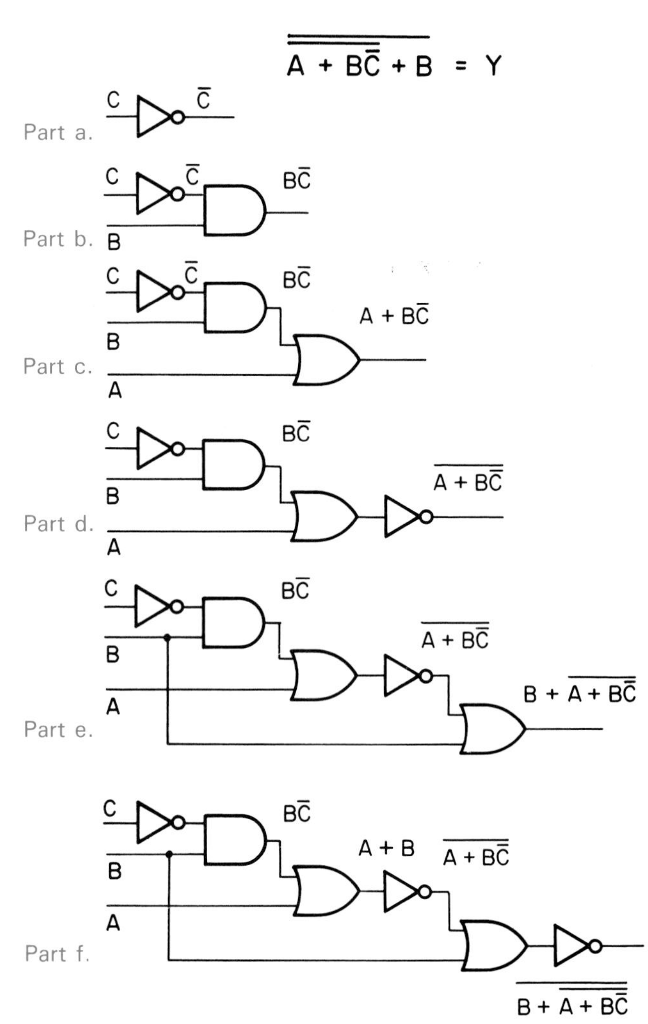

Fig. 4-22. This expression contains NOTs within NOTs.

SUM-OF-PRODUCTS METHOD

The easiest method of obtaining Boolean expressions from truth tables is called the SUM-OF-PRODUCTS method. It consists of writing a group of AND statements (products) based on an inspection of the truth table. These are then ORed (summed) to produce the expression.

$\overline{A}\overline{B} + AB = Y$

INPUTS		OUTPUT
A	B	Y
0	0	1
0	1	0
1	0	0
1	1	1

5

Fig. 4-23. Given a circuit's truth table, its Boolean expression can be obtained.

Example: Fig. 4-24 shows the truth table for a proposed circuit. It has three inputs and a single output. Under only two conditions will its output be 1. For all other input signal sets, Y is to equal 0.

Start with input set b (A = 1, B = 1, C = 1). If this were the only condition that resulted in Y = 1, the circuit would be an AND. Its expression would be:

$$ABC = Y$$

	INPUTS			OUTPUT
	A	B	C	Y
	0	0	0	0
	0	0	1	0
Part a.	0	1	0	1
	0	1	1	0
	1	0	0	0
	1	0	1	0
	1	1	0	0
Part b.	1	1	1	1

Part b. $ABC = 1 \cdot 1 \cdot 1 = 1$

Part a. $\overline{A}B\overline{C} = \overline{0} \cdot 1 \cdot \overline{0} = 1$

$ABC + \overline{A}B\overline{C} = Y$

Fig. 4-24. The sum-of-products method was used to write the Boolean expression for this truth table.

That is, 1 × 1 × 1 = 1. When any other input set is substituted into this expression, Y will equal 0.

However, there is a second input set (A = 0, B = 1, C = 0) that causes Y to equal 1. For this set, and this set only, the following expression will result in Y = 1:

$$\overline{A}B\overline{C} = Y$$
$$\overline{0} \times 1 \times \overline{0} = Y$$
$$1 = Y$$

The expression $\overline{A}B\overline{C}$ accounts for the input set at a. Note how it was obtained. If the value of a variable in a given set is 0, it is NOTed in the expression. If the value of a variable is 1 in the set, it appears unNOTed.

How should the two expressions be combined? A statement of the situation suggests the method. The output of this circuit is to be 1 when ABC = 1 OR when $\overline{A}B\overline{C}$ = 1. That is:

$$ABC + \overline{A}B\overline{C} = Y$$

Test this expression to see if it represents the truth table. That is, substitute input signal sets to show that Y equals 1 only for input sets a and b.

Remember, in sum-of-products, only input sets that result in the output being 1 are considered. Those that result in 0s can be ignored.

Fig. 4-25 is an implementation of the expression $ABC + \overline{A}B\overline{C} = Y$. If this circuit were built and tested in the laboratory, its truth table would be identical to that in Fig. 4-24.

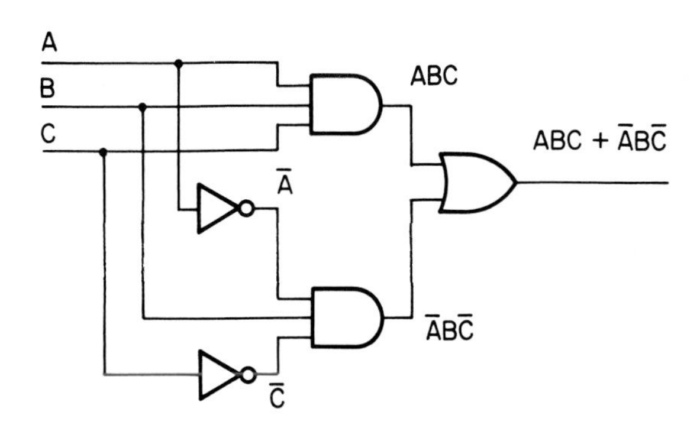

Fig. 4-25. With this implementation of the Boolean expression in Fig. 4-24, all three static descriptions have been completed.

Example: In the truth table in Fig. 4-26, Y = 1 for three input signal sets. These will be used to write a Boolean expression representing this table.

For set a, A is 0, so it will appear in NOTed form in the AND. That is:

$$\overline{A}BC$$

	INPUTS			OUTPUT	
	A	B	C	Y	
	0	0	0	0	
	0	0	1	0	
	0	1	0	0	
Part a.	0	1	1	1	$\bar{A}BC$
Part b.	1	0	0	1	$A\bar{B}\bar{C}$
Part c.	1	0	1	1	$A\bar{B}C$
	1	1	0	0	
	1	1	1	0	

$$\bar{A}BC + A\bar{B}\bar{C} + A\bar{B}C = Y$$

Fig. 4-26. Because three input sets result in Y = 1, the Boolean expression from this truth table contains three product terms.

For set b, both B and C are 0. The AND representing this input set becomes:

$$A\bar{B}\bar{C}$$

In set c, only B is 0, so the AND for this set is:

$$A\bar{B}C$$

ORing these three AND statements results in a Boolean expression representing the truth table.

$$\bar{A}BC + A\bar{B}\bar{C} + A\bar{B}C = Y$$

When implemented, the circuit in Fig. 4-27 results. To simplify the drawing, its inputs were treated as a bus. This does not change the meaning of the drawing.

To test this circuit, the next to the last input set in the table was applied. It should produce an output of 0. Note that it does. Satisfy yourself that this circuit represents the original truth table.

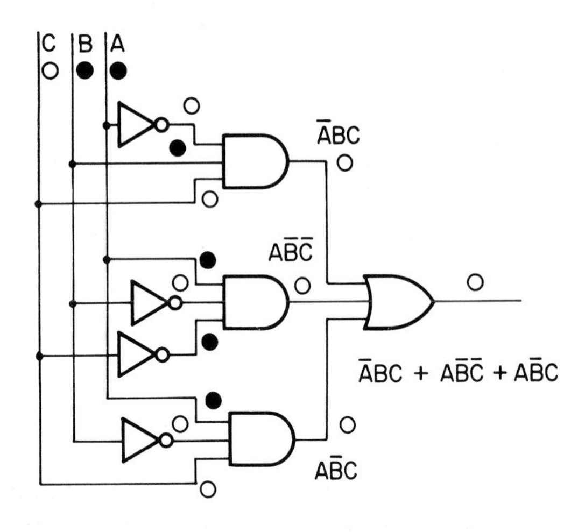

Fig. 4-27. The three product terms in the Boolean expression in Fig. 4-26 appear as ANDs in this circuit.

PRODUCT-OF-SUMS METHOD

For many workers in digital electronics, a knowledge of the sum-of-products method is sufficient. However, some workers should also be able to design circuits using the inverse of this process, the PRODUCT-OF-SUMS method.

The truth table in Fig. 4-28 suggests the advantage of the product-of-sums method. Most input sets result in Y = 1. In this case, the expression developed using the sum-of-products method would contain six ANDs. The product-of-sums method results in a simpler circuit.

The product-of-sums method is the inverse of the sum-of-products method in every way. For example:

1. Statements are written for input sets that result in Y = 0 (rather than Y = 1).
2. Input sets are represented by ORs (rather than by ANDs).
3. Variables that are 1 in the input set are NOTed in the OR; those that are 0 are not NOTed (the opposite of what is done in sum-of-products).
4. Statements representing individual input sets are ANDed to produce the final expression (again the opposite of what is done in sum-of-products).

	INPUTS			OUTPUT	
	A	B	C	Y	
	0	0	0	1	
	0	0	1	1	
	0	1	0	1	
Part a.	0	1	1	0	$A + \bar{B} + \bar{C}$
Part b.	1	0	0	0	$\bar{A} + B + C$
	1	0	1	1	
	1	1	0	1	
	1	1	1	1	

$$(A + \bar{B} + \bar{C})(\bar{A} + B + C) = Y$$

Fig. 4-28. The product-of-sums method was used to write the Boolean expression represented by this truth table.

Refer again to Fig. 4-28. Writing a Boolean expression from this truth table begins with any input set that results in Y = 0. For row a, the statement will be:

$$A + \bar{B} + \bar{C} = Y$$

If this input set is substituted, Y will equal 0. Substituting any other input set will result in Y = 1.

The statement for row b is:

$$\bar{A} + B + C = Y$$

When this input set is substituted, Y will equal 0.

To obtain the Boolean expression for this truth table, the two statements are ANDed. That is:

$$(A + \bar{B} + \bar{C})(\bar{A} + B + C) = Y$$

This Boolean expression is implemented and tested in Fig. 4-29. For the input set shown, Y should equal 1. Note that it does.

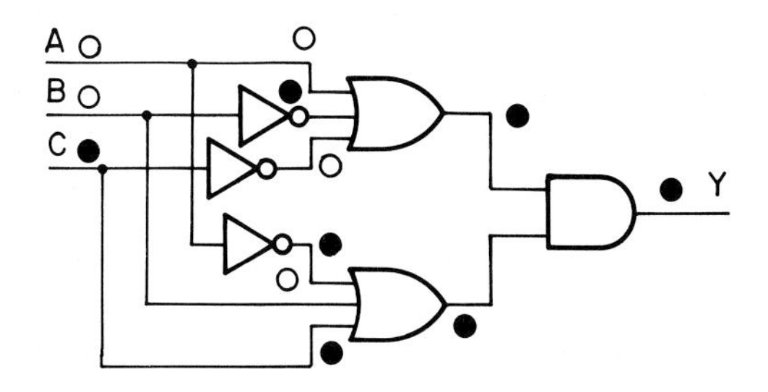

Fig. 4-29. This implementation of the Boolean expression, which is given in Fig. 4-28, contains two ORs.

EXAMPLE OF SUM-OF-PRODUCTS METHOD APPLIED TO CIRCUIT WITH TWO OUTPUTS

Part "a" of Fig. 4-30 shows a partial truth table for a proposed circuit. Only those input signal sets that result in an output of a 1 are of interest, so those resulting in 0s have been omitted. The sum-of-products method is used because there are fewer 1s than 0s in the truth table (other input sets result in Y = 0 and Z = 0).

The circuit has four inputs and two outputs. Two expressions are required, one for Y and one for Z. Input sets 1 and 3 will be used to develop the expression for Y; sets 2 and 3 will be used for Z.

The expression for Y is:

$$\bar{A}\bar{B}CD + ABC\bar{D} = Y$$

Review the truth table and convince yourself that this expression is valid.

The expression for Z is:

$$A\bar{B}CD + ABC\bar{D} = Z$$

Again check this against the truth table.

The implementation of these expressions is shown in part "b" of Fig. 4-30. $ABC\bar{D}$ appears in both expressions, but it does not need to be implemented twice. The output of the middle AND can drive both output ORs.

	INPUTS				OUTPUTS	
	A	B	C	D	Y	Z
1	0	0	1	1	1	0
2	1	0	1	1	0	1
3	1	1	0	0	1	1

Part a.

OTHER INPUT SETS RESULT IN Y=0 AND Z=0

$$\bar{A}\bar{B}CD + ABC\bar{D} = Y$$

$$A\bar{B}CD + ABC\bar{D} = Z$$

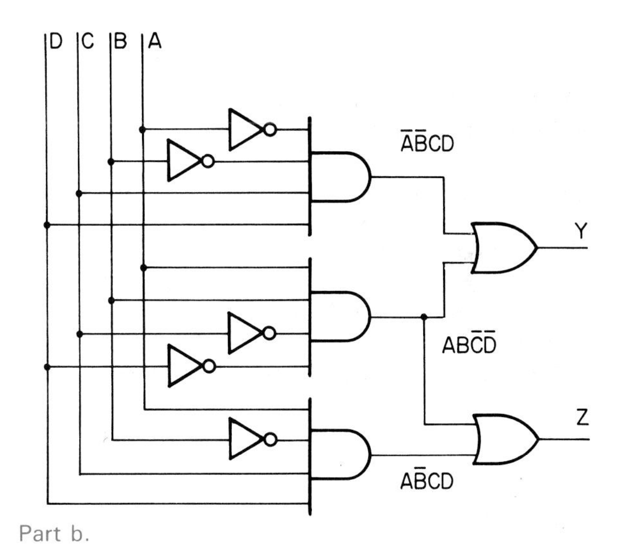

Part b.

Fig. 4-30. The sum-of-products method was used to write the Boolean expression for this truth table because there are fewer 1s then 0s in the output columns (other input sets result in Y = 0 and Z = 0).

SUMMARY

Logic diagrams, truth tables, and Boolean expressions can be used to describe a given combinational logic circuit. Each description contains the same information. Given any one of these, the other two can be determined.

TEST YOUR KNOWLEDGE

1. List three of the five types of transformations from one representation for the action of a circuit to another.
2. What logic element does the word "sum" represent in the term "sum-of-products?"
3. What logic element does the word "product" represent in the term "product-of-sums?"
4. When converting Boolean expressions to truth tables, there are rules stating the priority of each step. List in the proper order the letters in front of the following steps:
   A. Quantities under NOT signs are evaluated.
   B. ANDs are evaluated.
   C. Quantities within brackets are considered.
5. When more than one NOT is in a circuit, the __________ NOT must be evaluated first.
6. Does it make a difference whether NOTs are placed before or after a logic element?
7. Does a single test of the validity of a Boolean expression prove that it is correct?
8. The transformation from a truth table to a Boolean expression is often used to __________ a circuit.
9. The sum-of-products method looks for ________ (1s, 0s) in the output column of a truth table.
10. The product-of-sums method looks for ________ (1s, 0s) in the output column of a truth table.

STUDY PROBLEMS

1. Construct a truth table for the circuit shown in Fig. P4-1. The circuit has two inputs, so there are four possible input signal sets.

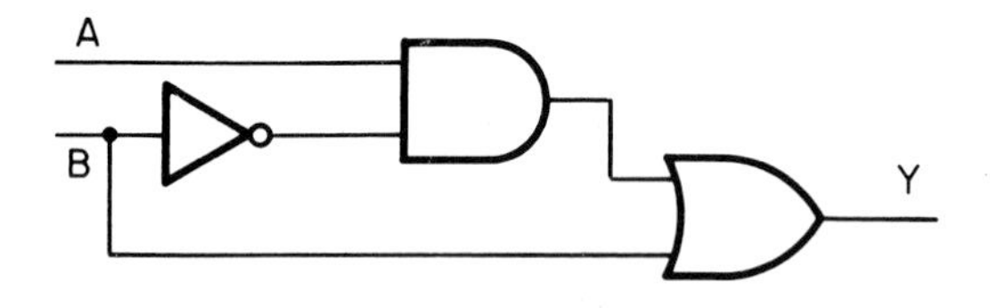

Fig. P4-1. Circuit with one AND, one OR, and one NOT is basis for truth table. Table has four lines.

2. Repeat Problem 1 for the circuit shown in Fig. P4-2. The truth table for the circuit will contain eight input sets.
3. Based on the following Boolean expression, construct a truth table. Again, it will contain eight input signal sets.

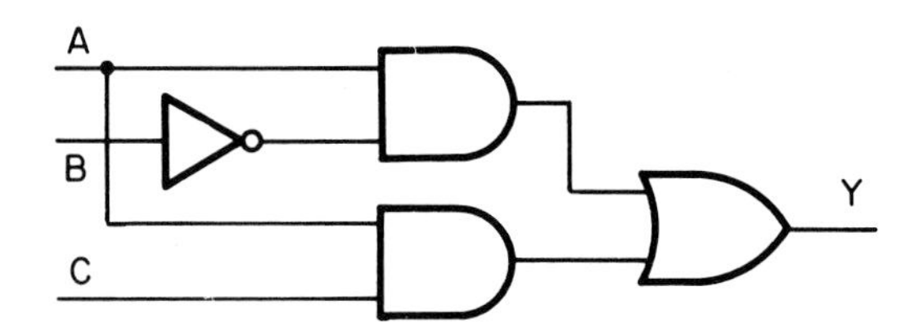

Fig. P4-2. Circuit with two ANDs, one OR, and one NOT is basis for truth table. Table has eight lines.

$$A(B + \overline{C}) = Y$$

4. Repeat Problem 3 for the following expression:

$$AB + \overline{A + B} = Y$$

5. Repeat Problem 3 for the following expression:

$$\overline{\overline{A} + B + AC} = Y$$

6. Write the Boolean expression for the circuit in Fig. P4-3. Do not attempt to simplify the expression.

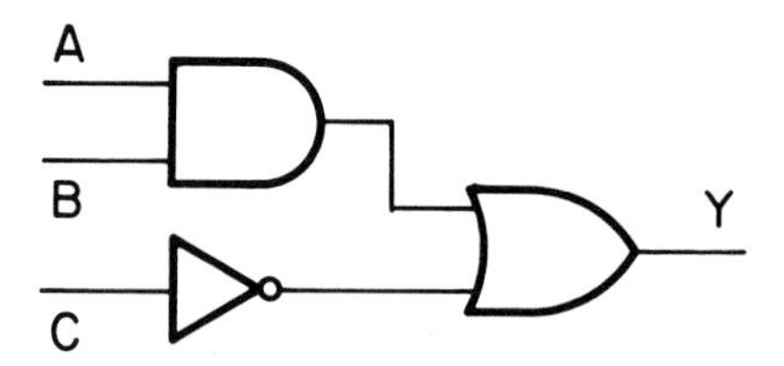

Fig. P4-3. Circuit with one AND, one OR, and one NOT is basis of Boolean expression.

7. Repeat Problem 6 for the circuit that is shown in Fig. P4-4.
8. Repeat Problem 6 for the circuit in Fig. P4-5.
9. Repeat Problem 6 for the circuit that is shown in Fig. P4-6.

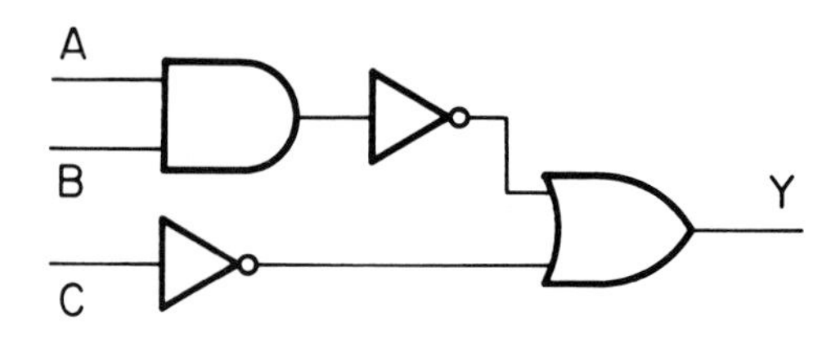

Fig. P4-4. Circuit with one AND, one OR, and two NOTs is basis of Boolean expression.

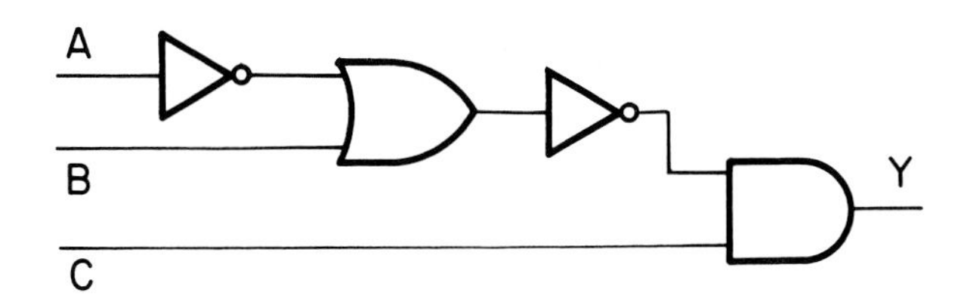

Fig. P4-5. Circuit with one OR, two NOTs, and an AND is basis for Boolean expression.

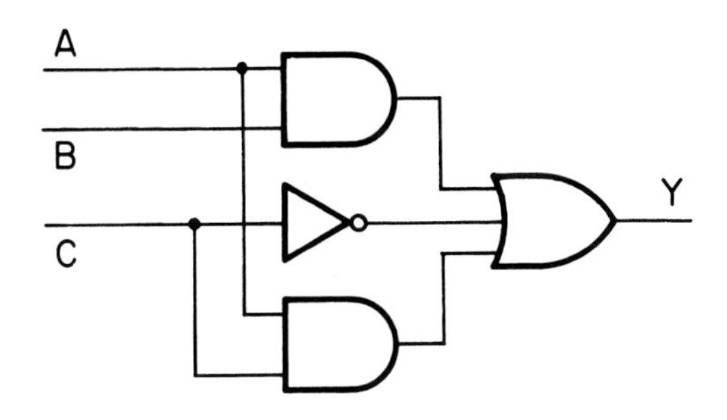

Fig. P4-6. Circuit with two ANDs, one NOT, and a three-input OR is basis for Boolean expression.

10. Repeat Problem 6 for the circuit in Fig. P4-7.
11. Carefully draw the logic diagram of the circuit represented by the following Boolean expression. Do not simplify. Use only AND, OR, and NOT elements.

$$AB + CD = Y$$

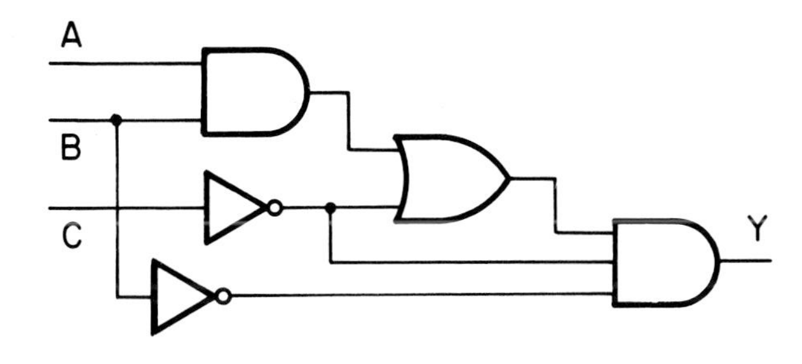

Fig. P4-7. Circuit with one two-input AND, one three-input AND, one two-input OR, and two NOTs is basis for Boolean expression.

12. Repeat Problem 11 for the following Boolean expression:

$$(A + C)A\overline{B} = Y$$

13. Repeat Problem 11 for the following Boolean expression:

$$\overline{AB} + C\overline{D} = Y$$

14. Repeat Problem 11 for the following Boolean expression:

$$\overline{\overline{AB} + \overline{C} + B} = Y$$

15. Repeat Problem 11 for the following Boolean expression:

$$AB(C + D) + A\overline{C}D = Y$$

16. Refer to the truth table in Fig. P4-8.
    1. Using sum-of-products, write a Boolean expression to represent the table. Do not simplify.
    2. Draw the logic diagram represented by your expression. Again, do not simplify.
    3. Using truth table circuit analysis, test your circuit to be sure it produces the action described in the truth table.

INPUTS		OUTPUT
A	B	Y
0	0	1
0	1	1
1	1	0
1	1	0

Fig. P4-8. Truth table provides basis for three student exercises.

INPUTS			OUTPUT
A	B	C	Y
0	0	0	0
0	0	1	0
0	1	0	1
0	1	1	0
1	0	0	1
1	0	1	0
1	1	0	1
1	1	1	1

Fig. P4-9. Truth table is basis for two student exercises.

17. Repeat parts 1 and 2 of Problem 16 for the truth table in Fig. P4-9. Test your circuit for any three input signal sets.

INPUTS				OUTPUTS	
A	B	C	D	Y	Z
0	1	1	1	1	1
1	0	0	1	0	1
0	0	0	0	0	0

OTHER INPUT SETS RESULT IN Y=0 AND Z=0

Fig. P4-10. Truth table with two outputs is basis for two student exercises. Three to four steps are required: use sum-of-products twice (for two outputs) and draw the logic diagram twice (or combine some of the circuits to eliminate a step).

INPUTS			OUTPUT
A	B	C	Y
0	0	0	1
0	0	1	1
0	1	0	1
0	1	1	0
1	0	0	0
1	0	1	0
1	1	0	1
1	1	1	1

Fig. P4-11. Truth table used as basis for sum-of-products exercise and product-of-sums exercise.

18. Repeat parts 1 and 2 of Problem 16 for the truth table in Fig. P4-10. The resulting circuit will have two outputs, so two expressions are needed. When implementing these expressions, note the duplication.
19. Using sum-of-products, write an expression for the circuit represented by the truth table in Fig. P4-11. Do not attempt to simplify.
20. Repeat Problem 19, but use product-of-sums on the truth table in Fig. P4-11. Although this expression differs from the one obtained in Problem 19, it has the same truth table.
21. Repeat Problem 19 for the truth table in Fig. P4-12. Use sum-of-products.
22. Repeat Problem 20 for the truth table in Fig. P4-12. Use product-of-sums. The expressions differ, but they have the same truth table.

INPUTS			OUTPUT
A	B	C	Y
0	0	0	1
0	0	1	0
0	1	0	1
0	1	1	1
1	0	0	1
1	0	1	0
1	1	0	1
1	1	1	1

Fig. P4-12. Truth table is basis for sum-of-products exercise and product-of-sums exercise.

Chapter 5

NAND, NOR, XOR Elements

LEARNING OBJECTIVES

After studying this chapter and completing lab assignments and study problems, you will be able to:

- Recognize symbols for NAND, NOR, and exclusive OR logic elements and explain their operations in terms of Boolean expressions, truth tables, and timing diagrams.
- Draw and construct NAND and NOR equivalents for AND, OR, and NOT elements.
- Express DeMorgan's theorem in terms of Boolean expressions and draw DeMorgan implementations for NORs and NANDs.
- Draw logic diagrams for expressions containing NAND, NOR, and exclusive OR elements. Also, be able to write Boolean expressions for circuits containing these elements.
- Do basic troubleshooting in circuits containing NAND, NOR, and exclusive OR elements.

Computers and other digital circuits can be assembled using only AND, OR, and NOT logic elements. However, other elements are available. These make designs simpler and may be less expensive than AND, OR, and NOT elements.

Three elements will be covered in this chapter. They are the NAND, NOR, and exclusive OR.

NAND LOGIC ELEMENT

Logic elements can have inverting and noninverting outputs. A noninverting output is a direct output. An inverting output provides the inverse of the usual output: a 1 becomes a 0 and a 0 becomes a 1.

The AND in part "a" of Fig. 5-1 is a noninverting element. When 1s are applied to its inputs, it provides an output of 1.

However, it is often less expensive to build IC chips with inverting outputs. That is, instead of outputting a 1 when the logic requirements of the element have been met, a 0 output is produced. Refer to the symbol shown in part "b" of Fig. 5-1.

The element in part "b" of Fig. 5-1 is called a NAND. The N stands for NOT. It is a NOTed AND. Only when 1s are applied to both inputs will this element output a 0. For any other input signal set, it outputs a 1.

The four methods used to describe AND, OR, and NOT elements will be used to describe the action of NANDs. To review, those methods are:

- Symbols.
- Truth tables.
- Boolean algebra expressions.
- Timing diagrams.

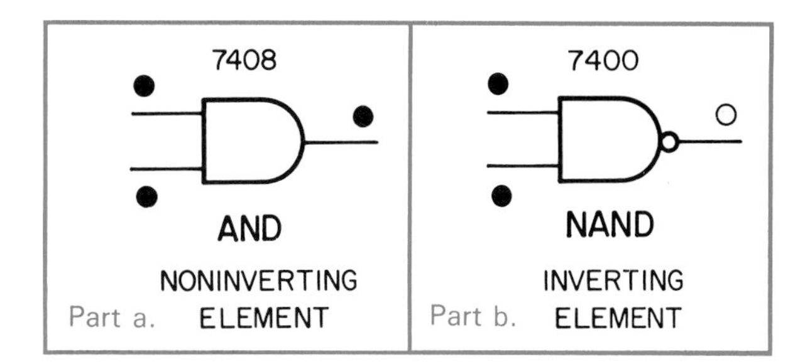

Fig. 5-1. A NAND is an AND with an inverted output.

NAND SYMBOL

The NAND symbol describes the basic action of the element. It is an AND with a NOTed output, Fig. 5-2.

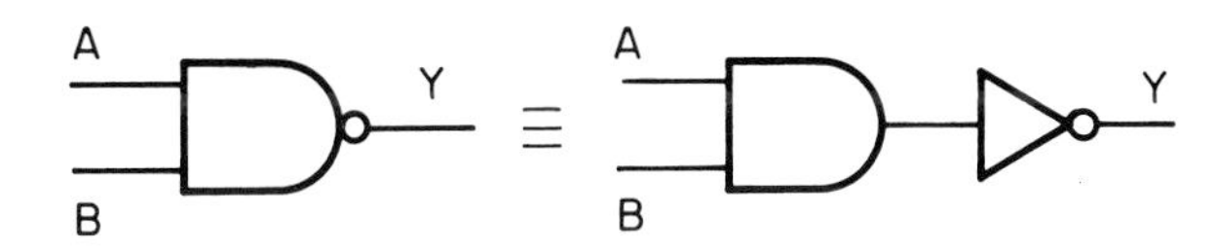

Fig. 5-2. The bubble at the output of a NAND indicates that its output is inverted. The triple line means "defined to be" or "identically equal to."

The bubble indicates the NOTing action that takes place within the chip.

NAND TRUTH TABLE

Fig. 5-3 shows truth tables for ANDs and NANDs. Note the differences in their outputs. The output of the NAND is the inverse of that for the AND.

AND	INPUTS A	INPUTS B	OUTPUT Y
	0	0	0
	0	1	0
	1	0	0
	1	1	1

NAND	INPUTS A	INPUTS B	OUTPUT Y
	0	0	1
	0	1	1
	1	0	1
	1	1	0

Fig. 5-3. Truth tables compare AND and NAND elements.

NAND BOOLEAN EXPRESSION

NAND action is described by the expression:

$$\overline{AB} = Y$$

The bar extends across both inputs. That is, A and B are ANDed and then NOTed.

NAND TIMING DIAGRAM

The timing diagram in Fig. 5-4 restates the action of a NAND in dynamic form. Only when A and B are both 1 will output Y be 0.

NAND elements are often referred to as *NAND GATES*. See Fig. 5-5. Input A is a series of pulses, and B acts as a gate or enable. When B is 0, output Y is forced to 1, and the pulses at A cannot get through. The gate is said to be closed.

When B goes to 1, the gate is open. Pulses at A are

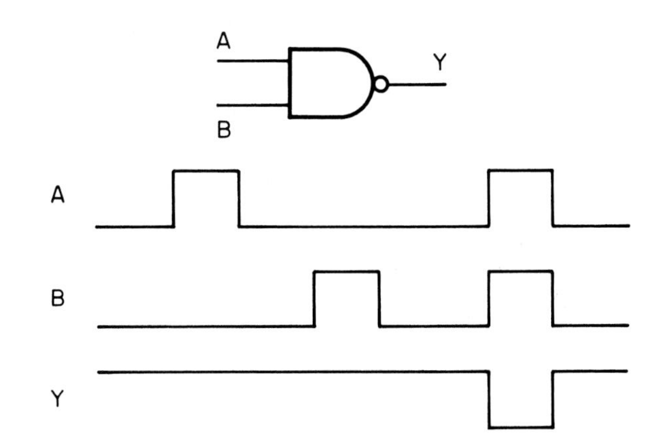

Fig. 5-4. A timing diagram restates the information in the NAND truth table.

ANDed with the 1 at B, and they appear at Y. Note the inversion. This results from the NOTing action of the NAND.

DE MORGAN'S THEOREM

Part "a" of Fig. 5-6 shows a second NAND symbol. To prove that it is a valid representation of a NAND, DeMorgan's theorem must be introduced.

Part "b" of Fig. 5-6 shows all possible input signal sets applied to a NAND. Only when A and B are both 1s will Y = 0. That is:

$$\overline{AB} = Y$$

Column c shows an OR with NOTed inputs. Note that its outputs are identical to those of the NAND. That is, this circuit could be substituted for the NAND. Its Boolean expression is:

$$\overline{A} + \overline{B} = Y$$

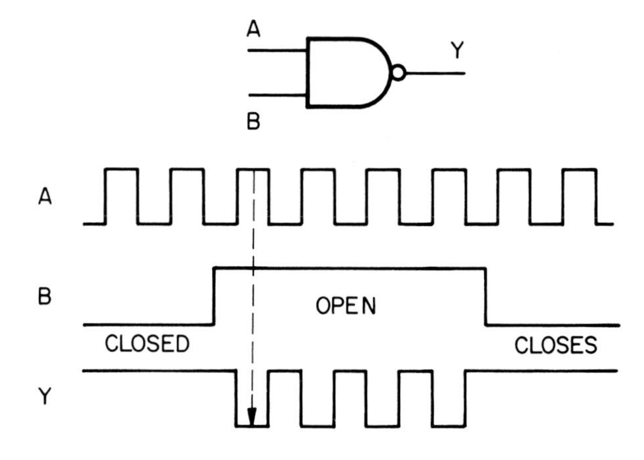

Fig. 5-5. A timing diagram emphasizes the gate action of NANDs.

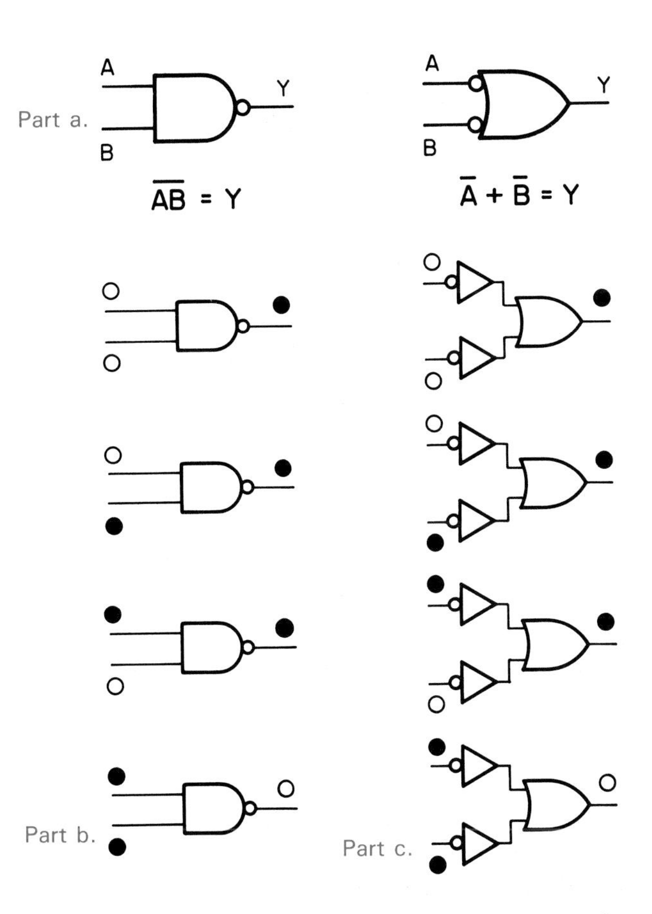

Fig. 5-6. Truth-table circuit analysis shows that DeMorgan's theorem is valid.

DeMorgan's theorem is a formal statement of this situation. That is:

$$\overline{AB} = \overline{A} + \overline{B}$$

Stated in words, an AND with a NOTed output (a NAND) has the same truth table as an OR with NOTed inputs. Either symbol can be used to describe a NAND. Workers in present-day digital electronics must be familiar with both.

ACTIVE HIGH AND ACTIVE LOW

The action of the two NAND symbols (the NOTed AND and the OR with NOTed inputs) can be described in terms of 1s and 0s. See the table in Fig. 5-7. Using this approach, bubbles are treated as NOTs.

However, bubbles may have an alternate meaning. The use of bubbles involves the concepts called *ACTIVE HIGH* (active when high) and *ACTIVE LOW* (active when low).

ACTIVE HIGH

Until this point, 1 has been assumed to be the active signal. Something happens when a 1 is output. For example, when L = 1, a lamp lights. Or the pressing of a button was represented as a 1.

ACTIVE LOW

In many computer and control circuits, however, it is desirable to use 0 (ground) as an active signal.

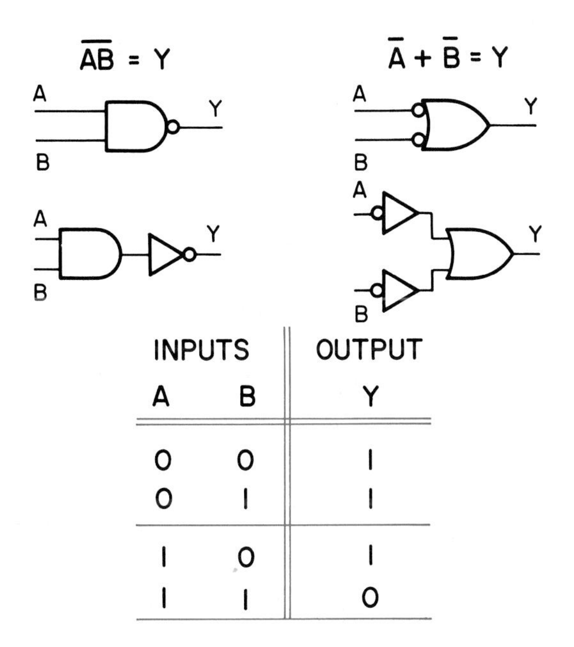

INPUTS A	INPUTS B	OUTPUT Y
0	0	1
0	1	1
1	0	1
1	1	0

Fig. 5-7. The two types of NAND symbols shown have the same truth table.

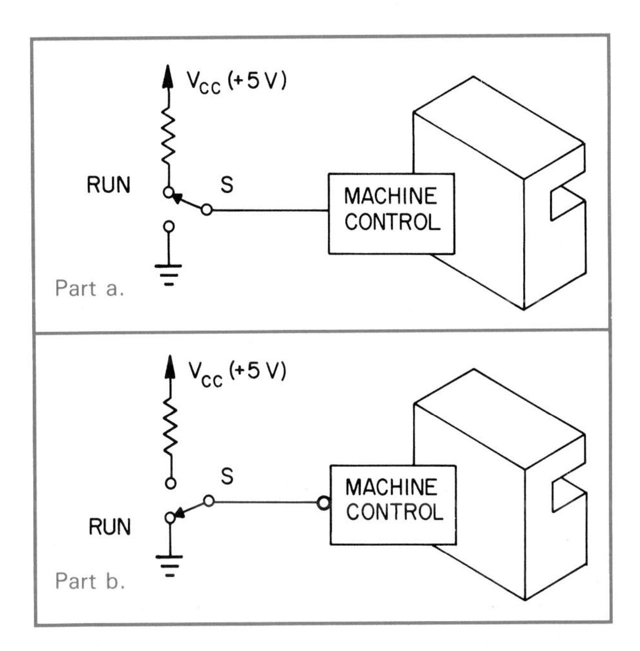

Fig. 5-8. For safety reasons, active low run signals are often used.

That is, things happen when 0s appear.

Fig. 5-8 is an example of an active low signal. The switch signals the start of the machine. In the upper circuit, 1 (+5 volts) is the active signal. When S is high, the machine starts. A similar circuit would be provided to signal the machine to stop.

If the control circuit contains TTL chips, there is a problem. If the lead between the switch and control were to open, the machine would start. This is a safety problem.

The lower circuit corrects the safety problem, since an active low start signal is used. This is indicated by the bubble at the control circuit input. The machine starts when S = 0. (It is stopped by another circuit that is not shown.) The machine will not start if leads to the control open.

APPLICATIONS OF ACTIVE HIGH AND ACTIVE LOW

Additional applications of active low are shown in Fig. 5-9. The first device has an input marked CLOCK. The bubble indicates that CLOCK is active low. That is, to operate this device, the input to CLOCK must go to 0.

The device at b has an input marked GATE. It operates much like CLOCK on the previous device. In this case, however, GATE is not bubbled, so it is active high. The circuit designer needed an active low device, so a NOT was added. Note the placement of the bubble on the NOT. This emphasizes that G is active low. That is, G must go to 0 to enable the device (to "enable" the device means to prepare the device for operation).

At c, input CL is bubbled; it is active low. Again, a NOT has been added, but this time the bubble is at the NOT's output. This emphasizes that CLEAR is active high.

The concept of active high and active low can be applied to NANDs. See Fig. 5-10. Both symbols represent NANDs. However, the AND-based symbol emphasizes row d in the truth table. Two active highs are needed at the inputs to produce an active low at its output.

	INPUTS A	INPUTS B	OUTPUT Y
Part a.	0	0	1
Part b.	0	1	1
Part c.	1	0	1
Part d.	1	1	0

Fig. 5-10. The OR-based NAND symbol emphasizes the first three rows of the truth table; the AND-based symbol emphasizes the last row.

The OR-based symbol in Fig. 5-10 emphasizes rows a, b, and c. Any combination of active lows at the element's inputs results in an active high output.

Both symbols may be used in the same diagram. See Fig. 5-11. Because A and B are active high, the AND-based symbol was used at the upper left. The OR-based symbol was used at the output, since it is driven by active low signals, but has an active high output.

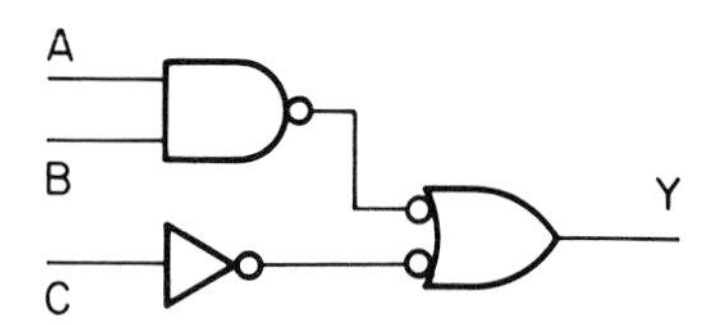

Fig. 5-11. AND-based and OR-based symbols can be used in the same logic diagram.

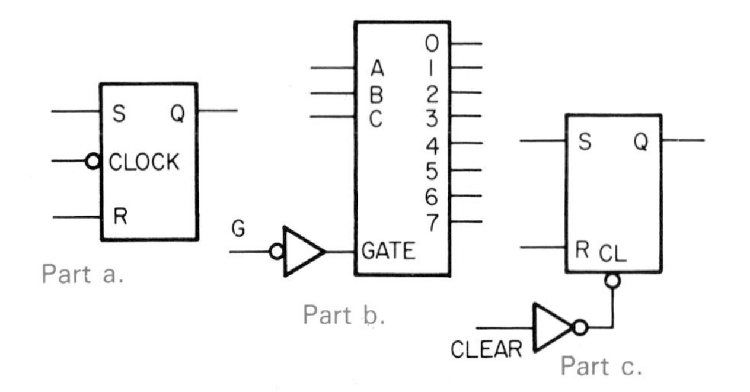

Fig. 5-9. Bubbles at CLOCK, G, and CL indicate active low inputs.

How many NAND symbols are there in Fig. 5-12? There are three (3, 4, and 6). Element 1 is not a NAND. It is called a NOR.

NAND APPLICATIONS AND NAND AS UNIVERSAL ELEMENT

NANDs are universal logic elements. Properly connected, NANDs can serve as ANDs, ORs, and NOTs. As a result, complete circuits can be constructed using only NANDs.

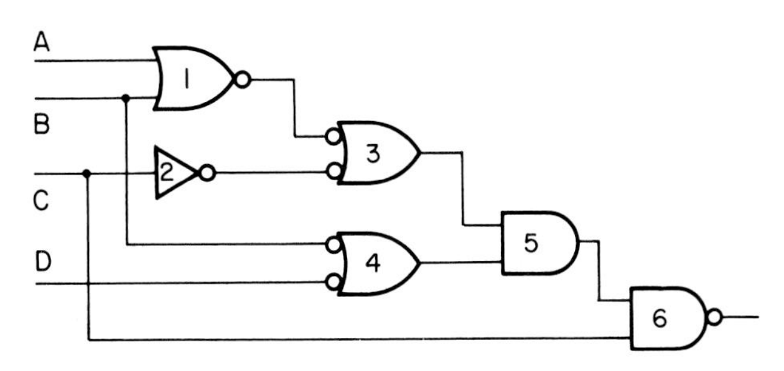

Fig. 5-12. Care is required to distinguish NAND symbols from those of other elements.

DIRECT USE OF NAND

Replacing AND-NOTs is an obvious use for NANDs, Fig. 5-13. The first circuit requires three chips (7404, 7408, 7432). When the AND-NOTs are replaced by NANDs, only two chips are required in the circuit (7400, 7432).

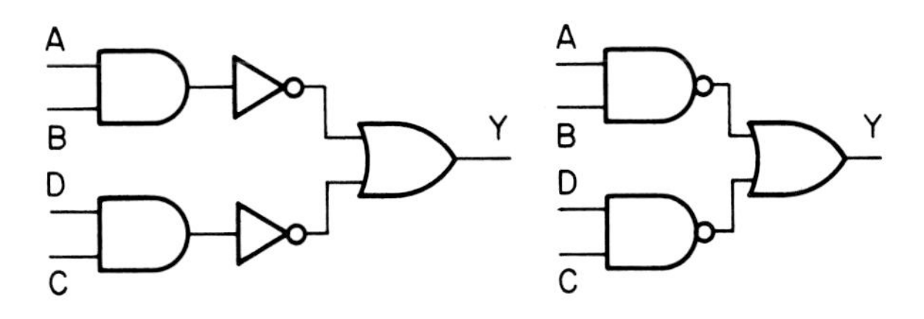

Fig. 5-13. NANDs can be used in place of AND-NOTs. Fewer chips are needed.

NAND USED AS NOT

NANDs can be used as NOTs. See Fig. 5-14. When inputs A and B are tied together, B = A. As a result, the element's truth table reduces to that of a NOT.

Three ways of representing a NAND-based NOT are shown in Fig. 5-15. Although the two symbols at the right have only one input lead, workers in digital electronics know that the inputs must be tied together. All three are actually connected as at the left.

Fig. 5-16 shows an application of NAND-based NOTs. The first circuit requires two chips (7400, 7404); the second requires only one (7400).

USING NOTS TO CONVERT NAND TO AND

When connected one after the other, the actions of NOTs cancel, Fig. 5-17. Output Y always equals A. When such NOTs appear in a circuit, they are said to be *REDUNDANT* (repetitious or being a duplication) and can usually be removed without changing circuit action. See Fig. 5-18.

Cancellation of redundant NOTs can be expressed

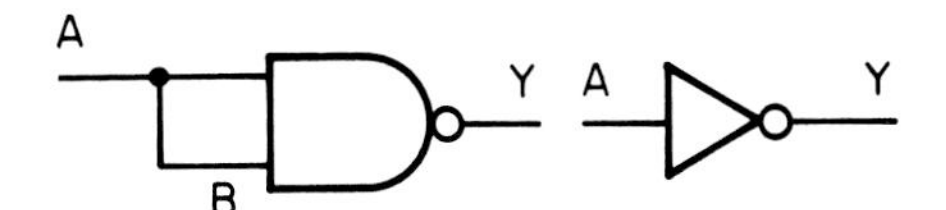

INPUTS A	INPUTS B=A	OUTPUT Y
0	0	1
0	0	1
1	1	0
1	1	0

=

INPUT A	OUTPUT Y
0	1
1	0

Fig. 5-14. NANDs can be connected to act like NOTs.

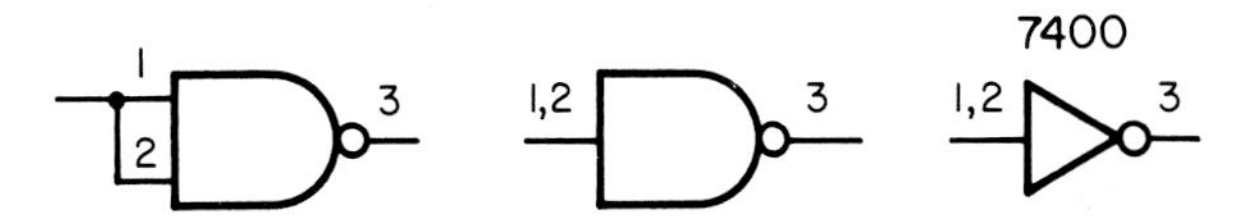

Fig. 5-15. Symbols such as the three symbols shown are used to represent NANDs connected as NOTs.

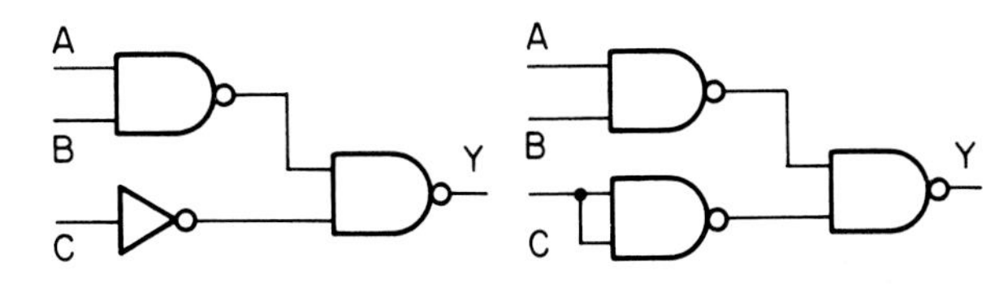

Fig. 5-16. In the circuit shown, a chip can be eliminated by using a NAND gate connected to be a NOT element.

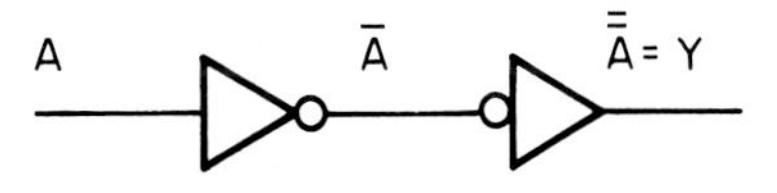

Fig. 5-17. When one NOT drives another NOT, their actions cancel.

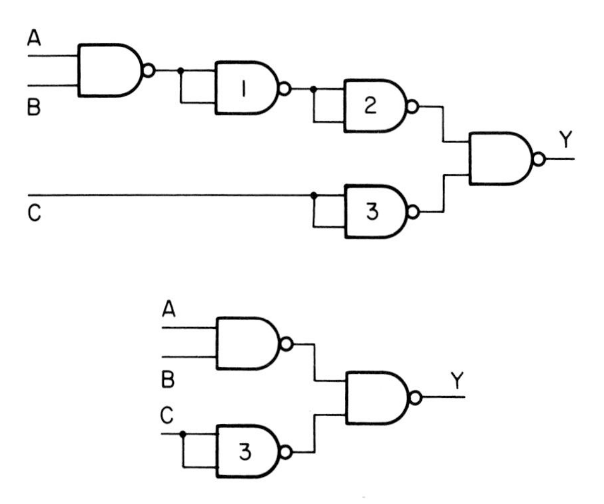

Fig. 5-18. Removal of redundant NOTs 1 and 2 does not change the action of this circuit.

by the Boolean expression:

$$\overline{\overline{A}} = A$$

This is read, "A NOT NOT equals A."

The ability of a second NOT to cancel the action of another NOT can be used to convert a NAND into an AND, Fig. 5-19. The external NOT cancels the action of the bubble.

Fig. 5-20 shows the use of NANDs to simplify a circuit. The upper circuit uses two chips (7404, 7408). The first NAND replaces the AND-NOT. The other two replace the AND. Both circuits require three logic elements, but the lower one uses only one chip (7400). Reliability is improved and cost is lowered.

NANDS USED AS AN OR

Properly connected, three NANDs can function as an OR, as shown in Fig. 5-21. Note the use of both OR- and AND-based NAND symbols (refer to Fig. 5-10). The NOTs cancel the actions of the input bubbles on the common NAND, so the circuit functions as an OR.

NAND AS UNIVERSAL ELEMENT

Because NANDs are universal elements, complete circuits can be constructed using only NANDs. The circuit in Fig. 5-22 is an example. The upper circuit implements the expression and uses three chips (7404, 7408, 7432). In the lower circuit, dashed lines highlight the substitutions of NAND-based elements. Six NANDs are needed (compared to four AND, four OR, and six NOT elements), but the lower circuit uses only two 7400 chips. The designer can cut costs and improve the product. Fewer NOTs are a benefit.

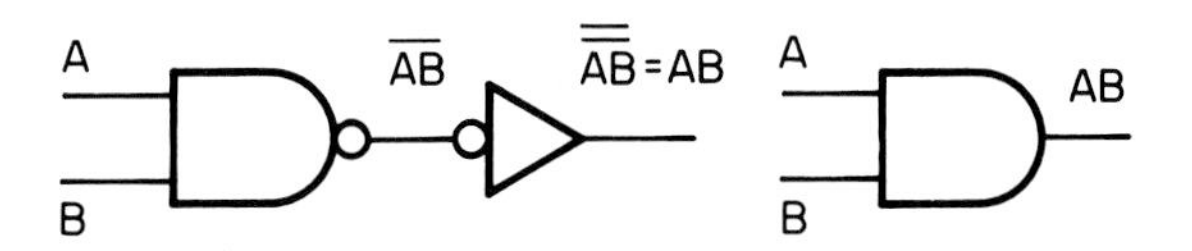

Fig. 5-19. A NAND-NOT acts like an AND.

$\overline{AB}C = Y$

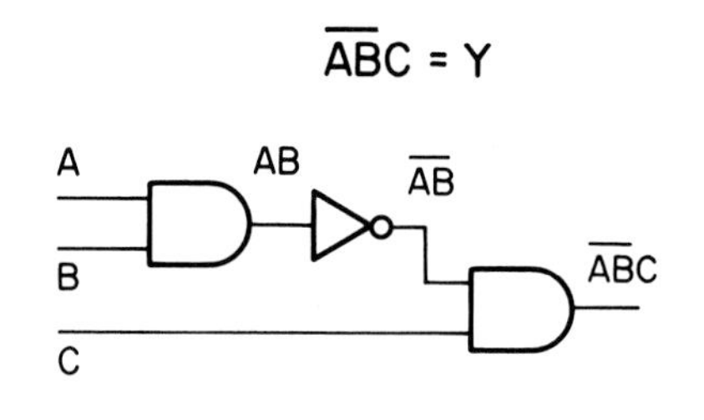

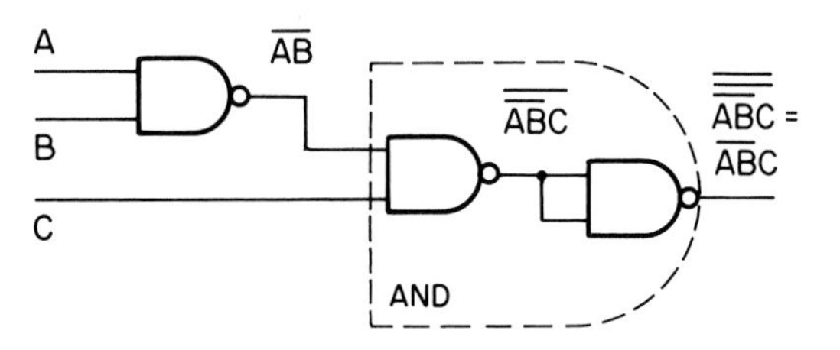

Fig. 5-20. The Boolean expression $\overline{AB}C = Y$ can be implemented using a quad NAND chip (chip with 4 NANDs on it). The 4th NAND is not used.

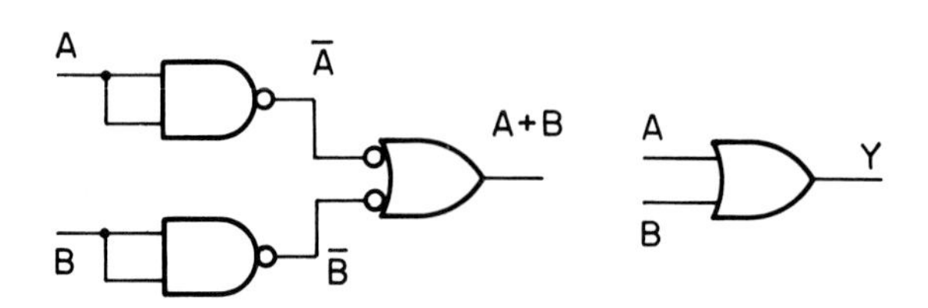

Fig. 5-21. Three NANDs can be connected to act like an OR.

NOR LOGIC ELEMENT

The second universal logic element is the NOR. It is an OR with a NOTed output. Like the NAND, it is widely used in computers and other digital circuits.

NOR SYMBOL

Fig. 5-23 shows the NOR symbol. The bubble indicates the NOT. If the concept of active high/active low is used, the bubble indicates that the output of the element is active low.

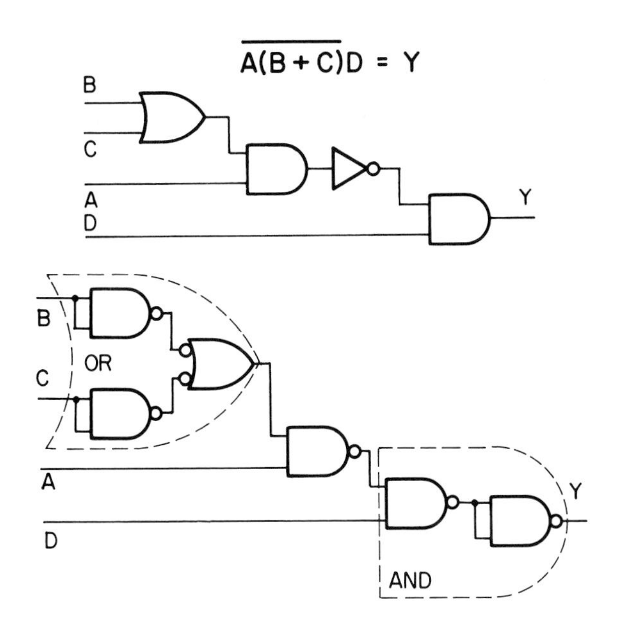

Fig. 5-22. Because the NAND is a universal element, this circuit can be implemented using only NANDs.

NOR BOOLEAN EXPRESSION

The basic NOR expression is:

$$\overline{A + B} = Y$$

A and B are first ORed, and the result is then NOTed.

DE MORGAN'S THEOREM FOR A NOR AND A NAND

A second form of DeMorgan's theorem applies to NORs. See Fig. 5-25. A NOR can be described as an OR with a NOTed output or as an AND with NOTed inputs. The OR-based symbol emphasizes rows b, c, and d in the NOR truth table; the AND-based symbol emphasizes row a.

The two forms of DeMorgan's theorem are next to each other in Fig. 5-26. In each case, all aspects of the symbols are inverted. The AND on the left of Fig. 5-26 becomes an OR. The OR on the left of Fig. 5-26 becomes an AND. ANDs become ORs; ORs become ANDs. Inputs and outputs that are bubbled lose their bubbles; those that are bare gain bubbles.

DeMorgan's theorem is a powerful tool that is widely used to explain the operation of digital circuits.

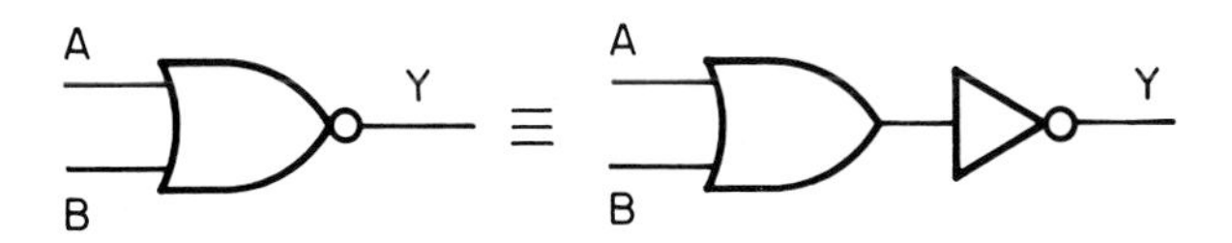

Fig. 5-23. A NOR is an OR-NOT.

NOR TRUTH TABLE

OR and NOR truth tables are shown in Fig. 5-24. Note that the output of the NOR is the inverse of the OR output.

OR

INPUTS A	INPUTS B	OUTPUT Y
0	0	0
0	1	1
1	0	1
1	1	1

NOR

INPUTS A	INPUTS B	OUTPUT Y
0	0	1
0	1	0
1	0	0
1	1	0

Fig. 5-24. Truth tables for OR and NOR elements. The output of the NOR is the inverse of the OR output.

	INPUTS A	INPUTS B	OUTPUT Y
Part a.	0	0	1
Part b.	0	1	0
Part c.	1	0	0
Part d.	1	1	0

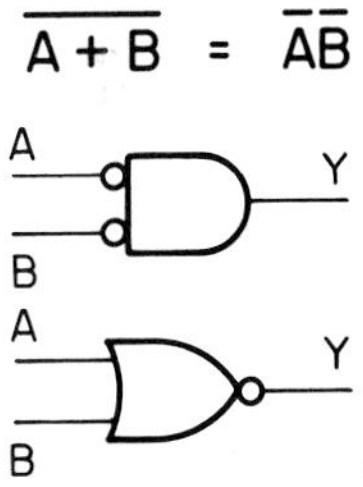

$$\overline{A+B} = \overline{A}\,\overline{B}$$

Fig. 5-25. DeMorgan's theorem permits two NOR symbols.

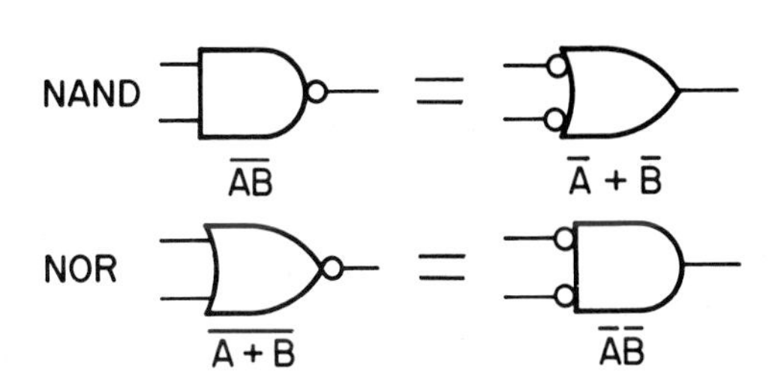

Fig. 5-26. One form of DeMorgan's theorem applies to NANDs; the other applies to NORs.

NOR TIMING DIAGRAM

The timing diagram in Fig. 5-27 is a dynamic description of the action of a NOR. It emphasizes that the output of a NOR element is 0 when either or both inputs are 1.

NORs are often described as NOR-GATES, Fig. 5-28. As long as B = 1, output Y is forced to 0. The pulses at A cannot appear at the output, and the gate is closed. When B changes to 0, the gate opens, and the pulses appear (in inverted form) at Y.

NOR APPLICATIONS

Like NANDs, NORs are universal logic elements. AND, OR, and NOT elements can be assembled from two or more NOR elements.

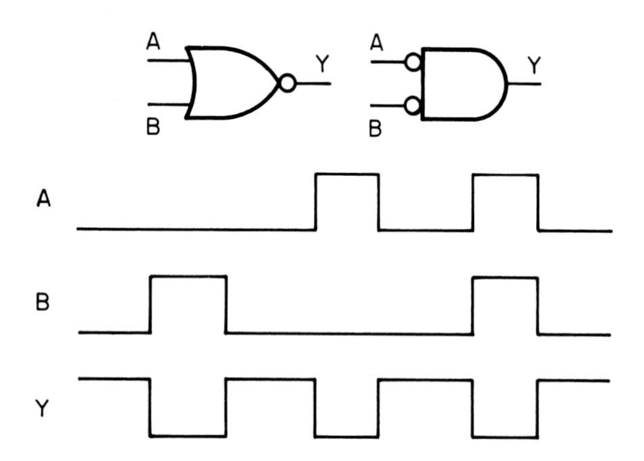

Fig. 5-27. This timing diagram is a dynamic representation of the NOR truth table.

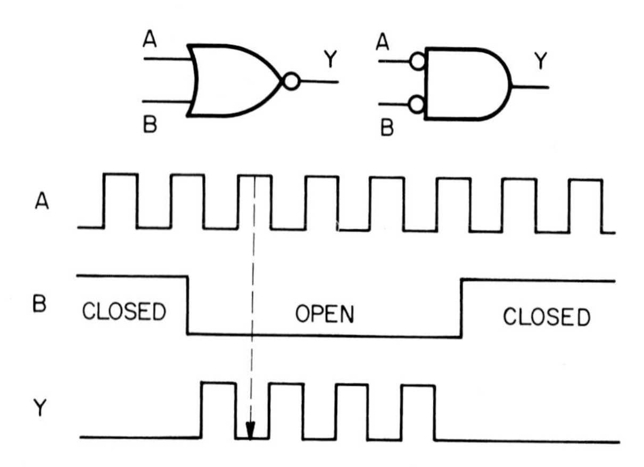

Fig. 5-28. This timing diagram shows the gate action of a NOR.

DIRECT USE OF NOR

NORs can replace OR-NOTs. See Fig. 5-29. The first circuit requires three chips (7404, 7408, 7432), while the second needs only two (7402, 7408).

ONE NOR USED AS A NOT

Like NANDs, NORs can be used as NOTs. See Fig. 5-30. Again, inputs are tied together even though the symbol may show only one input lead.

TWO NORS USED AS AN OR

Two NORs can be connected to produce the action of an OR, Fig. 5-31. The NOT cancels the action of the output bubble, and an OR results.

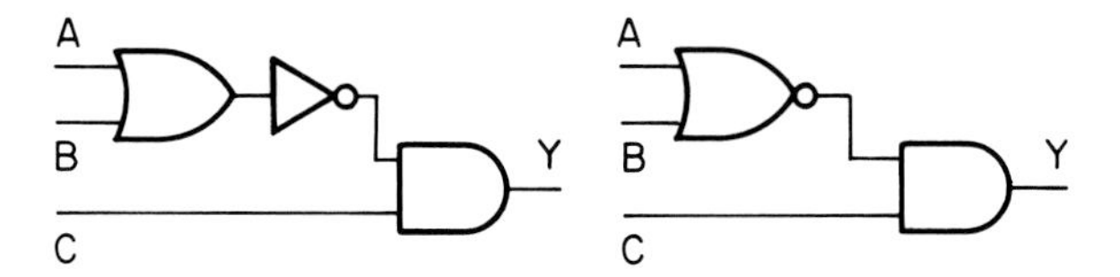

Fig. 5-29. NORs can be used in place of OR-NOTs.

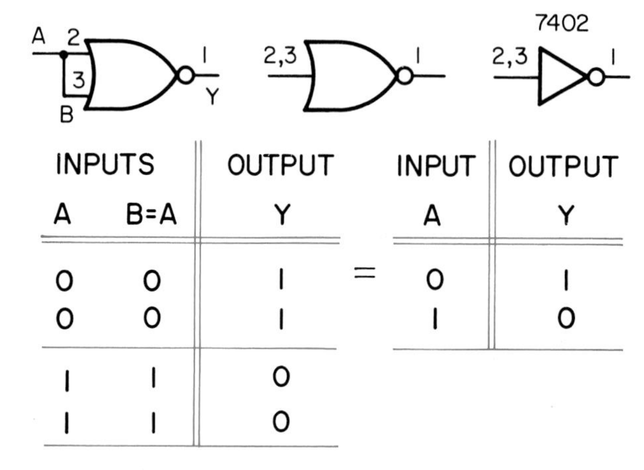

INPUTS A	INPUTS B=A	OUTPUT Y
0	0	1
0	0	1
1	1	0
1	1	0

=

INPUT A	OUTPUT Y
0	1
1	0

Fig. 5-30. NORs can be connected to act like NOTs.

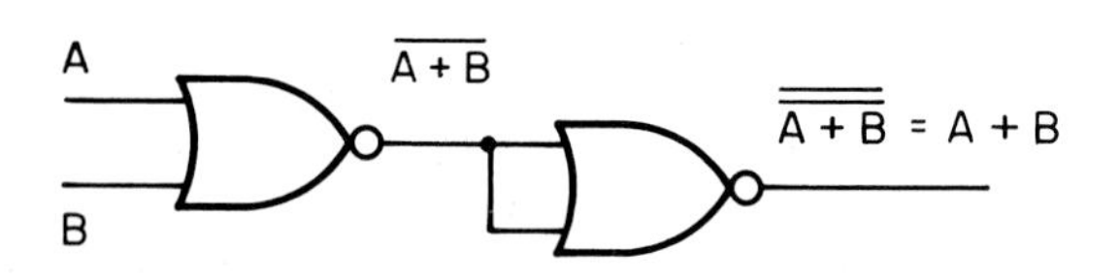

Fig. 5-31. NOR-NOTs act like ORs.

THREE NORS USED AS AN AND

Three NORs can be used to simulate an AND. See Fig. 5-32. The NOTs cancel the actions of the bubbles and the overall action is that of an AND.

NOR-BASED CIRCUITS AND NORS AS UNIVERSAL ELEMENTS

Fig. 5-33 shows the replacement of AND and OR elements by NORs. The first circuit requires two chips (7408, 7432). At b, five NORs are required. With four NORs per chip, two 7402s would be needed to implement this circuit. However, there are redundant NOTs in the upper path. If these are removed, this circuit requires only three NORs. One chip will suffice.

Fig. 5-34 emphasizes the universal nature of NANDs and NORs. The basic logic elements (AND, OR, and NOT) can be assembled using NANDs or NORs. As a result, complete circuits can be constructed using one type of logic element.

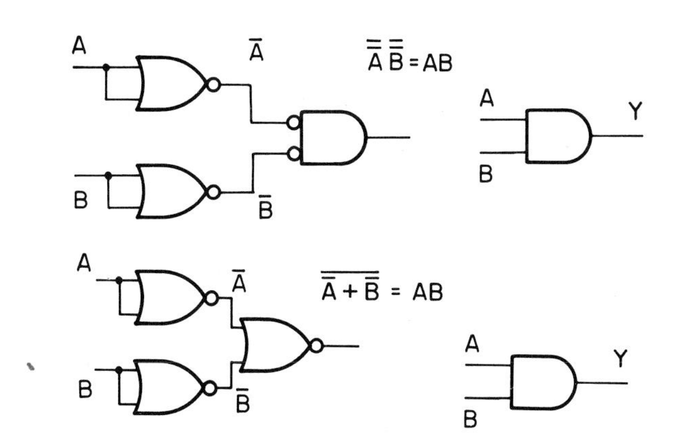

Fig. 5-32. Three NORs can be connected to act like an AND element.

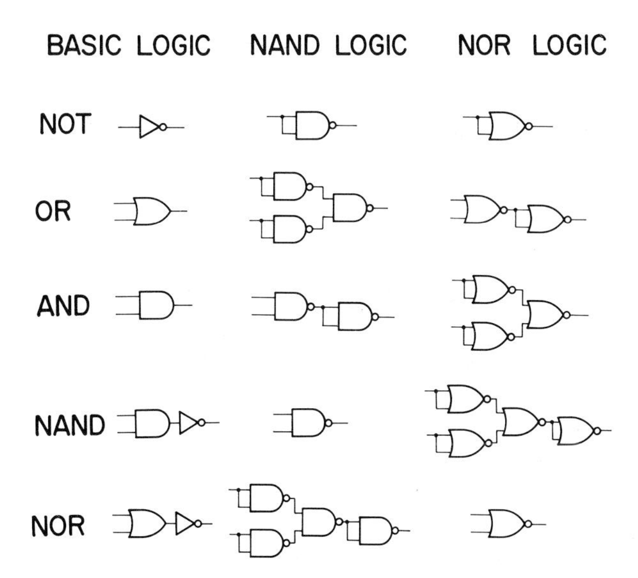

Fig. 5-34. Because NANDs and NORs are universal elements, they can be used to build the basic elements.

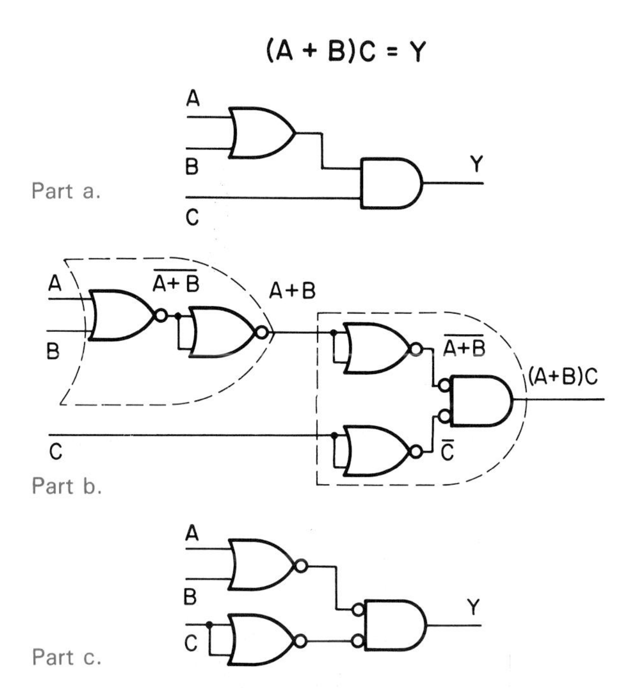

Fig. 5-33. The expression (A + B) C = Y can be implemented using three NORs.

EXCLUSIVE OR

Fig. 5-35 introduces the EXCLUSIVE OR (XOR). Like ordinary OR elements, this element outputs 1 when either input is 1. However, it outputs 0 when both inputs are 1. That is, the condition A = B = 1 is excluded. Compare the truth tables for the inclusive (ordinary) OR and exclusive OR.

XOR BOOLEAN EXPRESSION FROM SUM-OF-PRODUCTS AND PRODUCT-OF-SUMS

In Fig. 5-36, sum-of-products has been used to write the XOR expression. It is:

$$\overline{A}B + A\overline{B} = Y$$

Fig. 5-36 also shows the use of a circle around the + sign to indicate an XOR. While everyone in the digital electronics field understands this symbol, it has

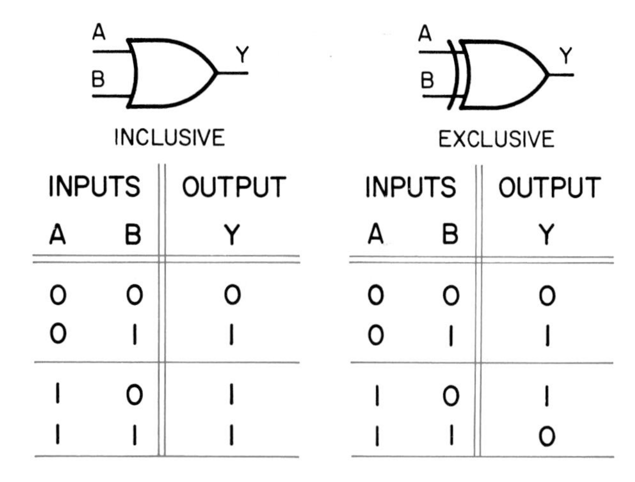

INPUTS A	INPUTS B	OUTPUT Y
0	0	0
0	1	1
1	0	1
1	1	1

INPUTS A	INPUTS B	OUTPUT Y
0	0	0
0	1	1
1	0	1
1	1	0

Fig. 5-35. Only the last rows of the OR and exclusive OR truth tables differ from each other.

INPUTS A	INPUTS B	OUTPUT Y
0	0	0
0	1	(1)
1	0	(1)
1	1	0

$$\bar{A}B + A\bar{B} = Y$$

$$A \oplus B = Y$$

Fig. 5-36. Sum-of-products was used to write the expression for an XOR.

Adders can be built many ways. The circuit in Fig. 5-40 uses two chips. The designer has room to decide which of the implementations is simpler, costs less, is faster in operation, or has some combination of simplicity, low cost, and speed.

XOR TIMING DIAGRAM

An XOR timing diagram is shown in Fig. 5-37. It restates the information in the element's truth table and Boolean expression in dynamic form.

Up to time t1, the XOR acts like an ordinary OR. However, between times t1 and t2, the nature of an XOR is emphasized. With 1s at both inputs, it outputs a 0.

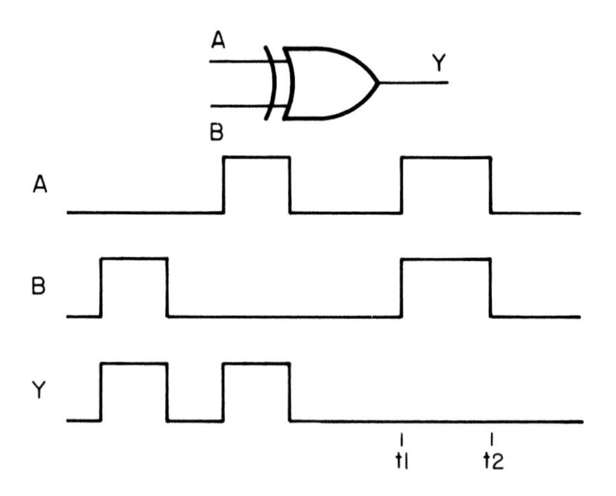

Fig. 5-37. This timing diagram restates the information in the XOR truth table.

problems. The XOR is not a fundamental logic element, so there are no Boolean rules for its manipulation. For example, if a circuit containing an XOR were to be simplified, the circled + form of notation could not be used in the circuit's expression. Rather, $\bar{A}B + A\bar{B}$ would be used to represent the exclusive OR element.

XOR BOOLEAN EXPRESSION FROM PRODUCT-OF-SUMS, WITH A COST ANALYSIS

A Boolean expression for an XOR can be written using product-of-sums. The Boolean expression is as follows:

$$(A + B)(\bar{A} + \bar{B}) = Y$$

The results of overlapping signals are shown in part "a" of Fig. 5-38. At times t1 and t2, the potential for glitch production is emphasized. It is almost impossible for the pulses at A and B to rise and fall at exactly the same time. As a result, glitches are likely if such pulses are applied to an XOR.

The race conditions at t1 and t2 in Fig. 5-38a are caused by the external circuit. Replacing the XOR will not correct the situation. Fig. 5-38b shows how serious the problem can be. Even with the same signal applied to A and B, small, unequal delays within the element result in glitches at Y.

XOR CIRCUITS

Various circuits are used in the construction of XORs. The most familiar uses ANDs, ORs, and NOTs. See part "a" of Fig. 5-39. This is the direct

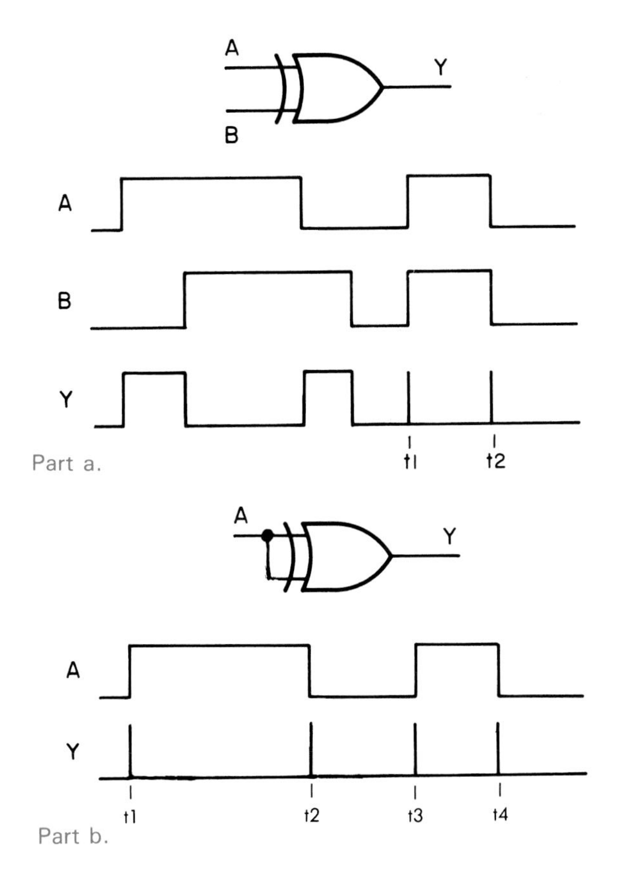

Fig. 5-38. When race conditions exist at the inputs of an XOR, glitches will appear at the element's output. Even when the same signal is applied to both inputs (as in part "b"), glitches are likely at Y. Unequal delays within the element produce these small spikes. A very high-speed oscilloscope is needed to see these glitches.

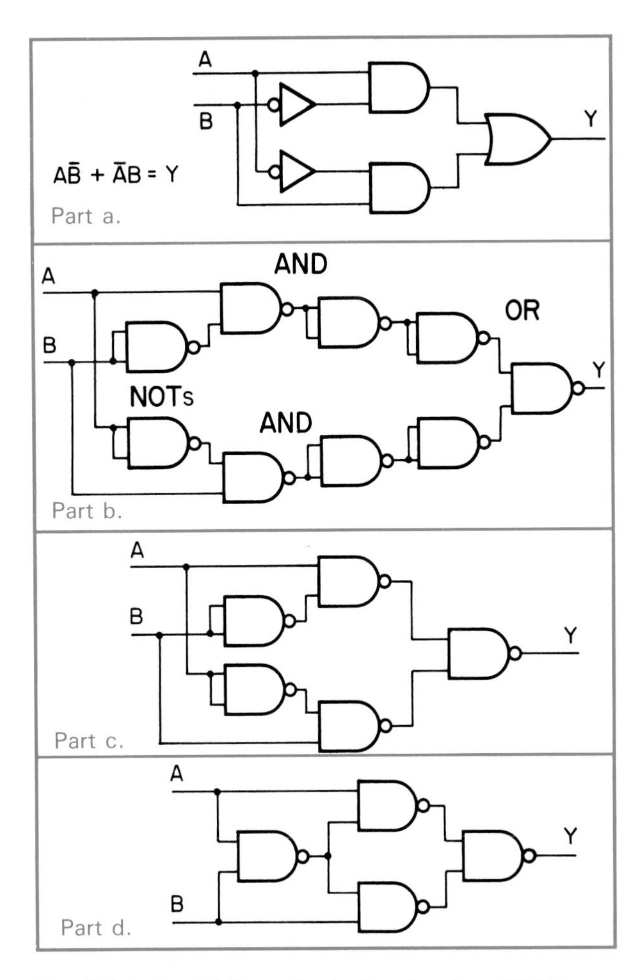

Fig. 5-39. An XOR can be built using four NANDs.

implementation of the expression $A\overline{B} + \overline{A}B$.

NANDs can also be the basis for an XOR. See part "b" of Fig. 5-39. NAND equivalents have replaced the ANDs, ORs, and NOTs. At c, redundant NOTs have been removed. At d, a trick was used to replace the remaining NOTs with a single NAND. Through truth table circuit analysis, it can be shown that this circuit is an XOR.

XOR APPLICATIONS

XORs are often found in the arithmetic units of computers. The adder will again be used as an example. The truth table for adding two binary bits is shown in Fig. 5-40. Sum-of-products was used to obtain expressions for the two outputs. As before, C was implemented by an AND. The expression for output S (sum) is identical to that of an XOR, so this element completes the circuit.

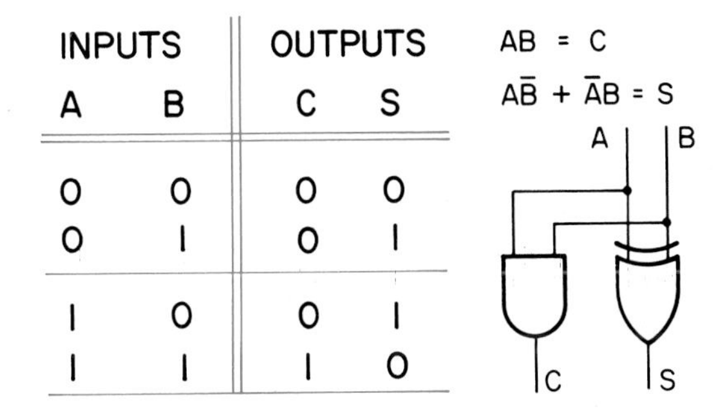

INPUTS		OUTPUTS	
A	B	C	S
0	0	0	0
0	1	0	1
1	0	0	1
1	1	1	0

Fig. 5-40. An XOR is used in the sum circuit of this adder. An AND provides the carry (C) output.

An XOR can be used as a NOT. See Fig. 5-41. The NOT action can be switched on and off. A system under complete computer control that must adjust to changing conditions can use a switchable NOT. The XOR is a versatile element.

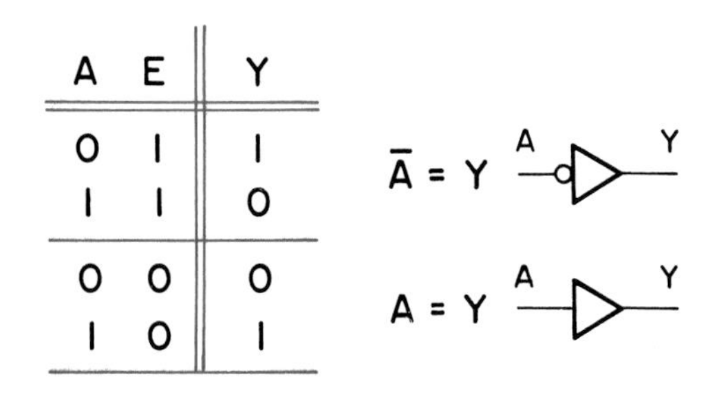

A	E	Y
0	1	1
1	1	0
0	0	0
1	0	1

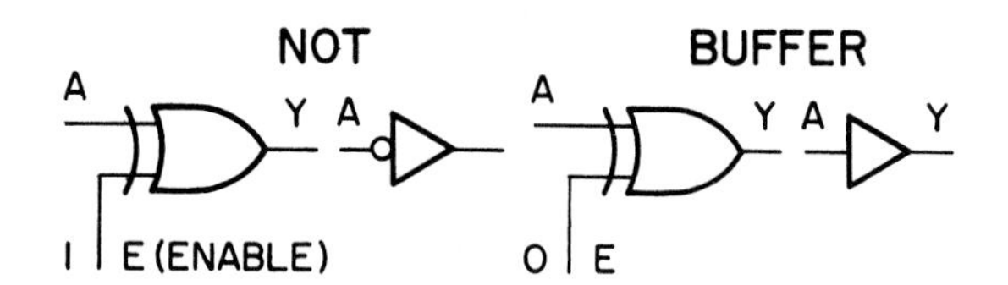

Fig. 5-41. An XOR can be used as a switchable NOT. The first two rows of the truth table show that Y is the complement of A when E = 1. The last two rows show that Y = A when E = 0. Depending on E (ENABLE), the NOT can be switched in or out of a circuit.

EXCLUSIVE NOR

Exclusive NORs are seldom used. They are, however, mentioned in the literature, so you should be familiar with this element. The three static descriptions of the XNOR are shown in Fig. 5-42. Using sum-of-products, the expression for this element is:

$$\bar{A}\,\bar{B} + AB = Y$$

XNORs might be described as equality circuits. When A equals B (both 1s or both 0s), Y is 1.

COMBINATIONAL LOGIC

Chapter 4 was concerned with combinational logic circuits composed of ANDs, ORs, and NOTs. The output of a combinational logic circuit is a combination of the input signals. In this chapter, NANDs, NORs, and XORs will be added.

INPUTS A	INPUTS B	OUTPUT Y
0	0	1
0	1	0
1	0	0
1	1	1

$\overline{A \oplus B} = Y$

$\bar{A}\bar{B} + AB = Y$

A

B

Y

Fig. 5-42. Sum-of-products was used to write the expression for this XNOR.

DIAGRAMS FROM EXPRESSIONS

Examples will be used to demonstrate the implementation of Boolean expressions. No attempt will be made to simplify expressions or circuits.

Example: NANDs and NORs can assist in the implementation of expressions containing NOTs. Refer to the expression in Fig. 5-43. Parentheses were added to emphasize that the two NORs are NANDed.

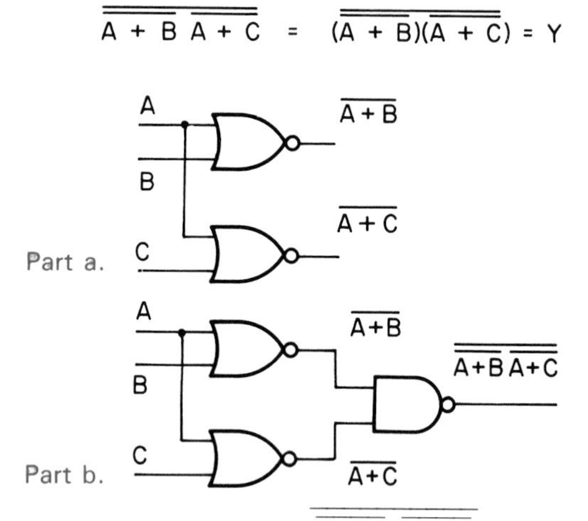

Fig. 5-43. The expression $\overline{\overline{A + B}\;\overline{A + C}} = Y$ has been implemented using NANDs and NORs.

Two steps are needed in Fig. 5-43. Part "a" of Fig. 5-43 involves the implementation of the NORs. In part "b," NANDing the outputs of the NORs completes the circuit.

Example: The expression in Fig. 5-44 is used. The quantity under the deepest set of NOTs in the expression in Fig. 5-44 will be implemented first. Four steps are needed:

1. In part "a" of Fig. 5-44, the NANDing of C and D begins the process.
2. In part "b," NORing the output of the NAND with E completes the more complex of the bracketed quantities.
3. In "c," an OR is used to create the quantity in the other set of brackets.
4. In "d," the ANDing of the bracketed quantities completes the implementation.

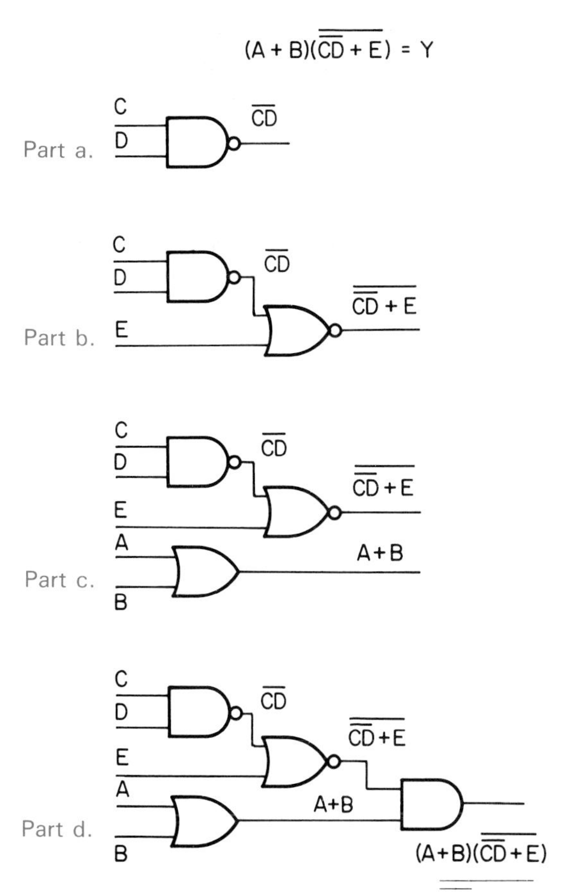

Fig. 5-44. For the expression $(A + B)(\overline{\overline{CD} + E}) = Y$, both inverting and noninverting elements were used.

Example: The Boolean expressions and circuits in Fig. 5-45 are identical. DeMorgan's theorem has merely been applied to the various elements.

Example: It is important to recognize the XOR in the expression in Fig. 5-46. The use of this element simplifies the implementation.

EXPRESSIONS FROM DIAGRAMS

Expressions for circuits containing NANDs, NORs, and XORs are obtained in the same way as those containing ANDs, ORs, and NOTs. Expressions are written for intermediate points in the circuit and then combined to produce the final expression.

Example: The 3-input NAND in Fig. 5-47 functions in the same way as a 2-input NAND. Its inputs are ANDed and then NOTed. The expression for this circuit is obtained by NORing C with the output of the 3-input NAND.

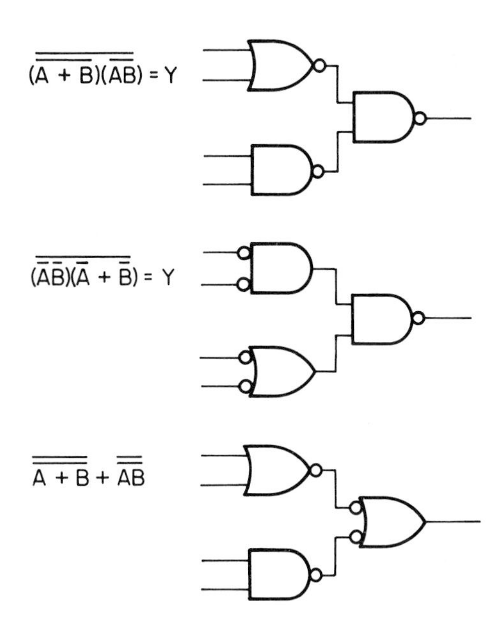

Fig. 5-45. These circuits are identical. Only the symbols have been changed by applying DeMorgan's theorem.

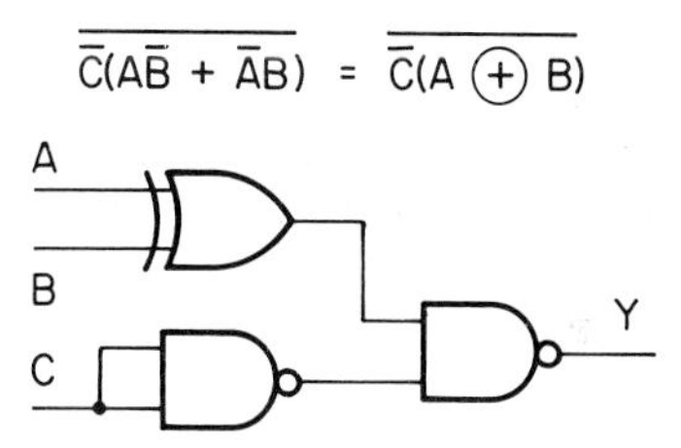

Fig. 5-46. The use of an XOR simplifies the implementation of the expression $\overline{C}\,\overline{(A\overline{B} + \overline{A}B)} = Y$.

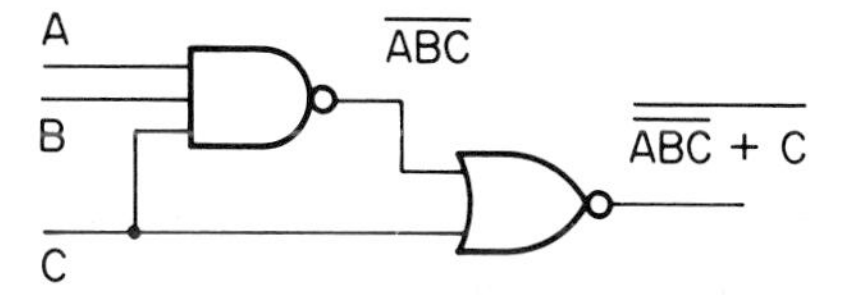

Fig. 5-47. The expression representing a circuit is obtained by writing expressions for all points in the circuit.

Example: The circuit in Fig. 5-48 contains an exclusive OR. While this makes the expression appear more complex, the writing process is similar to that of the example in Fig. 5-47.

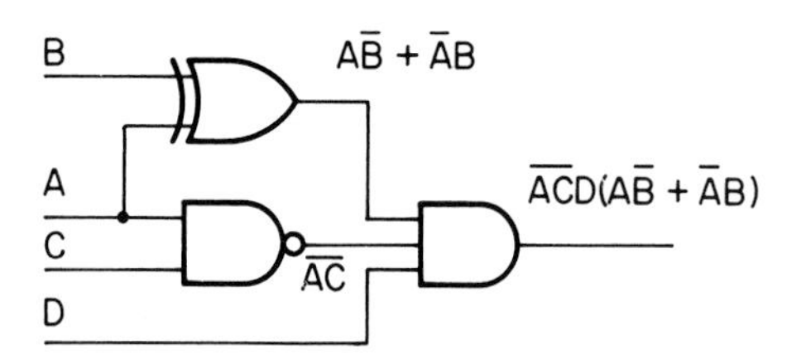

Fig. 5-48. The XOR makes the expression for this circuit appear complex.

Example: In Fig. 5-49, the input variables have been listed as D, E, and F. This was done to simplify the writing of the expression. It does not alter the results.

At b, the circled + notation has been used to represent the XOR. The output of this circuit is (DE) XORed with F. To represent this in Boolean algebra form, care must be taken to treat DE as one quantity. This is emphasized at c and d. In this expression, A = DE. When A is NOTed by the XOR, the whole quantity DE must be NOTed.

TROUBLESHOOTING

Troubleshooting of circuits containing NAND, NOR, and XOR elements follows the same pattern as those containing only AND, OR, and NOT elements. To review:

1. Input signals that produce incorrect outputs are applied.
2. Faults are localized.
3. Faults are isolated.
4. Faults are repaired.
5. Circuit is retested to insure that faults have been corrected.

Again, examples will be used.

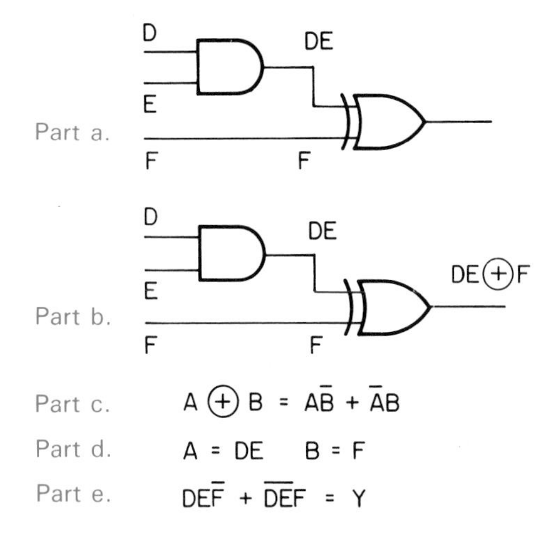

Fig. 5-49. D AND E is treated as one number by the XOR in the circuit.

Example: For the inputs shown, the circuit in Fig. 5-50 should output a 1. The measurements at b indicate that it outputs a 0. Comparing predicted and measured results suggests that the output of the first NOR is incorrect. The fault has been localized.

To isolate the fault, the lead between the NOR and NOT was opened. See the circuit at c. Based on the measured signals, the output of the NOR is s-a-1. Replacing this chip should correct the problem.

Example: Based on predicted and measured signals, the output of the XOR in Fig. 5-51 appears to be incorrect. This observation localizes the fault.

To isolate the fault, the lead between the XOR and the NAND was opened. See part "c" of Fig. 5-51. Note the output of the XOR. It changed to 0. The XOR appears to be working properly, so it is likely that the input to the NAND is s-a-1. Changing this chip should remove the fault.

USE OF DE MORGAN'S THEOREM ON AND, OR, NOT

DeMorgan's theorem has been applied to NANDs

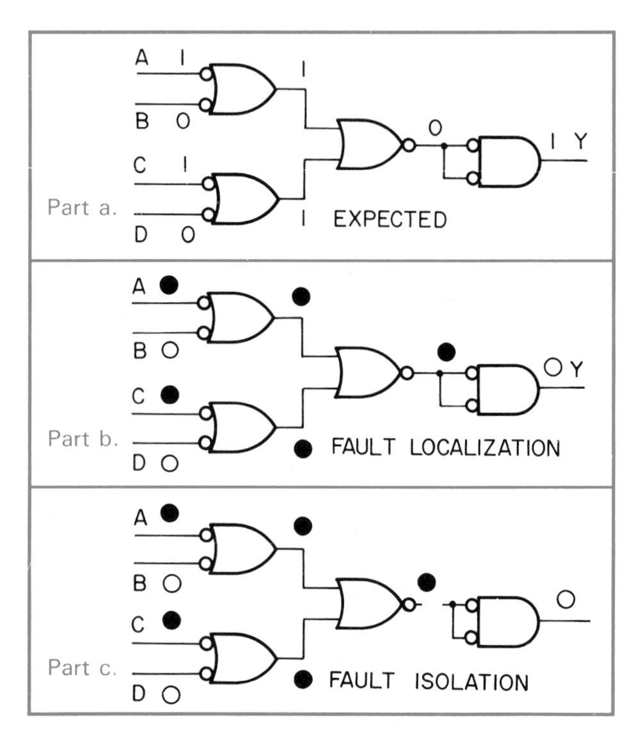

Fig. 5-50. Fault isolation in NAND/NOR circuits is identical to that used on circuits containing ANDs, ORs, and NOTs.

and NORs. It can also be applied to AND elements, OR elements, and NOT elements.

To review, DeMorgan's theorem inverts all aspects of a logic element. ANDs become ORs; ORs become ANDs. Inputs and outputs without bubbles become bubbled; those with bubbles lose them.

DE MORGAN IMPLEMENTATION OF AN AND

Fig. 5-52 shows the DeMorgan implementation of an AND. It becomes an active-low OR. Note the shaded portion of the truth table. When any input is low, the output of this device will be active low. Only when both inputs are inactive (high) will the output be inactive (high).

A Boolean algebra proof of the equivalency of the two symbols is shown at c. When $\overline{A} + \overline{B}$ is DeMorganed, it becomes $\overline{AB}$. And AB NOT NOT equals AB. The two symbols have the same truth table. The AND-based symbol emphasizes active high; the OR-based symbol emphasizes active low.

Fig. 5-53 shows a use for the DeMorganed AND. The circuits shown are identical. The action of the circuit at the left is a bit more difficult to follow. This is due to the bubbled outputs driving unbubbled inputs. This problem is solved at the right. The DeMorganed circuit emphasizes that output Y will be 0 when any input is 1. The circuit is a 4-input NOR.

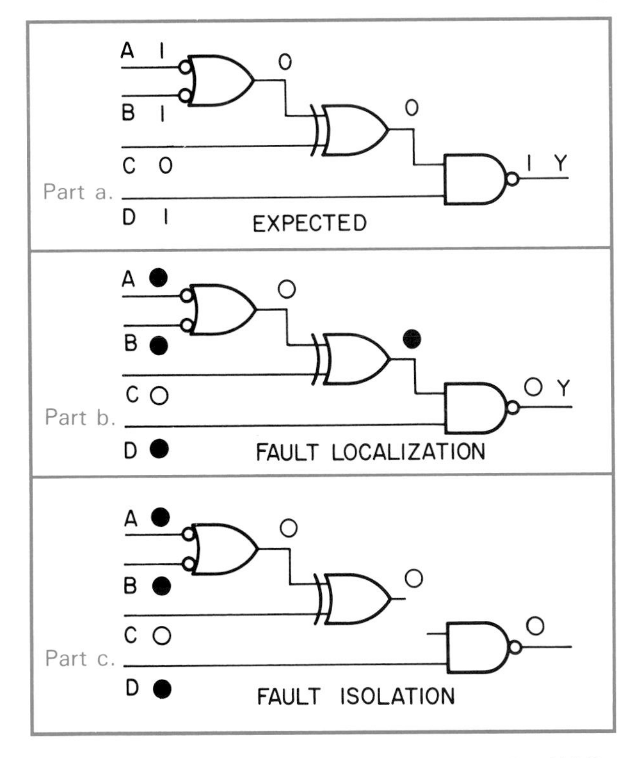

Fig. 5-51. Troubleshooting of circuits containing XORs follows the same methods used in other circuits.

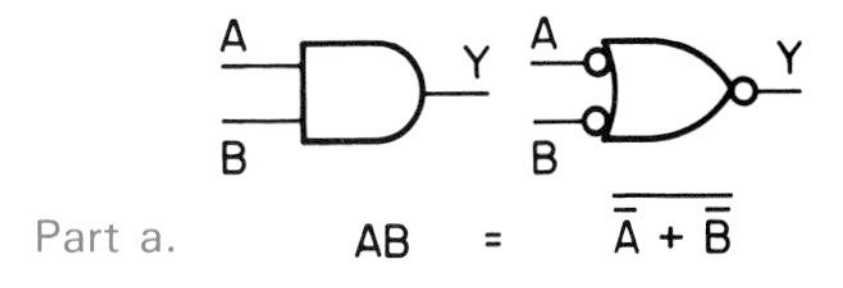

Part a. AB = $\overline{\overline{A} + \overline{B}}$

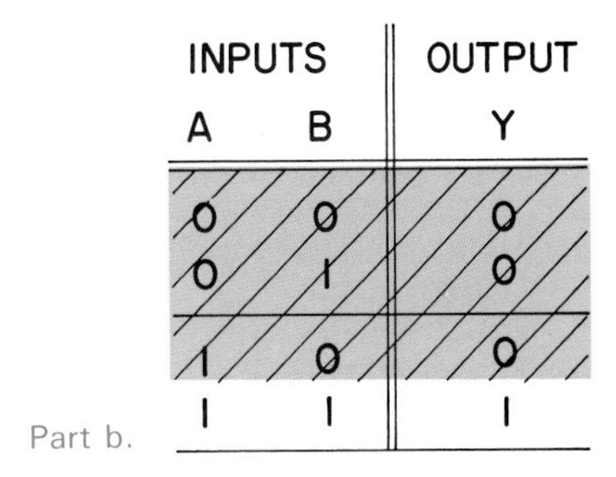

INPUTS		OUTPUT
A	B	Y
0	0	0
0	1	0
1	0	0
1	1	1

Part b.

$$AB = \overline{\overline{A} + \overline{B}}$$

$$AB = \overline{\overline{AB}}$$

Part c.

$$AB = AB$$

Fig. 5-52. DeMorgan's theorem can be applied to ANDs.

DE MORGAN IMPLEMENTATION OF AN OR

Fig. 5-54 shows the DeMorganing of an OR. An OR can be considered to be an active-low AND. Note the shaded portion of the truth table.

Fig. 5-55 shows an application of the DeMorganed OR symbol. In the circuit shown, both A and B must be low if C is to be active (low input). This is much more obvious in the drawing at the right.

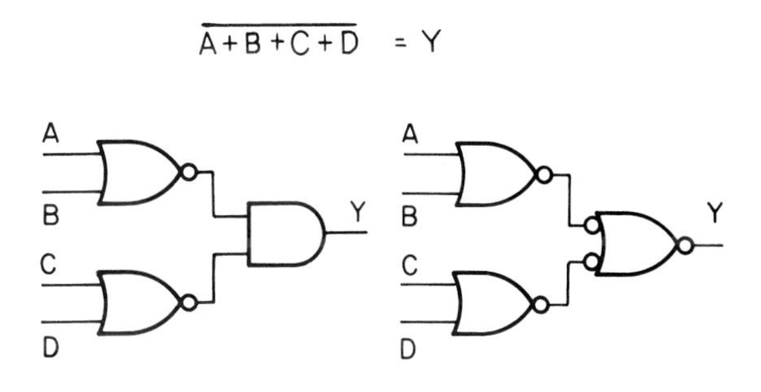

Fig. 5-53. The circuit containing the DeMorganed AND is easier to analyze.

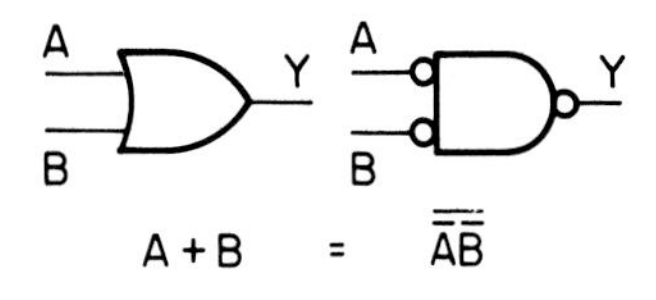

INPUTS		OUTPUT
A	B	Y
0	0	0
0	I	I
I	0	I
I	I	I

Fig. 5-54. DeMorgan's theorem can be applied to ORs.

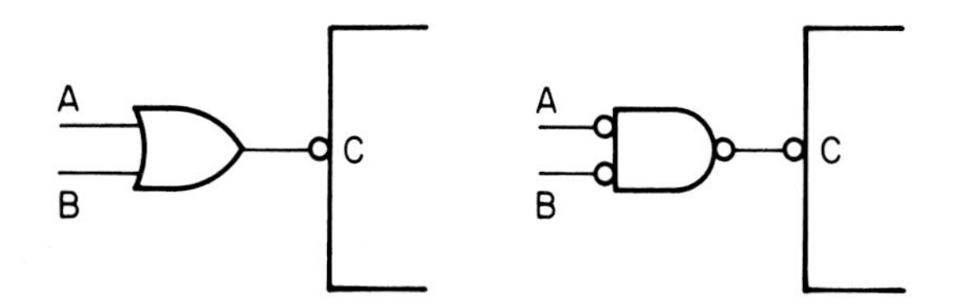

Fig. 5-55. The action of this circuit is easier to analyze when the OR has been DeMorganed.

DE MORGAN IMPLEMENTATION OF A NOT

Part "a" of Fig. 5-56 shows the DeMorgan implementation of a NOT. The only change is in the position of the bubble.

At b, the result of applying DeMorgan's theorem to a buffer (amplifier) is shown. Both symbols result in no logic signal change.

Fig. 5-57 summarizes the DeMorgan implementations. It emphasizes the rule that all aspects of a logic element are inverted when DeMorgan's theorem is applied.

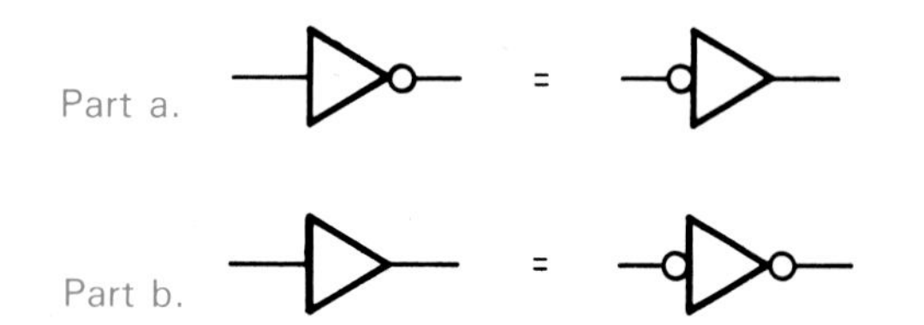

Fig. 5-56. DeMorgan's theorem can be applied to NOTs and buffers.

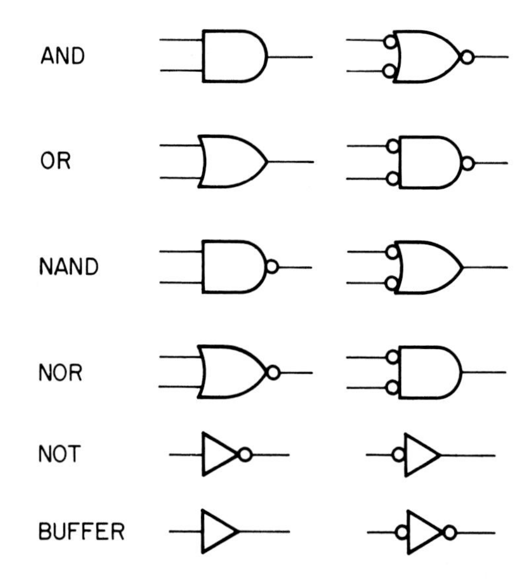

Fig. 5-57. In each case, either symbol can be used to represent the logic elements which are shown.

CIRCUIT ANALYSIS USING BUBBLES

When ANDs, NANDs, ORs, and NORs are used in a circuit, truth table circuit analysis is complicated by the number of truth tables that must be remembered. DeMorganing selected elements can speed analysis by encouraging the use of the concept of active-high/active-low. With this concept, analysis can be approached in terms of only ANDs and ORs.

Fig. 5-58 suggests the process.

Logic Element 1: This is an OR with active-low inputs. If an input is 0, it will output a 1. B is 0, so the output of element 1 is 1.

Logic Element 2: This is an AND with an active-low output. Its output will be 0 when 1s are applied to BOTH inputs. Both inputs are 1; its output is 0.

Logic Element 3: This is an OR with active-low inputs and output. If an input is 0, it will output a 0. One of its inputs is 0, so its output is 0.

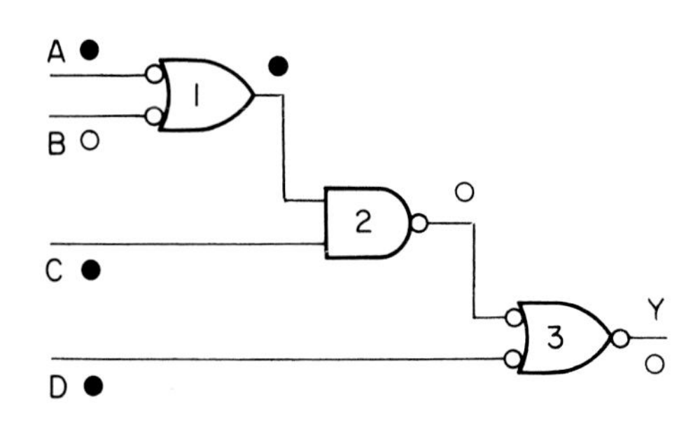

Fig. 5-58. Circuit analysis using bubbles is possible when the proper symbols are used.

Note that no mention was made of the fact that element 1 is a NAND and element 3 is an AND.

Trace Fig. 5-58 onto a piece of paper, and try to repeat the process for A = B = C = 0 and D = 1.

Logic Element 1: If an input of this element is 0, its output will be 1. Both are 0; its output is 1.

Logic Element 2: For its output to be 0, both inputs must be 1. They are not both 1; its output is 1.

Logic Element 3: If a 0 appears at an input of this element, it outputs a 0. There are no 0s at its inputs, so it outputs a 1. Y = 1.

RELATION OF NEGATIVE LOGIC TO DE MORGAN'S THEOREM

Most circuits use positive logic. That is, voltages representing logic 1s are more positive than those representing logic 0s. TTL logic is an example. As you are aware, logic 1s in this family are anywhere between +2 and +5 volts; logic 0s are between 0 and +0.8 volts.

However, there is no physical reason that this could not be reversed. The more positive voltage could represent logic 0; the less positive voltage could be taken as logic 1. Such an arrangement is called negative logic. See Fig. 5-59.

Based on DeMorgan's theorem, a positive-logic AND becomes a negative-logic OR. Also, a positive-logic OR changes into a negative-logic AND.

In the early days of digital electronics, complete families of logic elements were not available. If a circuit required large numbers of ANDs, but only high-speed ORs were available, designers might resort to negative logic. This turned the available ORs into negative-logic ANDs.

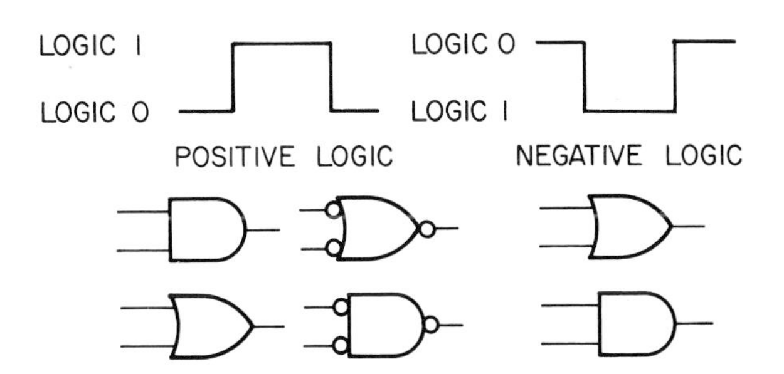

Fig. 5-59. In a negative-logic circuit, the more positive signal voltage represents 0; the less positive signal represents 1.

Negative logic is confusing for those who have worked only with positive logic. Little negative logic is used today. In those cases where negative logic would help explain the operation of a circuit, the concepts of active high and active low can be used. That is, the bubble notation of Fig. 5-57 can be substituted.

SUMMARY

NAND and NOR elements are NOTed ANDs and ORs. They are extensively used in digital circuits.

NANDs and NORs are universal logic elements. Complete circuits can be constructed using only NANDs or NORs. That is, they can be connected to simulate the actions of ANDs, ORs, and NOTs.

XORs may simplify certain circuits. They are often used in arithmetic sections of computers.

Boolean expressions can be written for circuits containing NAND, NOR, and XOR elements. Also, NANDs, NORs, and XORs can be used to implement such expressions. The methods for accomplishing these tasks are similar to those used on circuits containing only ANDs, ORs, and NOTs.

Truth table circuit analysis can be used to troubleshoot circuits containing NAND, NOR, and XOR elements.

IMPORTANT TERMS

Active high, Active low, Enable, Inverting, Negative logic, Redundant.

TEST YOUR KNOWLEDGE

1. What is an inverting output?
2. To make a NAND from an AND, are the inputs NOTed or is the output NOTed?
3. What is the Boolean expression for a 2-input NAND element?
4. For what reason is an active low signal used in the start/stop control of a machine?
5. What does the term "enable" mean?
6. How many NANDs are needed to build a NOT?
7. What is the meaning of a bubble in a logic diagram?
8. When DeMorgan's theorem is used on a logic element, ANDs become NANDS. (True or False?)
9. Does a DeMorgan process change all bubbled leads on a logic element to bare leads and all bare leads to bubbled leads?
10. A 2-input NOR used as a gate is open when one input is held at 1. (True or False?)
11. What is the output of an XOR element when both of its inputs are 1?
12. What is the minimum number of NANDs used to make an XOR? How many chips are used?
13. What is the most common Boolean expression for an XOR element?

14. How many NORs are used to produce an AND?
15. Can the circled plus sign be used in a Boolean expression?
16. Glitches caused by overlapping inputs in a timing diagram are a problem of what logic element?
17. Can an XOR be used as a NOT?
18. An XNOR is a NOR with a NOTed output. (True or False?)
19. An XNOR is an XOR with a NOTed output. (True or False?)
20. In negative logic, a negative voltage represents a logic 0. (True or False?)

STUDY PROBLEMS

1. Using the three static methods (symbol, truth table, and Boolean expression), describe the action of a 2-input NAND.
2. Trace the timing diagram in Fig. P5-1 onto a piece of paper and carefully construct the output signal of the 2-input NAND.

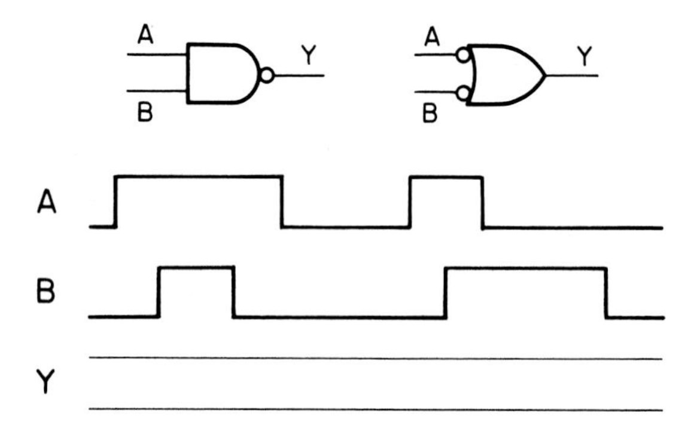

Fig. P5-1. Determine the shape of the output signal waveform for the 2-input NAND element and the input waves that are shown.

3. Repeat Problem 2 for the timing diagram and 3-input NAND in Fig. P5-2.
4. Draw the NAND equivalent of the following elements:
   a. NOT.
   b. AND.
   c. OR.
5. For a 2-input NAND, write the Boolean expression for DeMorgan's theorem.
6. Implement the expression in Fig. P5-3 using only NANDs. The brackets are not needed. They were used only to aid in solving this problem. Do not try to simplify the circuit. Practice using NANDs as NOTs and as universal elements.

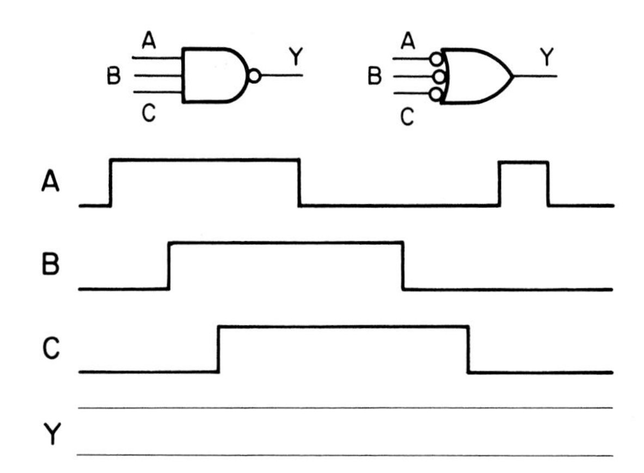

Fig. P5-2. Determine the output signal shape for the 3-input NAND and inputs shown.

$$(\overline{AB})C = Y$$

Fig. P5-3. NANDs can provide the circuit for the Boolean expression $(\overline{AB})\ C = Y$.

7. Implement the expression in Fig. P5-4 using only NAND elements.
8. Repeat Problem 7 for the expression in Fig. P5-5.

$$\overline{AB} + C = Y$$

Fig. P5-4. NANDs can describe the expression $\overline{AB} + C = Y$.

$$\overline{AB} + \overline{AC} = Y$$

Fig. P5-5. NANDs simplify $\overline{AB} + \overline{AC} = Y$.

9. Using the three static methods, describe the action of a 2-input NOR.
10. Trace the timing diagram in Fig. P5-6 onto a piece of paper and carefully construct the output signal of the 2-input NOR.
11. Repeat Problem 10 for the timing diagram and 3-input NOR shown in Fig. P5-7.
12. Implement the expression in Fig. P5-8 using only NOR elements.

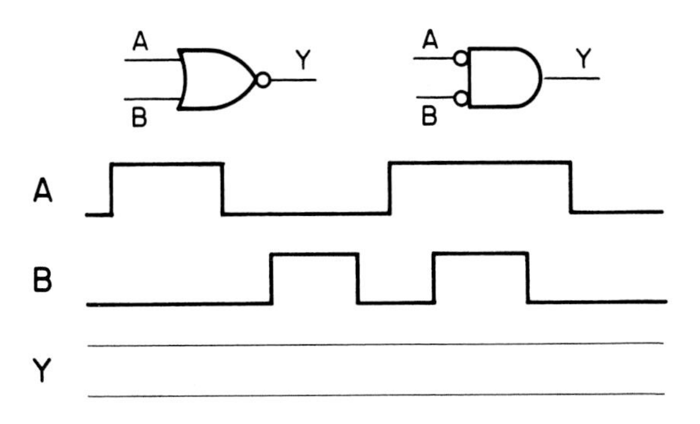

Fig. P5-6. Determine the output signal shape for the 2-input NOR and inputs shown.

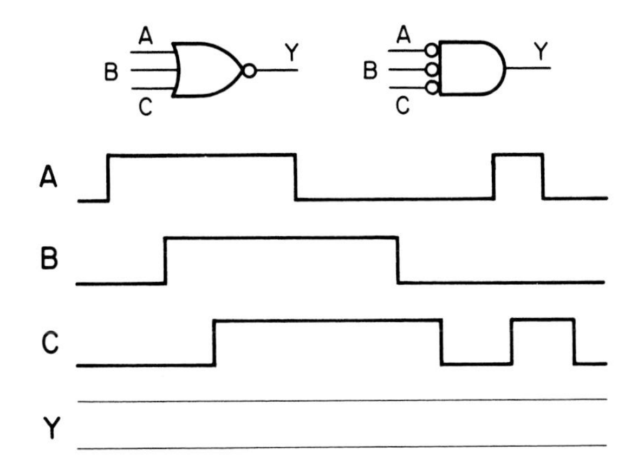

Fig. P5-7. Determine the output signal shape for the 3-input NOR and inputs shown.

$$\overline{\overline{A + B} + C} = Y$$

Fig. P5-8. NORs can provide the circuit for the expression $\overline{\overline{A + B} + C} = Y$. NORs simplify the expression.

13. Repeat Problem 12 for the Boolean expression given in Fig. P5-9.
14. Repeat Problem 12 for the Boolean expression given in Fig. P5-10.

$$\overline{A + B} + C = Y$$

Fig. P5-9. NORs can describe $\overline{A + B} + C = Y$.

$$\overline{A + B} + AC = Y$$

Fig. P5-10. NORs can describe $\overline{A + B} + AC = Y$.

15. Using the three static methods, describe the action of a 2-input XOR.
16. Trace the timing diagram in Fig. P5-11 onto a piece of paper and carefully construct the output signal of the XOR.
17. At which of the lettered points in Fig. P5-11 (the timing diagram of Problem 16) is a glitch likely to be produced?

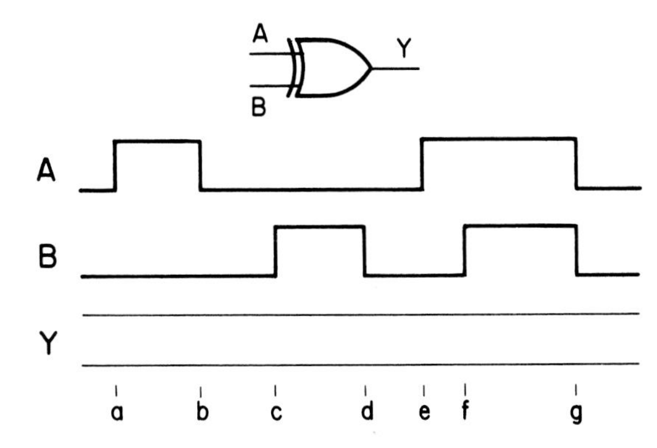

Fig. P5-11. Determine the output signal shape for the XOR and inputs shown. Note any problems with glitches.

18. Repeat Problem 16 for the XOR timing diagram that is depicted in Fig. P5-12.

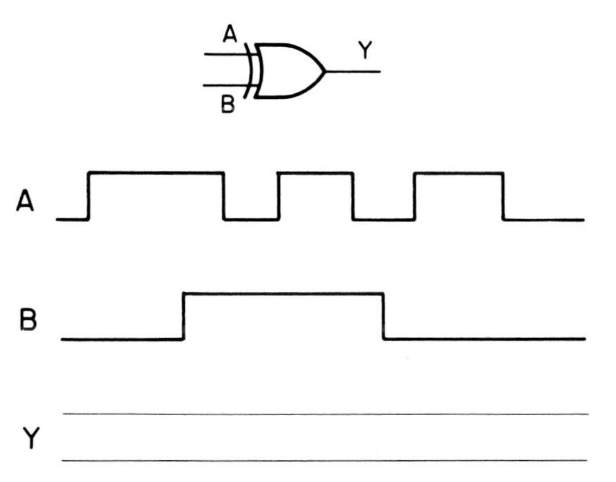

Fig. P5-12. Draw the output shape for the XOR and inputs shown. No glitches should occur.

19. Using only NANDs and XORs, implement the expression in Fig. P5-13.

$$\overline{(\bar{A}B + A\bar{B})C} = Y$$

Fig. P5-13. NANDs and XORs can provide the circuit for the expression $\overline{(\bar{A}B + A\bar{B})\,C} = Y$.

20. Repeat Problem 19 for the expression in Fig. P5-14. Again, use only NANDs and XORs.

$$A\bar{B} + C + \bar{A}B = Y$$

Fig. P5-14. NANDs and XORs can describe $A\bar{B} + C + \bar{A}B = Y$.

21. Write the simplest Boolean expression for the circuit in Fig. P5-15. Take the simplest expression from the circuit as it is drawn.

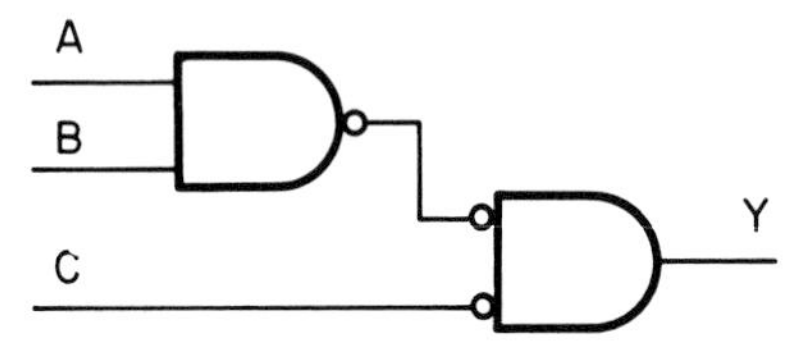

Fig. P5-15. There are four Boolean expressions for the NAND and the AND with NOTed inputs shown. Only the simplest of the four is to be written. The others involve DeMorgan processes.

22. Write the Boolean expression for the circuit in Fig. P5-16. Do not simplify.

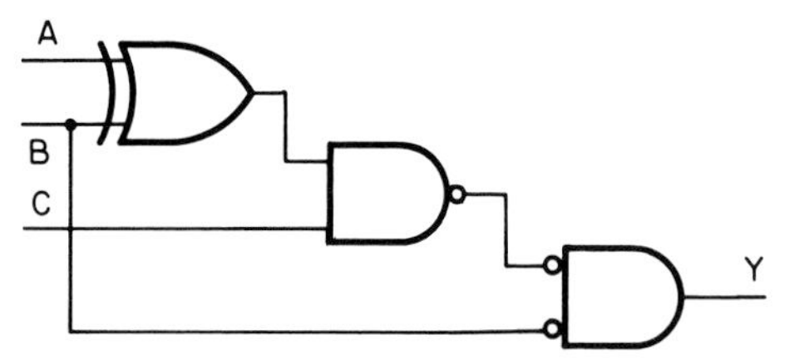

Fig. P5-16. The Boolean expression can be found for the XOR, NAND, and NOR shown.

23. Copy the circuit in Fig. P5-17 onto a piece of paper. Using dashed lines, identify NAND equivalents of an AND and an OR.

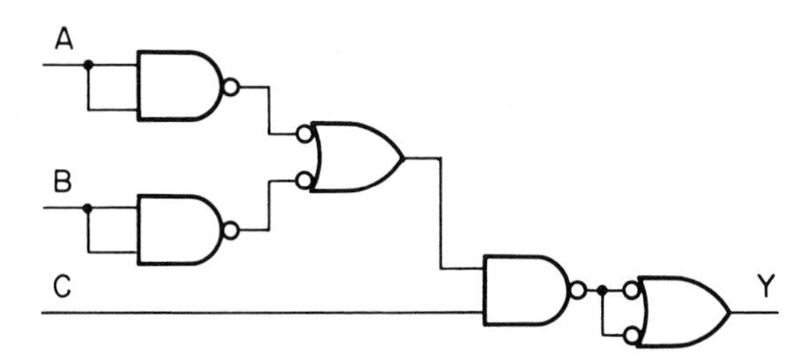

Fig. P5-17. An outline in dashed lines is needed to help identify an AND element and an OR element.

24. Evaluate (determine Y) the expression in Fig. P5-18 for A = 1, B = 0, and C = 1. The brackets are not needed. They were added to aid in solving the problem.

$$\overline{\overline{(A + B)}\,\overline{(A + C)}} = Y$$

Fig. P5-18. If A = 1, B = 0, and C = 1, what is the output of the Boolean expression $\overline{\overline{(A + B)}\,\overline{(A + C)}} = Y$?

25. Repeat Problem 24 for the Boolean expression given in Fig. P5-19.

$$\overline{\overline{(A + B)}C} + \overline{AB} = Y$$

Fig. P5-19. If A = 1, B = 0, and C = 1, what is the output of $\overline{\overline{(A + B)}\,C} + \overline{AB} = Y$?

26. For the inputs shown in Fig. P5-20, determine the output of this circuit.
27. Repeat Problem 26 for the circuit and input set in Fig. P5-21.
28. For the inputs shown, the output of the circuit in Fig. P5-22 is incorrect. At the output of which element does this error first appear?
29. Repeat Problem 28 for the circuit and measured signals in Fig. P5-23.

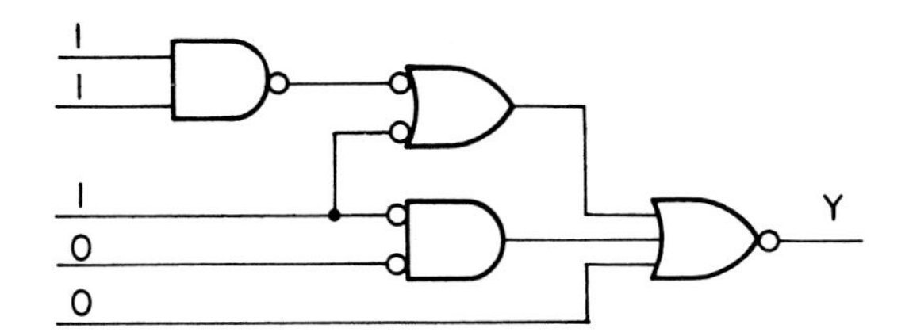

Fig. P5-20. Use of the active high/active low approach makes this circuit easy to analyze.

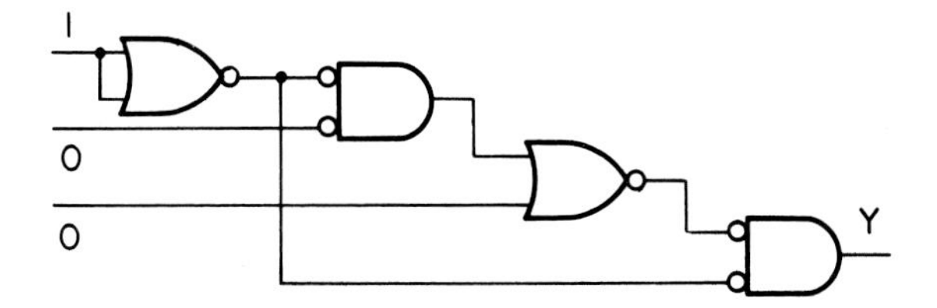

Fig. P5-21. What is the output if the inputs are 1,0,0?

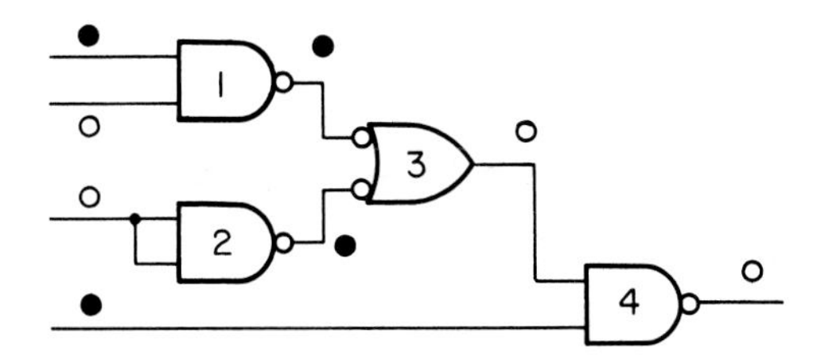

Fig. P5-22. Where does an error first appear in the circuit shown if the measured inputs are on, off, off, on?

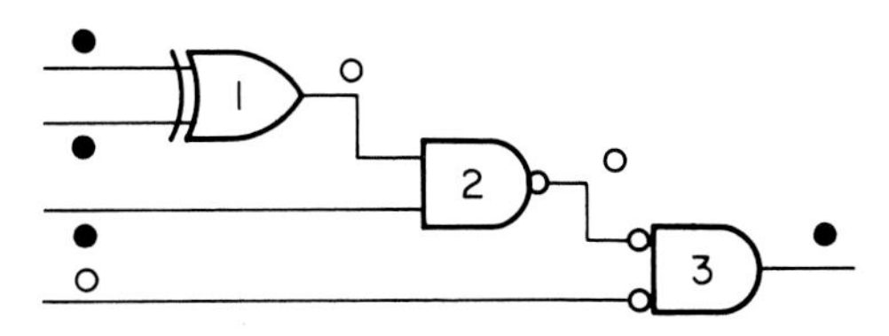

Fig. P5-23. Where does an error first appear in the circuit shown if the measured inputs are on, on, on, off?

30. Repeat Problem 28 for the circuit and measured signals in Fig. P5-24.

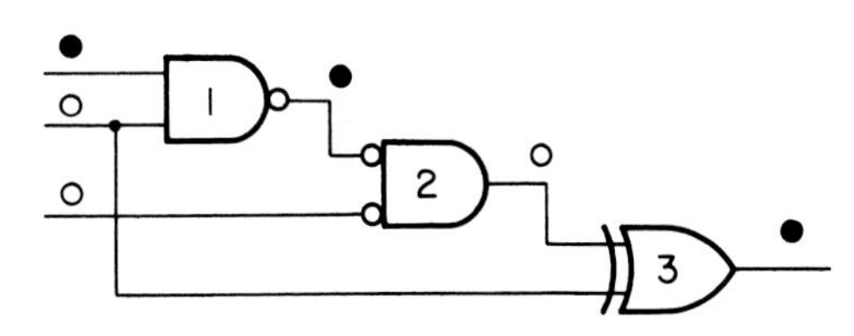

Fig. P5-24. Where does an error first appear in the circuit shown if the measured inputs are on, off, off?

For engineers and technicians, learning does not end with graduation. It continues in industry. (Hewlett-Packard Company)

Chapter 6

Design of Logic Circuits

LEARNING OBJECTIVES

After studying this chapter and completing lab assignments and study problems, you will be able to:

- Explain how to follow descriptions of the subjective approach to the design of digital circuits.
- Restate verbal descriptions of digital design problems in truth table form.
- Simplify expressions obtained from the sum-of-products method by using Boolean identities.
- Construct Karnaugh maps from truth tables for circuits with four or less inputs.
- Write Boolean expressions from Karnaugh maps.

The process of inventing digital circuits is called design. It begins with a detailed statement of a problem to be solved. The process ends with digital circuits that solve that problem.

Design is often considered an engineering activity. However, the need for help in circuit design by other workers in the digital field is growing. In addition, technical communication is more effective when team members are familiar with the work performed by co-workers.

DESIGN METHODS

There are two approaches to digital circuit design. One method depends heavily on the designer's skill and experience. The method is called subjective design. The other method uses truth tables and Boolean algebra as design tools. This method is called objective design.

SUBJECTIVE DESIGN

It is possible for a designer to go directly from a statement of a problem to a digital circuit. Such a process is called subjective design, since it occurs completely within the designer's mind.

The subjective approach has long been used to design switch-based circuits. It can also be used to design IC control circuits.

STATEMENT OF THE PROBLEM

All design processes begin with a detailed description of the problem to be solved. For example, a complete description of the cycle of a machine must be available and understood before controls can be designed.

DESIGN OF THE CIRCUIT

The second segment of the design process involves selecting a circuit that will solve the problem. The two-step nature of design is emphasized in the following example.

Example: Part "a" of Fig. 6-1 shows five switches involved in the starting of a machine. The problem is to design a digital circuit that signals when the machining cycle is to begin. A main cutoff is not shown. The subjective approach will be used.

The process begins with a statement of the problem:

1. In part "a" of Fig. 6-1, the cycle is to start when both START1 and START2 are pressed. A design must result in a safe machine, so safety is a consideration in all portions of the design process. However, this is an introduction to digital design. Attention to the details of safe machine design would unduly complicate the discussion. Even so, remember that safety is part of every good design.
2. However, the cycle must not start (even though both of the start switches have been pressed) if there is no part in the fixture. (A fixture is a vise or clamp used to hold parts for machining.) The presence of a part is sensed by the switch marked with the name PART.
3. Repair personnel must be able to cycle the machine without a part in the fixture by using the switch marked OVERRIDE. That is, either PART or OVERRIDE must be activated for the machine to start operating.
4. In addition, the cycle may start only if the spindle (rotating part of the machine) is in its uppermost

position. This is sensed by switch SPINDLE.

Assume the switches are active high. That is, a 1 indicates that a switch has been activated.

The second part of the process involves selecting circuits to meet the requirements.

1. The critical word in the description of the action of switches START1 and START2 is "both." It implies an AND. See part "b" of Fig. 6-1.
2. However, the machine must not start without a part in the fixture. That is, the output of the AND and switch PART must both be 1. See the AND in part "c" of Fig. 6-1.
3. Repair personnel must be able to cycle the machine without a part in the fixture, so either PART or OVERRIDE must be activated to start the cycle. The word either is important. It implies an OR. See part "d" of Fig. 6-1.
4. Finally, the switch that senses that the spindle is up enters into the problem. SPINDLE must be activated before the cycle can begin. Part "e" of Fig. 6-1 shows SPINDLE ANDed with the outputs of the previous elements.

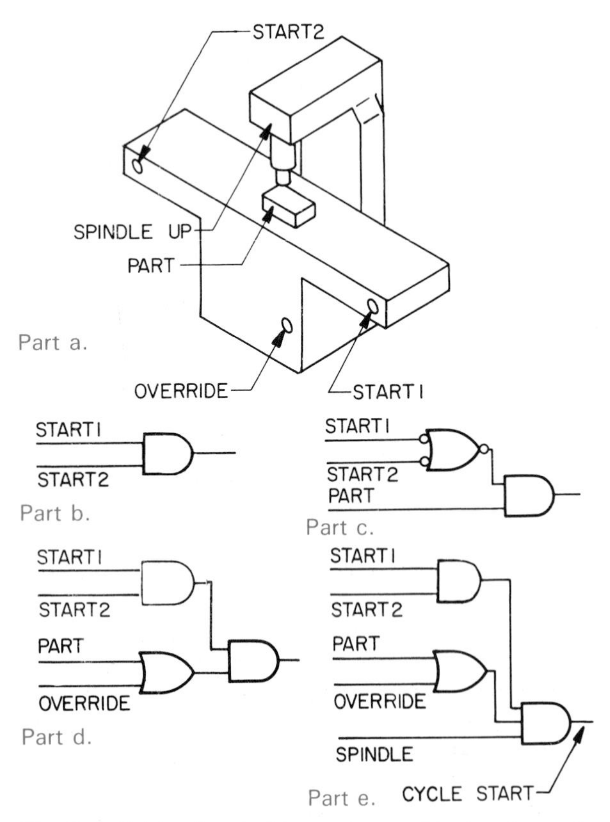

Fig. 6-1. Subjective design can be used to develop a circuit to control the start of a machine. A MAIN STOP, not shown, is a second override.

This process would continue until all aspects of the machine's operation have been accounted for. The resulting circuit would then be assembled and tested.

TRUTH-TABLE BASED DESIGN

The use of a truth table as the basis for the design of a circuit permits the use of Boolean algebra. As a result, this approach is highly objective and less dependent on the designer's skill and experience.

STEPS IN THE PROCESS

The steps in the truth-table based design process are as follows:

1. Problem is carefully stated.
2. Problem is restated in truth table form.
3. Truth table information is converted to Boolean expressions.
4. Expressions are simplified.
5. Simplified expressions are implemented (converted to logic diagrams).
6. Prototypes are constructed and tested.

The following are examples of steps 1 through 3. Simplification will be considered shortly.

Example (decoder for computer instructions): Design a circuit that will recognize three op codes.

Binary numbers are used to tell computers what to do. These are called *OPERATIONAL CODES* (OP CODES). The following op codes are to be used in this design problem:

Op Code	Operation
011	Add
100	Subtract
101	Memory write (store a word in memory)

The circuit is to have three inputs, I2, I1, and I0, for the op code. It will have three outputs, one for each operation. When the number 011 appears at the input, a 1 is to appear on the ADD lead. When 100 is applied, the ADD lead is to go to 0, and a 1 is to appear on SUBTRACT. Finally, when 101 is applied, the 1 is to move to MEMORY WRITE. When any other number appears on the input leads, the three outputs are to be 0.

Part "a" of Fig. 6-2 shows the truth table describing the above problem. Even a small computer would have more than three instructions, but for now, only add, subtract, and memory write will be considered.

Sum-of-products was used to write expressions b, c, and d. The resulting circuit is shown at e. It would be called a *DECODER*. This circuit decodes the op codes and places the correct signals on the appropriate control leads.

INPUTS			OUTPUTS		
I2	I1	I0	A	S	W
0	1	1	1	0	0
1	0	0	0	1	0
1	0	1	0	0	1
OTHER INPUT SETS NOT OF INTEREST			0	0	0
			0	0	0

Part a.

Part b. $(\overline{I2})(I1)(I0) = A$ (ADD)

Part c. $(I2)(\overline{I1})(\overline{I0}) = S$ (SUBTRACT)

Part d. $(I2)(\overline{I1})(I0) = W$ (WRITE)

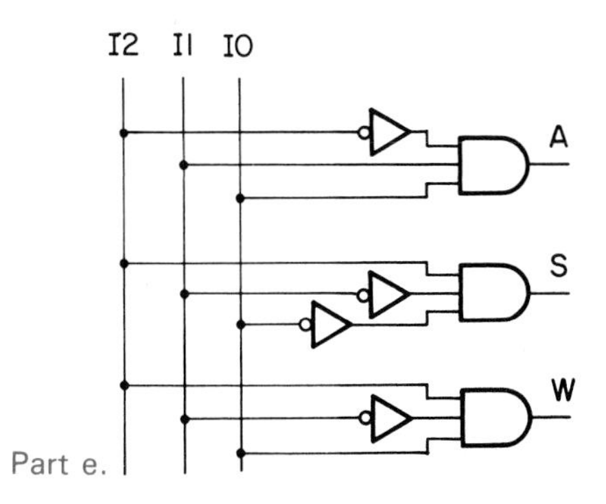

Part e.

Fig. 6-2. Sum-of-products was used in the design of this decoder. Only one set of products is used for each output, because each output is associated with only one input set.

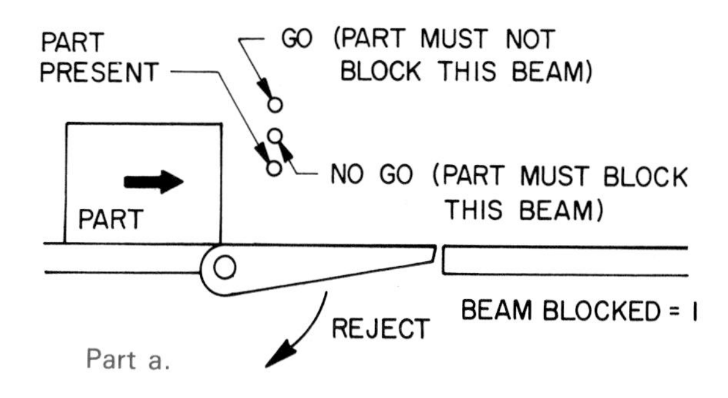

Part a.

	INPUTS			OUTPUT
	G	N	P	R
Row b.	1	1	1	1
Row c.	0	0	1	1
Row d.	0	1	1	0
Row e.	0	0	0	0

OTHER INPUT SETS NOT OF INTEREST

Part f. $GNP + \overline{G}\,\overline{N}P = R$

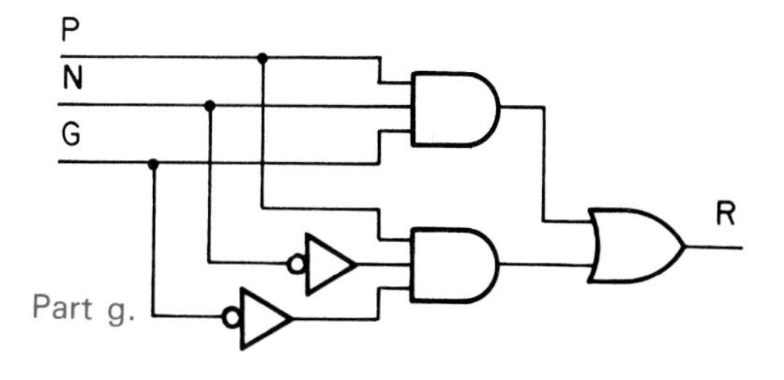

Part g.

Fig. 6-3. The design of this circuit begins with the construction of a truth table describing the action of the machine.

Example (automatic inspection): Design a circuit to reject components that are out of tolerance.

Part "a" of Fig. 6-3 shows an automatic gauging station. Parts to be inspected move from left to right. Light beams and photocells are used as detectors. Parts that are too tall break the upper beam, and they are rejected. Those that are too short will not break the second beam, and they are also rejected. The third beam detects the presence of a part.

In the truth table, G stands for GO. Acceptable parts should GO under the upper beam. N stands for NO GO. Acceptable parts should NOT GO under the second beam. P stands for PRESENT. A broken beam for any sensor means a logic 1.

At the output, R stands for REJECT. When R equals 1, the reject mechanism is activated.

In the truth table, row b represents a part that is too tall. All three beams are broken, giving 111, and therefore R = 1.

Row c represents a short part. Only beam P is broken. Again, R equals 1.

Row d represents a part that is within tolerance. It is neither too tall nor too short. The reject mechanism must not be activated, so R = 0.

Row e represents no part at the station. The reject mechanism should not be activated, so R again equals 0.

With three inputs, there are eight possible input signal sets. When the system is working properly, however, only the four conditions described will occur. For example, no part will ever break beam G without breaking N and P. That is, the input set 100 should not appear. These undefined input sets are referred to as *DON'T CARES*. For now, R will be set to 0 for the don't cares.

Once available, the truth table is the basis for the circuit's Boolean expression. Using sum-of-products, the expression at f results. It will not be simplified at this time.

The implementation of this expression is shown at g. Each time an out-of-tolerance part passes this gauging station, R will equal 1, and the gate will become open.

SIMPLIFICATION FOR SUBJECTIVE METHOD AND TRUTH TABLE METHOD

The circuits in Fig. 6-4 have the same truth table. Although they are physically different, their actions are identical.

The lower circuit is the simpler of the two, so it would probably be the one used to solve a given problem. The upper circuit was the one from the gauging station problem just considered. It was obtained by applying sum-of-products to a truth table. The process of obtaining the simpler circuit from the more complex is called simplification.

If only one or two circuits are to be built, simplification may not be worthwhile. Logic chips are relatively inexpensive, so the time spent on simplification may cost more than what is saved by eliminating chips.

If large numbers of circuits are to be built, simplification becomes important. In addition to direct savings, simplification leads to more reliable equipment.

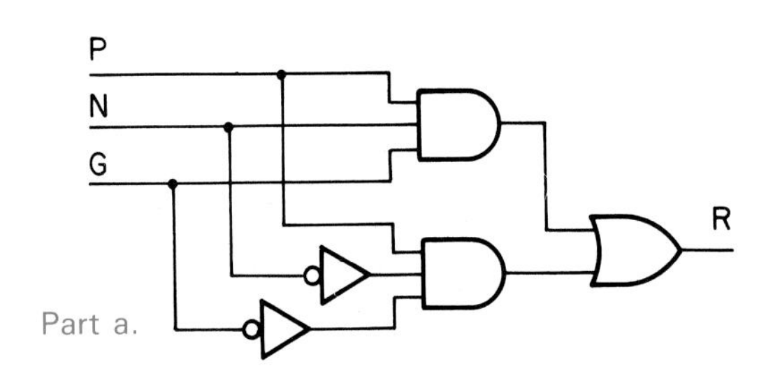

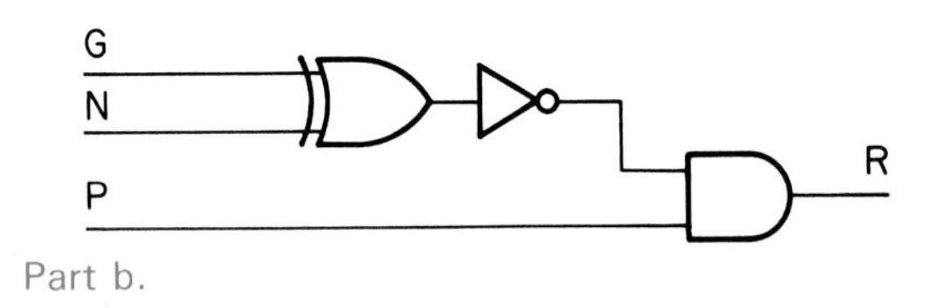

Fig. 6-4. Because these circuits have the same truth table, one could be substituted for the other.

SUBJECTIVE SIMPLIFICATION

An inspection of a logic diagram may suggest ways of simplifying a circuit. This is called subjective simplification.

This approach can be used to simplify the circuit in part "a" of Fig. 6-5. The redundant NOTs in the upper lead can be removed, and the AND-NOT in the lower lead can be replaced by a NAND. See part "b" of Fig. 6-5.

An experienced designer might see further simplifications. At c, the NOR has been DeMorganed. The bubbles on the inputs and outputs of the three elements cancel, and the circuit at d results. This is really a 4-input AND, so the original circuit has been simplified to a single logic element.

Subjective simplification depends on the skill and experience of the designer. Errors are likely, so careful testing of the simplified circuit is necessary.

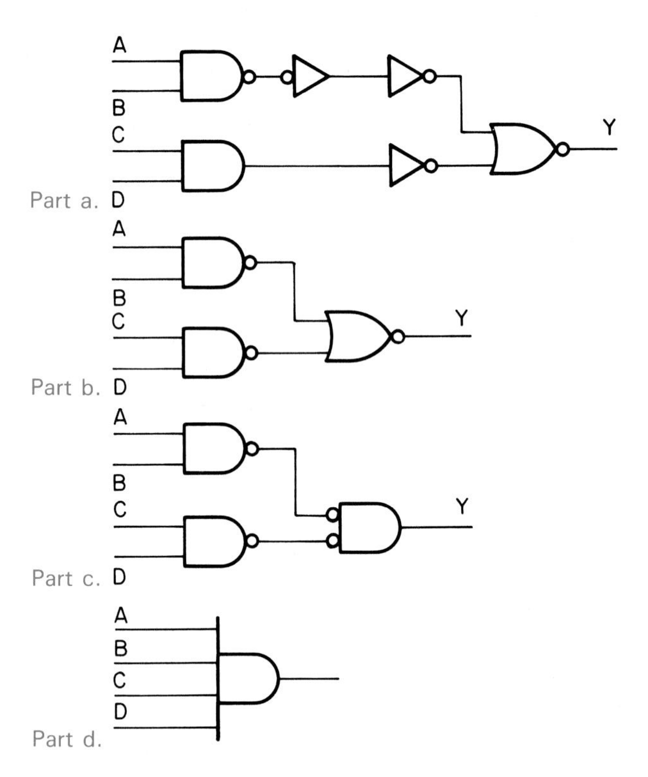

Fig. 6-5. Subjective simplification was used to simplify this circuit.

BOOLEAN SIMPLIFICATION

When complex circuits are to be simplified, the laws of Boolean algebra are usually used. For example, it would be difficult to predict that the circuit in part "a" of Fig. 6-6 simplifies to $\overline{B} = Y$. Yet, this is easily accomplished when the Boolean simplification process is used.

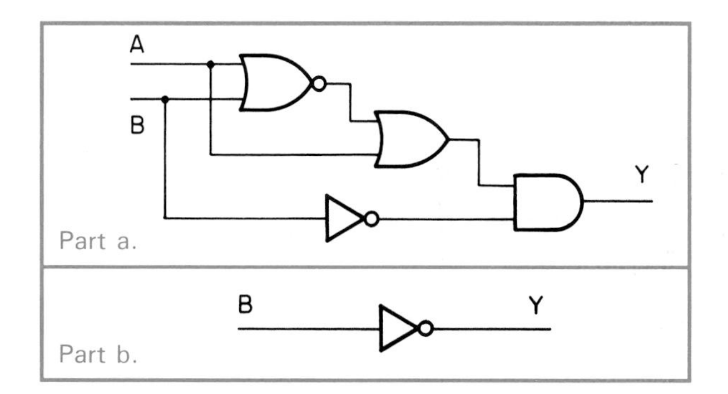

Fig. 6-6. This circuit simplifies to a single NOT. The value of A does not influence the value of Y.

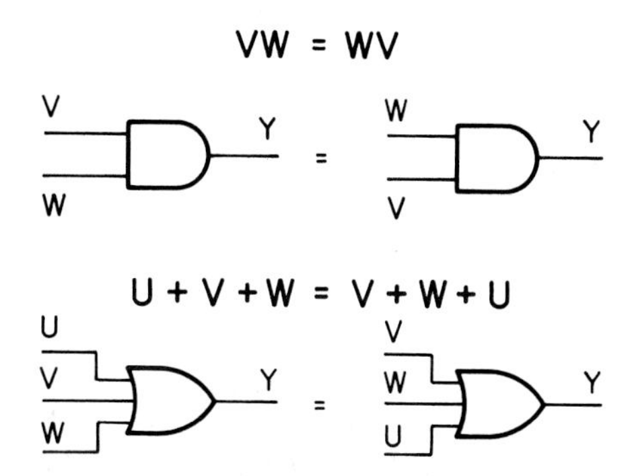

Fig. 6-7. Based on the commutative law, signals can be applied to logic elements in any order.

LAWS OF BOOLEAN ALGEBRA

Many of the laws of Boolean algebra are similar to those of ordinary algebra. The similarity makes the laws easier to remember.

COMMUTATIVE LAW

The commutative law states that signals can be applied to a logic element in any order, Fig. 6-7. That is: AB = BA.

This law applies to AND, OR, NAND, NOR, and XOR elements.

ASSOCIATIVE LAW

The associative law states that input signals to AND and OR elements can be grouped in any order. See Fig. 6-8. This law does not apply to NANDs and NORs. For NANDs and NORs, a bar over two or more letters determines the grouping.

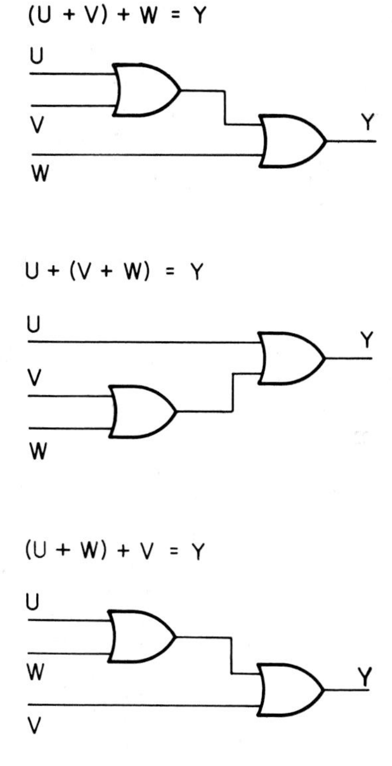

Fig. 6-8. The associative law permits AND and OR inputs to be grouped in any order.

DISTRIBUTIVE LAW

The distributive law is used to remove brackets. See Fig. 6-9. The process is similar to that of ordinary algebra. To remove the brackets in this expression, U is first ANDed with V. It is then ANDed with W. The two parts are then ORed.

The resulting circuit is more complex than the original. However, bracket removal is often used in preparation for simplification.

When two bracketed quantities are ANDed, the process is more complex. See Fig. 6-10. Note the similarity to ordinary algebra.

FACTORING

Factoring implies the removal of a common quantity from an expression. In the expression in Fig. 6-11, U is common to both sections. Factoring allows it to be placed outside the brackets.

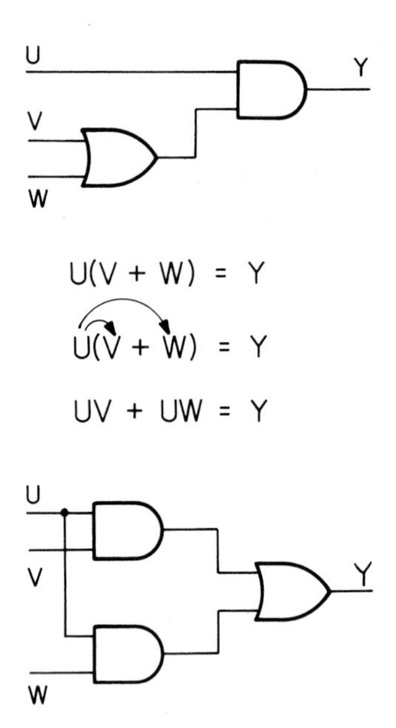

Fig. 6-9. The distributive law can be used to remove brackets.

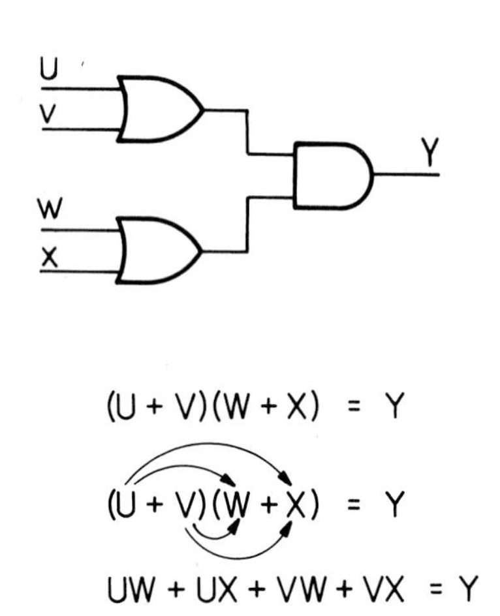

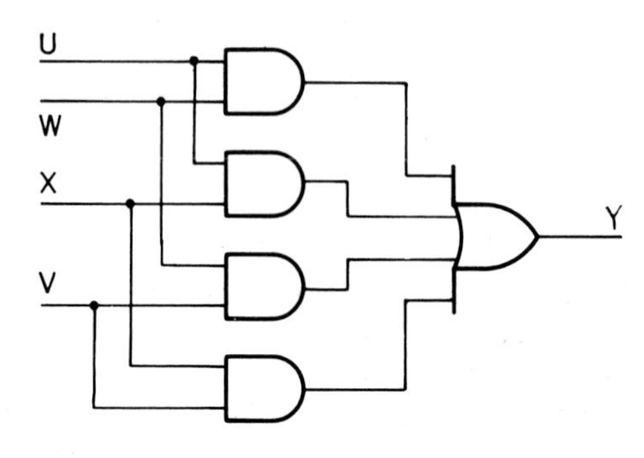

Fig. 6-10. The distributive law can be applied when two or more quantities are within each set of brackets.

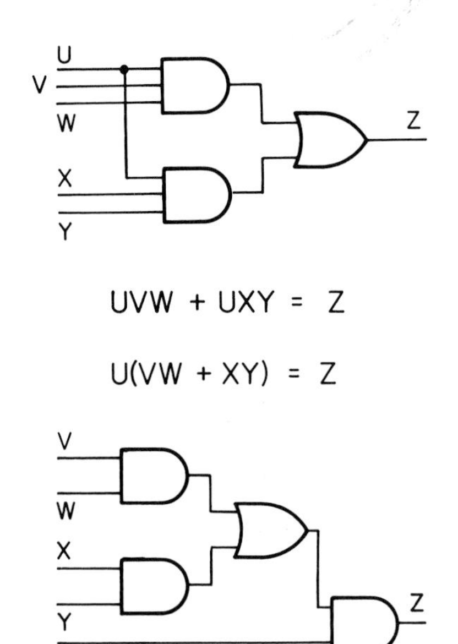

Fig. 6-11. Factoring in Boolean algebra is similar to that in ordinary algebra.

TAUTOLOGY

The term tautology means needless repetition. The rules of tautology are shown in Fig. 6-12. Based on your knowledge of AND and OR truth tables, convince yourself that each rule in Fig. 6-12 is valid. The following examples will show the use of these rules in circuit simplification.

Example: Simplify the circuit in Fig. 6-13.

Each simplification is unique. There are, however, a few general rules. With experience you will learn what leads to simplified circuits and what does not.

Step 1. Input A appears twice in this expression, so it is probably best to start with manipulations involving A. In this step the brackets are removed. That is, A is ANDed with A and B. The result is shown at 2.

Step 2. With the brackets removed, AA attracts attention. Applying Rule a, AA = A.

Step 3. Input A still appears twice in this expression. It can be factored from A + AB. The result is A(1 + B).

Step 4. When 1s and 0s appear in an expression, they attract attention. In this case, Rule d can be applied. 1 + B = 1. This removes B from the expression.

Step 5. Again a 1 appears. Rule c can be used. That is 1A = A.

BOOLEAN STATEMENT	LOGIC EQUIVALENT	SIMPLIFICATION
XX = X Part a.	X X	X X REMOVE AND
X + X = X Part b.	X X	X X REMOVE OR
1·X = X Part c.	1 X X	X X REMOVE AND
1 + X = 1 Part d.	1 X 1	Vcc 1 PERMANENT 1
0X = 0 Part e.	0 X 0	0 PERMANENT 0
0 + X = X Part f.	0 X X	X X REMOVE OR

Fig. 6-12. These Boolean statements can be used to simplify expressions. The right column shows the path that data takes or a graph of the effect of the data arrangement.

Step 6. The simplification is complete. Expressions that contain no 1s or 0s and no duplicate letters are usually in their simplest form.

The simplified circuit has the same truth table as the original. Note that B does not appear in the final expression. In the original expression and circuit, the value of B did not influence Y.

Example: Simplify the expression in Fig. 6-14.

This is a fairly complex simplification, so the process should be followed step by step.

Step 1. The brackets are removed first. The results are shown at 2.

Step 2. AA attracts attention. Applying Rule a, AA reduces to A.

Step 3. Input A appears in three sections of the expression and can be factored from all of them at one time.

Step 4. A 1 now appears inside the brackets. It permits application of Rule d. Anything ORed with 1 equals 1.

Step 5. Again the 1 is important. Rule c can be applied.

Step 6. This expression and circuit should have the same truth table as the original.

Accuracy is important. Even a simple copying error will cause completely incorrect results. It is best to record each step in the process.

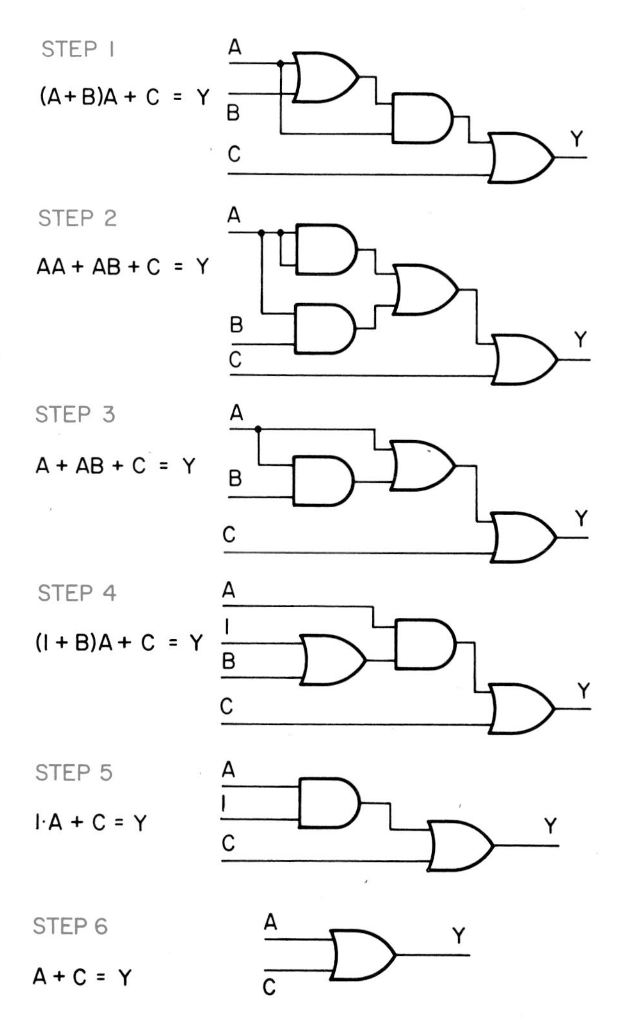

Fig. 6-13. All circuits in this illustration have the same truth table.

SIMPLIFYING USING NOT RULES

Fig. 6-15 shows rules used in simplifying circuits containing NOTs. Study the first three and convince yourself of their validity. The next two are DeMorgan's theorem. Then, XOR and XNOR expressions are listed. The last two are helpful in special cases (test them with X = 1 and with X = 0).

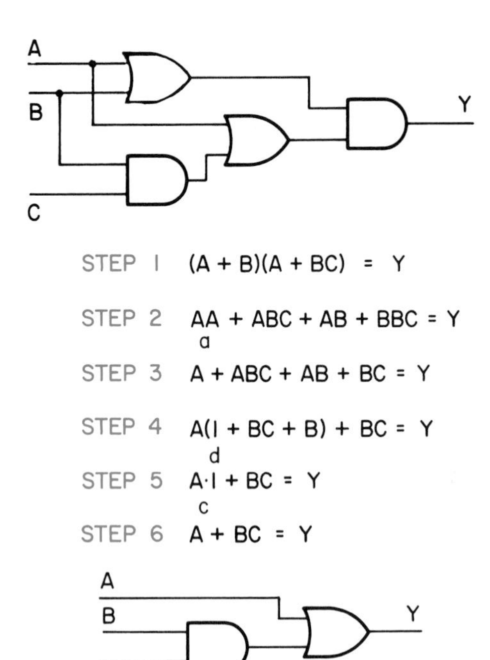

Fig. 6-14. This 4-element circuit can be simplified to one containing only two elements.

Example: Design a simple circuit based on the truth table in Fig. 6-16.

Step 1. Sum-of-products was used to obtain an expression. Four input signal sets result in Y = 1, so there are four ANDs in the expression.

Step 2. AB appears in the first two ANDs, and $\overline{A}\,\overline{B}$ appears in the other two. These can be factored.

Step 3. $C + \overline{C}$ is covered by Rule i. That is, $C + \overline{C} = 1$.

Step 4. The 1s can be removed by applying Rule c.

Step 5. This appears to be a complex expression, but Rule m indicates that it is an XNOR (an exclusive NOR).

Step 6. The expression can be implemented by an XNOR. In practice, it would probably be assembled using an XOR and a NOT. Note that C does not affect Y.

Example: Implement the expression in Fig. 6-17 using only NORs.

Step 1. The brackets are removed first.

Step 2. Using Rule h, $A\overline{A}$ becomes 0.

Step 3. The 0 can be removed by using Rule f.

Step 4. This expression contains two NORS. DeMorgan's theorem, Rule j, is applied to convert them to a more familiar form.

	BOOLEAN STATEMENT	LOGIC EQUIVALENT
Part g.	$\overline{\overline{X}} = X$	
Part h.	$X\overline{X} = 0$	
Part i.	$X + \overline{X} = 1$	
Part j.	$\overline{X + Y} = \overline{X}\,\overline{Y}$	
Part k.	$\overline{XY} = \overline{X} + \overline{Y}$	
Part l.	$X\overline{Y} + \overline{X}Y = X \oplus Y$	
Part m.	$XY + \overline{X}\,\overline{Y} = \overline{X \oplus Y}$	
Part n.	$X + \overline{X}Y = X + Y$	
Part o.	$\overline{X} + XY = \overline{X} + Y$	

Fig. 6-15. These Boolean statements are helpful when simplifying circuits containing NOTs.

Step 5. This expression contains two NORs and an OR. The OR can be implemented by using a NOR-NOT.

This is not the simplest possible circuit. It is, however the simplest that can be built using only NOR elements.

If there had been no limit on the logic type, what would the circuit look like? The simplification begins with Step 4. See Fig. 6-18.

Step 4a. $\overline{A}$ appears in both sections of the expression and can be factored.

Step 5a. $\overline{B} + \overline{C}$ can be DeMorganed using Rule k.

Step 6a. Again DeMorgan's theorem can be applied. Rule j can be used.

Step 7a. Implementation of this expression results in a circuit containing only two elements.

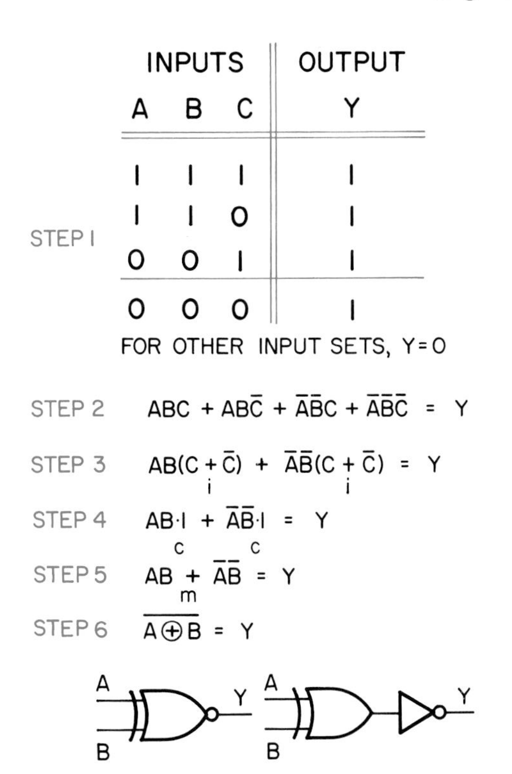

STEP 1

INPUTS			OUTPUT
A	B	C	Y
1	1	1	1
1	1	0	1
0	0	1	1
0	0	0	1

FOR OTHER INPUT SETS, Y=0

STEP 2 $ABC + AB\bar{C} + \bar{A}\bar{B}C + \bar{A}\bar{B}\bar{C} = Y$

STEP 3 $AB\underset{i}{(C + \bar{C})} + \bar{A}\bar{B}\underset{i}{(C + \bar{C})} = Y$

STEP 4 $\underset{c}{AB \cdot 1} + \underset{c}{\bar{A}\bar{B} \cdot 1} = Y$

STEP 5 $AB \underset{m}{+} \bar{A}\bar{B} = Y$

STEP 6 $\overline{A \oplus B} = Y$

Fig. 6-16. Although this truth table appears complex, it can be represented by a single XNOR.

STEP 1 $\bar{A}(A + \bar{B} + \bar{C}) = Y$

STEP 2 $\bar{A}A + \bar{A}\bar{B} + \bar{A}\bar{C} = Y$

STEP 3 $0 + \bar{A}\bar{B} + \bar{A}\bar{C} = Y$

STEP 4 $\bar{A}\bar{B} + \bar{A}\bar{C} = Y$

$\overline{A + B} + \overline{A + C} = Y$

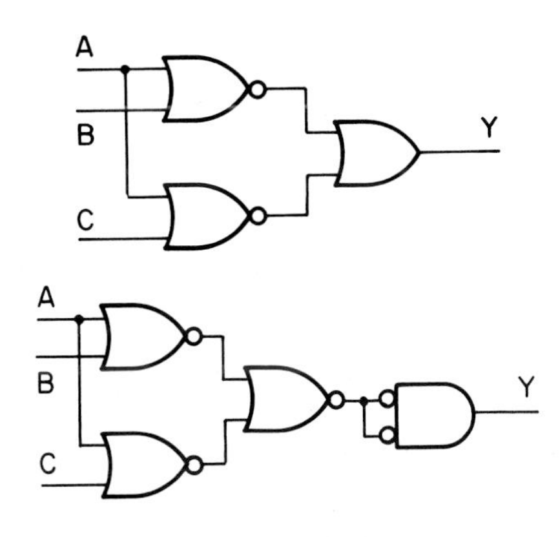

Fig. 6-17. Simplification methods were used to obtain a circuit that contains only NORs.

STEP 4a $\bar{A}\bar{B} + \bar{A}\bar{C} = Y$

STEP 5a $\bar{A}(\bar{B} + \bar{C}) = Y$

STEP 6a $\bar{A}(\overline{BC}) = Y$

STEP 7a $\overline{A + BC} = Y$

Fig. 6-18. If no limits are placed on the elements that can be used, the expression derived in Fig. 6-17 simplifies to a 2-element circuit.

KARNAUGH MAP

In the 1950s, Maurice Karnaugh (pronounced "car-no") introduced a method for simplifying Boolean expressions. It was based on a compact form of a truth table rather than on the standard column form. This truth table has the form of a matrix and is called a Karnaugh map. See part "a" of Fig. 6-19.

KARNAUGH MAP CONSTRUCTION

The circuit described in the truth table in part "b" of Fig. 6-19 has four inputs. Therefore, there are 16 rows in the table. Its Karnaugh map contains 16 boxes or cells.

Each cell represents one row of the truth table. Inputs A and B are plotted vertically. Inputs C and D are plotted horizontally. The 1s and 0s in the cells represent output Y.

Refer to input set ABCD = 0000 in the truth table. This is row 0 in the truth table and is represented by cell a in the map. The cell's coordinates are 00 (AB) and 00 (CD). The 0 in the cell indicates that input set 0000 results in Y = 0.

Row 1 of the truth table has coordinates of 00 (AB) and 01 (CD). Locate this cell on the map. It is cell b.

Care must be taken when mapping the next row (row 2 of the truth table). It is not next to cell b on the map. Its coordinates are 00, 10, and it is located at the upper right of the map. In Karnaugh maps, inputs are arranged so that adjacent cells represent reducible pairs. The details of this arrangement are not important here as long as the order 00, 01, 11, 10 is always used.

The following questions will aid in understanding

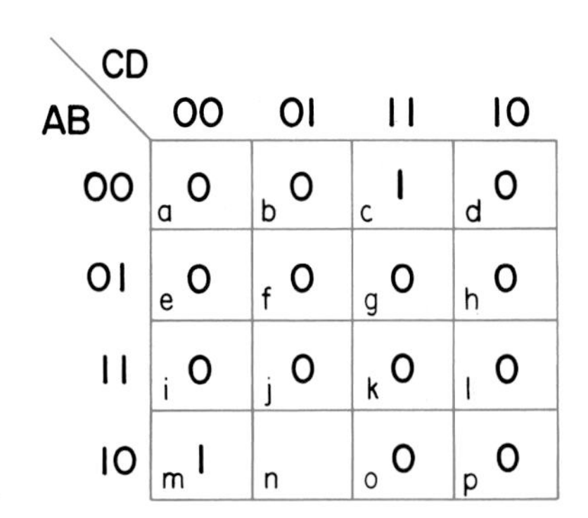

Part a.

	INPUTS A	B	C	D	OUTPUT Y
0	0	0	0	0	0
1	0	0	0	1	0
2	0	0	1	0	0
3	0	0	1	1	1
4	0	1	0	0	0
5	0	1	0	1	0
6	0	1	1	0	0
7	0	1	1	1	0
8	1	0	0	0	1
9	1	0	0	1	1
10	1	0	1	0	0
11	1	0	1	1	0
12	1	1	0	0	0
13	1	1	0	1	0
14	1	1	1	0	0
15	1	1	1	1	0

Part b.

Fig. 6-19. This Karnaugh map and truth table contain the same information.

the construction of Karnaugh maps. The questions refer to material in Fig. 6-19.

QUESTION: Which cell represents row 12? Its coordinates are 11 (AB) and 00 (CD). This maps to cell i.

QUESTION: Which row in the truth table does cell g represent? Its coordinates are 01 (AB) and 11 (CD), so it represents row 7.

QUESTION: What number (1 or 0) will appear in cell 10, 01? Input set 1001 is found in row 9 of the truth table. For this input set, Y = 1. A 1 will appear in cell n.

USE OF A KARNAUGH MAP

The Karnaugh map in Fig. 6-19 has been reproduced in Fig. 6-20. Letters representing the inputs have been added along its axes.

	$\bar{C}\bar{D}$ 00	$\bar{C}D$ 01	CD 11	$C\bar{D}$ 10
$\bar{A}\bar{B}$ 00	0	0	a 1	0
$\bar{A}B$ 01	0	0	0	0
AB 11	0	0	0	0
$A\bar{B}$ 10	b 1	c 1	0	0

STEP 1 (a, b, c) $\bar{A}\bar{B}CD + A\bar{B}\bar{C}\bar{D} + A\bar{B}\bar{C}D = Y$

STEP 2 $\bar{A}\bar{B}CD + A\bar{B}\bar{C}(D + \bar{D}) = Y$ (i and c)

STEP 3 $\bar{A}\bar{B}CD + A\bar{B}\bar{C} = Y$

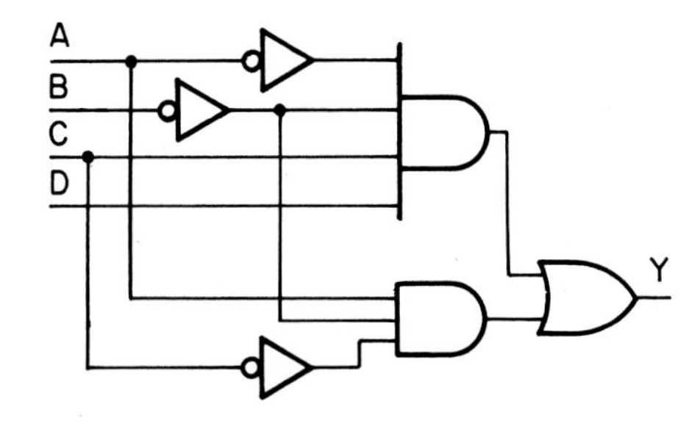

Fig. 6-20. Sum-of-products was applied to this Karnaugh map to obtain an expression for a circuit.

SUM-OF-PRODUCTS

Sum-of-products could be used to write an expression for the circuit represented by the Karnaugh map in Fig. 6-20. Three 1s appear, so there will be three ANDs. The following describes the writing and simplification of an expression based on this map.

Step 1. The coordinates of the three cells containing 1s become the three ANDs. Note that $A\overline{B}\,\overline{C}$ appears in both b and c, so it can be factored.

Step 2. Using Rule i, D + $\overline{D}$ can be replaced by 1. Applying Rule c eliminates the 1.

Step 3. $\overline{B}$ could be factored, but this would not simplify the circuit.

Nothing could be done to simplify $\overline{A}\,\overline{B}CD$ (other than factor $\overline{B}$). This leads to the first rule out of six rules of Karnaugh simplifications. The six rules are:

1. When a cell that contains a 1 is not adjacent to at least one other cell containing a 1, that entry in the final expression cannot be simplified.
2. When 1s appear in adjacent cells, only inputs common to those cells appear in the contribution made by those cells to the expression. Inputs in NOTed and unNOTed forms cancel.
3. Each loop must contain at least one cell not in any other loop.
4. Intersection areas of loops allow another step of simplification.
5. Each loop should contain the largest possible number of adjacent cells.
6. Loops may contain 1, 2, 4, or 8 adjacent cells. 6-cell loops are not allowed.

Refer to cell a. No simplification is possible, and its contribution to the final expression is merely its coordinates, $\overline{A}\,\overline{B}CD$.

Refer to cells b and c in Fig. 6-20. A, $\overline{B}$, and $\overline{C}$ are common to these cells. These inputs will appear in the final expression. However, $\overline{D}$ is associated with cell b and D is associated with cell c. D and $\overline{D}$ cancel, so D does not appear in the contribution made by cells b and c. This leads to the second Karnaugh map rule (when 1s appear in adjacent cells, only inputs common to those cells appear in the contribution made by those cells to the expression. Inputs in NOTed and unNOTed forms cancel.).

This suggests that the expression at Step 3 in Fig. 6-20 can be written directly from the Karnaugh map. The following example shows the process.

Example: Write the expression for the Karnaugh map in Fig. 6-21. Zeros have been omitted from the Karnaugh map to emphasize the 1s.

Notice the *LOOPS* (groups of 1s). There are three groups of 1s. The expression will contain three parts.

Step 1. Loop 1 contains a single cell, so its AND is easily written. It cannot be simplified. See part 1 of the expression.

Step 2. Loop 2 contains two cells. $\overline{A}$, $\overline{C}$, and D are common, so they will appear in the AND. B and $\overline{B}$ cancel. See part 2 of the expression.

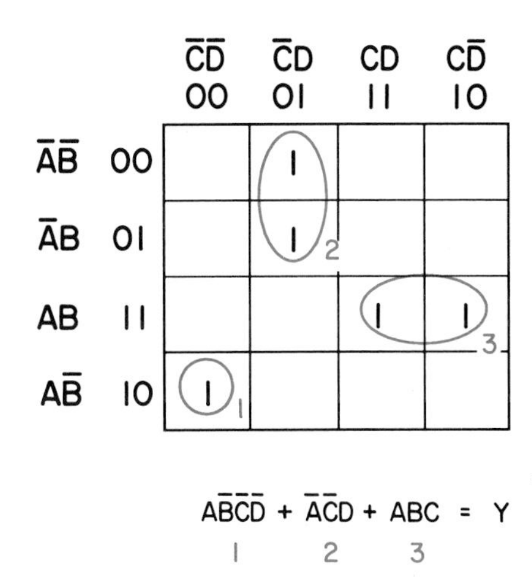

Fig. 6-21. Expressions can be written directly from Karnaugh maps. The terms involving B and $\overline{B}$ and D and $\overline{D}$ drop out.

Step 3. Loop 3 also has two cells. A, B, and C are common to these cells, but D and $\overline{D}$ cancel. The resulting contribution of these cells is shown at 3 in the illustration.

Except for some minor factoring, this is the simplest expression that could be written for this circuit.

Example: Write an expression for the map in Fig. 6-22. Two loops are possible, and the sharing of cells is permitted. However, Rule 3 is involved (each loop must contain at least one cell which is not in any other loop).

Step 1. In loop 1, $\overline{A}$, $\overline{C}$, and D are common (B and $\overline{B}$ cancel). In loop 2, $\overline{A}$, B, and D are common (C and $\overline{C}$ cancel). See parts 1 and 2 of Fig. 6-22.

Step 2. The intersection (common area) of loops 1 and 2 is called loop 3. In loop 3, $\overline{A}$ and D are common. The term $\overline{A}D$ can be factored out. The final expression is shown in part 3 of Fig. 6-22. The factoring process reduces the number of logic components needed from 5 to 4. The process suggests Rule 4 (intersection areas of loops allow another step of simplification).

What would have happened if loop 2 had circled only one cell as shown in Fig. 6-23? The resulting expression (shown below the map) is valid. That is, a circuit based on this expression would work. However, it is not in its simplest form. When this expression is compared with that obtained earlier, the C in the 4-input AND is found not to be needed.

This suggests Rule 5 (to obtain the simplest

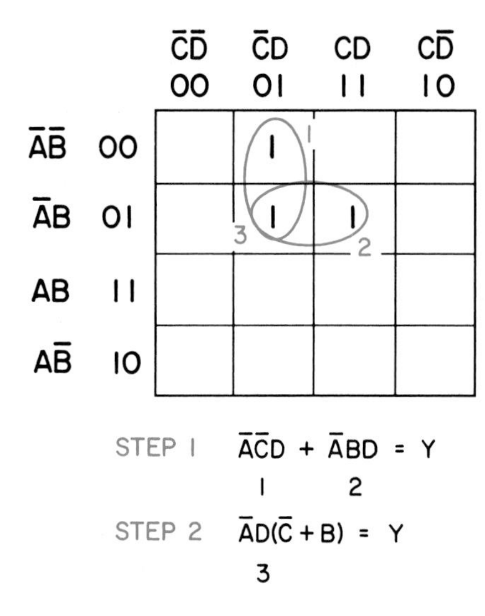

Fig. 6-22. Ones can be used in more than one loop as long as each loop contains at least a single 1 which is not used in other loops.

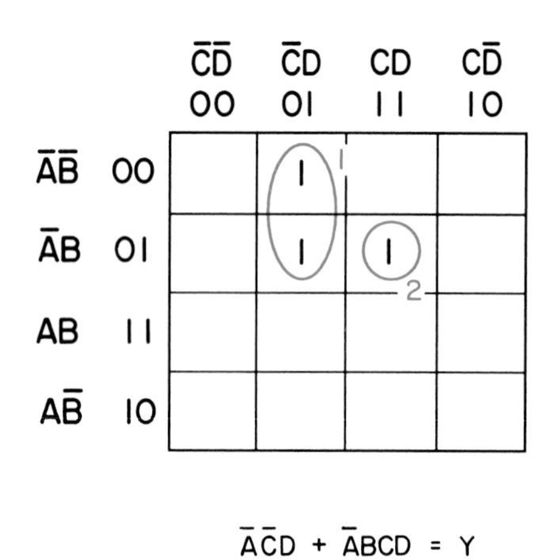

Fig. 6-23. Because loop 2 contains a lone 1, no simplification results.

expression, each loop should contain the largest possible number of adjacent cells. [Remember, only cells containing 1s are of interest.]).

Example: Write an expression for the map which is shown in Fig. 6-24.

It is the order of the inputs at the top and side of a Karnaugh map that is important. Where the series begins is not important. That is, the column headings on the map in Fig. 6-24 are:

$\overline{C}\overline{D}$ $\overline{C}D$ CD $C\overline{D}$

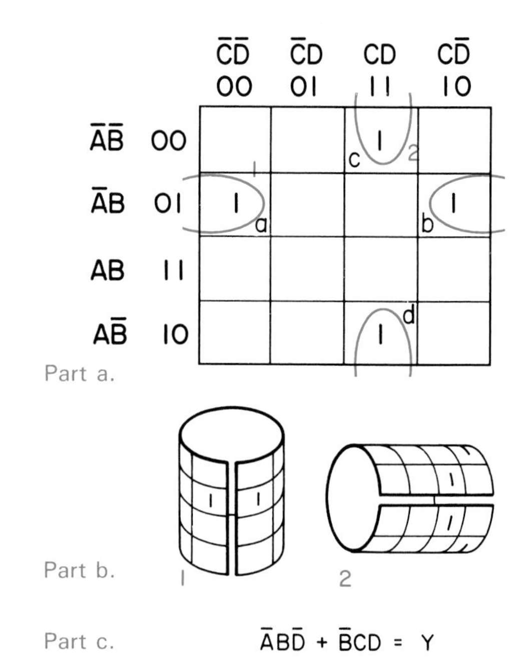

Fig. 6-24. Because Karnaugh maps can be wrapped, cells containing 1s at their edges can be combined.

They could be shifted one place to the left without harming the concept of the map. That is:

$\overline{C}D$ CD $C\overline{D}$ $\overline{C}\overline{D}$

Because this is true, the map has no beginning or end. It can be wrapped back on itself either vertically or horizontally. See the cylinders at b.

Because maps can be wrapped, cells a and b can be combined into one loop. Cells c and d can also be combined into one loop.

For loop 1, $\overline{A}$, B, and $\overline{D}$ are common (C and $\overline{C}$ cancel). For loop 2, $\overline{B}$, C, and D are common (A and $\overline{A}$ cancel). The resulting expression is shown at the bottom of Fig. 6-24.

Example: Write the expression for the map which is shown in Fig. 6-25.

Without looking at the expression below the map, attempt to write the expression for this map. The three loops have been outlined.

The intersection area indicates that some factoring is possible. $\overline{B}D$ can be factored, but 6 components are required in either case.

To work with more complex maps, rule 6 is needed (loops may contain 1, 2, 4, or 8 adjacent cells. 6-cell loops, for example, are not allowed.).

Fig. 6-26 shows several 4-cell loops. In the first map, B and $\overline{C}$ are common to all four cells. A and $\overline{A}$ and D and $\overline{D}$ cancel. The resulting expression for this loop is simply $B\overline{C} = Y$.

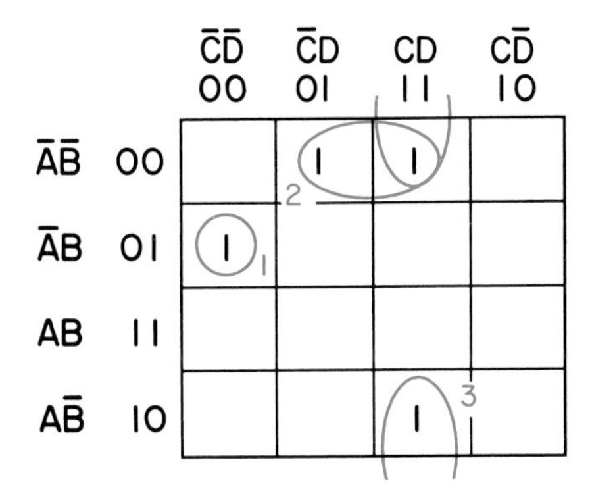

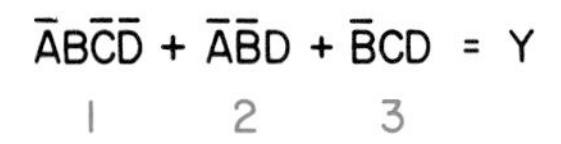

Fig. 6-25. The 1 in the bottom row can be wrapped around and grouped with the 1 in $\overline{AB}$, CD to eliminate A from product 3.

As indicated by Rule 5, the larger the number of cells in a loop, the greater the simplification. The 8-cell loops which appear in Fig. 6-27 result in entries containing only one input.

What happens when there are only three inputs? The matrix is no longer square. See Fig. 6-28. However, the map is handled in the same manner as those for 4-input circuits. Attempt to write the expression for the map in Fig. 6-29 without looking at the answer.

2-input circuits have simple Karnaugh maps. See Fig. 6-30. In loop 1, B and $\overline{B}$ cancel leaving only $\overline{A}$. In loop 2, A and $\overline{A}$ cancel leaving only B.

DON'T CARES CAN SIMPLIFY A KARNAUGH MAP

The truth table for the automatic inspection station in Fig. 6-3 has been reproduced in Fig. 6-31. Remember that only four input signal sets (of the eight possible sets) were used in the operation of that station. That is, only the four starred numbers in the map at b were defined. The other four cells are "don't cares." The don't cares are marked D.

DON'T CARES AS 0s

At b, the don't cares have been set to 0. The resulting expression is shown beside the map. Because each loop contained only one cell, no simplification resulted. The expression is the same as the one obtained from sum-of-products.

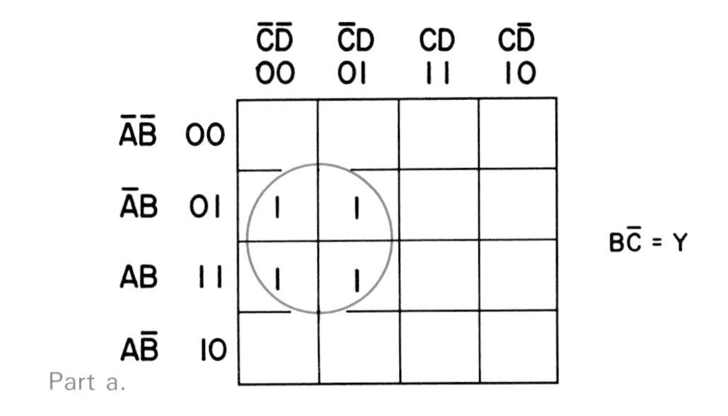

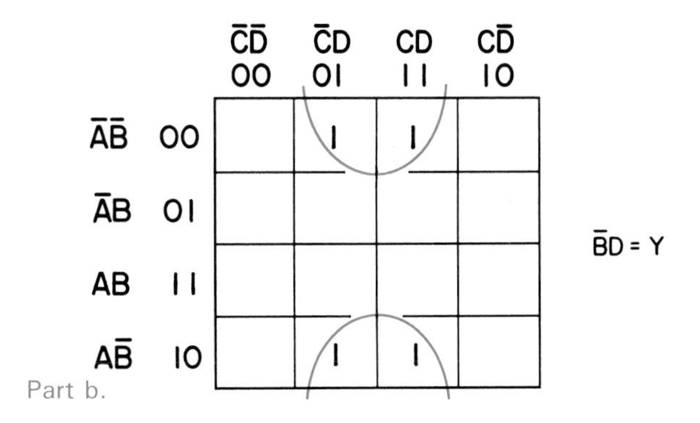

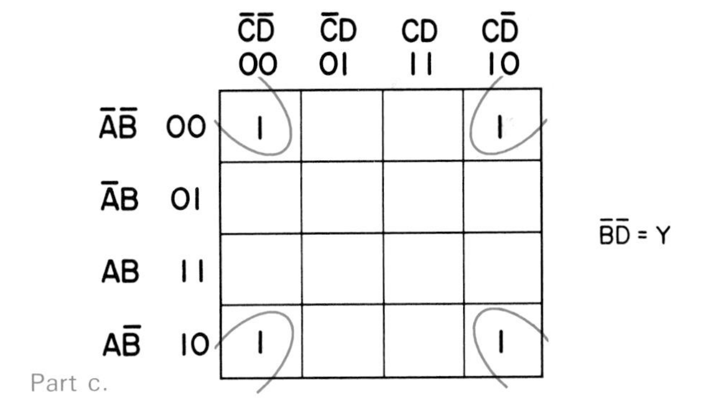

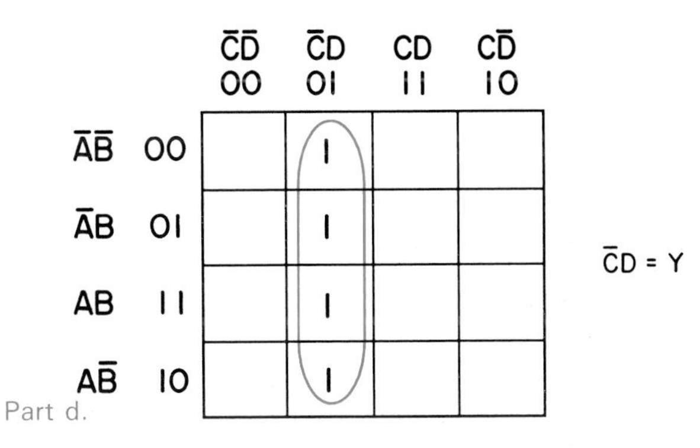

Fig. 6-26. Simplification from 4 sets of 4-input elements to one 2-input element results when four 1s appear in the same loop.

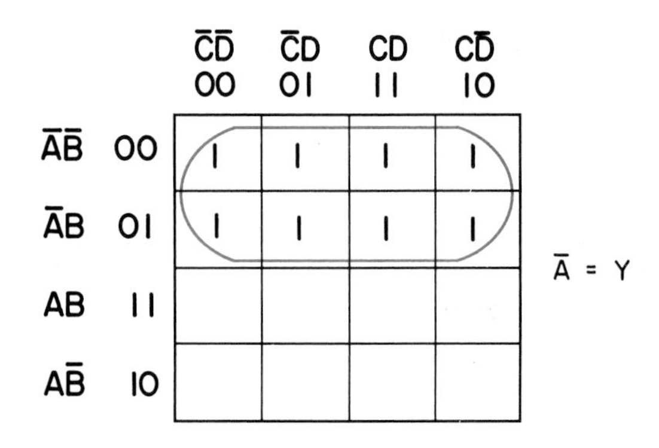

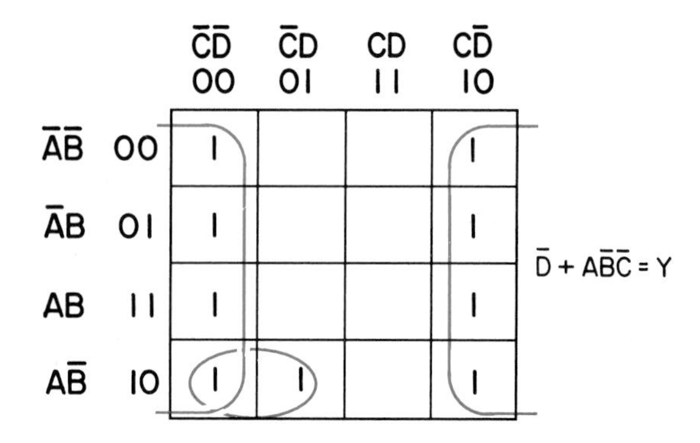

Fig. 6-27. An 8-entry loop simplifies to just one variable.

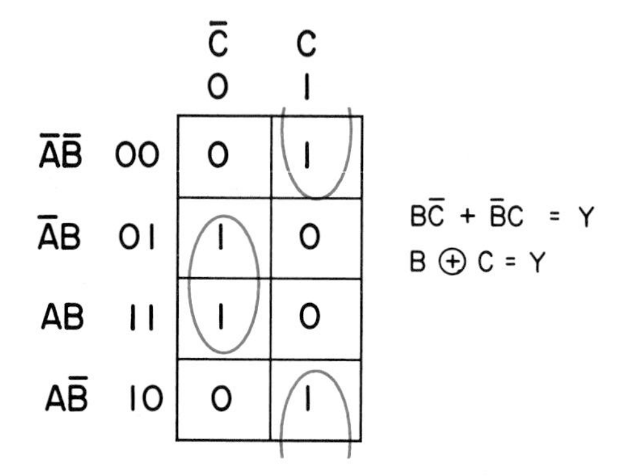

Fig. 6-28. Karnaugh maps for 3-input circuits are used in the same manner as those for 4-input circuits.

DON'T CARE AS 1 OR 0

At c, 1s were placed in three selected cells. This is permissible because these input signal sets will not appear in the normal operation of the machine.

Note the result. Both loops are now multi-celled, so simplification occurs.

Caution is needed. The starred cells are not don't cares. They must remain 0, since the values of Y for

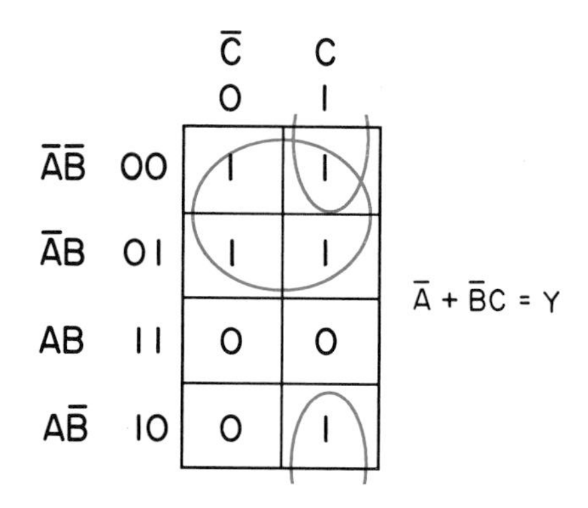

Fig. 6-29. Loops containing four 1s greatly simplify expressions for 3-input circuits.

these cells were established in the original truth table. Also, the cell at 01, 0 should remain 0. Placing a 1 here would add a third AND to the expression.

DE MORGAN/KARNAUGH MAP FOR PRODUCT-OF-SUMS METHOD

To review, expressions written directly from truth tables are seldom in their simplest form. Boolean algebra is often used to simplify such expressions. To obtain the simplest starting point, sum-of-products is usually used if there are few 1s in the table. If there are few 0s, product-of-sums is used.

When a Karnaugh map is used to obtain an expression, the choice is less important. Karnaugh maps produce simplified expressions automatically, so both sum-of-products and product-of-sums produce equally simple expressions. For example, Fig. 6-32 contains more 1s than 0s, yet the expression obtained using sum-of-products (from the 1s) is very simple. In most cases, sum-of-products is all you will need, but for the sake of completeness, product-of-sums will be described.

The map in Fig. 6-32 has been redrawn in Fig. 6-33. Dashed loops have been drawn around the 0s. The first expression was written using the same methods used for 1s. Note, however, that Y is NOTed. When zeros are used, the resulting expression is the complement of the one obtained using 1s.

To undo the complement (to find Y rather than $\overline{Y}$), both sides of the expression at "a" must be complemented (NOTed). In Boolean algebra, an expression can be complemented merely by changing all terms and operations to their complement. ANDs become ORs; ORs become ANDs. UnNOTed variables are NOTed (A becomes $\overline{A}$). NOTed variables are unNOTed ($\overline{A}$ becomes A). The expression at "b" results when these operations have been performed. Note that this is a product-of-sums expression. It is

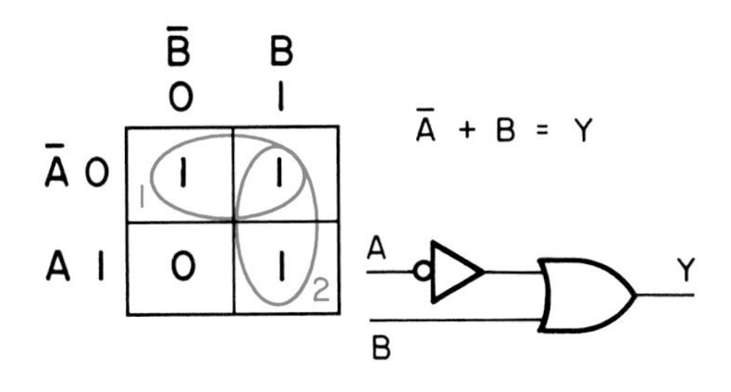

Fig. 6-30. The 2x2 Karnaugh map describes a 2-input circuit.

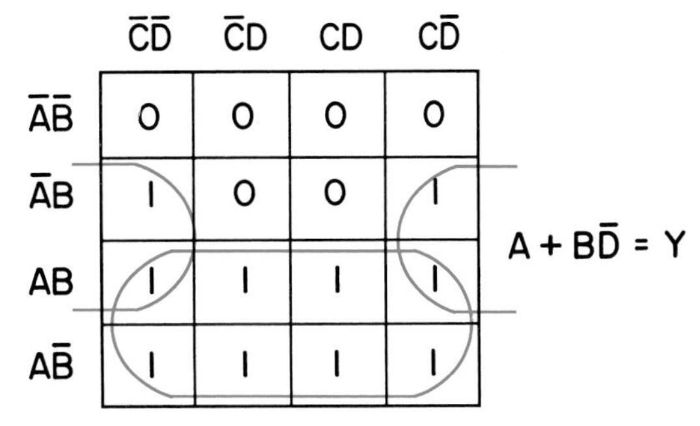

Fig. 6-32. Even though there are more 1s than 0s in this Karnaugh map, sum-of-products results in a simple expression.

INPUTS			OUTPUT
G	N	P	R
1	1	1	1
0	0	1	1
0	1	1	0
0	0	0	0

OTHER INPUT SETS NOT OF INTEREST

Part a.

		$\bar{P}$ 0	P 1
$\bar{G}\bar{N}$	00	* 0	* (1)
$\bar{G}N$	01	D 0	* 0
GN	11	D 0	* (1)
$G\bar{N}$	10	D 0	D 0

$\bar{G}\bar{N}P + GNP = R$

Part b.

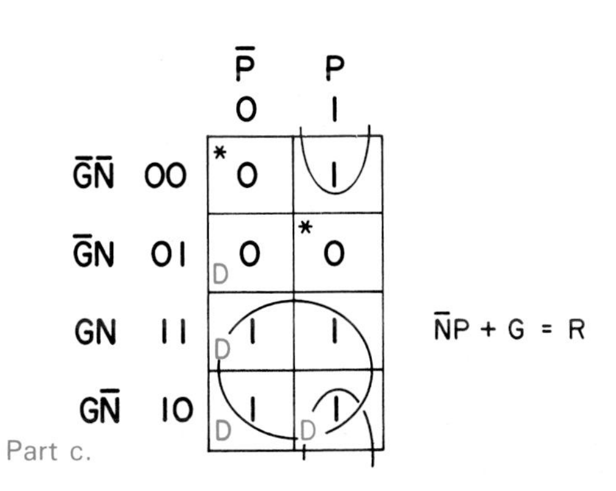

Part c.

Fig. 6-31. Don't cares can be used to obtain simpler expressions. The don't cares are marked D.

almost as simple as the one obtained by drawing loops around 1s. Boolean algebra can be used to show that the expressions in Fig. 6-32 and Fig. 6-33 are equivalent.

Fig. 6-34 is another example of obtaining a product-of-sums expression from a Karnaugh map. Remember, the expression obtained from loops drawn around 0s is for the complement of Y. When it is complemented, the product-of-sums expression results.

It turns out that the complementing can be accomplished before or after the writing of the expression based on 0s. In Fig. 6-35, product-of-sums labels have been added within the dashed rectangles. They were obtained by complementing the sum-of-product labels. For example, $\bar{A}\bar{B}$ becomes A + B. Using these labels, product-of-sums expressions can be written directly from the map. The expression in the first set

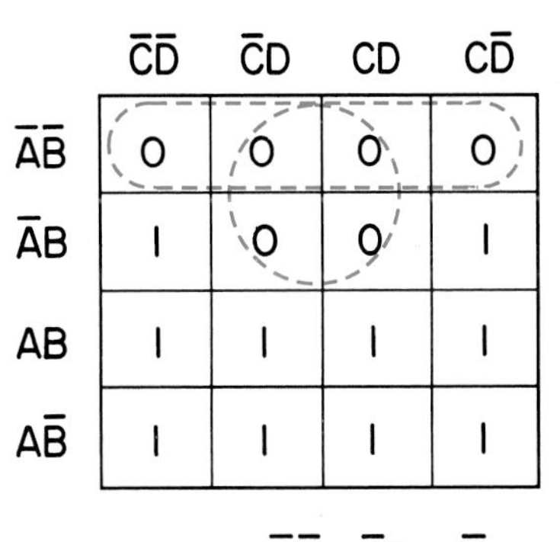

Part a. $\bar{A}\bar{B} + \bar{A}D = \bar{Y}$

Part b. $(A+B)(A+\bar{D}) = Y$

Fig. 6-33. If 0s are used, the resulting expression is the complement of Y. If this expression is then complemented, a product-of-sums expression results.

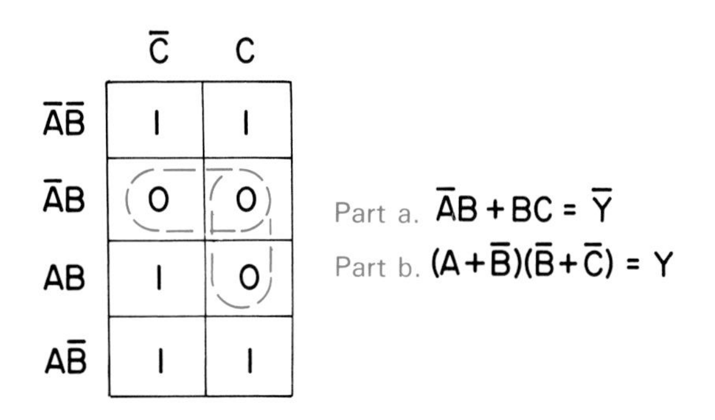

Fig. 6-34. A product-of-sums expression can be obtained from this 3-input Karnaugh map.

of brackets was obtained from the horizontal loop. C and $\overline{C}$ cancel, so this row represents $A + \overline{B}$. The second set of brackets represents the vertical loop. A and $\overline{A}$ cancel, so the expression within the brackets is $\overline{B} + \overline{C}$.

SUMMARY

An example is used to summarize this chapter.

Example: Design a circuit to output a 1 if the sum

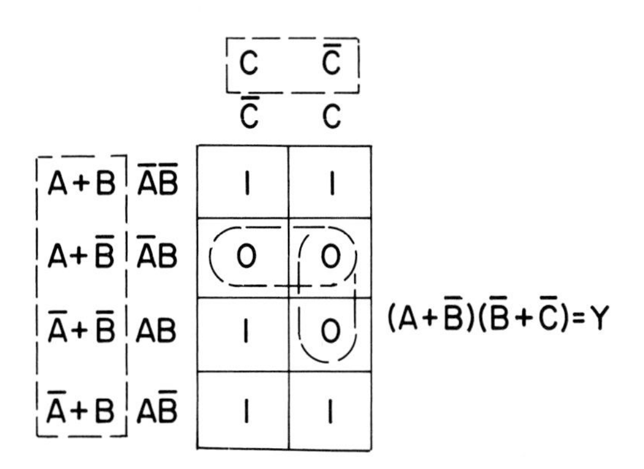

Fig. 6-35. By adding complements of the labels to this Karnaugh map, product-of-sums expressions can be written directly from this map.

of three coins equals or exceeds 15¢. Use sum-of-products. Then use product-of-sums.

The input is to be any combination of one quarter, one dime, and one nickel, Fig. 6-37. There is to be no provision for making change. The circuit is to be as simple as possible. To check the design, both truth table and Karnaugh map methods are to be used.

Step 1. The truth table restates the problem. Using sum-of-products, an expression can be written. It will have five ANDs since there are five input sets that result in Y = 1.

Step 2. Simplification begins with the observation that BC can be factored from two ANDs, and $A\overline{B}$ can be factored from two others.

Step 3. Rules i and c are used to eliminate $\overline{A} + A$ and eliminate $\overline{C} + C$.

Step 4. A can be factored from the last two ANDs.

Step 5. Rule o permits the removal of B inside the brackets.

Step 6. The quantity in the brackets can be DeMorganed.

Step 7. Rule n can be applied.

Step 8. The expression is in its simplest form. Its implementation requires only an AND element and an OR element.

Based on the truth table, two Karnaugh maps were constructed. The first contains two sum-of-products loops. The resulting expression is identical to that obtained from the truth table. The product-of-sum expression was obtained from the second Karnaugh map. Implementations of these expressions are shown in Fig. 6-36.

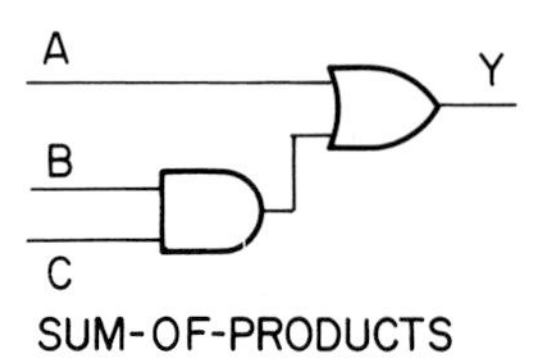

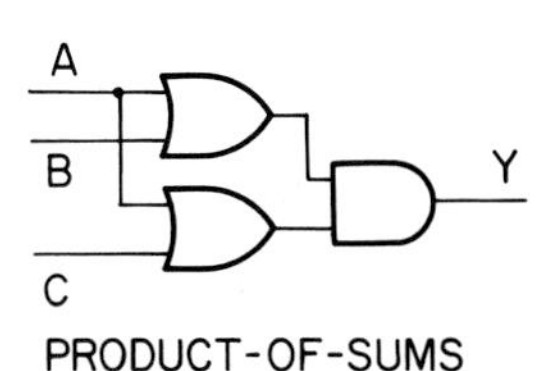

Fig. 6-36. These circuits are implementations of the expressions in Fig. 6-37.

TEST YOUR KNOWLEDGE

1. What is a fixture?
2. Can "don't cares" be set to zero?
3. When is simplification not worthwhile?
4. Simplification resulting in fewer components leads to more ________ equipment.

STEP 1

INPUTS 25 A	10 B	5 C	SUM ≥15 Y
0	0	0	0
0	0	1	0
0	1	0	0
0	1	1	1
1	0	0	1
1	0	1	1
1	1	0	1
1	1	1	1

STEP 2 $\bar{A}BC + A\bar{B}\bar{C} + A\bar{B}C + AB\bar{C} + ABC = Y$

STEP 3 $BC(\bar{A} + A) + A\bar{B}(\bar{C} + C) + AB\bar{C} = Y$

STEP 4 $BC + A\bar{B} + AB\bar{C} = Y$

STEP 5 $BC + A(\bar{B} + B\bar{C}) = Y$

STEP 6 $BC + A(\bar{B} + \bar{C}) = Y$

STEP 7 $BC + A\overline{BC} = Y$

STEP 8 $BC + A = Y$

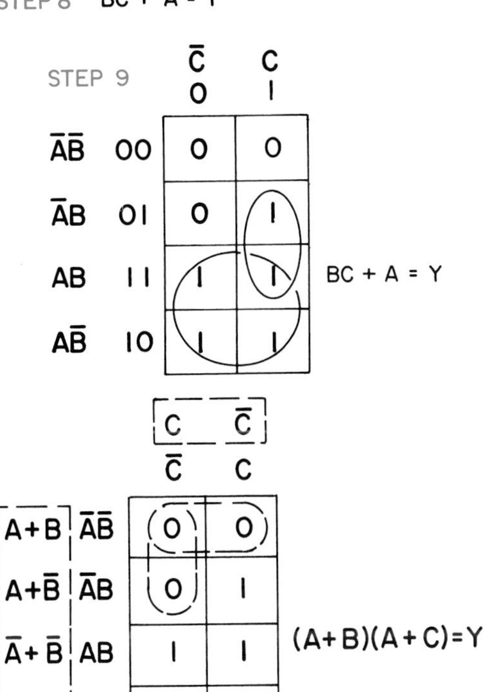

Fig. 6-37. Sum-of-products and product-of-sums were used in the design of the circuits in Fig. 6-36.

5. The ________ law of Boolean algebra states that signals can be applied to a logic element in any order.
6. One can often factor if a common quantity is in two terms of a statement. (True or False?)
7. To wrap a Karnaugh map means to connect the top to the left side. (True or False?)
8. Is the expression $ABC + \bar{A}BC = Y$ in its simplest form?
9. How many 1s must appear together in a Karnaugh map for the simplified expression to be like the expression $A = Y$?
10. A Karnaugh map for 3 inputs is used much the same as a map for 4 inputs. (True or False?)

STUDY PROBLEMS

1. Rewrite the following list to show the order of steps in the truth-table-based design process:
   1. Simplification.
   2. Construct truth table.
   3. Precise description of problem.
   4. Write expression.
   5. Construct and test circuit.
   6. Draw logic diagram.
2. Complete these identities.
   1. $1A =$
   2. $0A =$
   3. $AA =$
   4. $A + A =$
   5. $1 + A =$
   6. $0 + A =$
3. Complete these identities.
   1. $A\bar{A} =$
   2. $A + A =$
   3. $A + \bar{A} =$
4. Which of the following is a valid statement of DeMorgan's theorem?
   1. $A + \bar{A}B = A + B$
   2. $\bar{A}B + A\bar{B} =$ XOR
   3. $\bar{A} + \bar{B} = \overline{AB}$
5. Copy the truth table in Fig. P6-1 onto a sheet of paper, and complete column Y. It is to show the results of voting by A, B, and C. A yes vote is indicated by a 1 in an input column. If the majority votes yes, a 1 is to appear at output Y.
6. Lay a piece of paper over the Y column of the truth table in Fig. P6-2 and mark dividing lines on it to match those placed between every third input set. Then complete column Y on this piece of paper.

   This device is a COMPARATOR. Its inputs are two, 2-bit binary numbers A1A0 and B1B0. Output Y is 1 when A1A0 is larger than B1B0. When A1A0 equals or is smaller than B1B0, Y equals 0.

INPUTS			OUTPUT
A	B	C	Y
0	0	0	
0	0	1	
0	1	0	
0	1	1	
1	0	0	
1	0	1	
1	1	0	
1	1	1	

Fig. P6-1. Truth table for voting. A majority of voters is 2 or more.

7. Remove the brackets from these expressions. Do not simplify.
   1. A(B + C) = Y
   2. AB(C + EF) = Y
   3. (A + B) (C + DE) = Y
   4. A(B + C) (ED + F) = Y
8. Factor the largest possible number of variables from these expressions. Do not simplify.
   1. AB + AC = Y
   2. AB + BC + BD = Y
   3. ABD + ACD + ADE = Y
   4. ABC + ABD + CD = Y
9. Using Boolean algebra, simplify the expression AB(A + B) in Fig. P6-3. Show all steps. Implement (draw a circuit representing) the simplified expression.
10. Repeat Problem 9 for the expression in Fig. P6-4.
11. Repeat Problem 9 for the expression in Fig. P6-5.
12. Repeat Problem 9 for the expression in Fig. P6-6.
13. Repeat Problem 9 for the expression in Fig. P6-7.
14. Repeat Problem 9 for the expression in Fig. P6-8.
15. Copy the Karnaugh map at the bottom of Fig. P6-9 onto a sheet of paper. Based on the truth table, complete the Karnaugh map.
16. Repeat Problem 15 for the truth table and Karnaugh map in Fig. P6-10.
17. Write the Boolean expression that best represents the map in Fig. P6-11. Call the output Y. Zeros have been omitted from the Karnaugh map to emphasize cells containing 1s.
18. Repeat Problem 17 for the map in Fig. P6-12.
19. Repeat Problem 17 for the map in Fig. P6-13.
20. Repeat Problem 17 for the map in Fig. P6-14. Two loops have been indicated, so your expres-

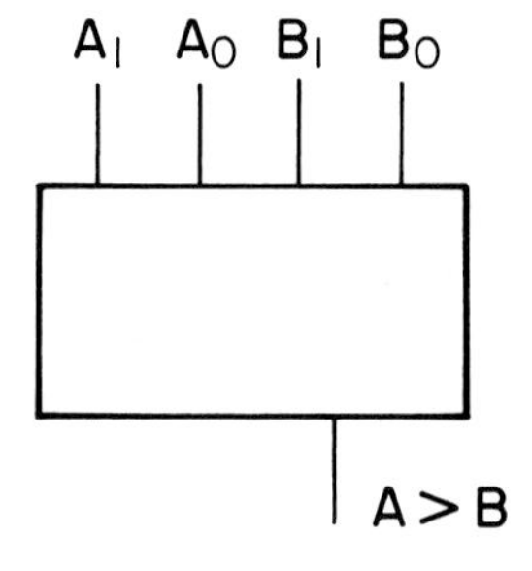

INPUTS				OUTPUT
A_1	A_0	B_1	B_0	A>B
0	0	0	0	
0	0	0	1	
0	0	1	0	
0	0	1	1	
0	1	0	0	
0	1	0	1	
0	1	1	0	
0	1	1	1	
1	0	0	0	
1	0	0	1	
1	0	1	0	
1	0	1	1	
1	1	0	0	
1	1	0	1	
1	1	1	0	
1	1	1	1	

Fig. P6-2. Comparator logic element. The symbol ">" means "greater than." "A>B" means "A is greater than B."

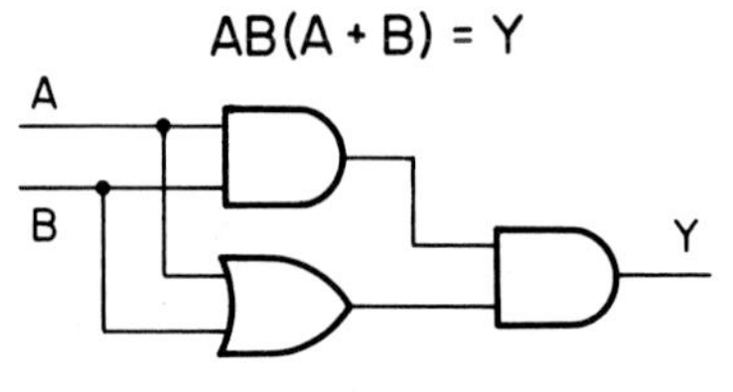

Fig. P6-3. The circuit having 2 ANDs and an OR can be simplified. Multiplication within the Boolean expression is the first step. The simplified Boolean expression can be implemented.

$(A + B)(A + C) = Y$

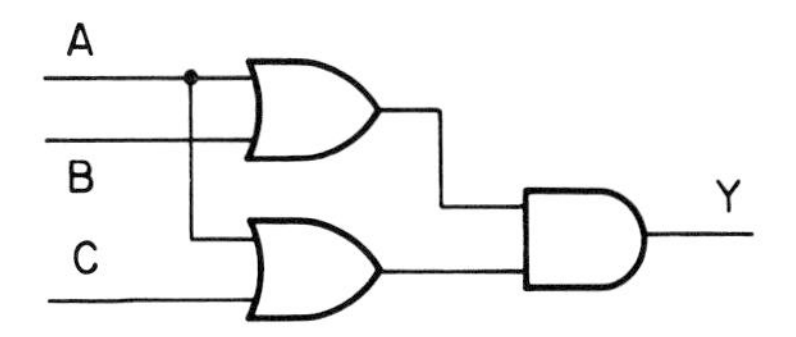

Fig. P6-4. The circuit with 2 ORs and an AND can be simplified.

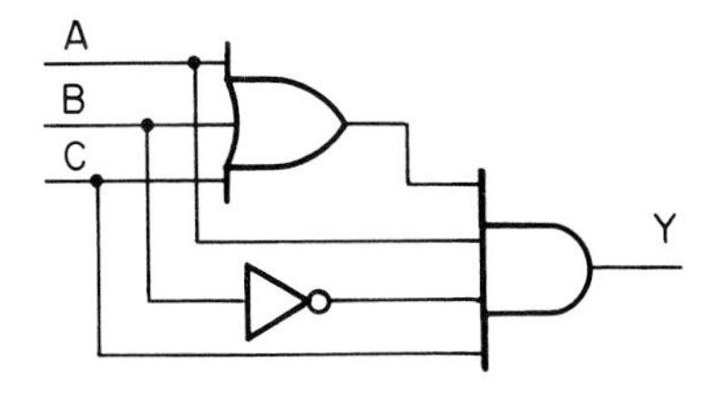

Fig. P6-5. The circuit with a 3-input OR, a NOT, and a 4-input AND can be simplified.

$A\bar{B}(\bar{A} + \bar{B}) + C = Y$

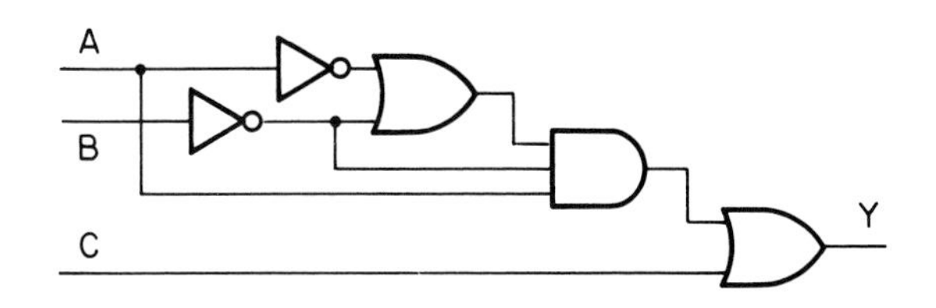

Fig. P6-6. The circuit with 2 NOTs at the inputs of an OR can be simplified.

$\bar{A}(\overline{A + B}) + \bar{B} = Y$

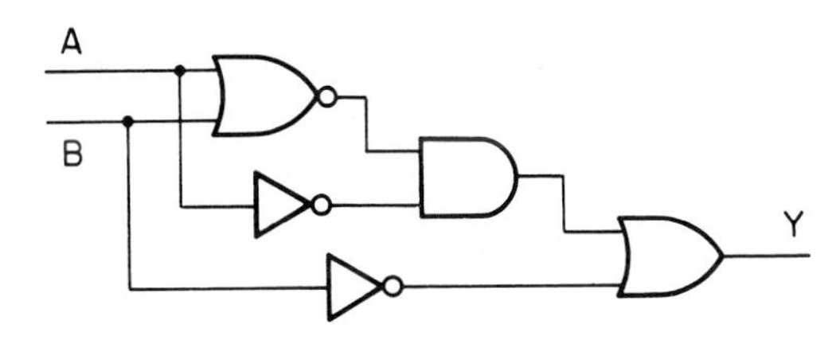

Fig. P6-7. The circuit with a NOR, an AND, an OR, and two NOTs can be simplified.

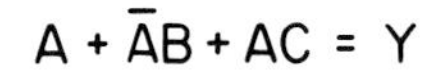

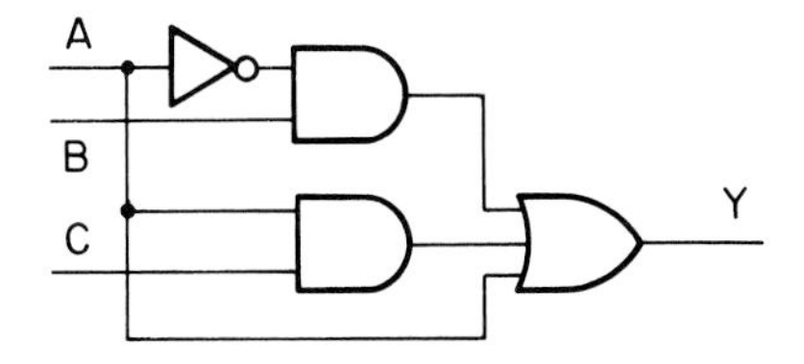

Fig. P6-8. The circuit with two 2-input ANDs, a 3-input OR, and a NOT can be simplified.

INPUTS			OUTPUT
A	B	C	Y
0	0	0	0
0	0	1	0
0	1	0	0
0	1	1	1
1	0	0	1
1	0	1	1
1	1	0	0
1	1	1	0

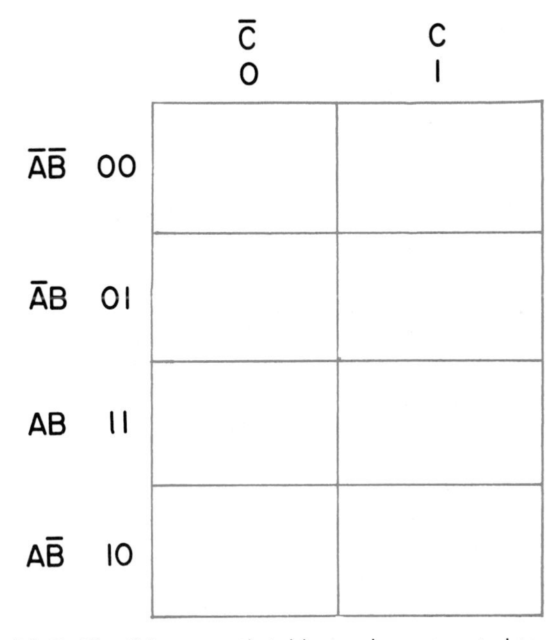

Fig. P6-9. The 3-input truth table can be converted to a 2-column Karnaugh map.

INPUTS				OUTPUT
A	B	C	D	Y
1	1	1	1	1
1	1	0	0	1
0	1	0	1	1
1	1	1	0	1

OTHER INPUT SETS RESULT IN Y = 0

	$\bar{C}\bar{D}$ 00	$\bar{C}D$ 01	CD 11	$C\bar{D}$ 10
$\bar{A}\bar{B}$ 00		1		
$\bar{A}B$ 01		1		
AB 11			1	1
$A\bar{B}$ 10				

Fig. P6-12. A Boolean expression with 4 ANDs is not the simplest expression taken from the Karnaugh map.

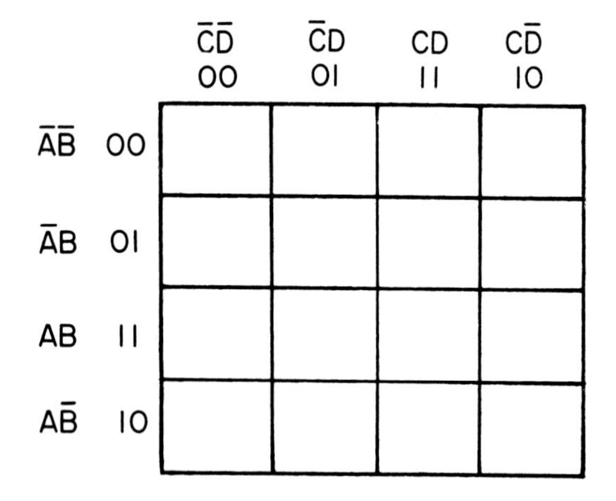

Fig. P6-10. The 4-input truth table can be converted to a 4-column Karnaugh map.

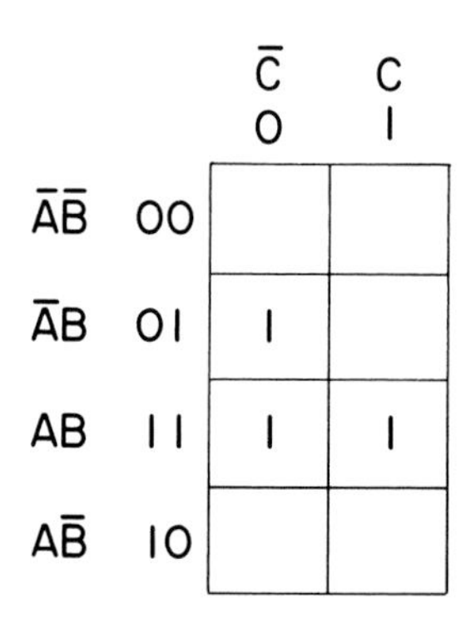

Fig. P6-13. The Boolean expression uses less than 3 ANDs.

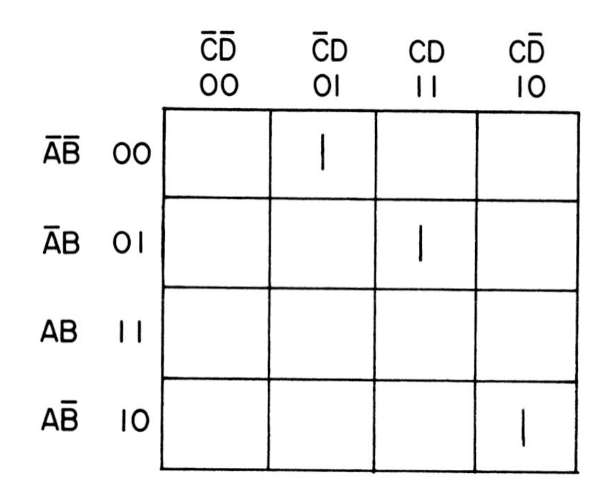

Fig. P6-11. A Boolean expression with 3 ANDs can be written from the Karnaugh map.

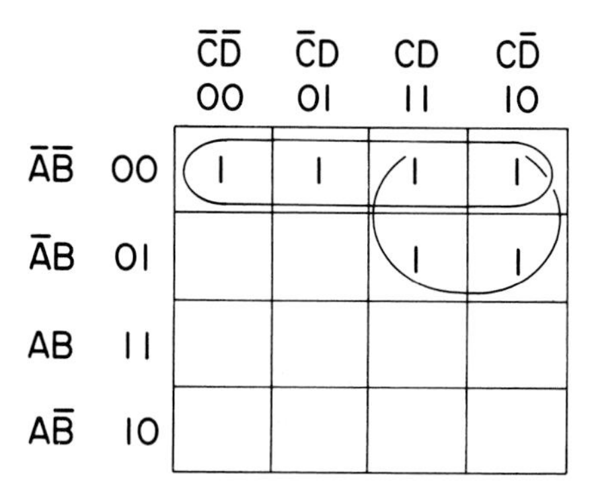

Fig. P6-14. The Boolean expression reduces 6 cells to 2 AND elements.

sion will contain two ANDs.

21. Repeat Problem 17 for the map in Fig. P6-15. Use two loops.

		$\bar{C}\bar{D}$ 00	$\bar{C}D$ 01	CD 11	$C\bar{D}$ 10
$\bar{A}\bar{B}$	00		1	1	
$\bar{A}B$	01	1			1
AB	11				
$A\bar{B}$	10		1	1	

Fig. P6-15. The Boolean expression has 2 ANDs.

22. Use sum-of-products to write the expression for the truth table in Fig. P6-16. Then simplify.

INPUTS			OUTPUT
A	B	C	Y
0	0	0	1
0	0	1	1
0	1	0	0
0	1	1	0
1	0	0	0
1	0	1	1
1	1	0	0
1	1	1	0

Fig. P6-16. A Boolean expression can be written using sum-of-products. Rearrangement of the terms can simplify the resulting Boolean expression.

23. Copy the Karnaugh map in Fig. P6-17 onto a sheet of paper, and complete it using data from the truth table in Problem 22. Write the expression represented by this map, and compare it with the one obtained using sum-of-products.

24. Copy the Karnaugh map in Fig. P6-18 onto a piece of paper and do the following:
   a. Add product-of-sums labels within the dashed rectangles.
   b. Using these labels, write the product-of-sums expression representing this map.
   c. Implement (draw the circuit diagram) the expression.
   d. The sum-of-products expression representing this map is $\bar{A}\bar{B} + \bar{B}C = Y$. Use Boolean simplification to show that your expression is its equivalent.

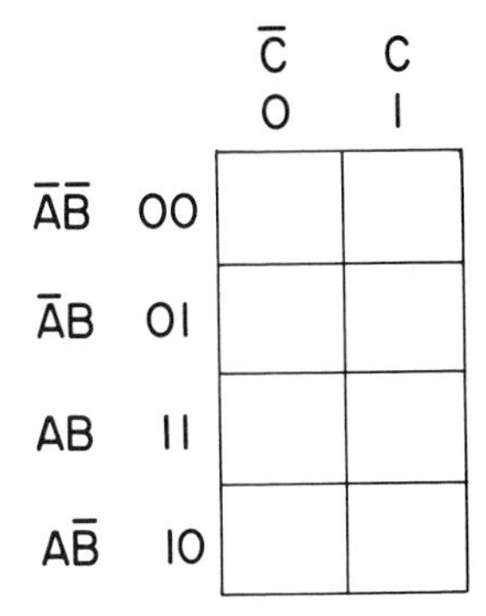

Fig. P6-17. The Karnaugh map uses data from Fig. P6-16 (Problem 22).

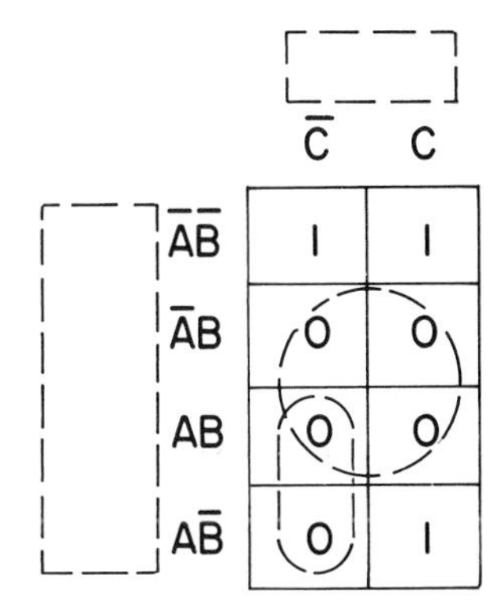

Fig. P6-18. If product-of-sums labels are added to this Karnaugh map, a product-of-sums can be written directly from this map.

Complex setups are needed to test the operation of modern large-scale integration chips. (Tektronix, Inc. 1996©)

Chapter 7

Sequential Logic

LEARNING OBJECTIVES

After studying this chapter and completing lab assignments and study problems, you will be able to:

- Recognize relay-based memory circuits and explain their operation.
- Recognize the symbol for an R-S flip-flop and describe its action using truth tables and timing diagrams.
- Explain the operation of a simple register based on the R-S flip-flop.
- Explain the operation of a NAND-based R-S flip-flop.
- Draw circuit diagrams for R-S flip-flops using either NANDs or NORs.
- Complete the truth table for an R-S flip-flop with active-low inputs.
- List the basic fault modes in NAND-based R-S flip-flops.

When a pocket calculator is used to add two numbers, say 4 and 6, the numbers are entered one after the other. If the 4 is entered first, it must be stored within the circuit until the second number has been entered. Only then does the addition take place. The resulting sum must also be stored so it can be continuously displayed.

Combinational-logic circuits (the type discussed in Chapters 2, 4, 5, and 6) are used to do the addition. Another group of circuits, called *SEQUENTIAL-LOGIC* circuits, accomplish the storage. This chapter introduces the elements used in such circuits.

RELAY-BASED MEMORY ELEMENT

When pushbutton switches are used to start and stop electric motors, a memory element must be provided. The circuit must remember whether START or STOP was pushed last.

The following describes the conversion of a combinational-logic circuit into a sequential-logic circuit. That is, a circuit without memory will be converted to one that can remember which switch, START or STOP, was pressed last.

COMBINATIONAL-LOGIC CIRCUIT

Part "a" of Fig. 7-1 shows the parts in a motor-starting circuit. A lamp has been substituted for the motor to make the circuit look more like the one used in Chapter 2.

The device marked 1CR is of special interest. It is a control relay. A relay is defined as an electrically operated switch.

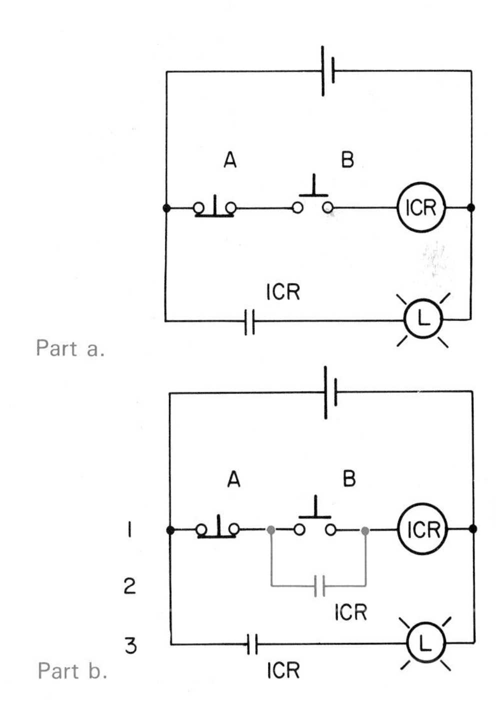

Fig. 7-1. The upper circuit is a combinational-logic circuit. It can be changed into a sequential logic circuit by adding the contacts on line 2 of the lower drawing.

RELAY ACTION

Part "a" of Fig. 7-2 shows the construction of a relay. Its input is a coil (coil of wire). When current flows through this coil, an electromagnetic field is created. This field in turn attracts the iron armature, and it moves upward. Through a simple mechanical linkage (see the dashed line), this motion operates the contacts.

To restate the action of a relay, when current flows in its coil, the electrical contacts are activated. When the current stops, the contacts are deactivated.

RELAY SYMBOL

Part "b" of Fig. 7-2 shows the relay symbol. Its coil (input) is represented by a circle. Parallel bars are used for contacts (outputs). A single coil can operate several sets of contacts. Some contacts may be NO (normally open); some may be NC (normally closed). As with switches, NO contacts close when activated; NC contacts open when activated. A bar across a set of contacts indicates NC.

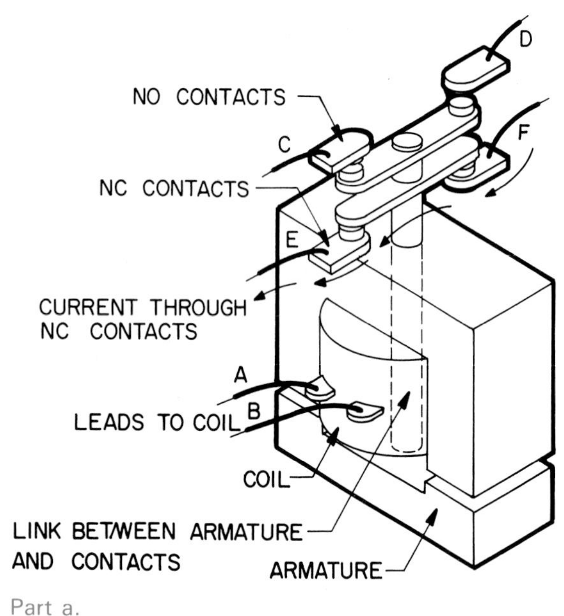

Part a.

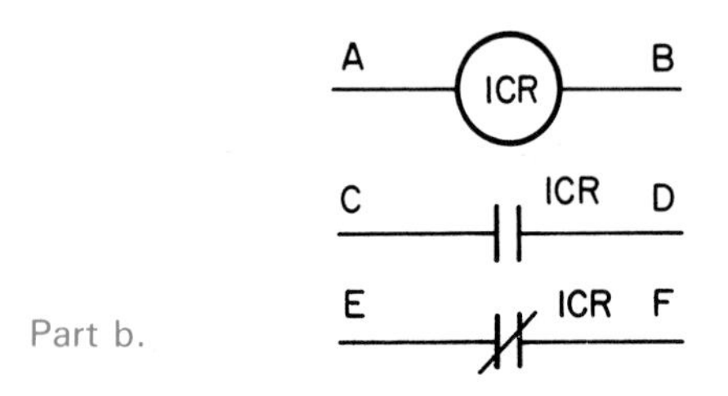

Part b.

Fig. 7-2. When current flows in the coil of the relay, its contacts are activated.

COMBINATIONAL-LOGIC CIRCUIT ACTION

Refer again to part "a" of Fig. 7-1. When B is pressed, the upper loop is completed, current flows through the relay coil, and the contact on line 2 closes. The lamp lights.

When B is released, the lamp goes out. This is the expected action for a combinational-logic circuit. The condition of its output depends directly and immediately on the combination of its input signals. In this case, that combination is:

$$\overline{A}B = L$$

SEQUENTIAL-LOGIC CIRCUIT

The addition of one set of relay contacts changes this circuit to sequential logic. See part "b" of Fig. 7-1.

START CIRCUIT

When B is pressed, current flows through the relay coil, and both sets of contacts close. The lower set lights the lamp.

RUN CIRCUIT

The effect of the second set of contacts becomes obvious when B is released. Although B is open, current continues to flow in the relay coil. Its path is through the closed set of contacts. The lamp remains lit, since the circuit remembers that B was pressed last.

The contacts on line 2 are called *HOLDING* or *SEALING CONTACTS.* They may also be referred to as *AUXILIARY CONTACTS.*

STOP CIRCUIT

Switch A is used to turn off the lamp. When pressed, A opens and stops the flow of current through the relay coil. This results in the opening of both sets of contacts.

When A is released, the circuit will not re-energize, since both B and the holding contacts in the upper loop are open. The circuit remembers that A was pressed last.

SEQUENTIAL-LOGIC RELAY TRUTH TABLE

Fig. 7-3 describes the action of the sequential-logic relay memory circuit. A and B are inputs. Ones in these columns indicate the pressing of a switch. A 1 in the output column indicates that the lamp is lit. The symbol Q means output.

The heading of the output column has the

suffix t + 1. A suffix is a symbol or group of symbols that follows a symbol. The symbol t + 1 follows the symbol Q. (Here, + is read as plus.) The t + 1 implies a small time after some action has taken place.

The first row states that shortly after B has been pressed, the lamp will light. The second row indicates that the lamp will be out shortly after A is pressed. The next row is important, for it indicates the memory ability of the circuit. It states that shortly after a button is released (be it A or B), the lamp will be in the same condition it was just before that release.

The result of pressing and releasing both buttons at the same time cannot be predicted. The lamp may stay off or come on depending on the order of release of the switches.

As a result, this situation is not defined and should be avoided. Refer to the last row in the truth table.

INPUTS A	INPUTS B	OUTPUT Q_{t+1}
0	1	1
1	0	0
0	0	Q_t
1	1	*

* NOT DEFINED

Fig. 7-3. Truth table for the relay-based circuit which is shown in part "b" of Fig. 7-1.

APPLICATION OF RELAY-BASED MEMORY

Fig. 7-4 depicts a common use of relay-based memories. This circuit starts and stops a large industrial motor.

Power flows to the motor through the upper three lines of this diagram. The line voltage is 240 volts or more. Current flows through a disconnect switch (1SW), fuses (FU), motor starter contacts (M), and overload protection heaters (OL). Only the motor starter contacts will be of interest here. The other parts will be considered later.

The lower portion of the diagram is the control circuit. Power is supplied by the transformer (T), and the circuit is fused (4FU). Coil M is part of the motor starter. This device is an oversize relay. Its contacts are capable of carrying the large currents demanded by the motor. Overload contacts (OL) are normally closed. If the motor draws too much current, they open and stop the machine.

The action of the circuit is as follows. When 2PB is pressed, coil M is energized. All four contacts close. Those in the power circuit supply current to the motor. Those in the control circuit are holding contacts. When 2PB is released, the holding contacts maintain current through coil M.

Pressing 1PB breaks the circuit supplying current to M. All four contacts open and the motor stops.

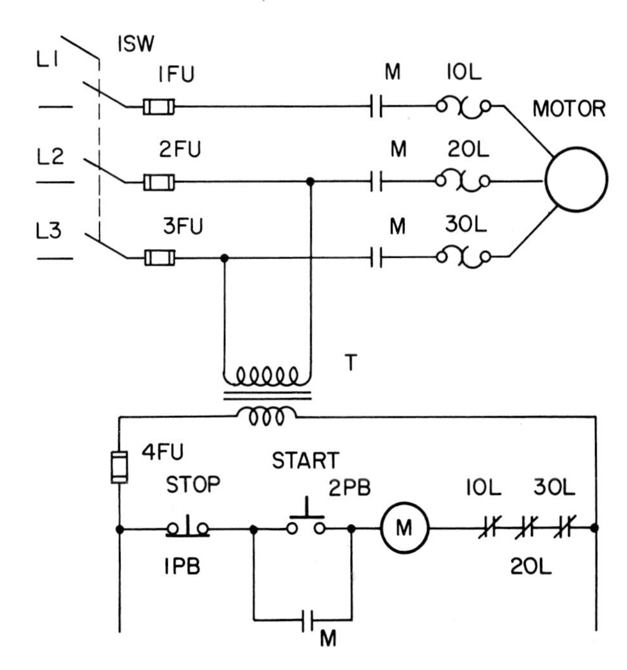

Fig. 7-4. Note the use of holding contacts in this motor starting circuit. The holding contacts are at bottom center.

IC MEMORY ELEMENT

The formal name for the IC memory element is *BISTABLE MULTIVIBRATOR*. It is usually called a *FLIP-FLOP*. It duplicates the action of relays that have holding contacts. The flip-flop has two states.

FLIP-FLOP SYMBOL

Part "a" of Fig. 7-5 shows the flip-flop symbol. The device has two inputs. S stands for SET; R stands for RESET. As a result, this device is usually called an R-S flip-flop.

The letter Q is usually used at the flip-flop's output. Power and ground connections are required, but these are seldom shown on the symbol.

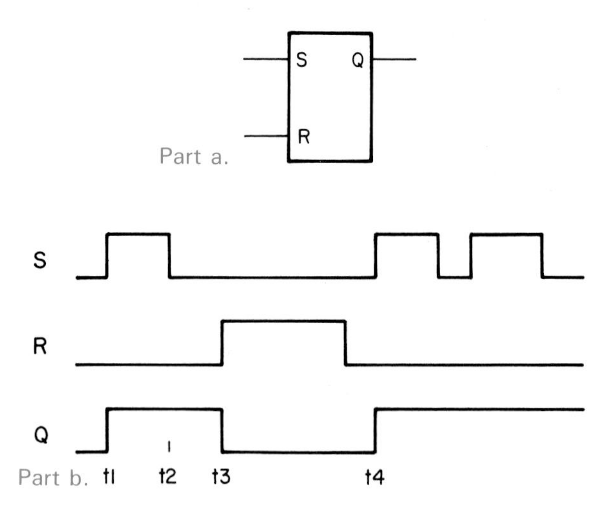

Fig. 7-5. The symbol for an R-S flip-flop and a timing diagram for an R-S flip-flop.

FLIP-FLOP ACTION

Flip-flop operation is identical to that of relay-based memory elements. When a 1 is applied to input S, output Q goes to 1. When S returns to 0, Q remains at 1.

To store a 0, a 1 is applied to input R. Q resets and remains at 0 even when R has returned to 0. That is, the flip-flop remembers where the last 1 was applied.

FLIP-FLOP TRUTH TABLE

The truth table in Fig. 7-6 is similar to the truth table for relay-based elements. It differs only in the identifying letters used.

Note the last row. A condition in which R and S are 1 at the same time is not defined. Designers must avoid this situation when using R-S flip-flops. If 1s appear at both inputs of such devices, a fault probably exists in the driving circuits.

FLIP-FLOP TIMING DIAGRAM

Flip-flops are sequential-logic devices, so the order of arrival of input signals influences observed outputs. As a result, timing diagrams are often used to describe flip-flop action.

Part "b" of Fig. 7-5 is the timing diagram for an R-S flip-flop. At time t0, Q = 0. At t1 a SET signal was applied to S, so Q changed to 1. At t2, S returned to 0, but Q remained 1. At t3, the flip-flop was reset by the application of a 1 to input R.

Note the lengths of the SET and RESET pulses. The fact that one is shorter than the other does not affect the operation of the device.

Also notice that two SET pulses appear after time t4. Because Q is already 1, the second pulse has no effect on the output.

INPUTS		OUTPUT
S	R	Q_{t+1}
1	0	1
0	1	0
0	0	Q_t
1	1	*

*NOT DEFINED

Fig. 7-6. Truth table for R-S flip-flop. For the input set 1, 1, the output is not defined. The input set 1, 1 is not allowed.

FLIP-FLOP APPLICATION

In computers and other digital circuits, numbers must be stored temporarily. Circuits used for this purpose are called *REGISTERS.*

Fig. 7-7 shows a 3-bit register. Numbers enter at the bottom (A2A1A0); stored numbers are available at the top (Q2Q1Q0).

To store a number in this register, WRITE ENABLE must go to 1. This opens the AND gates. If the number on bus lead A2 is 1, the first AND has 1s applied to both its inputs, so S2 is 1. At the same time, the NOT applies a 0 to one of the inputs of the second AND. R2 will be 0. As a result, the first flip-flop will be set (Q2 = 1).

If A1 = 0, just the opposite happens at Q1. The 0 from A1 is applied to the first AND, so S1 is 0. Due to the NOT, the second AND has 1s applied to both inputs, so R1 equals 1. The flip-flop in the middle of the register is reset (Q1 = 0).

When the number has been stored, WRITE ENABLE is returned to 0. With 0s applied to the ANDs, all S and R inputs are 0. The stored numbers are retained until WRITE ENABLE is again activated.

NAND FLIP-FLOP

The circuit in Fig. 7-8 contains two NANDs and two NOTs, yet it is not a combinational-logic circuit.

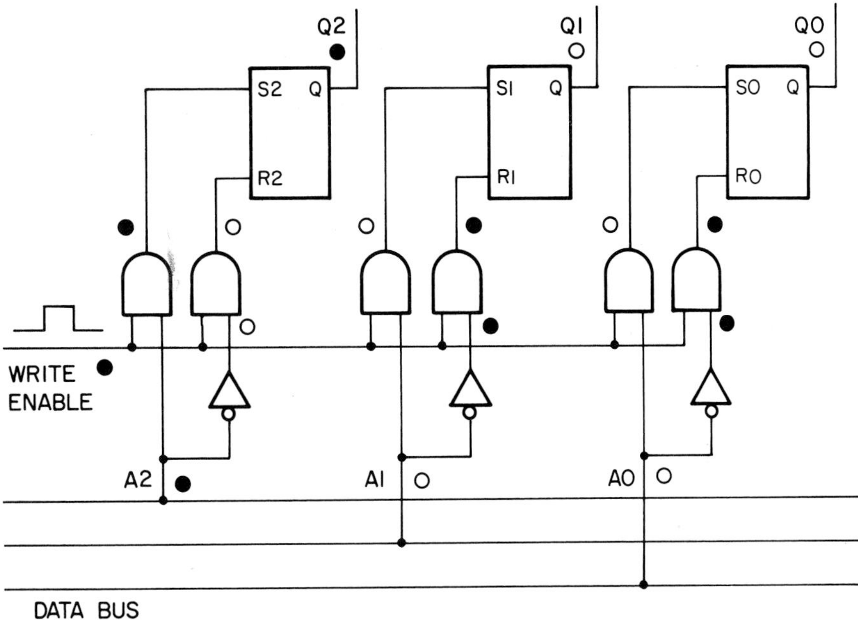

Fig. 7-7. R-S flip-flops are the memory devices in this 3-bit register.

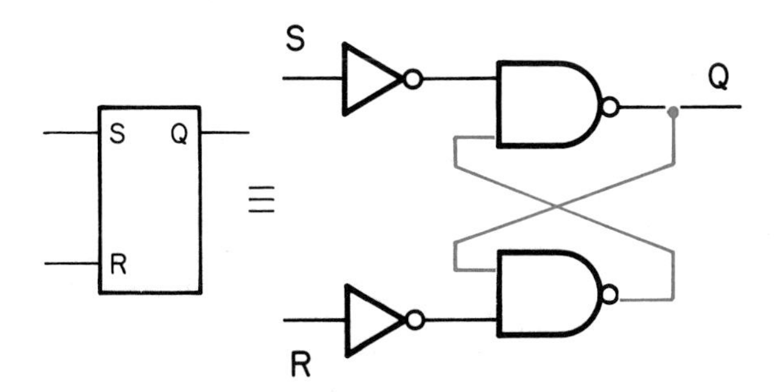

Fig. 7-8. The cross-coupling of the outputs back to the inputs changes this combinational-logic circuit into an R-S flip-flop.

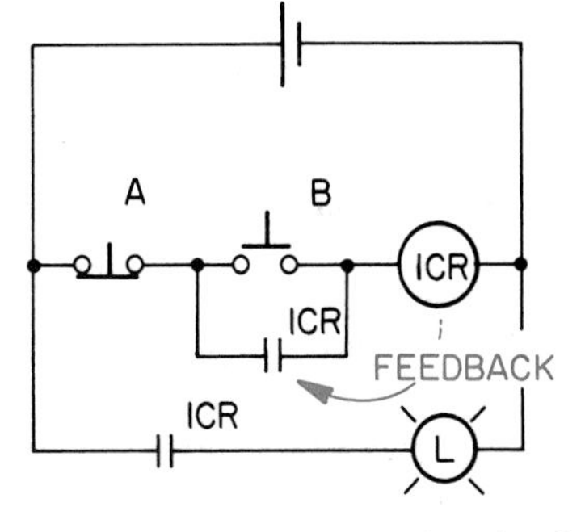

Fig. 7-9. For this relay-based logic circuit to have memory capabilities, feedback is required.

The crossed leads cause it to function as a sequential-logic circuit. It is a simple R-S flip-flop.

FEEDBACK

Memory elements must contain feedback. That is, output signals are returned to the circuit's input. This can be seen in the relay-based memory circuit in Fig. 7-9. Relay 1CR is associated with the circuit's output. However, one set of contacts is in parallel with input B. This paralleling of input and output elements represents feedback.

The crossed leads in Fig. 7-8 also provide feedback. Output Q is returned to the input of the lower AND.

REVIEW OF NAND

A brief review of NAND action is appropriate. Fig. 7-10 shows both NAND symbols. The first emphasizes that the application of two 1s to a NAND results in an output of 0. The second points out that a 0 applied to either input results in the output of a 1.

NAND FLIP-FLOP ACTION

The circuit and timing diagram in Fig. 7-11 is used to describe the action of the NAND flip-flop. The NOTs do not enter directly into the operation of the circuit. They merely make inputs S and R active

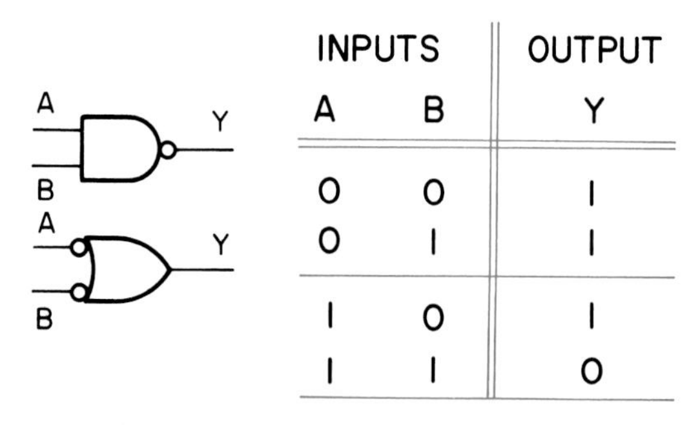

INPUTS		OUTPUT
A	B	Y
0	0	1
0	1	1
1	0	1
1	1	0

Fig. 7-10. A review of NAND action emphasizes the two symbols used to represent the NAND element.

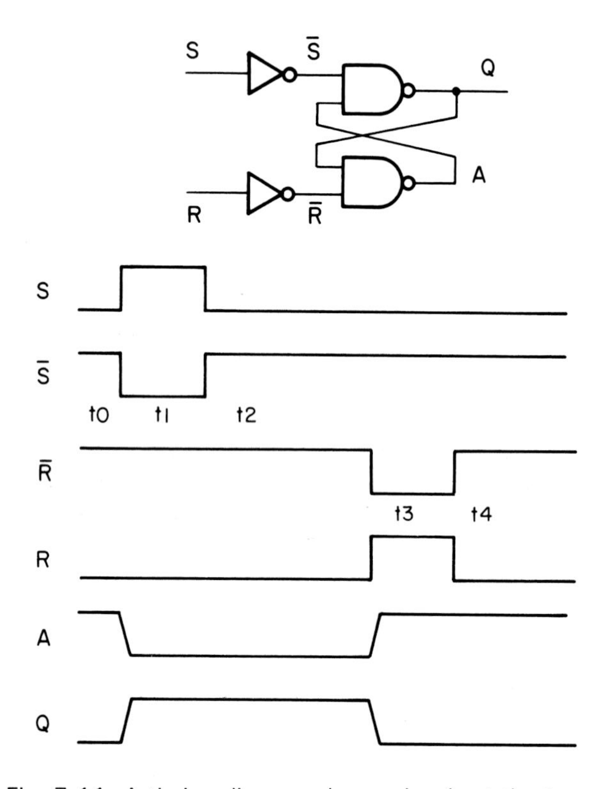

Fig. 7-11. A timing diagram shows signals at the terminals and within the NAND-based flip-flop.

high. (The circuit is SET and RESET by 1s.)

Except during the change from one stored number to another, the circuit must be stable. That is, signals fed back by the crossed leads must hold output Q in whatever state it is in. If Q = 1, the signal fed around the figure-eight path must force Q to 1. This will be tested for the five situations in the timing diagram.

1. TIME t0. At time t0, S and R are both 0, and Q = 0. That is, at some time in the past, a 0 was stored in the flip-flop.

The problem will be one of proving that the feedback signal will hold Q at 0. Fig. 7-12 will be used in the proof. Note that both NAND symbols have been used. This is acceptable, since they are both valid representations of NANDs.

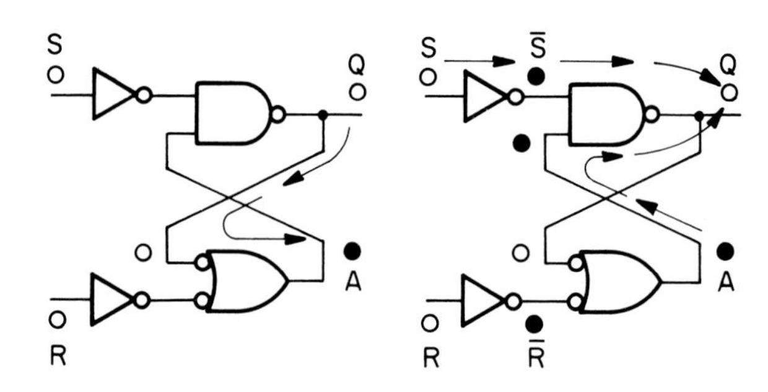

Fig. 7-12. At time t0 in the timing diagram in Fig. 7-11, the stored 0 is not changed by the presence of 0s at R and S.

The first diagram shows the 0 from Q being fed to an input of the lower NAND. If either input of a NAND is 0, its output will be 1 (A = 1).

The diagram at the right shows the 1 at A being fed to one of the inputs of the upper NAND. Here it combines with the NOTed 0 from S. With two 1s at its input, this NAND outputs a 0. That is, the signal fed around the figure-eight path supports the 0 that was assumed at Q. The circuit is stable. It will remain in this state until a signal is applied to S or until the power is removed.

2. TIME t1. Refer to time t1 in the timing diagram of Fig. 7-11. Just before t1, S changed to 1. As a result, the output of the upper NOT changed to 0. To follow the resulting action, refer to Fig. 7-13. (The change of symbols does not alter the circuit's action.)

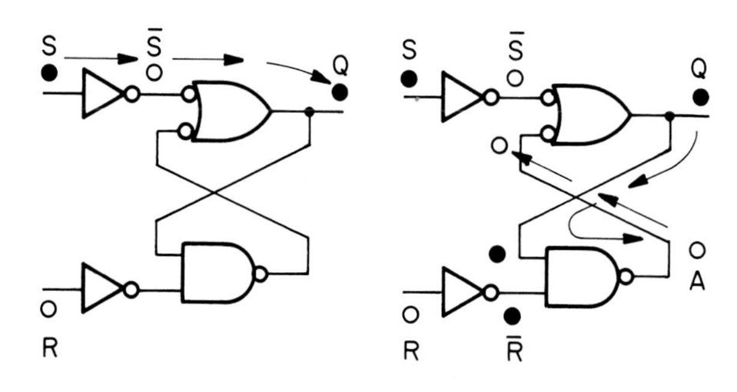

Fig. 7-13. When S goes to 1 at time t1, output Q changes to 1.

With S = 1, a 0 is applied to the upper NAND. With either input a 0, this NAND outputs a 1, so Q changes to 1.

In the diagram at the right, this 1 is shown being fed to the lower NAND. With both inputs at 1, this NAND outputs a 0 (A changes to 0).

This 0 is then applied to the upper NAND. However, it does not alter Q, since the output of a NAND is 1 whether one or both inputs are 0. The circuit is stable in this state.

3. TIME t2. At t2, the SET signal at S has been removed. Will Q remain 1?

   Fig. 7-14 shows the 1 at Q being fed to the input of the lower NAND. With 1s at both inputs, this NAND outputs a 0.

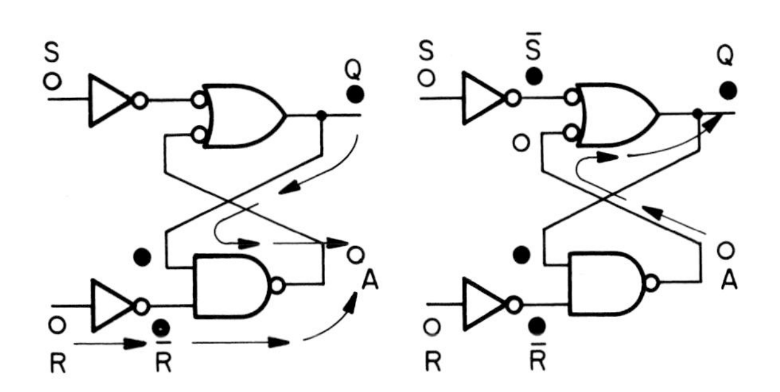

Fig. 7-14. When S returns to 0, Q continues to display the stored 1.

In the diagram at the right, the 0 at A is fed to the upper NAND. With one input at 0, its output will be 1. The circuit is indeed stable in this state.

4. TIME t3. The storing of a 0 is similar to the storing of a 1. The action is much like that in Fig. 7-13, except that the change of state starts with the lower NAND. Using Fig. 7-15, attempt to follow the action of this circuit from R = 1 to A and then around the loop back to A. The fed back signal should support A = 1.
5. TIME t4. At t4, the circuit is again in the state described in Fig. 7-12. Its inputs are both 0, and a 0 has been stored.

NOR FLIP-FLOP

NORs can be used to build R-S flip-flops. See Fig. 7-16. Such circuits operate in much the same manner as NAND flip-flops. Note that Q is taken from the lower NOR element.

OTHER R-S FLIP-FLOPS

Many types of R-S flip-flops are available. Only two will be described at this time:
1. With dual output.
2. With active-low inputs.

DUAL OUTPUT FROM R-S FLIP-FLOP

Refer again to the timing diagram in Fig. 7-11. Note that A is always the inverse of Q. This means that Q and $\overline{Q}$ are available in the circuit.

In circuit design, the availability of $\overline{Q}$ can be helpful, so both Q and $\overline{Q}$ are often brought out, Fig. 7-17. The upper symbol is most often used. However, the bubbled symbol has the same meaning.

$\overline{Q}$ is available in both NAND- and NOR-based circuits. The output leads on the NOR flip-flop must be crossed, Fig. 7-17, to match the standard positions of Q and $\overline{Q}$ on the general symbol.

ACTIVE-LOW INPUTS INTO NAND R-S FLIP-FLOP

Omitting the NOT logic elements at the inputs of a NAND flip-flop results in a memory with active-low inputs, Fig. 7-18. To set this flip-flop, S must go

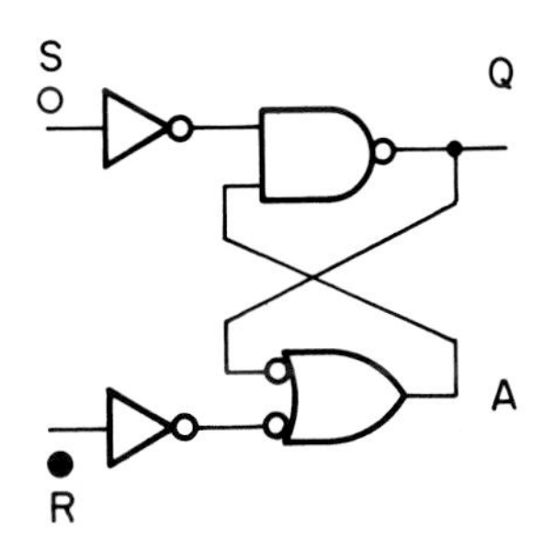

Fig. 7-15. This illustration is a self test. See if you can trace the signal from R to A and then from A around the loop back to A.

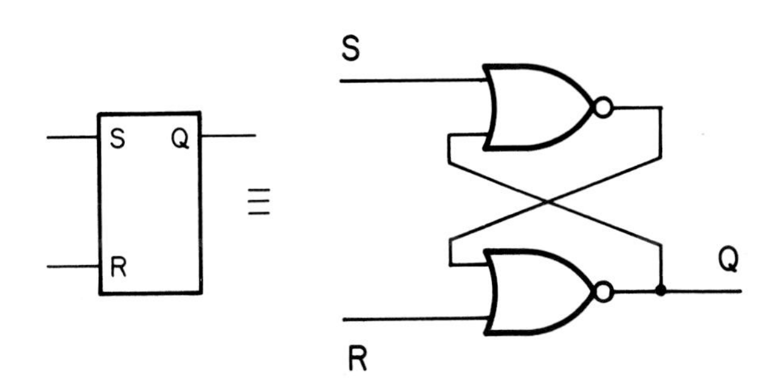

Fig. 7-16. NORs can be used to build R-S flip-flops.

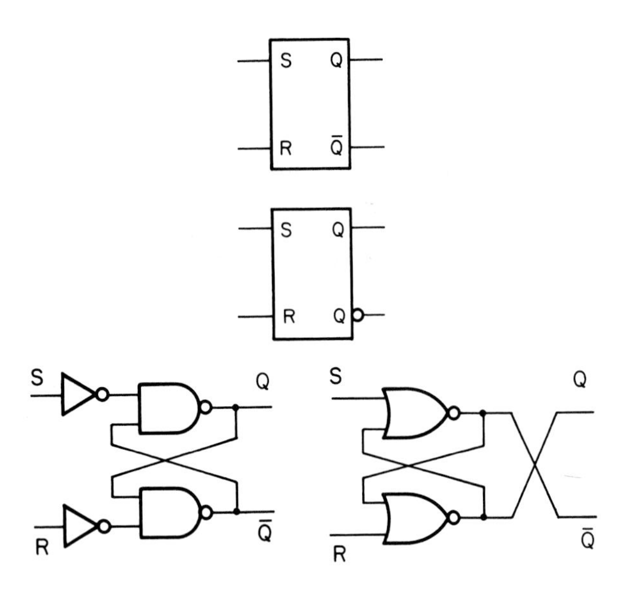

Fig. 7-17. Because it is available in the R-S flip-flop circuit, $\bar{Q}$ is often brought out.

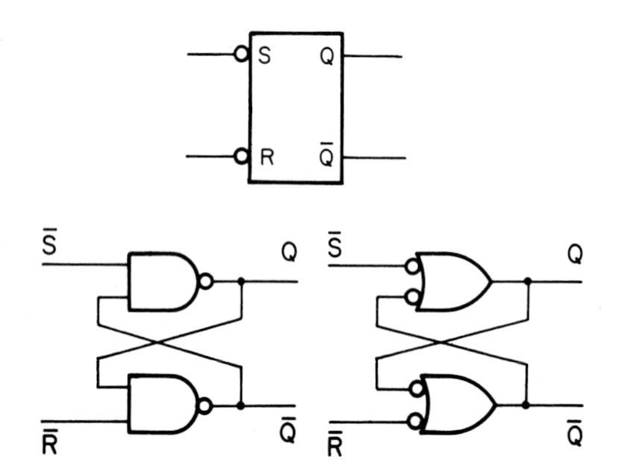

Fig. 7-18. To emphasize its active-low input, the R-S flip-flop circuit has been redrawn using the alternative NAND symbol.

INPUTS		OUTPUT
S	R	Q_{t+1}
0	1	1
1	0	0
1	1	Q_t
0	0	*

* NOT DEFINED

Fig. 7-19. Truth table for R-S flip-flop that has active-low inputs. The input set 0, 0 is not allowed.

to 0. To reset, R must go to 0. During the time that a number is stored, S and R must both be 1. See the truth table shown in Fig. 7-19.

Fig. 7-20 shows a timing diagram for an active-low flip-flop. The action in the circuit takes place when an input goes low.

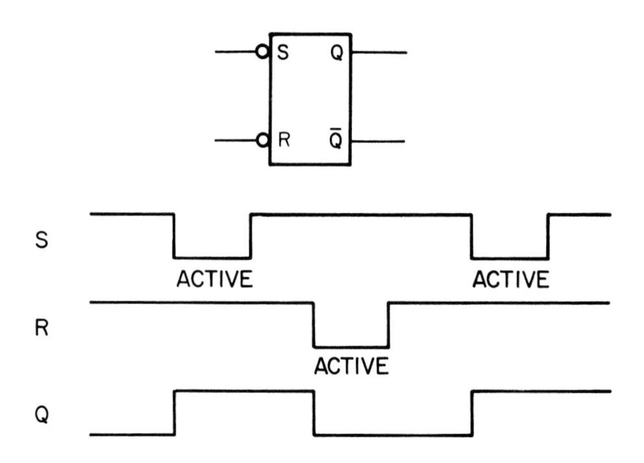

Fig. 7-20. Timing diagram for an R-S flip-flop that has active-low inputs.

TROUBLESHOOTING

The troubleshooting methods used with combinational-logic circuits can be applied to sequential-logic circuits. However, the order of arrival of signals is important in sequential circuits, so dynamic measurements are often necessary.

TROUBLESHOOTING A STUCK-AT-0 FAULT

Most faults in sequential-logic circuits are of the type called "stuck-at." Care must be taken to distinguish between real and apparent stuck-at faults.

TRUE S-A-0 FAULT

Output Q in part "a" of Fig. 7-21 is s-a-0. This 0 is fed to the lower portion of the circuit, and $\bar{Q}$ is held at 1. The SET and RESET signals have no effect on the outputs. Replacing the flip-flop is likely to correct the fault.

APPARENT S-A-0 FAULT

The outputs in part "b" of Fig. 7-21 are identical to those in the upper timing diagram. In this case, however, no SET signal is present. The flip-flop has

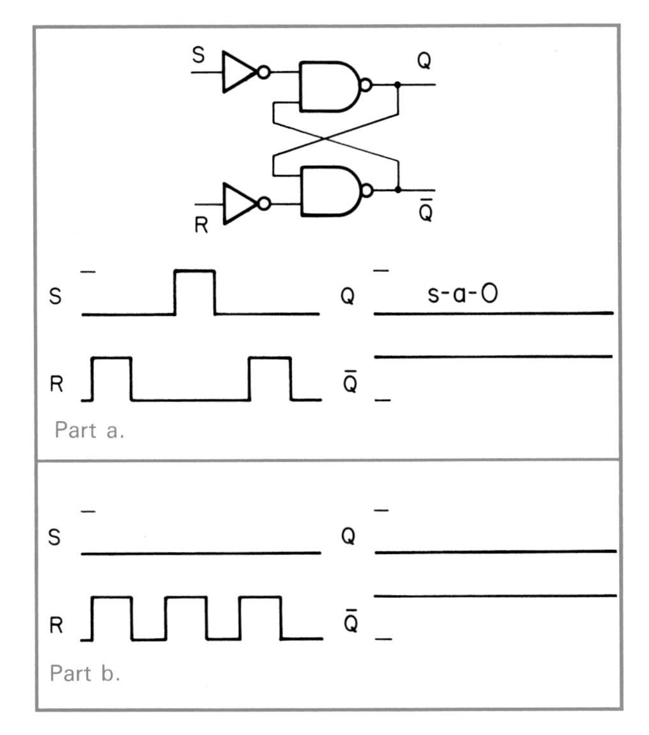

Fig. 7-21. The upper timing diagram represents a true s-a-0 fault at Q. The lower diagram shows only an apparent s-a-0 fault.

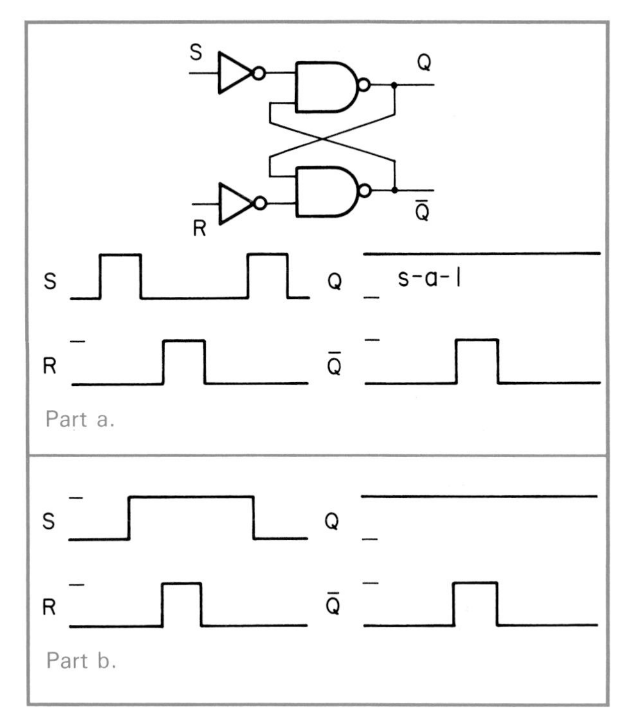

Fig. 7-22. The upper timing diagram shows a true s-a-1 fault at Q. The lower diagram shows an apparent s-a-1 fault.

been RESET in the past. The signal at R is not calling for any change. The flip-flop remains in the RESET state.

If the lack of signal at S is due to an internal short to ground, replacing the flip-flop is likely to correct the fault. If the fault is in the circuit driving S, this action will not correct the fault. Circuits ahead of the flip-flop need to be inspected.

TROUBLESHOOTING A STUCK-AT-1 FAULT

When the output of an R-S flip-flop is s-a-1, interesting output signal patterns result. The patterns help determine the fault.

TRUE S-A-1 FAULT

Output Q of the flip-flop in part "a" of Fig. 7-22 is s-a-1. Due to the feedback, this 1 is permanently applied to the lower NAND. With a 1 at one input, this NAND acts as if it were a NOT. Its action cancels that of the NOT in lead R, and input R appears at $\overline{Q}$.

When Q and $\overline{Q}$ are observed at the same time, this fault is obvious. $\overline{Q}$ must always be the inverse of Q for a properly operating flip-flop. Replacing the flip-flop is likely to correct the fault.

APPARENT S-A-1 FAULT

The outputs in part "b" of Fig. 7-22 look like those of the upper circuit. However, the flip-flop is not faulty. The problem is in the applied signals. S and R are 1 at the same time. In the R-S flip-flop truth table, this is a disallowed condition. Changing the flip-flop would not correct the situation, since the fault is in the driving circuit.

GLITCH-INDUCED ERRORS

The output of the flip-flop in Fig. 7-23 is incorrect. The device was SET by the pulse on S, but then, for no apparent reason, it RESET at t2.

Such action suggests a glitch on R. A high speed oscilloscope or logic probe with pulse capabilities could be used to inspect lead R. It is likely that a glitch exists at t2. It may be small and may not last long, but flip-flops are sensitive to such disturbances.

The fault is not in the flip-flop, and replacing it would not correct the problem. Rather, there is a design fault in the circuit driving the flip-flop. Two signals arriving at the same time in the driving circuit may have caused a race condition at the output of the driving circuit. The race condition must be found and corrected.

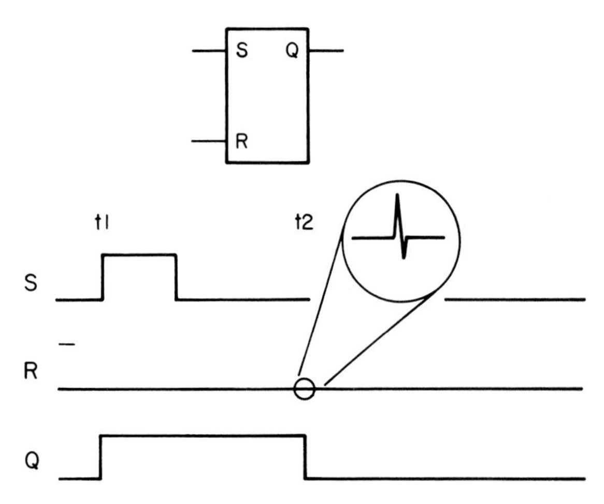

Fig. 7-23. A glitch occurs at time t2 on lead R of the R-S flip-flop shown. Because of the glitch on lead R, the R-S flip-flop resets at time t2.

LOSS OF POWER

When supply voltage is removed from a flip-flop, the stored number is lost. Such memory elements are said to be *VOLATILE.*

When power is reapplied, there is no way of knowing what number will appear at Q. It might be a 1; it might be a 0.

SUMMARY

Unlike combinational-logic circuits, the outputs of sequential-logic circuits depend on past as well as present input signals. Such circuits can remember.

Both relay-based and IC memory elements are available. NANDs and NORs can be used to construct R-S flip-flops. When an active signal is applied to the SET input of such a memory element, output Q goes to 1. It stays in this state until an active signal is applied to RESET. Then Q goes to 0.

Faults in flip-flops are similar to those in combinational-logic circuits. They tend to be s-a-0 and s-a-1. Glitches also cause errors in flip-flop outputs.

TEST YOUR KNOWLEDGE

1. Which type of logic circuit, combinational or sequential, is used to store numbers?
2. What letters are used to mark relay contacts that are normally open?
3. What type of contacts (load, holding, or primary) convert a combinational-logic relay circuit to a sequential-logic circuit?
4. The result of pressing and releasing both buttons in a sequential-logic relay circuit at the same time cannot be predicted. (True or False?)
5. What is another name for flip-flop?
6. In an R-S flip-flop with active-high inputs, what will output Q be when S = 1 and R = 0?
7. Is the following set of inputs for an R-S flip-flop with active-high inputs allowed: R = 1, S = 1?
8. At what levels must the inputs to an R-S flip-flop be to "remember" the last state achieved?
9. A register is made up of one or more flip-flops. (True or False?)
10. Memory elements must contain ________ (active-high inputs, bias, feedback).
11. In Fig. 7-15, what signals (1s and 0s) will appear at point A and point Q, and are these stable outputs? Are the signals at point A and point Q inverses of each other?
12. The two outputs of an R-S flip-flop are inverses of each other. (True or False?)
13. Qt + 1 means immediately. (True or False?)
14. To set an active-low R-S flip-flop, S must go to a level of 1. (True or False?)
15. In part "b" of Fig. 7-21, which of the following actions would remove the apparent s-a-0?
    a. Apply a 1 to input S.
    b. Apply a 1 to input R.
16. Which of the following give an apparent s-a-1 fault (as in Fig. 7-22)?:
    a. S = 0, R = 1.
    b. S = 1, R = 1.
17. A true s-a-1 fault is seen by checking if Q and $\overline{Q}$ are inverses of each other (see Fig. 7-22). (True or False?)
18. A true s-a-1 fault is seen by checking if Q and $\overline{Q}$ are inverses at the same time that S = 0 and R = 1. (True or False?)
19. What two instruments can find a glitch on lead R?
20. Can the number stored in a flip-flop remain when the supply voltage is removed?

STUDY PROBLEMS

1. Draw a circuit diagram for a relay-based memory element. Use two pushbutton switches as inputs, a control relay with two sets of contacts, and a lamp as an output. Use a battery as a power source for the circuit.
2. Fig. P7-1 depicts a starting circuit for a small motor. Which set of contacts (refer to the letters a through g) would most likely be described as holding contacts?
3. Draw the symbol for an R-S flip-flop. It is to have active-high inputs and a single output. Also, construct a truth table to describe the action of the R-S flip-flop.

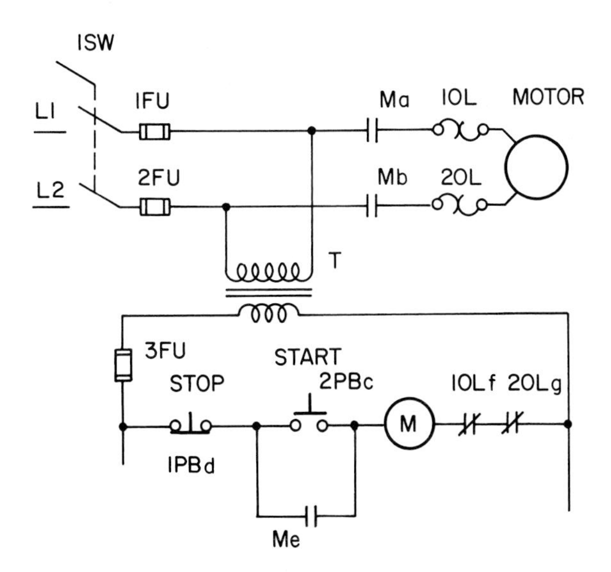

Fig. P7-1. Sequential logic relay circuit for starting motor has holding contacts.

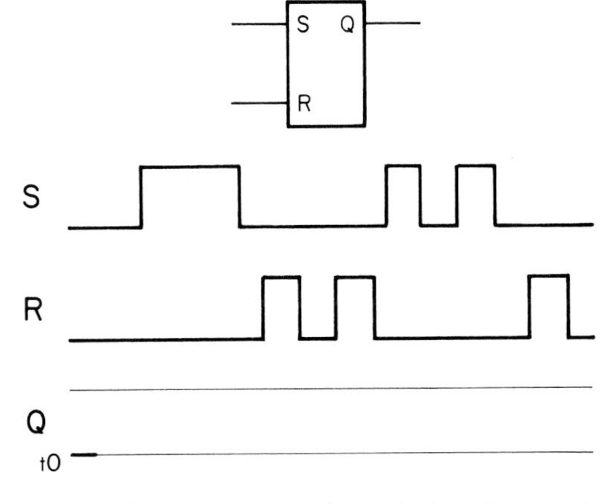

Fig. P7-3. Q can be found from timing diagram for R-S flip-flop. Q = 0 at time t0.

4. Copy the timing diagram in Fig. P7-2 onto a piece of paper, and construct the output signal. Assume Q = 0 at time t0.
5. Repeat Problem 4 for the set of input signals that are shown in Fig. P7-3.
6. In Fig. P7-4, why is it impossible to determine the value of Q at time t1?
7. Fig. P7-5 shows the circuit for a 1-bit register. Using the letters A, B, and C, match the following functions with the leads of the circuit.
   1. Data input.
   2. Data output.
   3. Write enable lead.

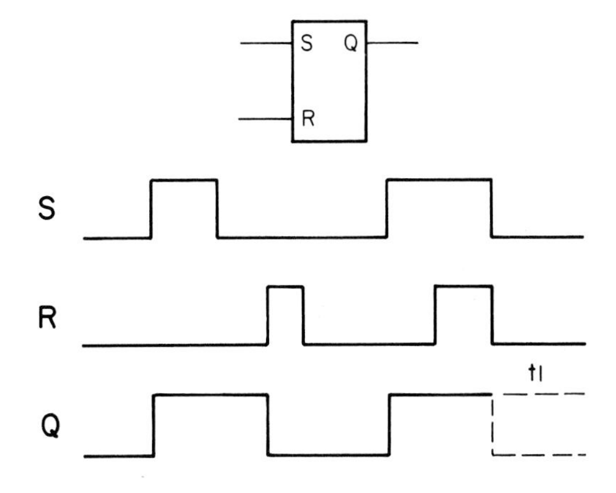

Fig. P7-4. Q is not defined for all possible inputs. Note the case at time t1.

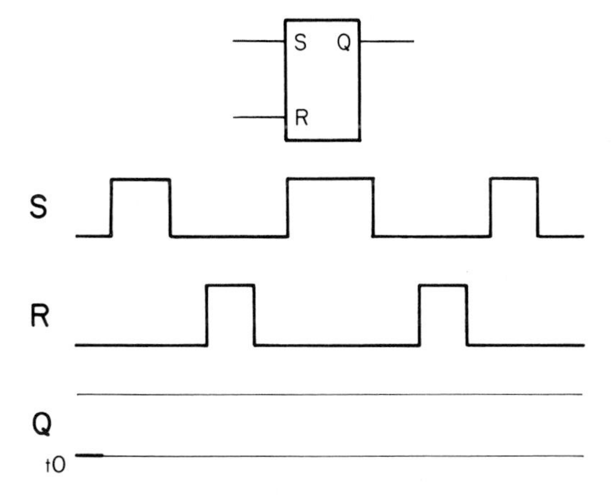

Fig. P7-2. Q can be found from timing diagram for R-S flip-flop. Q = 0 at time t0.

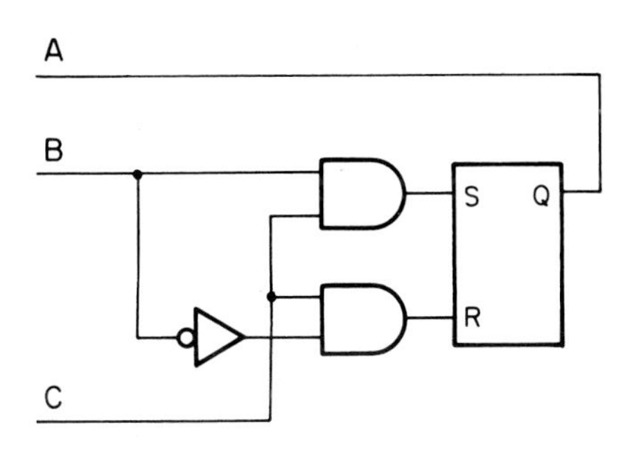

Fig. P7-5. Leads A, B, and C can be identified for 1-bit register.

8. Using NAND elements, draw the logic diagram for an R-S flip-flop with active-high inputs. Use NANDs for the NOTs. Identify the input leads and the output leads with the letters S, R, and Q.
9. Repeat Problem 8 for a NOR-based flip-flop.
10. Each of the circuits in Fig. P7-6 represent a NOR-based flip-flop. In each, certain signals have been indicated; others are unknown. In each case, determine the values of the missing signals. Record your answers in the form:

Aa = ________
Qa = ________
Ab = ________
etc.

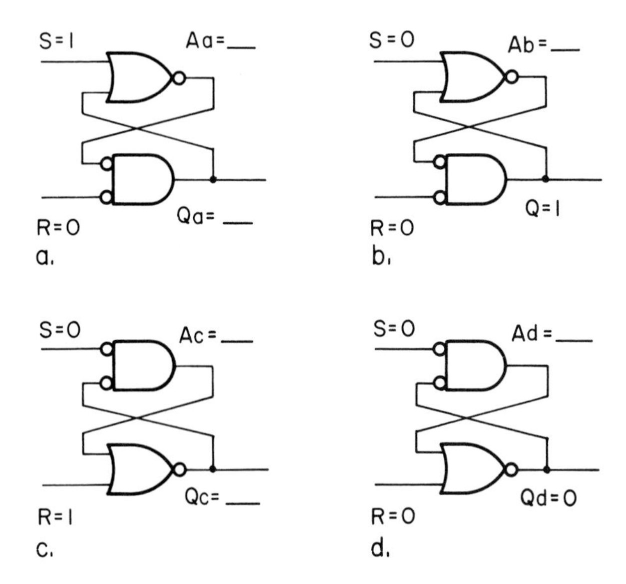

Fig. P7-6. Outputs can be found from inputs for the four circuits shown. Q is given in the cases with R = 0 and S = 0.

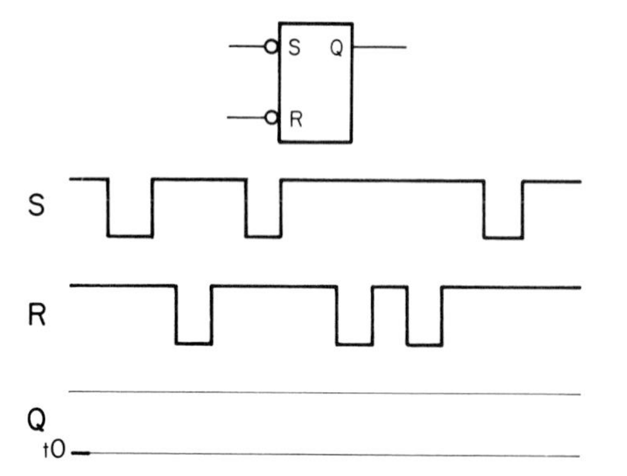

Fig. P7-7. Q can be found from timing diagram for R-S flip-flop with active-low inputs. Q = 0 at time t0.

11. Draw the symbol and construct the truth table for an R-S flip-flop with active-low inputs.
12. Trace the timing diagram in Fig. P7-7 onto a piece of paper and carefully construct the output signal. Assume Q = 0 at time t0.
13. Show how the flip-flop in Fig. P7-8 could be modified so it has active-high inputs. The bubbles cannot be removed, since they are part of the internal circuit.

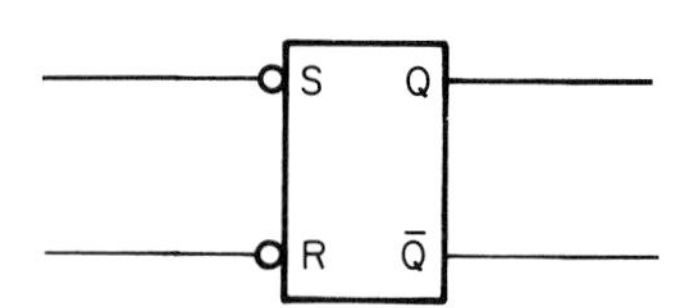

Fig. P7-8. An R-S flip-flop with active-low inputs can be converted to one with active-high inputs.

14. Fig. P7-9 shows the pinout for a 74279. Are its inputs active-high or active-low?

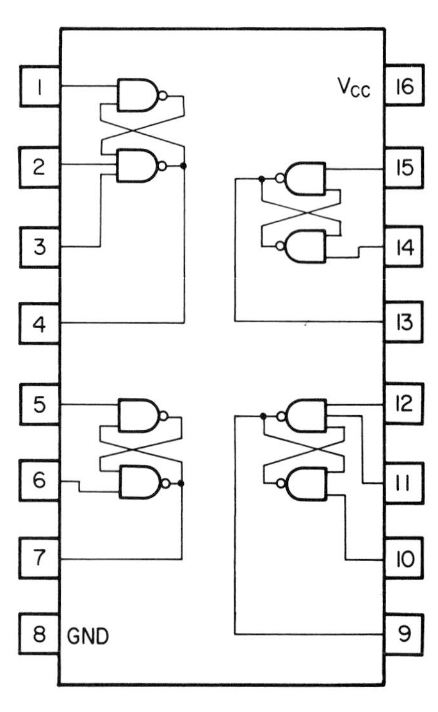

Fig. P7-9. Pin outline diagram for a 74279. Whether its inputs are active-high or active-low can be determined.

15. The outputs of the three flip-flops in Fig. P7-10 appear to be either s-a-1 or s-a-0. However, only

one is faulty. Based on the input signals, the other two are working properly. Which one of the circuits shown is probably faulty?

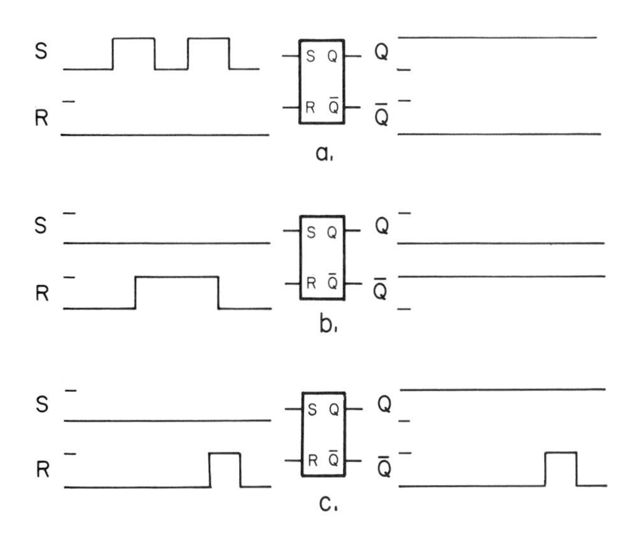

Fig. P7-10. Correct operation of R-S flip-flops can be distinguished from incorrect operation. Three cases aid in the study of the R-S flip-flop.

16. The output of the flip-flop in Fig. P7-11 is incorrect. There is a chance that a glitch caused this error. At what point in time and on which input would you look for a glitch? Draw the desired graph for Q.

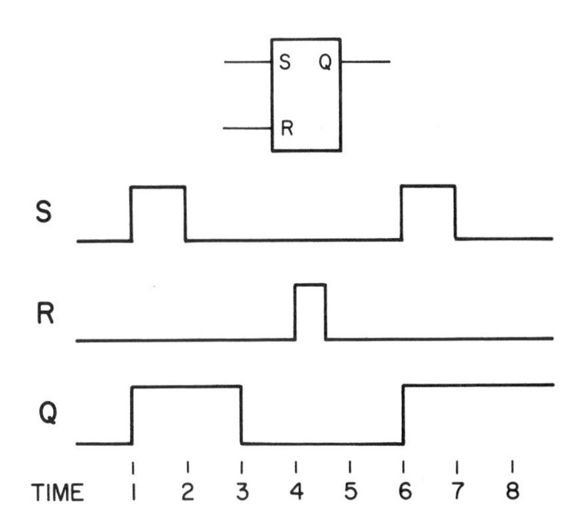

Fig. P7-11. A glitch or other fault can be found by comparing inputs with the output Q.

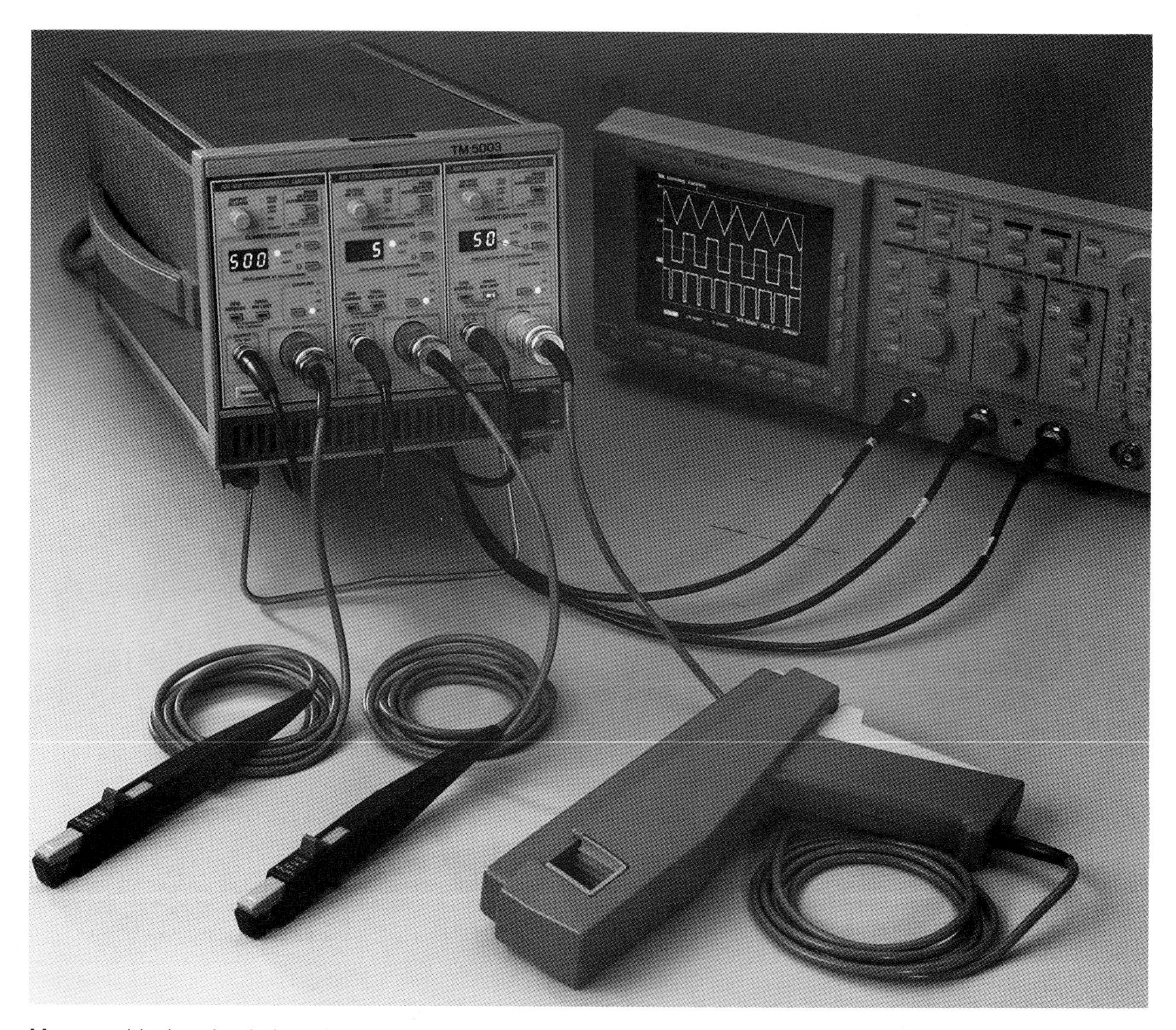

Most troubleshooting is based on voltage measurements. In some cases, however, the measurement of current is helpful. The probes illustrated above, clamp around leads and the current flow is displayed. (Tektronix, Inc. 1996©)

Chapter 8

Clocked Flip-Flops

LEARNING OBJECTIVES

After studying this chapter and completing lab assignments and study problems, you will be able to:

- ► Recognize the symbol for a clocked R-S flip-flop and use truth tables and timing diagrams to describe its action.
- ► Draw the circuit diagram for a NAND-based clocked R-S flip-flop.
- ► Draw and explain the action of a register using clocked R-S flip-flops as the memory elements.
- ► Recognize the symbol for a data latch and use truth tables and timing diagrams to explain its action.
- ► Draw the circuit diagram for a data latch which is built from NAND elements.
- ► Explain the action of a memory that uses data latches as memory elements.

A flip-flop is more useful if an enable lead is available. Such inputs are usually called *CLOCK* leads. When a clock lead is active, new numbers can enter and be stored. When inactive, stored numbers cannot be changed. This chapter introduces clocked flip-flops and describes several applications.

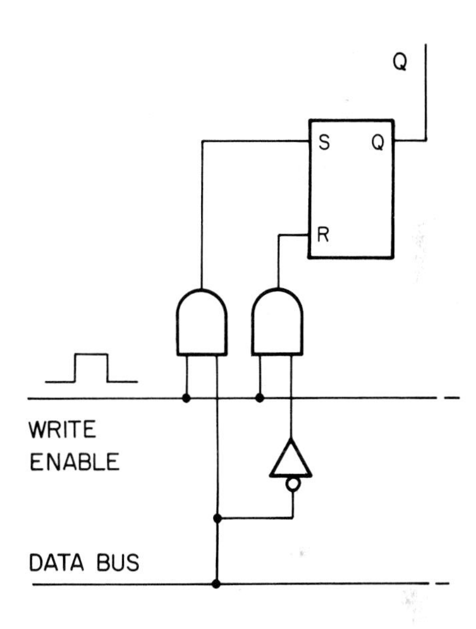

Fig. 8-1. A section for one bit of a register that uses R-S flip-flops as storage elements.

ENABLE CIRCUIT

A section for one bit of the 3-bit register described in the last chapter is reproduced in Fig. 8-1. Data enters through the lead at the bottom, is stored in the flip-flop, and is available at output Q.

Of the many numbers that travel back and forth on the bus, only a few will be stored by this circuit. To answer this need, an enable circuit (composed of two ANDs) has been provided. When a number to be stored appears on the bus, WRITE ENABLE is made active (it goes to 1). This opens the AND gates and permits the number to reach the flip-flop inputs. When the number has been stored, WRITE ENABLE is returned to its inactive state. This disconnects the flip-flop from the bus, so numbers being transported to other locations cannot alter the stored number.

CLOCKED R-S FLIP-FLOP

Enable circuits are so often used with flip-flops that chips are available with built-in enables. Part "a" of Fig. 8-2 shows a functional diagram of such a circuit. It is called a clocked R-S flip-flop. Part "b" of Fig. 8-2 shows its symbol.

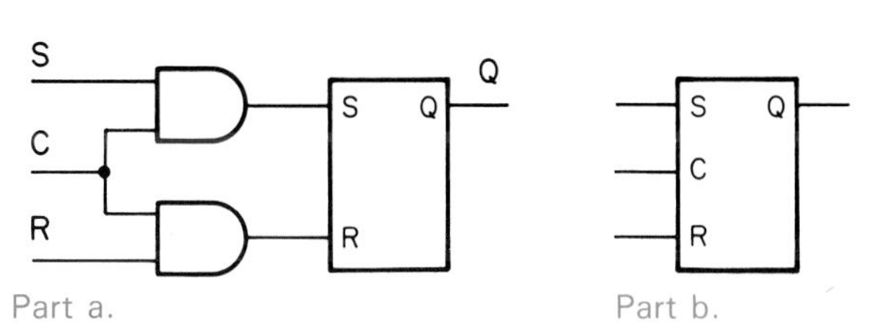

Fig. 8-2. The enable circuit at the left is part of the clocked R-S flip-flop at the right.

The term *CLOCK* (C) is from the time of early computers. A clock is a device that outputs a series of pulses. These are used to time the various functions within computers. In some applications, clock signals were applied to the flip-flop enables, so this input was called the CLOCK INPUT. The terms CLOCK and CLOCK INPUT are still used.

CLOCKED R-S FLIP-FLOP ACTION

When C of a clocked flip-flop is active (high), the circuit acts like an ordinary R-S flip-flop. Its truth table and timing diagram are identical to those studied in the last chapter.

When C is inactive (low), the inputs are disconnected from the memory portion of the device. New numbers cannot be stored, and the output retains the previous number.

CLOCKED R-S FLIP-FLOP TRUTH TABLE

Fig. 8-3 shows the truth table for a clocked R-S flip-flop. It is identical to that of an ordinary R-S flip-flop except that the definition of Qt + 1 has been modified. Here, t + 1 indicates that the clock has gone active and returned to its inactive state.

CLOCKED R-S FLIP-FLOP TIMING DIAGRAM

Timing diagrams for clocked R-S flip-flops appear complex. However, they can be made more understandable if you remember that nothing can happen (no matter what signals are at the device's inputs) if C is inactive. When C is active, the device acts like an ordinary R-S flip-flop.

INACTIVE C

Inactive C is made clear in Fig. 8-4. At times t1 and t2, signals are at inputs S and R, yet Q is unchanged.

INPUTS S	INPUTS R	OUTPUT Q_{t+1}
1	0	1
0	1	0
0	0	Q_t
1	1	*

*NOT DEFINED

Fig. 8-3. Truth table for clocked R-S flip-flop.

The graph of the clock suggests the reason for this inactivity. With C = 0, new numbers cannot enter the flip-flop.

This is also true at t3. Here, an interesting use for the clock is suggested. Both S and R are 1 at the same time. If these signals were applied to an ordinary R-S flip-flop, a fault would result, since this is an undefined input set. In this case, the output is unaffected, since the clock is inactive.

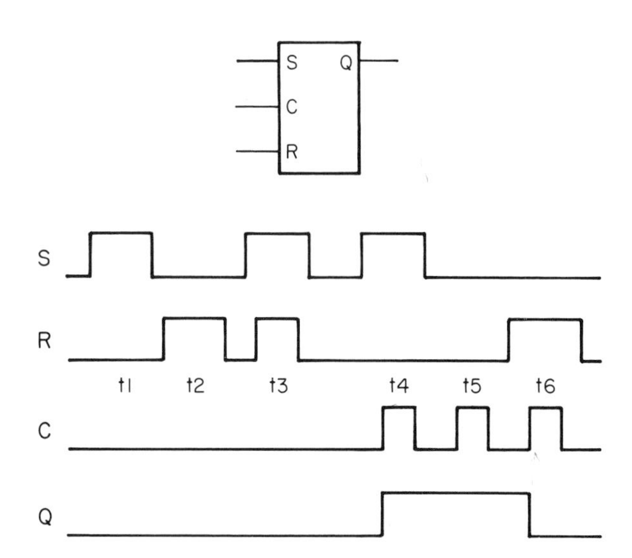

Fig. 8-4. New numbers can be stored in a clocked R-S flip-flop only at a time when the clock becomes active.

ACTIVE C

Active C is clarified in Fig. 8-4. At t4, C is active, S = 1, and R = 0. As a result of the input set 1, 0, a 1 is stored (Q = 1).

At t5, C is again active, but Q did not change. The graphs for R and S indicate the reason. Based on the third row of Fig. 8-3, Qt + 1 equals Qt, when 0s appear at both S and R.

At t6, C is active, S = 0, and R = 1. As a result of the input set 0, 1, a 0 is stored (Q = 0). The input R = 1 has allowed the flip-flop to be reset. The state called RESET is the storage of a 0.

ACTIVE-LOW CLOCK

The clock in Fig. 8-5 is bubbled. This means that it is active low. The action of the flip-flop is identical to that of an active-high clocked flip-flop except that changes occur when the clock is low. See the timing diagram in Fig. 8-5.

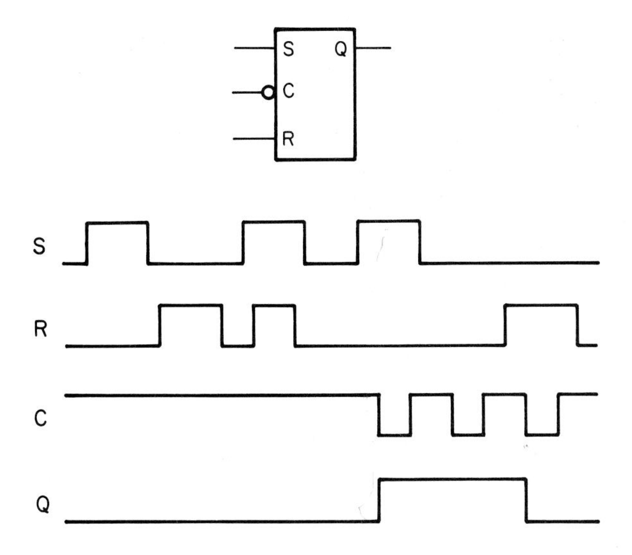

Fig. 8-5. The flip-flop shown has an active-low clock.

LEVEL-TRIGGERED CLOCK

The clock inputs described so far are level triggered. (Edge-triggered clocks will be introduced in the next chapter.) This means that S and R are connected to the flip-flop portion of the circuit during the time that the clock is at its active level. If the clock pulse is short, the process of enabling the inputs may not cause problems. See time t1 in Fig. 8-6.

If the clock is active for a long time, however, input signals may change while the inputs are connected to the flip-flop. See times t2 and t3. With the flip-flop directly connected to the inputs, the output follows changes in input signals. The number at Q at the time the clock returns to its inactive level is the one that is stored. See the point marked "stored."

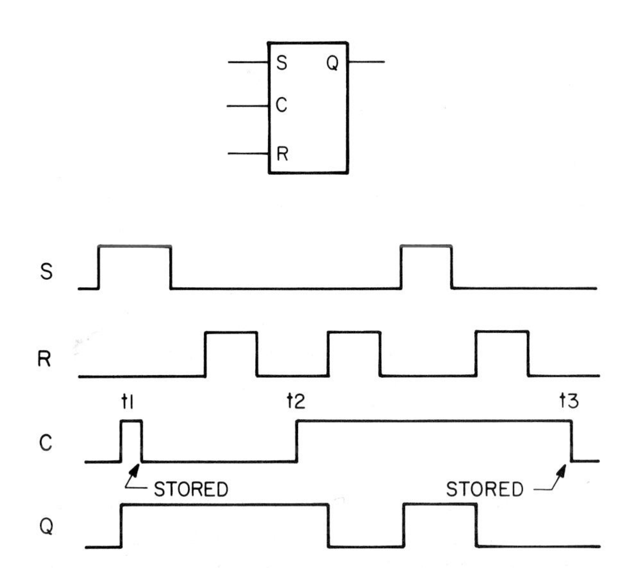

Fig. 8-6. During the time that its clock is active, the clocked flip-flop acts like an unclocked element.

In some applications, the changes in Q that occur between t2 and t3 cause no problem. In other applications, long clock pulses may force the use of other triggering methods.

CLOCKED R-S FLIP-FLOP CIRCUIT

Fig. 8-7 shows how a widely used circuit for clocked R-S flip-flops is designed. At b, the enable circuit has been brought out. At c, a NAND-based circuit has been substituted for the R-S flip-flop. At d, the AND-NOTs have been replaced by NANDs. This suggests that a clocked R-S flip-flop can be constructed using four NANDs.

Fig. 8-8 carries the design one step further. The symbol for an R-S flip-flop with active-low inputs has been substituted. Active-low inputs on the R-S flip-flop allow the use of NANDs (on the left) with the

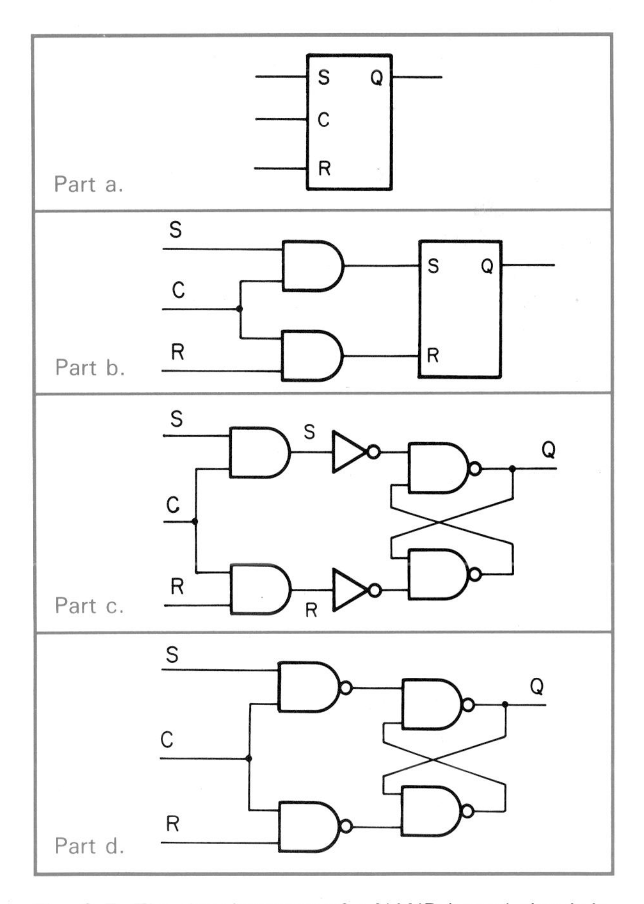

Fig. 8-7. The development of a NAND-based circuit is traced from the element's symbol to its circuit diagram.

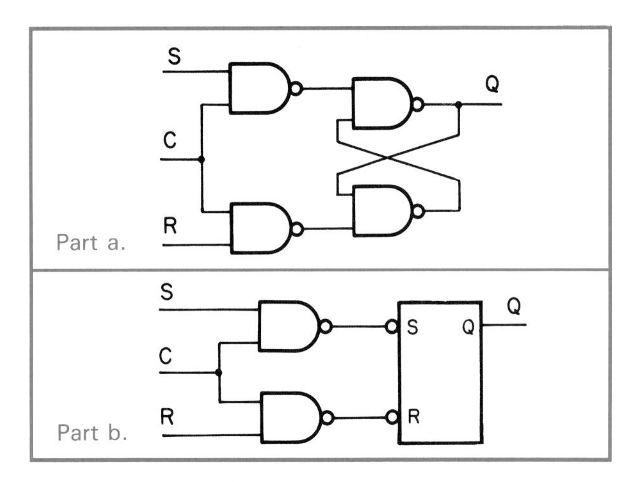

Fig. 8-8. The flip-flop portion of the circuit in Fig. 8-7 has active-low inputs.

flip-flop (on the right) that is most available. Designs using NANDs are simple.

CLOCKED R-S FLIP-FLOP APPLICATION

The register in Fig. 8-9 is similar to the one described in the last chapter. Here, however, clocked R-S flip-flops have been used. As a result, the enable circuits of the earlier circuit are not needed. When the number on the bus is to be stored, CLOCK is made active. When the clock input lead returns to its inactive state, the number is stored.

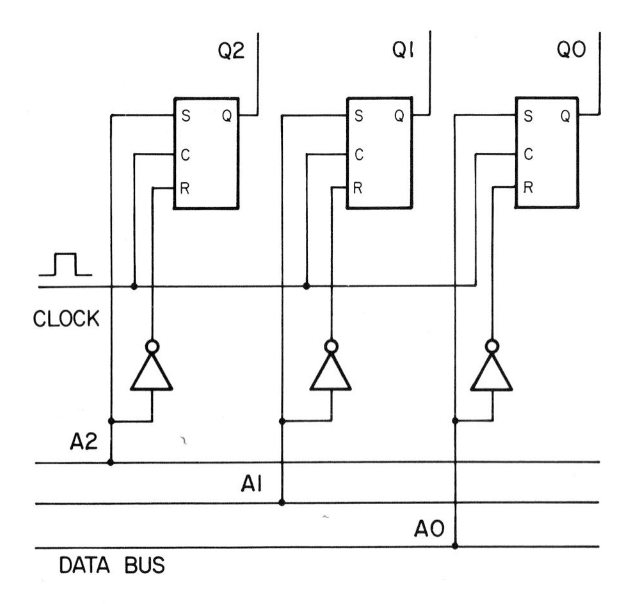

Fig. 8-9. Because clocked flip-flops are used, this circuit is less complex than that of register in Chapter 7. See Fig. 7-7. Fewer ANDs are needed.

DATA LATCH

Part "a" of Fig. 8-10 shows a section for one bit of the above register. It suggests another useful flip-flop circuit. If the NOT in lead R were made part of the flip-flop, a memory element with but one input would result. See parts "b" and "c" of the illustration. Such a flip-flop is called a *DATA LATCH.*

The term LATCH means to catch or hold. A data latch catches and stores signals available at its single input. The timing of that storage is determined by the circuit's clock input.

A data latch must have a clock. Fig. 8-11 shows that a data latch without a clock has no practical value. Because $S = D$ and $R = \overline{D}$, the output of this circuit merely follows the changes in D. An amplifier would do the same thing with less cost and complexity. Data latches must have a clock to indicate when a number is to be stored.

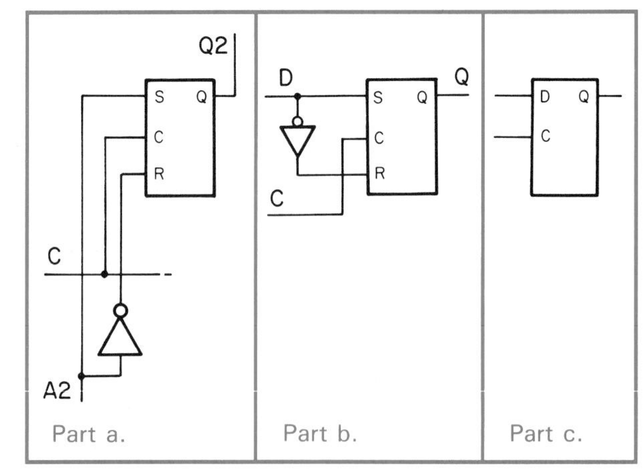

Fig. 8-10. If the NOT element is placed within the clocked R-S flip-flop, a data latch results.

DATA LATCH ACTION

When C is active (high), the number at D in Fig. 8-10 enters the flip-flop and appears at Q. When C returns to its inactive level, the number at Q is stored. Q remains unchanged until C is again active.

DATA LATCH TRUTH TABLE

Because NOTs and data latches have but one input, their truth tables are simple. See Fig. 8-12. As before, $Qt+1$ implies that Q takes on the indicated value when the clock has gone active and the clock has returned to its inactive state.

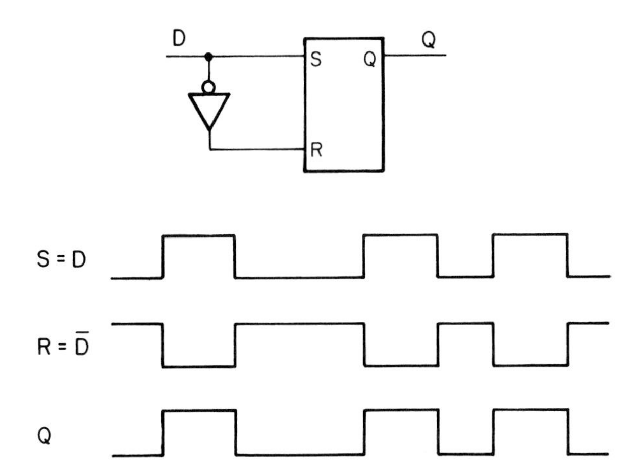

Fig. 8-11. An unclocked data latch has no practical value, since its output merely reproduces its input.

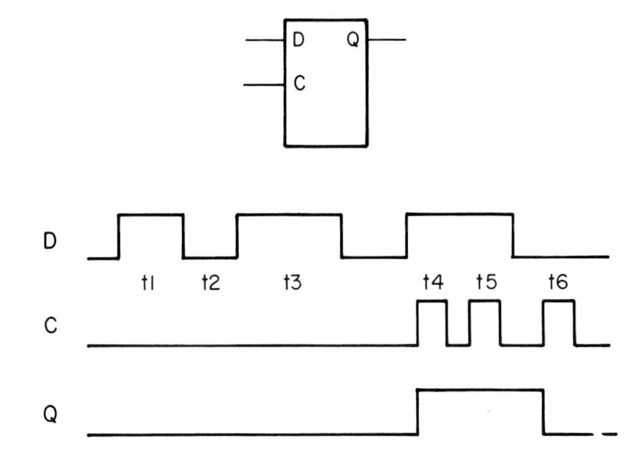

Fig. 8-13. Numbers that appear on data input lead D can enter this data latch only when C is active.

INPUT D	OUTPUT Q_{t+1}
1	1
0	0

Fig. 8-12. Truth table for data latch.

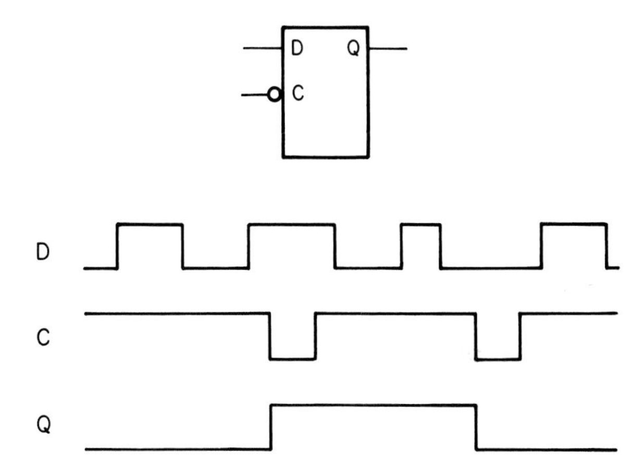

Fig. 8-14. The data latch shown has an active-low clock, so numbers that appear on data input lead D enter when C = 0.

DATA LATCH TIMING DIAGRAM

The timing diagram of a data latch reflects its basic action. When the clock is inactive, the stored number remains unchanged no matter what data is presented to input D. When C is active, the number at D enters the flip-flop. When C returns to its inactive state, the last value of Q is stored.

ACTIVE-HIGH CLOCK

Fig. 8-13 is the timing diagram for a data latch with an active-high clock. At times t1, t2, and t3, data was applied to input D. Because C was inactive, this data was ignored, and Q was unchanged.

At t4, C is active, and the 1 at D was stored. At t5, C was again active, but Q did not change. In this case, Q was already 1, so the storing of another 1 resulted in no change. At t6 a 0 was stored.

ACTIVE-LOW CLOCK

Fig. 8-14 is the timing diagram for a data latch with an active-low clock. Except for the inverted clock input, its action is identical to that of a data latch with an active-high clock.

LEVEL-TRIGGERED CLOCK

Fig. 8-15 emphasizes the action of a data latch with a level-triggered clock. When the clock is active output Q follows input D. When the clock returns to its inactive level, the number at Q at that instant is stored.

DATA LATCH CIRCUIT

A data latch can be assembled by placing a NOT across the inputs of an R-S flip-flop, Fig. 8-16. In a NAND-based circuit, a NAND element might be used in place of the NOT element.

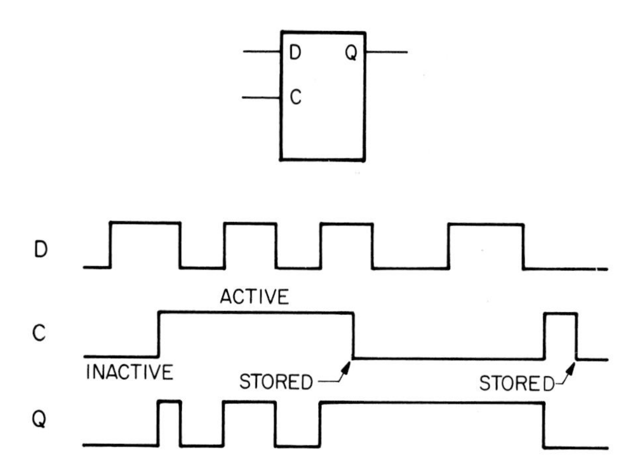

Fig. 8-15. While C is active, output Q equals D. The number present when C goes inactive is the number stored.

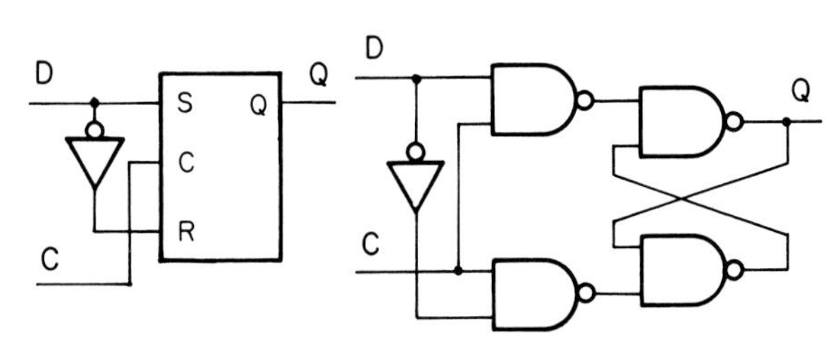

Fig. 8-16. A circuit for a data latch can be obtained by placing a NOT across the input of the clocked R-S flip-flop developed earlier.

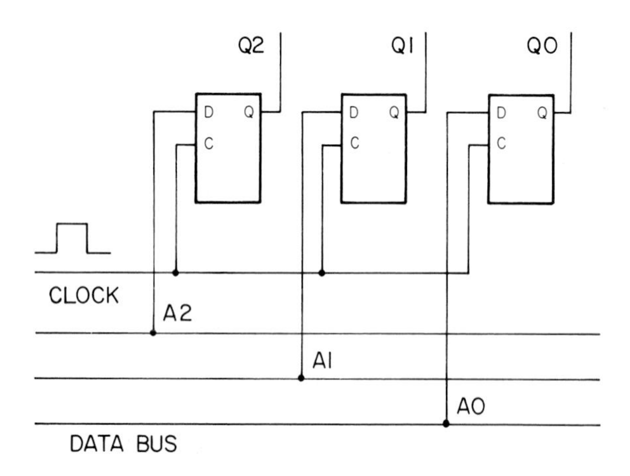

Fig. 8-17. With most of the circuitry built into the flip-flops, registers based on data latches have simple circuits.

DATA LATCH APPLICATION

A register is used as an application of a data latch. See Fig. 8-17. A register which is assembled from data latches is a simple system.

MEMORY USING DATA LATCHES

The study of individual elements is basic to understanding digital circuits. Yet, such knowledge is not in itself sufficient. Skill must be developed in the analysis of relatively complex circuits. To lay a foundation for that skill, an analysis of the circuit in Fig. 8-18 is described in detail.

The circuit in Fig. 8-18 is complex, yet you understand the operation of all its parts and subcircuits. The problem will be one of dividing the circuit into easily understood sections, analyzing the actions of those sections, and noting their contributions to the action of the whole.

Here are some hints on how to approach such a circuit:

1. Know what the circuit does.
2. Consider it in small, recognizable subcircuits.
3. Start with subcircuits that are easily analyzed; then proceed to the more difficult.

FUNCTION OF CIRCUIT

The circuit in Fig. 8-18 is a memory. It is a group of registers. Each register is called a *MEMORY LOCATION*. Each register can store one multi-bit binary number. Such multi-bit binary numbers are often referred to as *WORDS*.

This is a 3 x 2 memory. That is, it can store three words. Each word is two bits long.

An examination of the functions of the leads into and out of the circuit often aids in understanding the operation of a circuit. It is helpful to examine the functions of the following leads:

1. Data leads (input and output).
2. Address leads.
3. A READ/WRITE lead.

DATA LEADS

The word length is two, so there are two input data leads (D1D0). Compare Fig. 8-19 with Fig. 8-18. There are also two output leads (Q1Q0).

ADDRESS LEADS

A number may be stored in any one of the three memory locations available in this circuit. Leads A1 and A0 are called *ADDRESS LEADS* and the

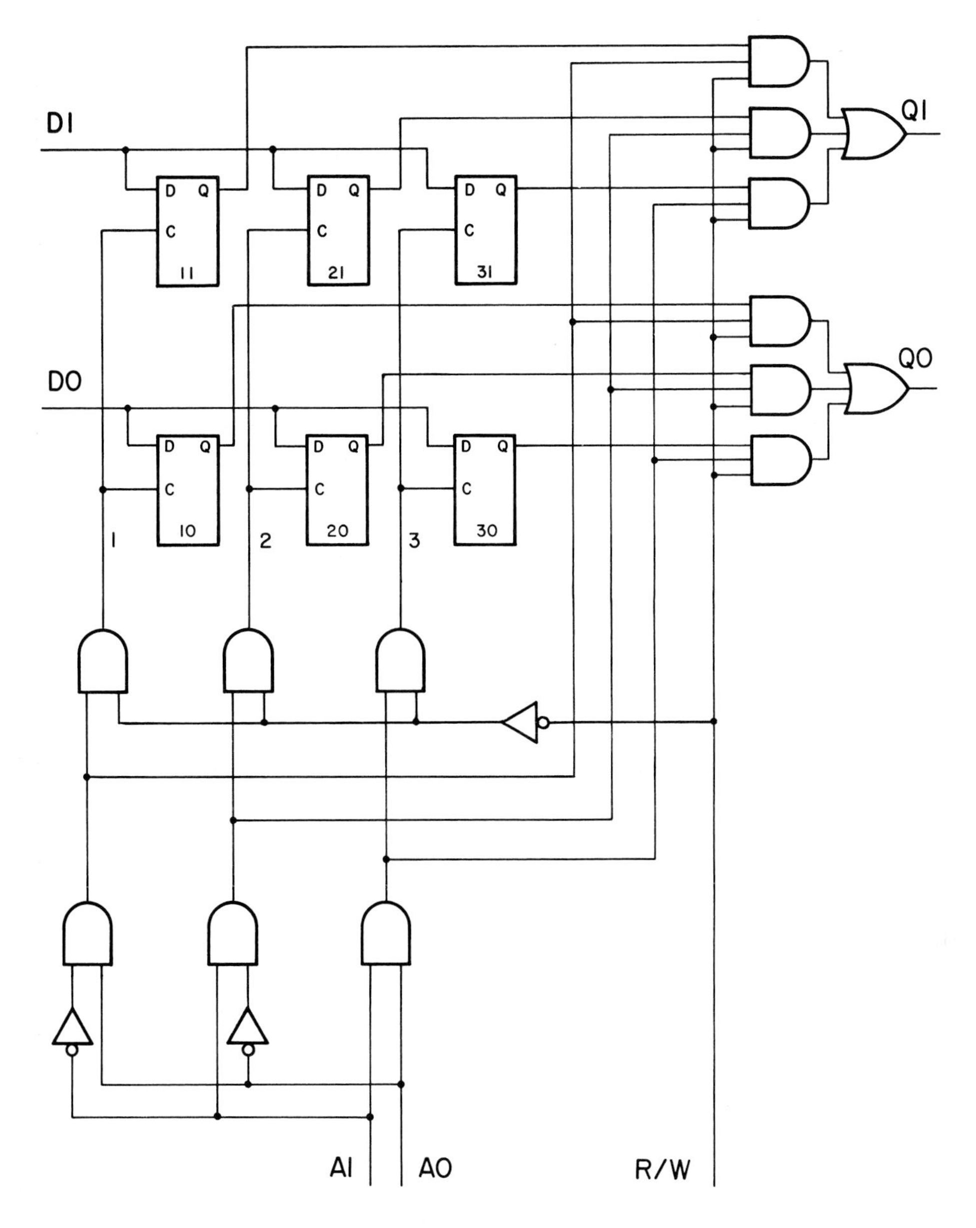

Fig. 8-18. Data latches are used as memory elements in the small memory shown.

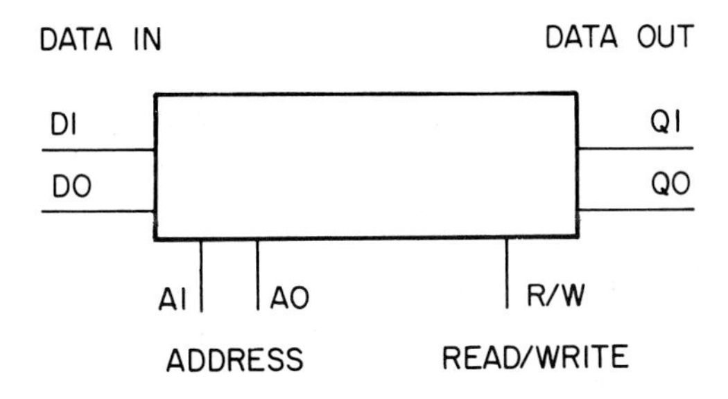

Fig. 8-19. Because this memory has a 2-bit word, it has two input and two output data leads.

numbers on the address leads indicate the memory location which is involved in a given read or write operation.

If a number at D1D0 is to be stored in location 2, A1A0 will equal 10. See part "a" of Fig. 8-20. If a number is to be stored at location 3, A1A0 would equal 11. See part "b" of Fig. 8-20.

The same address leads are used to indicate which number is to be output. For example, if the number in location 3 were to be output, A1A0 would be set to 11. See part "a" of Fig. 8-21.

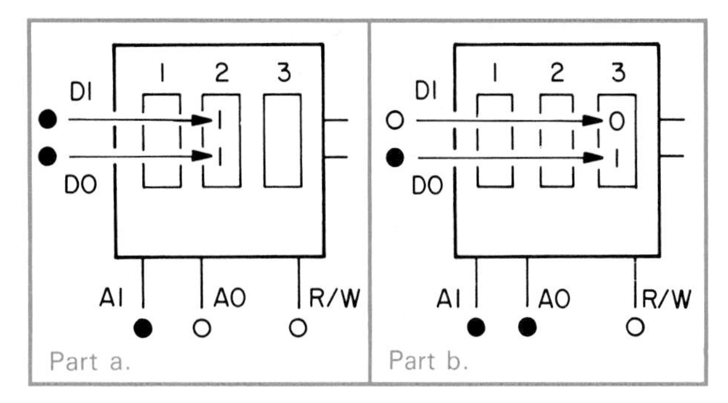

Fig. 8-20. Address leads are used to direct input data to the place where it is to be stored. The address leads are marked A1 and A0.

READ/WRITE LEAD

R/W stands for *READ / WRITE.* This is a control lead. Based on standard notation, R/W means that a 1 on this lead will result in a read (the outputting of a stored number). A 0 will result in a write (the storing of a number).

To review the action of a memory circuit, refer to part "b" of Fig. 8-21. For the data, address, and control signals shown, the number 11 will be stored in location 1. That is, the 10 presently in the location will be replaced by 11.

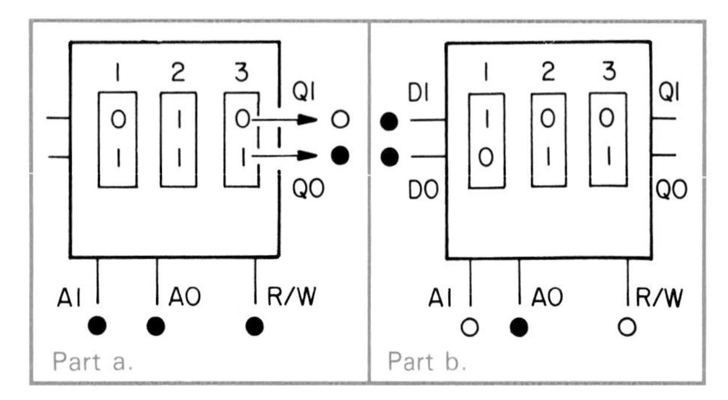

Fig. 8-21. The signal on R/W determines whether an input or output occurs. At b, 11 replaces 10 in cell 1.

SUBCIRCUITS

Before detailed analysis can begin, the circuit in Fig. 8-18 should be divided into recognizable subcircuits. See Fig. 8-22. The order of analysis is not critical, but the letters suggest an appropriate sequence. Memory elements are basic to the circuit's operation, so they will be considered first.

Before going on, locate the memory array in the circuit. See Fig. 8-18 and Fig. 8-22.

MEMORY CELLS

Memory elements are called *CELLS.* The six memory elements or cells are data latches. The memory cells are shown separated from the rest of the circuit in Fig. 8-23.

High order bits of the three stored numbers are held in the upper data latches. All numbers input through D1 are delivered at the same time to the three D inputs. Whether a given number is stored or not, and where it is stored is determined by leads 1, 2, and 3 at the bottom of the circuit. If a number is to be stored in location 2, a 1 must appear on lead 2. This causes the clock on data latch 21 to go active and store the number present at its D input. Low order bits are stored in the lower three data latches.

As soon as a number enters a data latch, it is available at its output. As indicated at the right of the circuit in Fig. 8-23, these outputs are delivered to the output circuit for further processing.

ADDRESS DECODER

Subcircuit B of Fig. 8-22 is the address decoder. Fig. 8-22 has been redrawn in Fig. 8-24. Addresses enter the circuit in binary form and must be decoded. Depending on the binary number input, a 1 will appear on one of the three decoded address leads.

Note that there is no decoder for A1A0 = 00. When this address is applied, the circuit is disabled. Neither a read nor write can take place because no enable signal is available.

READ/WRITE

Again refer to Fig. 8-22. Subcircuit C determines whether this memory is in its read or write mode. The circuit has been redrawn in Fig. 8-25.

Write mode

When R/W = 0, a number is to be stored. The NOT inverts this 0, and the resulting 1 is applied to the three AND gates at the left. In the memory, these gates are located between the address decoder and the clock inputs on the data latches. Refer to Fig. 8-22.

When these gates open, the decoded address can reach the clock inputs on the two data latches where the number is to be stored. The C inputs go active, and the number is stored.

When R/W = 1 (the read mode), the NOT applies a 0 to the three AND gates on the left in Fig. 8-25. The decoded address cannot reach the clock inputs, so none are active. New numbers cannot be stored during a read operation. Because new numbers are restricted, the data being output is protected.

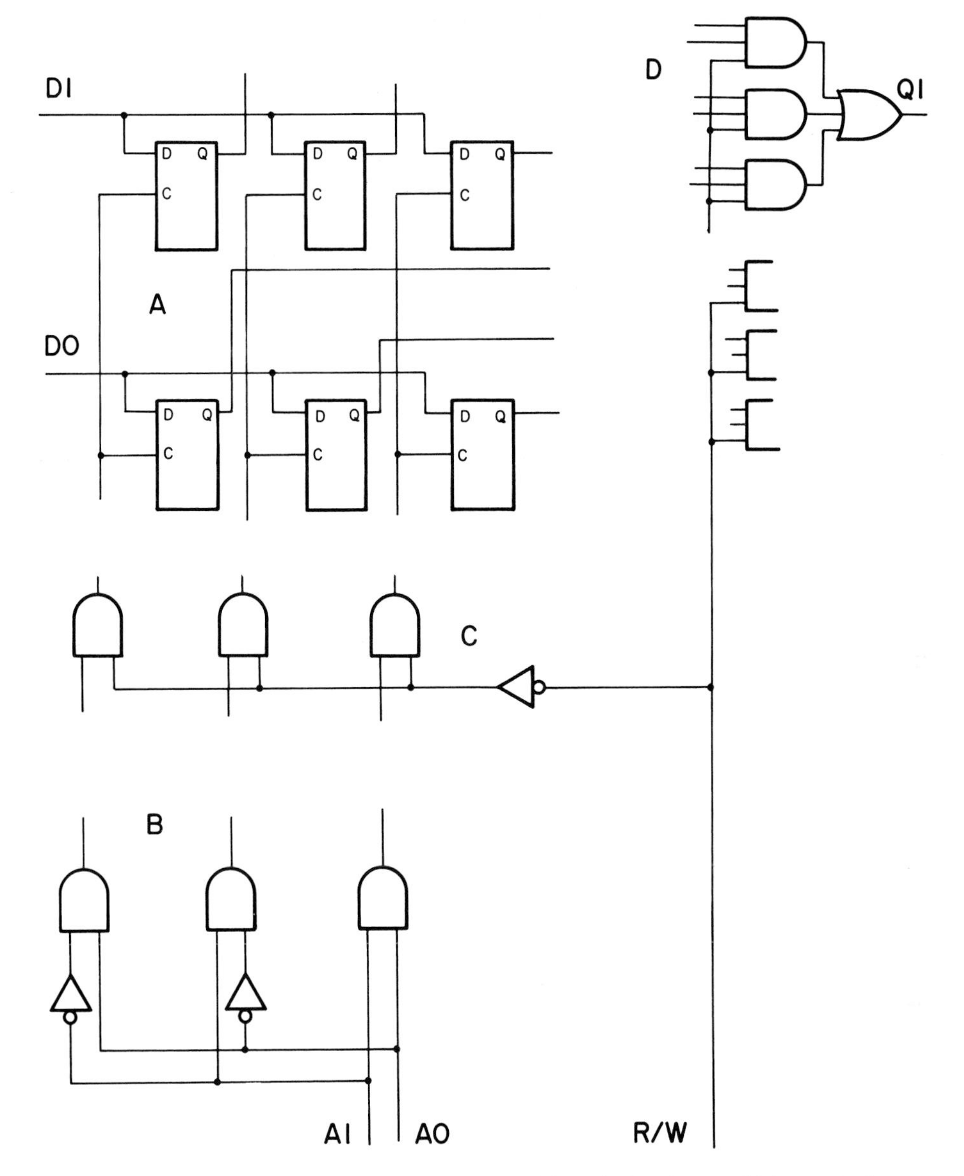

Fig. 8-22. The circuit is broken into recognizable subcircuits to aid in analysis.

Read mode

When R/W = 1, the AND gates in the output portion of the circuit are opened. Numbers from the output leads of the data latches have a path to the output lead at Q1 and the output lead at Q0.

When R/W = 0, these six gates are closed. The memory cannot output when it is in its write mode.

OUTPUT

Subcircuit D is shown in Fig. 8-22 and Fig. 8-26. The action of the circuit for output Q0 is identical to that for Q1.

Of the three gates in this portion of the output circuit, only one will be open at any given time. Even that one gate will be open only when R/W equals 1.

R/W lead

R/W is applied to the lower inputs of the three AND gates in Fig. 8-26. When the memory is in its write mode (R/W = 0), all three of the gates will be closed, and 0s will appear at their outputs. With 0s at its inputs, the OR will output a 0 (Q1 = 0).

When R/W = 1 (read mode), all three AND gates will have 1s at their lower inputs. This permits them to open, and an output can be obtained.

Outputs are well defined and are useful. No special interface circuit is needed.

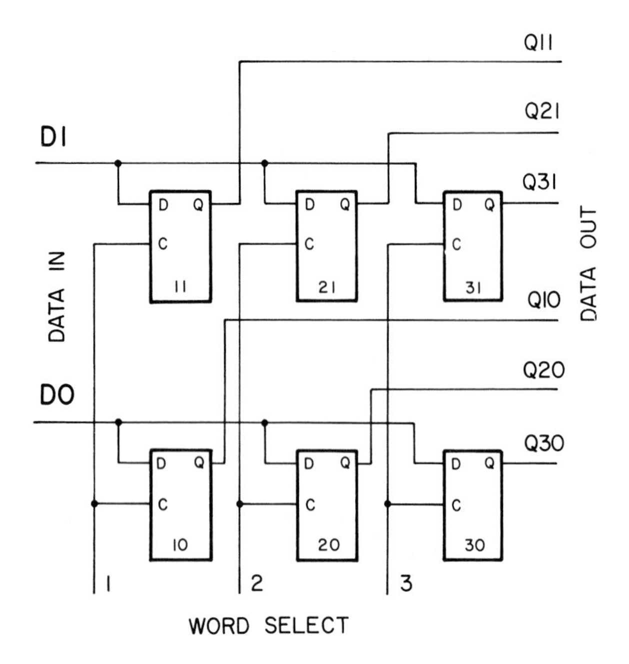

Fig. 8-23. The action of the memory cell array is analyzed first.

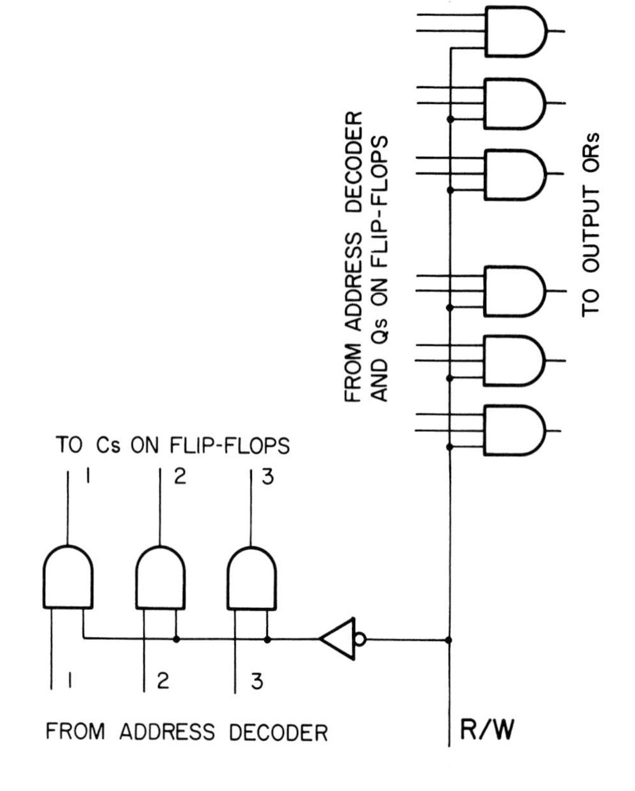

Fig. 8-25. The control circuit shown determines whether a read or write will occur.

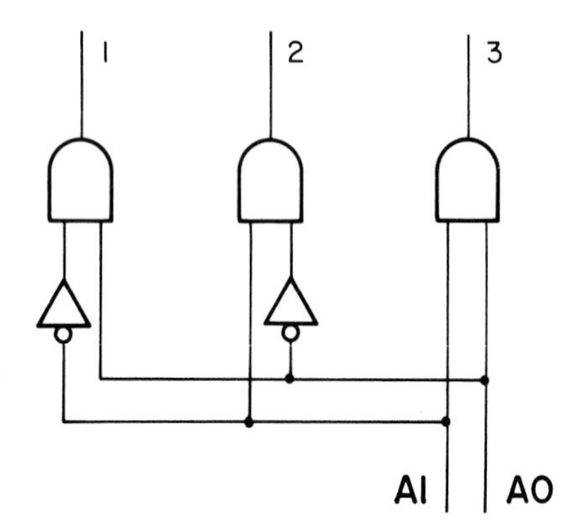

Fig. 8-24. The circuit shown decodes the address.

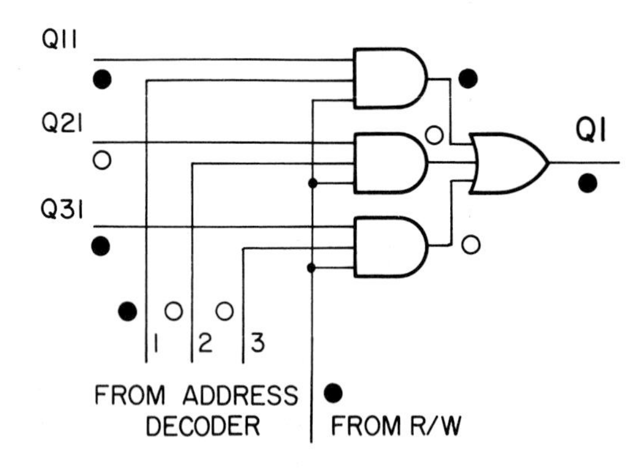

Fig. 8-26. The circuit shown selects the proper bit from the correct memory cell and outputs the bit at Q1.

Decoded-address lead

The middle input of each AND gate is connected to an output of the address decoder. In normal operation, a 1 will appear on only one of these leads. As a result, only the gate associated with the memory location being addressed will open. The number (1 or 0) stored at that location will be delivered to the OR and then to output Q1.

The measured signals in Fig. 8-26 show the outputting of data stored in location 1. Decoded-address lead 1 and R/W are both 1, so the upper AND gate is open. (The other gates are closed, because 0s appear on their decoded-address leads.) The 1 stored at Q11 is applied to the OR and appears at Q1.

Fig. 8-27 shows a stored 0 being output. It is stored at location 2, so the second AND gate is opened. When the stored 0 reaches the input of the OR, it com-

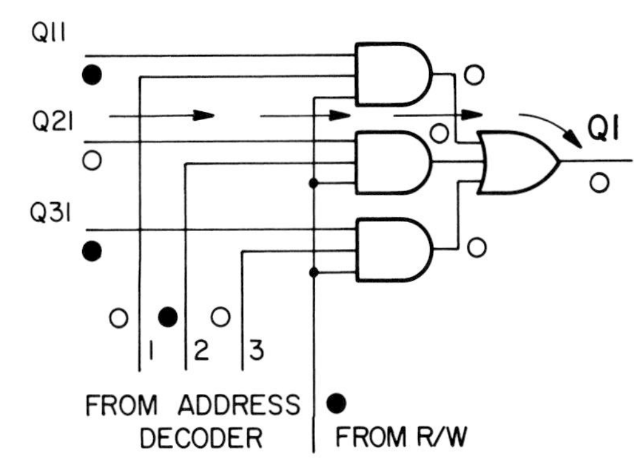

Fig. 8-27. For the control and address signals shown, the 0 stored in fip-flop 21 will be output.

bines with the 0s from the three closed AND gates to the left of the OR and results in Q1 = 0. A 0,0,1 (on address leads) lets out the 1 at Q31.

REVIEW OF PROCESS

Return to Fig. 8-18. The circuit shown may appear less complex than it did when first encountered. When you look at it, visualize the four subcircuits. It is only by dividing it into recognizable sections that its overall action can be understood.

When analyzing a circuit:

1. Know what the circuit does.
2. Consider it in small, recognizable subcircuits.
3. Start with subcircuits that are easily analyzed. Then proceed to the more difficult subcircuits. Complete the analysis by reviewing the types of interactions between the simple subcircuits.

SUMMARY

A clock input increases the usefulness of R-S flip-flops, since the clock input acts like an enable. Such a flip-flop can be constructed using four NANDs.

A data latch is a clocked flip-flop with only one input. A clocked R-S flip-flop can be converted into a data latch by adding a NOT across its inputs.

Complex digital circuits are best analyzed by dividing them into recognizable subcircuits.

TEST YOUR KNOWLEDGE

1. A clock input on a clocked R-S flip-flop is used as a/an __________ lead.
2. When C of an active-high clocked flip-flop is active __________ (high, low), the circuit acts like an ordinary flip-flop.
3. How many NAND elements are used to build a clocked R-S flip-flop circuit?
4. A data latch can be obtained from a clocked R-S flip-flop by placing a __________ element across the R-S input of the R-S flip-flop.
5. A register circuit does not use an address code for its operation. (True or False?)
6. A number is stored in a register after an active-high clock goes from the inactive state to the active state. (True or False?)
7. A memory circuit with a word length of 3 bits and a word storage capacity of 4 words has 4 data input leads. (True or False?)
8. To understand a complex circuit, break it into two or more __________ circuits.
9. Memory elements are called __________ (lattices, cells, intersections, planes, repeaters).
10. A memory has a lead called R/W. What signal (1 or 0) should be on the lead to write into the memory?

STUDY PROBLEMS

1. Trace the timing diagram in Fig. P8-1 onto a sheet of paper and carefully construct the expected output. Assume Q = 0 at t = 0.

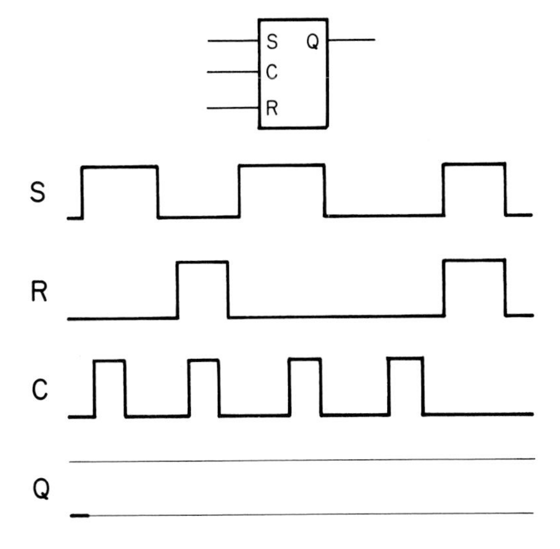

Fig. P8-1. The output timing diagram for the clocked R-S flip-flop can be found from its inputs.

2. Repeat Problem 1 for the timing diagram and circuit symbol that is shown in Fig. P8-2.
3. Repeat Problem 1 for the timing diagram and flip-flop with active-low clock shown in Fig. P8-3.
4. Repeat Problem 1 for the timing diagram in Fig. P8-4. In this case, the clock is active-high. Rather than being a short pulse, the clock signal remains active for a long time.

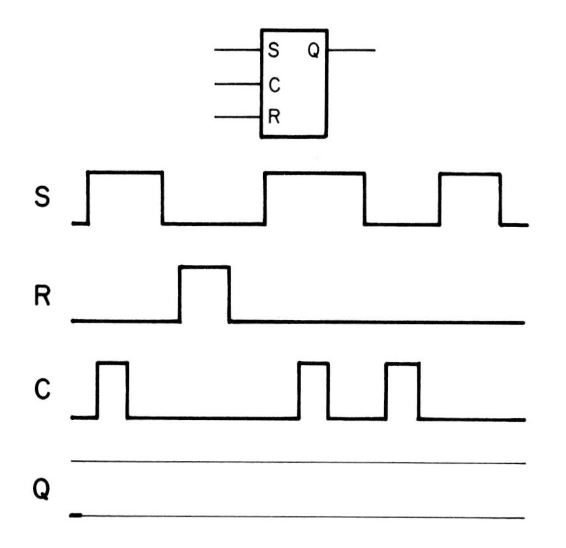

Fig. P8-2. A number can be input to a clocked R-S flip-flop only when the clock is active.

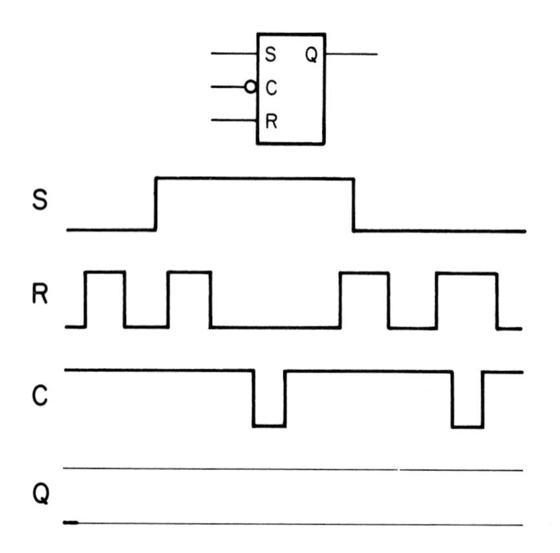

Fig. P8-3. Of the many combinations of 1s and 0s on S and R, only signals present when the clock goes active (low) are stored.

5. Using four NANDs, draw a logic diagram for a clocked R-S flip-flop. Label the leads of the clocked R-S flip-flop S, R, C, and Q.
6. For the input signals shown in Fig. P8-5, what signals (1s and 0s) are expected at points 1, 2, 3, and 4 in the circuit?
7. For the inputs and output shown in Fig. P8-6, what signals (1s and 0s) are expected at points 1, 2, and 4 in the circuit?

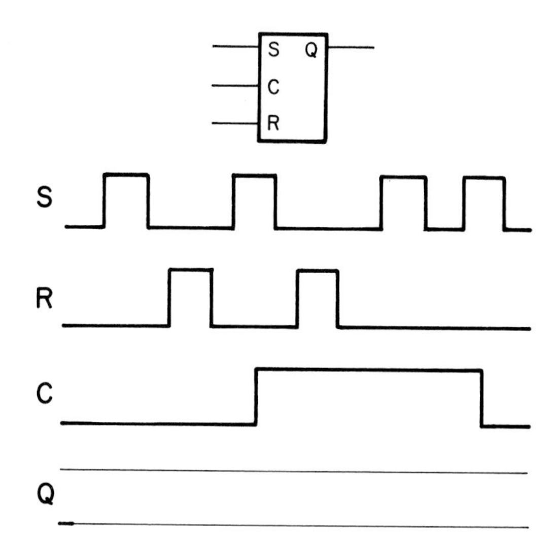

Fig. P8-4. A long clock pulse for a clocked R-S flip-flop with active-high clock requires careful study to obtain the correct output timing diagram.

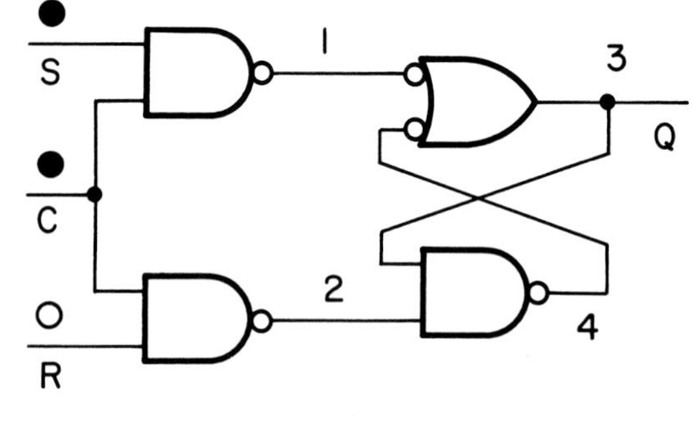

Fig. P8-5. Measured logic levels at the inputs and outputs of the NANDs can be found for the clocked R-S flip-flop shown.

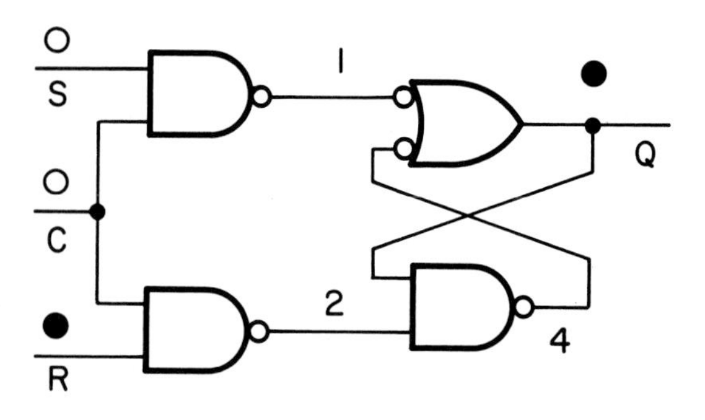

Fig. P8-6. Measured logic levels for the NANDs of a clocked R-S flip-flop can be found from the inputs and one output shown.

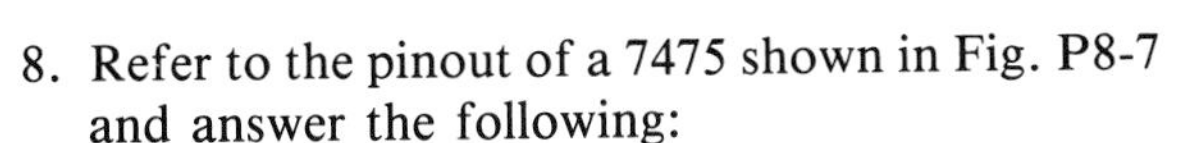

8. Refer to the pinout of a 7475 shown in Fig. P8-7 and answer the following:
   1. How many data latches are in a 7475?
   2. What letter is used to indicate the clock input?
   3. How many clock leads are brought out of the DIP package that is shown?
   4. Is the clock active-high or active-low?

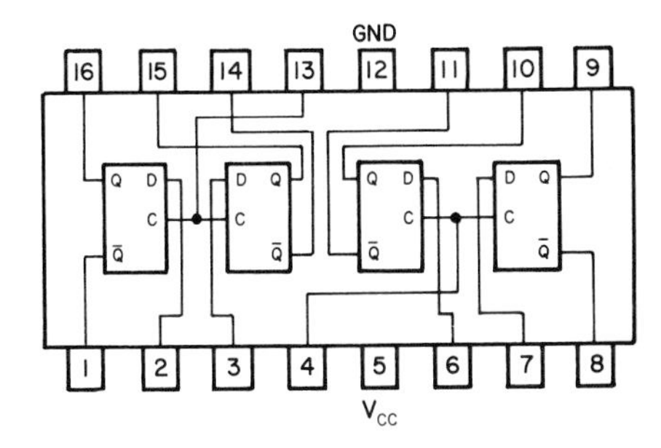

Fig. P8-7. Specifications can be read from the pin outline (pinout) diagram of a 7475 dual in-line package (DIP).

9. Trace the timing diagram in Fig. P8-8 onto a sheet of paper and carefully construct the expected output. Assume Q = 0 at t = 0.

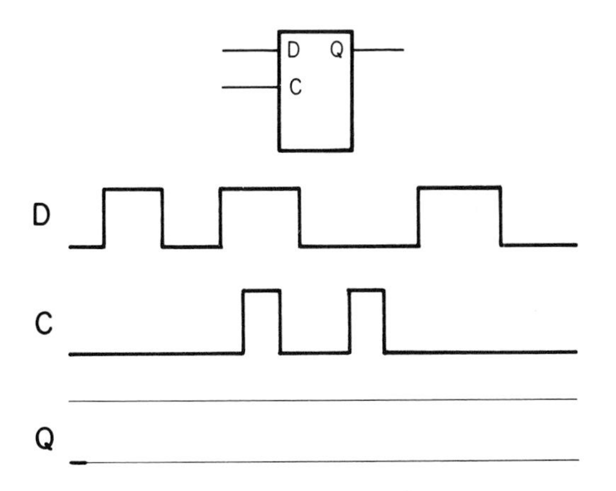

Fig. P8-8. The output timing diagram can be found for the data latch shown. Assume Q = 0 at time t = 0.

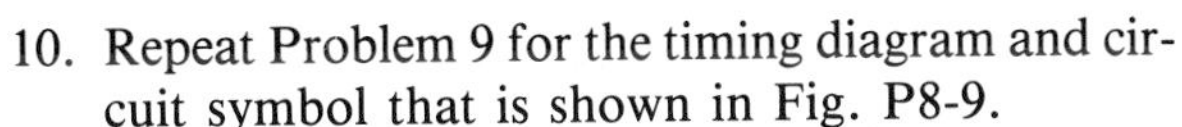

10. Repeat Problem 9 for the timing diagram and circuit symbol that is shown in Fig. P8-9.
11. Repeat Problem 9 for the timing diagram in Fig. P8-10. This data latch has an active-low clock.
12. Repeat Problem 9 for the timing diagram in Fig. P8-11. The clock on this data latch is active-high. Note the length of the clock pulse.

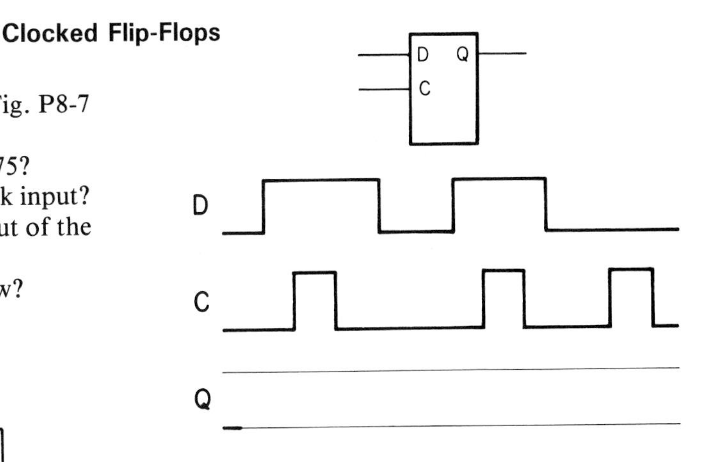

Fig. P8-9. Output of a data latch can be analyzed with the help of timing diagrams.

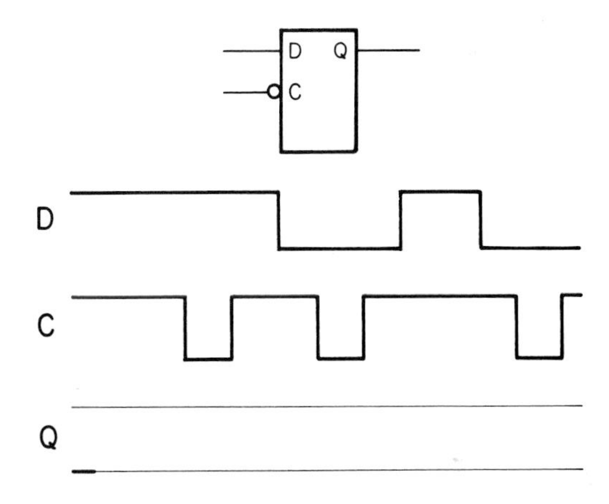

Fig. P8-10. Output of a data latch with active-low clock can be analyzed from input timing diagrams.

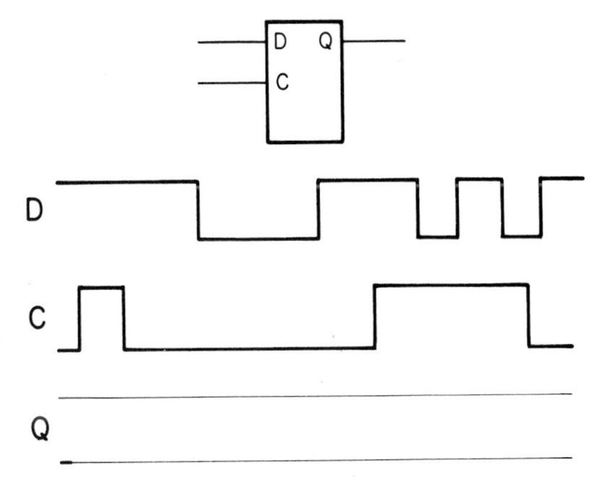

Fig. P8-11. Output for data latch with active-high clock is found from input timing diagrams. A short clock pulse and a clock pulse of medium length are shown. Assume the output Q is 0 at time t = 0.

13. Use five NANDs and draw the logic diagram for a data latch. Label its leads D, C, Q, and $\overline{Q}$.
14. Using data latches, draw a logic diagram for a 4-bit register. Its inputs are to be A3, A2, A1, and A0. Its outputs are to be Q3, Q2, Q1, and Q0. Label its enable lead WRITE ENABLE. It is to have an active-high enable.
15. Based on the diagram of a 3-bit register shown in Fig. P8-12, answer the following questions.
    1. When the control lead is 1, is this circuit in its read (outputting a stored number) or its write (storing a new number) mode?
    2. When in its write mode, which of the following best describes the number at the outputs that are marked Y2, Y1, and Y0?
    (1). Three 0s will appear at these leads.
    (2). The previously stored number will be output.
    (3). The new number input through A2, A1, and A0 will immediately appear at the output.

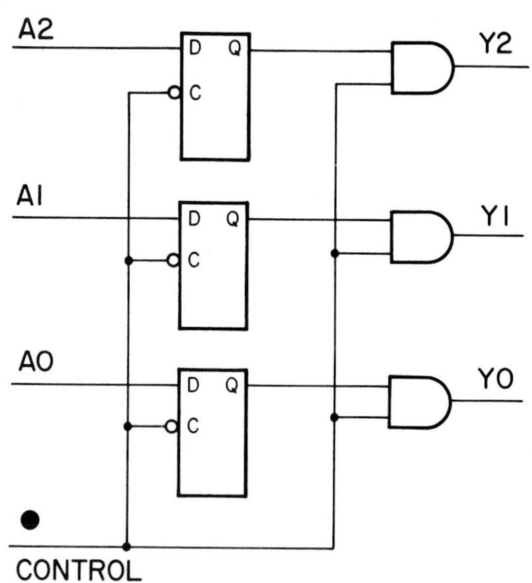

Fig. P8-12. A 3-bit register provides practice in determining outputs from data inputs and control lead inputs.

Computers and testing equipment are based on digital circuit design.

Master-Slave Flip-Flops

Chapter 9

LEARNING OBJECTIVES

After studying this chapter and completing lab assignments and study problems, you will be able to:

- Draw circuit diagrams for master-slave data latches and R-S flip-flops.
- Describe the actions of master-slave flip-flops using truth tables and timing diagrams.
- Draw circuit diagrams for basic shift registers and ring counters.
- Complete timing diagrams for shift registers and ring counters.
- Recognize, list, and match symbol names with the correct symbols for level-triggered and edge-triggered flip-flops and use timing diagrams to describe the actions of the circuits.

The flip-flops studied so far contain only one cross-coupled memory element. In such devices, newly stored numbers immediately appear at the circuit's output. Old (previously stored) numbers are lost as soon as the storage process begins.

In some flip-flop applications, it is necessary to read the old number while a new number is being stored. To do this, flip-flops called *MASTER-SLAVE* flip-flops are used. These memory elements and their applications are described in this chapter.

MASTER-SLAVE DATA LATCH

Part "a" of Fig. 9-1 shows the timing diagram for an ordinary data latch. When its clock goes active, the number (1 or 0) at D immediately appears at Q. The old (previously stored) number is lost as soon as C goes active.

The timing diagram for a master-slave data latch is shown at b. When its clock goes active at t1, the number at D enters. However, it does not appear at output Q until t2, the time when its clock returns to its inactive level. The old number can be read from the output of this data latch at the same time that a new number is being stored.

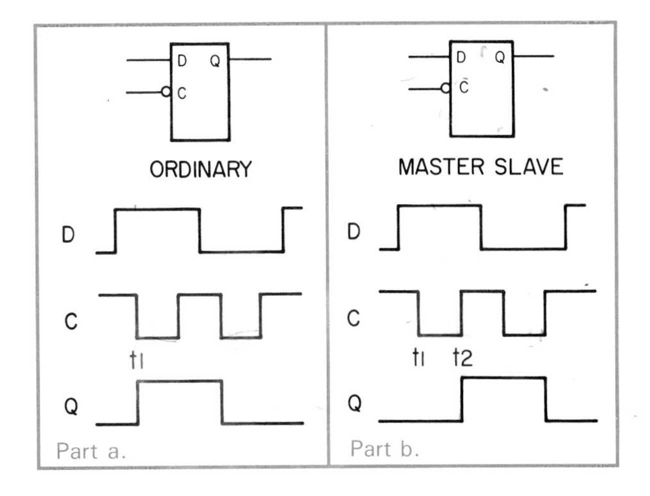

Fig. 9-1. The previously stored number is available at the output of the master-slave flip-flop during the time that a new number is being stored. This is not true of ordinary flip-flops.

NEED FOR MASTER-SLAVE FLIP-FLOP

Part "a" of Fig. 9-2 shows a *SHIFT REGISTER*. Each time SHIFT goes active (low), numbers at Q0, Q1, and Q2 move one place to the right. The number at IN appears at Q0. The stored number is said to *SHIFT RIGHT*.

MASTER-SLAVE CIRCUIT FOR SHIFT REGISTER

The timing diagram in part "b" of Fig. 9-2 shows the circuit's action when master-slave data latches are used. All outputs are assumed to be 0 at time t0. At time t1, SHIFT is active, so the 1s and 0s at the three D inputs are accepted. A 1 (from IN) enters the first data latch. Zeros (from Q0 and Q1) enter the other two. That is, the old numbers at Q0 and Q1 are available even though numbers are being input. This could not happen if ordinary flip-flops were used.

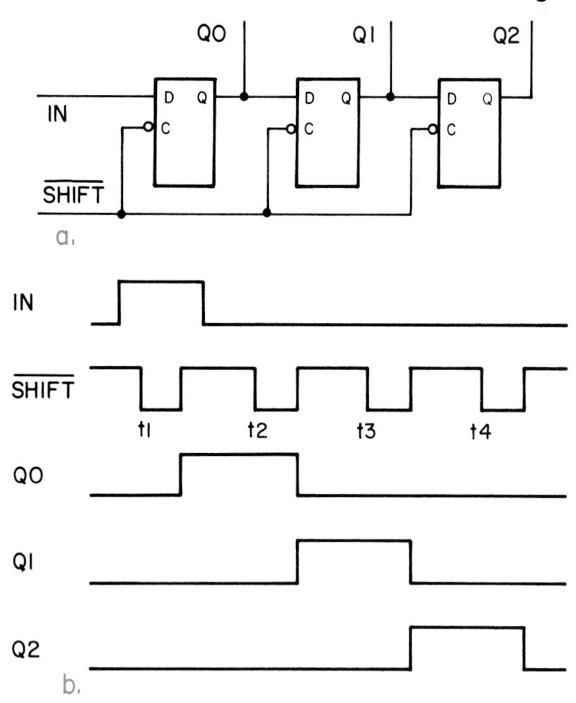

Fig. 9-2. The flip-flops in the shift register shown must be of the master-slave type, since their outputs must be read during the time that new numbers are being stored. All outputs are assumed to be 0 at time t0.

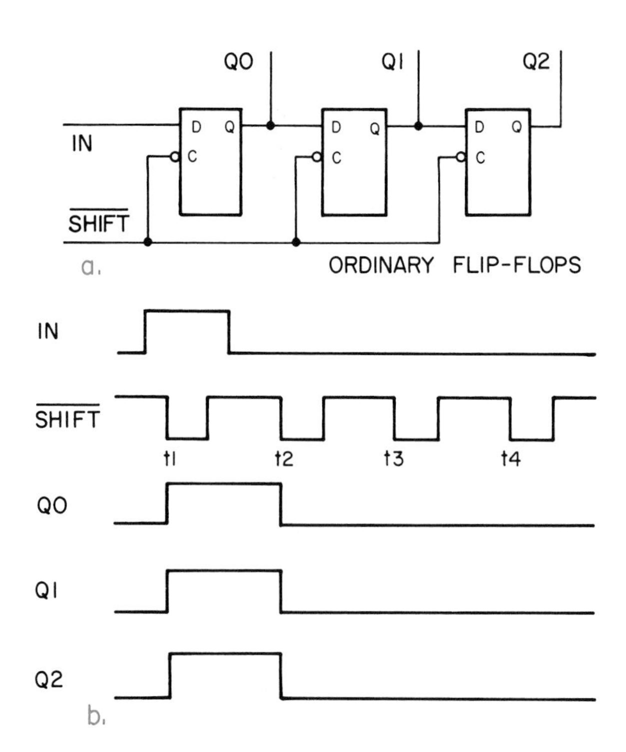

Fig. 9-3. This timing diagram shows what happens when ordinary (non-master-slave) flip-flops are used in a shift register.

When SHIFT returns to 1 (inactive), the new numbers appear at the three outputs. The act of SHIFT returning to 1 has a special effect on the second half (the output half) of the master-slave flip-flop. The most recent numbers input at the D inputs become readable at each one of the Q outputs.

At t2, SHIFT is again active. This time, IN = 0, Q0 = 1, and Q1 = 0. These numbers are input by the flip-flops, and appear at their outputs when SHIFT returns to its inactive state. The 1 entered during the first shift cycle has moved to output Q1.

The process is repeated at t3 and t4. At t3 the 1 moves to Q2. At t4 it is said to have fallen off the end of the register. That is, 0s appeared at the inputs of the three flip-flops. When SHIFT returns to its inactive state, 0s appear at the three outputs.

In a properly operating shift register, numbers input through IN shift from one output to another under the control of SHIFT.

ORDINARY FLIP-FLOP CIRCUIT

Ordinary flip-flops are shown in Fig. 9-3. At time t1, SHIFT is active, and numbers enter the three data latches. Because these are not master-slave units, the 1 from IN immediately appears at Q0. The clock of the second flip-flop is active, so that the 1 at Q0 is accepted and appears at Q1. The same thing happens at the last flip-flop. As a result, all three outputs take on the value of IN.

At t2, IN equals 0, so this number appears at all outputs. This is not the action of a shift register. To function as a shift register, master-slave flip-flops must be used. That is, elements that output old numbers at the same time new numbers are being input are needed.

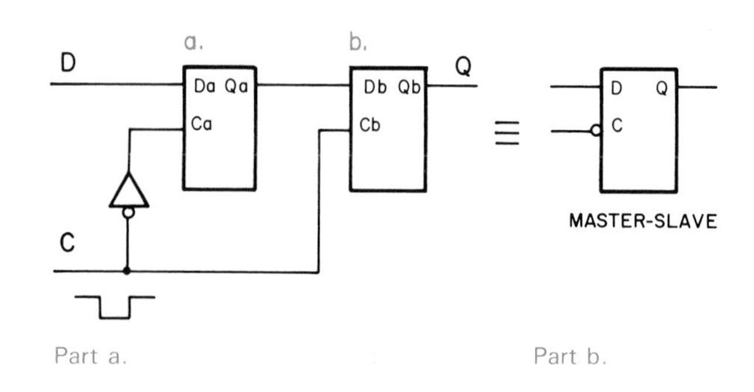

Fig. 9-4. A master-slave flip-flop can be constructed using two ordinary data latches.

MASTER-SLAVE FLIP-FLOP CIRCUIT

To obtain master-slave action, two flip-flops are connected one after the other, as shown in part "a" of Fig. 9-4. Clock input C drives both Ca and Cb. The NOT insures that only one flip-flop is active at a given time.

The symbol at b indicates that this circuit has an active-low clock. The symbol does not distinguish between ordinary and master-slave flip-flops. Only their applications and identifying numbers indicate the difference.

MASTER-SLAVE FLIP-FLOP CIRCUIT ACTION

To emphasize their actions, clocks Ca and Cb have been shown as switches in Fig. 9-5. All outputs are assumed to be 0 before time 1.

Using Fig. 9-5, master-slave flip-flop circuit action is described as follows:

1. Time t1: At time t1, C = 1 (inactive). The NOT inverts this 1 and applies the resulting 0 to Ca. This opens the switch, so new numbers cannot enter.
   The 1 at C is also applied to Cb. This switch is closed, and the number stored in the first flip-flop (in this case a 0) is applied to the second. There it is stored and appears at output Q.
2. Time t2: Here, C = 0 (active). This 0 opens switch Cb. The flip-flops are isolated from each other, so the 0 stored in the second element continues to appear at Q.
   The NOT inverts C and applies the resulting 1 to Ca. This closes the switch and permits the new number (in this case a 1) to enter the input flip-flop. Remember, it cannot reach the other element, since Cb is inactive.
3. Time t3: C has returned to 1, and the circuit is again inactive. Ca is open, so new numbers cannot enter. Cb is closed, and the transfer of the new number has been made.

To review, new numbers enter master-slave flip-flops while C is active. These new numbers are delivered to the second flip-flop and the output when C returns to its inactive level.

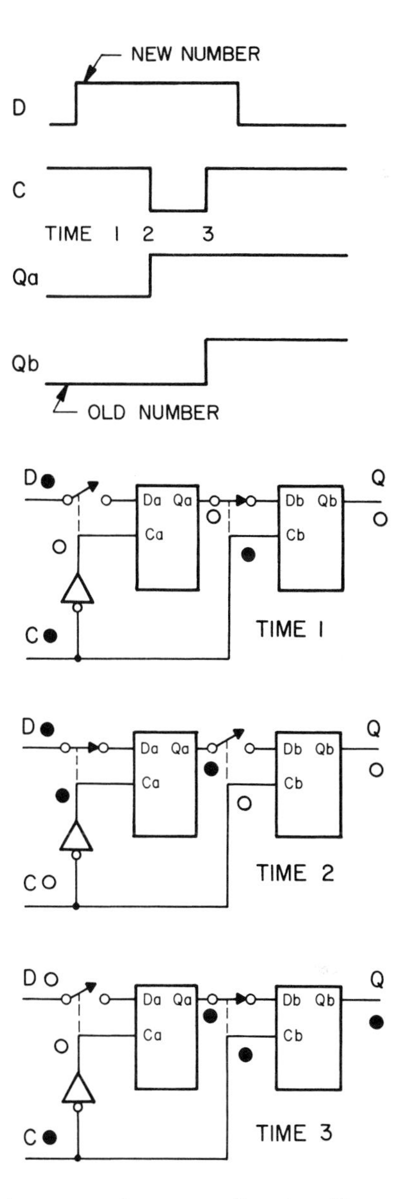

Fig. 9-5. A new number enters Da when C goes active. It is transferred to Db when C returns to its inactive state. All outputs are assumed to be 0 before time 1.

PLACEMENT OF A NOT ON DATA LATCHES FOR MASTER-SLAVE FLIP-FLOP

The NOT may be placed in series with either Ca or Cb, Fig. 9-6. However, the upper circuit is usually preferred.

To operate properly, the output flip-flop must be disconnected before the input flip-flop accepts a new number. Because it takes a short time for signals to pass through the NOT, its placement in the lead to Ca insures that Cb will be inactive before Ca becomes active. This placement results in input C being bubbled. Proper timing for an active-high clock is achieved as shown in part "c" of Fig. 9-6.

NAND-BASED DATA LATCH IN MASTER-SLAVE FLIP-FLOP

Fig. 9-7 shows a NAND-based master-slave level-triggered active-low data latch. It consists of two ordinary NAND-based data latches connected one after the other.

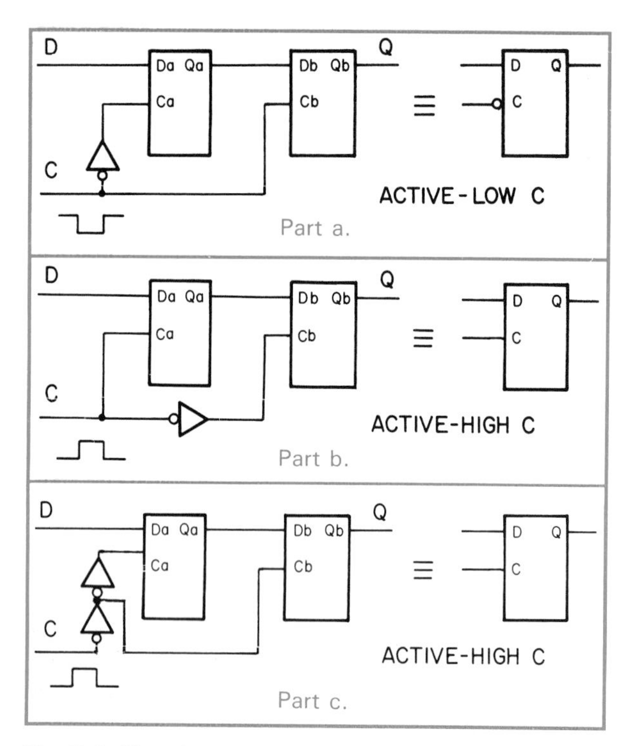

Fig. 9-6. The placement of the NOT on data latches for a master-slave flip-flop determines whether C will be active high or active low.

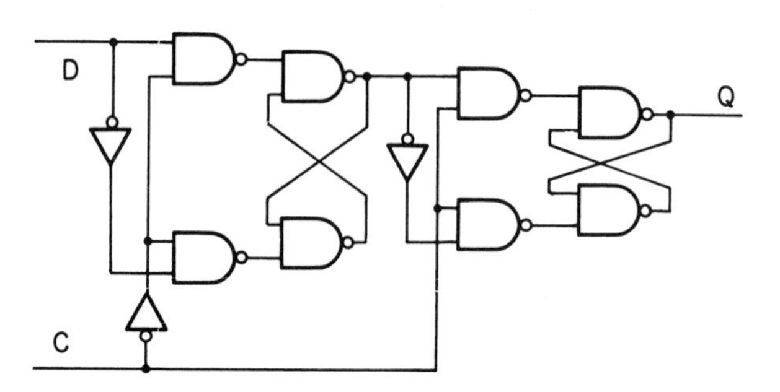

Fig. 9-7. NANDs are used in the construction of a master-slave data latch. The clock input, which is marked C, is of the active-low type.

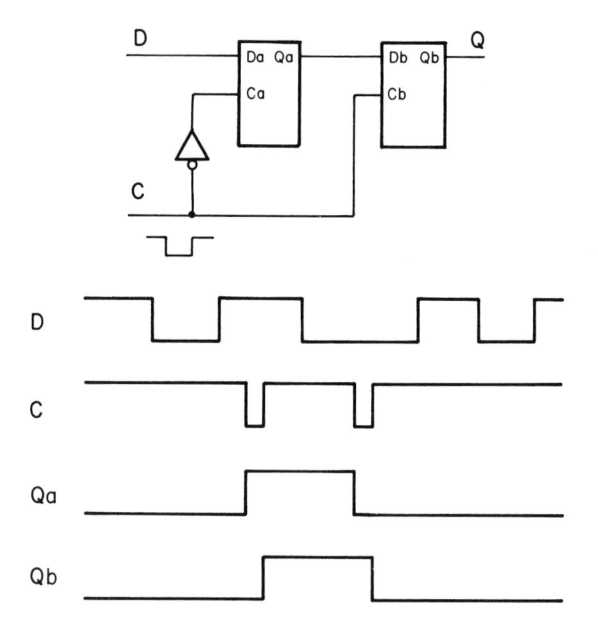

Fig. 9-8. Clock pulses in the timing diagram shown are short compared to the signals at D.

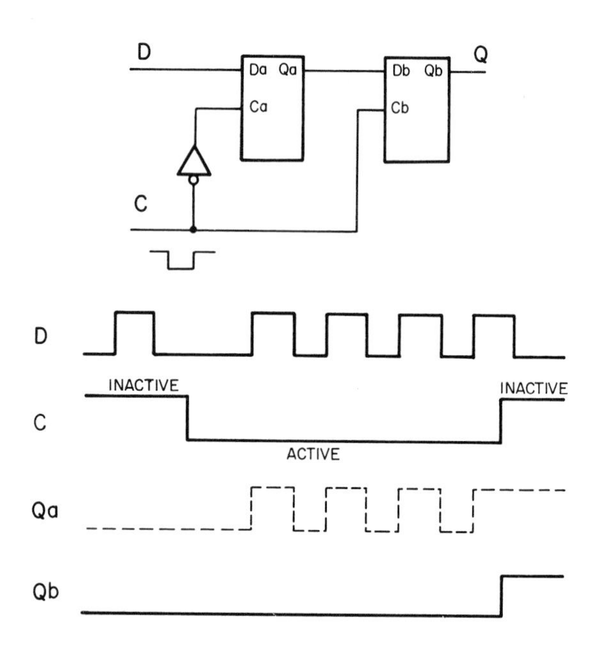

Fig. 9-9. When the clock pulse is long, the action of the circuit is more complex.

LEVEL-TRIGGERED ACTION

The clock pulses shown in Fig. 9-8 are relatively short. Numbers enter when C becomes active; they appear at output D when C returns inactive. The level-triggered nature of this circuit causes no problem.

The long clock pulses shown in Fig. 9-9 emphasize the level-triggered nature of this circuit. While the clock is active, the input flip-flop accepts whatever numbers are applied to D. See graph Qa. Dashed lines have been used to emphasize that these numbers are not transferred to the output flip-flop, since its clock is inactive.

When C goes inactive, the number in the first flip-flop is stored, applied to the output flip-flop, and appears at Q.

MASTER-SLAVE R-S FLIP-FLOP

Master-slave R-S flip-flops can also be constructed. See Fig. 9-10. Numbers enter when the clock is active; they appear at the circuit's output when the clock returns to inactive.

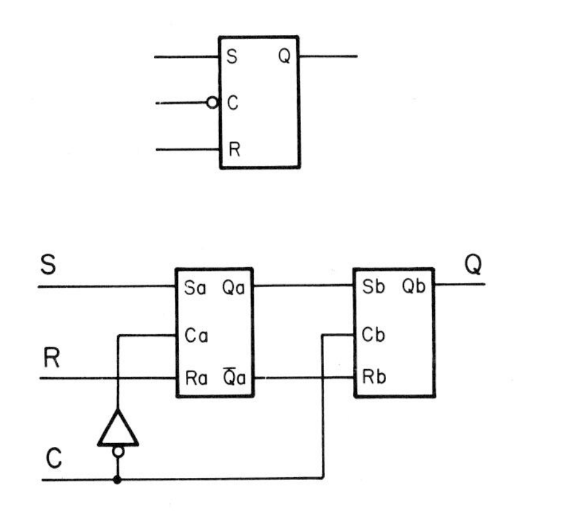

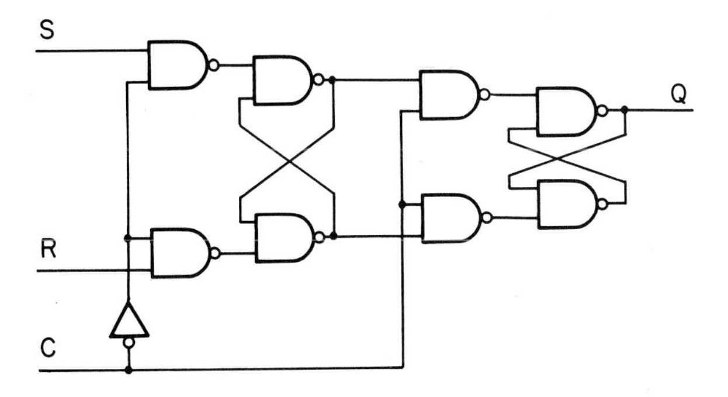

Fig. 9-10. Symbol and circuit for a master-slave R-S flip-flop.

MASTER-SLAVE APPLICATIONS

In most applications, master-slave memory elements may be used in place of ordinary flip-flop or data latch elements. For example, master-slave data latches could be used in the register in Fig. 9-11. Its action is similar to that of the register described in the last chapter. The only difference is found in the timing of its output. When ordinary flip-flops are used, changes in output signals occur when ENABLE goes active. When master-slave elements are used, new numbers are not output until ENABLE returns to its inactive state.

As an aside, a bar has been placed over ENABLE. This notation is often used to emphasize that ENABLE is active low. It should not be considered a Boolean notation.

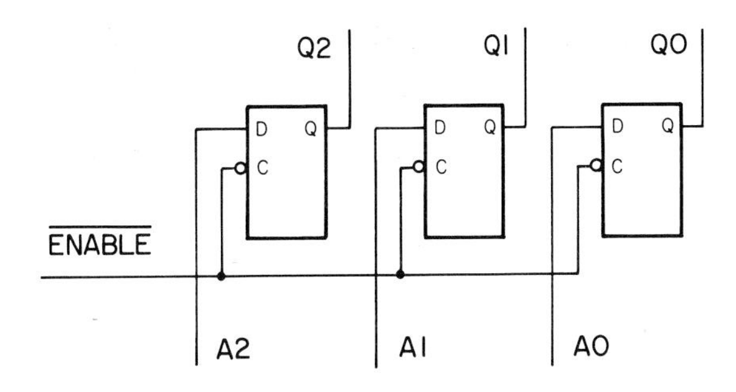

Fig. 9-11. Although it uses master-slave flip-flops, the register shown looks and functions like one using ordinary flip-flops.

MASTER-SLAVE SHIFT REGISTER APPLICATION TYPES

As stated earlier, flip-flops used in shift registers must be of the master-slave type. That is, previously stored numbers must be available during the storing of new numbers.

There are several types and uses of master-slave shift registers. The following are the uses for shift registers:

1. Serial-to-parallel conversion.
2. Serial-to-serial storage.
3. Parallel-to-serial conversion.

SERIAL-TO-PARALLEL CONVERSION

When multi-bit binary numbers are transmitted on a single pair of wires, they must be in *SERIAL* form. That is, one bit is sent at a time. A telegraph signal is an example.

Fig. 9-12 shows the number D3D2D1D0 = 1011 represented in serial form. In this graph, time is plotted from left to right. D3 is transmitted first. D2, D1, and D0 follow.

Although it is convenient to transmit data in serial form, it is usually used in *PARALLEL* form. That is, all bits of a number must be available at the same

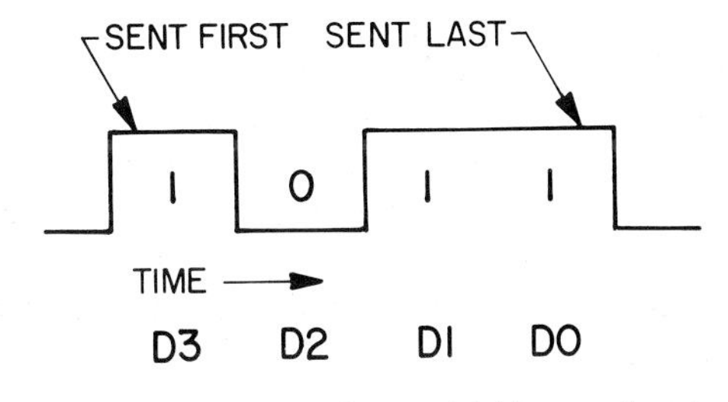

Fig. 9-12. Graph of a serial binary signal.

time on parallel leads. The conversion from serial to parallel can be accomplished by a shift register.

In the timing diagram in Fig. 9-13, time is plotted horizontally. The bits available at the parallel outputs, Q3, Q2, Q1, and Q0, are plotted vertically. To emphasize this relationship, place a piece of paper over the graph. Move it to the right as the following description progresses. Start with the edge of the paper on the dashed line at t0. The circuit action is as follows:

1. Time t0: At t0, SHIFT is inactive and all inputs and outputs are 0. Note that D = 0. The graph shows that the four outputs are 0. The register is said to be *CLEARED.*
2. Time t1: Two things have happened. Data bit D3 (a 1) has appeared at input D, and SHIFT has gone active. The 1 has entered the first flip-flop; 0s have entered the other three.
3. Time t2: SHIFT is inactive, so the numbers input at time t1 now appear at Q0, Q1, Q2, and Q3. Data bit D3 is at output Q0.
4. Time t3: SHIFT is again active, and new numbers enter the flip-flops. The 0 at D is accepted by the first fip-flop. The 1 at Q0 is accepted by the second. The last two accept 0s from Q1 and Q2.
5. Time t4: SHIFT is inactive, so the new numbers appear at the circuit's outputs. Data bits D3 (1) and D2 (0) have been entered, but they are not yet in their correct positions.
6. Time 5: SHIFT is again active, and data bit D1 (1) enters.
7. Time t6: With the clock inactive, data bits D3, D2, and D1 are in the shift register. The circuit's output (Q3Q2Q1Q0) is 0101. Read upward from the bottom of the second last column.
8. Time t7: The last bit of the serial number (D0 = 1) has been input. Q0 is already 1, so no change is needed. Q1 needs to make its output match the 1 inside it.
9. Time t8: All of the bits inside the latches are placed at the outputs. The conversion from serial to parallel is complete. The Qs appear in reverse order in the circuit. This does not change the stored number. Q3Q2Q1Q0 = 1011.

Timing diagrams can provide much information that can be obtained no other way. Timing diagrams that are displayed on an oscilloscope describe circuit operations well. See part "b" of Fig. 9-13.

SHIFT must go active at the proper times. It must go active for data to enter. That is, data timing and SHIFT must be SYNCHRONOUS.

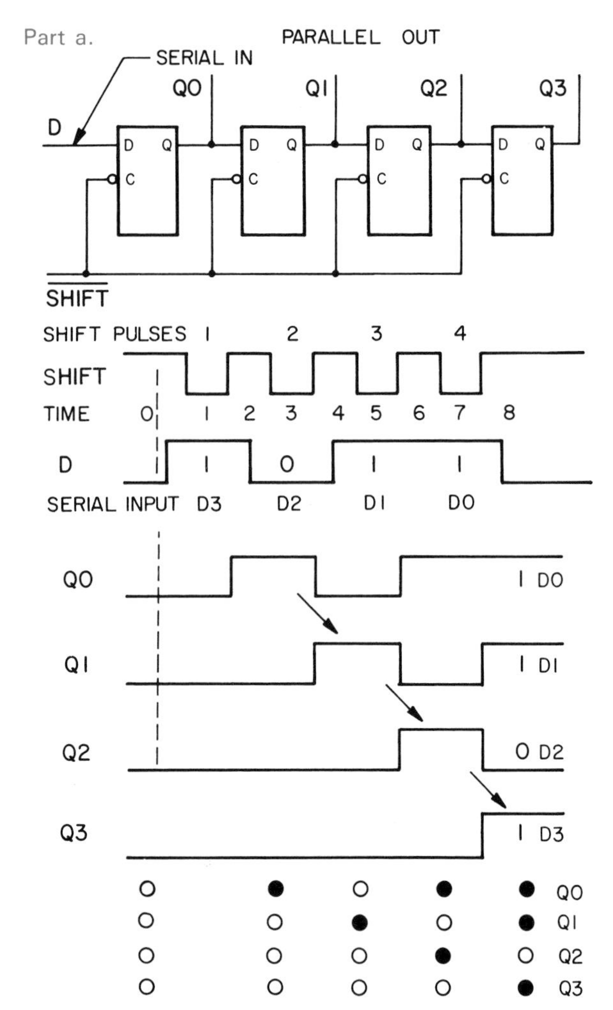

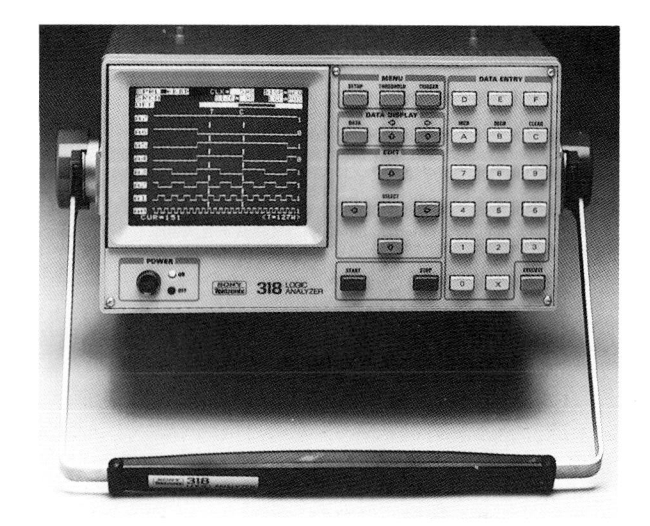

Fig. 9-13. In part "a," shift registers can convert serial data to parallel data. In part "b," a similar timing diagram display is shown. Designers, engineers, and technicians are aided by these diagrams. (Tektronix)

SERIAL-TO-SERIAL STORAGE

In some applications, serial data must be stored and

then output in serial form. Shift registers can accomplish this task.

To do the serial-to-serial conversion, the serial number is input through D as described in the serial-to-parallel conversion. At time t1 in Fig. 9-14, the number D3D2D1D0 = 1011 is shown stored in a shift register.

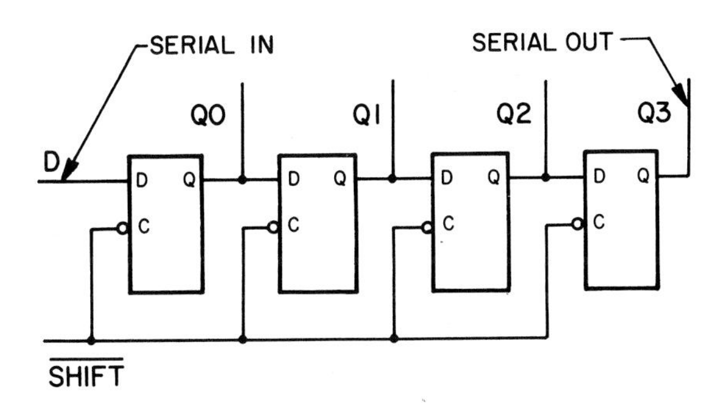

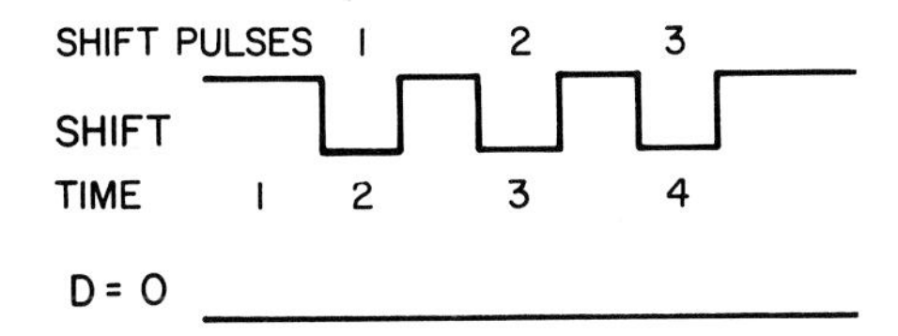

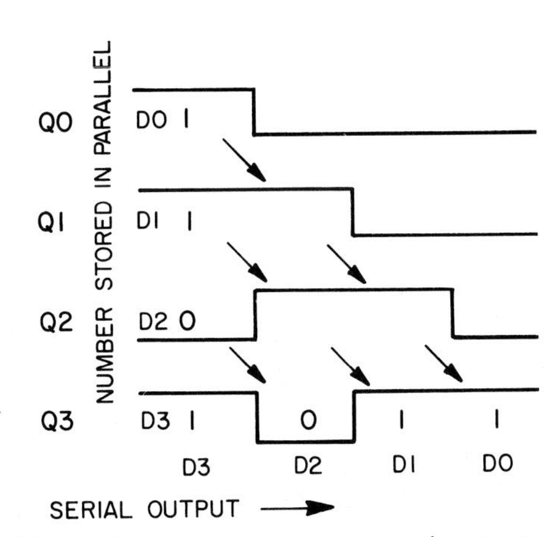

Fig. 9-14. Shift registers can store and output serial data.

Q3 will be used as the serial output. Note that data bit D3 is already available at Q3. Activating SHIFT will move data bit D2 to Q3. And activating it twice more will move D1 and D0 to this serial output. In other words, pulsing SHIFT three times will deliver the number to Q3 in serial form. In Fig. 9-14, D = 0 because no new number is being input while the old stored number is being read out. After reading out, the outputs are in the sequence 0,0,0,1.

To do a serial-to-serial conversion, SHIFT is pulsed four times to enter the serial number. After an appropriate delay, it is pulsed three more times to output the stored number in serial form.

UNIVERSAL SHIFT REGISTER

The basic shift register just described can make two conversions. They are:

1. Serial-in to parallel-out.
2. Serial-in to serial-out.

By modifying the flip-flops, two additional conversions are possible. They are:

3. Parallel-in to parallel-out.
4. Parallel-in to serial-out.

Flip-flops with PRESET and CLEAR inputs are often used to construct devices called *UNIVERSAL SHIFT REGISTERS.* Universal shift registers are capable of the four conversions.

PRESET/CLEAR

The master-slave data latch symbol in part "a" of Fig. 9-15 has two new leads. PR stands for PRESET (sometimes referred to merely as SET); CL stands for CLEAR (sometimes called RESET).

PRESET and CLEAR are bubbled inputs. When a 0 is applied to PR, a 1 is stored (Q = 1). When a 0 is applied to CL, a 0 is stored (Q = 0).

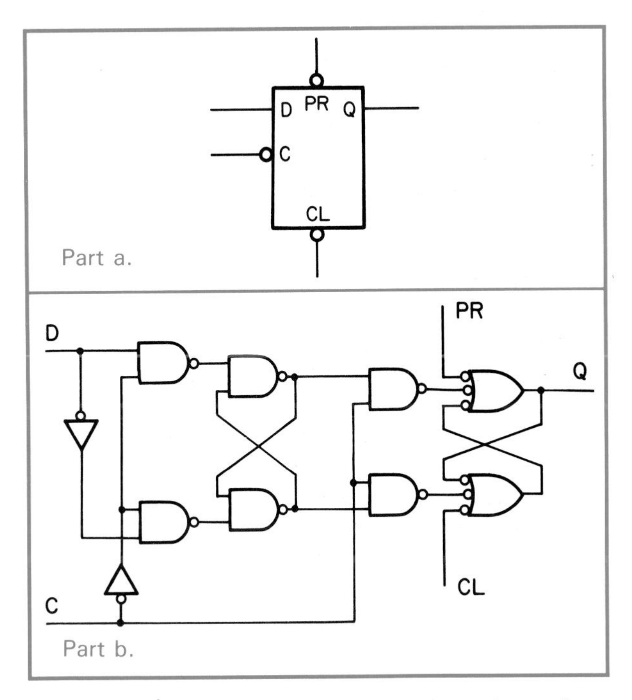

Fig. 9-15. The R-S master-slave flip-flop shown has preset and clear leads.

Input D (and R and S on R-S flip-flops) are said to be *SYNCHRONOUS* inputs. Synchronous means the inputs operate only when the clock is active. PR and CL are *ASYNCHRONOUS* (not synchronous).

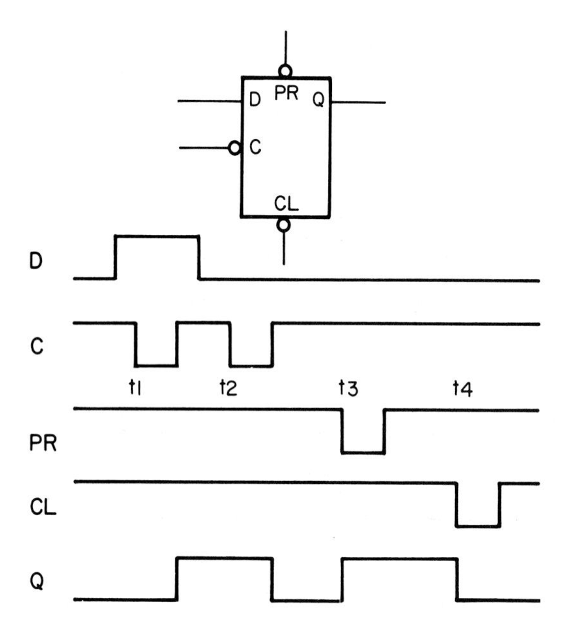

Fig. 9-16. Numbers can be input through synchronous input D or through asynchronous inputs PR and CL.

The clock need not be active for numbers to be entered through these inputs. In most flip-flops, PR and CL override the synchronous inputs. That is, if a 1 is being input through D, and a 0 is being input through CL, the output will go to 0.

Part "b" of Fig. 9-15 shows a NAND-based master-slave data latch with PR and CL. When a 0 is applied to PR, the output of the upper NAND is forced to 1, and the flip-flop is set. If a 0 is applied to CL, the output of the lower NAND is forced to 1 and Q goes to 0. The flip-flop is cleared.

Fig. 9-16 shows a timing diagram for a data latch with PR and CL. At t1 and t2 numbers enter through D. Because the flip-flop is a master-slave type and is level-triggered, these numbers appear at the output when the clock returns to its inactive state.

At t3 and t4, numbers enter through PR and CL. This portion of the circuit may be thought of as an unclocked R-S flip-flop. Numbers appear at Q as soon as they enter.

PARALLEL INPUT IN UNIVERSAL SHIFT REGISTER WITH PRESET AND CLEAR

PR and CL can be used to add a parallel input to a shift register. See Fig. 9-17. To input a parallel number, ENABLE is set to 1.

Input B0 shows the result of inputting a 0. With

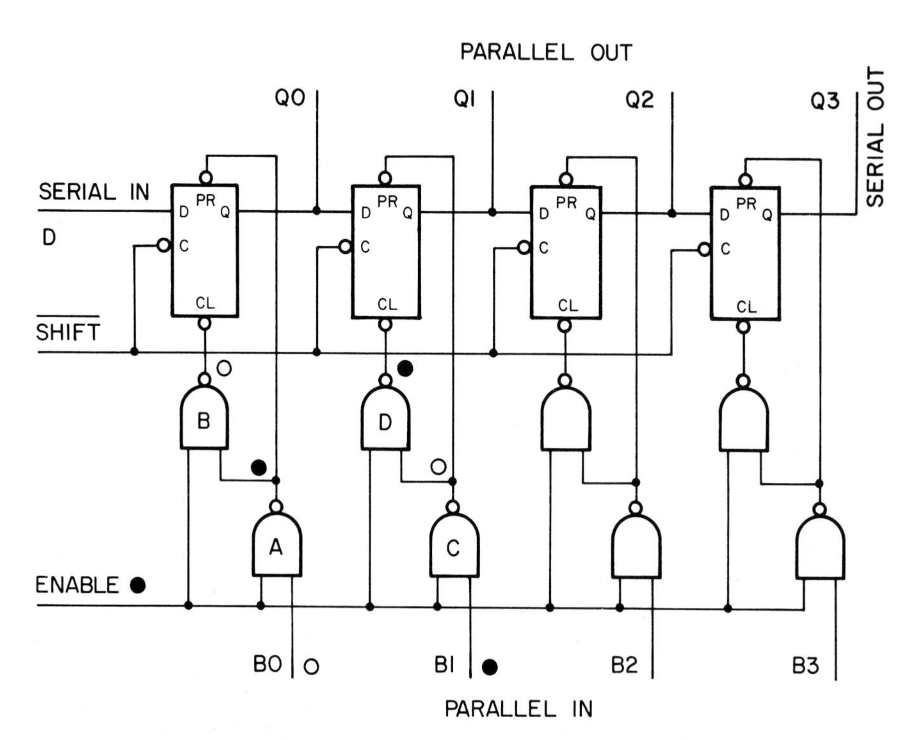

Fig. 9-17. A circuit with 4 master-slave data latches, parallel inputs (the inputs that are labeled B0 through B3), enable, and shift can perform the 4 series-parallel conversions.

a 1 (from ENABLE) and a 0 (from B0) at its inputs, NAND A outputs a 1. Because PR is active low, this 1 does not preset the flip-flop. However, with 1s at both of its inputs, NAND B outputs a 0. This causes CL to be active, and a 0 is entered into the first data latch circuit.

Input B1 shows the result of inputting a 1. With 1s at both inputs, NAND C outputs a 0. Applied to PR, this 0 causes a 1 to be stored in the second flip-flop.

When ENABLE is low, all eight NANDs have 0s at at least one input, so they all output 1s. Because PR and CL are active low, this turns off the parallel input circuit paths.

Numbers input through the parallel input immediately appear at the circuit's output. As a result, this circuit can be used as a parallel-to-parallel register.

PARALLEL-TO-SERIAL CONVERSION IN UNIVERSAL SHIFT REGISTER

The universal shift register circuit shown in Fig. 9-17 can convert parallel numbers to serial numbers. The parallel input is used to enter the number under the control of ENABLE. To output the number in serial form, three pulses are then applied to SHIFT. The four bits will shift to the right and appear in serial form at Q3.

RING COUNTER

You are probably familiar with radios called police scanners. Such receivers automatically scan a group of channels searching for one that is active. Often, lamps are used to indicate the channel being sampled, and the light sweeps the line of lamps again and again. The voltages necessary to accomplish this scan can be generated by a device called a *RING COUNTER*.

A ring counter is a shift register with feedback, Fig. 9-18. When operating normally, a 1 will appear at only one output. Each time C is pulsed, this 1 will move one place to the right. When it reaches Qd, it will jump back to Qa.

If the 1 is at Qa, Qb, or Qc, the output of the NOR will be 0. As the 1 moves to the right, 0s will fill in behind it.

When the 1 reaches Qd, outputs Qa, Qb, and Qc will be 0. The NOR will output a 1 to the input D. When the clock becomes inactive, Qa will become 1. The timing diagram in Fig. 9-18 describes the circuit's action in detail.

The ring counter circuit in Fig. 9-18 is self-clearing. The 3-input NOR performs the clearing process. When first turned on, any combination of 1s and 0s

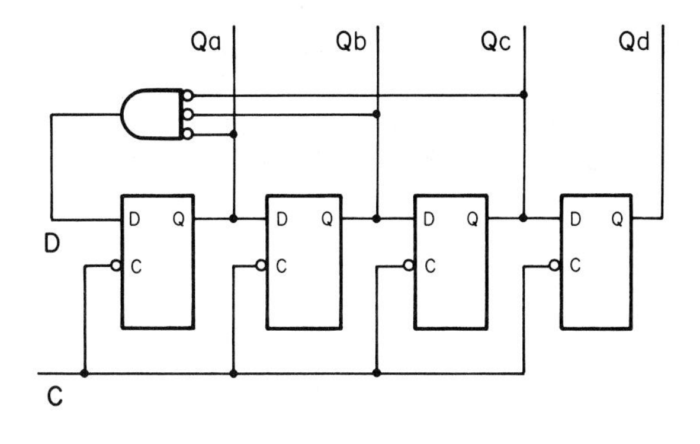

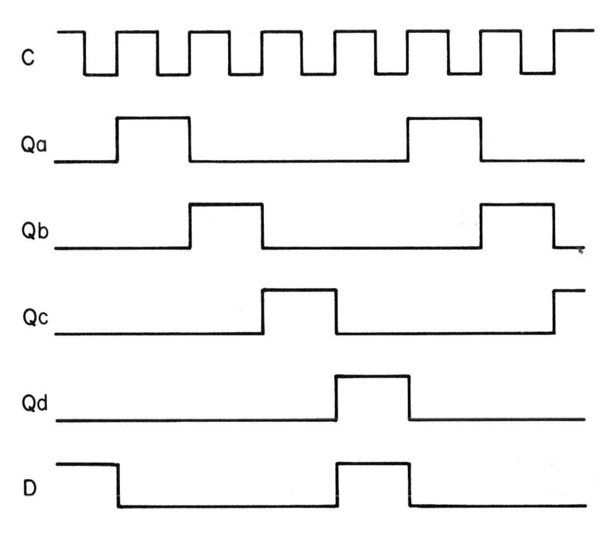

Fig. 9-18. When feedback is added, a shift register becomes a ring counter. The unit is self-clearing. The 3-input NOR puts out 0s until the inputs Qa, Qb, and Qc are all 0.

may appear. Once the clock is started, however, these will shift to the right and fall off the end of the counter. After it has cleared itself, a single 1 will circulate.

LEVEL-TRIGGERED VERSUS EDGE-TRIGGERED CLOCKS IN TYPES OF MASTER-SLAVE DATA LATCHES

Part "a" of Fig. 9-19 shows the action of a level-triggered data latch. To review, a number is stored when the clock returns to its inactive level. For short clock pulses (see time t1), this causes no great problem. For the longer pulse starting at t2, care must be taken. It is the number present at D when the clock goes inactive that is stored.

It is not the leading edge of the clock pulse that is critical; it is the trailing edge. As a result, the length of a clock pulse becomes important. If the length of

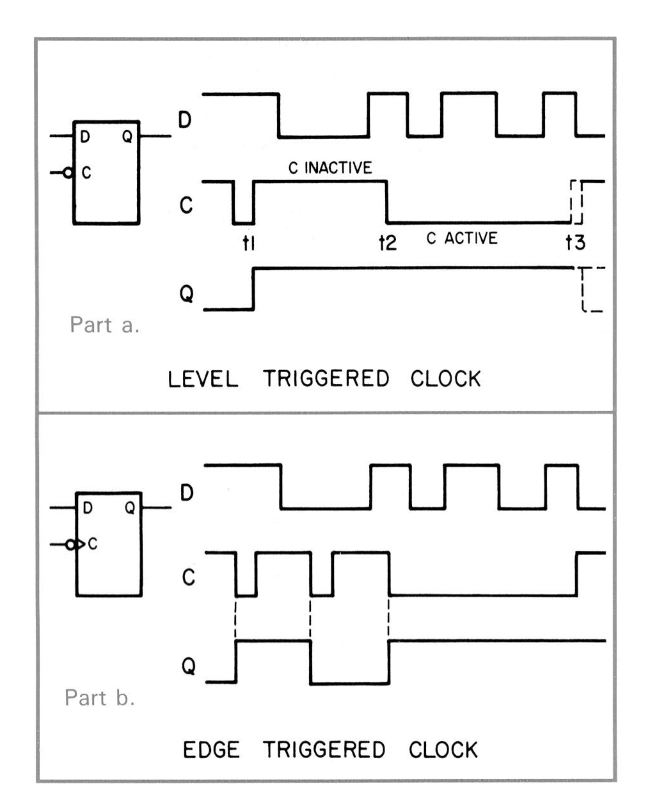

Fig. 9-19. Clocks in master-slave flip-flops may be level-triggered or edge-triggered. Level-triggered clocks require great care in design to avoid timing problems.

the pulse that starts at t2 changes, the stored number might also change. Note in part "a" of Fig. 9-19 the problem that can occur at t3, which depends on the length of the pulse that starts at t2.

To overcome such problems, clocks called *EDGE-TRIGGERED* clocks are available. The small triangle at C in part "b" of Fig. 9-19 indicates an edge-triggered clock. Because a bubble is shown, the negative-going edge of the clock pulse will cause the number at D to be stored. Refer to the timing diagram.

In part "a" of Fig. 9-20, arrows have been used to emphasize the negative-going edges. Although the numbers seem to appear at the output the instant the clock goes active, there is a slight delay. See part "b" of Fig. 9-20. This is a master-slave flip-flop, so a number can be read from its output at the same time a new number is being input. Therefore, shift registers and ring counters can be constructed using edge-triggered flip-flops.

A circuit for an edge-triggered data latch is shown in Fig. 9-21. It is complex, since there are four cross-couplings. Three are in flip-flops. The fourth is between flip-flops (see leads A and B).

An analysis of this circuit is not necessary. For the

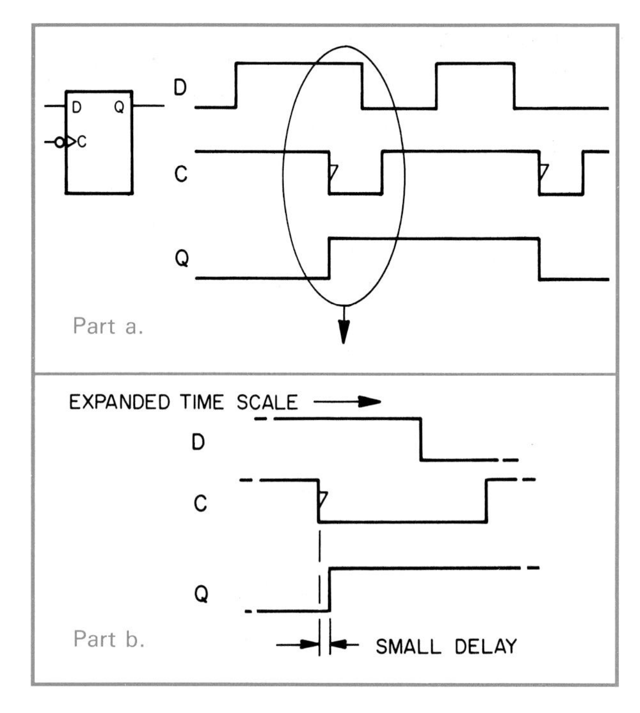

Fig. 9-20. The output of the edge-triggered flip-flop shown can be read at the same time a new number is being input.

sake of completeness, however, one approach will be outlined. To suggest the method of analysis, the first three steps will be described. It is helpful to examine the detailed operation of the circuit.

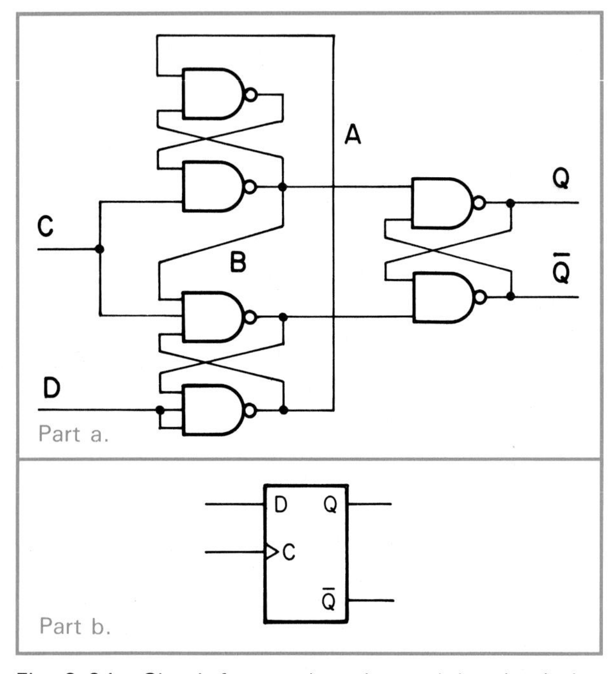

Fig. 9-21. Circuit for an edge-triggered data latch. Its symbol lacks a bubble, so it is triggered by a positive-going edge.

To simplify the circuit, RS flip-flop symbols will be substituted for the three NAND-based flip-flops. Part "a" of Fig. 9-22 reviews this substitution. Note that inputs R and S are active-low.

The analysis begins with the signals shown in part "a" of Fig. 9-23. Signals on the input and output leads are examined first. At the circuit's outputs, Q = 0 and $\overline{Q}$ = 1. A 0 must have been stored at some time in the past. Lead $\overline{Q}$ was drawn with a heavy line to emphasize the presence of a 1.

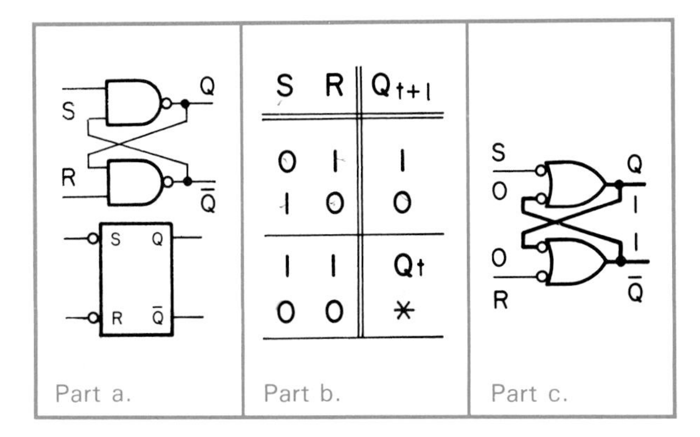

Fig. 9-22. This drawing provides a review of the action of an unclocked RS flip-flop with active-low inputs.

At the circuit's inputs, small timing diagrams have been drawn to show the history of the signals on these leads. Note that D was 0 when the clock last went active (refer to the positive-going edge of the pulse). At that instant, a 0 was stored. The clock then returned to 0.

Next, consider signals within the circuit. The 0 at C is applied to inputs on both flip-flops at the left (see inputs S2 and RB1). With 0s at their active-low inputs, 1s appear at the corresponding outputs (Q2 = 1 and $\overline{Q}$1 = 1). These 1s are, in turn, applied to R and S of flip-flop 3. To refresh your memory, see the truth table in part "b" of Fig. 9-22. With 1s at both active-low inputs, the number stored in the output flip-flop remains unchanged. This flip-flop might be thought of as being in standby.

A closer look at the circuit in part "a" of Fig. 9-23 suggests a problem. With C = 0 and D = 0, there are 0s at both inputs of flip-flop 1 (RB1 = 0, S1 = 0). The last line of the truth table describes this as undefined. Normally this input set is avoided, but this is one of the few circuits used in industry that permits 0s to appear at both active-low inputs at the same time.

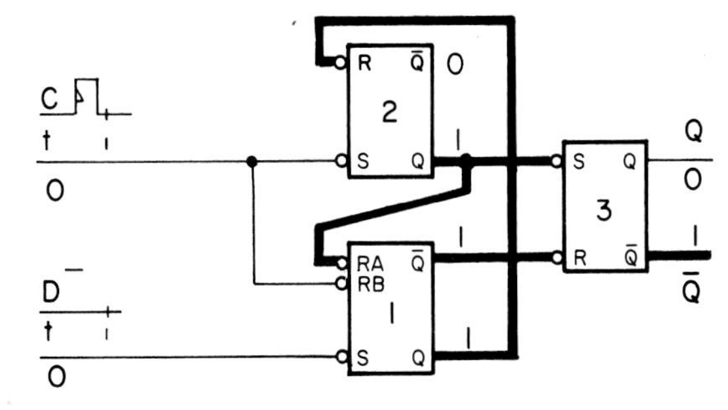

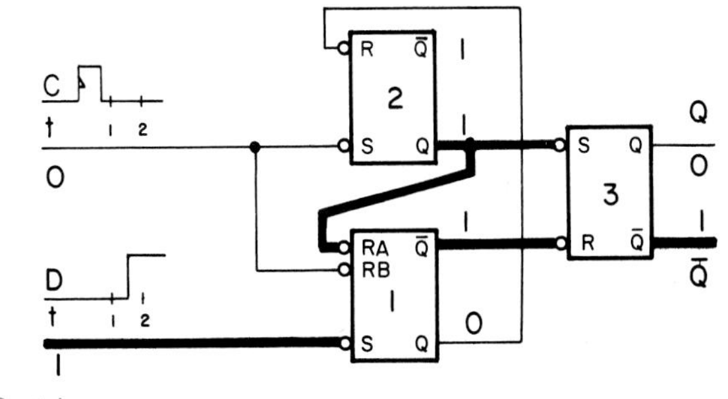

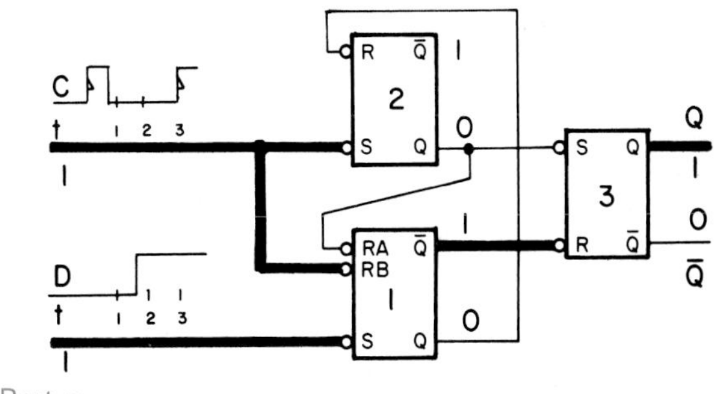

Fig. 9-23. To aid in the analysis of this edge-triggered data latch, the NAND-based flip-flops in Fig. 9-21 have been replaced with RS flip-flop symbols.

To determine the signals at Q1 and $\overline{Q}$1 of flip-flop 1, the NAND-based circuit is DeMorganed. See part "c" of Fig. 9-22. With 0s at R and S, both Q and $\overline{Q}$ output 1s. Normally this is unacceptable, but in this case, it does not harm the circuit's action.

In preparation for storing at 1, D is 1 at time t2 in part "b" of Fig. 9-23. Because C is still 0, 1s continue to appear at the inputs to flip-flop 3. The stored number cannot change.

When D changed to 1 (that is, S1 = 1), the undefined condition was removed from flip-flop 1. With RB1 = 0 and S1 = 1, this active-low flip-flop output a 1 at $\overline{Q}$1 and a 0 at Q1. However, the undefined condition moved to flip-flop 3. Again, this does not hurt the circuit's action.

Part "c" of Fig. 9-23 shows storing of a 1. At time t3, C went from 0 to 1. Both inputs to flip-flop 1 became 1s, so the number stored there cannot change (it moved to its standby mode). The 0 from Q1 of the lower flip-flop continues to appear at R2 of the upper flip-flop. As a result, flip-flop 2 reset (R2 = 0, S2 = 1, so Q2 = 0). The 0 at Q2 is applied to S (active-low) of the output flip-flop, so a 1 is stored. If this analysis were continued, it would show that no new number can enter this circuit until C goes from 0 to 1 again.

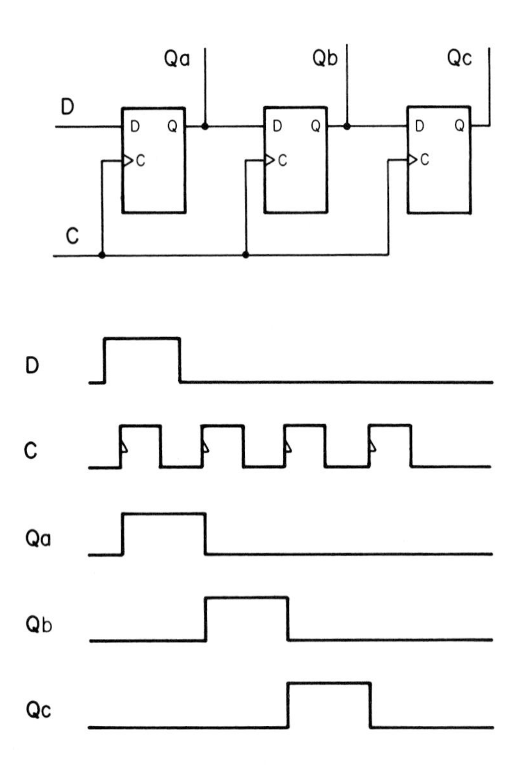

Fig. 9-24. The shift register shown uses edge-triggered data latches that respond to positive-going edges. Note the second flag and check the edge with a straightedge to compare it with Qa and Qb.

EDGE-TRIGGERED SHIFT REGISTER

Fig. 9-24 shows a shift register using edge-triggered data latches. The latches respond to a positive-going edge. Remember that these are master-slave flip-flops. It appears that Qa goes to 0 (at the right end of the pulse at Qa) at the same time that the clock goes active. However, there is enough delay that the 1 at Qa acts as an input for the next flip-flop. Check the edges with a straightedge to compare the flag section with Qa and Qb.

EDGE DETECTOR

The circuit in Fig. 9-25 might be used to detect negative-going edges. Note that the clock input is bubbled and edge-triggered. The flip-flop shown in Fig. 9-25 is triggered by negative-going edges.

Also note that D is permanently connected to a logic 1. Whenever C goes negative, a 1 will be input.

At t1 in the timing diagram, the negative-going edge at C stores a 1. The edge has been detected. At t2, CL was used to clear the flip-flop and prepare it for the next edge. The next edge arrived at t3. At t3, a 1 is stored (it reverses the state left by CLEAR). At t4, the flip-flop was not cleared before the next edge arrived, so the edge at t4 goes undetected.

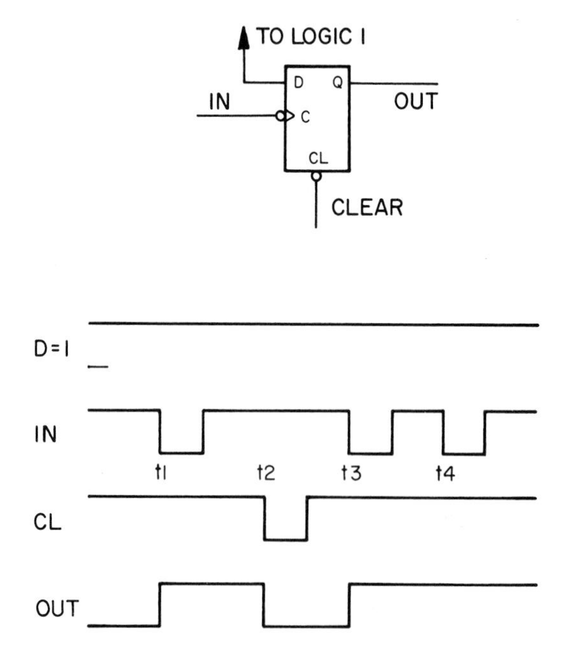

Fig. 9-25. The circuit shown detects negative-going edges on lead IN. The CLOCK is marked "IN." CLEAR is marked "CL." The clock at IN only operates on the negative-going edges when IN goes to active-low.

SUMMARY

Master-slave flip-flops permit the reading of old numbers at the same time new numbers are being input. Master-slave data latches and R-S flip-flops are used in shift registers and ring counters.

Preset and clear inputs are available on master-slave flip-flops. Also, their clocks may be level- or edge-triggered.

IMPORTANT TERMS

Asynchronous, Edge-triggered, Level-triggered, Parallel, Serial, Synchronous.

TEST YOUR KNOWLEDGE

1. Is the old number in a master-slave flip-flop available while a new number is being stored?
2. A number goes to a level-triggered master-slave data latch output when clock goes active (low). (True or False?)
3. A _________ register requires the use of master-slave data latches.
4. To obtain master-slave action, two _________ are connected one after the other.
5. For a clock of a master-slave flip-flop, both flip-flops are active at the same time. (True or False?)
6. List the application types for master-slave shift registers. See page 147.
7. Although it is convenient to transmit data in serial form, it is usually used in _________ form.
8. In serial-to-serial conversion using 4 latches, pulsing SHIFT _________(1, 2, 3, 4) times will output the stored number.
9. In serial-to-serial conversion using 4 latches, a number is entered after _________ (1, 2, 3, 4) SHIFT pulses.
10. A universal shift register can perform how many types of conversions?
11. _________ and CLEAR can be used to add a parallel input to a shift register.
12. In parallel-to-serial conversion, a parallel input on a shift register with PR and CL is used to enter a number under the control of _________.
13. A ring counter is a shift register with _________.
14. The ring counter circuit in Fig. 9-18 is a self-clearing circuit. (True or False?)
15. With level-triggered master-slave latches, which edge of the active half of clock pulse (leading, trailing) is more important?
16. In part "c" of Fig. 9-22, can both Q and $\overline{Q}$ be 1 at the same time?
17. Edge delay allows old numbers to be read from an edge-triggered data latch while new numbers are being input. (True or False?)
18. An edge-triggered data latch cannot be used in a shift register. (True or False?)
19. An edge (following a 1 at an input) in an edge detector goes undetected if the last state was not _________.
20. Edge-triggered data latches (see Fig. 9-24) can input data twice each clock cycle—once when their clocks go from 0 to 1 and again when they go from 1 back to 0. (True or False?)

STUDY PROBLEMS

1. Draw the circuit diagram for a NAND-based, master-slave data latch. Use eleven NANDs. Three are to be connected as NOTs. Place the NOT in the clock circuit so the clock will be active low. Include a $\overline{Q}$ output. Label the leads with the symbols D, C, Q and $\overline{Q}$.
2. Repeat Problem 1 for an R-S flip-flop. Use nine NANDs. Label the leads S, R, C, Q and $\overline{Q}$.
3. Trace the timing diagram in Fig. P9-1 onto a sheet of paper and construct the expected output. This is an ordinary data latch (not master-slave). Assume Q = 0 at t = 0.
4. For a second time, trace the timing diagram in Fig. P9-1 (from Problem 3) onto a sheet of paper and again construct the expected output. This time the data latch is to be master-slave and level-triggered.

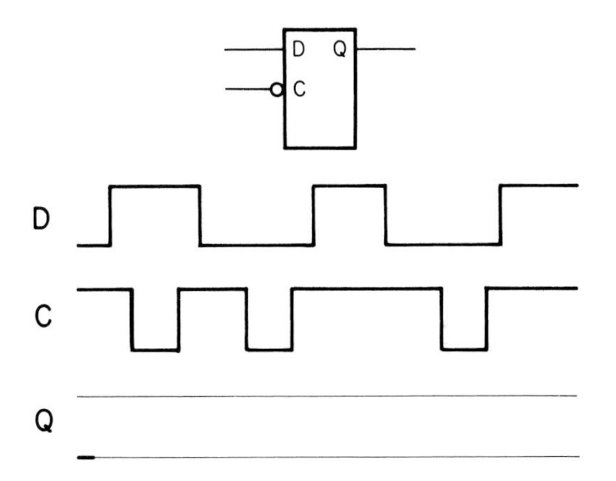

Fig. P9-1. Circuit helps compare the action of ordinary data latch with that of level-triggered master-slave data latch.

5. Trace the timing diagram in Fig. P9-2 onto a sheet of paper and construct its expected output. This is a master-slave, level-triggered data latch. Assume Q = 0 at t = 0.
6. Repeat Problem 5 for the ordinary R-S flip-flop shown in Fig. P9-3.
7. Trace the timing diagram in Fig. P9-4 onto a piece of paper and construct the expected signals at Qa and Q. The two flip-flops that make up this master-slave circuit are ordinary data latches. Assume Q = 0 at t = 0.
8. Trace the timing diagram in Fig. P9-5 onto a piece of paper and construct the expected output. The data latch is master-slave and level-triggered.

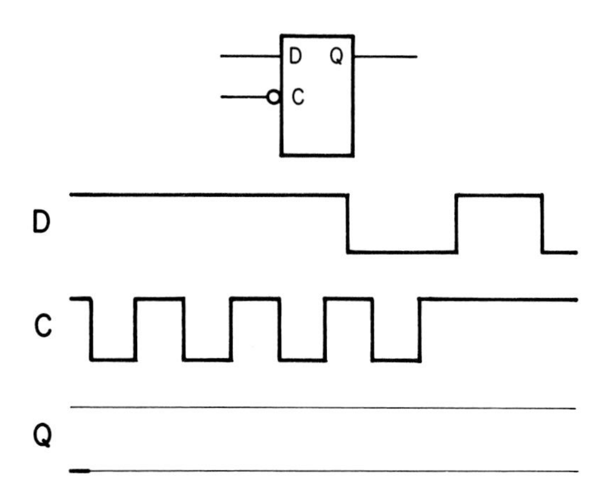

Fig. P9-2. Master-slave level-triggered data latch. Assume Q = 0 at t = 0.

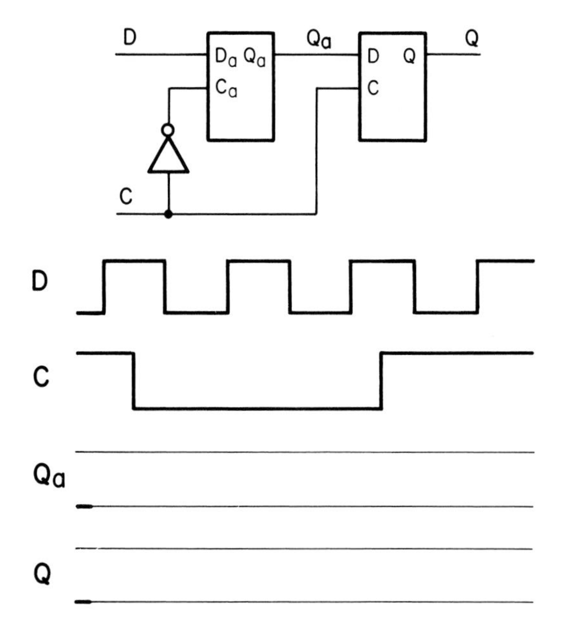

Fig. P9-4. The two outputs Qa and Q help show the action of the master-slave circuit made up of two ordinary data latches. Assume Q = 0 at t = 0.

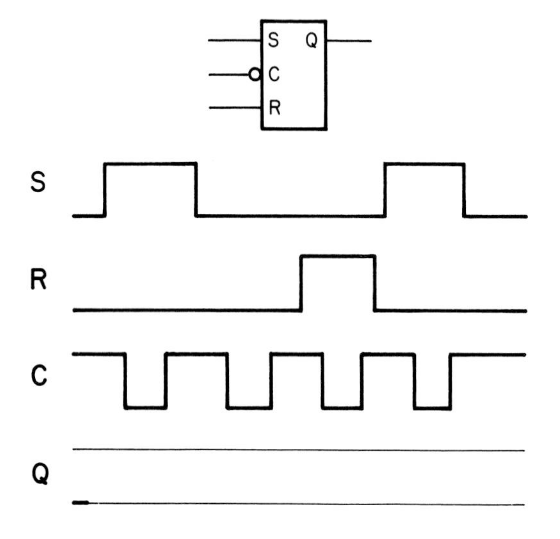

Fig. P9-3. Ordinary R-S flip-flop.

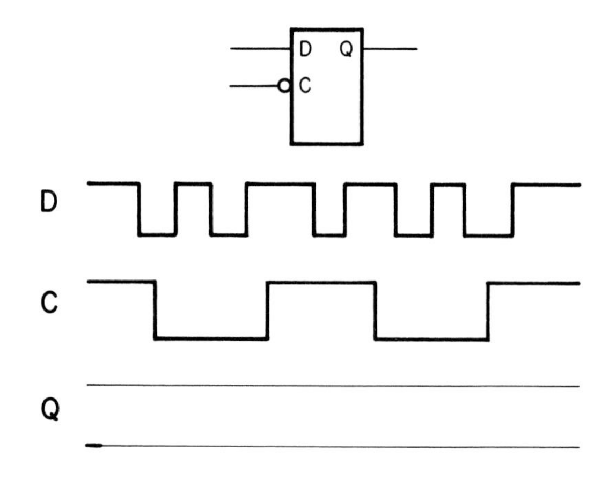

Fig. P9-5. Output can be drawn for the master-slave level-triggered data latch shown. Assume Q = 0 at t = 0. It may help to sketch the internal output of the first flip-flop of the master-slave unit.

Note that the clock is active for long periods.

9. Draw the circuit diagram for a 4-bit shift register. Use R-S flip-flops with active low clocks. Add a NOT to the first flip-flop to convert it to a data latch. Label the leads with the symbols IN, SHIFT, Q0, Q1, Q2, and Q3.
10. Trace the timing diagram in Fig. P9-6 onto a sheet of paper and construct the expected outputs. Assume all outputs are 0 at t = 0.
11. Repeat Problem 10 for the shift register and timing diagram shown in Fig. P9-7.
12. Trace the timing diagram in Fig. P9-8 onto a sheet of paper and construct the expected output.
13. A universal shift register is shown in Fig. P9-9.

Match the following functions with the letters on its leads. If more than one lead is needed to accomplish a given function, list all appropriate leads. A given lead may serve more than one function. Example: 1. UXY

1. Shift control.
2. Parallel input.
3. Serial input.

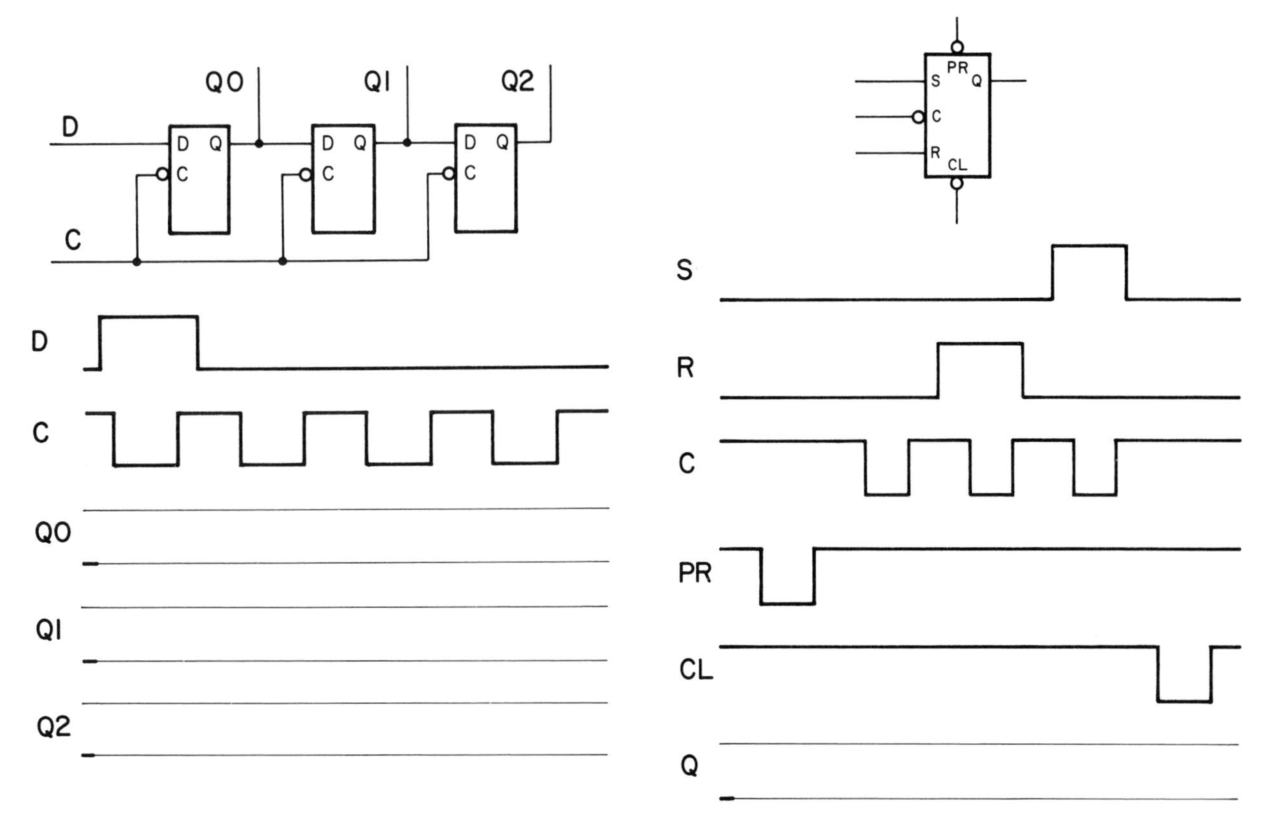

Fig. P9-6. Output can be drawn for 3-bit shift register shown.

Fig. P9-8. Output can be drawn for master-slave R-S flip-flop with preset and clear.

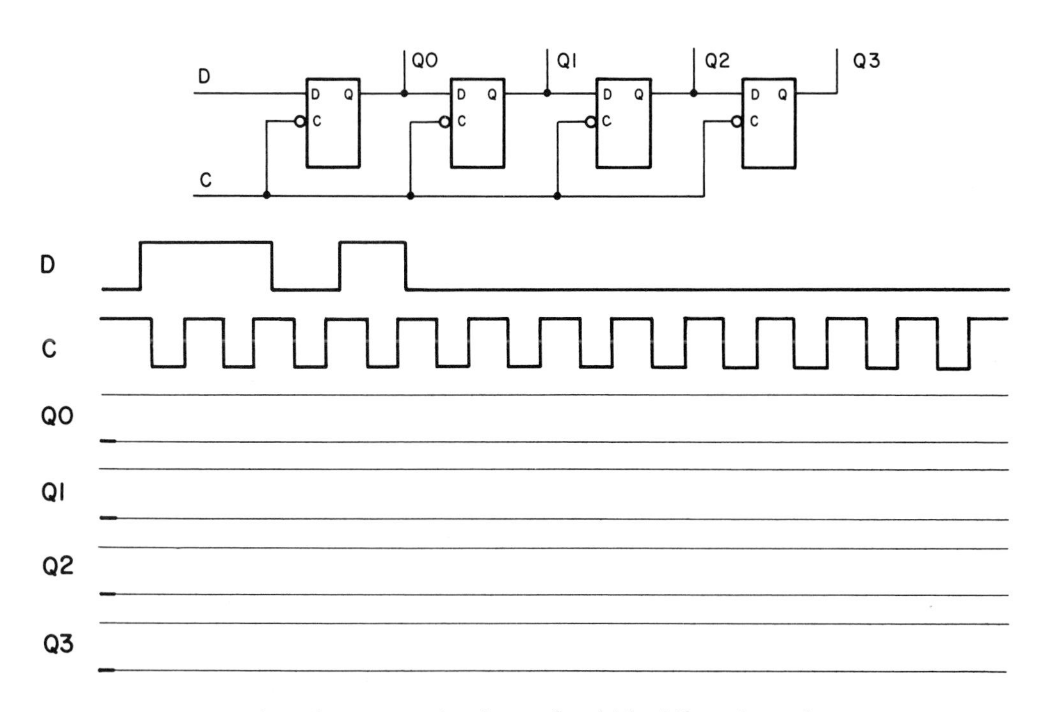

Fig. P9-7. Output can be drawn for 4-bit shift register shown.

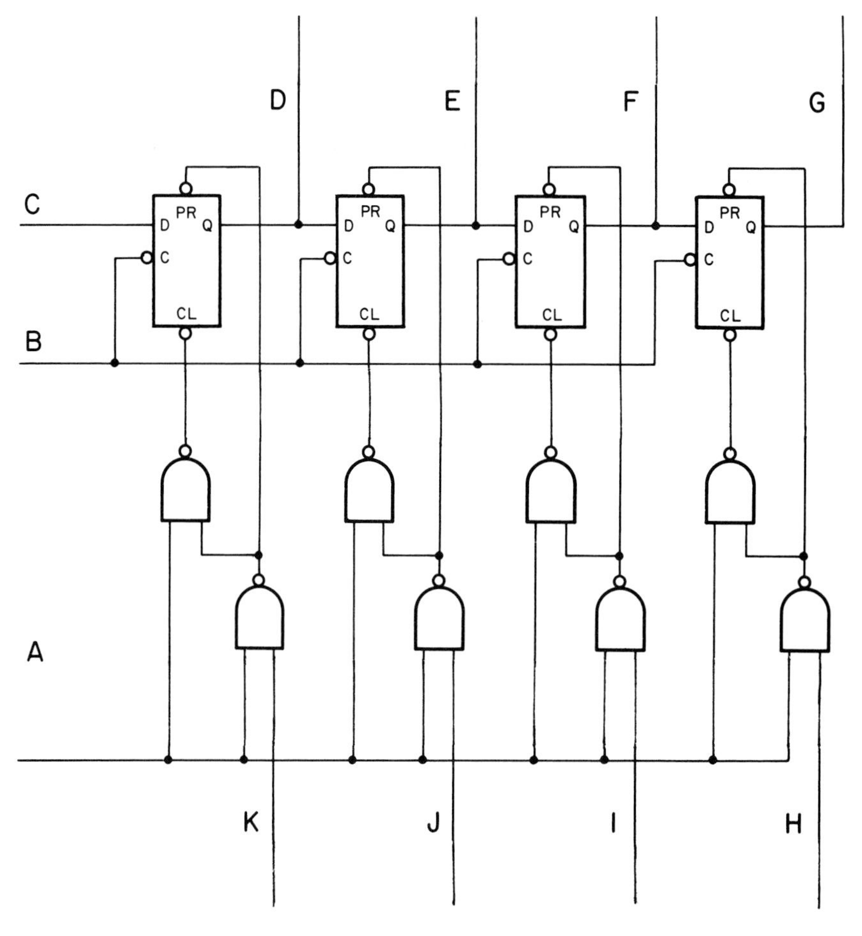

Fig. P9-9. Functions can be identified for leads in the universal shift register shown.

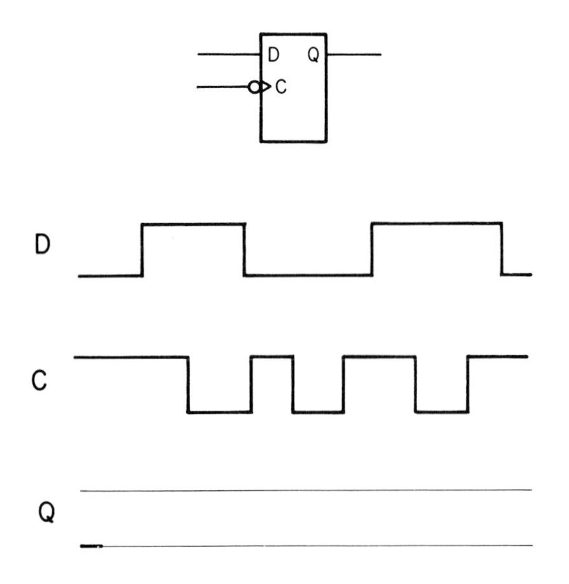

Fig. P9-10. Output can be drawn for edge-triggered data latch that responds to negative-going edges.

4. Parallel output.
5. Serial output.

14. Trace the timing diagram in Fig. P9-10 onto a sheet of paper and construct its expected output. The flip-flop is an edge-triggered data latch that responds to negative-going edges.
15. Trace the timing diagram of a shift register shown in Fig. P9-11 onto a sheet of paper and construct its expected outputs. Assume all outputs of the circuit are 0 at t = 0.
16. Trace the timing diagram for the ring counter shown in Fig. P9-12 onto a sheet of paper and construct its expected outputs. Assume all inputs to the 3-input NOR (Q0, Q1, and Q2) are 0 when the time is at t = 0.
17. Match the following circuits with the diagrams shown in Fig. P9-13. Example: 1. E
    1. Register.
    2. Ring counter.
    3. Shift register.

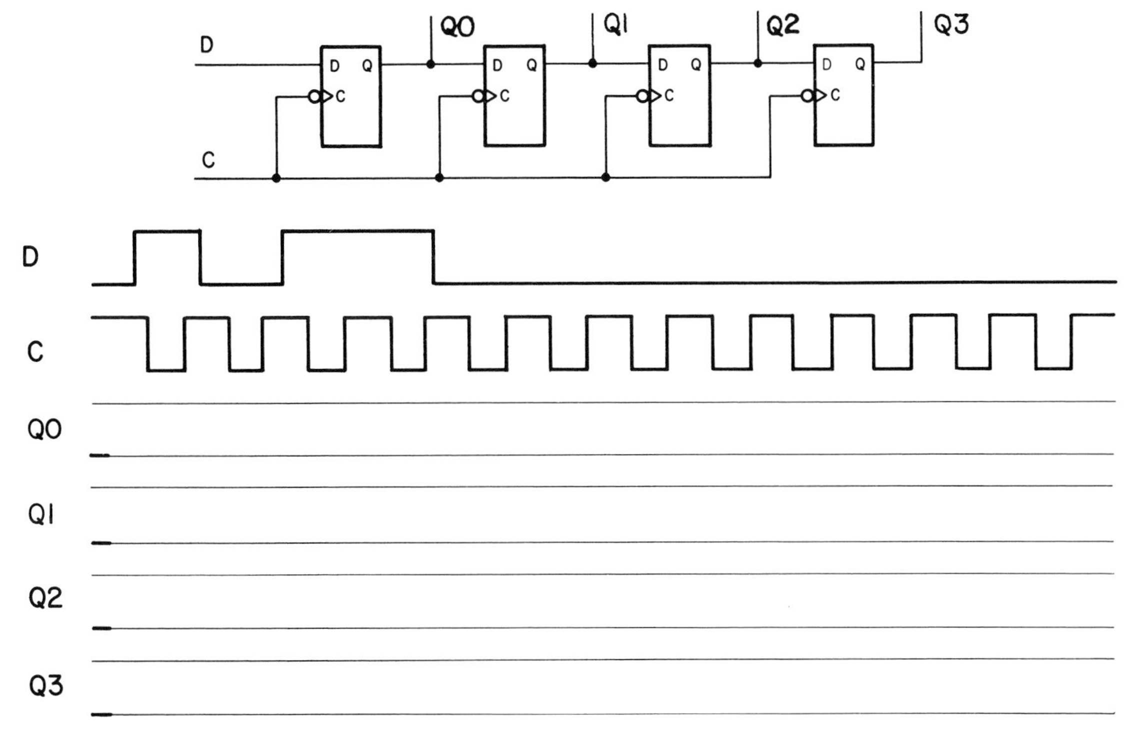

Fig. P9-11. Output can be drawn for 4-bit shift register shown.

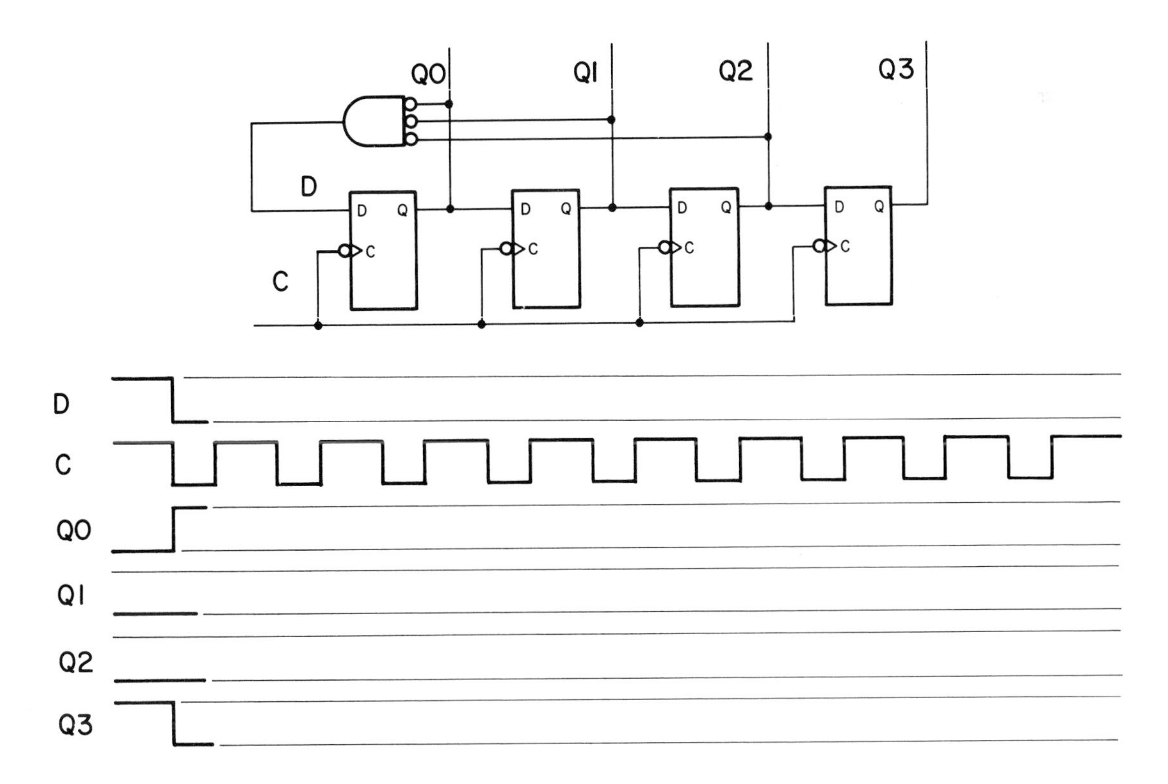

Fig. P9-12. Output can be drawn for self-clearing ring counter shown. Assume all inputs to the 3-input NOR (Q0, Q1, and Q2) are 0 at t = 0.

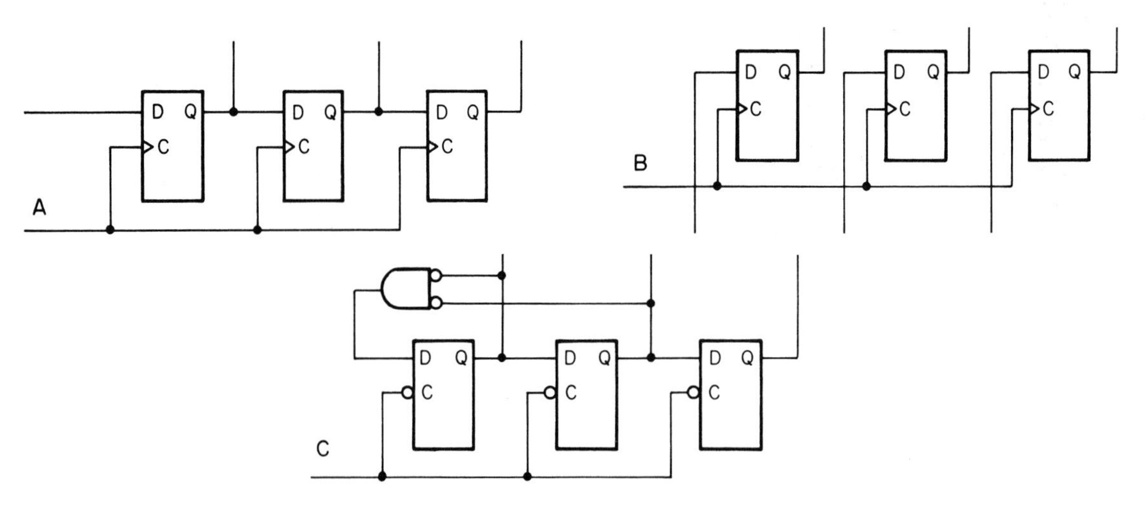

Fig. P9-13. Matching of circuits to their names helps clarify their uses.

Toggling Flip-Flops

Chapter 10

LEARNING OBJECTIVES

After studying this chapter and completing lab assignments and study problems, you will be able to:

- Draw the symbol and construct the truth table for J-K flip-flops.
- Draw the circuit for a NAND-based master-slave level-triggered J-K flip-flop.
- Complete timing diagrams for level- and edge-triggered J-K flip-flops.
- Draw the symbol and complete the timing diagram for a toggle flip-flop.
- Draw the circuits and predict the maximum counts and frequency divisions for ripple counters.
- Follow and explain descriptions of the actions of preset and up/down counters.
- Recognize and list differences in the outputs of synchronous and asynchronous counters.
- Describe problems created by bounce and draw a circuit for debouncing a mechanical switch.
- Draw the symbol and describe the action of a Schmitt trigger.

The term *TOGGLE* implies a device with two stable states. In the general sense, all flip-flops are toggling devices. In digital circuits, however, toggling flip-flops are those with the ability to *COMPLEMENT* their outputs on command. The word complement means that if the output of a toggling flip-flop is 1, it will become a 0 when pulsed; if it is 0, it will become a 1.

This chapter deals with the actions and applications of toggling flip-flops. The usefulness of toggling flip-flops in the design of circuits without noise is emphasized.

J-K FLIP-FLOP

The J-K flip-flop is a universal flip-flop, since it can be converted into almost any other type of flip-flop. It is an R-S flip-flop with toggling capabilities.

REVIEW OF R-S FLIP-FLOP

Fig. 10-1 reviews the action of a clocked R-S flip-flop. Qt + 1 implies the value of Q after the clock has cycled. Qt is the value of Q just before the clock was applied. With 1s at both inputs, the output of the device cannot be predicted. It is up to circuit designers to avoid this input set.

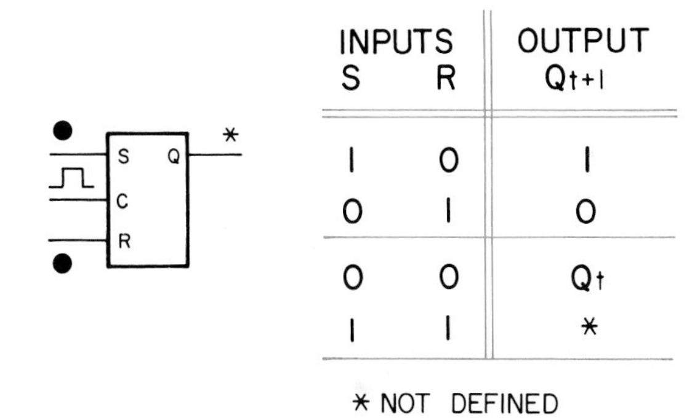

INPUTS S	INPUTS R	OUTPUT Q_{t+1}
1	0	1
0	1	0
0	0	Q_t
1	1	*

* NOT DEFINED

Fig. 10-1. When the clock of an R-S flip-flop is active, 1s should not be applied to both inputs. The result is an undefined output.

J-K FLIP-FLOP ACTION

Fig. 10-2 shows the symbol and truth table for a J-K flip-flop. The first three rows are identical to those of an R-S flip-flop. The difference is found in the last row. It is permissible to apply 1s to both inputs of a J-K flip-flop.

When the clock is activated and 1s are at J and K, the output of this device toggles (complements itself). That is, $Qt+1 = \overline{Q}t$. If Q = 1 just before the clock goes active, it will be 0 when the clock returns

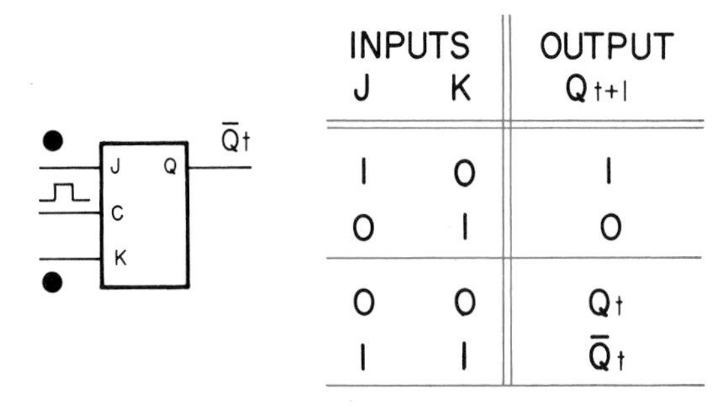

INPUTS J	INPUTS K	OUTPUT Q_{t+1}
1	0	1
0	1	0
0	0	Q_t
1	1	$\bar{Q}_t$

Fig. 10-2. A J-K flip-flop toggles when 1s are applied to both inputs.

to its inactive state; if Q = 0, it will be 1 when the clock returns.

J-K FLIP-FLOP CIRCUIT

Fig. 10-3 shows a NAND-based circuit for a J-K flip-flop. It is a master-slave level-triggered R-S flip-flop with feedback added. Refer to the cross-coupled leads marked A and B.

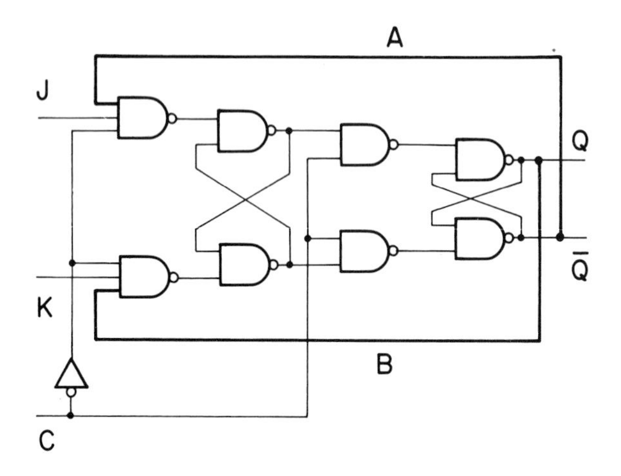

Fig. 10-3. When feedback leads A and B are added to an R-S master-slave flip-flop, a J-K flip-flop results.

Fig. 10-4 shows the R-S equivalent of a J-K flip-flop. With 1s applied to J and K, the circuit is in its toggling mode.

The 0 at $\bar{Q}$ is applied to the upper AND, so S must equal 0. The 1 at Q is applied to the lower AND. This 1 combines with the 1 at K, so R = 1. When the clock goes active (to 0), the 1 at R enters the master

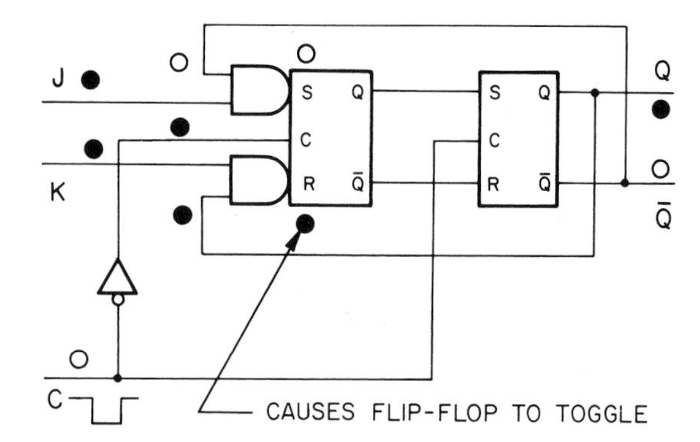

Fig. 10-4. Two R-S flip-flops can be used to construct a J-K flip-flop.

flip-flop. When the clock returns to its inactive state, this stored number is passed to the slave and the circuit's outputs. Q will go to 0; $\bar{Q}$ will go to 1. The outputs have toggled.

The action when Q = 0 is similar. Due to the cross coupling, the 0 will be applied to the lower AND, and the 1 at $\bar{Q}$ will appear at the upper AND. When the clock is activated, the flip-flop will set (Qt + 1 will equal 1). When the clock goes inactive, Q of the slave flip-flop goes to 1. The device has again toggled.

J-K FLIP-FLOP CLOCK TRIGGERING MODES

J-K flip-flops are available with level- and edge-triggered clocks. Signals can be output either when the clock returns to inactive or just after the clock goes active.

LEVEL-TRIGGERED FLIP-FLOPS

The NAND-based circuit just described is level triggered. For the position of the NOT shown, it is active-low.

Fig. 10-5 shows the symbol and timing diagram for such a device. For the clock pulses at t1, t2, and t3, its action is identical to that of a level-triggered, R-S flip-flop. Numbers enter when the clock goes active; they appear at the output when the clock returns to its inactive level.

At t4 and t5, the device toggles. This is the action of a J-K flip-flop with 1s at J and K. Note that the result of the toggling appears at the output when the clock returns to its inactive level.

EDGE-TRIGGERED FLIP-FLOPS

Fig. 10-6 shows the symbol and timing diagram of an edge-triggered J-K flip-flop. Except for the tim-

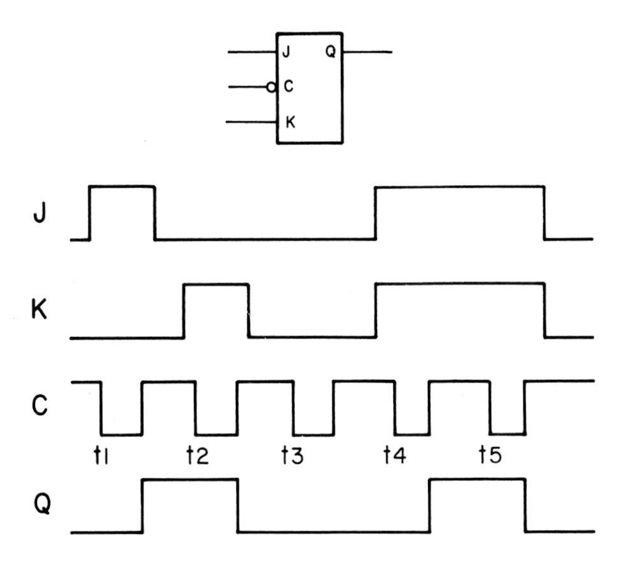

Fig. 10-5. Data enters this level-triggered flip-flop when its clock goes active. That data appears at the output when the clock returns to its inactive state.

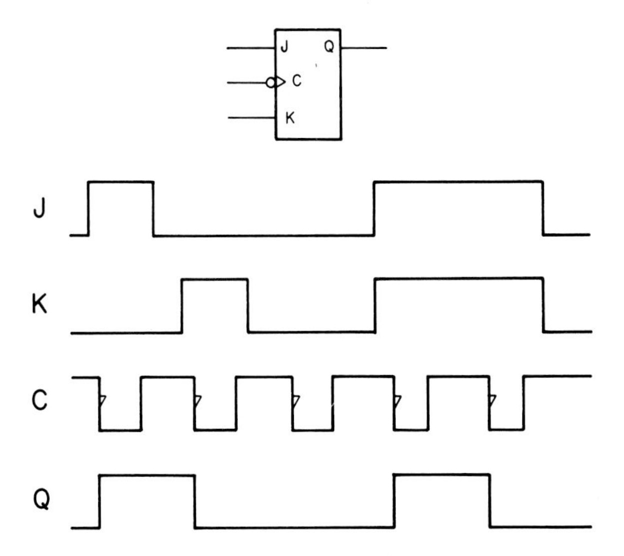

Fig. 10-6. Data enters and appears at the output of this edge-triggered flip-flop at the negative-going edge of the clock signal.

ing of the output, this diagram is identical to that of the level-triggered device. Rather than waiting for the clock to return to its inactive level, new numbers appear after a very short delay.

Although the output appears to occur at the same time the clock goes active (at its negative-going edge), this is a master-slave device. Numbers can be read at its output at the same time new numbers are entered.

TOGGLE FLIP-FLOP

In many applications, only the toggling ability of J-K flip-flops is used. The symbol in Fig. 10-7 emphasizes this situation. It is assumed that 1s are permanently applied to J and K, and C becomes the only input. C is renamed T on such symbols.

Each time T goes active, Q toggles. Refer to the timing diagram in Fig. 10-7.

J-K flip-flops can be changed into toggle flip-flops by tying J and K to logic 1s. See Fig. 10-8. If TTL chips are used, permitting these inputs to float is often sufficient.

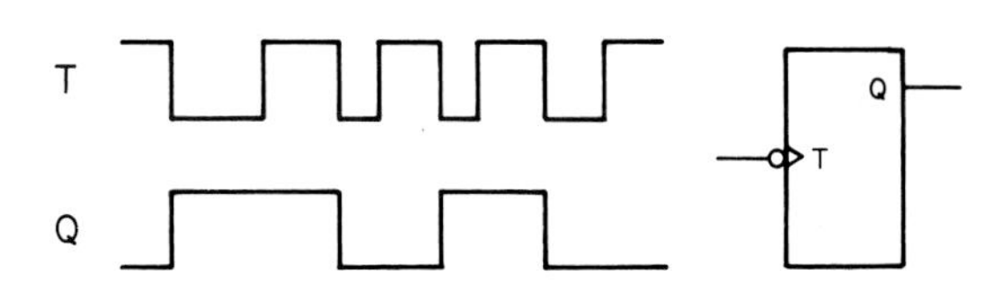

Fig. 10-7. Each time the clock (T) of this flip-flop goes active, its output toggles.

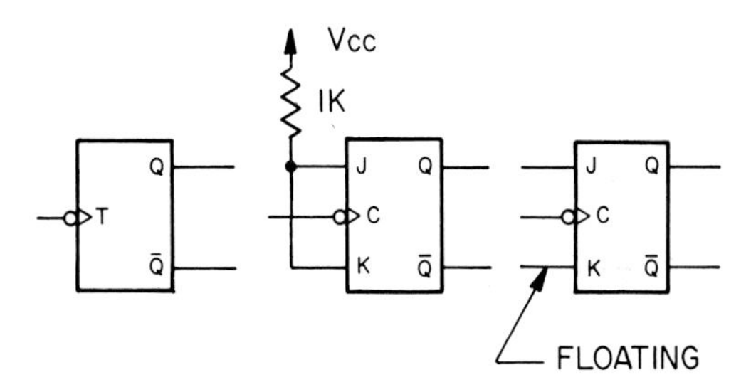

Fig. 10-8. This illustration shows three ways of obtaining toggle action.

J-K FLIP-FLOP APPLICATIONS

Applications of J-K flip-flops may be divided into two groups:

1. Being a universal flip-flop, J-K flip-flops can be used in place of data latches and R-S flip-flops.
2. Their ability to toggle permits their use in unique circuits.

J-K FLIP-FLOPS USED AS DATA LATCH AND R-S FLIP-FLOP

In a properly designed digital circuit, 1s will normally not appear at both inputs of an R-S flip-flop at the same time. See the last line of the R-S flip-flop table in Fig. 10-9. Because this signal set is avoided in circuits containing R-S flip-flops, J-K flip-flops can often serve as substitutes for R-S units. That is, the signal set that causes J-K flip-flops to toggle are just not present in R-S based circuits. Three equivalencies of J-K flip-flops to circuits based on R-S flip-flops are shown in Fig. 10-9. In Fig. 10-10, J-K flip-flops have been subtituted for R-S flip-flops in a shift register.

While J-K flip-flops can often substitute for R-S elements, the opposite is often not acceptable. Circuits using J-K flip-flops often utilize the toggling ability of these flip-flops. R-S flip-flops cannot toggle.

J-K flip-flops can also be substituted for data latches. If a NOT is placed across the inputs of a J-K flip-flop, 1s cannot appear at J and K at the same time, so the flip-flop's toggling feature will never function.

S	R	Q_{t+1}
1	0	1
0	1	0
0	0	Q_t
1	1	*

J	K	Q_{t+1}
1	0	1
0	1	0
0	0	Q_t
1	1	$\bar{Q}_t$

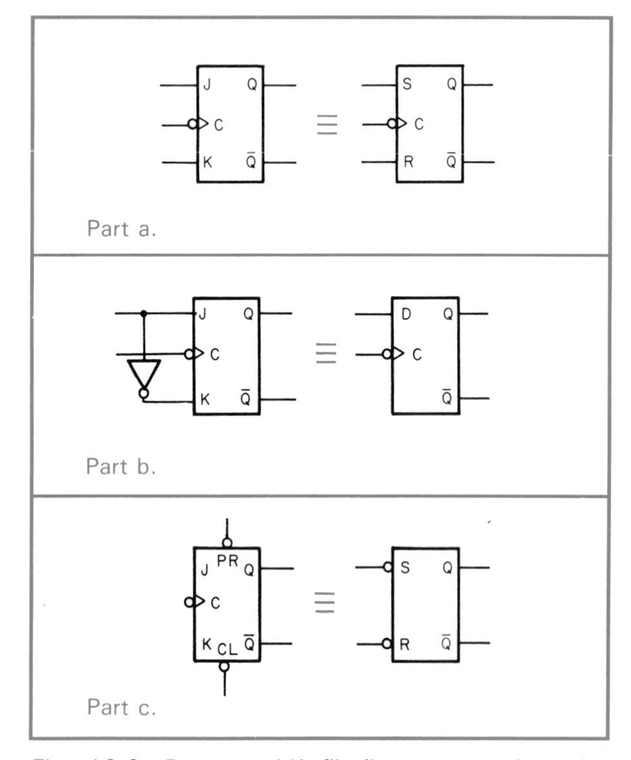

Fig. 10-9. Because J-K flip-flops can replace the elements shown, they are called universal flip-flops.

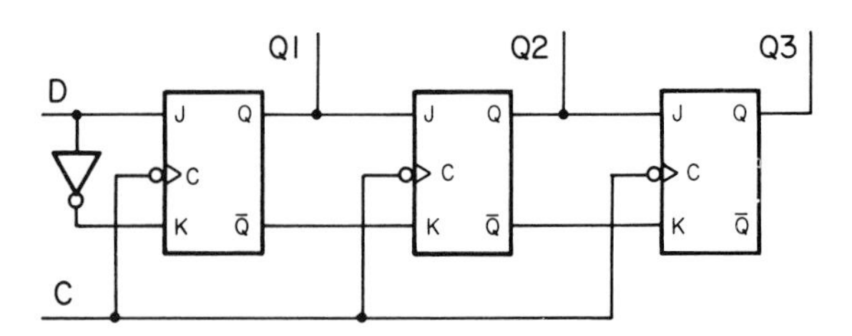

Fig. 10-10. J-K flip-flops have been used in place of R-S units in the shift register shown. Note that the circuit design does not allow J and K to be 1 at the same time (especially note J and K for the second and third flip-flops).

TOGGLE FLIP-FLOPS IN RIPPLE COUNTER

Some circuits are useful for counting. Fig. 10-11 shows a type of circuit that is called a *RIPPLE COUNTER*. Pulses applied to IN are tallied, and the total appears in binary form at the group of three outputs for the circuit.

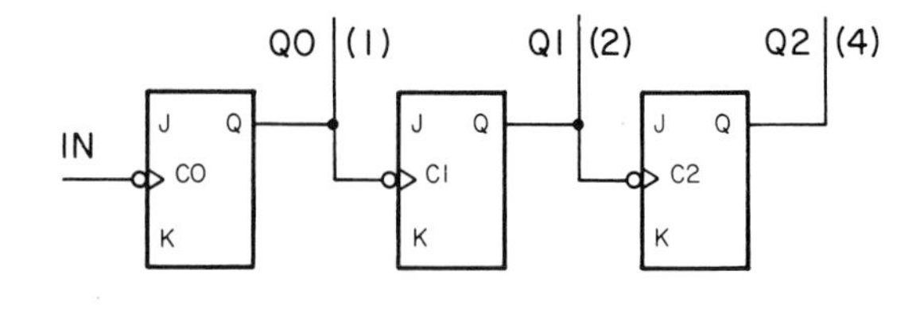

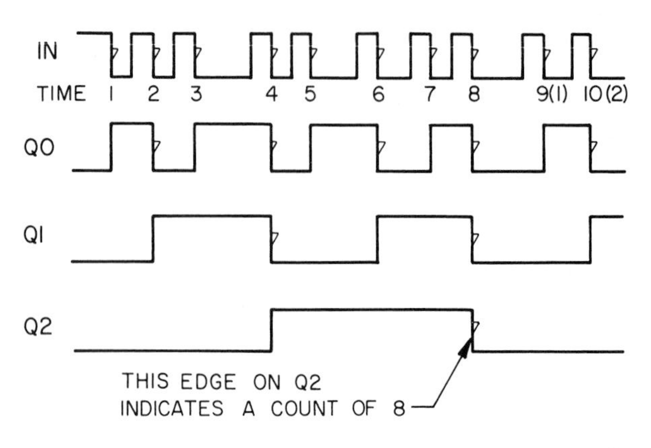

Fig. 10-11. The ripple counter shown can count to 8. The count begins at Q2Q1Q0 = 000 and goes to 111 (decimal 7). On count 8, it again outputs 000.

Ripple counters are widely used. They can be used to count the number of parts produced by a machine. Ripple counters are so fast that they are used to count the number of cycles per second which occur in high frequency ac waves.

Ripple counters use the toggling abilities of J-K flip-flops and toggle flip-flops. As a result, the signal path shown in Fig. 10-11 is from Q to C (or T). Each time a signal at input C goes negative, a flip-flop toggles.

RIPPLE COUNTER CIRCUIT ACTION

In the timing diagram shown in Fig. 10-11, the graph of IN represents a series of pulses to be counted. The flip-flops have negative-going, edge-triggered clocks, so the circuit really counts trailing edges. See the arrows.

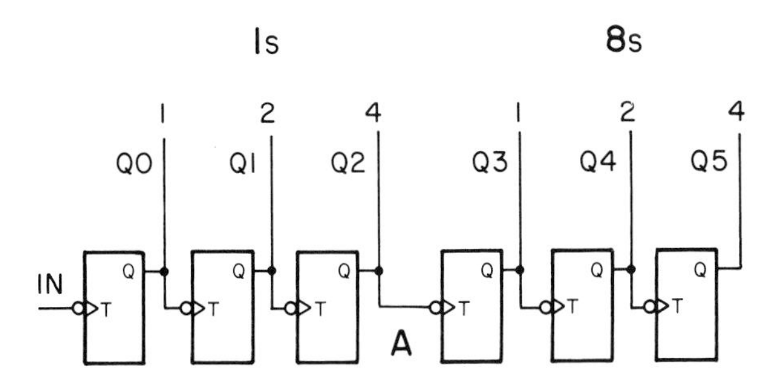

Fig. 10-12. The second three bits of this ripple counter count the negative-going edges at point A.

When the first negative-going edge is applied to C0, this flip-flop toggles. A 1 appears at output Q0.

At t2, a second negative-going edge is applied to C0, so the first flip-flop again toggles. This time, its output goes from 1 to 0, so a negative-going edge is applied to C1. This causes the second flip-flop to toggle (Q1 goes from 0 to 1).

At t3, Q0 again toggles. Remember, these flip-flops have negative edge-triggered clocks. The transition of Q0 from 0 to 1 does not activate C1. Three negative edges have been applied, and the circuit's output is Q2Q1Q0 = 011 (or decimal 3).

At t4 the fourth negative-going edge is applied. Q0 toggles from 1 to 0. The resulting edge causes Q1 to toggle from 1 to 0. This negative-going edge causes Q2 to toggle, and the circuit's output becomes Q2Q1Q0 = 100 (a decimal 4).

Follow the action of the counter until t8 is reached. An orderly progression of pulses occurs.

RIPPLE ACTION

Just before t8 in Fig. 10-11, 1s appear at all outputs (111 binary equals 7 decimal). The negative-going edge at t8 causes the first flip-flop to toggle. When its output goes from 1 to 0, another negative-going edge is created. It causes the second flip-flop to toggle from 1 to 0. This edge causes the last one to toggle. The result is Q2Q1Q0 = 000.

Note how the toggle action ripples through the circuit. The ripple action that occurs in the circuit is the source of the term ripple counter. It also slows the counter's action. Each count must pass through all flip-flops before a correct count can be output.

Some theory is needed on the following:

1. Octal counter.
2. Hexadecimal counter.
3. Maximum count.

The first topic tells how to recognize a complete cycle, as in the following:

1. Octal counter: The term *OCTAL* implies 8. Because it can count to binary 8, this circuit is often called an octal counter.

Just before t8 in Fig. 10-11, 1s appear at all outputs (111 binary equals 7 decimal). On the next count, Q2Q1Q0 = 000. This suggests that an output of 000 can be thought of as a 0 or an 8.

On the eighth count, Q2 outputs a negative-going edge. This edge may also be used to represent a total of 8. That negative edge could cause some action to be taken or it could drive another counter. See Fig. 10-12. Each time the first counter (first group of three) reaches 8, a negative-going edge appears at A. That edge causes the second counter (second group of three) to add 1 to its total. Counter one counts up to 8; counter two keeps track of the number of eights.

2. Hexadecimal counter: Fig. 10-13 shows a 4-digit ripple counter. When the eighth count arrives, it does not clear. Rather, a 1 appears at Q3. Sixteen pulses are needed to go from 0000 to 0000. The counting system using 16 as a base is called *HEXADECIMAL*.

On the sixteenth count all outputs go to 0. The negative-going edge at Q3 marks this maximum count.

3. Maximum count: Because ripple counters display their counts in binary, place values of outputs are easily determined. To determine the maximum count possible with a given counter, 1 is added to the sum of the available place values. See part "a" of Fig. 10-14. With 1s at its three outputs, the count is 111 (binary) or 7 (decimal). Because 000 is a valid number, the maximum count occurs when the counter returns to 000. That is, 111 + 1 = 1000 (binary) or 7 + 1 = 8 (decimal).

Part "b" of Fig. 10-14 shows the maximum count method applied to a 6-bit counter. Here, 63 + 1 = 64. The maximum count is determined to be 64 from the method.

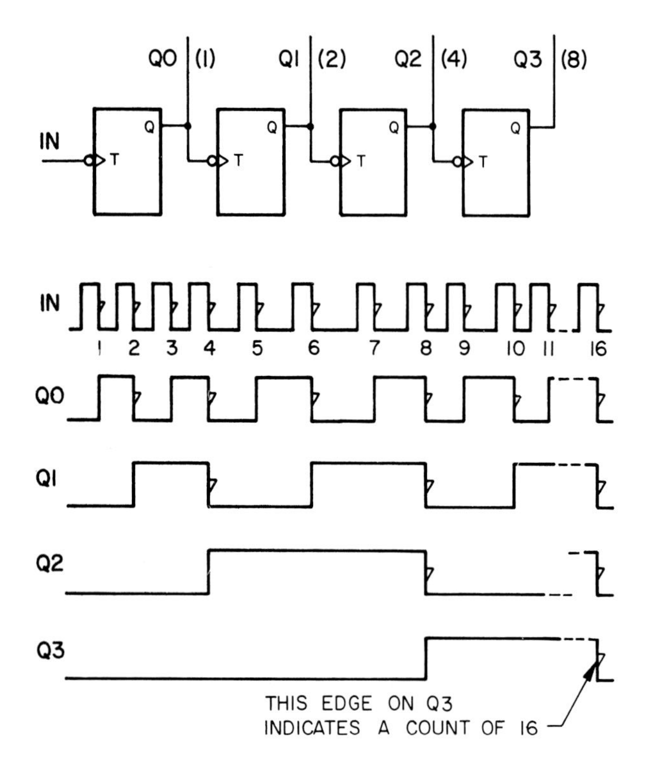

Fig. 10-13. The use of four toggle flip-flops results in a maximum count of 16.

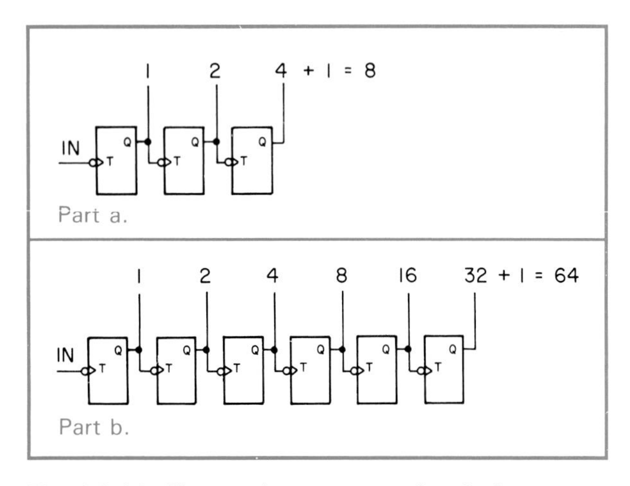

Fig. 10-14. The maximum count of a ripple counter equals the sum of its place values plus 1.

Part "a" of Fig. 10-15 shows a quicker way of finding maximum count. When the 4-bit circuit in part "a" of Fig. 10-15 reaches its maximum count, Q2 outputs a negative-going edge. If you imagine one more flip-flop (see the dashed lines), it will toggle in response to this edge. The place value of this imaginary flip-flop is the maximum count of the circuit.

Determine the maximum count of the circuit in part

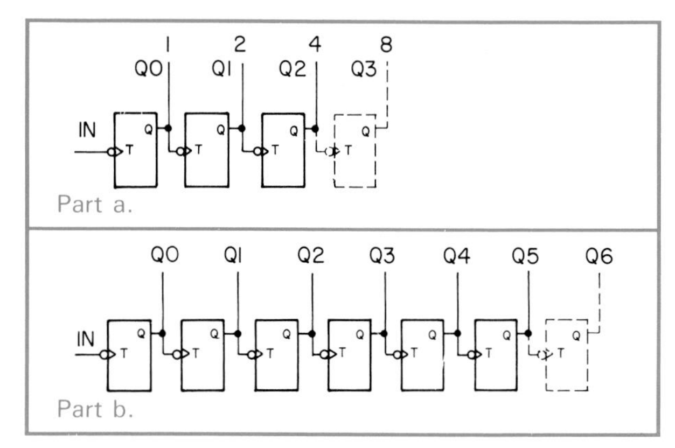

Fig. 10-15. The maximum count of a ripple counter can be found by imagining an additional bit. The maximum count of the counter equals the place value of that added bit.

"b" of Fig. 10-15. If you find the maximum count to be 64, you are correct.

The counters in Fig. 10-16 can be used to practice reading counter outputs. You may find the process easier if you start at the MSB and work toward the LSB. The correct counts are indicated at the right of each counter circuit drawing.

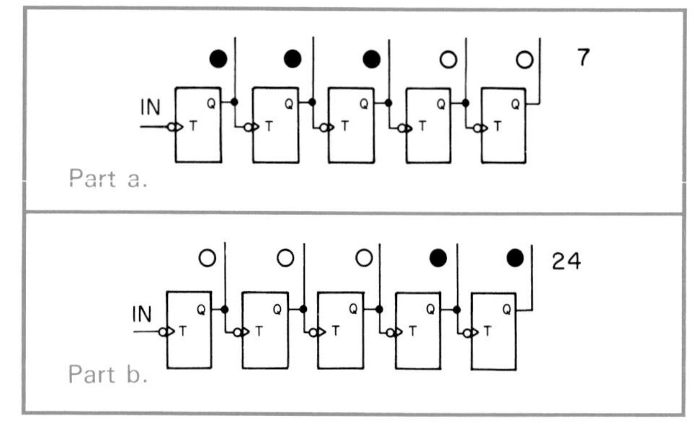

Fig. 10-16. These counters are used for practice in reading ripple counters.

TROUBLESHOOTING COUNTERS

Like other logic elements, J-K flip-flops and toggle flip-flops tend to fail in the s-a-0 and s-a-1 conditions. Ripple counters containing such faults will usually fail to count above some fixed number. That number is helpful in localizing and isolating the fault.

Counters also suffer from a unique problem called *BOUNCE.* This problem is not in the counter, but in its driving circuit.

The ripple counter shown in Fig. 10-17 counts the closings of the switch. Each time the switch grounds input T, a negative-going edge results. The counter counts these edges.

The timing diagram suggests a problem. At t1, the circuit should have output a 1. Instead it indicated a count of 3. At t2 it made an even greater mistake. This is a negative-going, edge-triggered circuit. Yet it added 1 to its count on a positive-going edge. Bounce in the mechanical switch caused this problem.

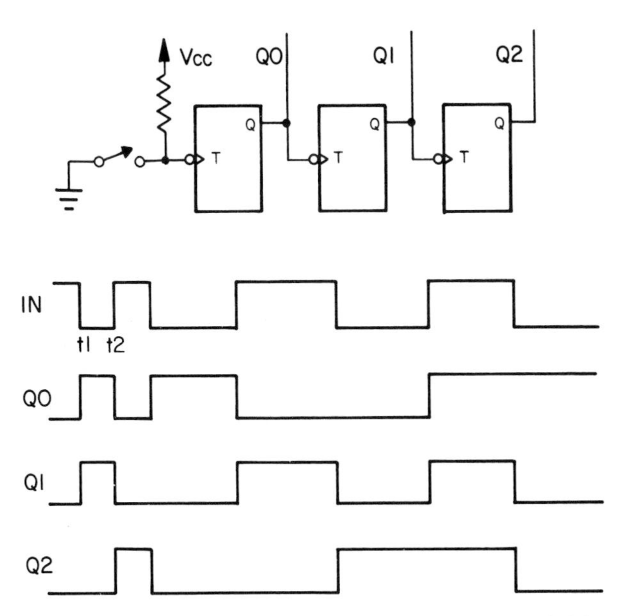

Fig. 10-17. Due to contact bounce, the output of this counter is meaningless.

BOUNCE IN SWITCHES

When mechanical switches close, they appear to make firm and permanent contact. In reality most switch contacts bounce. Contact is made, and current starts to flow. Then the contacts bounce, and the circuit is opened for an instant.

Switch bounce is too fast to be seen without the aid of high-speed motion pictures. In most applications it causes little difficulty. However, modern digital counters are so fast that they can count individual bounces. If, upon closing, a switch bounces three times, a counter will record a count of 3 rather than the intended single count.

Fig. 10-18 shows the electrical result of switch bounce. Instead of a single negative-going edge, the closing of the switch produced three.

Electrical noise is also produced when a switch opens, Fig. 10-18. Note the negative-going edge when the contacts open.

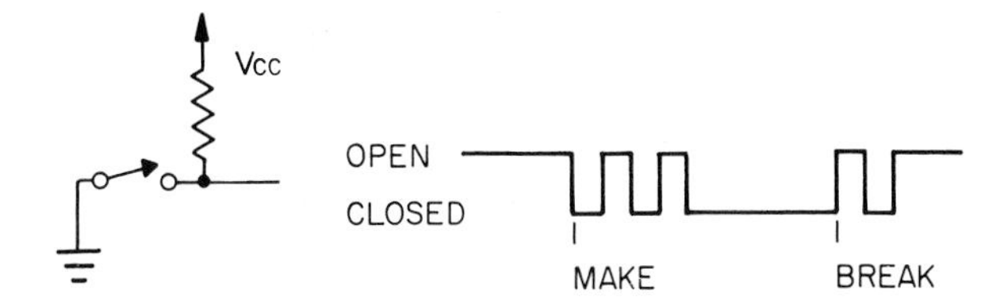

Fig. 10-18. Contact bounce occurs on both make and break.

DEBOUNCING METHODS

Errors resulting from switch bounce cannot be corrected by changing chips. Because most mechanical switches bounce, changing switches is seldom effective. Usually debouncing circuits must be placed between switches and counters.

R-S FLIP-FLOP DEBOUNCER

An unclocked R-S flip-flop is the basis for a widely used debouncer, part "a" of Fig. 10-19. Note the bubbled inputs. Also, the switch is a single-pole, double-throw (SPDT) type of switch.

The timing diagram shows the result of moving the switch blade from R to S. As the blade moves away from R, there is bounce. See graph R. However, Q

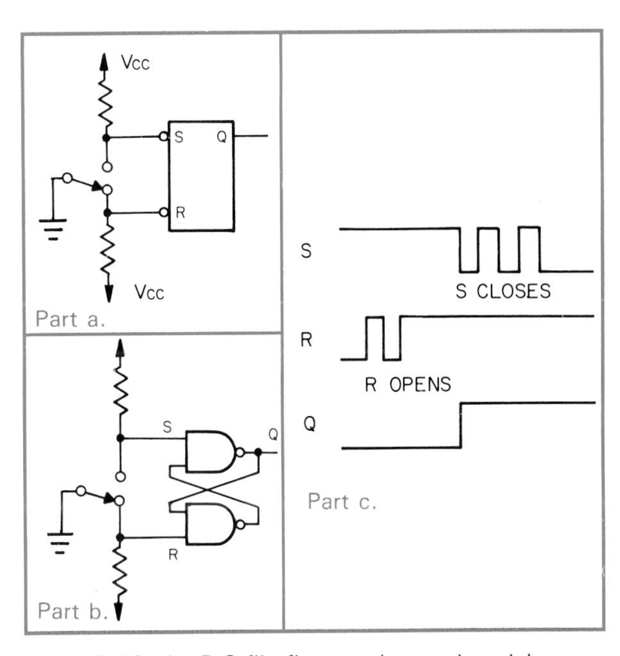

Fig. 10-19. An R-S flip-flop can be used to debounce an SPDT mechanical switch.

is already 0, so the regrounding of R does not change the output.

When the blade reaches and grounds S, Q becomes 1. Again the regrounding of S does not change the output. The output of the flip-flop is a clean transition from 0 to 1.

Although it is widely used, the circuit shown in Fig. 10-19 has a disadvantage. It requires an SPDT switch. The following circuit permits the use of an SPST (single-pole, single-throw) switch.

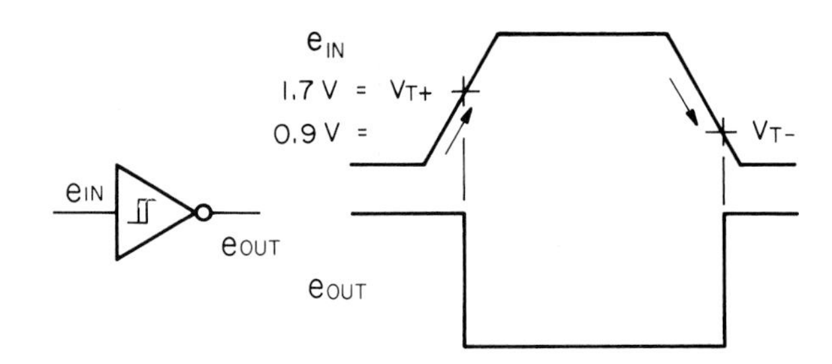

Fig. 10-21. The level at which a Schmitt trigger switches depends on whether the input signal is increasing or decreasing.

SCHMITT-TRIGGER DEBOUNCER

The circuit in Fig. 10-20 uses two new components:
1. A Schmitt trigger.
2. A capacitor.

In the Schmitt-trigger debouncer circuit, the Schmitt trigger is a NOT, but it has the characteristic of quickly going from one logic level to the other when certain signal levels are exceeded. The capacitor has the ability to temporarily store electricity.

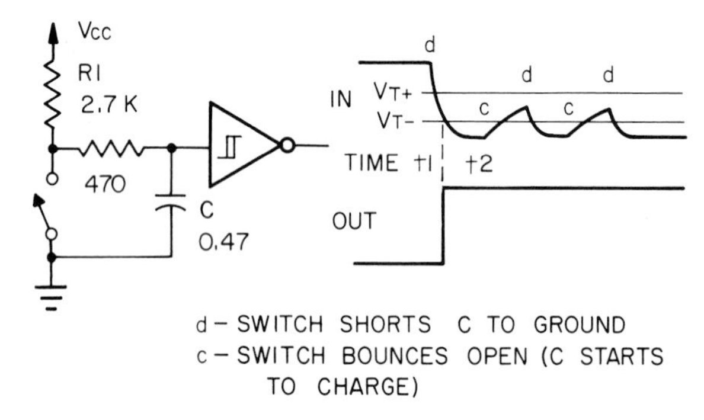

Fig. 10-20. A Schmitt trigger can be used to debounce an SPST mechanical switch.

Fig. 10-21 shows the characteristic response curve of a Schmitt-trigger circuit. As the input voltage increases, its output remains unchanged until a point called the *THRESHOLD* is reached. Then the output quickly changes. The threshold for increasing voltages is usually at about 1.7 volts for TTL elements and is designated VT+.

For negative-going input signals, there is a different threshold. It is usually at about 0.9 volts and is often designated VT−.

Because VT+ and VT− differ, a Schmitt trigger is said to have *HYSTERESIS*. That is, it follows different curves when coming and going. A gear train that has backlash is said to have hysteresis. When you reverse such a gear train, you have to take out the backlash before the gears will turn in the opposite direction. The little curve in the logic symbol of a Schmitt trigger represents a hysteresis curve.

Return to Fig. 10-20. Component C is a capacitor. It can temporarily store electricity. As with any storage container, it takes time to fill or empty a capacitor.

When the switch shorts the input of the Schmitt trigger to ground at t1, the charge on the capacitor is quickly drained through the 470 ohm resistor to ground. This lowers the voltage at the input of the Schmitt trigger, and its output goes to 1.

At t2, the switch has bounced open, and the capacitor starts to charge. Because a fairly large resistor (2,700 ohms) is in series with Vcc, this takes a comparatively long time. Before the capacitor can charge to the Schmitt trigger's threshold, the switch again closes and returns the voltage to 0. As a result, the output of the circuit is a clean transition from a signal level of 0 to a signal level of 1.

Schmitt triggers can be used to "clean up" noisy pulses. See Fig. 10-22. When the input voltage reaches VT+, the output switches. It remains a 1 until the input voltage falls below VT−. Note that at A, the input voltage fell below VT+, the positive-going

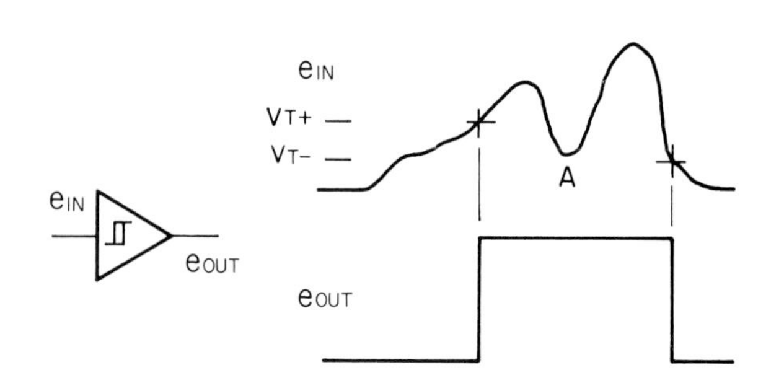

Fig. 10-22. Schmitt triggers can be used to remove noise from signals. The voltage at A can be less than VT+ but not less than VT−.

threshold. This did not change the output, since the input voltage must fall below the negative-going threshold for the output to switch back to 0. Note in Fig. 10-22 that no bubble is used on the Schmitt trigger symbol. Both kinds of Schmitt trigger are available. An external NOT can be used on the Schmitt trigger of Fig. 10-20 and Fig. 10-21.

Schmitt triggers can be used to convert sine waves into pulses, Fig. 10-23. The output of the circuit is rectangular rather than square. That is, it is off longer than it is on, but in most applications this imbalance is not a problem.

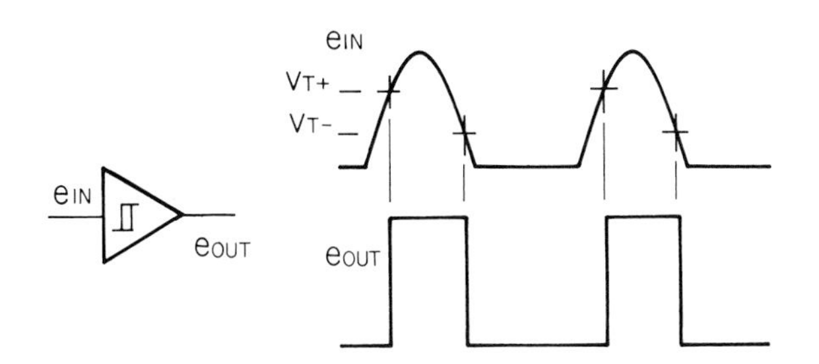

Fig. 10-23. Schmitt triggers can be used to change sine waves into pulses. The sine wave is first put through a half-wave rectifier.

PRESET COUNTER

When the circuit in Fig. 10-24 reaches its maximum count (decimal 16), Q3 outputs a negative-going edge. That edge could be used to cause some action. For example, when enough parts have been placed in a box, that edge could signal the need to remove the box and position an empty box.

The counters studied so far have a major limitation. They can count only to 2, 4, 8, 16, 32, etc. To permit other maximum counts, they must be modified.

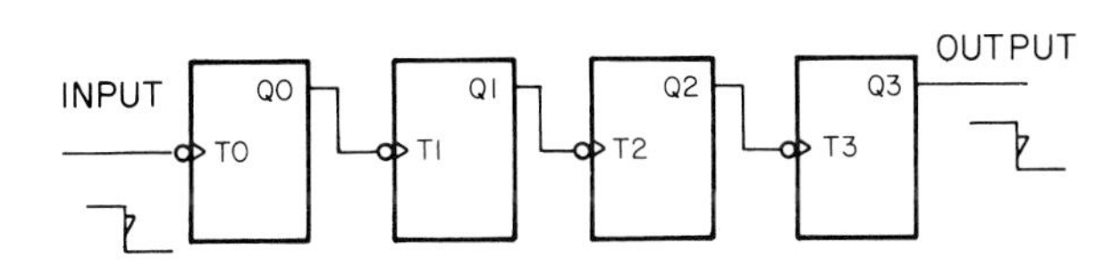

Fig. 10-24. On the sixteenth input pulse, the circuit made of four toggle flip-flops outputs a negative-going edge.

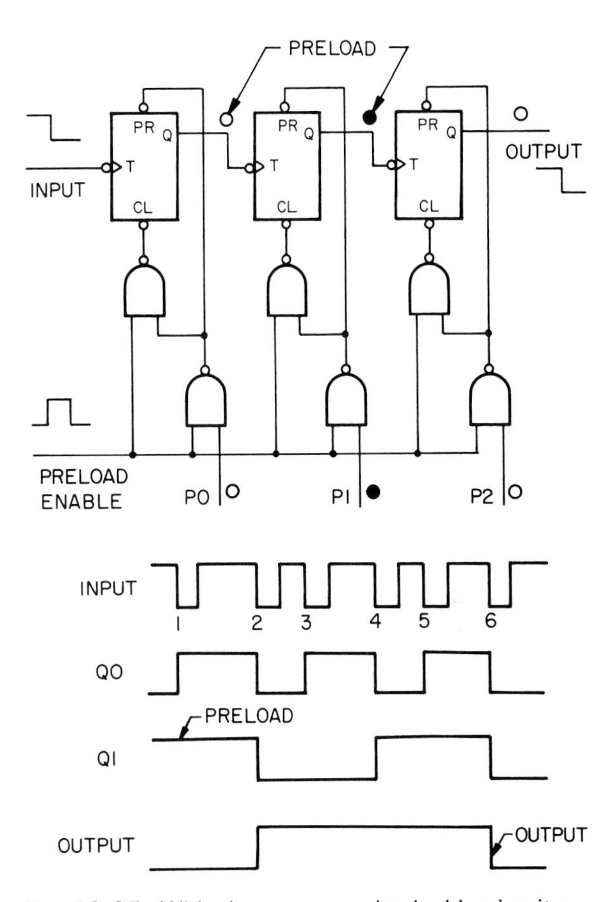

Fig. 10-25. With the proper preload, this circuit can count to any number between 1 and 8.

PRESETTING A COUNTER

By inputting the proper number through P2P1P0, the counter in Fig. 10-25 can have any maximum count between 1 and 8. The count is accomplished by preloading the difference between the desired count and 8. For example, if a count of 6 is required, 2 would be loaded. Before the count starts, Q2Q1Q0 would equal 010. As a result, only six counts are needed to bring it to 000 and produce a negative-going edge at Q2.

The fact that the difference between the desired count and the maximum count must be computed and entered as a preset is a disadvantage. The following circuit begins to overcome this problem, but a MOD counter is often a solution.

DOWN COUNTER

When a pulse is applied to an ordinary counter, 1 is added to the stored number. When a down counter is used, each pulse subtracts 1. To accomplish this, T inputs in down counters are connected to $\overline{Q}$ rather than Q. See Fig. 10-26.

DOWN COUNTER CIRCUIT ACTION

The action of the circuit for the down counter in Fig. 10-26 is best analyzed in two steps. First, the timing diagram based on $\overline{Q}$ is constructed. Then, the signals at Q2, Q1, and Q0 are found by complementing $\overline{Q}2$, $\overline{Q}1$, and $\overline{Q}0$.

Assume the counter is cleared at t = 0. That is, Q2Q1Q0 = 000 and $\overline{Q}2\overline{Q}1\overline{Q}0$ = 111. Refer to the timing diagram in Fig. 10-26.

When the first negative-going edge is applied to T0, $\overline{Q}0$ toggles (goes from 1 to 0). This creates a negative-going edge and $\overline{Q}1$ toggles (goes from 1 to 0). This causes the last flip-flop to toggle ($\overline{Q}2$ goes from a level of 1 to a level of 0).

The rest of the timing diagram for the NOTed outputs can be developed like that of any ripple counter. Watch the following:

1. The action of the negative edges from the output of $\overline{Q}0$ on the input at T1.
2. The action of negative edges from $\overline{Q}1$ on the input at T2.

Toggling occurs at negative edges, as expected. The resulting changes in the outputs allow a sequence of counting down in three digits.

The graphs for Q2, Q1, and Q0 are easily determined. They are merely the complements of $\overline{Q}2$, $\overline{Q}1$, and $\overline{Q}0$. At t = 0, Q2Q1Q0 equals 000. This can represent a count of 0 or 8. On the first count, Q2Q1Q0 changes to 111 (decimal 7). That is, 000 − 1 = 111, because 000 can equal 8. On the next count, Q2Q1Q0 = 110 (decimal 6). Again 1 has been subtracted from the stored number. Additional pulses will count the circuit down to 000.

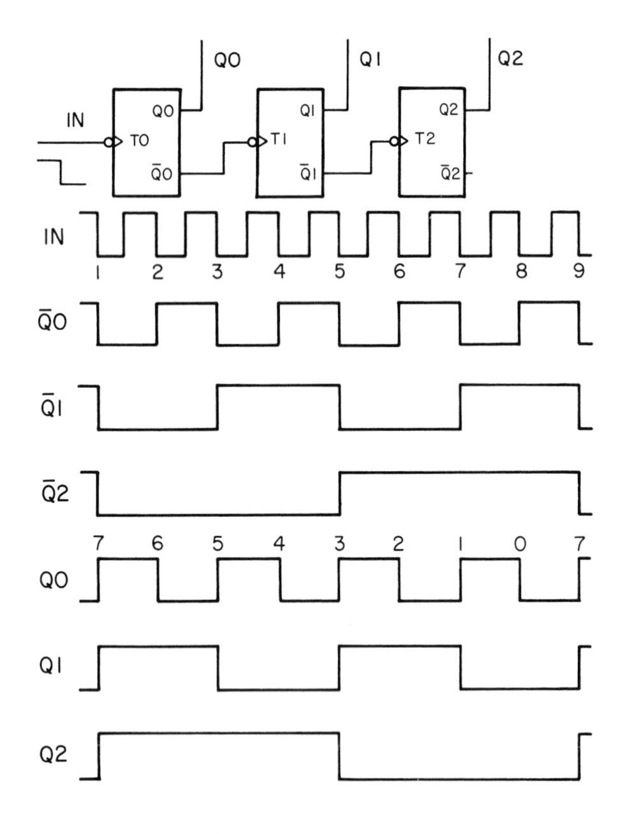

Fig. 10-26. Because its T inputs are connected to $\overline{Q}$s, the circuit shown is a down counter. Notice the output numbers above Q0. Starting at Q2Q1Q0 = 000, input is active 8 times to count down from 000 through 111, 110, 101, 100, 011, 010, 001, to 000. More stages can be connected at $\overline{Q}2$.

DOWN COUNTER WITH SPECIAL OUTPUT CIRCUIT

If the down counter in Fig. 10-26 is to be cascaded with other counters (see Fig. 10-12), it must output a negative-going edge when it reaches 000. That edge acts as the input to the next counter. In circuits considered up to this point, this edge could be obtained from Q2. In this circuit, however, Q2 is already 0 when 000 is reached. Q0 outputs a negative-going edge at 000, but it outputs many such edges and cannot be used. The solution is shown in Fig. 10-27. When Q2Q1Q0 goes to 000, the OR will output the required negative-going edge.

DOWN COUNTER WITH PRELOAD

Fig. 10-28 shows a down counter with a preload circuit. It counts down by starting from any number between 1 and 8.

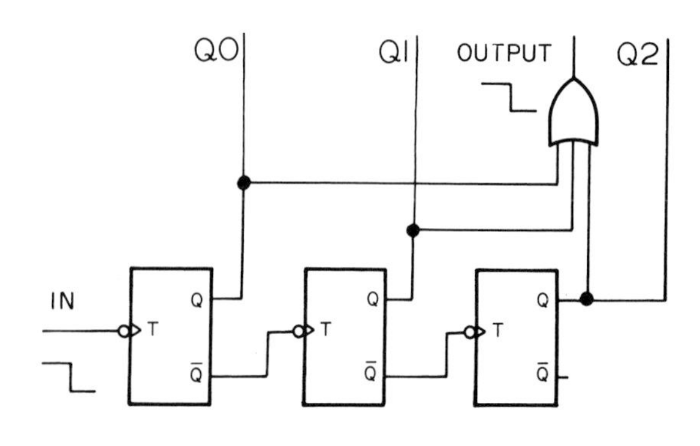

Fig. 10-27. When this down counter reaches 000, it outputs a negative-going edge.

UP/DOWN COUNTER

Fig. 10-29 shows an UP/DOWN counter. Note the coupling between the flip-flops. When the upper AND-gates are open (a 0 applied to DOWN/UP), it is an up counter. When the lower AND-gates are open (a 1 is applied to DOWN/UP), it is a down counter.

The notation at the DOWN/UP control lead suggests the signals that must be applied to obtain each function. Because DOWN is above the line, a 1 results in a down counter. With UP below the line, a 0 changes this circuit into an up counter.

A CLEAR lead has been provided. Note the NOT. When CLEAR is 1, 0s are applied to the CL (bubbled) leads, and all outputs are forced to 0.

ADVANTAGES AND DISADVANTAGES OF PRESET COMPARED TO MOD COUNTER

The use of preset or preloads to determine the maximum count of a counter has advantages and disadvantages. The preset arrangement permits the count to be changed easily. Merely changing the preset changes the maximum count. On the other hand, the preset must be re-entered before each count begins. For repetitive counts, the preset entry may be very inconvenient (60 presets per second, for example).

If the maximum count is fixed, other counters are usually used. A MOD counter is most often used and avoids the error-prone situation of entering presets.

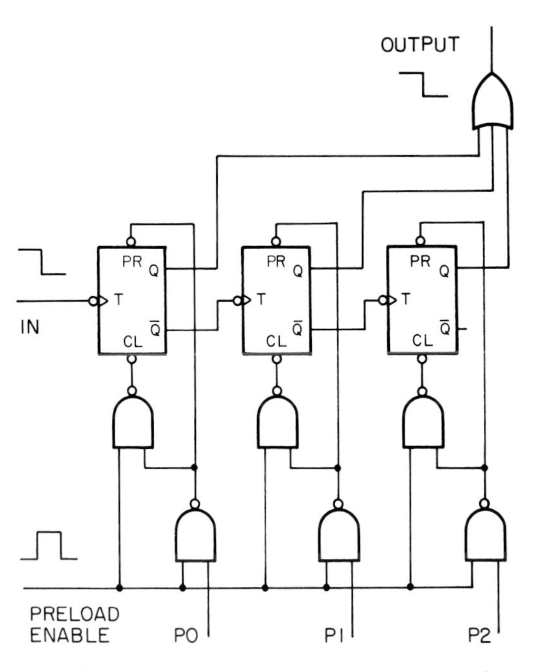

Fig. 10-28. This down counter can be preset for any count between 1 and 8. Preset is sometimes called preload.

MOD COUNTER

The number of counts required for a counter to go from 0 to 0 is referred to as its *MODULUS* (mod). As indicated, the moduli (plural of modulus) of basic ripple counters must be 2, 4, 8, 16, etc. Counters with other moduli are often referred to as mod counters. Mod counters are used when the maximum count is a fixed (unchanging) number.

DECADE COUNTER

The output of the counter in Fig. 10-30 is in binary form. However, when it reaches decimal 10, it clears. That is, on the count after Q3Q2Q1Q0 = 1001, its output goes to 0000. It is called a mod-10 or DECADE counter.

The circuit is a ripple counter with feedback added. This causes the circuit to clear when the count reaches decimal 10. Note the use of the CL leads.

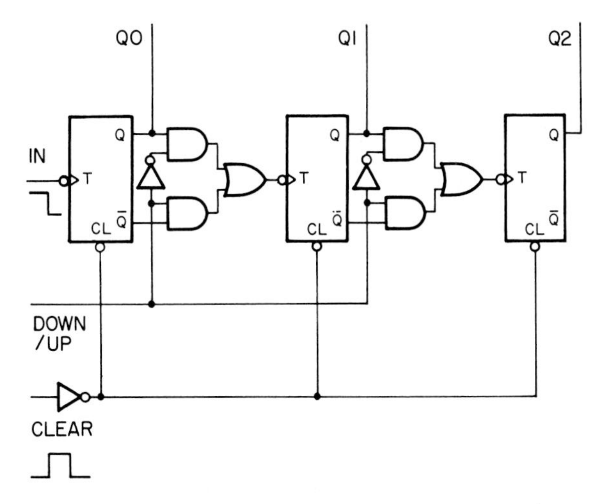

Fig. 10-29. The circuit shown can count up or down.

DECADE COUNTER CIRCUIT ACTION

Up to count 9, the decade counter circuit shown in Fig. 10-30 functions like an ordinary ripple counter. During these counts, Q1 and Q3 are never 1 at the same time. Therefore, the output of the NAND stays 1, and the CLs are not activated.

On the tenth count, however, Q1 and Q3 are both 1 for a very short time. See graphs Q1 and Q3. This

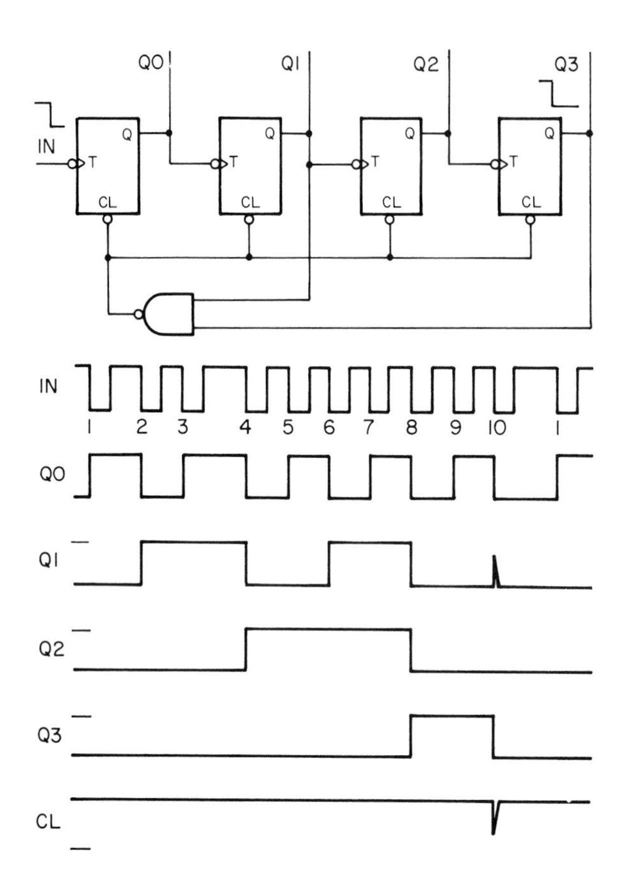

Fig. 10-30. When this counter reaches 10 (decimal), it automatically clears. Clearing occurs when Q1 = Q3 = 1. Negative-going signal on CLEAR does the clearing process. Because of the clear, Q1 drops down.

causes the output of the NAND to go to 0 and clear the flip-flops. Refer to the graph of CL.

Note the spike on the graph of Q1. As soon as Q1 goes to 1, this flip-flop is immediately cleared by CL. This spike is so fast that it may be difficult to find even with a high-speed oscilloscope. Even so, this is a major disadvantage of the decade counter circuit. The spike may function as a glitch and cause problems in reading data, writing data, or timing a circuit.

COUNTS HIGHER THAN TEN

To count higher than ten, additional decades can be added to a decade counter. See Fig. 10-31. Each time the ones counter reaches ten, it outputs a negative-going edge. These are counted by the tens counter. In turn, the tens counter outputs negative-going edges when it reaches ten counts. These are recorded by the hundreds counter.

Each decade outputs a binary number between 0 and 9. As a result, it is easy to read this total, since

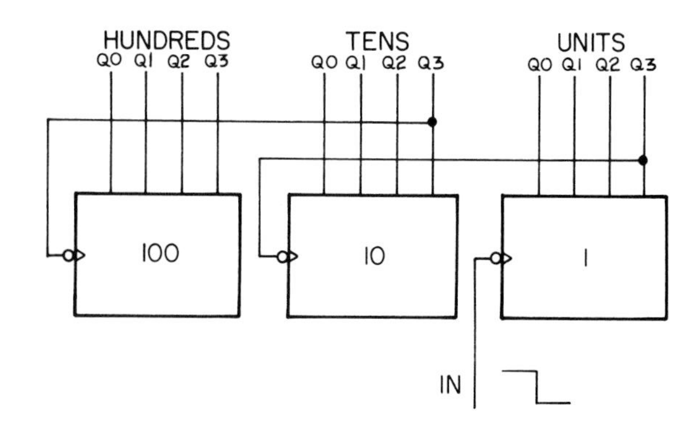

Fig. 10-31. Decade counters can be cascaded. Note the output of each decade counter, which is the position of the old Q3 lead.

each set of four outputs can easily be converted to decimal. This system of numbers is called *BINARY CODED DECIMAL* (BCD). BCD will be covered in detail later.

7493 CHIP USED IN A DECADE COUNTER

The 7493 chip is a general purpose counter chip. It contains four toggle flip-flops and a clear circuit, Fig. 10-32. Coupling between the first two flip-flops is brought out. An external jumper must be used when this connection is needed. The others are internally connected.

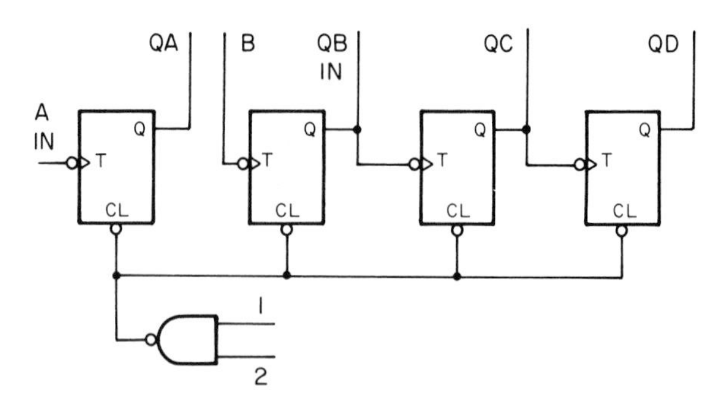

Fig. 10-32. The 7493 chip contains four toggle flip-flops and a feedback circuit. The chip is easy to connect into various counter circuits.

The 7493 chip can be connected to form a decade counter. Note the ease of connecting the 7493 chip leads as shown in Fig. 10-33. The modulus is determined by the NAND element.

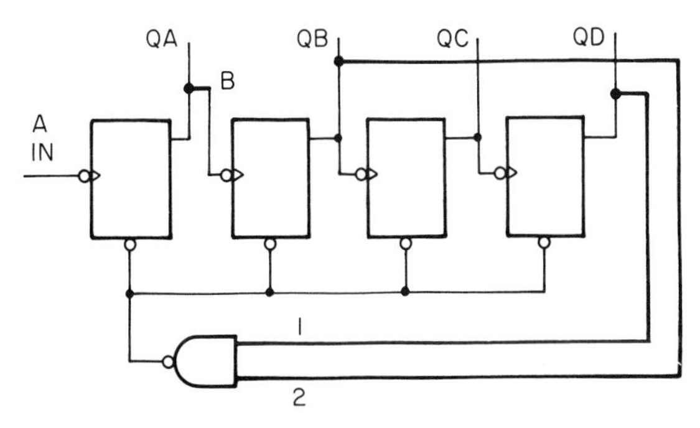

Fig. 10-33. Only three external connections are needed to make a 7493 chip into a decade counter.

COUNTERS WITH MODULI OTHER THAN 10

Counters with almost any maximum count can be constructed. Fig. 10-34 shows the 7493 chip connected for maximum counts of 2 through 9. Sometimes a NAND with more than two inputs is needed. An external AND can provide extra inputs.

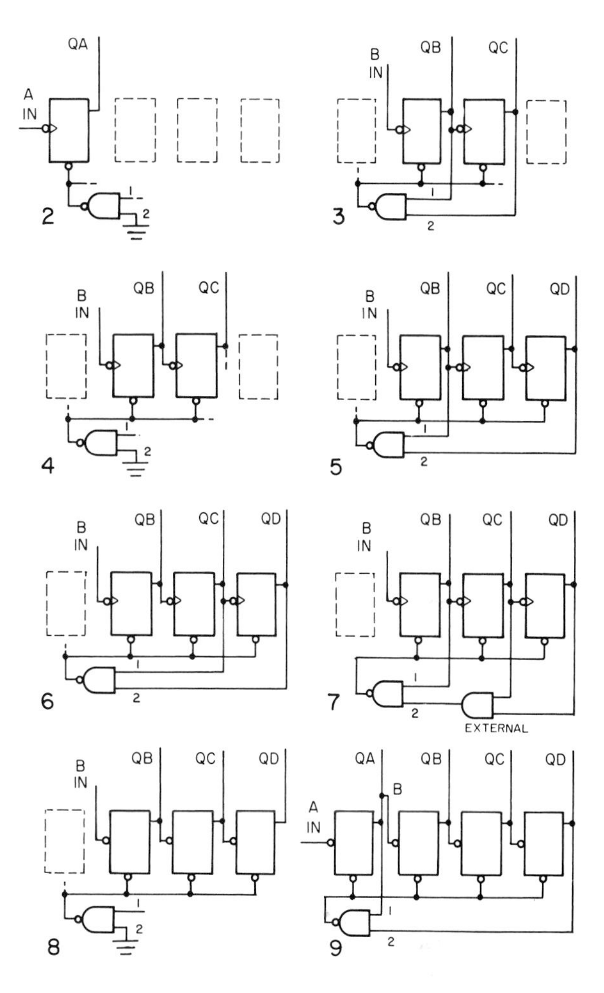

Fig. 10-34. Using the 7493 chip, a variety of maximum counts can be obtained.

SYNCHRONOUS COUNTER

Counters described up to this point have been *ASYNCHRONOUS* (non-synchronized). That is, as the effects of a count ripple through a circuit, its outputs do not change at the same time.

In most applications, the small delay between the change in the first and the last output can be ignored. In high-speed circuits or when high counts are required, such delays may be a problem. The solution to this problem is the use of devices which are called *SYNCHRONOUS* counters. In such circuits, all outputs change at the same time.

Fig. 10-35 shows a synchronous counter. Its action can be best described by making three preliminary observations.

First, high-order counter outputs change from 0 to 1 only when all lower-order bits are 1. For example, the output of a counter might be Q3Q2Q1Q0 = 0011. On the next count, 0011 + 1 = 0100. Output Q2 changed to 1 because the lower-order bits were all 1s. Q3 will change to 1 on the count after 0111.

The second observation concerns the input lead in Fig. 10-35. COUNT goes to the clocks of the four flip-flops. They are activated at the same time.

Third, J-K inputs are not permanently connected to 1s in this circuit. (Except for the first element, toggle flip-flops cannot be used.) Before a flip-flop can toggle, 1s must appear at J and K.

The design is based on knowing what the outputs will be before they are created. The system is fast.

SYNCHRONOUS COUNTER CIRCUIT ACTION

For the output of an AND to be 1, all lower-order outputs must be 1. For example, flip-flop number 3 in Fig. 10-35 will not toggle until Q2, Q1, and Q0 are all 1. Synchronous counters can be faster than ripple counters. In synchronous counters such as this, all flip-flops that are supposed to toggle on a given count do so at the same time. In ripple counters, each count must ripple down the line of flip-flops. If there are many flip-flops in such a counter, this can be time consuming.

SYNCHRONOUS MOD COUNTERS

Synchronous counters with moduli other than 2, 4, 8, etc. are possible. A mod-3 counter and a mod-5 counter are shown in Fig. 10-36.

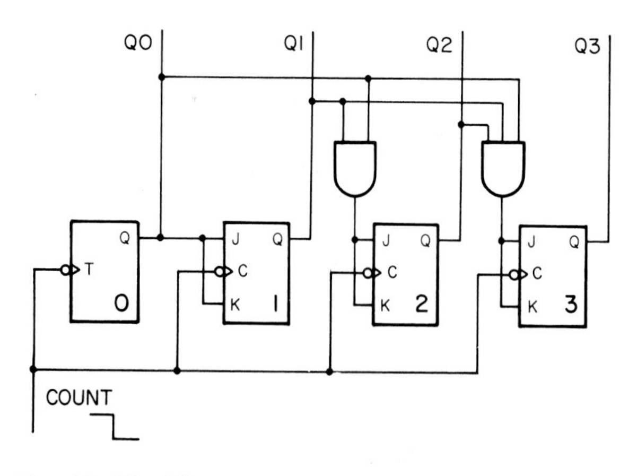

Fig. 10-35. All outputs of the synchronous counter change at the same time.

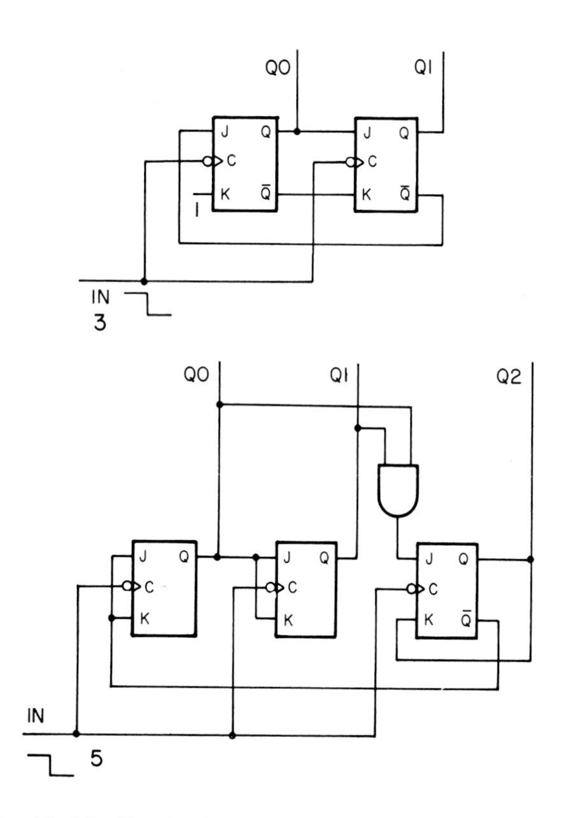

Fig. 10-36. Circuits for synchronous mod counters are shown. The top circuit is a mod-3 synchronous counter and the bottom circuit is a mod-5 synchronous counter.

Synchronous counters are more difficult to design than are ripple counters. Because synchronous counters tend to contain many more elements than ripple counters, they are generally more expensive and should be used only when their special features are needed.

The analysis of the actions of such counters is time consuming and will not be considered here. Circuits for most moduli are available. Catalogs and other literature describe circuits to fit each need. Many types of feedback (and sometimes use of feed-forward) allow the choice of modulus.

FREQUENCY DIVIDER

The circuit in Fig. 10-37 is identical to that of a ripple counter. Yet, it is called a frequency divider. The difference is found in circuit applications.

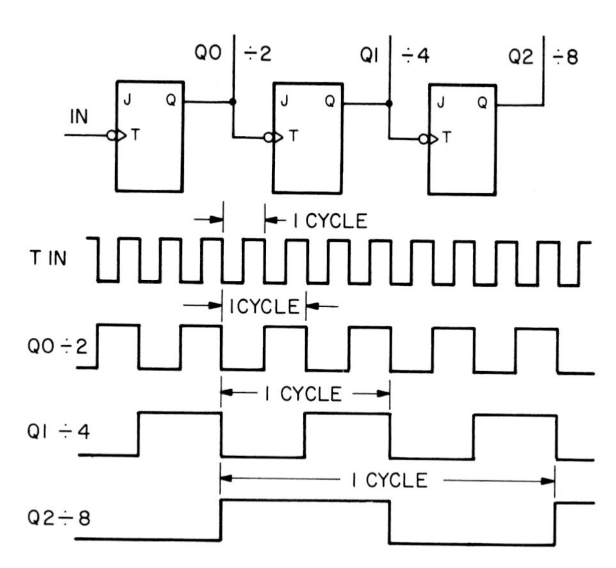

Fig. 10-37. This circuit divides the frequency of the incoming signal by 2, 4, and 8.

Note the uniformity of the input pulses in the frequency divider's timing diagram. Rather than representing individual events to be counted, these pulses form an alternating-current wave. Such waves are made up of trains of identical *CYCLES* (a cycle is a complete wave). See Fig. 10-38. The time required for one cycle is its *PERIOD*. And the number of cycles in one second is the wave's *FREQUENCY*. The name HERTZ is given to the dimension which represents cycles-per-second.

If the period of a wave is known, its frequency can be determined from the equation:

$$f = \frac{1}{P}$$

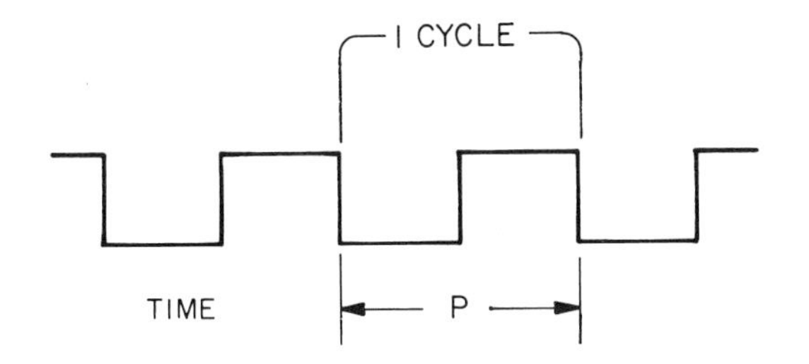

Fig. 10-38. The time required for one cycle of a wave is called its period.

where f is the frequency in Hertz (cycles-per-second), and P is the period in seconds per cycle. Note that the longer the period, the lower the frequency.

NEED FOR FREQUENCY DIVISION

In digital circuits, it is often desirable to lower the frequency of a signal by a fixed amount. Digital watches offer an example. A simple watch might consist of a series of counters. Starting with a 1 Hertz (1 cycle-per-second) signal, a mod-60 counter could be used to keep track of the seconds. The output of the seconds counter could drive another mod-60 counter. It could count the minutes. Its output could then drive a mod-12 counter to keep track of the hours, etc.

The accuracy of such a watch depends directly upon the accuracy of the 1 Hertz signal. However, it is difficult to build a small, accurate signal source at such a low frequency. As a result, a digital watch usually contains a high frequency signal source. Its frequency is then divided down to the required low frequency. The low frequency signal is as accurate as the original high frequency signal was.

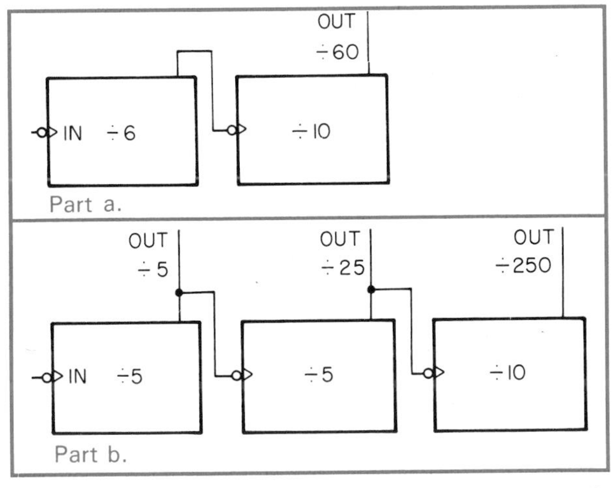

Fig. 10-39. Frequency dividing circuits can be cascaded. The output frequency ratio is the product of the individual divider ratios.

FREQUENCY DIVIDER CIRCUIT ACTION

Refer again to Fig. 10-37. The time between negative-going edges of the wave in graph T represents one cycle of the input signal. Output Q0 toggles on each negative-going edge of T, so it takes two cycles of the input signal to produce one cycle at Q0. Because its period is twice as long, the frequency of the signal at Q0 is half that of the signal at input T.

Each flip-flop divides the frequency by 2. The 3-bit divider shown in Fig. 10-37 divides the input frequency by 8. If higher ratios are needed, additional flip-flops can be added. Ratios other than 2, 4, 8, etc. can be obtained by using counters with other moduli.

CASCADE FREQUENCY DIVIDERS

Part "a" of Fig. 10-39 shows two frequency dividers in cascade (one after the other). For this divide-by-6 circuit and "÷"-by-10 circuit, the overall ratio is 1/60. The final ratio is the product of the individual ratios.

The second circuit shows more than two dividers cascaded. Intermediate ratios can be output along the circuit. TV camera vertical and horizontal circuits use cascaded dividers having several "taps" (at intermediate frequencies). Waves with most equally spaced positive-going edges (or most equally spaced negative-going edges) are in the flip-flops farthest from input flip-flop (due to averaging over time).

SUMMARY

The difference between J-K and R-S flip-flops is found in the last lines of their truth tables. When both J and K are 1, a clock pulse will cause the output of a J-K flip-flop to toggle. A toggle flip-flop is a J-K flip-flop with permanent 1s applied to J and K.

Counters are an important use for toggling flip-flops. The ripple counter is the simplest counter. It is an asynchronous circuit. Without feedback, available moduli are 2, 4, 8, 16, etc. If feedback is added, other moduli can be obtained.

Counters can be used as frequency dividers. By cascading dividers, almost any ratio of input to output frequency can be obtained.

When mechanical switches are used with counters, care must be taken to overcome bounce. Several circuits are available for debouncing switches.

IMPORTANT TERMS

Binary coded decimal, Cascade, Complement, Decade counter, Feedback, Frequency divider, Hexadecimal, Hysteresis, Modulus, Octal, Period, Ripple counter, Switch bounce, Toggle.

TEST YOUR KNOWLEDGE

1. The term toggle implies a device with three stable states. (True or False?)
2. The term complement means to __________ (fortify, add, reverse).
3. With 1s applied to both J and K of a J-K flip-flop, is the result an undefined state?
4. With J = 1 and K = 1, will the flip-flop toggle when the clock is pulsed?
5. A J-K flip-flop is a master-slave level-triggered R-S flip-flop with __________ added.
6. If a TTL toggling flip-flop is desired, J and K can be allowed to float. (True or False?)
7. In which type of flip-flop, edge-triggered or level-triggered, is the newly stored number output just after clock goes active?
8. Can a J-K flip-flop replace an R-S flip-flop?
9. In a ripple counter, all outputs occur at the same time. (True or False?)
10. In decade counters in Fig. 10-30 and Fig. 10-31, circuit clears if CLEAR has negative-going signal on it. At clear, does Q1 stay up or drop down? At clear, does Q3 go up or down? At clear, had Q3 just gone high or had it been high a while? Does clear need to wait for Q1 or for Q3?
11. If flip-flops are arranged in groups of three, with the second group of three keeping track of the number of 8s, a third group of flip-flops keeps track of the number of __________ (32s, 64s, 512s) in the count.
12. To get maximum count of a ripple counter, add __________ (1, 2, 4) to sum of the place values.
13. Ripple counters with s-a-0 or s-a-1 faults will usually __________ (1. fail to count, 2. continue to count, 3. start to count, 4. reverse the count) above some fixed number.
14. Counters are sensitive to a problem called switch contact __________.
15. What two circuits debounce a switch signal?
16. A Schmitt-trigger debouncer uses a Schmitt trigger and a/an __________.
17. What is an SPST switch?
18. What is an SPDT switch?
19. A Schmitt-trigger response curve demonstrates __________ (ringing, noise, memory, hysteresis, polarization).
20. The term __________ refers to at least two points on the response curve of a Schmitt trigger which determine when the trigger will respond.
21. Schmitt triggers can be used to convert __________ waves into square wave pulses or pulses which have a rectangular shape.
22. Schmitt triggers "__________" noisy pulses.
23. Presetting or preloading a counter allows it to count only to 2, 4, or 8 etc. (True or False?)
24. In a down counter, the input to the second and third flip-flops comes from which of the following leads?: Q, T, $\overline{Q}$, $\overline{T}$.
25. A 3-bit down counter with __________ counts down from any number between 1 and 8.
26. A/an __________ counter can reverse its count direction.
27. A __________ counter avoids some of the problems with entering presets when counts other than 2, 4, 8, etc. are desired.
28. What is the plural form of the word modulus?
29. A mod-10 type of counter is often called a __________ counter.
30. A decade counter clears on the count after Q3Q2Q1Q0 = 1000. (True or False?)
31. A second group of counters following a decade counter keeps track of the __________ (tens, hundreds).
32. The abbreviation BCD stands for what?
33. In a __________ counter, all of the outputs change at the same time.
34. The last flip-flop in a 4-bit counter toggles on the count after the previous flip-flops output the following: __________ (000, 101, 111, 110).
35. Is a synchronous counter (Fig. 10-35) more complex than a similar asynchronous unit? Extra __________ (ANDs, ORs, NANDs, NORs) are used in synchronous counters.
36. Is a uniform wave more likely at the input of a frequency divider or a counter?
37. The name __________ is given to the dimension cycles-per-second.
38. Frequency f equals 1 __________ (times, divided by) the period P.
39. The formula for the period P in terms of the frequency f is as follows:
    1. P = f
    2. P = 1/f.
    3. P = f^2.
40. In a cascaded frequency divider, the resulting frequency ratio is the __________ (sum, product) of the individual frequency ratios.

STUDY PROBLEMS

1. Draw the symbol for a J-K flip-flop. Use a level-triggered active-low clock. Include both Q and $\overline{Q}$. Also, show preset and clear inputs. Label the leads J, K, C, Q, $\overline{Q}$, PR, and CL.
2. Construct the truth table for the flip-flop in Problem 1 for inputs J and K. Label the output column Qt + 1.
3. Draw the NAND-based circuit for the flip-flop in Problem 1. Do not show the preset and clear.
4. Draw the symbol for a toggle flip-flop. Its clock is to trigger on the negative-going edge of the in-

put signal. Include preset and clear. Label its leads T, PR, CL, Q, and $\overline{Q}$.

5. Trace the timing diagram in Fig. P10-1 onto a sheet of paper and construct the expected output. The clock is level-triggered. Because it is a J-K flip-flop, it is a master-slave unit.

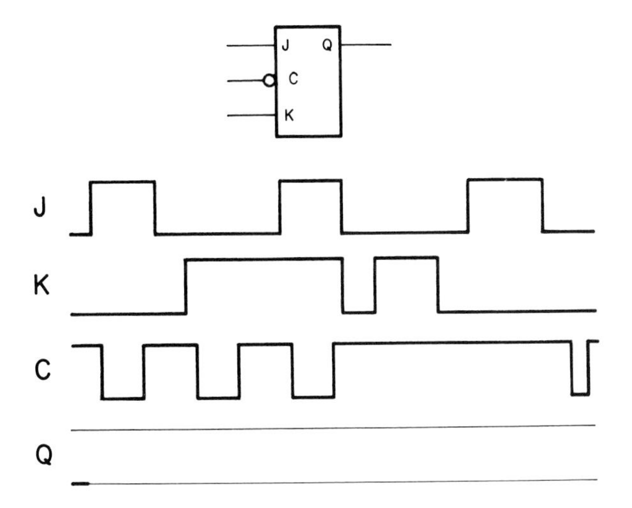

Fig. P10-1. A level-triggered master-slave J-K flip-flop can be examined for all combinations of J, K, and C.

6. Repeat Problem 5 for the flip-flop and timing diagram shown in Fig. P10-2. The clock is edge-triggered.

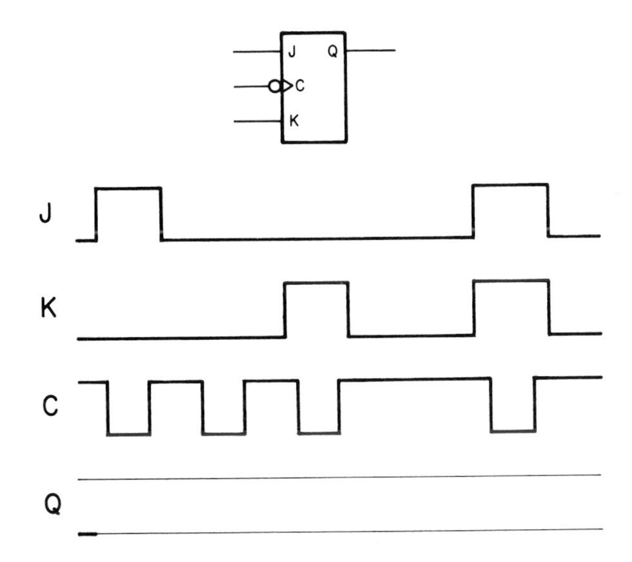

Fig. P10-2. An edge-triggered master-slave J-K flip-flop can be examined for all combinations of J, K, and C.

7. Repeat Problem 5 for the toggle flip-flop and timing diagram shown in Fig. P10-3.

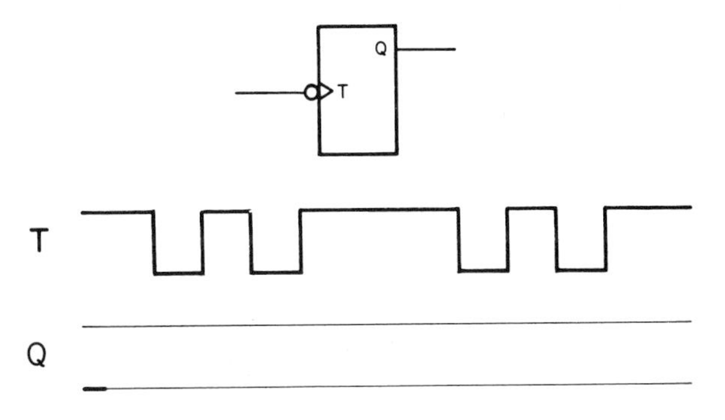

Fig. P10-3. Examine the output of a toggle flip-flop.

8. Using toggle flip-flops, draw a 4-bit ripple counter. Label its leads T, Q0, Q1, Q2, and Q3. Q0 is to be the LSB. The T inputs are to be negative-going edge-triggered.
9. Repeat Problem 8, but make the circuit a down counter.
10. Trace the timing diagram shown in Fig. P10-4 onto a sheet of paper and construct the expected outputs. Assume all outputs are 0 at t = 0.
11. Refer to the completed timing diagram from Problem 10. How many counts (negative-going edges) were needed for this counter to go from 000 to 000?
12. Assume that a 3 has been preloaded into the counter shown in part "a" of Fig. P10-5. Trace the timing diagram in part "b" of Fig. P10-5 onto a sheet of paper and construct the expected outputs.
13. For each of the following basic ripple counters (no feedback), indicate their maximum counts.
    1. Three-bit (three flip-flops).
    2. Four-bit.
    3. Five-bit.
    4. Six-bit.
14. Describe the difference between the output signals of synchronous and asynchronous counters.
15. Draw circuit diagrams to show how a 7493 might be connected to produce the following moduli.
    1. Mod-4.
    2. Mod-8.
    3. Mod-10.
    4. Mod-12.
16. Draw the circuit for an R-S flip-flop debouncing circuit. Use two NANDs and an SPDT switch. Include the resistor symbols shown in the text.
17. Draw the symbol for a Schmitt-trigger NOT.

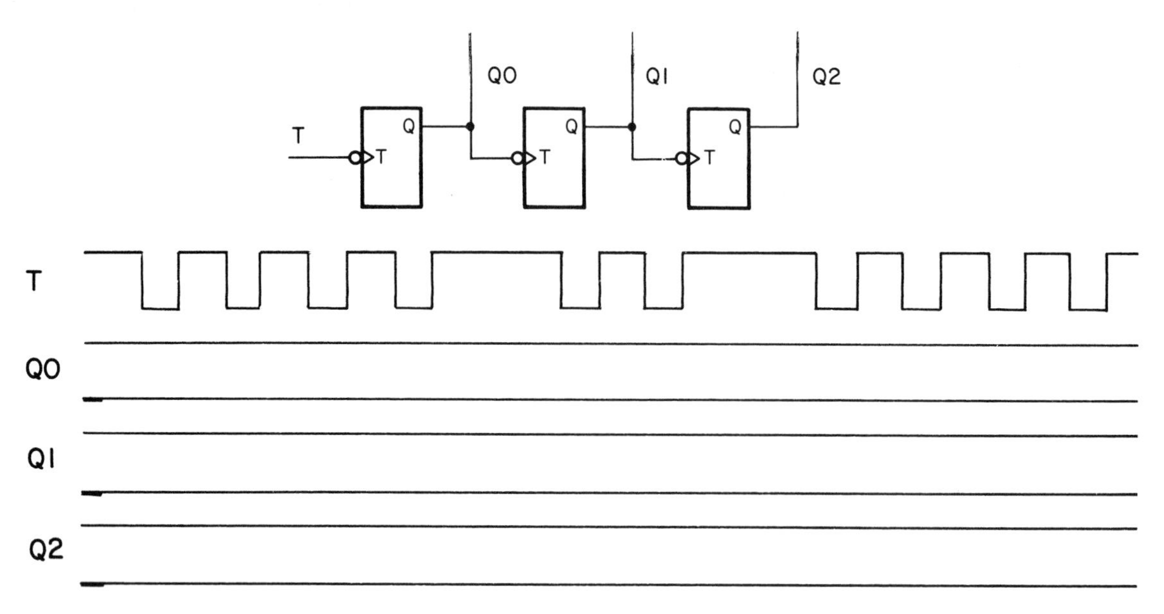

Fig. P10-4. Examine the outputs of a ripple counter.

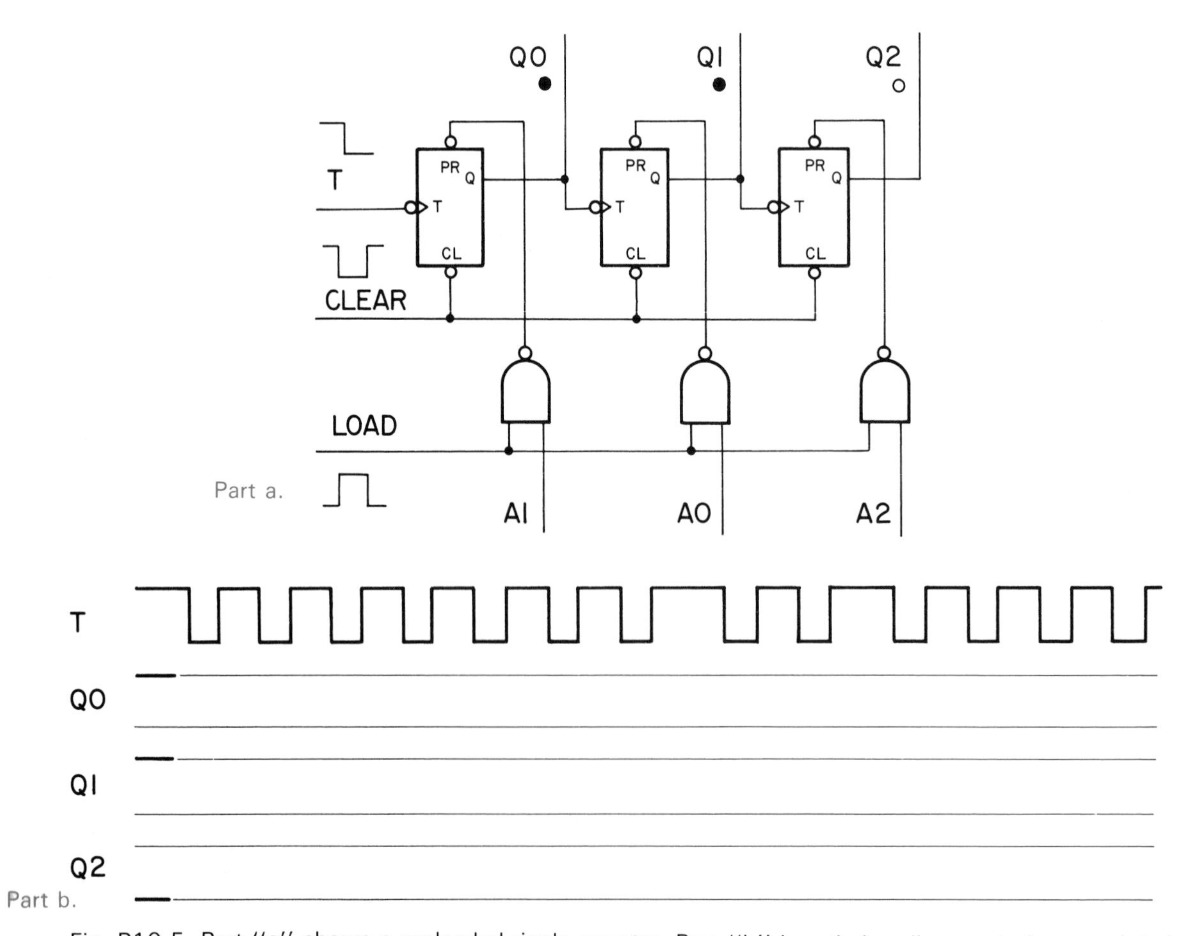

Fig. P10-5. Part "a" shows a preloaded ripple counter. Part "b" is a timing diagram to be completed.

Binary Number Shortcuts and Arithmetic Processing Methods

Chapter 11

LEARNING OBJECTIVES

After studying this chapter and completing lab assignments and study problems, you will be able to:

- Count using the binary numbering system.
- Compute place values for the digits of binary numbers.
- Convert binary numbers to decimal and decimal numbers to binary.
- Add unsigned binary numbers.
- Determine the two's complement of binary numbers and use the two's complement method to do binary subtraction.
- Add signed binary numbers.
- Determine if overflows have occurred when binary numbers have been added.

Binary is the numbering system used within computers. As a result, workers in the digital electronics field must be completely familiar with this system. They should be able to convert back and forth between binary and decimal. To design and repair arithmetic circuits, they must understand binary arithmetic.

This chapter reviews and expands on the introduction to the binary number system presented in Chapter 1. Many examples are provided to aid in understanding number systems and arithmetic circuits.

REVIEW OF BINARY NUMBERS

You are already familiar with counting in binary numbers. You can also convert binary numbers to decimal. However, a review of these topics is appropriate before additional concepts are introduced. To understand some shortcuts made possible by the two's complement method of adding and subtracting, it is important to be skilled in binary counting and binary error identification. Two methods of subtracting can be compared.

REVIEW OF READING A BINARY NUMBER

The binary number 01,111,010 is read, "Zero, one (pause), one, one, one (pause), zero, one, zero." It should not be read, "One million one hundred eleven thousand ten." Such notation implies a decimal number.

REVIEW OF BINARY COUNTING

The radix or base of the binary system is two, so its symbol set contains two symbols. The symbols are 1 and 0.

No matter what base is used, counting proceeds in the same manner. Symbols in a system's symbol set are used first. On the count after the symbol set has been exhausted, a 0 results, and a carry is generated.

Counting in binary starts with 0. The result of the first count is:

```
COUNT 1:        0
              + 1
              ---
                1
```

This is identical to the first count in any system. (Because this is binary arithmetic rather than Boolean algebra, the + sign is read as plus.)

On the second count, a problem arises. The symbol set has been depleted. The solution is identical to that of other numbering systems. When 1 is added to the last symbol of the set, a 0 results, and a carry is generated. That is:

```
COUNT 2:      c1
                1
            +   1
            -----
               10
```

The c1 indicates a carry. Binary 10 equals two counts.

The next count is simple. A 1 is merely added to the 0.

```
COUNT 3:       10
            +   1
            -----
               11
```

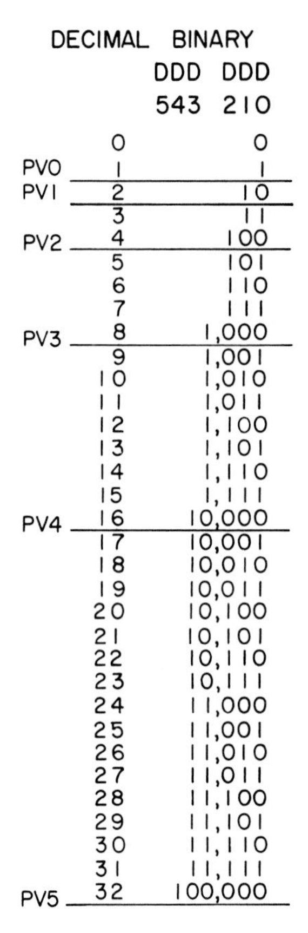

	DECIMAL	BINARY DDD 543	DDD 210
	0		0
PV0	1		1
PV1	2		10
	3		11
PV2	4		100
	5		101
	6		110
	7		111
PV3	8	1,	000
	9	1,	001
	10	1,	010
	11	1,	011
	12	1,	100
	13	1,	101
	14	1,	110
	15	1,	111
PV4	16	10,	000
	17	10,	001
	18	10,	010
	19	10,	011
	20	10,	100
	21	10,	101
	22	10,	110
	23	10,	111
	24	11,	000
	25	11,	001
	26	11,	010
	27	11,	011
	28	11,	100
	29	11,	101
	30	11,	110
	31	11,	111
PV5	32	100,	000

Fig. 11-1. Counting in decimal and binary. PV stands for place value.

On the fourth count, two carries result.

COUNT 4:

```
 c11
  11
+  1
 100
```

The table in Fig. 11-1 continues the count to decimal 32. To review counting in binary, set up a table and count to decimal 20 in binary. Check your results against the table.

REVIEW OF BIT NOTATION

To permit easy reference to individual bits of a binary number, the following notation is used:

Dn . . . D5 D4 D3 D2 D1 D0

In the number 111,010, D2 is 0, and D1 is 1. The symbol Dn implies that the notation can be extended to any number of bits.

REVIEW OF PLACE VALUES

The contribution of a bit to the total count of a number depends on its value (1 or 0) and its position in the number. This is demonstrated by the number D1D0 = 11 (base 2). The 1 at DO contributes a single count. The 1 at D1 represents two counts.

The amount a bit contributes is called its *PLACE VALUE.* The place values of the first six bits are:

D5	D4	D3	D2	D1	D0
32	16	8	4	2	1

Based on this progression, place values for higher-order bits can be predicted. Each is twice the previous value, giving the following:

D11	D10	D9	D8	D7	D6
2048	1024	512	256	128	64

REVIEW OF BINARY-TO-DECIMAL CONVERSION

The binary system is a completely valid numbering system. Any computation that can be done in decimal can also be done in binary. However, people tend to think in terms of decimal, so conversions between binary (the numbering system of computers) and decimal are often made.

BINARY-TO-DECIMAL EQUATION

The binary-to-decimal conversion process described in the first chapter can be expressed in equation form. The equation is as follows:

$$Dn(Pn) + . . . + D4(16) + D3(8) + D2(4) + D1(2) + D0(1) = N$$

Here, N is in base 10. The bracketed numbers are place values. The Ds represent the bits (1s and 0s) that make up the binary number. Dn(Pn) indicates that the equation can be extended to any number of bits.

USE OF BINARY-TO-DECIMAL EQUATION

To convert a binary number to decimal, the 1s and 0s that make up the number are substituted for the Ds. After the indicated multiplications have been made, the sum of the products is the value of the number in decimal.

Example: Convert 10,110 (base 2) to decimal.

Substituting into the equation results in:

$$D4(16) + D3(8) + D2(4) + D1(2) + D0(1) = N$$
$$1(16) + 0(8) + 1(4) + 1(2) + 0(1) = N$$
$$16 + 0 + 4 + 2 + 0 = N$$
$$22 = N$$

The binary number 10,110 equals 22 (base 10). The equation formalizes the process you have been using to convert from binary to decimal.

DECIMAL-TO-BINARY CONVERSION

Converting decimal numbers to binary is a bit more

difficult than the binary-to-decimal process. It involves finding the group of binary place values that exactly equals the decimal number.

Two commonly used methods will be described:
1. MSB method.
2. LSB method.

Both methods of converting decimal numbers to binary numbers use the equation just introduced.

MSB METHOD

As its name suggests, the MSB method determines the MSB (most significant bit) of the binary number first. An example will be used to describe the process.

Example: Convert 25 (base 10) to binary using the MSB method.

The highest-order, non-zero bit of the resulting binary number is determined first. This is most easily accomplished by writing binary place values until one is found that is larger than the decimal number to be converted. For example:

D5	D4	D3	D2	D1	D0
32	16	8	4	2	1

The place value of D5 is larger than the number to be converted. It cannot be part of the equivalent binary number. D4 will be the highest-order, non-zero bit in the number.

Computation: D4 will be 1, but a simple computation will be made to demonstrate the process. If 16 (the place value of D4) is divided into 25, the quotient will be 1 and a remainder of 9 will result. See part "a" of Fig. 11-2.

With D4 = 1, this bit contributes 16 counts to the number. Bits D3, D2, D1, and D0 will contribute the remaining 9 counts. See part "b" of Fig. 11-2.

Computation: To determine if D3 is 1 or 0, this bit's place value (8) is divided into the remainder (9). See part "a" and part "b" of Fig. 11-3.

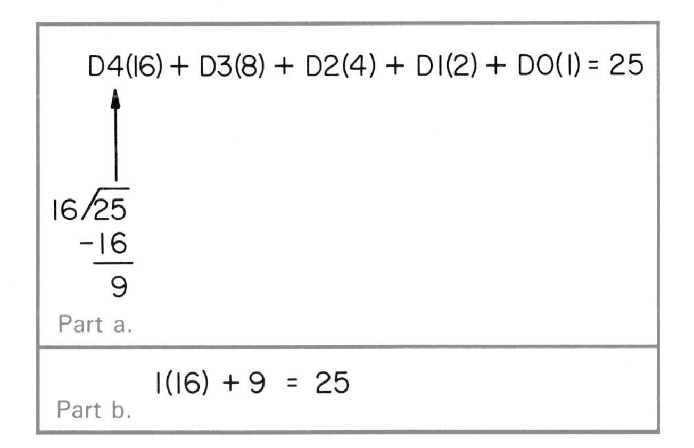

Fig. 11-2. First step of computation used to convert decimal 25 to binary. The MSB method is used.

Based on the indicated division, D3 = 1, and there is a remainder of 1.

Computation: To determine D2, its place value (4) is divided into the remainder from the previous step. Four will not go into 1, so D2 equals 0. See part "c" and part "d" of Fig. 11-3. The remainder stays at 1, as it was in the previous computation.

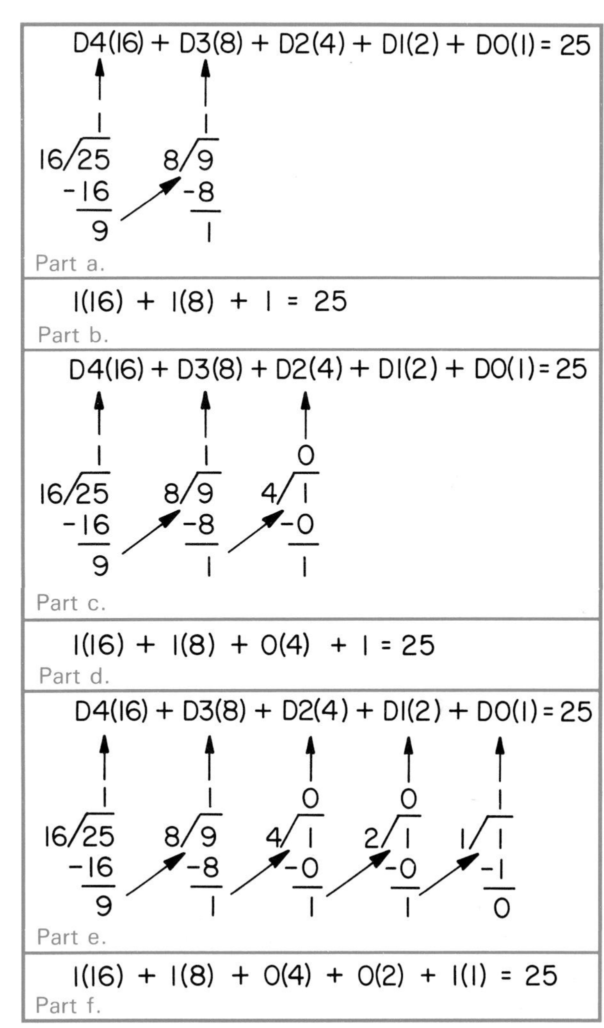

Fig. 11-3. Second, third, and fourth steps used to convert decimal 25 to binary.

Computation: Part "e" and part "f" of Fig. 11-3 show the last two steps in the conversion. D1 equals 0, since its place value is larger than the remainder. D0 = 1, since its place value can be divided into the remainder. The result is: 25 (base 10) equals 11,001 (base 2).

Example: Convert the decimal number 45 to binary. Use the MSB method.

The place value of D6 (64) is larger than the number to be converted, so D5 will be the highest-order, nonzero bit. The rest of the computations are shown in Fig. 11-4.

The conversion of decimal 45 into binary is easily accomplished on a calculator or by hand. Fig. 11-5 shows a simplified form for the conversion. Starting with the MSB, if its place value can be subtracted, that bit is 1. If it cannot be subtracted (that is, the place value is larger than the remainder), that bit is 0.

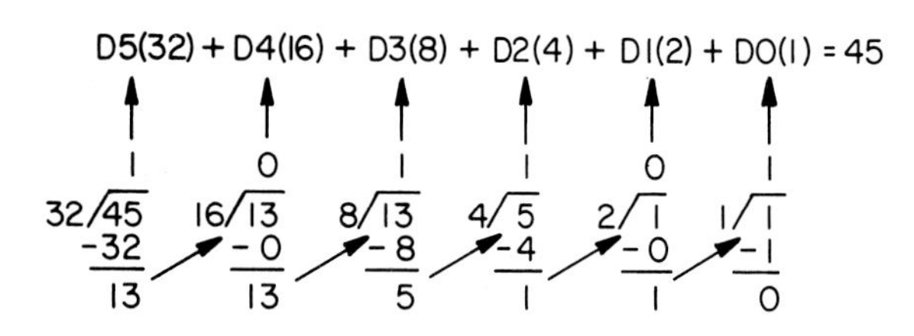

1(32) + 0(16) + 1(8) + 1(4) + 0(2) + 1(1) = 45

Fig. 11-4. Steps in the conversion of decimal 45 to binary.

LSB METHOD

A second widely used method of converting from decimal to binary starts with the LSB of the binary number. It involves repeated division by 2.

Example: Convert 23 (base 10) to binary. Use the LSB method.

PV	REMAINDERS	BITS
	45	
32	−32	D5 = 1
	13	
16	−0	D4 = 0
	13	
8	−8	D3 = 1
	5	
4	−4	D2 = 1
	1	
2	−0	D1 = 0
	1	
1	−1	D0 = 1
	0	

Fig. 11-5. Method used to recognize a place value which contributes to the total of 45 in decimal form.

Computation: The process begins by dividing the decimal number 23 by 2. See part "a" of Fig. 11-6. The remainder will be either 1 or 0, and it becomes the value of D0. In this case, D0 = 1.
Computation: To determine D1, the quotient from the result of the previous division, (11) is divided by 2. See part "c." The remainder is 1, so D1 is 1.
Computation: The process is repeated for D2. The

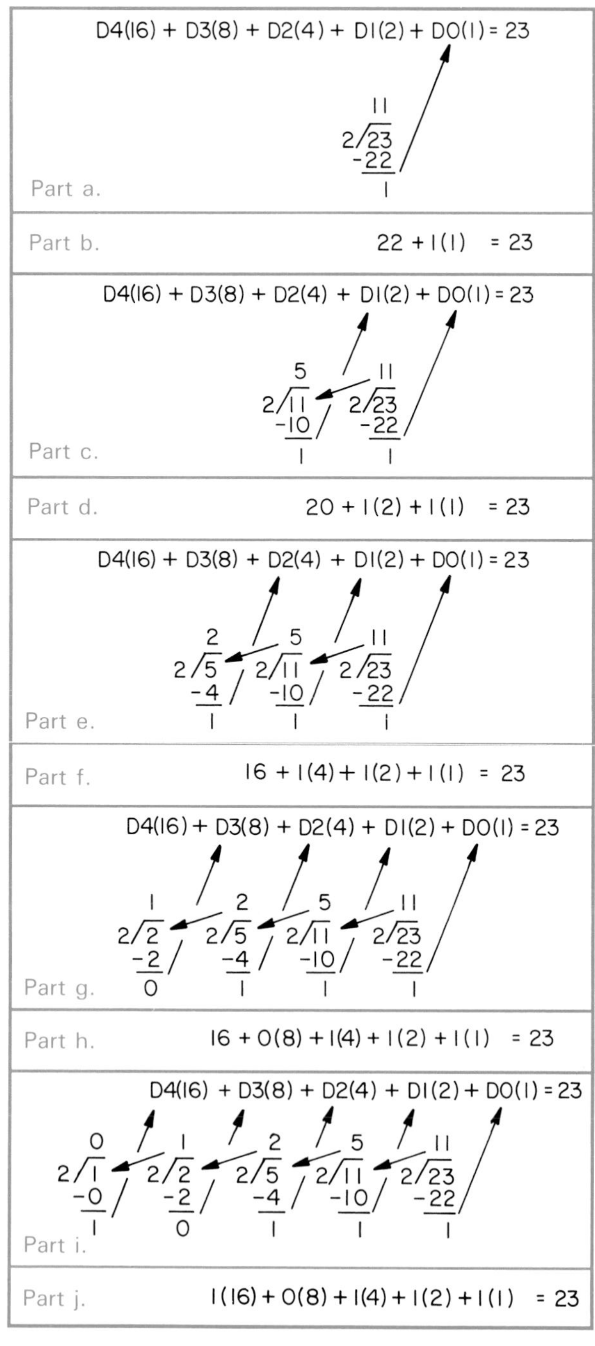

Fig. 11-6. Steps in the conversion of decimal 23 to binary. The LSB method is used.

previous quotient (5) is divided by 2. See part "e" of Fig. 11-6. The remainder is 1, so D2 is 1.
Computation: In the step at "g" in Fig. 11-6, 2 is divided into 2. The remainder is 0, so D3 = 0.
Computation: The conversion ends at "i" when 1 is divided by 2. The remainder is 1 so D4 is 1.

Example: Convert the decimal number 19 to binary using the LSB method.

Because the remainder is always 1 or 0, the form shown in Fig. 11-7 can be used. If the quotient from a division has a remainder of 1, the corresponding bit is 1. If the quotient is a whole number (no remainder), the associated bit is 0.

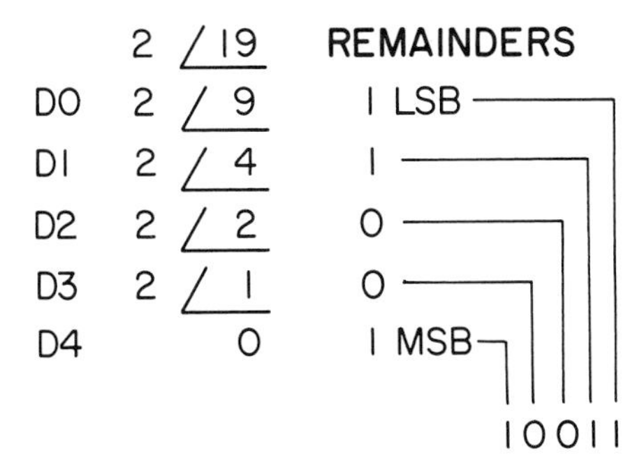

Fig. 11-7. Converting decimal 19 to binary using the LSB method. Remainders are always 1 or 0.

THEORY BEHIND LSB METHOD

You can effectively use the LSB method without understanding its theory. Therefore, this section was included only for the sake of completeness.

Its LSB (D0) determines whether a binary number is ODD or EVEN. If D0 = 1, the number is odd; if D0 = 0, it is even. You need look only at D0 to know that 10,111 (decimal 23) is odd and 1,110 (decimal 14) is even. To explain this, remember that the place values above D0 are all even (2, 4, 8, etc.). Their sums will always be even. See the numbers over the dashed line at "a" in Fig. 11-8. If the complete number (the part under the solid line) is odd, D0 must be 1. If the number is even, D0 must be 0.

This concept can be used to find D0. If a decimal number is odd, D0 in the binary number will be 1; if it is even, D0 will be 0. To test if the number is odd or even, it is divided by 2. See part "aa." (Ignore the dots.) If the remainder is 1, the number is odd and D0 = 1. If it is 0, D0 = 0.

This approach can be used to find D1. Refer to part "b" of Fig. 11-8. Because the place values beyond D1 (those over the dashed line) are multiples of 2, their sum always contains an even number of 2s. If the sum of the place values including D1 (those under the solid line) has an odd number of 2s, D1 must be 1. The computation at "bb" determines if the number of 2s is odd or even. The division is by 4. There is a remainder (2), so D1 is 1.

Place values above D2 are multiples of 4, so D2 can be determined. See parts "c" and "cc." There is no remainder (the number of 4s above D2 is even), so D2 must be 0. The process is repeated for D3 and D4.

Part "f" shows a simplification of the computations. Each dotted number has been divided by its bit's place value. This is acceptable, since the ratio of the dividends and divisors is unchanged. Note that the method described in Fig. 11-6 is the result.

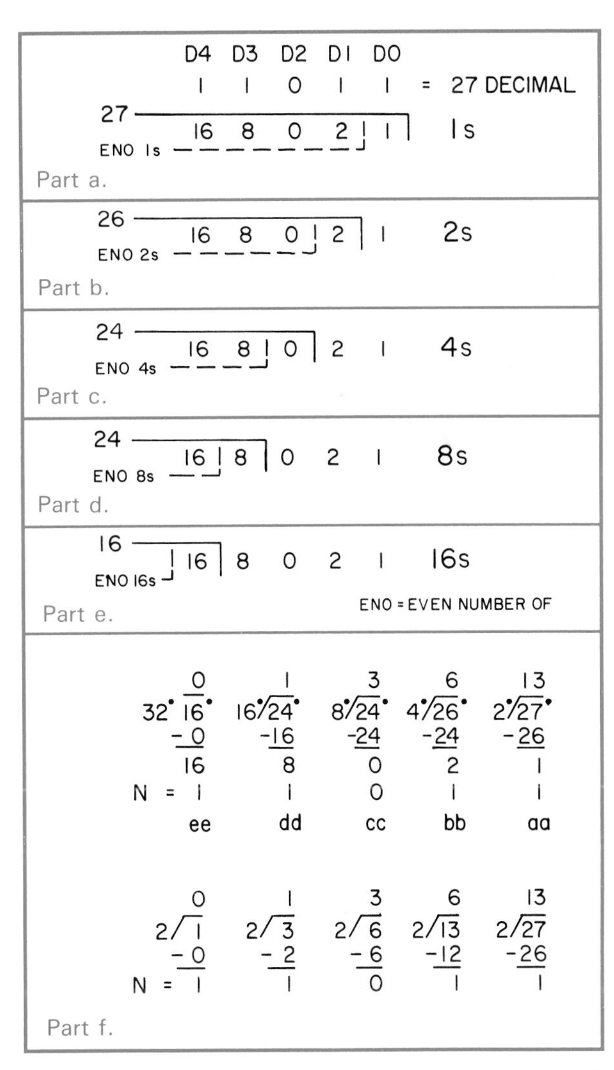

Fig. 11-8. The LSB method for converting from decimal to binary numbers is based on determining whether a given place value appears an odd or even number of times.

When using the LSB method, remember that the first bit evaluated is D0. D0 is used instead of D1 because 2 raised to the 0 power is 1.

BINARY NUMBERS LESS THAN 1

Binary numbers which are less than 1 are called fractional binary numbers. The number 1/2 is a fraction which can be converted to binary as in the following list.

decimal 1/2 = decimal 0.5 = binary 0.1
decimal 1/4 = binary 0.01
decimal 1/8 = binary 0.001
decimal 1/16 = binary 0.0001

The list can be continued. Any fraction in decimal can be converted to binary as a suitable sum of the powers of 1/2. For example, decimal 3/4 equals 1/2 + 1/4, so that decimal 3/4 equals 0.110 in binary. Decimal 7/8 = 1/2 + 1/4 + 1/8 = 0.1110 in binary. Decimal 5/8 = 1/2 + 1/8 = 0.1010 in binary. Decimal 15/16 = 0.11110 in binary.

Note that numbers like 0.1111 in binary are nearly equal to 1. A fractional binary number with an endless string of 1s would be equal to 1. That is:

$0.11111111\overline{1}$ in binary equals 1.

The bar over the last 1 means that it repeats forever. The usual mathematical statement of the property is as follows:

$$\lim_{x \to \infty} \sum_{n=1}^{x} \frac{1}{2^n} = 1$$

That is, if the powers of 1/2 go on forever while being summed with the $\sum$ sign, the result equals 1.

Errors can result by calling the dot to the left of the 1s a "decimal point." This mixes decimal concepts with binary concepts. A better term to use for the dot is "place point."

UNSIGNED NUMBERS

The symbols + and − have two meanings. They can indicate addition and subtraction (3 + 9 = 12 and 5 − 3 = 2). They can also indicate positive and negative (+4 and −16).

An *UNSIGNED* number is one without a positive or negative sign. For example, 11,101 and 43 are unsigned numbers. They are assumed to be positive. They can be added, subtracted, multiplied, or divided, but the result of such computations must be positive numbers. In unsigned numbers there is no way of indicating negative numbers.

ADDITION OF UNSIGNED NUMBERS

You are already familiar with the rules for adding unsigned numbers. Addition of unsigned decimal numbers is familiar. Addition of 1 digit unsigned binary numbers is less familiar. A review is appropriate. A comparison can help.

RULES OF BINARY ADDITION

The rules of binary addition are different from those for decimal addition. The rules are different enough to require a statement of their four fundamental properties. The four rules of addition are:

Rule 1 0 + 0 = 0
Rule 2 1 + 0 = 1
Rule 3 1 + 1 = 10
Rule 4 1 + 1 + 1 = 11

The following example demonstrates the use of these rules in multi-bit addition. An example can be checked one digit at a time.

Example: Add the binary numbers 1110 and 1100.

The LSBs of both numbers are 0, so D0 of the sum will be 0. This is an application of Rule 1.

```
  1110
+ 1100
------
     0
```

The addition of the next two bits demonstrates Rule 2.

```
  1110
+ 1100
------
    10
```

Rule 3 is used to do the next addition.

```
   c1
  1110
+ 1100
------
   010
```

Rule 4 is used to complete the process.

```
  c11
   1110
+  1100
-------
  11010
```

To test the result, the three numbers can be converted to decimal. A check for a consistent pattern using a comparison of the decimal and binary systems is a good proof.

```
   1110 (base 2)  =   14 (base 10)
+  1100           = + 12
  -----              ---
  11010           =   26
```

Example: Add the binary numbers 11011 and 01011.

The carries and sum that result from this addition are shown below.

```
c11 11
  11011
  01011
 ------
 100110
```

LEADING ZEROS

A 1 or 0 must appear in every bit of a computer's word. There can be no blanks. For example, if the number 1011 were placed in an 8-bit memory, the stored number would appear as 00,001,011. Any of the 0s which are in front of the highest-order 1 are called *LEADING ZEROS.*

A printout of binary numbers stored in an 8-bit computer might take the form:

01011011
00010111
00100001
00000101

Each number contains eight bits. Each bit is either a 1 or a 0. This is the form in which the numbers are stored and manipulated within the computer.

Note, however, that it is difficult to pick out the largest number. The leading zeros cause confusion. Without them, the list would take the form:

1011011
10111
100001
101

Some computer programs can suppress leading zeros when data is printed. Within the computer, however, all bits must be filled. Workers in the digital electronics field should be skilled in reading numbers with leading zeros.

OVERFLOW

The size of a computer's word determines the largest number that can be stored, moved, and manipulated at one time. An *OVERFLOW* occurs when the result of an addition or other computation is larger than the number that can be represented by that number of bits. When an overflow occurs, the number reported by the computer is incorrect. If the incorrect number is used in future computations, additional errors will be present.

NUMBER SIZE

Large computers have 32 or more bits in their words. Microprocessors usually have 8- or 16-bit words. The maximum numbers that can be represented in such words are shown below:

Word Size	Binary	Decimal
8-bit	1111,1111	255
16-bit	1111,1111,1111,1111	65,535

If the sum resulting from an addition exceeds the above numbers, an overflow results. The concept of overflow will be covered next.

NATURE OF OVERFLOW

Overflow most often occurs in addition, rather than in subtraction. An example is useful. The addition of the following 8-bit numbers results in an overflow:

```
   c11  1   1
     11,011,001 (base 2) =    217 (base 10)
  +  01,010,101             +  85
     00,101,110                46
```

This sum is incorrect. Binary 00,101,110 equates to decimal 46, and 46 is certainly not the sum of 217 and 85. With only an 8-bit word, the largest number that can be manipulated is 255 (the sum of 217 and 85 is 302), so an overflow has occurred. A 9-bit word is required to hold 100,101,110 (the binary representation of 302), so the MSB bit of the sum has been lost. Unless steps are taken in the program, this error will go undetected.

DETECTING OVERFLOW IN UNSIGNED NUMBERS

In many computers, the sum appears in a register called the *ACCUMULATOR* or A register. A flip-flop is usually placed just beyond the MSB of this register. When a carry occurs out of the MSB of the sum, this flip-flop is set (Q = 1). When no carry occurs, it is reset (Q = 0). This flip-flop is called the *CARRY FLAG.*

When unsigned numbers are added, the CARRY flag should not be set. That is, there should be no carry out of the MSB.

A flag can be *TESTED* (checked to see if it is set or cleared). Programming methods are available for testing the CARRY flag. Programs should be written such that this flag is tested after each addition. If the CARRY flag is found to be set, a subroutine (a small program within a larger program) should be available to output an error message and possibly stop the computer.

TROUBLESHOOTING

If tests for overflow are not built into programs, overflow can create difficult troubleshooting problems. Overflow is the responsibility of programmers, but the errors that result from overflow may occur at almost random intervals. As a result, they may look much like intermittent circuit faults. Personnel working with hardware may spend much time looking for a circuit fault when the problem is really in the software. As a result, hardware personnel should be aware of this source of difficulty.

These additions resulted in incorrect sums. One was

caused by overflow; the other is the result of a faulty adder. Which is which?

```
a.   10,110,011        b.   00,111,001
     01,101,001             10,100,001
  c1 00,011,100          c1 11,000,010
```

The incorrect sum at the left resulted from overflow. This is a software problem. The incorrect sum at the right resulted from a faulty adder. This can be corrected by hardware repair methods.

SUBTRACTION OF UNSIGNED NUMBERS

The concept of carry is basic to addition. Its counterpart, *BORROWING,* is basic to subtraction. A borrow is necessary when a large digit is subtracted from a smaller digit. For example:

```
  32
 - 5
  27
```

The 5 is larger than the 2, so 10 must be borrowed from the 30. The 5 is then subtracted from the resulting 12.

Due to the limited symbol set, many carries are generated when binary numbers are added. The same is true of borrow when binary numbers are subtracted. To avoid this situation and simplify arithmetic circuits, most computers use a system called *TWO'S COMPLEMENT* to subtract.

DEVELOPING TWO'S COMPLEMENT

To review, to complement means to invert a bit. If it is a 1, it becomes a 0; if it is a 0, it becomes a 1. Two concepts are needed:

1. One's complement.
2. Two's complement.

One's complement is defined as follows:

1. One's complement: The one's complement of a multi-bit number is obtained by complementing all bits. For example:

   Number = 00,011,010

   $\bar{1}$s (one's complement) = 11,100,101

Note that leading 0s become 1s.

2. Two's complement: The two's complement of a number is obtained by adding 1 to its one's complement. For example:

```
Number = 00,010,110
1s     = 11,101,001
              +  1
2s     = 11,101,010
```

TWO'S COMPLEMENT SUBTRACTION

To accomplish a subtraction, the two's complement of the subtrahend (the number to be subtracted) is merely added to the minuend (the number from which it is to be subtracted). An example will show that this does result in a subtraction.

Example: Using two's complement, do the following subtraction: 00,011,010 − 00,001,111 (26 − 15 = 11 base 10).

Step 1: Compute the two's complement of the subtrahend.

```
Subtrahend = 00,001,111
1s         = 11,110,000
                   + 1
2s         = 11,110,001
```

Step 2: Add the minuend and the two's complement of the subtrahend. The result is the difference.

```
                        c111
Minuend            =      00,011,010  =   26
2s of subtrahend   =  +   11,110,001  = − 15
Difference             c1 00,001,011  =   11
```

The carry is ignored. It merely indicates that a borrow occurred during the subtraction. The process is somewhat like "casting out nines" in the decimal subtraction process.

Example: Using two's complement, do the following subtraction: 00,010,011 − 00,000,111 (19 − 7 = 12 base 10).

When doing two's complement subtraction, the order of addition is not important. That is, the two's complement can be found and directly added to the other number. For example:

```
Subtrahend =      00,000,111  =    7
1s         =      11,111,000
                         + 1
2s                11,111,001  =  − 7
Minuend         + 00,010,011  =   19
Difference     c1 00,001,100  =   12
```

Again the carry is ignored. Because these are unsigned numbers, all numbers must be positive. There is no way of indicating negative results. The minuend must be larger than subtrahend.

SIGNED NUMBERS

The numbers +455 and −22 are signed numbers. The first is larger than 0; the second is smaller than 0.

In computers, there is no direct way for keeping track of the signs of numbers. With only 1s and 0s, there is no unique symbol for + and −. To overcome this problem, one bit of a signed number is reserved for sign.

SIGN BIT

A standard method of indicating sign is to use the MSB as a SIGN BIT. If it is 0, the number is positive;

if it is 1, the number is negative. For example:

00,001,110 is positive
01,110,001 is positive
10,001,110 is negative

SIGN AND WORD SIZE

When one bit is used for the sign, the largest number that can be represented must be reduced. In 16-bit and larger machines, the loss of one bit may not be important. In an 8-bit machine, however, this loss might be critical. Maximum counts for signed numbers are:

Word Size	Binary	Decimal
8-bit	S111,1111	127
16-bit	S111,1111,1111,1111	32,767

S = sign bit

The decision to use signed numbers is made by the programmer. The loss of word size must be considered.

TWO'S COMPLEMENT AND SIGNED NEGATIVE NUMBERS

Two's complement is often used to represent negative numbers within computers. When a negative number is placed in a computer, it is usually immediately converted to its two's complement. It is then stored and used in this form. Note that the sign bit is automatically 1, so the use of a 1 in the sign bit to represent a negative number is compatible with two's complement arithmetic. For example:

Decimal number input	−18
Binary number	00,010,010
2s complement	11,101,110

The bottom number would appear inside most computers to represent −18 (base 10). Its sign bit is 1 (indicating that it is negative), and it is in two's complement form.

INVERSE OF TWO'S COMPLEMENT FOR NEGATIVE NUMBERS

When a negative number is to be output, it must be converted to its original form. That is, the two's complement process must be reversed. The reversal of the two's complement process is accomplished by again taking the two's complement. For example, the first number below is −18 in two's complement form. To convert it so the binary number can be read as 10,010, the two's complement is taken a second time.

Number in computer	=	11,101,110
1s	=	00,010,001
		+ 1
2s (binary 18)	=	00,010,010

Example: Determine the decimal value of the signed number 11,011,010.

The sign bit is 1, so it is a negative number. It is in two's complement form and must be converted before its decimal value can be read.

Number in computer	=	11,011,010
1s	=	00,100,101
		+ 1
2s	=	00,100,110
Decimal	=	−38

SIGNED ARITHMETIC

The basic computer operation is addition. Signed numbers are added like unsigned numbers. The difference is in the interpretation of the resulting sum. Sign bits can be added like any other digit. The resulting sign bit takes care of itself.

SUM OF POSITIVE AND NEGATIVE NUMBERS

Positive and negative numbers can be added in many combinations. Depending on the sizes of the numbers, the sum of a positive and a negative number may be positive or negative.

Example (Positive sum): The lamps shown in Fig. 11-8A show the content of two registers. When added, the sum appears in register A. These are signed numbers. What is their sum in base 10?

Adding the numbers as they appear, the sum is:

```
      c111 111
A =     11,101,110
B =     00,010,100
A = c1  00,000,010(sum)
```

The carry is ignored. The sum is positive and equals 2 (base 10). To test this result, A and B will be changed to decimal. Because A is negative, it must be converted before being changed to decimal.

A	=	11,101,110
1s	=	00,010,001
		+ 1
2s	=	− 00,010,010
A (base 10)	=	−18

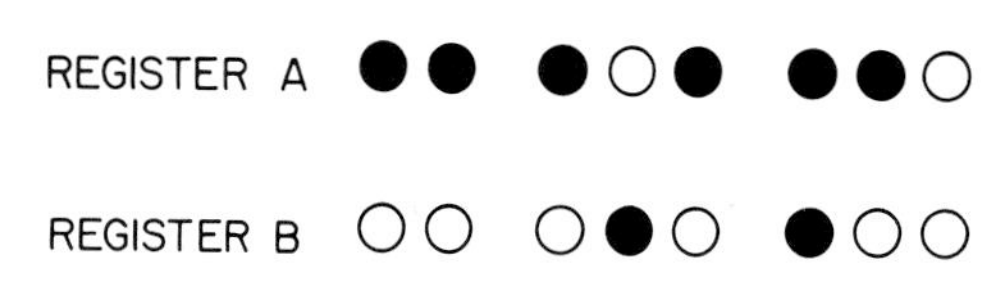

Fig. 11-8A. Registers contain two signed numbers to be added. The number in register A is negative; the number in register B is positive. The sum is put into register A. The sum is positive.

In base 10:

A = – 18
B = 20
A = 2 (sum)

Example (Negative sum): The lamps shown in Fig. 11-9 show the content of registers A and B. Determine the sum of these numbers and convert it to decimal.

A = 11,001,111
B = 00,010,000
A = 11,011,111 (sum)

The sign bit is 1, so the sum is negative. To test this answer, both sum and A must be converted to ordinary binary.

A = 11,001,111
$\bar{1}$s = 00,110,000
+ 1
$\bar{2}$s = – 00,110,001
A = – 49
Sum = 11,011,111
$\bar{1}$s = 00,100,000
+ 1
$\bar{2}$s = – 00,100,001
Sum = – 33

In base 10, the addition takes the form:

A = – 49
B = 16
A = – 33 (sum)

REGISTER A ●● ○○● ●●●

REGISTER B ○○ ○●○ ○○○

Fig. 11-9. Registers contain two signed numbers to be added. The sum is negative.

SUM OF TWO NEGATIVE NUMBERS

When two negative numbers are added, their sum should be negative. If their sum is positive, an overflow has occurred.

Example: Determine the sum of the numbers displayed in Fig. 11-10. Do the addition in binary. Then test the results using decimal.

A = 11,100,000
B = 11,001,000
A = c1 10,101,000 (sum)

The carry is ignored. The sum is indeed negative. To test the results, all three numbers must be converted to decimal. This means that the two's complements of all three numbers must be taken.

$\bar{2}$sA = – 00,100,000 = – 32
$\bar{2}$sB = – 00,111,000 = – 56
$\bar{2}$sA = – 01,011,000 = – 88 (sum)

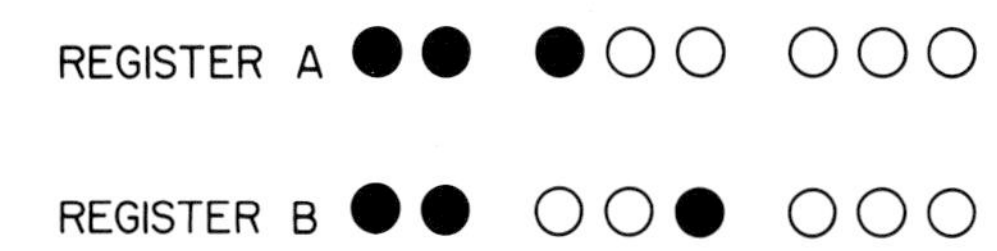

Fig. 11-10. If the sum of these negative numbers is positive, an overflow has occurred. If an overflow has not occurred, the sum of the two negative numbers is negative.

SUBTRACTING A NEGATIVE NUMBER

A negative number is subtracted in the same manner as a positive number. Its two's complement is added to the other number.

Example: Do the following subtraction using two's complements: A – B = 10,100,010 – 11,100,000. Note that both A and B are negative (both have a sign bit equal to 1).

B = 11,100,000
$\bar{1}$s = 00,011,111
+ 1
$\bar{2}$s = 00,100,000
A = + 10,100,010
Dif = 11,000,010

To test the results, all three numbers are converted to decimal.

$\bar{2}$sA = – 01,011,110 = – 94
$\bar{2}$sB = – 00,100,000 = – (– 32)
$\bar{2}$sDif = – 00,111,110 = – 62

OVERFLOW

When signed numbers are added, the carry flag may be set when borrows occur. This means that the carry flag cannot be used as an overflow indicator as it was for unsigned numbers. Rather, the sign bit must be tested to determine if an overflow has occurred. Two rules are important. These rules are:

1. When numbers with opposite signs are added, no overflow will occur.
2. When numbers with the same sign are added, an overflow has occurred when the sign of the sum differs from that of the original numbers.

When numbers with opposite signs are added, the absolute value of the sum will always be smaller than the larger of the original numbers. This can be shown by adding decimal numbers with opposite signs.

```
     9          (− 9)
+ (− 4)      +    4
-------      -------
     5          − 5
```

The size of the sum (whether positive or negative) is smaller than the larger of the original numbers. It can be assumed that the original numbers are within the capacity of the computer.

To apply rule 1 to signed numbers, the sign bits of the numbers to be added are inspected. If they differ, an overflow cannot occur. For example:

```
     01,100,011
+    10,110,100
---------------
c1   00,010,111
```

The setting of the carry flag is unimportant, since this is signed arithmetic. The original numbers have opposite signs, so there is no overflow. Rule 2 of the two rules is important (when numbers with the same sign are added, an overflow has occurred when the sign of the sum differs from that of the original numbers).

The sum of positive numbers should be positive; the sum of negative numbers should be negative. If the sign of the sum does not match the sign of the numbers added, an overflow has occurred. For example:

```
  01,111,101 =   125
+ 00,100,001 = +  33
--------------------
  10,011,110    − 98
```

The sum is negative. When converted to decimal (its two's complement was taken before it was converted), it was found to equal − 98. This is certainly not the sum of 125 + 33. This error can be detected by inspecting the sign bits.

Overflow involving negative numbers has the same symptom. For example:

```
     10,000,100 =   (− 124)
+    11,000,101 = + ( − 59)
--------------------------
c1   01,001,001 =       73
```

Ignore the carry, since this is signed arithmetic. The change in the sign bit (D7) indictes an overflow.

SUMMARY

Personnel working with hardware must be familiar with binary numbers and arithmetic. They should be able to quickly make binary-to-decimal and decimal-to-binary conversions.

Two's complement is often used to do binary subtraction. Two's complement is also used to represent negative numbers in signed arithmetic. The MSB is usually the sign bit. If it is 0, the number is positive; if it is 1, the number is negative.

Overflow is a programming problem. However, personnel working with hardware should be aware of this source of error. In unsigned arithmetic, the carry flag can be used to indicate overflow. In signed arithmetic, the sign bit is used.

IMPORTANT TERMS

Leading zeros, One's complement, Overflow, Sign bit, Two's complement.

TEST YOUR KNOWLEDGE

1. On the count after the symbol set has been exhausted, a 0 results, and a __________ is generated.
2. PV stands for __________ __________.
3. The number 32 is the place value for __________ (D3, D4, D5, D6).
4. The binary-to-decimal conversion process described in the first chapter can be expressed in __________ form.
5. Converting a number from decimal to binary uses the __________ method or the LSB method.
6. If two negative numbers are added together and the resulting sum is positive, an overflow has occurred. (True or False?)
7. If two negative numbers are added together and the resulting sum is negative, no overflow has occurred. (True or False?)
8. The two numbers resulting from the division process are the quotient and the __________.
9. LSB stands for ____________________________.
10. Which is larger, the MSB or the LSB?
11. In the LSB method of decimal-to-binary conversion, the bit is a 1 or a 0 depending on whether the __________ (quotient, remainder) is 1 or 0.
12. The number −6 is a/an __________ (signed, unsigned) number.
13. The decimal number 1/2 is written in binary as __________ (0.5, 0.1, 0.01, 0.05).
14. The number $0.1111111\overline{1}$ in the binary numbering system equals ______ (1/2, 2, 1).
15. The point which separates whole binary numbers from fractional binary numbers is called the __________ __________.
16. In a binary number which has no blanks in the number, the 0s which are in front of the highest-order 1 are called ________ ________.
17. What is rule 4 of binary addition?
18. In binary, 1 + 1 = __________.
19. Some computer programs __________ leading zeros when data is printed.
20. Using unsigned numbers in an 8-bit microprocessor, what is the maximum number represented (by all digits equal to 1)?

21. The flip-flop in a circuit built for unsigned numbers that stores the carry out is called the __________ __________.
22. In subtraction, the number to be subtracted is the __________, while the number from which it is to be subtracted is the minuend.
23. To accomplish a subtraction, the __________ __________ of the subtrahend is added to the minuend.
24. Is a carry in subtraction ignored?
25. To reverse the process of taking the two's complement, do the following:
    a. Take the one's complement.
    b. Invert the sign bit.
    c. Take the two's complement.
    d. Take the one's complement and subtract 1.

STUDY PROBLEMS

1. Using binary, count from 0 to 20 (base 10).
2. Convert the following binary numbers to decimal.
    a. 1,011
    b. 101,000
    c. 1,001,100
    d. 11,100,001
3. Using the MSB method, convert these decimal numbers to binary.
    a. 19
    b. 100
    c. 300
    d. 1200
4. Using the LSB method, convert these decimal numbers to binary.
    a. 19
    b. 100
    c. 200
    d. 109
5. Do the following binary additions. Give your answers in binary.

```
a.    101          d.     101,100
   +    1             +    10,101

b.    1,011        e.  11,010,110
   +      1           +  1,001,101

c.   11,011        f.  11,110,000
   +     11           +     10,010
```

6. Write the one's and two's complements of these numbers.
    a. 00,010,011
    b. 00,110,110
    c. 10,110,000
    d. 11,111,011
7. Using two's complements, do the following subtractions.

```
a.    00,111,000       c.    00,101,001
    − 00,000,111           − 00,010,001

b.    10,001,110       d.    00,001,000
    − 01,100,100           − 00,000,101
```

8. The following are signed numbers. Which are positive and which are negative?
    a. 00,111,001
    b. 10,000,011
    c. 11,010,110
9. Convert these signed numbers to base 10. Indicate the signs of the base 10 numbers.
    a. 01,011,011
    b. 11,100,011
    c. 11,010,100
10. These are unsigned numbers. Do the indicated additions and indicate which result in overflows.

```
a.    00,110,110       c.    10,100,000
    + 00,010,010           + 01,110,000

b.    10,110,101
    + 01,101,000
```

11. Both of the following additions indicated overflows. One is a true overflow (the sum is too large); the other was caused by a faulty adder. Which was caused by the faulty adder? These are unsigned numbers.

```
a.      01,001,111     b.      01,001,111
      + 01,010,001           + 11,110,001
   c1   00,011,110        c1   01,000,000
```

12. These are signed numbers. Do the following additions and indicate which result in overflows.

```
a.    10,001,011       d.    10,001,001
    + 01,111,000           + 11,111,000

b.    00,101,001       e.    00,100,000
    + 01,110,000           + 01,100,000

c.    10,100,000       f.    10,000,111
    + 01,110,000           + 10,000,001
```

Chapter 12

Numbering Systems and Codes Other Than Binary System

LEARNING OBJECTIVES

After studying this chapter and completing lab assignments and study problems, you will be able to:

- Count using the octal and hexadecimal numbering systems.
- Convert between binary and octal numbers and between binary and hexadecimal numbers.
- Convert between decimal and octal numbers and between decimal and hexadecimal numbers.
- Convert between decimal and binary-coded decimal numbers.
- Describe the nature of the gray code and its most important application.
- State the meaning of the term ASCII.

Binary numbers are difficult to remember, express in words, and write without error. Converting binary numbers to decimal numbers overcomes these problems. However, the conversion process requires computations and increases the possibility of error.

Two numbering systems, OCTAL and HEXADECIMAL, have an interesting relationship to binary. Binary numbers can be converted to octal or hexadecimal by inspection (without computations). As a result, binary numbers are often expressed in the octal or hexadecimal system.

NUMBERS IN OCTAL SYSTEM

The term octal implies eight. The octal numbering system has a base or radix of eight, so its symbol set consists of 0, 1, 2, 3, 4, 5, 6, and 7. The symbol 8 does not appear in any octal number. The combination symbol 10 is used to represent a quantity of 8 in the octal numbering system.

COUNTING IN THE OCTAL SYSTEM

As in the decimal and binary systems, counting in the octal system proceeds until all of the symbols have been used. Then a 0 results and a carry is generated. Refer to Fig. 12-1 for a summary of different counting system procedures.

DECIMAL	BINARY	OCTAL
0	0	0
1	1	1
2	10	2
3	11	3
4	100	4
5	101	5
6	110	6
7	111	7
8	1,000	10 *
9	1,001	11
10	1,010	12
11	1,011	13
12	1,100	14
13	1,101	15
14	1,110	16
15	1,111	17
16	10,000	20 *
17	10,001	21
18	10,010	22
19	10,011	23
20	10,100	24
21	10,101	25
22	10,110	26
23	10,111	27
24	11,000	30 *
25	11,001	31
26	11,010	32
27	11,011	33
28	11,100	34
29	11,101	35
30	11,110	36
31	11,111	37
32	100,000	40 *

* CARRY

Fig. 12-1. Counting in the decimal system, binary system, and octal system. In each system, counting proceeds until the symbol set is exhausted. Then a 0 results and a carry is generated. Notice that the carry events in the octal numbering system occur at several of the carry events for the binary numbering system. This makes conversions between the two systems easy.

USE OF OCTAL NUMBERING SYSTEM

The relation between the octal system and binary system permits easy conversion between these systems. The first three bits of a binary number represent counts between 0 and 7. This corresponds to the range of the octal symbol set. Note the range of symbols for the binary system and octal system in the following:

Binary	Octal
000	0
001	1
010	2
011	3
100	4
101	5
110	6
111	7

On the next count (count 8 in decimal), a carry is generated in both systems. In the binary system, that carry goes into the next group of three bits. In the octal system, the carry goes into the next higher digit. That is:

Binary	Octal
1,000	10

The groups of three in the binary number and the digits in the octal number build together. As a result, binary numbers can be converted to octal numbers merely by separating the digits into 3-bit groups and converting them according to the previous table. For example:

101,011 (binary) = 53 (octal)

BINARY-TO-OCTAL CONVERSION

Part "a" of Fig. 12-2 shows a 9-bit monitor. To aid in reading, its lamps have been arranged in groups of three. The binary number can be read as the octal number 263. The advantage of this number (263) over its binary equivalent (010,110,011) is easily seen.

A word of caution is needed. Although 263 appears to be a decimal number, it is not. It is octal. For example, 263 + 5 does not equal 268. It equals 270 (octal). There is no symbol 8 in the octal system. When 5 (octal) is added to 3 (octal), the sum is the combination symbol 10 (octal).

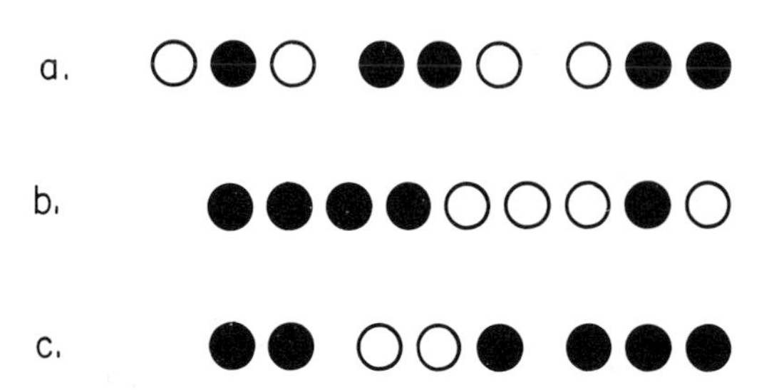

Fig. 12-2. Reading binary numbers directly in octal is aided by arranging the bits in groups of three.

The term BINARY-CODED-OCTAL (BCO) may be used to describe the lamp arrangement in part "a" of Fig. 12-2. As far as the circuit is concerned, however, the data is presented in binary form (no conversion is needed to output in BCO). It is the user that converts the number from binary to octal.

While the groups of three aid in reading the monitor in part "a" of Fig. 12-2 in octal, such grouping is not needed. See part "b" of Fig. 12-2. The 9-bit monitor shown can be read as 742 (octal) even though the lamps are equally spaced.

Part "c" of Fig. 12-2 shows an 8-bit monitor. With only two lamps in the highest-order group, the third digit of the octal number cannot be larger than 3. This monitor is displaying the octal number 317.

OCTAL-TO-BINARY CONVERSION

Octal numbers are easily converted to binary. For example, 245 (octal) equals 10,100,101 (binary). As a result, binary numbers are often written as octal in the literature. For example, the op code for a given operation might be 11,100,000. In the instruction manual, it would be written as 340 (octal).

OCTAL AND DECIMAL CONVERSIONS AND EQUATIONS

The primary use of octal is the representation of binary numbers in condensed form. However, there are times when conversions between the octal system and the decimal system are necessary. Examples can be checked one place value at a time. A fundamental equation can show the relation between the digits of two or more numbering systems.

OCTAL-TO-DECIMAL CONVERSION

Conversion from octal to decimal is similar to the conversion from binary to decimal. It is based on place values. Two topics must be examined. They are:

1. Octal place values.
2. Octal-to-decimal conversion process.

The following look at the first topic is a review of place values:

1. Octal place values: In the last chapter, an equation for expressing binary numbers in decimal was used. It has the form:

$$Dn(Pn) + \ldots + D3(8) + D2(4) + D1(2) + D0(1) = N \text{ (base 10)}$$

The place values (numbers in the brackets) have an interesting relation to the system's base or radix. They represent that radix raised to the power of the position. That is:

$$Dn(2^n) + \ldots + D3(2^3) + D2(2^2) + D1(2^1) + D0(2^0) = N$$

Any number raised to the 0 power equals 1. Any number raised to the 1 power equals itself. The rest of the place values can be found by multiplying 2 times itself the indicated number of times. For example, the place value of D3 is 2 x 2 x 2 = 8.

Also note the match between the exponents and the suffixes for the Ds. A suffix is a symbol following another symbol. The numbers following the Ds are suffixes. To find the place value of any digit, the radix is raised to the power of its suffix. For example, the place value of D5 is 2 raised to the fifth power (2 x 2 x 2 x 2 x 2 = 32).

This equation explains why 0 was used to designate the unity bit. To have a match between the suffixes and powers, the series had to start with D0.

The above equation can be generalized (made to work for any base) by substituting R for the 2s in the exponential terms. That is:

$$Dn(R^n) + \ldots + D3(R^3) + D2(R^2) + D1(R^1) + D0(R^0) = N$$

By substituting the radix of the octal system (8) for R, the place values for the octal system can be found.

$$Dn(8^n) + \ldots + D3(8^3) + D2(8^2) + D1(8^1) + D0(8^0) = N$$

The place values are found by evaluating the exponential parts. Remember that any number raised to the 0 power is 1; any number raised to the 1 power is that number.

$$Dn(8^n) + \ldots + D3(512) + D2(64) + D1(8) + D0(1) = N$$

Note that each place value is 8 times the previous value. Higher-order values are easily obtained.

D6	D5	D4
262,144	32,768	4096

2. Octal-to-decimal conversion process: Conversion from octal to decimal is accomplished in the same way as binary-to-decimal conversion. Each digit is multiplied by its place value. The sum of the resulting products is the decimal value.

Example: Convert 362 (base 8) to decimal.

The digits of the octal number (D2 = 3, D1 = 6, and D0 = 2) are multiplied by their place values. The sum of the resulting products is the decimal equivalent.

$$D2(8^2) + D1(8^1) + D0(8^0) = N \text{ (base 10)}$$
$$3(64) + 6(8) + 2(1) = N$$
$$242 = N$$

That is, 362 (base 8) equals 242 (base 10).

DECIMAL-TO-OCTAL CONVERSION

The two methods used to obtain binary numbers from decimal numbers will be used to convert decimal numbers to octal numbers. These methods are:

1. The MSD (most significant digit) method.
2. The LSD (least significant digit) method.

The first topic is simple, as shown in the following:

1. MSD method: The term bit does not apply to octal numbers, so the highest-order digit in an octal number is designated MSD (most significant digit). In this method, the highest-order, non-zero digit is found first.

Example: Convert the decimal number 356 to octal.
Computation: Refer to part "a" of Fig. 12-3. As

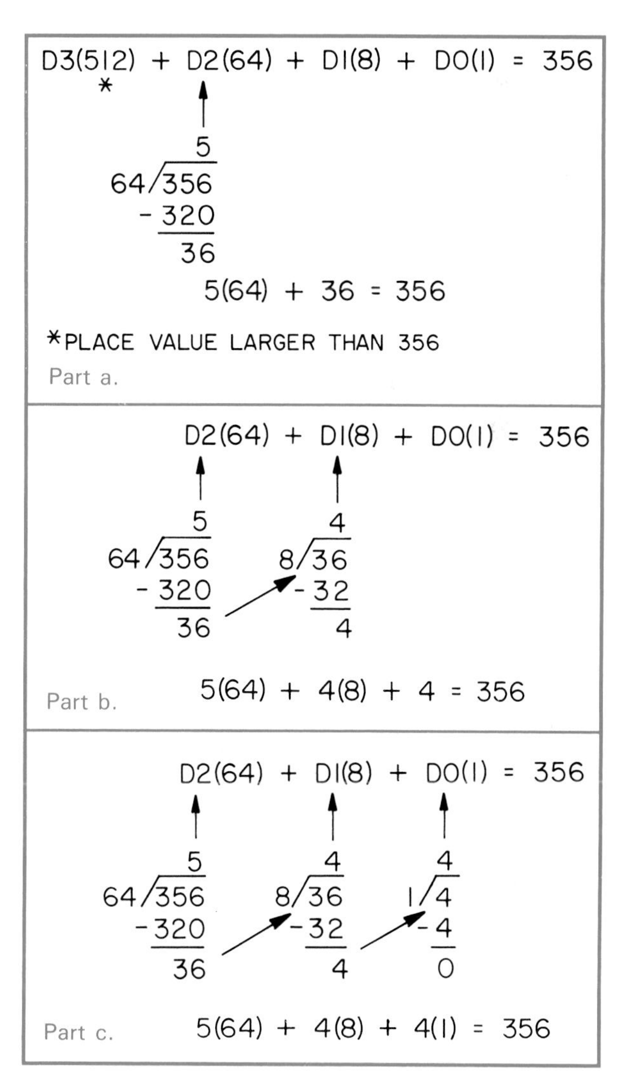

Fig. 12-3. Steps in the conversion of the decimal number 356 to octal using the MSD method. A procedure is used that is like that used to convert decimal numbers to the equivalent binary numbers.

before, place values are written until one is found that is larger than the number being converted. See the first row in Fig. 12-3. The place value of D3 is larger than 356, so the MSD of the octal number will be in the position represented by D2.

To determine D2, its place value is divided into the decimal number. The portion of the resulting product that is a whole number is D2. D2 equals 5.

Computation: To determine D1, the remainder from the last computation is divided by the place value of D1. See part "b" of Fig. 12-3. The whole portion of the resulting quotient is D1 (D1 = 4).

Computation: The value for the digit D0 is found next. In part "c" of Fig. 12-3, a step can be saved if the procedure is well thought out. The remainder from the previous computation is the last digit of the octal number. D0 = 4.

The result is checked by reversing the conversion. Based on this conversion, the number 544 (base 8) equals the number 356 (base 10).

2. LSD method: The LSD or successive-division method can be used to convert from decimal to octal. The process is like that for decimal to binary. As before, the LSD is found first.

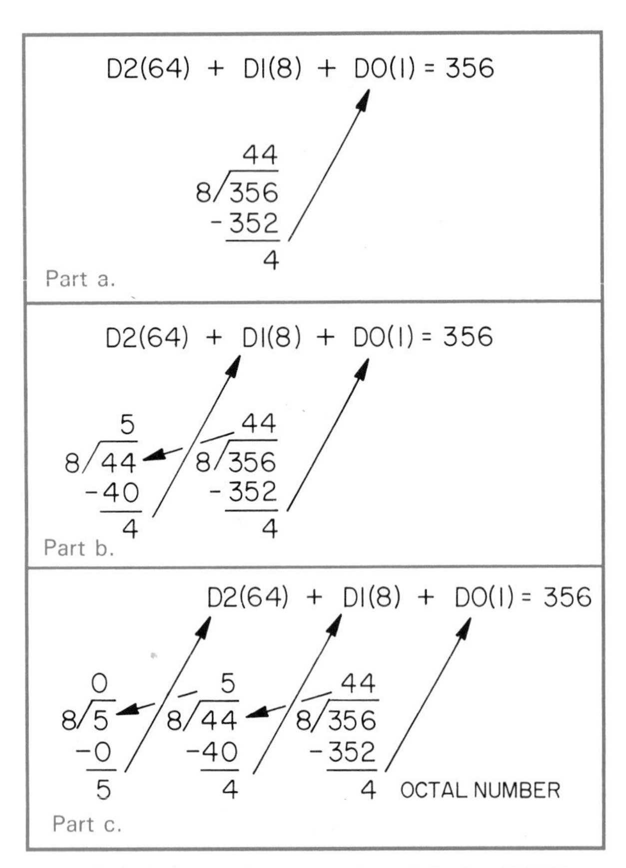

Fig. 12-4. Steps in the conversion of decimal 356 into octal using the LSD method. The procedure is like that for decimal to binary.

Example: Use the LSD method to convert 356 (base 10) to octal. Note the similarities with converting from decimal to binary by using the LSB method.

Computation: The base of the system (8) is divided into the decimal number. See part "a" of Fig. 12-4. The resulting remainder is D0. D0 = 4.

Computation: To determine D1, 8 is divided into the whole portion of the quotient of the previous division. 8 is divided into 44. See part "b" of Fig. 12-4. The remainder of this division is D1. D1 = 4.

Computation: The whole portion of this quotient (5) is less than 8, so it is D2, the MSD of the new octal number. See part "c" of Fig. 12-4. D2 = 5.

Remember, the LSD is found first when this method is used. The result is that 356 (base 10) equals 544 (base 8).

HEXADECIMAL NUMBERS

Hex or hexa means six. Decimal implies ten. When the term hexa is combined with the term decimal, the resulting term implies sixteen. The hexadecimal system has a base of 16, so there are 16 symbols in its symbol set. They are 0, 1, 2, 3, 4, 5, 6, 7, 8, 9, A, B, C, D, E, and F. Some letters are used as numbers. Any symbols could have been used beyond 9. For example, 0, 1, 2, 3, 4, 5, 6, 7, 8, 9, @, &, $, ¢, #, and * would have been acceptable.

COUNTING IN HEXADECIMAL

As in decimal, binary, and octal, counting in hexadecimal proceeds until all symbols have been used. Then a 0 results and a carry is generated. See the table in Fig. 12-5. The set is depleted at the count of F (15 in base 10). The next count (16 in base 10) is represented by the number 10 (base 16).

USE OF HEXADECIMAL

The properties of hexadecimal are similar to those of octal. The following two situations are true:

1. As with the conversion from binary to octal, the conversion from binary to hexadecimal can be made by inspection.
2. As with the conversion from octal to binary, the conversion from hexadecimal to binary can be made by inspection.

The first topic is clarified as follows:

1. Binary to hexadecimal: Sixteen counts are needed to generate a carry in hexadecimal. Four bits are needed to represent this count in binary. As a result, only two hexadecimal digits are needed to represent an 8-bit binary number. For example, the binary number displayed in part "a" of Fig. 12-6

DECIMAL	BINARY	OCTAL	HEX.
0	0	0	0
1	1	1	1
2	10	2	2
3	11	3	3
4	100	4	4
5	101	5	5
6	110	6	6
7	111	7	7
8	1,000	10 *	8
9	1,001	11	9
10	1,010	12	A
11	1,011	13	B
12	1,100	14	C
13	1,101	15	D
14	1,110	16	E
15	1,111	17	F
16	10,000	20 *	10 *
17	10,001	21	11
18	10,010	22	12
19	10,011	23	13
20	10,100	24	14
21	10,101	25	15
22	10,110	26	16
23	10,111	27	17
24	11,000	30 *	18
25	11,001	31	19
26	11,010	32	1A
27	11,011	33	1B
28	11,100	34	1C
29	11,101	35	1D
30	11,110	36	1E
31	11,111	37	1F
32	100,000	40 *	20 *

* CARRY

Fig. 12-5. Counting in decimal, binary, octal, and hexadecimal.

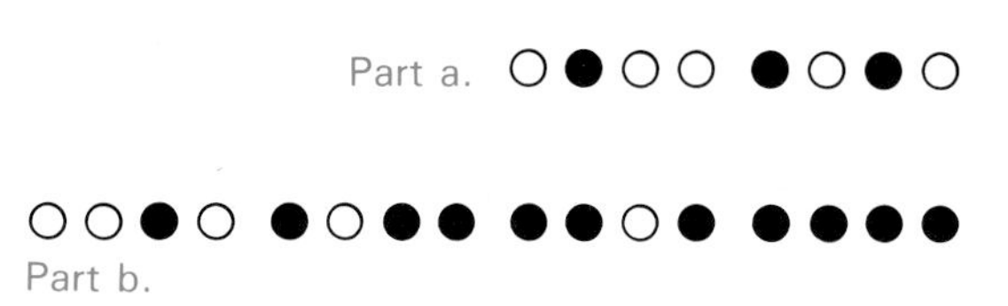

Fig. 12-6. Arranging bits in groups of four aids when binary numbers are to be read directly in the hexadecimal system.

can be directly read in hexadecimal.

0100,1010 (base 2) = 4A (base 16)

The term *BINARY-CODED-HEXADECIMAL* (BCH) is used to describe the grouping of lamps to aid in reading binary numbers in hexadecimal. Remember, the lamps are really displaying binary numbers. They are merely read in hexadecimal.

Hexadecimal has advantages over octal. Only two hexadecimal digits are needed to describe an 8-bit binary number. For example:

10111001 base 2
271 base 8
B9 base 16

The advantage is greater when 16-bit numbers are considered, as shown in part "b" of Fig. 12-6. This number converts to the number 2BDF (base 16) or the number 025737 (base 8).

Learning the new symbols (those above 9) is a disadvantage. However, as you work with these numbers, they will become as familiar as the symbols 0 through 9. Fig. 12-7 shows lamp patterns for the counts 9 through F.

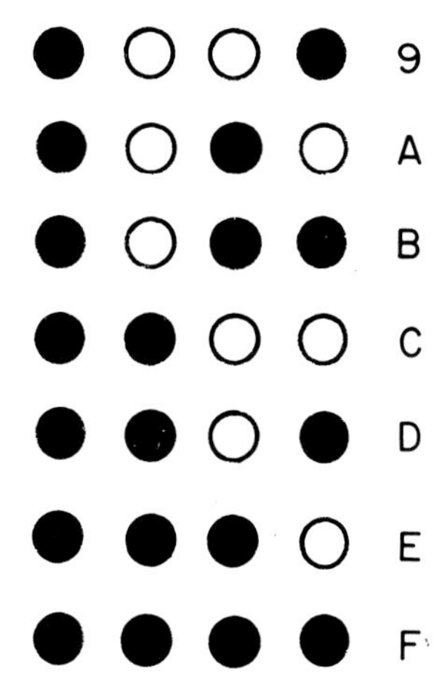

Fig. 12-7. Most people quickly become familiar with lamp patterns for hexadecimal numbers A through F.

2. Hexadecimal to binary: Conversion of hexadecimal numbers to binary is easily accomplished. For example, A5 (base 16) can be read as 1010,0101 (base 2).

In control applications, computer inputs and outputs are often in hexadecimal. Fig. 12-8 shows a keypad from a microprocessor-based computer. The numbers run from 0 through F. A simple 1- or 2-chip circuit converts the keypad's output to binary.

Fig. 12-8. Keys A through F on this key pad do not represent letters. They are the hexadecimal numbers A (10 in base 10) through F (15 in base 10).

Fig. 12-9. In the display shown, the symbols A and F are hexadecimal numbers. b and d are lower case letters.

HEXADECIMAL AND DECIMAL CONVERSIONS AND EQUATIONS

Fig. 12-9 shows a 7-segment display from a small computer. It is outputting the hexadecimal number 03AF. (The leading zero has not been supressed.) If this is output data, it would probably be converted to decimal before being used. The processes for converting a hexadecimal number to a decimal number and converting a decimal number to a hexadecimal number follow.

HEXADECIMAL-TO-DECIMAL CONVERSION

Conversion of hexadecimal to decimal is similar to octal-to-decimal conversion. It is based on place values. Two topics must be examined. They are:

1. Hexadecimal place values.
2. Hexadecimal-to-decimal conversion process.

The first topic can be organized as follows:

1. Hexadecimal place values: The general equation for expressing numbers in decimal can be used to determine hexadecimal place values. A change of the base from 2 or 8 to 16 is used as follows.

$$Dn(16^n) + \ldots + D3(16^3) + D2(16^2) + D1(16^1) + D0(16^0) = N$$

Evaluating the exponential terms results in the following.

$$Dn(16^n) + \ldots + D3(4096) + D2(256) + D1(16) + D0(1) = N$$

The numbers in the brackets are the place values. Starting at the number 1, each place value is 16 times the previous place value.

2. Hexadecimal-to-decimal conversion process: To convert a hexadecimal number to decimal, each digit is multiplied by its place value. The sum of the resulting products is the decimal equivalent.

Example: Convert the hexadecimal number 3AF to decimal. Recall the hexadecimal number 03AF from Fig. 12-9. The leading zero is not suppressed. It can be suppressed in the following calculation. The digits 3, A, and F are used. "A" equals 10 (base 10) and "F" equals 15 (base 10). Substituting into the equation:

$$3(256) + 10(16) + 15(1) = N \quad \text{(base 10)}$$
$$943 = N$$

DECIMAL-TO-HEXADECIMAL CONVERSION

The two conversion methods used for binary and octal can be used for hexadecimal. These methods are:

1. The MSD method.
2. The LSD method.

The first method is easy to use, as in the following:

1. MSD method: As before, the MSD is found first. The method takes a minimum of steps.

Example: Convert 713 (base 10) to hexadecimal using the MSD method. Note the similarities with converting from decimal to binary by using the MSB method.

Computation: To determine the highest-order, nonzero digit in the hexadecimal number, place values are written until one is found that is larger than the decimal number. See part "a" of Fig. 12-10. In this case, the place value of D3 is larger than N, so D2 is the MSD.

To find D2, its place value is divided into 713. The whole portion of its quotient is D2 (D2 = 2).

Computation: The remainder from the previous division is used to find D1. That remainder is divided by the place value of D1. See part "b" of Fig. 12-10. The whole portion of the resulting quotient is D1. It equals 12 (base 10) or C (base 16).

Computation: D0 will equal the remainder from the last division. See part "c" of Fig. 12-10.

D3(4096) + D2(256) + D1(16) + D0(1) = 713

256/713 = 2
−512
201

2(256) + 201 = 713

*PLACE VALUE LARGER THAN 713

Part a.

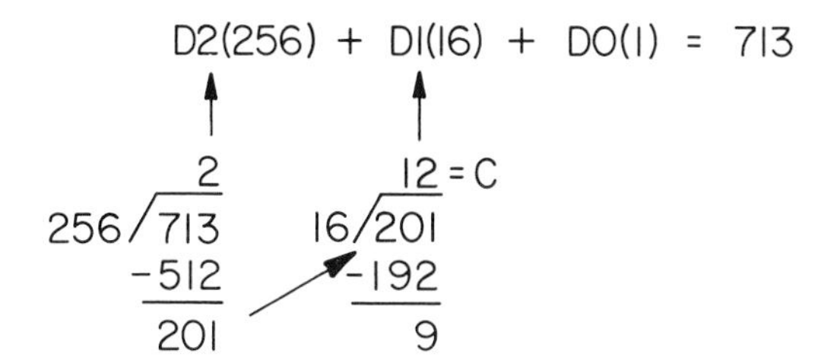

Part b. 2(256) + 12(16) + 9 = 713

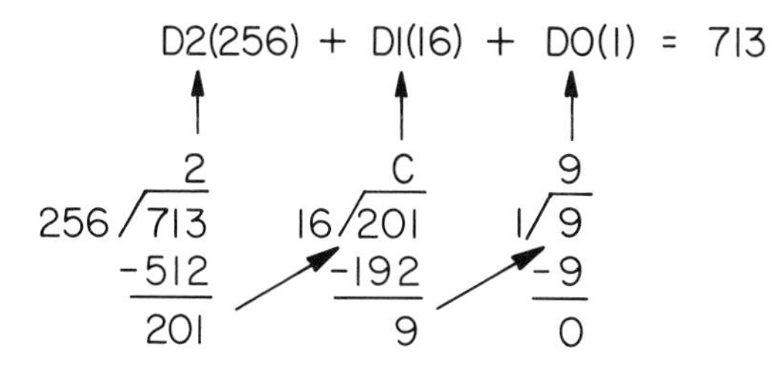

Part c. 2(256) + 12(16) + 9(1) = 713

Fig. 12-10. Steps in the conversion of the decimal number 713 to a hexadecimal number using the MSD method.

The decimal number 713 equals 2C9 (base 16). The symbol "C" is treated like any other digit in the hexadecimal number system.

2. LSD method: The LSD or successive division method for converting a decimal number to a hexadecimal number can be used. Sometimes more steps are used with the LSD method than for the MSD method.

Example: Convert 4000 (base 10) to hexadecimal. Use the LSD method. Note the similarity with decimal-to-binary conversion using the LSB method. In a similar way, the process is identical to that of the decimal-to-octal conversion except 16 is used instead of 8.

Computation: The decimal number is first divided by the radix of the system (16). See part "a" of Fig. 12-11. The remainder from this division is D0. In this case, the remainder is 0, so D0 = 0 (base 16).

Computation: Higher-order digits are found by repeating the division by 16. The whole number portion of the quotient is used in each of these divisions. The remainder becomes the value of the previous digit. See part "b" of Fig. 12-11.

The hexadecimal equivalent of 4000 (base 10) is FA0. To test this result, it can be converted back to decimal.

15(256) + 10(16) + 0(1) = 4000

D2(256) + D1(16) + D0(1) = 4000

16/4000 = 250
−4000
0

Part a.

D2(256) + D1(16) + D0(1) = 4000

16/15 = 0, −0, 15 = F
16/250 = 15, −240, 10 = A
16/4000 = 250, −4000, 0

HEX. NUMBER

Part b.

Fig. 12-11. Steps in the conversion of the decimal number 4000 to a hexadecimal number using the LSD method.

CONVERSIONS BETWEEN HEXADECIMAL AND OCTAL SYSTEMS

The easiest way to convert between hexadecimal and octal is through binary. The number to be converted is written in binary. The binary number is then converted to the desired system. The following is an example of a hexadecimal-to-octal conversion.

Base 16	Base 2	Base 8
B8 =	1011,1000	
	10,111,000	= 270

Converting a number from octal to hexadecimal is similar to that for hexadecimal to octal. The following is an example.

Base 8	Base 2	Base 16
307 =	11,000,111	
	1100,0111	= C7

RELATION OF PLACE VALUES TO NUMBER OF DIGITS

It is useful to know how many digits are in a number for one base compared to another base. It is appropriate to review the concept of place values. The place value of D0 is 1 in all systems. See the table shown in Fig. 12-12. The place value of D1 equals the system's base or radix.

The larger the symbol set, the larger the place values of a numbering system. Because each symbol contains more information, systems with large bases require fewer digits to express a given number. For example:

0110,1100 (base 2) = 6C (base 16)

As a result, numbers with large bases are easier to remember, write, and describe verbally.

PLACE VALUES IN BASE 10

SYSTEM	D4	D3	D2	D1	D0
BASE 16	65536	4096	256	16	1
BASE 10	10000	1000	100	10	1
BASE 8	4096	512	64	8	1
BASE 2	16	8	4	2	1

Fig. 12-12. Place values of the first five digits of the numbering systems that are most often used.

DECIMAL	BCD
0	0
1	1
2	10
3	11
4	100
5	101
6	110
7	111
8	1000
9	1001
10	1,0000
11	1,0001
12	1,0010
13	1,0011
14	1,0100
15	1,0101
16	1,0110
17	1,0111
18	1,1000
19	1,1001
20	10,0000

Fig. 12-13. Counting in decimal and BCD (binary-coded-decimal). Note the similarity to the binary system for the first 9 digits. The system does not follow the binary pattern for decimal numbers equal to or larger than 10 (base 10).

BINARY-CODED-DECIMAL SYSTEM

A code is a system of symbols and rules for representing information. BCO (binary-coded-octal) and BCH (binary-coded-hexadecimal) can be thought of as natural codes. Conversion from binary into these codes comes directly from the representation of binary numbers. Bits only need to be placed in groups of three or four to aid in the direct conversion of a binary number to an octal or hexadecimal number.

This direct conversion is not possible between binary and decimal. To overcome this limitation, the *BINARY-CODED-DECIMAL* system was developed. It is not a natural code, since it is not a true binary system.

COUNTING IN BCD

The table in Fig. 12-13 shows the method of counting in BCD. From 0 to 9 (base 10), the count follows the rules of binary. However, on the count after 1001, the low-order group of four bits goes to 0000, and a carry is generated. The BCD number 1,0000 equals decimal 10.

When the first four bits again reach 1001, the action is repeated. That is, the first four bits cannot count higher than 9 (base 10).

The action of the next group of four bits is identical to that of the first four. When the number 1001,1001 (99 in base 10) is reached, both groups clear, and the number 1,0000,0000 (100 in base 10) results from the counting process.

AN ADVANTAGE OF BCD IS RELATED TO SIMPLICITY

BCD allows the direct conversion of numbers expressed as 1s and 0s to decimal. As a result, the primary use of the BCD numbering system is at the

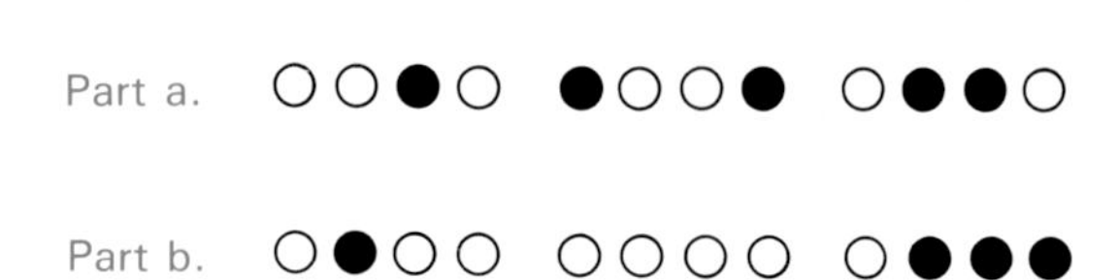

Fig. 12-14. The indicators shown are displaying BCD numbers, so no group of four lamps can output a binary number larger than 9 (base 10). The examples can be used to convert BCD numbers to decimal.

interface between a computer and the human operator.

Only a minimum knowledge of binary numbers is needed to convert the BCD number in part "a" of Fig. 12-14 to decimal. It reads 296.

Each group of four bits converts to a decimal digit. The number that is displayed in part "b" of Fig. 12-14 is 407 (base 10).

DISADVANTAGES OF BCD

Within a computer, BCD is usually an inconvenient way to represent numbers. Binary is usually better. Choices sometimes have to be made between speed and efficiency. Three problems are:

1. Unused bit combinations with BCD numbering system.
2. Special arithmetic is needed with BCD numbers.
3. BCD numbering system is nibble oriented, in opposition to computer structure, which is byte oriented.

Examine the first of the topics for comparison with binary as follows:

1. Unused bit combinations with BCD numbering system: In BCD, no 4-bit group can count higher than 1001 (9 in base 10). This means that bit combinations 1010, 1011, 1100, 1101, 1110, and 1111 are never used. As a result, a counter using a 4-bit binary number can count to 15 (base 10); a counter using a 4-bit BCD number can count only to 9 (base 10). Because of unused bit combinations, BCD numbers require more space in memory than equivalent binary numbers.
2. Special arithmetic is needed with BCD numbers: While BCD numbers look like binary, they are not. The addition of the BCD numbers 0011 and 1000 cannot be accomplished in an ordinary binary adder. The sum would be 1011, which is incorrect. It should be 0001,0001 (BCD).

Special circuits or programming must be used with BCD. The result is increased cost.

3. BCD numbering system is nibble oriented, in opposition to computer structure, which is byte oriented: Terms used to describe the parts of a word are shown in Fig. 12-15. To review, a BIT is a single digit of a number made up of 1s and 0s. A group of four bits is a *NIBBLE*. A *BYTE* is an 8-bit group.

Computers tend to be byte oriented. For example, word size for a computer is usually a multiple of eight bits (8, 16, 32, etc.)

BCD is nibble oriented. Even in small computers, the storing of a single 4-bit BCD digit in an 8- or 16-bit memory location results in inefficient use of memory.

To overcome this problem, a process called *PACKING* is often used. Digits from more than one BCD number are placed in each word. For example, two BCD digits are placed in an 8-bit word; four can be packed into 16-bits, etc. However, packing is not without cost. Programming steps are needed to pack and unpack such words.

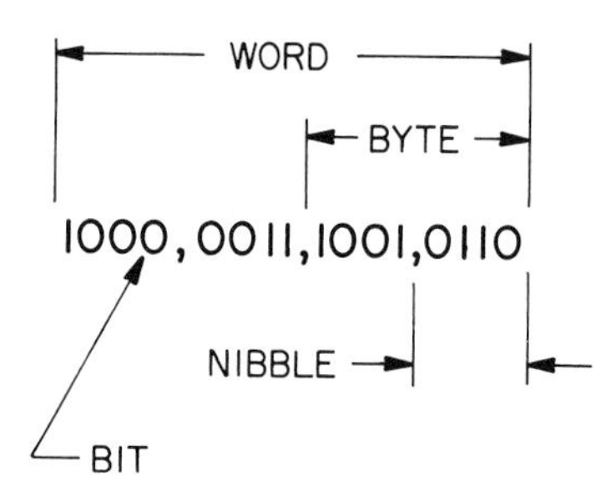

Fig. 12-15. Four bits are often referred to as a *NIBBLE*. BCD is a nibble-oriented numbering system.

TROUBLESHOOTING BCD

When lamps are placed in groups of four, should they be read in BCD or BCH? When working with equipment, reference should be made to instruction manuals to determine the numbering system in use.

A BCD monitor is shown in Fig. 12-16. Its output is faulty, since its second digit reads 11 (base 10). In a properly operating BCD circuit, this is not possible. No digit greater than 9 (base 10) is allowed. The fault is probably in the driving circuit, since lamps are seldom in the s-a-1 fault mode. Lamps usually fault in the s-a-0 mode (they burn out).

Fig. 12-16. The display shown is outputting an incorrect number. Note that the second decimal digit is outputting the number 11. No number greater than 9 (base 10) is allowed.

BCD CIRCUITS

Some circuits output BCD naturally. Fig. 12-17 shows the decade counter described in Chapter 10, Fig. 10-30 and Fig. 10-31. The decade counter in Fig. 12-17 outputs BCD.

Circuits that use the BCD numbering system tend to be a little slower than circuits using the binary numbering system. An output must wait until at least

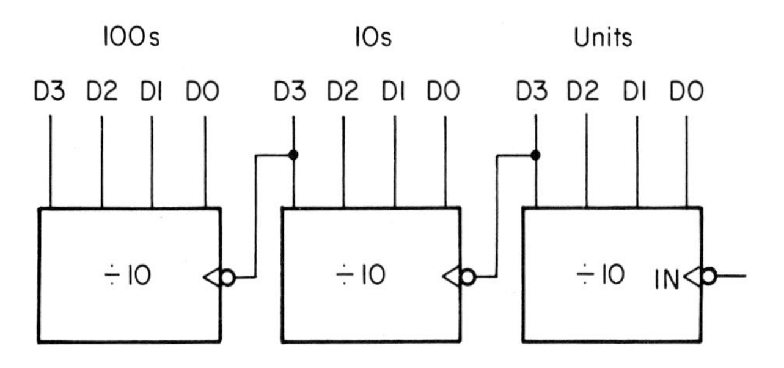

Fig. 12-17. The decade counters in this circuit automatically output BCD numbers. The leads D3, D2, D1, and D0 represent 1/2 byte. Each section clears when the counter in that section reaches 10 (it is 10 for a very short time).

4 flip-flops have settled down. Binary circuits add a delay of 1 or 2 flip-flops.

Buffer circuits between BCD units and binary units are expensive. Less compatibility is the result.

GRAY CODE

Part "a" of Fig. 12-18 shows a shaft position encoder. As the shaft turns, light passes through the transparent areas of the disk and excites the photocells. The 1s and 0s output by these cells represent the position of the disk and shaft.

The obvious pattern on the disk would be one that produced a binary output, part "b" of Fig. 12-18. However, when the output goes from 1 (0001) to 2 (0010) a problem is likely to exist. It is almost impossible for the light to switch from bit D0 to D1 without overlap. During the transition, D0 and D1 are likely to be 1 at the same time. For at least a short period, the encoder will output a 3 (0011) when it should be outputting a 1 or 2.

To avoid this problem, GRAY CODE is often used. In this counting system, only one bit changes at a time. See the table in Fig. 12-19 and refer to part "c" of Fig. 12-18. With only one bit changing at a time, there can be no overlap.

Although the gray code uses 1s and 0s, it is not binary. It does not follow the rules of binary. Once the position of the shaft has been encoded using gray code, it can be converted to binary. A simple decoding circuit can do this.

ASCII CODE

Within computers, only 1s and 0s can be stored and manipulated. As a result, letters and other symbols must be converted to codes based on 1s and 0s. ASCII is such a code. These letters stand for American

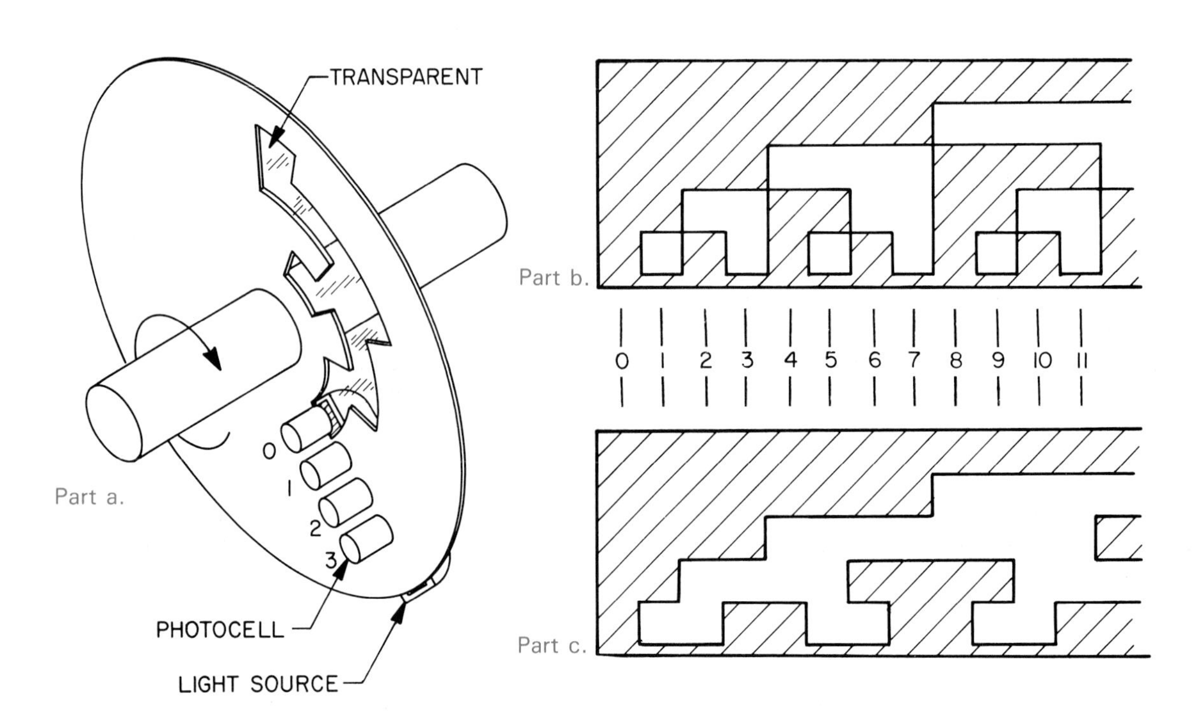

Fig. 12-18. Gray code is often used on shaft position encoders.

DECIMAL	GRAY CODE
0	0000
1	0001
2	0011
3	0010
4	0110
5	0111
6	0101
7	0100
8	1100
9	1101
10	1111
11	1110
12	1010
13	1011
14	1001
15	1000

Fig. 12-19. Counting in decimal and gray code.

Standard Code for Information Interchange. ASCII is usually read as as-kee-two.

This widely used code is said to represent information that is called *ALPHANUMERIC* information. That is, both letters and numbers can be encoded. For example, capital A has the 7-bit code 100,0001. The number 6 is 011,0110. The control signal for returning the carriage of a teletype machine is 000,1101.

Most keyboards output ASCII signals. Also, most printers respond to ASCII. Details of this code are widely available in the literature.

Because ASCII codes require 7 bits, only one character (one letter, number, punctuation mark, etc.) can be stored in an 8-bit word. As a result, the storage of *STRINGS* (continuous groups of alphanumeric characters) requires a great deal of memory space. For example, the simple sentence, INPUT YOUR NAME, requires at least 16 memory locations in an 8-bit machine or microprocessor.

In machines which have larger words, packing can be incorporated to more efficiently use memory. As mentioned before, however, packing requires additional programming steps and is costly.

SUMMARY

Binary numbers are difficult to remember, write down, and express in words. Converting binary numbers to decimal solves these problems, but the conversion process is time consuming and the conversion introduces errors.

Binary numbers are, however, easily converted to octal and hexadecimal. As a result, these numbering systems are widely used.

BCD permits numbers composed of 1s and 0s to be easily converted to decimal. BCD and binary are not the same numbering systems, so a group of special circuits or a type of special programming is needed to create BCD numbers.

Gray code is often used in shaft and position encoders. Only one digit of this numbering system changes with each count.

ASCII is used to store and communicate alphanumeric information. It is widely used in the field of digital electronics.

IMPORTANT TERMS

Alphanumeric, Binary-coded-hexadecimal (BCH), Byte, Leading zero, Nibble, Packing.

TEST YOUR KNOWLEDGE

1. What does the term octal imply?
2. What is a radix?
3. In any numbering system, counting proceeds until all ________ have been used.
4. Binary numbers can be converted to octal numbers by first separating the digits into ________ groups.
5. Any number that is raised to the 0 power equals the number ________.
6. A ________ is a symbol following another symbol.
7. What does MSD mean? What does LSD mean?
8. The number 7A5 is a ________ number.
9. Hex or hexa means ________.
10. The letter E in the hexadecimal numbering system stands for what number in the decimal numbering system?
11. Only two hexadecimal digits are needed to represent a/an ________ -bit binary number.
12. Will a lamp have a fault that is s-a-1 or will the fault be s-a-0?
13. In the number 08B9, the leading 0 on the number has not been ________.
14. Numbers with ________ (large, small) bases are easier to remember, write, and describe verbally.
15. What does the term BCH mean?
16. To convert a binary number to a hexadecimal number, the first step is to arrange the digits in groups of ________ (2, 3, 4, 8).
17. The first 4 bits of a BCD number cannot count higher than ________ (base 10).
18. The primary use of the BCD numbering system is at the ________ between a computer and the human operator.

19. In the BCD numbering system, how many bit combinations are not used?
20. What is a nibble? What is a byte?
21. Computers tend to be ________ (nibble, byte) oriented.
22. The process called packing is used to save ________ (time, electricity, memory space, programming space).
23. The counter in Fig. 10-31 uses three decade counters of the type in Fig. 10-30. In what form are the outputs of these counters?
    a. Binary.
    b. Decimal.
    c. Octal.
    d. BCD.
24. One use of the ________ code is in a shaft position encoder circuit.
25. ________ (Asymmetric, Alphanumeric, Asynchronous) characters are encoded in the American Standard Code system called ________.

STUDY PROBLEMS

1. Set up a table and count to twenty in both decimal and octal.
2. Convert the following binary numbers to octal.
    a. 110,010
    b. 101,011,111
    c. 10,011,100
    d. 11011011
3. Read the numbers displayed on the lamp monitors in Fig. P12-1 in octal.
4. Convert the following octal numbers to binary. Do not show leading zeros.
    a. 42
    b. 426
    c. 521
    d. 6634

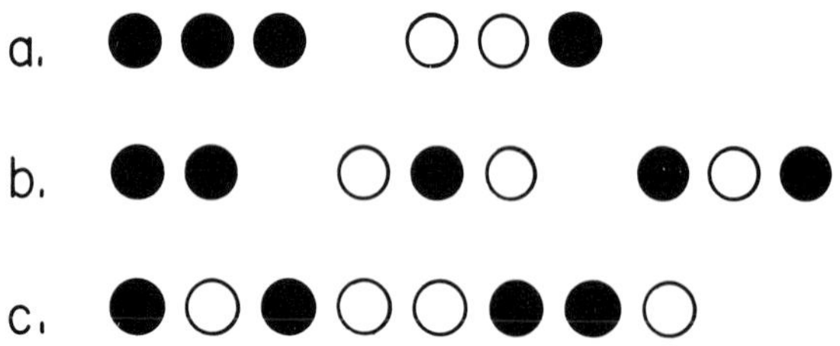

a. ●●● ○○●

b. ●● ○●○ ●○●

c. ●○●○○●●○

Fig. P12-1. Lamp combinations to be read in octal.

5. Convert these octal numbers to decimal.
    a. 34
    b. 825
    c. 725
    d. 4362
6. Convert these decimal numbers to octal. Use the MSD method.
    a. 234
    b. 1879
7. Repeat Problem 6 using the LSD approach.
8. The number 3962 is not an octal number. How can you be sure that this is true?
9. Set up a table and count to twenty in both decimal and hexadecimal.
10. Convert the following binary numbers to hexadecimal.
    a. 1000,0011
    b. 0111,1011
    c. 1100,0110,1010
    d. 0010111
11. Read the numbers displayed by the lamps in Fig. P12-2 in hexadecimal.

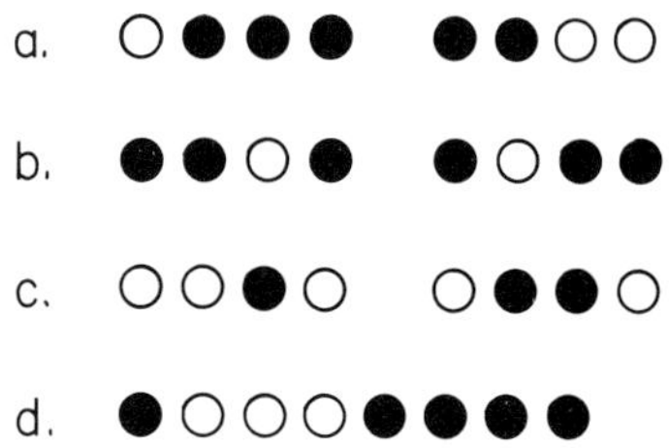

a. ○●●● ●●○○

b. ●●○● ●○●●

c. ○○●○ ○●●○ ●○●○

d. ●○○○●●●●

Fig. P12-2. Lamp combinations to be read in hexadecimal.

12. Convert the following hexadecimal numbers to binary. Do not show leading zeros.
    a. 4A
    b. EDA
    c. E6
    d. 4DB
13. Convert these hexadecimal numbers to decimal.
    a. 46
    b. 3B2
    c. DFE
    d. 94B3
14. Convert the following decimal numbers to hexadecimal. Use the MSD approach.
    a. 932
    b. 44255
15. Repeat Problem 14 using the LSD approach.
16. Any of the following numbers could be hexadecimal. Which one is most likely hexadecimal?
    a. 110010
    b. 4DA
    c. 38571
    d. 5616
    e. 27777

17. Convert the following octal numbers to hexadecimal numbers.
    a. 352
    b. 73
18. Convert the following hexadecimal numbers to octal numbers.
    a. A5
    b. 94
19. The numbers displayed on the monitors in Fig. P12-3 are BCD numbers. Indicate their values in decimal number form.
20. Write the following decimal numbers in BCD.
    a. 48
    b. 926
21. Write the following binary numbers in BCD. Should you convert directly from the binary system or should you first convert to the decimal system?
    a. 10,101
    b. 1011,1001

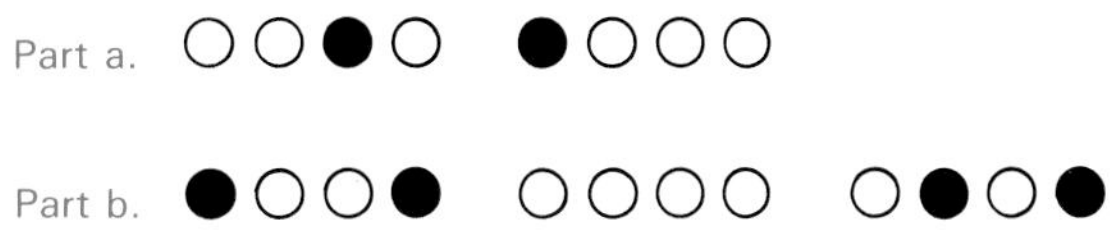

Part a.

Part b.

Fig. P12-3. Lamp combinations to be read in decimal number form.

In the past, digital and analog circuits were often thought of as separate technologies. Today, however, many analog signals are digitized for processing and analysis. As a result, those specializing in digital electronics should have a firm knowledge of analog circuits. (Tektronix, Inc. 1996©)

Arithmetic/Logic Circuits

Chapter 13

LEARNING OBJECTIVES

After studying this chapter and completing lab assignments and study problems, you will be able to:

- Recognize and name the parts of the circuit for a basic multi-bit adder and describe its action.
- Describe the action of an XOR complementing circuit.
- Follow a description of the action of a two's complement subtractor and contrast it with that of an adder.
- Recognize and name the parts of circuits for registers that can shift and rotate.
- Recognize and name the parts of circuits for registers that can increment and decrement.
- Describe the actions of typical carry, zero, sign, and parity flags.
- Predict the contents of a flag register based on the results of mathematical and logic operations.

In computers, manipulation of data takes place in sections called *ARITHMETIC/LOGIC UNITS* (ALUs). Among the basic functions of ALUs, a unit does the following:

1. Add and subtract.
2. Complement.
3. Compare.
4. Shift and rotate.
5. Increment and decrement.
6. AND, OR, and XOR.

In addition to the results of the above arithmetic and logic operations, other information is needed. Is the result of a subtraction positive or negative? Did an overflow result when two large numbers were added together?

Such information is supplied by the flip-flops that make up the flag register. Each flip-flop or flag describes some characteristic of the results of a computation. The flags to be studied in this chapter are:

1. Carry.
2. Sign.
3. Zero.
4. Parity.

A given computer may have between four and ten flags. However, the above are the most common. A flag is helpful in checking the logic path that produced an error. A technician measures the signal of a flag flip-flop to see if it has been set. A series of lamps make circuit tracing easier.

ADDERS

In computers, the command called ADD is a basic instruction. Most ALUs (arithmetic/logic units) contain a circuit for accomplishing this task. The circuit can have as few as three elements or as many as 100 elements.

A simple adder was described in an earlier chapter. That circuit will be reviewed and advanced adders will be introduced.

HALF ADDER

An adder with two inputs and two outputs is called a *HALF ADDER*. It is considered less than a complete adder because it cannot accept carries from lower-order bits.

To review, the table in Fig. 13-1 shows the truth

INPUTS		OUTPUTS	
A	B	Co	S
0	0	0	0
0	1	0	1
1	0	0	1
1	1	1	0

Co = CARRY OUT S = SUM

Fig. 13-1. Truth table for half adder.

table for a half adder. It represents the following rules of binary addition:

0	0	1	1
+ 0	+ 1	+ 0	+ 1
0	1	1	c1 0

To continue the review, the sum-of-products method of designing a circuit can be used to write Boolean expressions for the two outputs. See Fig. 13-2 for Boolean expressions and a circuit.

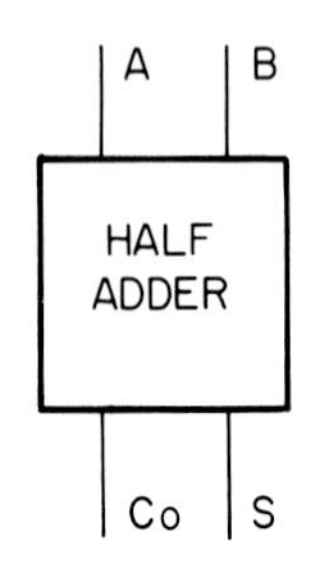

$$\bar{A}B + A\bar{B} = S$$

$$AB = Co$$

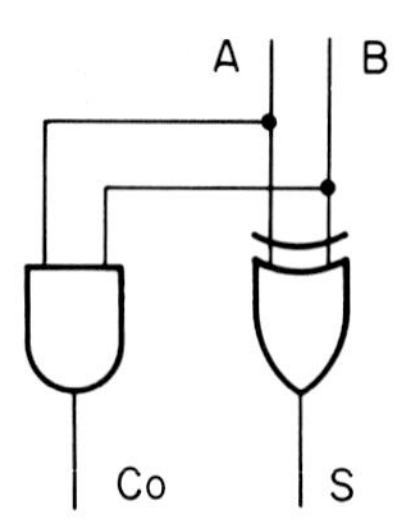

Fig. 13-2. Only two logic elements are needed to implement the expression for the half adder shown.

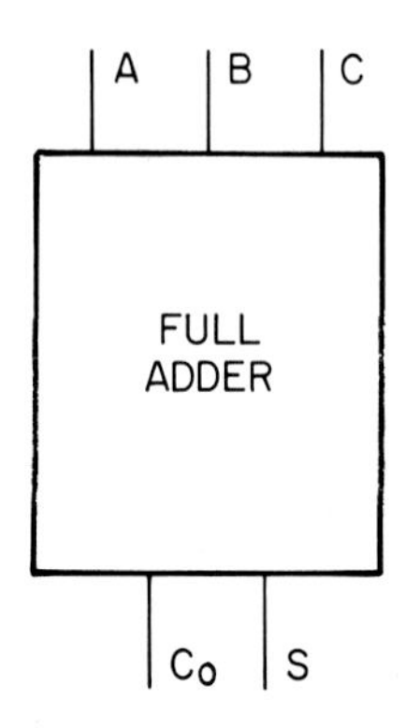

Fig. 13-3. A full adder has three input leads—two for the bits to be added and one for the carry from the previous bit.

Only two logic elements are needed to implement the expressions in Fig. 13-2. For the sum, an XOR can be used; for the carry, an AND will suffice.

FULL ADDER

When multi-bit numbers are to be added, circuits called *FULL ADDERS* are needed. A full adder has three inputs and two outputs. See Fig. 13-3. A and B are the bits to be added. C is the carry from the previous bit.

The table in Fig. 13-4 is the truth table for a full adder. To review, the rules of addition are:

0	0	0	1
+ 0	+ 0	+ 1	+ 1
+ 0	+ 1	+ 1	+ 1
0	1	c1 0	c1 1

There are 8 rules. Only 4 of the eight are shown. Other combinations result from exchanging an A with a B, or a B with a C, etc. A C is given as much weight in the formulas as that given to a B or an A.

Example: Design a full adder circuit based on the table in Fig. 13-4. Simplify the resulting expression before implementing. The circuit which is used to sum numbers will be designed first.

INPUTS			OUTPUTS	
A	B	C	Co	S
0	0	0	0	0
0	0	1	0	1
0	1	0	0	1
0	1	1	1	0
1	0	0	0	1
1	0	1	1	0
1	1	0	1	0
1	1	1	1	1

C = CARRY IN   Co = CARRY OUT

Fig. 13-4. Truth table for full adder.

Step 1. Using sum-of-products, the expression for S is obtained. See part "a" of Fig. 13-5. C is factored from the first and last products. $\bar{C}$ is factored from the two middle products.

Step 2. The expressions within the brackets are special. The first represents an XNOR (exclusive NOR). The second is an XOR.

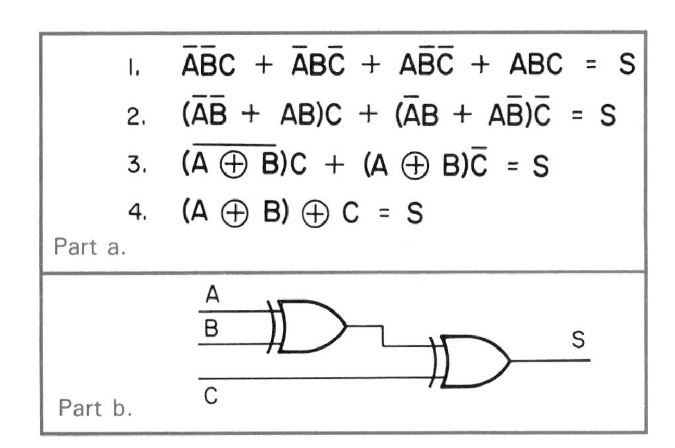

Fig. 13-5. The simplification of the sum-of-products expression shown can be implemented by two XORs.

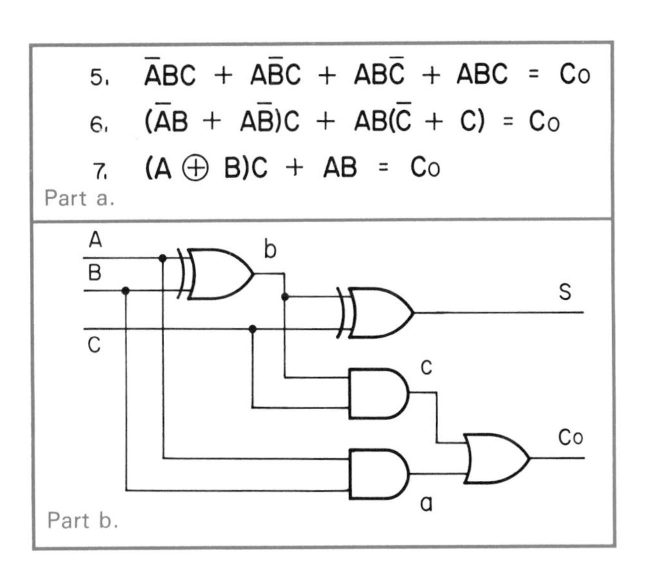

Fig. 13-6. Two ANDs, an XOR, and an OR are used in the carry portion of the full adder shown.

Step 3. Although the expression shown appears to be complex, it is merely an XOR. Let X = (A XOR B), and Y = C. Then $\bar{X}Y + X\bar{Y} = S$ can be recognized as an XOR.

Step 4. The statement shown is the simplified expression for the sum circuit. When implemented, it consists of two XORS. See the circuit at the bottom of Fig. 13-5.

The writing and simplification of the expression for the carry is shown in Fig. 13-6. Terms in the expression can cancel each other.

Step 5. Sum-of-products was used to write the expression for carry. C can be factored from the first two terms. AB can be factored from the last two.

Step 6. The expression in the first set of brackets is an XOR. In the second set, $\bar{C}$ cancels C.

Step 7. The implementation of the expression shown completes the task. The term A XOR B was implemented as part of the sum circuit, so it does not need to be repeated in the carry portion of the circuit.

The action of the carry circuit is as follows. When A and B are both 1, a carry is generated. See the AND marked a in Fig. 13-6. When ABC = 101 or ABC = 011, a carry is also generated. In either case, the output of the XOR marked b is 1. This output is ANDed with C at c and results in a carry output (called a carry out).

MULTI-BIT ADDER

Fig. 13-7 shows a 3-bit adder. Bit 0 is a half adder; the other two are full adders.

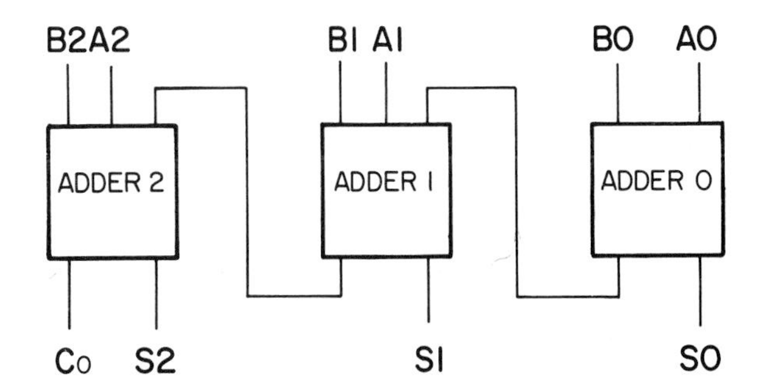

Fig. 13-7. The circuit shown can add two 3-bit binary numbers.

Part "a" of Fig. 13-8 shows the following addition:

A2A1A0 =		011
B2B1B0 =		+ 010
S2S1S0 =	(Co = 0)	101

It is important to check the measured outputs for an overflow. If the numbers in part "a" of Fig. 13-8 (the previous list of "A2A1A0," etc.) are unsigned numbers, no overflow occurred. There was no carry out of the MSB. If, however, these are signed numbers, an overflow did occur. A and B are both positive; the sum is negative.

In part "b" of Fig. 13-8, the following addition is shown by solid dots and by empty circles:

A2A1A0 =		111
B2B1B0 =		+ 011
S2S1S0 =	(Co = 1)	010

If these are unsigned numbers, an overflow occurred. This is indicated by the carry out of the MSB. If A and B are signed, there is no overflow. When one number is positive and the other is negative, there can be no overflow.

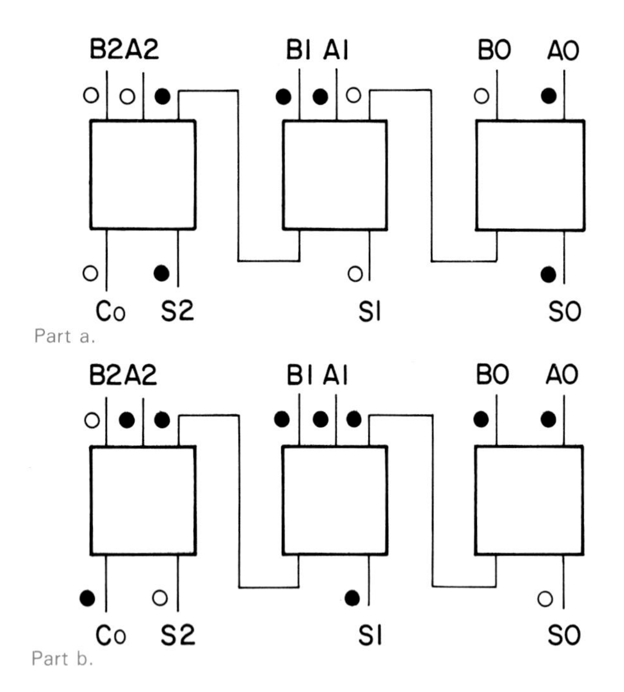

Fig. 13-8. The drawing with two 3-bit adders and measured signals shows typical addition problems being solved by a binary adder.

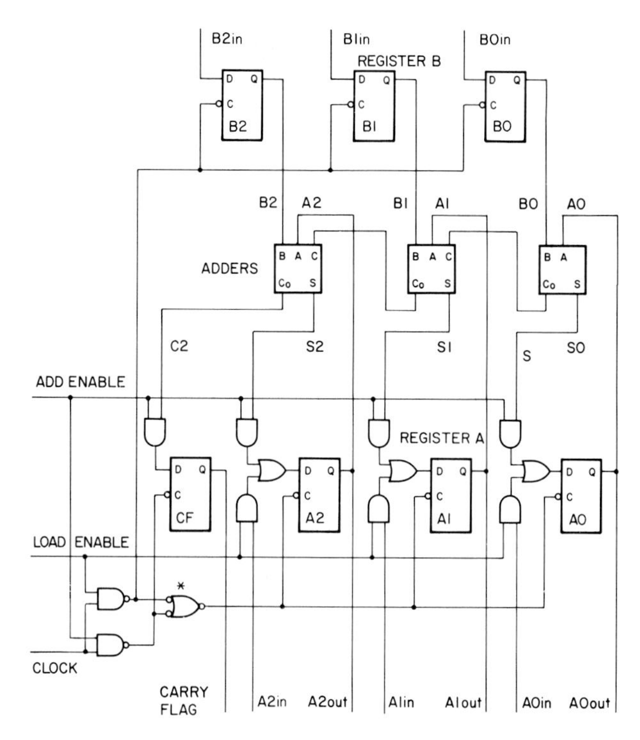

Fig. 13-9. The circuit shown is a 3-bit adder. Numbers to be added are stored in Registers A and B. The sum is returned to Register A.

ADDER AND REGISTERS

In a computer, numbers to be added are placed in registers adjacent to the adder. Such a circuit is shown in Fig. 13-9. The action of this circuit will be analyzed in detail.

To review, helpful hints for analyzing complex circuits are:

- Know what the circuit does.
- Consider it in small, recognizable subcircuits.
- Start with a subcircuit that is easily analyzed. Then proceed to the more complex.

Eight topics need to be discussed. These topics are as follows:

- Function of circuit.
- Leads.
- Adders.
- Loading register B.
- Load register A from A2A1A0.
- Store sum in register A.
- Need for master-slave flip-flops.
- Carry flag.

The first of the topics gives an overview as follows:

1. Function of circuit: The circuit in Fig. 13-9 is a 3-bit adder with a carry flag. Numbers to be added are stored in two registers, A and B. Register A is an accumulator. That is, the sum that results from the addition is returned to this register.
2. Leads: Numbers to be added are input through A2A1A0 (in) and B2B1B0 (in). See Fig. 13-10. The resulting sum appears on leads A2A1A0 (out). The carry out is on lead CARRY FLAG.

There are three control leads. When new numbers are to be entered, LOAD ENABLE must be high (1), and CLOCK must be active. To do an add, ADD ENABLE must be high, and again, CLOCK must be in the active state.

Fig. 13-11 shows the sections of the circuit that will be analyzed. The numbers indicate the order of the analysis.

3. Adders: Three adders are used, Fig. 13-12. Based on the number of input leads, the first adder is a half adder. The other two are full adders.

The adders shown in Fig. 13-12 are not clocked or enabled. Whatever numbers appear at their inputs are immediately added.

4. Loading register B: The number to be stored in register B enters through B2B1B0 (in). See Fig. 13-13. The memory elements are data latches. Their clocks are shown as active-low and level-triggered. Negative-going, edge-triggered units could be substituted without changing the circuit action and behavior.

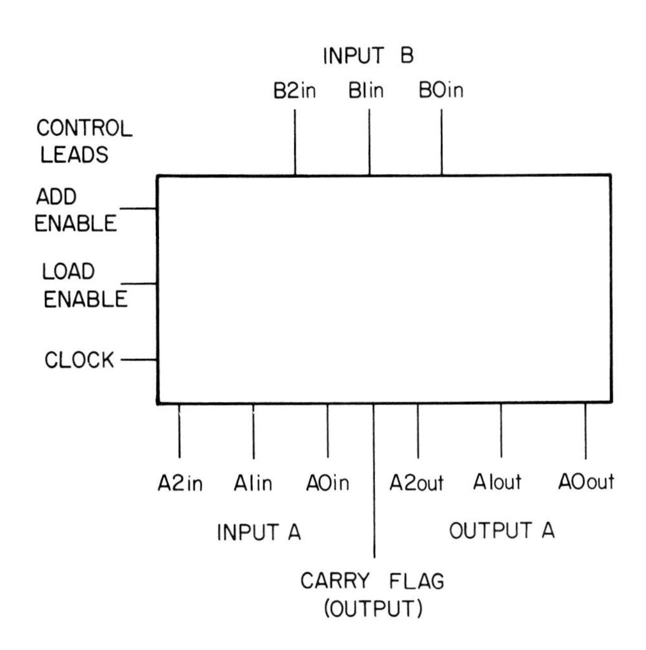

Fig. 13-10. The drawing of a circuit symbol with leads emphasizes the functions of the leads of the circuit in Fig. 13-9.

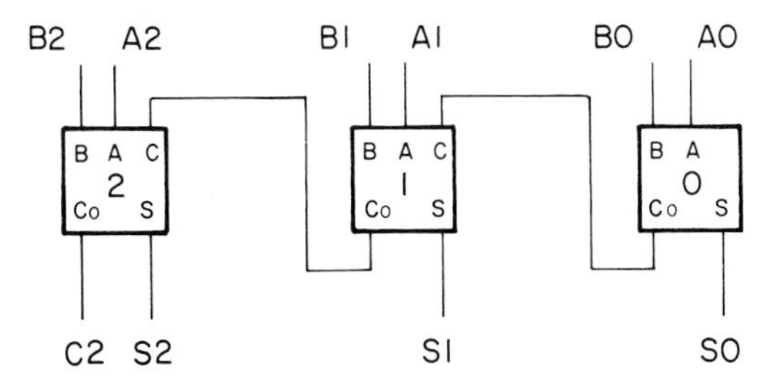

Fig. 13-12. The adder portion of the circuit shown is identical to the one in Fig. 13-7 and Fig. 13-8.

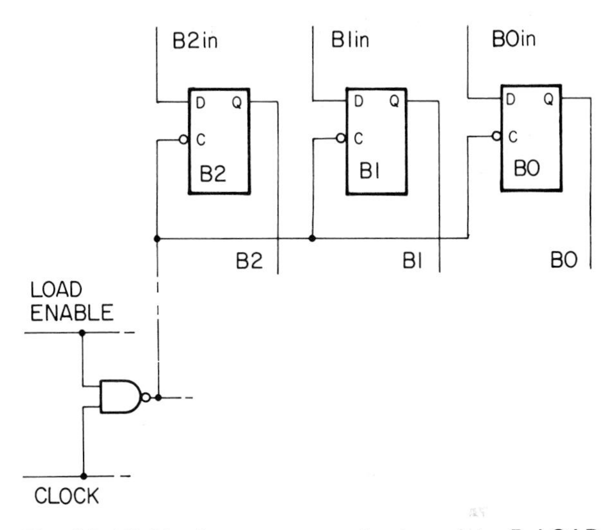

Fig. 13-13. To place a new number in register B, LOAD ENABLE and CLOCK must be active at the same time. The data latch clocks shown are active-low and level-triggered.

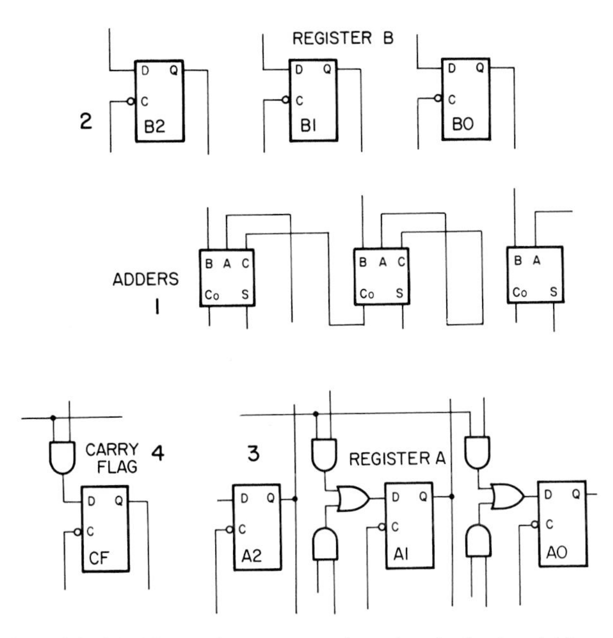

Fig. 13-11. To analyze a complex circuit, it should be divided into recognizable subcircuits.

To store a new number, both LOAD ENABLE and CLOCK must be active (high). Note the NAND in the control circuit at the lower left.

5. Load register A from A2A1A0: Register A can be loaded from two directions. New numbers can enter through A2A1A0 (in) or from the adders (S2S1S0). This description concerns A2A1A0 (in).

   Refer to Fig. 13-14. For a number at A2A1A0 (in) to reach the D inputs of the data latches, the three AND gates in a horizontal line must be open. For this to happen, LOAD ENABLE must be active. With these gates open, numbers from A2A1A0 (in) reach the ORs and pass through to the D inputs.

   In addition to having a new number at the D inputs, the clocks of the data latches must be active. The control circuit at the lower left provides this signal.
6. Store sum in register A: To review, as soon as numbers appear at the outputs of the registers, they are added. Their sum appears at S2S1S0. If a carry out is generated, it appears at C2.

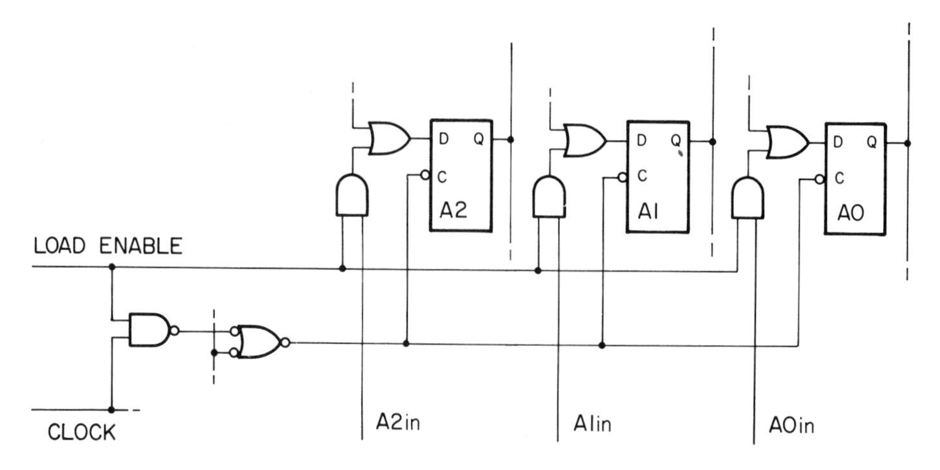

Fig. 13-14. As with register B, LOAD ENABLE and CLOCK must be active at the same time to place a new number in register A.

Fig. 13-15 shows one bit of the adder and register A. To store a sum in A, ADD ENABLE must go high. This opens the AND gate between the adder and input D of the data latch. Data can leave the adder at S2S1S0 and enter register A. However, the storage is not completed until the clock on the data latch goes active. That is, ADD ENABLE and CLOCK must be active at the same time. Refer to the NAND at the lower left of Fig. 13-15. The NAND is not the same one used for LOAD ENABLE in Fig. 13-14.

Control leads ADD ENABLE and LOAD ENABLE must not be active at the same time. If they are (and CLOCK is active), numbers from the adders (S2S1S0) and A2A1A0 (in) will reach the flip-flops at the same time. The resulting stored number would be meaningless.

7. Need for master-slave flip-flops: The data latches in register A, Fig. 13-15 and Fig. 13-9, must be of the master-slave type. Numbers stored in register A are added to those from register B. The sum is returned to A. That is, A is read from and written into at the same time. If master-slave flip-flops were not used, the sum would appear at the output and be added to the number in B. A new (and faulty) sum would then be returned to A. See the arrows in Fig. 13-15.

When master-slave data latches are used, the new sum enters the master portion of the flip-flops when the clock goes active. It does not appear at the outputs until the clock goes inactive.

8. Carry flag: Fig. 13-16 shows the carry flag. When an add is accomplished, the carry from the MSB of the adder (C2 in Fig. 13-9) is stored in this flip-flop. In unsigned arithmetic, this flip-flop can be used to indicate an overflow.

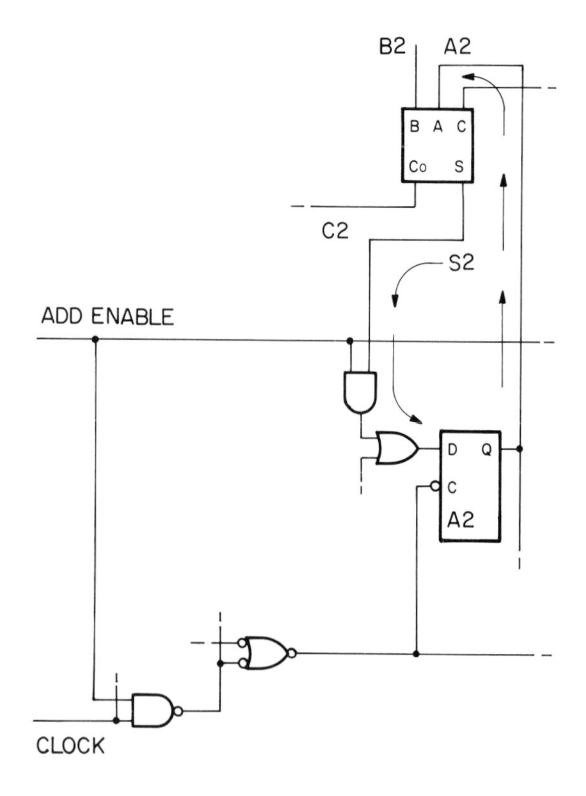

Fig. 13-15. The storage of a sum is emphasized in the drawing shown. Because flip-flop A2 is read and written into at the same time, it must be of the master-slave type. Note the data path on the vertical line at the right.

What happens to the carry flip-flop when a new number is placed in register A? In most computers,

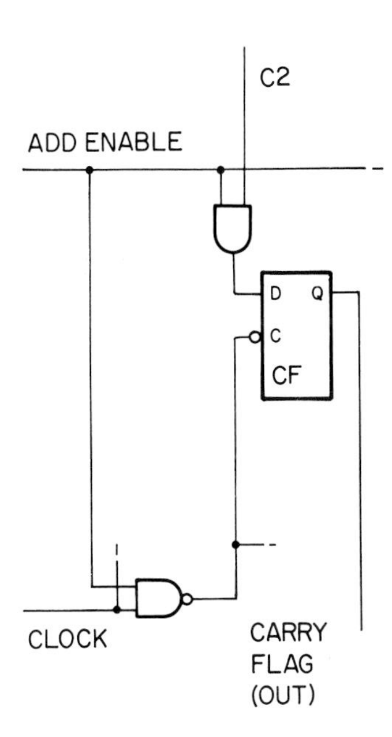

Fig. 13-16. The carry flip-flop changes only when an add is accomplished. Refer to Fig. 13-9.

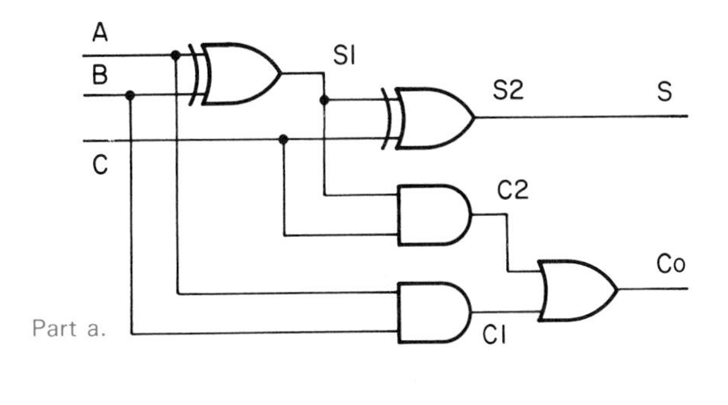

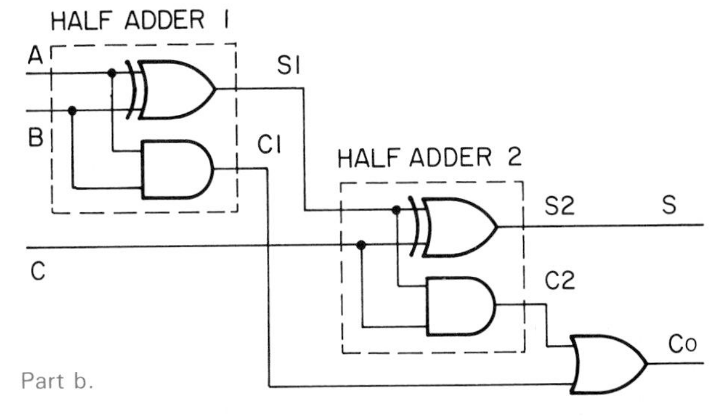

Fig. 13-17. The symbols in the full adder shown have been rearranged to show that it consists of two half adders. Refer to Fig. 13-6.

the carry flag does not change. Refer to Fig. 13-9. Only an arithmetic operation (in this case, an addition) will change this flag.

VARIATIONS IN THE DESIGN OF ADDER CIRCUITS

The adder in part "a" of Fig. 13-17 is a direct implementation of the expression obtained from the full adder truth table, Fig. 13-4. There are, however, many variations of this basic circuit. The following three topics are described:

1. Two half adders.
2. Full adder made only of ANDs and ORs.
3. Ripple and look-ahead carries compared.

The first topic looks for simplifications as follows:

1. Two half adders: The presence of two XORs and two ANDs suggest that a full adder might be built from two half adders. See part "b" of Fig. 13-17. This is the same circuit as in part "a" of Fig. 13-17. It has merely been redrawn. A full adder is indeed two half adders with an OR added.
2. Full adder made only of ANDs and ORs: The previous circuit is not as simple as it appears. XORs are not basic elements. Each contains at least four NANDs or NORs.

The circuit in part "a" of Fig. 13-18 uses about the same number of logic elements as the circuit with XORs. It has the advantage of a fast carry. That is, input signals pass through only two elements before generating a carry out. Many special adder circuits have been designed. Each has some special feature that is useful in specific applications.

To suggest the validity of the circuit in part "a" of Fig. 13-18, the Karnaugh map for its carry is shown in part "b" of Fig. 13-18. The map shown contains three 2-cell loops. The resulting expression contains three 2-input ANDs. The carry circuit is the implementation of this expression.

3. Ripple and look-ahead carries compared: Part "a" of Fig. 13-19 shows 8 bits of a 32-bit adder. Note that the carry from a given bit serves as an input to the next bit. This type of linear adder circuit is said to be a *RIPPLE CARRY.*

The outputs of any bit (say S5 and C5 of bit 5) cannot be correct until a correct carry has been received from the previous bit (that is, C4 must be correct). This means that high-order sums must wait until the carries have passed (rippled) through all previous bits before a correct output can result.

The delay per bit is small. In computers with short word lengths (say 8 or so), such delays can often be ignored. In machines with long words (32-bits, for example), the sum of such delays becomes a problem. Such computers must wait until the carry has rippled the full length of the adder before a correct sum can be output.

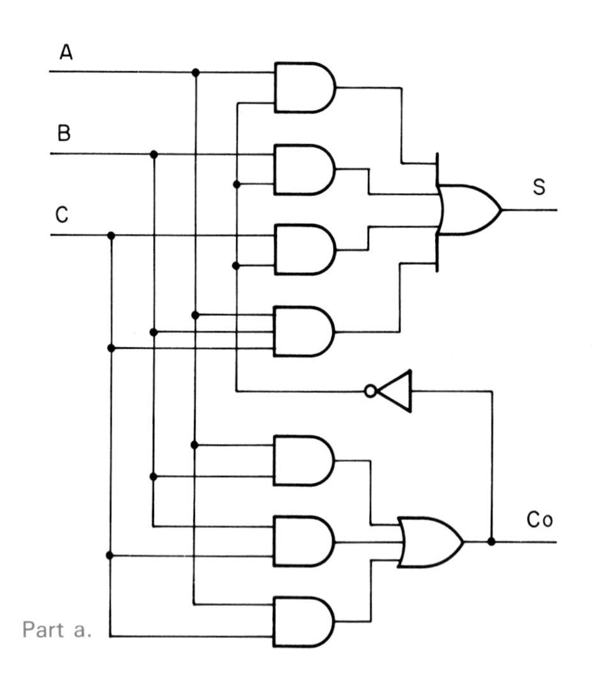

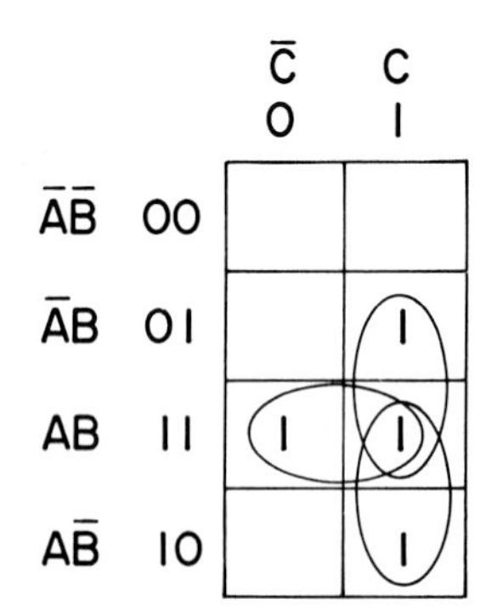

AB + BC + AC = Co

Fig. 13-18. The full adder circuit shown has a fast carry. That is, signals pass through only two elements before a valid carry is generated.

The circuit in part "b" of Fig. 13-19 reduces the delay in the carry signal. It contains a circuit called a *LOOK-AHEAD CARRY*. The first four bits of the number are applied to both the adders and a carry circuit. While the bits are being added, the look-ahead circuit determines if C3 will be 1 or 0. Rather than waiting for the carry to be generated at the starred output, C3 is immediately supplied to the next group of four bits. This permits computation to begin there before the normal carry from the previous four bits is available. There is still some delay, but it is smaller than with a ripple carry.

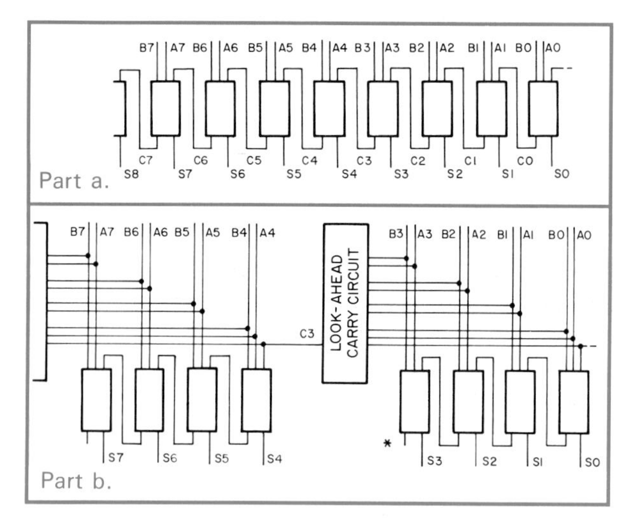

Fig. 13-19. In ripple counters (upper circuit), high-order carries cannot be generated until all lower orders are correct. The look-ahead circuit overcomes this problem.

COMPLEMENTING CIRCUIT

Most computers can execute (perform, carry out) a complement instruction. Part "a" of Fig. 13-20 shows a circuit that can accomplish this task. The XOR truth table in part "b" of Fig. 13-20 explains the circuit's action.

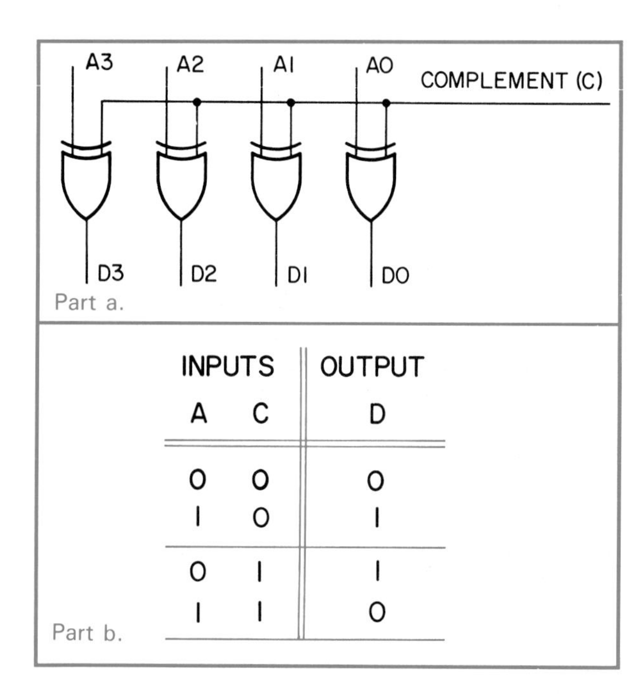

INPUTS		OUTPUT
A	C	D
0	0	0
1	0	1
0	1	1
1	1	0

Fig. 13-20. A complementing circuit can be built using XORs. The circuit is a switchable NOT. See Fig. 5-41.

When COMPLEMENT ENABLE (lead C) is 0, output D equals input A. When lead C is 1, D is the complement of A. That is:

$$\bar{A} = D \text{ when } C = 1$$
$$A = D \text{ when } C = 0$$

SUBTRACTOR/ADDER

In most computers, the two's complement method is used to accomplish subtraction. To review, the two's complement of the subtrahend is added to the minuend. The sum is the difference.

Fig. 13-21 shows a circuit that can add and subtract. Its mode is controlled by the SUBTRACT/ADD lead.

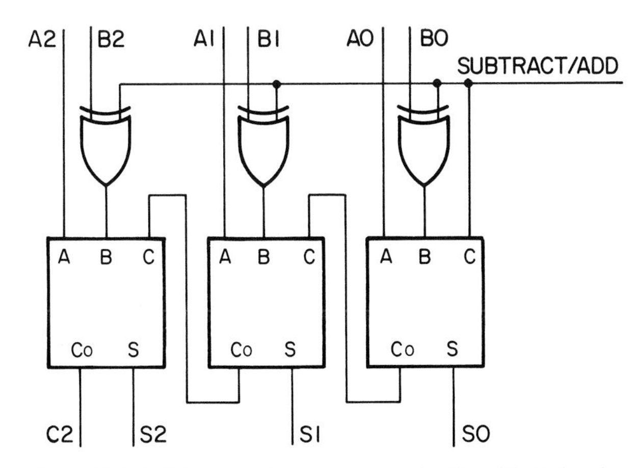

Fig. 13-21. The circuit shown can both add and subtract. The one's complement is provided by the complement circuit made of XORs. The two's complement is provided by adding 1 through C of the first adder.

ADD MODE

When the SUBTRACT/ADD lead in Fig. 13-21 is 0, A and B will be added. With 0s applied to the XORs, B will not be complemented. Also, when C of the first adder is 0, it acts like a half adder. Addition proceeds in the usual way.

SUBTRACT MODE

When the SUBTRACT/ADD lead in Fig. 13-21 is 1, B is subtracted from A. The one's complement of B is provided by the XORs. The additional 1 that is needed to convert its one's complement to two's complement is provided at C of the first adder. C is automatically added to the one's complement of B2B1B0. The number at S2S1S0 is the difference between the two input numbers.

COMPARATOR

The function of a comparator is to indicate when two numbers are equal. If A equals B, an active output results. Circuits can be built for the following common uses:

1. One-bit comparator.
2. Multi-bit comparator.

Examine the first type of comparator as follows:

1. One-bit comparator: Part "a" of Fig. 13-22 shows a 1-bit comparator. It has an active-low output. When A = B, it outputs a 0.

Part "b" through part "e" of Fig. 13-22 suggest that an XOR can be used as a comparator. When A and B are both 1s or both 0s, it will output a 0.

2. Multi-bit comparator: Fig. 13-23 shows a multi-bit comparator. Because the individual XORs output 0s when a match is achieved, a NOR is used to combine the outputs. As a result, the output of the circuit is active high.

COMPARATOR CHECKS IF INPUT A IS GREATER THAN INPUT B

In addition to detecting equality, comparators can be used to indicate when A is larger than B. See Fig. 13-24 for a circuit with NOTs, ANDs, and an OR.

In Fig. 13-24, suppose A1A0 = 10 and B1B0 = 01. Then the ANDs in a row will output 1, 0, and 0. The OR outputs a 1, meaning A1A0 is greater than B1B0.

Suppose A1A0 = 11 and B1B0 = 10. Then the ANDs output 0, 0, 1. The OR outputs a 1.

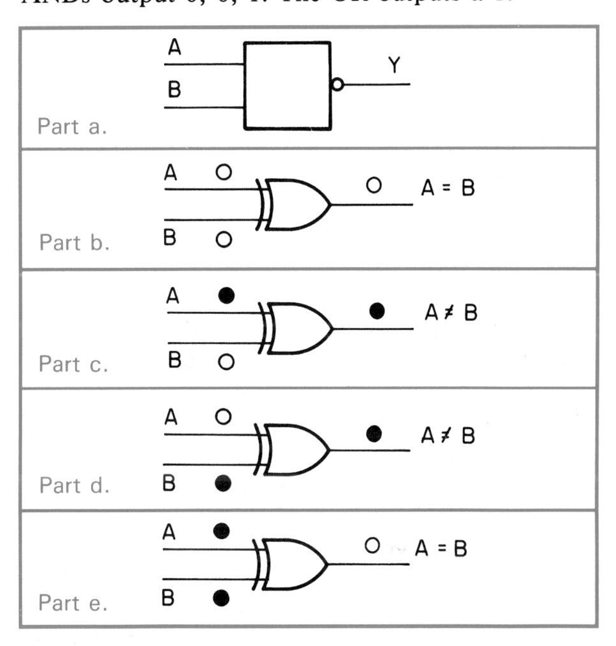

Fig. 13-22. An XOR can be used to compare two bits.

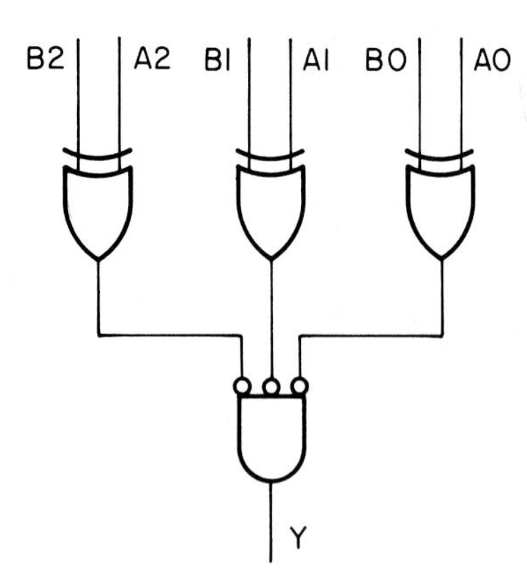

Fig. 13-23. The circuit shown compares two 3-bit binary numbers.

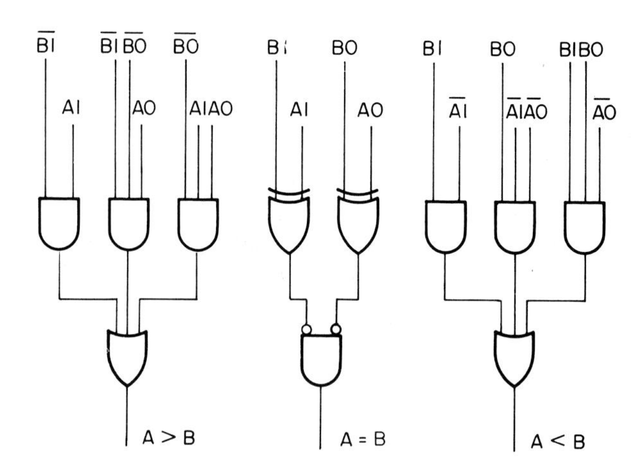

Fig. 13-25. The circuit shown compares A1A0 and B1B0 and indicates whether they are equal or not. If they are not equal, it indicates which is larger.

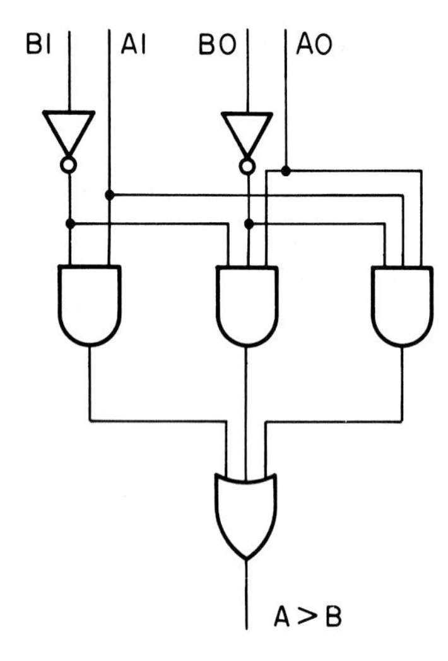

Fig. 13-24. The circuit shown outputs a 1 when A1A0 is larger than B1B0. The output is active high.

Suppose A1A0 = 01 and B1B0 = 10. The ANDs are 0, 0, 0. The OR is 0.

Suppose A1A0 = 11 and B1B0 = 11. The ANDs are 0, 0, 0. The OR is 0.

A similar circuit could be used to produce an active output when A1A0 is smaller than B1B0. Combining the three circuits results in the magnitude comparator in Fig. 13-25.

Note the method of showing input signals. The necessary NOTs have been omitted. These connections have been left to those who do the wiring. This technique is often used on complex circuits that have many inputs. The designer can choose a notation to save design steps or designing time.

SHIFT AND ROTATE

The term SHIFT means to move the bits of a number one or more places to the right or left. See part "a" of Fig. 13-26. In this case, the shift is to the

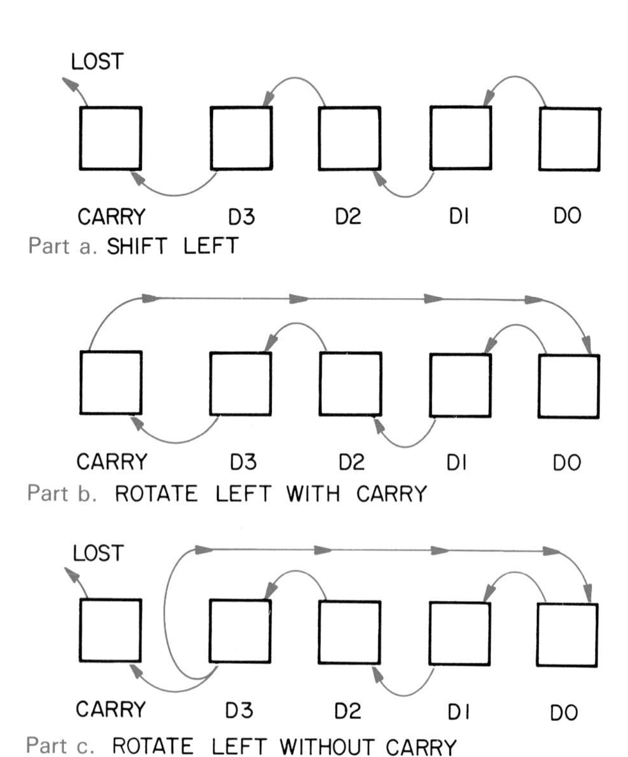

Fig. 13-26. In computers, registers may be capable of shifting and rotating numbers. Typical forms of shift and rotate processes are shown.

left. When D0 is vacated, a 0 is usually inserted. As is shown, the carry flag may be included. The bit originally in the carry flip-flop is lost. Shift instructions are used in binary multiplication and for testing individual bits.

There is a term called *ROTATE.* Like shift, ROTATE involves moving the bits of a number to the right or left. However, individual bits are not lost by this action. When a bit reaches the end of a register, it is returned to the other end. See b and c in Fig. 13-26. The carry flip-flop may or may not be within the loop.

ROTATE/SHIFT CIRCUIT

Fig. 13-27 is a review of shift registers. Here bits move to the right. The number to be shifted is brought in through A3A2A1A0 under the control of PARALLEL LOAD. The selection of rotate or shift is determined by signal on ROTATE/SHIFT. When operation has been completed, number is output through Q3Q2Q1Q0. Two topics will be discussed:

1. Rotate mode.
2. Shift mode.

The circuit action involved with the first topic is as follows:

1. Rotate mode: When a 1 is placed on ROTATE/SHIFT, the AND gate at the left opens. The number at Q0 is applied to input D of the first flip-flop. When SHIFT goes active, this number enters, and a rotate has been accomplished.
2. Shift mode: When ROTATE/SHIFT is 0, the AND gate is closed, so it outputs a 0. When a shift command is given, the stored number moves to the right. The bit at Q3 is replaced by a 0, and a shift has been accomplished.

BIDIRECTIONAL SHIFT REGISTER

Fig. 13-28 shows a register that can shift in either direction. The parallel-load circuit has been omitted for clarity. Also, outputs have been shown below the circuit. The following topics are discussed:

1. Shift-right mode.
2. Shift-left mode.
3. Rotate right and left.

The first shift mode is examined first as follows:

1. Shift-right mode: When a 1 is applied to control lead R/L, the circuit will shift to the right. The AND gates that permit signals to move from left to right are open. See part "a" of Fig. 13-29. The AND at the left will output a 0. As the number moves to the right, it is replaced by 0s.
2. Shift-left mode: With a 0 on R/L, the shift will be to the left. See part "b" of Fig. 13-29. The AND at the right will output a 0, so the vacated bits will

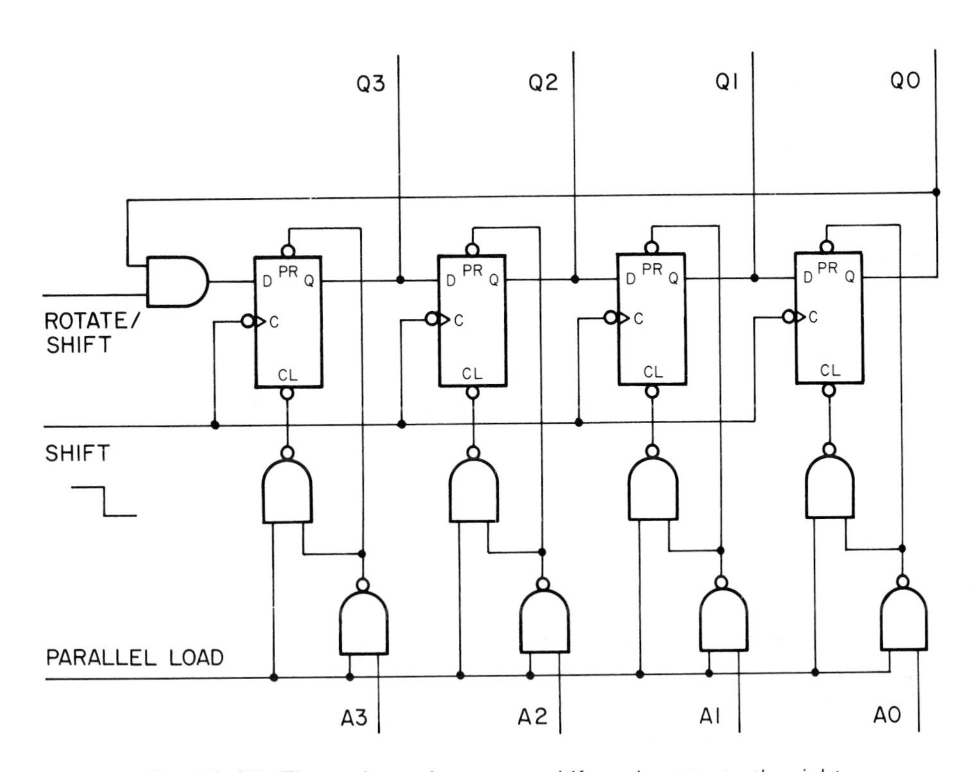

Fig. 13-27. The register shown can shift and rotate to the right.

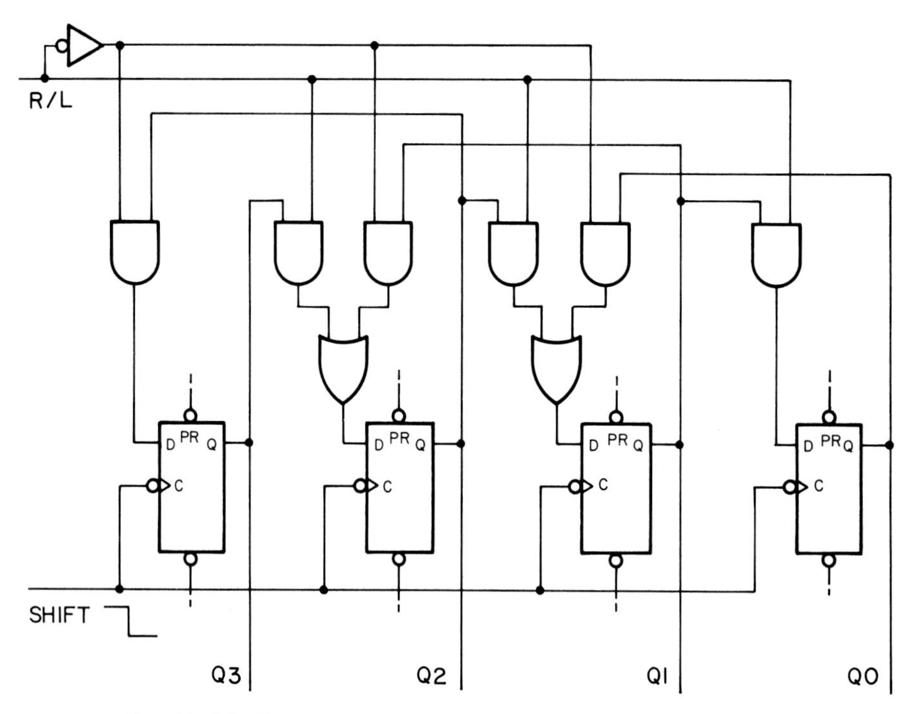

Fig. 13-28. The register shown can shift to the right or left.

be filled with 0s.

3. Rotate right and left: To emphasize that shift and rotate are similar operations, two ANDs were added to the shift register in Fig. 13-28 to produce this register that can rotate numbers. The control lead was relabeled, ROTATE R/L.

INCREMENT AND DECREMENT

Adding 1 to a number is called *INCREMENTING.* Subtracting 1 is called *DECREMENTING.* In computer programming, these are important operations. A register, called a program counter, is used to keep track of where a computer is in a program. As each step is completed, the program counter increments. Also, in computer programs it is easier to count down to zero than up to a prescribed number. As a result, decrement is widely used.

INCREMENTING REGISTER

The logic circuit for incrementing is an up counter, Fig. 13-31. This is a 4-bit register. Numbers are entered in parallel through A3A2A1A0. The input and output leads on the diagram are in reverse order to simplify the appearance of the circuit.

To increment a stored number, a negative-going edge is applied to INCREMENT. This adds the necessary count to the number.

INCREMENT/DECREMENT REGISTER

Fig. 13-32 shows a circuit that can both increment and decrement. The control depends on whether the toggles receive data from the previous Q or from the previous $\bar{Q}$. If the toggle receives from Q, the counter will increment. If the toggle receives from $\bar{Q}$, the counter will decrement.

INCREMENT MODE

When control lead I/D is 1, the upper AND gates are opened. This connects the Q outputs to the T inputs, and the circuit functions as an up counter. When I/D ENABLE goes active, 1 is added to the number stored in the register.

DECREMENT MODE

When I/D is 0, the lower AND gates open. The $\bar{Q}$ outputs are connected to the T inputs, and the circuit functions as a down counter. A negative-going edge at I/D ENABLE subtracts 1 from the number stored in the register.

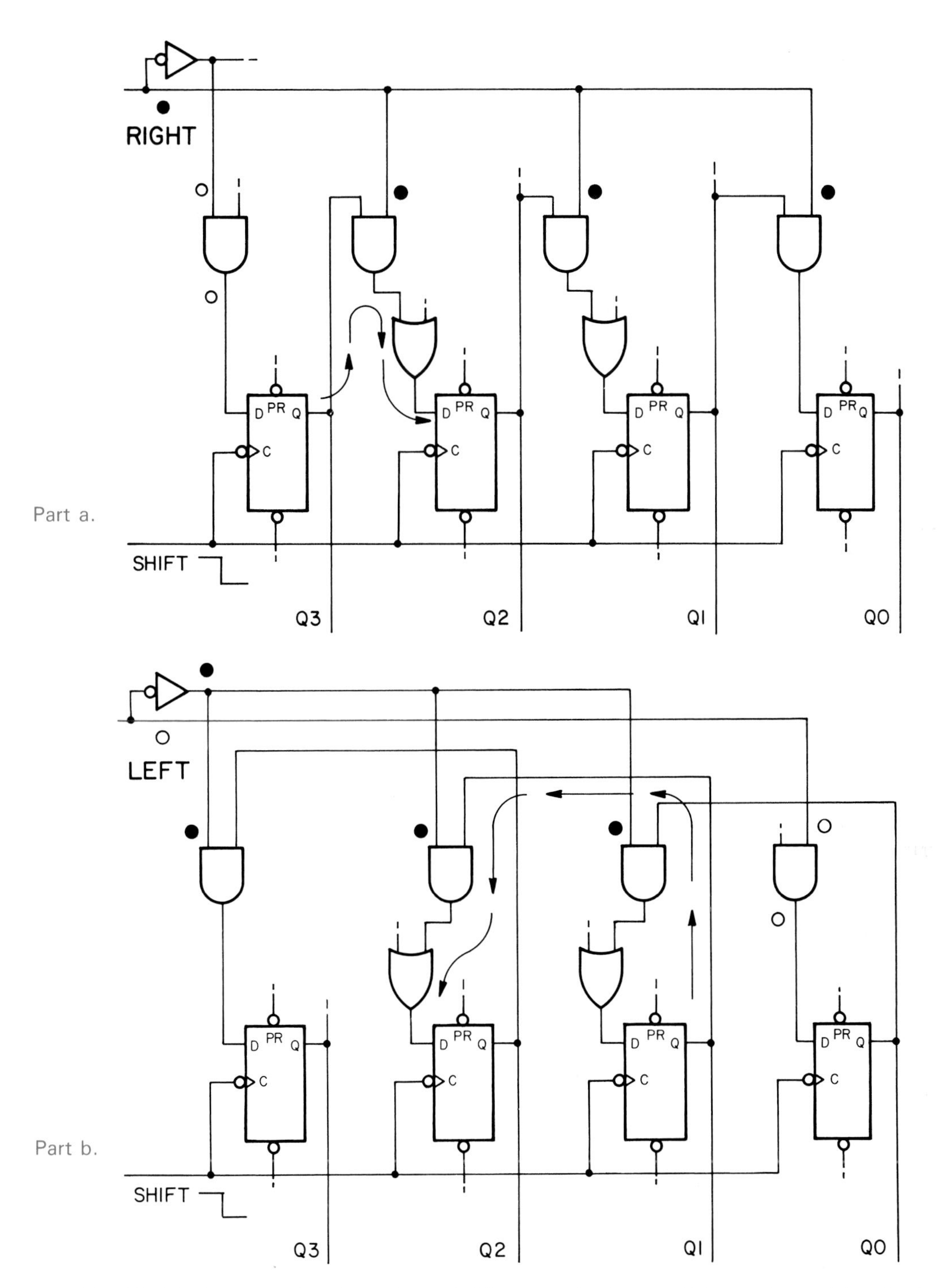

Fig. 13-29. To emphasize the actions of shifting to the right and left, portions of the circuit in Fig. 13-28 have been omitted.

FLAGS

Flags are closely associated with the operation of the ALU. As indicated earlier, they consist of flip-flops that are set or cleared depending on the outcomes of arithmetic and logic operations.

The carry flag has been described in some detail. Several other flags will be considered. The following

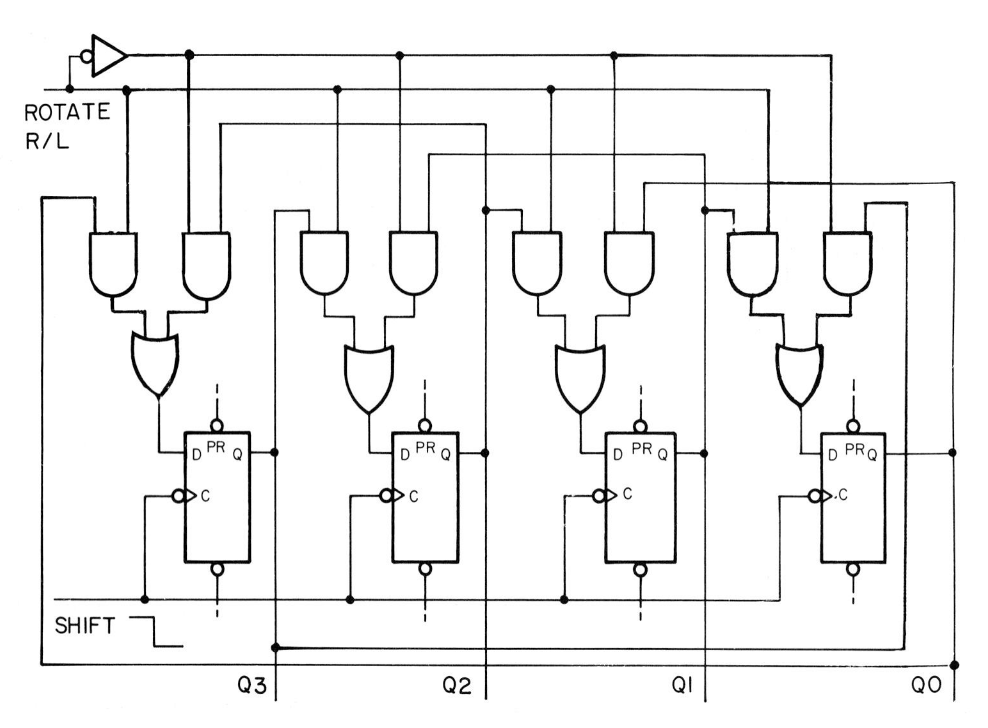

Fig. 13-30. Circuit for rotating right and left.

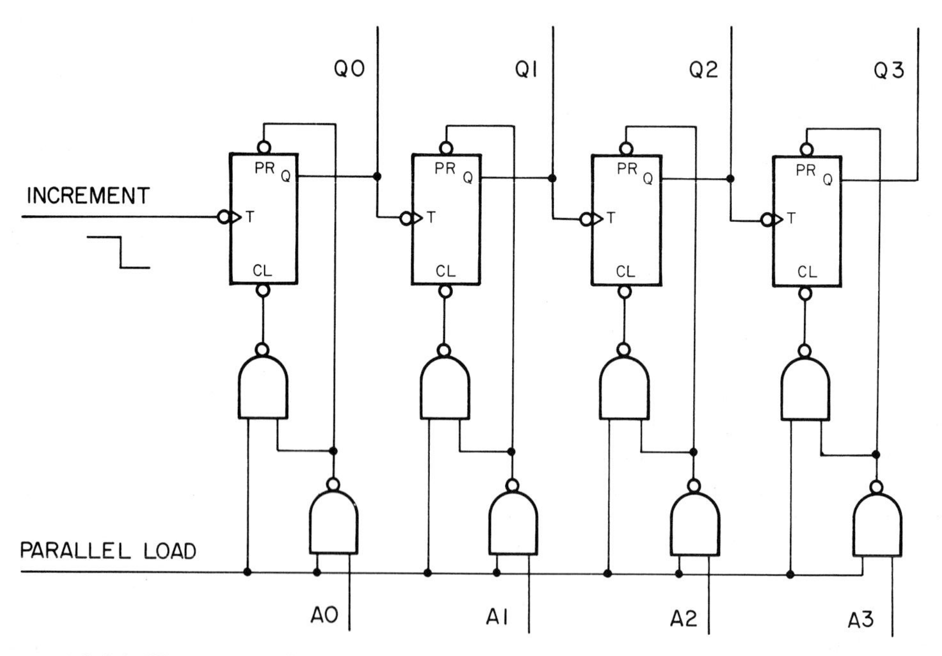

Fig. 13-31. The register shown can be incremented. It consists of a ripple counter with parallel inputs.

topics are discussed:

1. Zero flag.
2. Sign flag.
3. Parity flag.
4. Flag register.
5. Predicting flag register content.

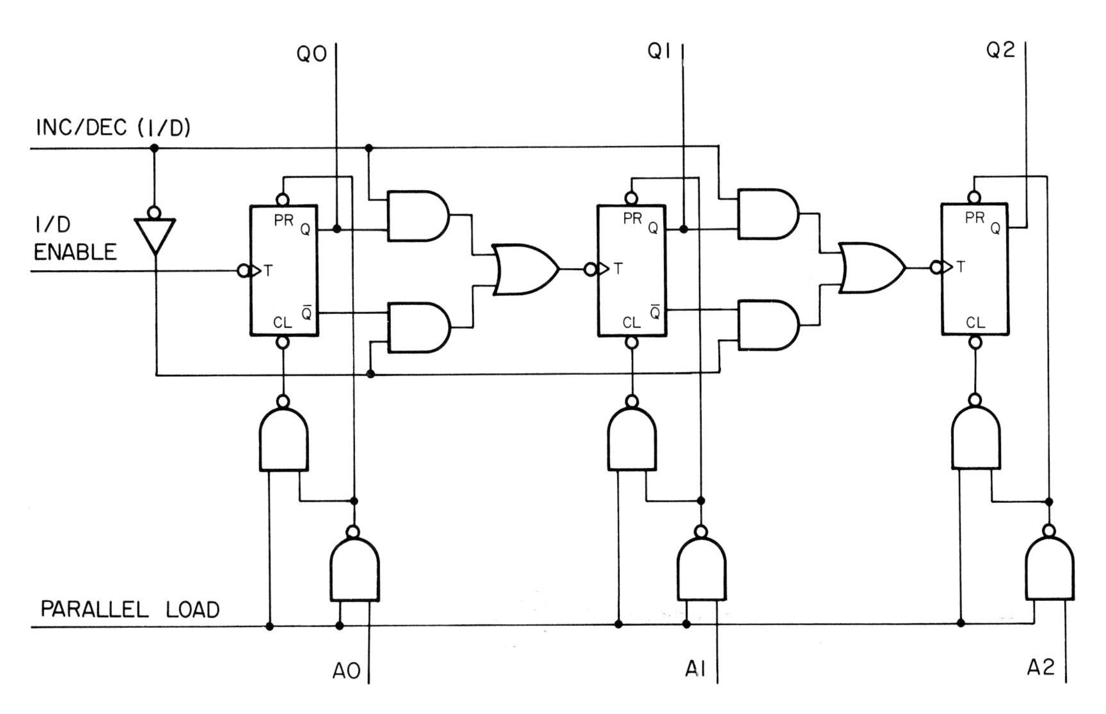

Fig. 13-32. The register shown can be both incremented and decremented. The toggle inputs receive outputs from the previous Q to do an increment step, and the toggle inputs receive from $\bar{Q}$ to do a decrement step.

6. Reading flag register.
7. Other flags.

The zero flag is simple, as shown in the following:

1. Zero flag: When the results of an arithmetic or logic operation are zero, the ZERO FLAG is set (goes to 1). If the results are anything but zero (all bits 0), the flag is cleared (goes to 0).

Fig. 13-33 shows a simple circuit that might be used. In the term "Z-flag," Z stands for "zero." The circuit has two sections. The NOR detects the presence of a zero answer. The flip-flop stores the information.

The zero flag can be confusing. When this flag is set (Z = 1), the result of the computation was zero. Z = 1 implies that the answer was 0000. If the zero flag is reset (Z = 0), the result was nonzero.

The zero flag has many uses. For example, a programmer might want to go through a given segment of a program a number of times. This is called looping (a program loops back on itself). Let us assume that it is necessary to go around a loop five times. A five would be placed in a register. That register would be called a counter.

Near the end of the loop, the programmer would use an instruction to decrement the counter. Then the zero flag would be tested. The instruction would say, "If the zero flag is cleared, go back to the beginning of the loop."

Because the counter contains a 4 (5 − 1 = 4), the program would jump back and repeat the loop. Each time, the counter would be decremented and the zero flag would be tested.

On the fifth pass through the loop, the counter would be decremented to zero, and the flag would be set. When tested, it would be found to equal 1, so

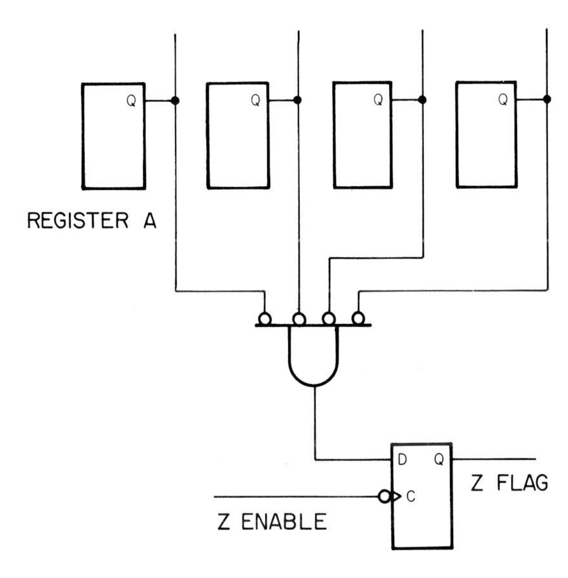

Fig. 13-33. The Z-flag flip-flop will contain a 1 (be set) only when all the flip-flops in register A contain 0s.

the jump would not be made. The program would then continue with the next step that is outside of the repeated loop.

2. Sign flag: The SIGN FLAG merely stores the MSB of the result of an arithmetic or logic operation. In signed numbers, this would be the sign bit. See the left side and lower center of Fig. 13-34.

The most obvious use for the sign flag is the determination of the sign of a number. However, it has many uses. A programmer might want to know if D5 of a number is 1 or 0. The number would be placed in register A and shifted to the left twice. That is, D5 would be moved to D7. In an 8-bit machine, this would cause the sign flag to be set or cleared. Testing this flag would determine whether D5 was a 1 or 0.

3. Parity flag: In mathematics, parity implies a certain odd-even relationship. If two numbers are both even or both odd, they are said to have the same parity. If one is odd and the other is even, they have different parity.

In computers and digital circuits, the parity of a number is found by counting the number of 1s it contains. If the number of 1s is even, the number is said to have an even parity. If the count for the 1s is odd, the parity of the number is odd.

The first three numbers in the following list have even parity. The last three have odd parity. Note that zero is considered to have even parity.

Even parity: 01,101,001
11,001,111
00,000,000
Odd parity: 00,111,110
00,000,010
10,111,001

An XOR can be used to determine the parity of a 2-bit number, Fig. 13-35. If A1A0 equals 00 or 11, its parity is even. If it equals 01 or 10, its parity is odd. Note that the exclusive OR outputs a 0 for even parity of a number.

Fig. 13-36 shows a parity flag circuit. The XORs determine the parity of the number in the register. The flip-flop (the parity flag) holds the results of that determination.

Note the NOT. In this circuit, Q = 1 implies even parity. For A3A2A1A0 = 1101, the parity flag would output a 0 (odd parity).

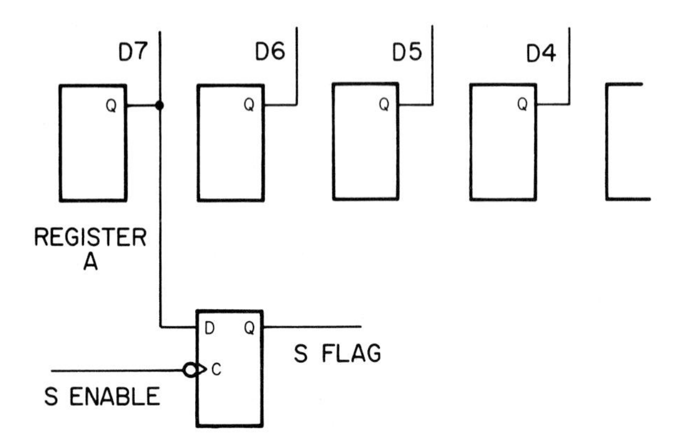

Fig. 13-34. After an arithmetic or logic operation, the S-flag flip-flop and MSB flip-flop contain the same number.

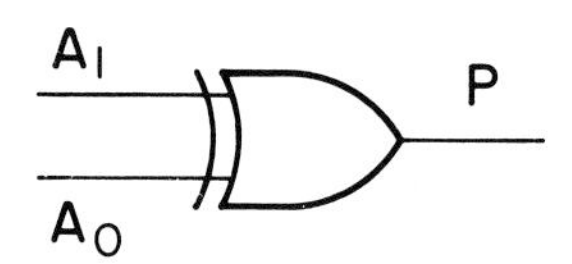

INPUTS		OUTPUT	
A_1	A_0	P	
0	0	0	EVEN
1	1	0	
0	1	1	ODD
1	0	1	

Fig. 13-35. An XOR can be used to determine the parity of a pair of binary digits.

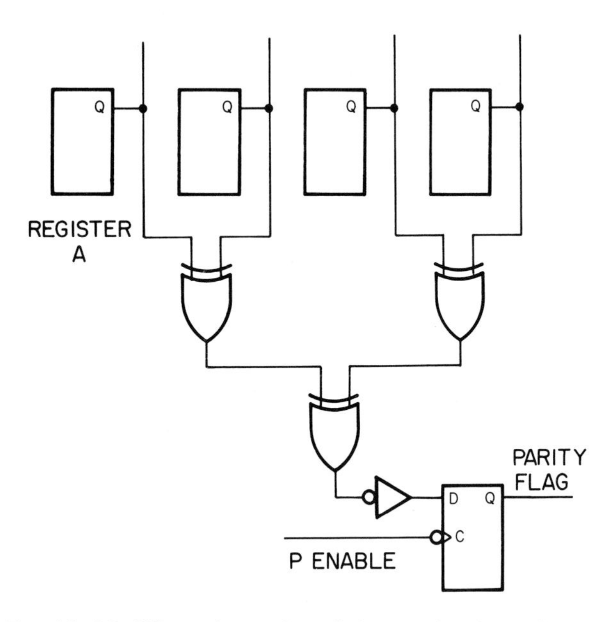

Fig. 13-36. When the parity of the number in register A is even, the circuit shown outputs a 1.

Parity is often used to detect errors in numbers that have been stored or transmitted from place to place. For example, a number might be transmitted from a teletype machine to a computer. Some noise on the telephone line might change a 1 to a 0. The parity can be used to detect such errors.

The process involves converting all numbers to be stored or transmitted to a specified parity. This is done by using a parity bit (usually the MSB). A 1 or 0 is placed in that bit such that the parity of the resulting number is always the same. In the following example, D7 is set so the parity is always even:

Number	Number with parity
__0,001,011	10,001,011
__1,001,111	11,001,111
__0,011,101	00,011,101

Computer programs can insert the proper parity bit, or a circuit like the one in Fig. 13-37 might be used. If the parity of the 4-bit number is even, D4 is 0. If the parity of the original number is odd, the circuit makes D4 a 1. The parity of the 5-bit number is always arranged to be even.

When a number and its parity bit are brought out of storage or received at a distant location, the parity of the number is checked. If it has changed, at least one bit has changed from 1 to 0 or 0 to 1. The number is known to be faulty.

Parity cannot detect errors that occur in pairs. If two bits change, the parity of the number would remain unchanged. In most cases, however, errors involve only one bit. As a result, parity is an effective error detecting method.

4. Flag register: Registers are usually thought of as places for temporarily storing numbers. Mathematical operations are often performed on the numbers stored in registers.

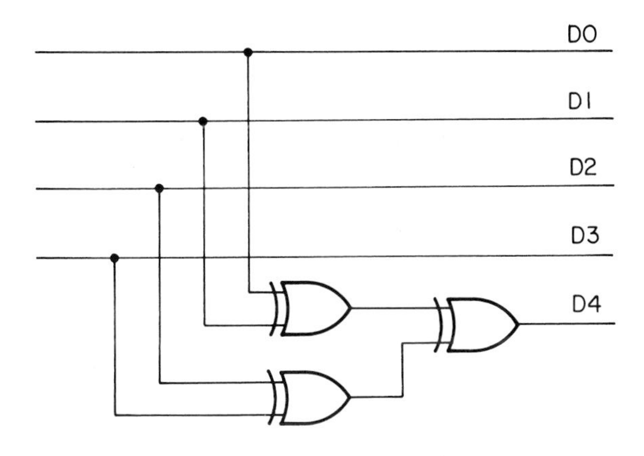

Fig. 13-37. The parity generator shown outputs a bit (D4) such that the parity of the resulting number is always even.

In the usual sense, the flag register is not a register. Rather, it is a collection of flags. Although individual flags can be set and cleared under program control, numbers cannot be stored in the flag register. Its content cannot directly enter into mathematical operations. However, the content of the flag register can be stored in memory.

The flag register in Fig. 13-38 contains four flags (four flip-flops). Unused bits are permanently filled with 1s. Many systems of storing the flag bits in an orderly way are possible.

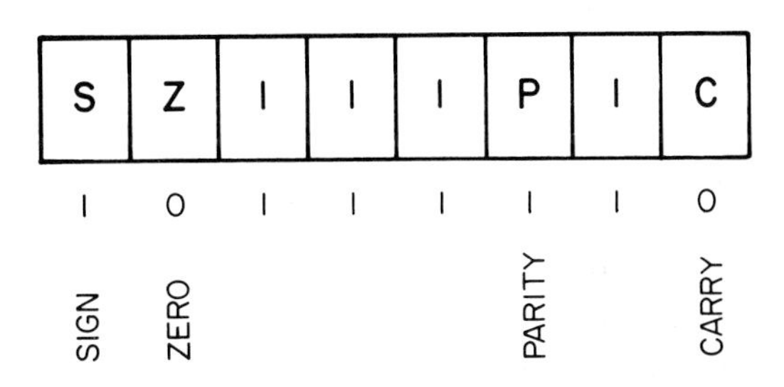

Fig. 13-38. The row of symbols shows one of the many possible arrangements of an 8-bit flag register.

5. Predicting flag register content: Based on a given arithmetic operation, the content of the flag register can be predicted. Assume 01,001,001 is added to 01,101,000 in signed arithmetic. What will be the content of the flag register?

```
      01,001,001
    + 01,101,000
   c0 10,110,001
```

An overflow has occurred (the sign bit has changed from 0 to 1). Sometimes an overflow flag is used. It would be set to 1.

Sign Flag: The sign flag reproduces the sign bit, D7. In this case, the sign flag is set (1).

Zero Flag: The sum is obviously not zero, so this flag will be cleared (0).

Parity Flag: There are four 1s in the sum. Its parity is even, so the parity flag is set (1).

Carry Flag: There was no carry out of D7, so this flag will be cleared (0).

Based on the previous computation, and the row of symbols in Fig. 13-38, the flag register will contain 10,111,110. Note that the four unused bits are set to 1.

Suppose 00,101,001 is added to 01,001,000. What is the content of the flag register?

```
      00,101,001
    + 01,001,000
   c0 01,110,001
```

No overflow has occurred. Overflow = 0. The flag

register contains the following: 00,111,110.

6. Reading flag register: In certain troubleshooting tasks, a knowledge of the content of a given flag can be helpful. For this purpose, most computers provide methods for outputting the flag register. This is usually in hexadecimal or octal.

To determine the content of a given flag, this number must be converted to binary. For example, if a small computer were to output 7E for its flag register, what is the condition of its zero flag? Refer to Fig. 13-39. When the hexadecimal number is converted to binary, the Z flag can be read. In this case it is set (made equal to 1). The result of the last computation was 0000,0000.

7. Other flags: Only the four most common flags have been described. Others are available. For example, some computers have an overflow flag. When an overflow occurs, this flag is set to 1.

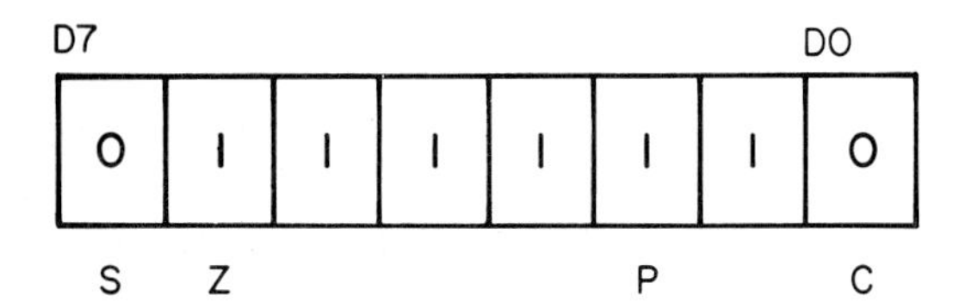

Fig. 13-39. The number in the flag register shown is 7E. As a result, the Z flag is set (1). A conversion of the hexadecimal number 7E in the flag register to a binary number is the first step. The second digit (Z) is then read.

SUMMARY

Registers, arithmetic/logic units, and flags are circuits involved in calculations and decision-making in computers. Some of the functions accomplished by typical ALU sections are:

1. Addition and subtraction.
2. Complementing.
3. Comparison.
4. Shifting and rotating.
5. Incrementing and decrementing.
6. ANDing, ORing, and XORing.

Circuits for accomplishing these tasks are constructed using the combinational- and sequential-logic circuits studied in the early chapters of this book. Master-slave data latches and flip-flops make reading and storing of data easy.

Decision-making in computers is based on testing flags. Flags are flip-flops that record certain properties of the results of mathematical and logic operations. The flags studied in this chapter are:

1. Carry.
2. Zero.
3. Sign.
4. Parity.

IMPORTANT TERMS

ALU, Comparator, Decrement, Fast carry, Flag, Full adder, Half adder, Increment, Look-ahead carry, Parity, Rotate.

TEST YOUR KNOWLEDGE

1. What does the term ALU mean?
2. A(n) __________ (adder, shift, flag) register helps a technician trace errors.
3. Can a full adder be constructed from only two half adders?
4. A full adder has __________ (one, two, three, four, five) input leads.
5. A full adder has __________(one, two, three, four) output leads.
6. The sum from adder elements is often stored in a(n) __________ register.
7. In the adder with fast carry in Fig. 13-18, input signals pass through only __________ (one, two, four) elements before generating a carry out.
8. A look-ahead carry circuit is faster than a __________ carry circuit.
9. A(n) __________ element can be used as a complementing circuit.
10. Checking if the number A equals the number B is the function of a __________.
11. Can a comparator check if A is greater than B?
12. A magnitude comparator consists of __________ (one, two, three, four, five) sections.
13. When a __________ process to the right or left is carried out, data bits are lost.
14. When a __________ process is carried out, no bits are lost.
15. The bidirectional shift register in Fig. 13-28 uses __________ (ANDs, NANDs), ORs, and a NOT to control whether a shift is to the right or to the left.
16. Adding 1 to a number is called __________.
17. __________ consist of flip-flops that are set or cleared depending on the outcomes of arithmetic and logic operations.
18. Does the number 11,011,001 have even parity or odd parity?
19. The parity of a number helps to trace if a(n) __________ has occurred in storing or transmitting a number.
20. A parity-generating and parity-checking process cannot detect errors that occur in __________.

STUDY PROBLEMS

1. Construct the truth table for a half adder. Use A and B for inputs and S (sum) and Co (carry out) for outputs. Then use sum-of-products to write the expression for this circuit. Do not simplify.
2. For the inputs shown at the left in Fig. P13-1, determine the sum and carry outputs of the half adder circuit shown.

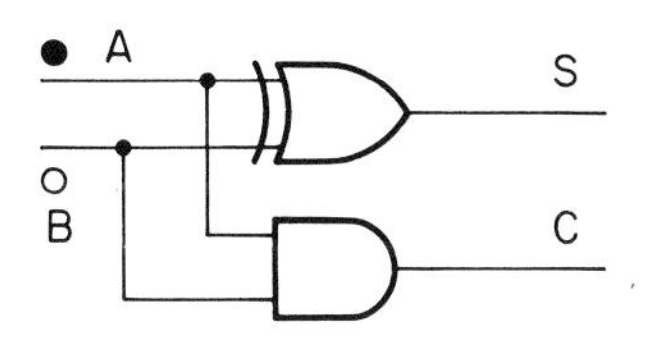

Fig. P13-1. Half adder circuit with input of 1,0 can be easily analyzed for its output. The states 1,0 and 0,1 produce the same output.

3. Construct the truth table for a full adder. Use A, B, and C (carry in) as inputs and S (sum) and Co (carry out) as outputs. Then use sum-of-products to write the expressions for this circuit. Do not simplify.
4. Copy the circuit in Fig. P13-2 for a full adder onto a sheet of paper. For the inputs shown, determine the expected signals at the indicated points (points a through h).

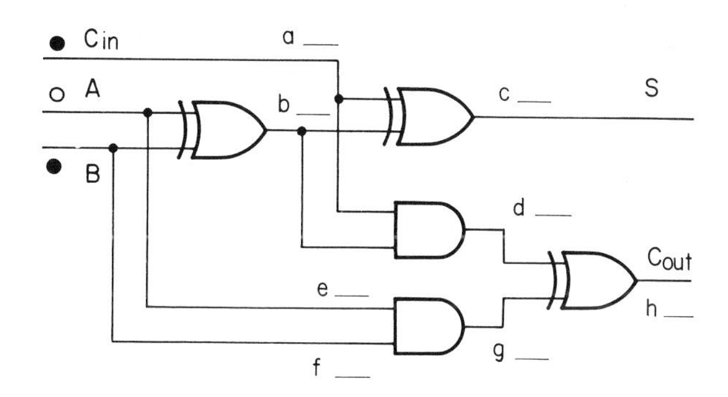

Fig. P13-2. Eight signal levels for full adder circuit can be filled in. Inputs are shown at left.

5. Repeat Problem 4 for the inputs which are as follows: C = 1, A = 1, and B = 1.
6. A 3-bit adder is shown in Fig. P13-3. List the three individual adders (Adder 2, Adder 1, and Adder 0), and indicate which are half adders and which are full adders.
7. For the inputs shown in Fig. P13-3, determine the expected outputs (Co, S2, S1, and S0).

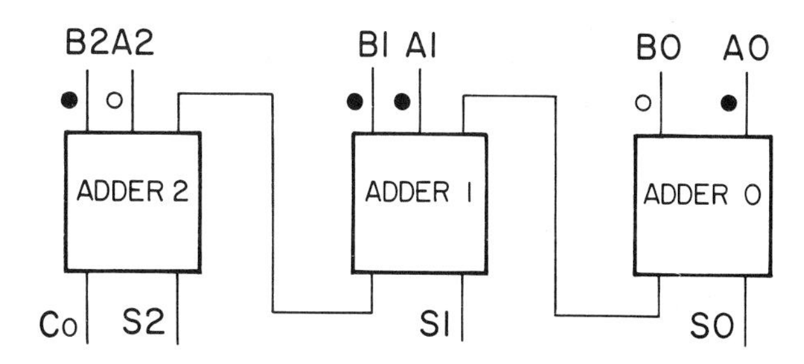

Fig. P13-3. Parts can be labeled for 3-bit adder. Expected outputs (Co, S2, S1, and S0) can be determined.

8. A full adder is shown in Fig. P13-4. For the inputs shown, determine the signals at the inputs to the ORs (points a through g) and at S and Co.
9. The circuit in Problem 8 is said to have a fast carry. In which of the following adder circuits would this be a major advantage?

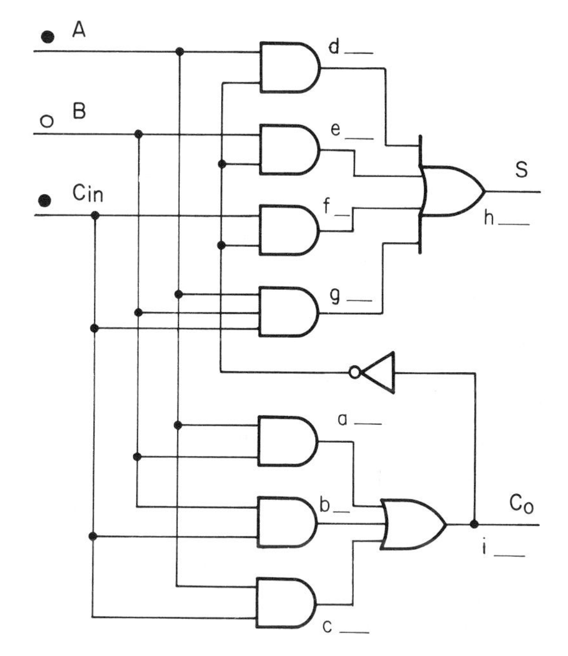

Fig. P13-4. Adder has a fast carry. Nine signal levels can be filled in. Advantages for ripple-carry adder and look-ahead-carry adder can be compared.

a. Ripple-carry adder.
b. Look-ahead-carry adder.

10. Copy the circuit in Fig. P13-5 onto a piece of paper. For the inputs shown, determine the outputs. What relation does the number D2D1D0 have to the number A2A1A0?

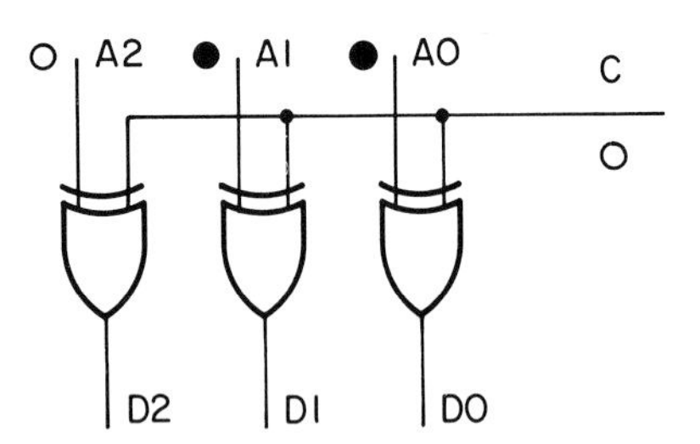

Fig. P13-5. Circuit with three separate XORs has a special property. Check outputs when C = 0 and when C = 1.

11. Repeat Problem 10 with C changed to 1.
12. The circuit shown in Fig. P13-6 can add and subtract. List the three individual adders (Adder 2, Adder 1, and Adder 0) and indicate which are full adders and which are half adders.

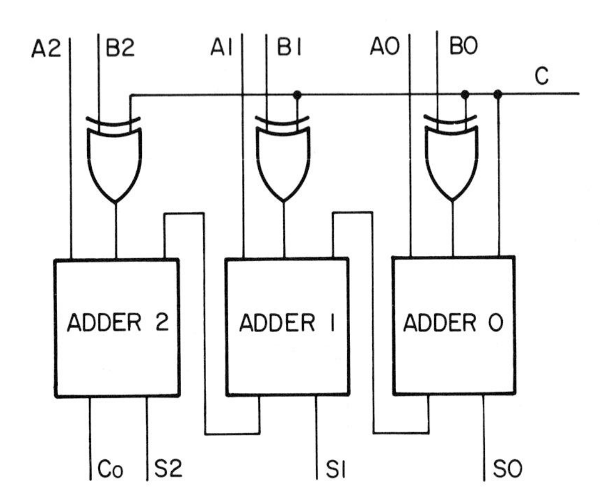

Fig. P13-6. Parts can be labeled for 3-bit subtractor/adder. Signal on lead C that provides an add process can be determined.

13. For the circuit in Fig. P13-6 to accomplish an add, what signal (1 or 0) must be placed on control lead C at the right?
14. Copy the comparator circuit in Fig. P13-7 onto a sheet of paper. For the inputs shown, determine the expected signals at the indicated points.

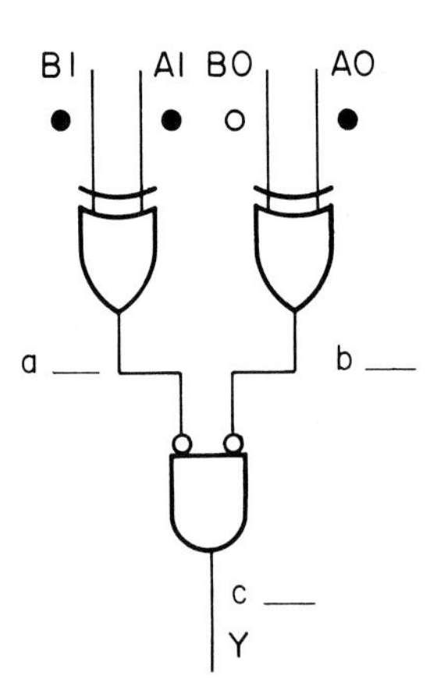

Fig. P13-7. Three signal levels for comparator can be filled in. Inputs are shown at top.

15. Between times t1 and t2, which operation was performed on the number shown below?
    a. Shift left.
    b. Rotate left.
    t1 A = 10,100,100
    t2 A = 01,001,001
16. Insert a 1 or 0 in bit D7 of each of these numbers to produce numbers with even parity.
    a. ____1,001,110
    b. ____0,001,100
    c. ____0,111,110
    d. ____1,110,111
17. For the input signals shown in Fig. P13-8, determine the expected signals at the indicated points. What is the parity of the input signal (D3D2D1D0)? What is the parity of the output signal (D4D3D2D1D0)?
18. Do the following addition and determine the

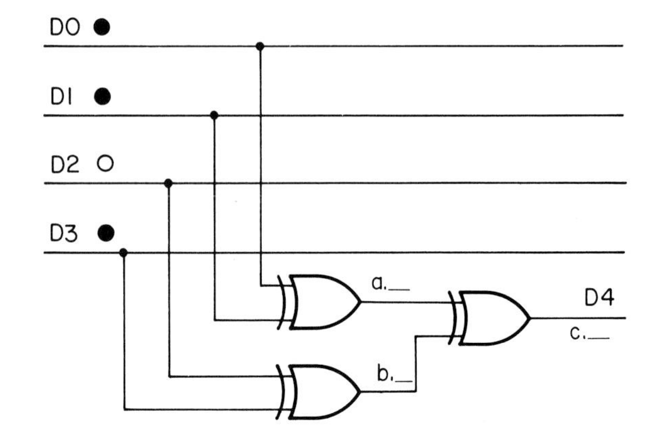

Fig. P13-8. Three signal levels can be filled in. Inputs are at left.

values (1s and 0s) of the zero flag, sign flag, parity flag, and carry flag. Assume they are all active high and that the parity flag indicates even parity. The numbers to be added are signed numbers.

$$\begin{array}{r} 10,100,001 \\ +\ 01,100,101 \\ \hline \end{array}$$

19. For the flag register shown in Fig. P13-9, determine the value of each flag (1 or 0) when this register contains the number BE (base 16).

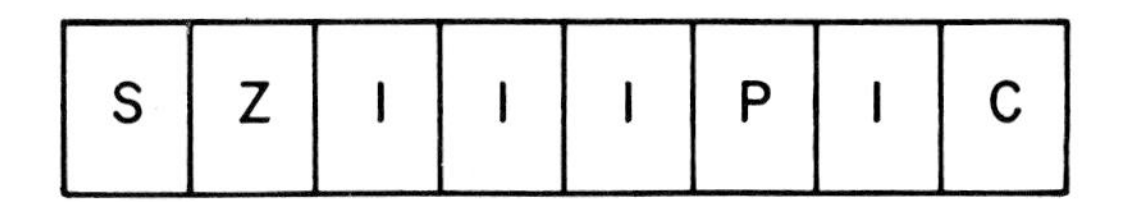

Fig. P13-9. Blanks in the flag register can be filled in. Number in register is BE (base 16).

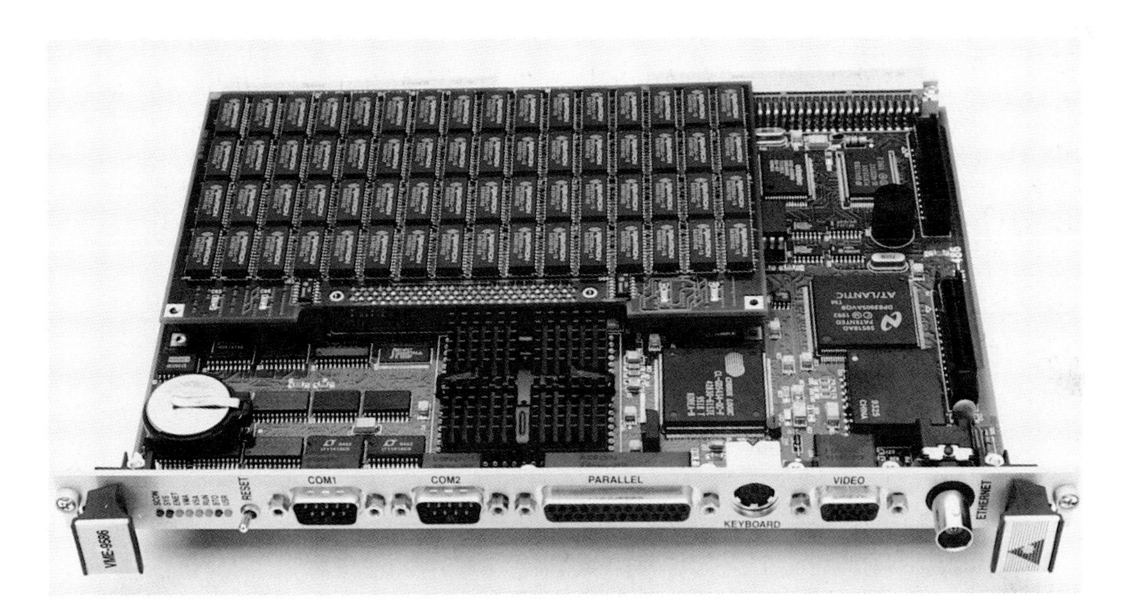

This is a single-board personal computer. It mounts vertically in a rack. Because such devices are available, it is relatively easy to design computers into machine and process controls. (Logic Design Group, Inc.)

Pin spacing can be so narrow that special adapters are needed to attach test probes. (Tektronic, Inc. 1996©)

Chapter 14

Buses

LEARNING OBJECTIVES

After studying this chapter and completing lab assignments and study problems, you will be able to:

- ► List the three buses commonly used in computers.
- ► Describe the action of wired ANDs and explain the problems associated with their use.
- ► Describe the action of open-collector elements when connected in wired ANDs.
- ► Explain the need for and operation of pullup resistors.
- ► Describe the action of three-state elements and their use as bus drivers.
- ► Define the term fanout and describe at least three ways of increasing fanout.
- ► Define the term fanin and describe a method of decreasing the value of fanin.
- ► Recognize and list the uses for any buffer/drivers used on bidirectional buses.

Within a computer, paths must be provided between sections for the exchange of data, addresses, and control signals. If direct, section-to-section electrical connections were used, the complex network shown in part "a" of Fig. 14-1 would result. Each interconnecting line represents between 8 and 40 parallel conductors. The resulting printed circuits and cables would be impossibly complex. Troubleshooting would be difficult.

To overcome the problems of point-to-point connections, most computers use bus systems, part "b" of Fig. 14-1. Computer sections communicate over a common set of conductors. The bus functions somewhat like a rural telephone line. All sections can listen to data, addresses, and control signals. However, care must be taken to insure that only one section places a signal on a bus at a given time.

This chapter deals with problems associated with buses. Special logic elements used to drive buses are described.

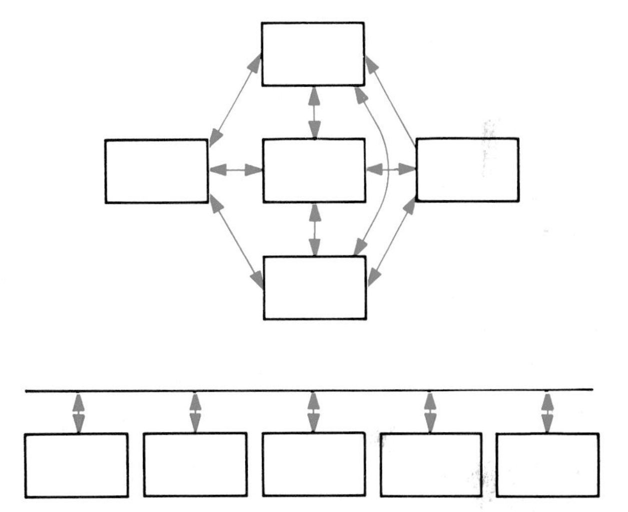

Fig. 14-1. Two methods for interconnecting computer sections are shown above. The top figure illustrates point-to-point wiring and the bottom figure illustrates the bus approach.

BUS STRUCTURES

Fig. 14-2 shows one typical block diagram for a simple computer. The block at the left contains the arithmetic/logic unit, registers, and flags. The control unit is next. Everything that happens within this computer is under the control of the circuits in this block. Memory and input and output circuits complete the diagram. Except for the small number of leads, the bus system shown is typical of those used in most computers today.

On the control bus, data usually flows in only one direction from the control block to other sections of the computer. However, these control leads are often grouped together. The control leads are usually referred to as control buses.

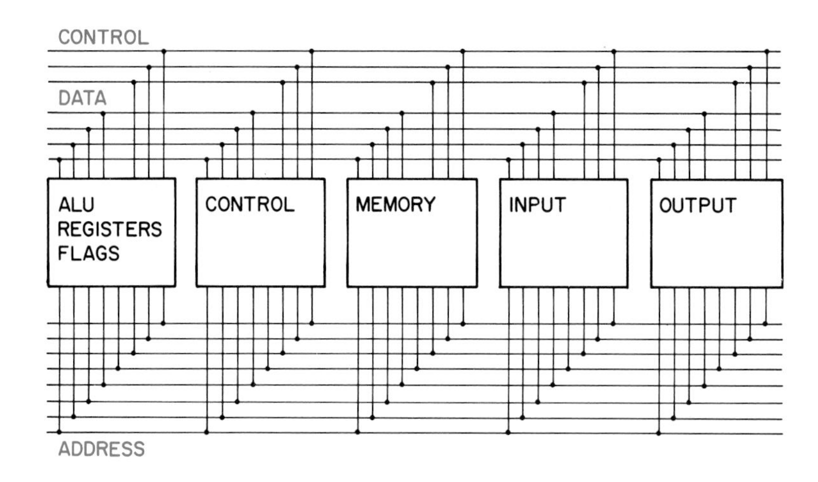

Fig. 14-2. Typically, computers have three buses: control, data, and address.

The three types of buses need to be discussed. These are the data bus, the control bus, and the address bus.

DATA BUS

The data bus transfers data between sections of the computer in Fig. 14-2. For example, the data bus would be used to move a number from memory to the ALU. After the number had been operated upon, it might be sent (again by way of the data bus) to the output portion of the computer.

CONTROL BUS

To move data from memory to the ALU, the memory must be told that it is to supply the data. The ALU must be informed that it is to take that data from the bus. The control bus distributes control information throughout the computer.

ADDRESS BUS

Even in the simple computer in Fig. 14-2, 256 words can be stored in memory. When a word is to be read from memory (or stored there), the address bus is used to indicate where that word is in memory.

ONE LEAD OF A BUS

Because they contain many leads, buses appear complex. However, the study of buses can be simplified by dealing with one lead at a time. For example, one lead from the data bus in Fig. 14-2 might be studied. However, it must be kept in mind that any circuits considered would be repeated for each lead in the bus.

Fig. 14-3 shows several input/output circuits connected to this lead. Each unit can output data to the bus; each can accept data from the bus. OUT ENABLE (OE) and IN ENABLE (INE) control the direction of data flow. If neither enable is active, the unit is disconnected from the bus.

This circuit appears capable of doing the required task. However, it has a major problem. This involves the connection of multiple outputs to a common lead.

PARALLEL CONNECTED OUTPUTS

Consider the transfer of data from unit A to unit C in Fig. 14-3. To place data on the bus, OEA (OUT ENABLE A) would be activated. The first AND gate would be opened.

To permit C to accept the number on the common lead, DINC (DATA IN C) would be made active. All other enables would be inactive (0).

But there is a problem. With 0s on their enable leads, the output ANDs at B, C, and D would output 0s. While the output AND at A is attempting to drive the common lead to 1, the other three ANDs are forcing it to 0. Will the 1 at the active output or the 0s at the inactive outputs prevail? The answer is found in the concept of WIRED ANDs.

DESCRIPTION OF WIRED AND

Fig. 14-4 shows the switch equivalent of an OR and an AND. Their outputs drive a common lead. For the switch positions shown, the OR is attempting to place a 1 on the lead; the AND is attempting to drive it to 0.

Note that the circuit from Vcc to ground is complete. The arrows show the resulting current. (Electron current flows in the opposite direction.)

Due to the low resistance of the switches, the common lead is grounded by the AND. That is, a logic 0 will appear on the lead. Most of Vcc will be dropped across the resistor in the OR.

WIRED AND ACTION

Part "a" of Fig. 14-5 shows the four possible output signal combinations. Remember, when the output

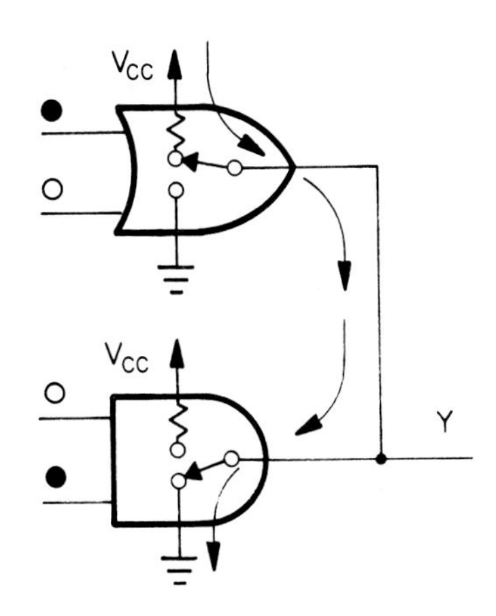

Fig. 14-4. The outputs of the elements shown are driving a common lead. When they attempt to output opposite numbers, a short circuit results.

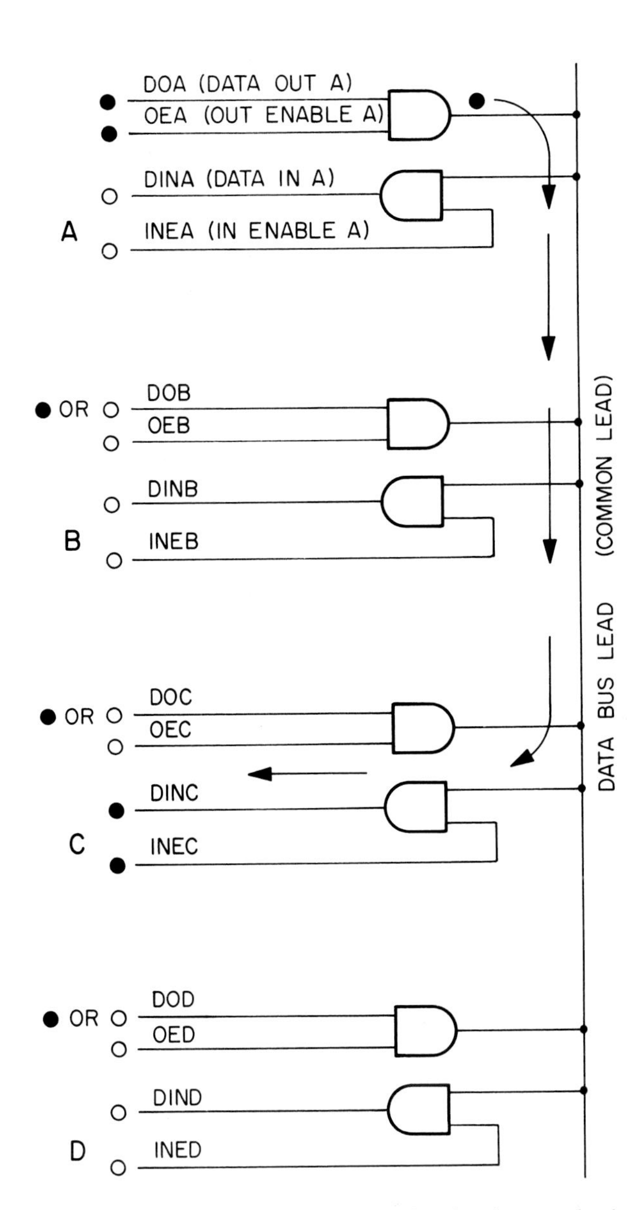

Fig. 14-3. Although it appears that the simplest method for placing numbers on a bus lead will work, it has serious problems.

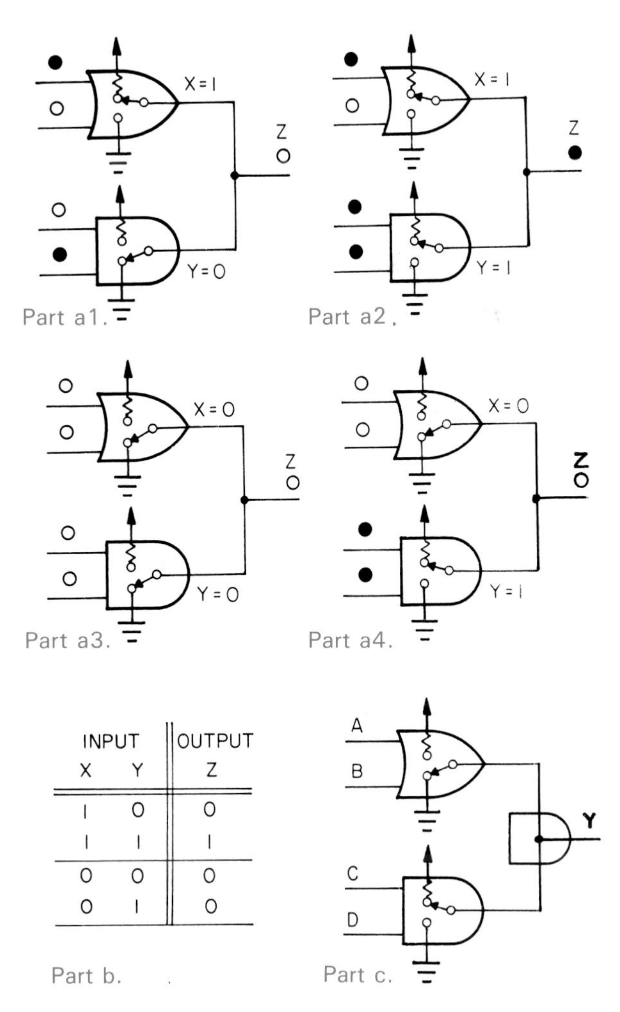

INPUT X	INPUT Y	OUTPUT Z
1	0	0
1	1	1
0	0	0
0	1	0

Fig. 14-5. When two or more outputs are connected to a common lead, a WIRED AND results.

switches are in opposite positions, the grounded switch prevails.

The results of these four conditions are described in the truth table at b. This is the truth table of an AND. That is, when the outputs of TTL logic elements are connected to a common lead, an unseen logic element results. The term WIRED AND is appropriate. To emphasize this action, the symbol at c is often added to logic diagrams containing wired ANDs. The expression for the circuit is as follows:

$$(A + B)(CD) = Y$$

DISADVANTAGES OF WIRED AND

A careful analysis of wired ANDs can reveal properties that the technician and the designer should be aware of. Caution is needed. Because wired ANDs appear to give something for nothing, there is a temptation to use them. However, they have several disadvantages. Two topics are important. These are:

1. Overload.
2. Troubleshooting.

The first of the topics is critical to designing for burnout protection, as described in the following:

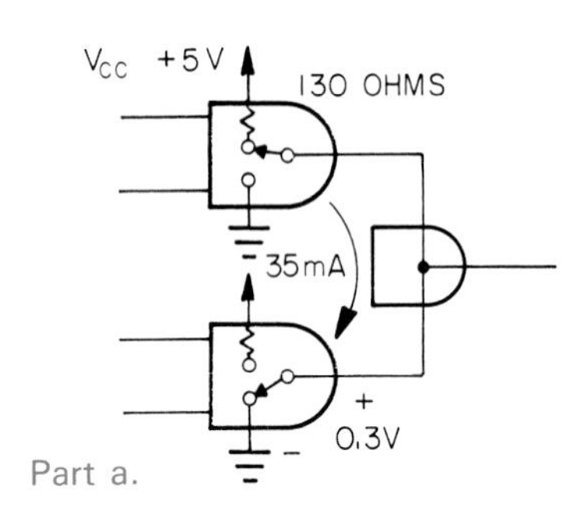

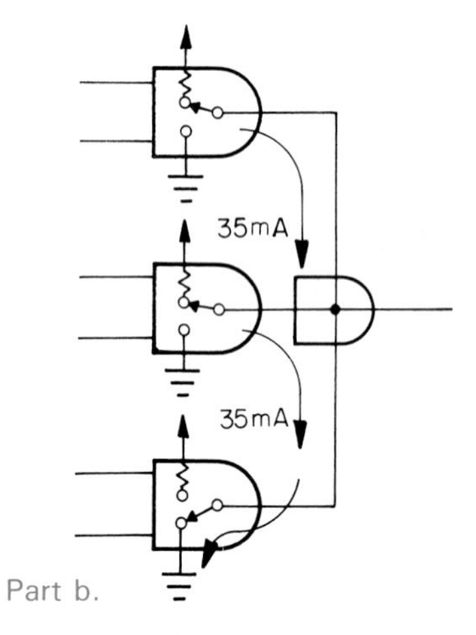

Fig. 14-6. Current flow in wired ANDs can be high enough to damage logic elements. The voltage drop, 4.7 V, is divided by 130 Ω to give approximately 36 mA.

1. Overload: Part "a" of Fig. 14-6 shows typical voltages and resistance values for a TTL wired AND. The resulting current is likely to be near 35 mA (0.035 Amperes). This is enough to cause borderline chips to fail by burning out. Even if chip damage does not result, this current is higher than the 16 mA (0.016 Ampere) rating of TTL outputs. Therefore, the signal voltage on the common lead may be higher than the 0.4 V upper limit recommended for a valid 0.

Part "b" of Fig. 14-6 shows that a wired AND with more than two driving elements must be avoided. In this 3-element circuit, the current through the lower AND could be over 70 mA. This current is likely to damage the chip.

2. Troubleshooting: Another disadvantage of wired ANDs is shown in Fig. 14-7. The dots show the results of measurements made on this circuit. Based on its inputs, the OR should output a 1. It is a 0. Is the OR faulty? In this case, it is operating properly. The wired AND has forced the output of the OR to 0.

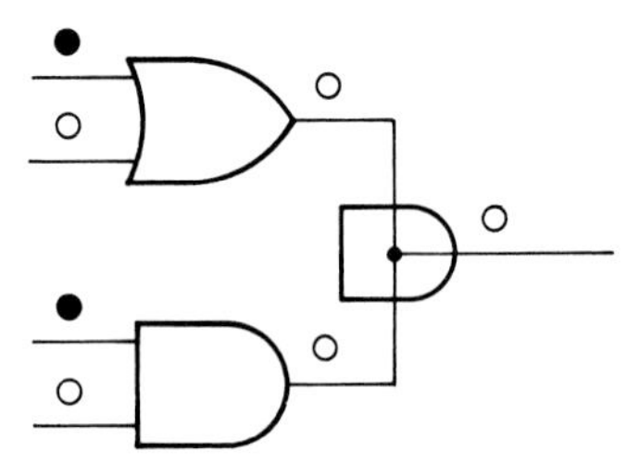

Fig. 14-7. Wired ANDs may cause troubleshooting problems.

If repair personnel are not aware that a wired AND is present, this could cause confusion. As a result, wired ANDs should be avoided in logic circuits.

OPEN-COLLECTOR DEVICES

In logic circuits, wired ANDs can be avoided. However, they must be used when buses are involved. Because buses are basically wired ANDs, methods have been developed to overcome the problems of this configuration. One of these methods uses logic elements with output leads called *OPEN COLLECTORS*.

Part "a" of Fig. 14-8 shows the switch equivalent of a 7408, an ordinary AND. When it outputs a 1, lead Y is connected to Vcc through a resistor. At the right is a 7409, an open-collector AND. The resistor

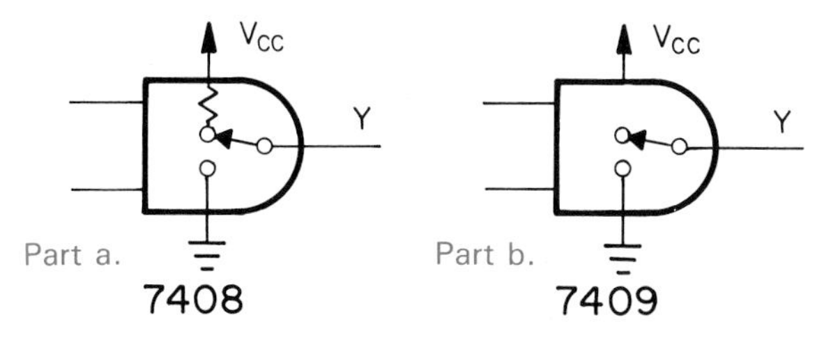

Fig. 14-8. The AND at the right is of the open-collector type. It outputs an open circuit when the ordinary AND outputs a logic 1.

that connected the switch to Vcc has been omitted. When it outputs a 1, lead Y floats.

OPEN-COLLECTOR ACTION

Part "a" of Fig. 14-9 shows an open-collector wired AND. The upper AND is outputting a 1; the lower is outputting a 0. The common lead is grounded through the lower AND.

Due to the open-collector lead in the upper element, no current flows from the upper AND. The problem of one output shorting another has been solved.

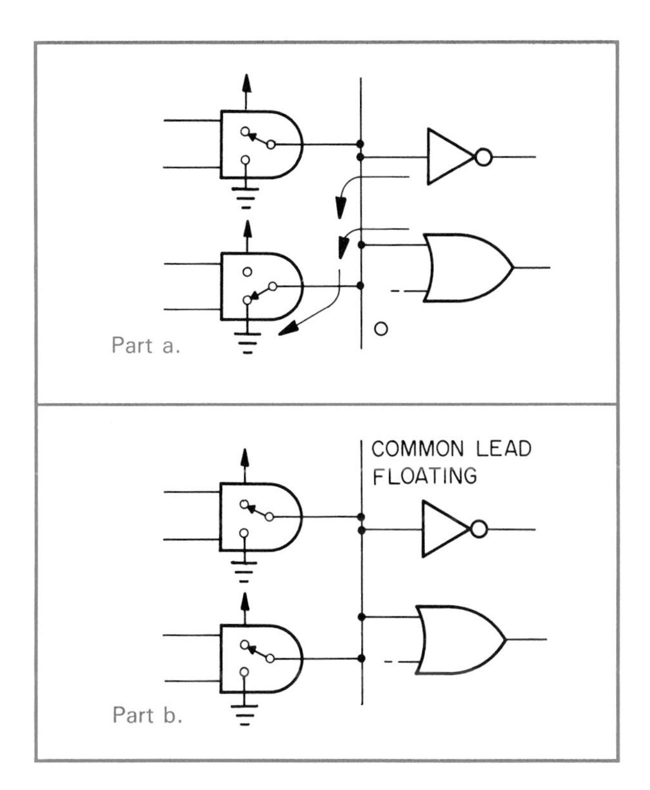

Fig. 14-9. When open-collector elements drive a common lead, that lead floats when all driving elements output 1s.

There is still a small current flowing from each input. Whenever a 0 is applied to the input of a TTL element, a current flows from Vcc, through the internal circuit of the chip, and out its input. The driving element must provide a path to ground for this current. As a result, TTL is said to be a *CURRENT-SINKING* family. The technician must get used to the fact that current can come out of an input.

When both ANDs output 1s, the common lead floats. See part "b" of Fig. 14-9. With no path to ground, current can no longer flow from the inputs of the driven elements. The floating lead is treated as a logic 1.

OPEN-COLLECTOR PROBLEMS

The use of open-collector elements solves the problem of one element shorting another. However, the floating common lead creates other problems. There are problems with the following:

1. Noise.
2. Reduced speed.

The problem is external to a chip, as seen in the following:

1. Noise: Strong electric and magnetic fields are present in most industrial plants. These fields can induce glitches in floating leads. A lead can act as an antenna, while long looped leads can act as pickup coils. As a result, floating leads are to be avoided.
2. Reduced speed: Floating leads also slow the action of logic circuits. That is, it takes longer for such circuits to go from logic 0 to logic 1. Electrical capacitance, present in all logic elements, contributes to this problem.

To review, capacitors temporarily store electricity. They consist of two conducting plates separated by an insulator.

Because conductors must be used in the construction of electronic devices, a small amount of capacitance is unintentionally built into all logic elements. It is called *STRAY CAPACITANCE* or *STRAY CAPACITY.*

Stray capacitance from the inputs, outputs, and wiring of a circuit is usually gathered together and represented as a single capacitor, part "a" of Fig. 14-10. A dashed symbol is used to indicate that it is the sum of the stray capacities.

As described earlier, it takes time to fill storage devices. For example, if a small hose is used, it takes a long time to fill or empty a barrel. If a large hose is used, it can be filled or emptied quickly.

The same is true of capacitors. If the current is small, it will take a long time to charge or discharge a capacitor. If the current is large, it can be charged

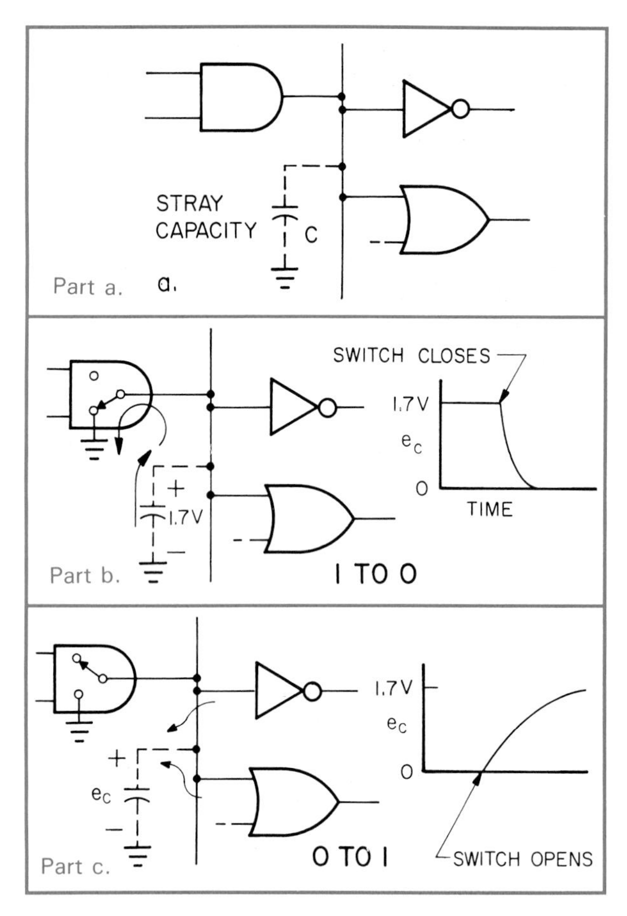

Fig. 14-10. When open-collector drivers are used, it takes a long time for the voltage on the common lead to go from 0 to 1.7 Volts. This is caused by the small current available to charge the stray capacity.

and discharged quickly.

Part "b" of Fig. 14-10 shows the rapid discharge of the stray capacity when the signal goes from 1 to 0. Note the steep edge in the curve at the right. The current is high because the resistance of the internal switch is low.

Part "c" of Fig. 14-10 shows the signal change from 0 to 1. Note the slow rise of the signal. Due to the open collector, current cannot flow from the output of the driving element. It must come from the inputs of the driven elements. This current is small, so it takes a long time to charge the stray capacity. Such a slowly changing signal could greatly reduce the speed of a computer or other digital circuit.

PULLUP RESISTOR

To overcome open-collector problems, components called *PULLUP RESISTORS* are often used. Resistor R in part "a" of Fig. 14-11 is a pullup resistor. A more consistent signal can be obtained. Erroneous signals are easier to trace and to check. Three topics are important:

1. Noise problem improvement.
2. Speed improvement.
3. Pullup resistor selection.

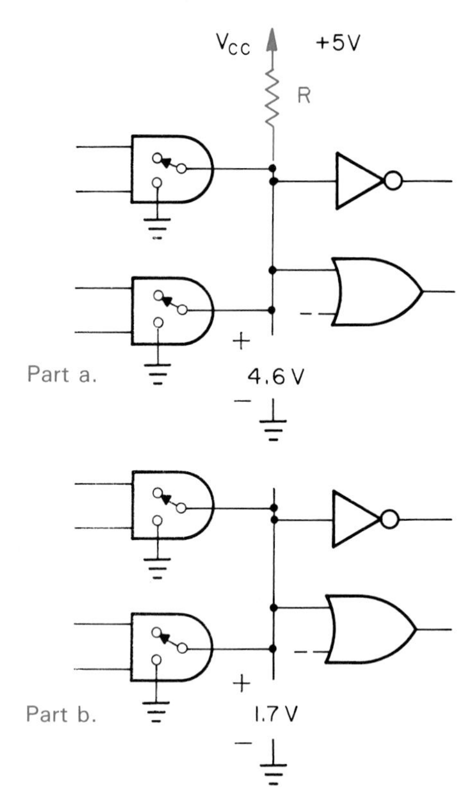

Fig. 14-11. Adding a pullup resistor (labeled R) solves some of the problems of open-collector circuits.

The first of the topics compares a floating input voltage with a "solid" input voltage as follows:

1. Noise problem improvement: In part "b" of Fig. 14-11, the common lead floats when both drivers output 1s. The voltage between the common lead and ground will be about 1.7 V. This is not a solid logic 1 (although it is treated as a 1 by the TTL inputs). As indicated earlier, this circuit is sensitive to noise. A noise voltage of only 1 V can reduce the 1.7 V to 0.7 V, which could be interpreted by the input as a logic 0 when it should be a logic 1.

When a pullup resistor is added, the common lead no longer floats. It is connected to Vcc through the resistor. The voltage between the common lead and ground will be about 4.6 volts and will represent a solid logic 1. A very large negative-going noise pulse would be needed to convert this voltage into a logic 0.

2. Speed improvement: Fig. 14-12 shows that a pullup resistor can improve the charging rate of the stray capacitance. Current flowing through the resistor adds to that from the inputs of the driven elements.
3. Pullup resistor selection: To reduce noise problems and obtain high-speed operation, pullup resistors should be as small as possible (permit high current). However, if this resistor is too small, problems result.

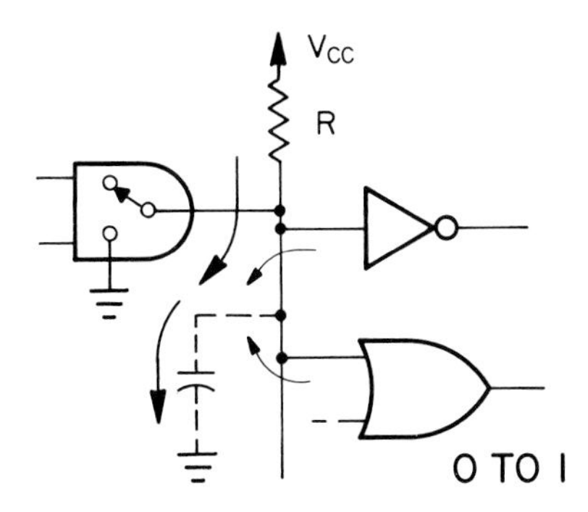

Fig. 14-12. Because a pullup resistor adds to the charging current for the stray capacitance, the circuit will go from 0 to 1 faster than an open-collector circuit without a pullup resistor will.

Fig. 14-13 shows the flow of current when a driving element outputs a 0. In addition to sinking the currents from the inputs, the element must also conduct the pullup resistor current to ground. If R is too small, the total current may exceed the rating for the output of the AND.

In most applications, a 1,000 Ω resistor results in acceptable operation. This is small enough to insure solid logic 1s and permit relatively fast action, yet it does not result in excessive current when logic 0s are on the common lead.

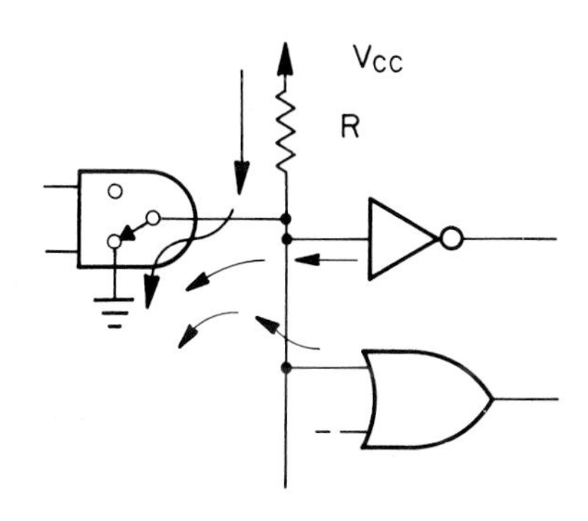

Fig. 14-13. Current from a pullup resistor adds to the current the driving element must sink. As a result, fewer inputs can be driven.

BUSES DRIVEN BY OPEN-COLLECTORS

Fig. 14-14 shows a bus driven by an open-collector. Note the use of NANDs. Due to the NOTed outputs, data appears on the bus in complemented form. That is, if a 1 is sent, a 0 appears on the bus. The NOTs at each input correct this situation.

Assume data is to be sent from A to C. Ones would be placed on OUT ENABLE A and IN ENABLE C. All other control leads would be 0.

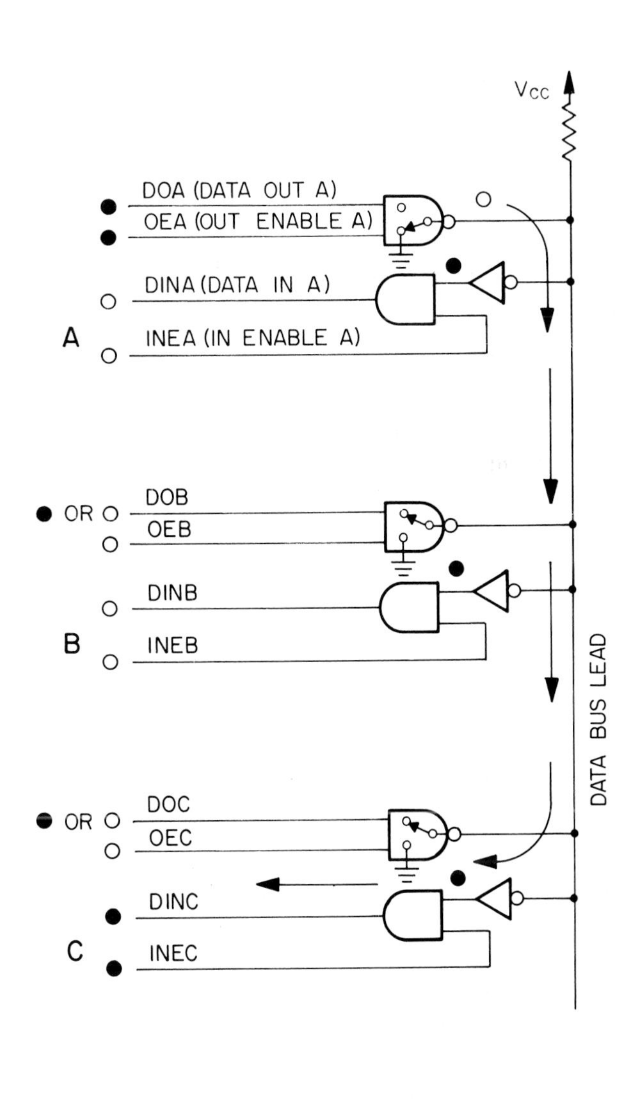

Fig. 14-14. When OUTPUT ENABLE A and INPUT ENABLE C are active, a bit is transferred from Block A to Block C. Only the driver at A is active. The other outputs are open circuits.

TRANSMISSION OF A 1

If Aout = 1, the switch of the upper open-collector NAND would be in the position shown. (All other output switches would be in the open-circuit position.) This would ground the bus and transmit a 0.

The NOT at unit C would invert the 0 on the bus and present a 1 to the AND. With a 1 on IN ENABLE C, this 1 would appear at DINC.

TRANSMISSION OF A 0

If Aout = 0, the switch in the upper NAND would be in its open position. With all outputs in this position, little current would flow through the pullup resistor, and a 1 would appear on the bus.

The NOT at C would again invert the number on the bus. With AND gate C open, DINC will equal 0.

THREE-STATE DEVICES

The use of open-collector elements and pullup resistors is an acceptable method for driving buses. Another widely used method has components called *THREE-STATE DEVICES.* Part "a" of Fig. 14-15 shows the symbol for a three-state NOT.

Part "b" of Fig. 14-15 shows the switch equivalent of a three-state NOT device. The lead on the bottom side of the symbol is its output enable. When a 0 is placed on this lead, switch B opens. Its output floats. When it is enabled (a 1 is placed on this lead), switch B closes, and the element functions as an ordinary NOT. That is, its output is either a 1 or 0 depending on the input signal. In this state, the output is driven in both directions. The bus is either connected to ground or to Vcc (through the internal resistance).

The term three-state comes from the three states of the output. They are:

Logic 0: Output grounded through switches B and A.
Logic 1: Output connected to Vcc through switches labeled B and A.
Floating: Switch B is open.

The floating state is often referred to as the high impedance state. High impedance implies that little or no current can flow into or out of the element.

Many logic elements are available in three-state form. Also, active-high and active-low output enables are available. See Fig. 14-16.

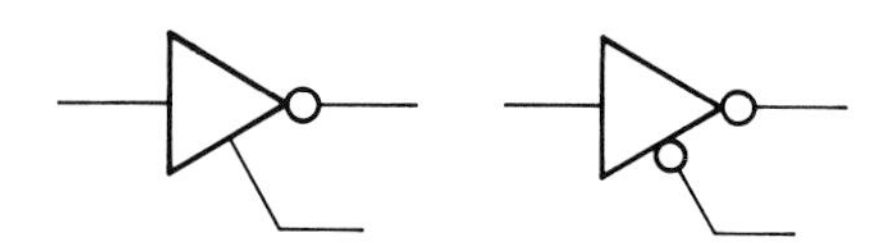

Fig. 14-16. Output enables on three-state elements may be active-high or active-low.

Fig. 14-17 shows a lead from a bus driven by three-state elements. At any given time, only one OUT ENABLE will be active. That device will be connected to the bus. The other drivers will be in their high impedance state.

In present-day computers, most buses are driven by three-state elements. The devices have a consistent operation and a proven design.

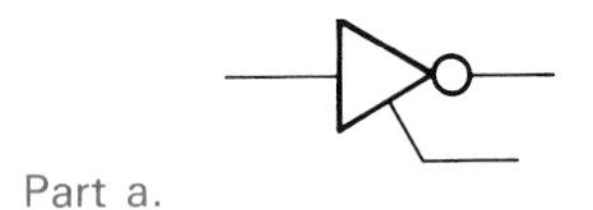

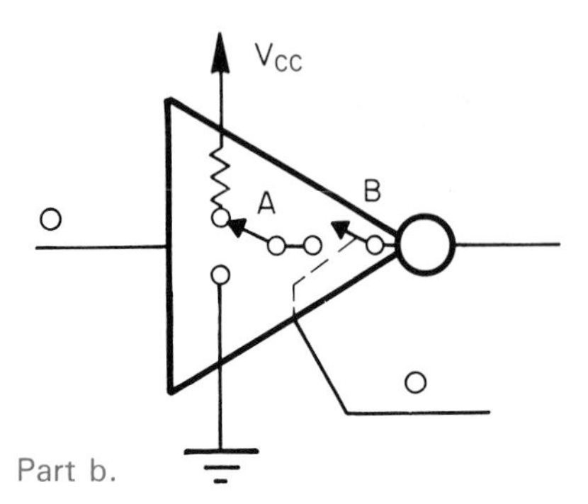

Fig. 14-15. The three-state NOT shown can output a 1, a 0, and an open circuit.

MULTIPLE INPUTS

Open-collector and three-state elements solve the problems resulting from multiple bus drivers. However, problems also result when too many inputs are connected to a common lead or bus. The designer must keep track of the currents used. Current ratings must not be exceeded.

CURRENT RATINGS

Fig. 14-18 emphasizes that current flows in TTL inputs when signals are applied. If the signal is 0, current flows out of the inputs. The driving element must provide a path to ground for the sum of these currents. If the signal number is a 1, current flows into the inputs. Again, the driving element must supply the sum of these currents. Two topics are important. These are:

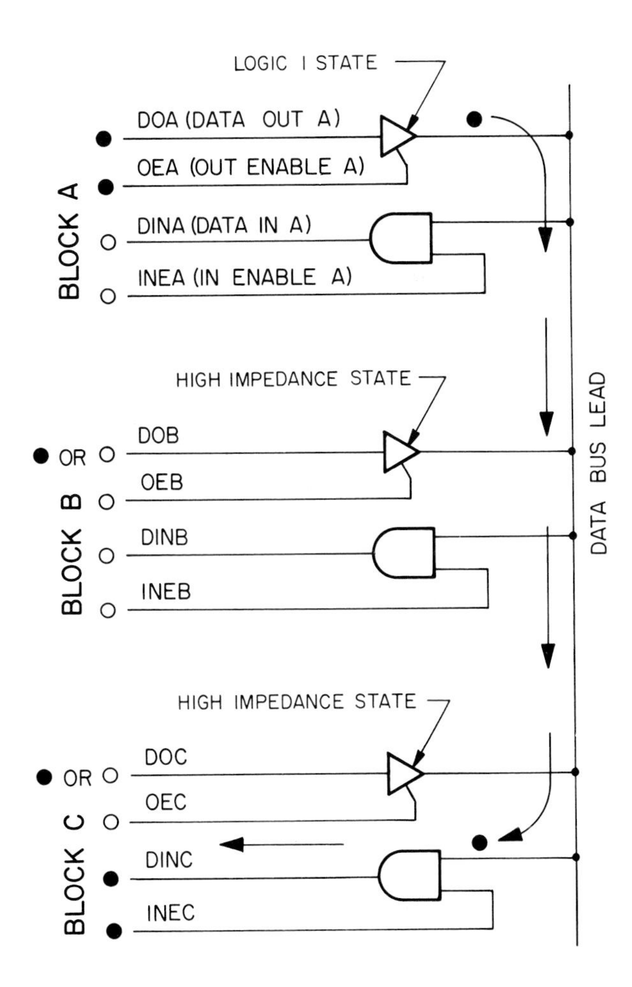

Fig. 14-17. Three-state buffers drive the bus in the circuit shown. Only one output enable is active at a given time.

1. Input ratings.
2. Output ratings.

The first topic shows typical input ratings and prepares for a fanout calculation, as in the following first step:

1. Input ratings: The current from a single TTL input can be as high as 1.6 mA when a 0 is applied. The current flows out of the input.

   When a 1 is applied, the current can be as high as 0.04 mA. The current flows into the input.
2. Output ratings: When outputting a 0, most TTL outputs can sink (conduct to ground) 16 mA. When outputting a 1, their rating is 0.4 mA. If these currents are exceeded, signal voltages on the bus or common lead may not be within acceptable limits for 0s and 1s. If currents considerably higher than these flow in a TTL output, the chip could overheat and be damaged.

FANOUT

Based on the current ratings of TTL inputs and outputs, one output can drive ten inputs. This ratio (10 to 1) is said to be the family's *FANOUT*.

Other families have different fanouts. For example, a CMOS output can drive 50 or more CMOS inputs. This family's fanout is 50.

FANOUT PROBLEMS

Because many inputs may be connected to a single lead of a bus, the fanout of ordinary elements may not be sufficient. The resulting problems are especially difficult when a low-power chip (such as CMOS) drives elements from a high-power family (such as TTL). Solutions to fanout problems are described within the listings of the next few headings.

BUFFER/DRIVERS

In electronics, to *BUFFER* means to isolate. A DRIVER implies a high-power circuit. In digital circuits, a BUFFER/DRIVER is an element capable

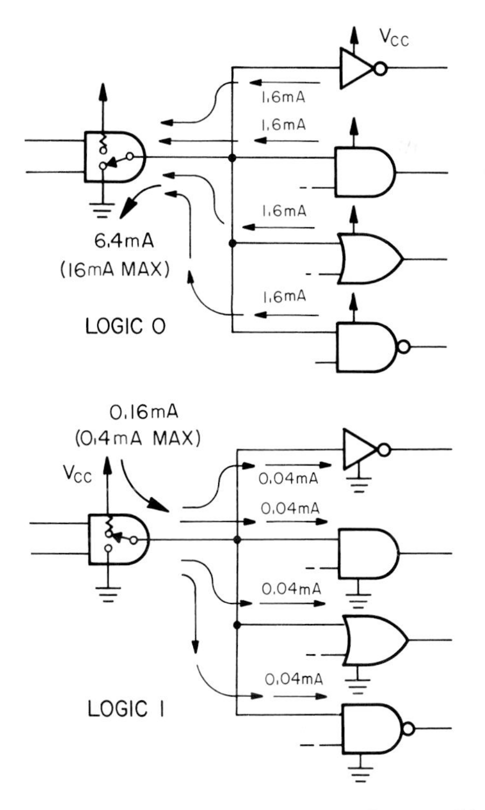

Fig. 14-18. The driving element must be capable of passing the sum of the input currents of the driven elements.

of sinking higher than normal currents. Buffer/drivers are often used to increase the fanout of low-fanout devices.

The 7437 is very much like the 7400 except it has a current rating of 48 mA. (The rating of the 7400 is 16 mA.) At 1.6 mA per TTL input, a 7437 can drive 30 inputs. That is, the use of the 7437 chip increases the fanout of the TTL family from 10 to 30.

LOW-POWER FAMILIES

Some component families require little input current. If two families are compatible (they can operate in the same circuit without special interfacing), elements from the higher-power family can drive those of the lower-power family. For example, one TTL output can drive 75 CMOS inputs.

There are also low-power versions of the TTL family. One such family is designated by the letters LS. For example, the 74LS08 is a low-power 7408. With a logic 0 at its input, the signal current of a 74LS08 is only 0.36 mA (compared to 1.6 mA for a standard 7408). A standard TTL output can drive forty 74LS08 inputs. Fig. 14-19 shows a mix of standard and low-power chips driven by one TTL output. By keeping track of the total input current, one TTL output can drive more than 10 inputs.

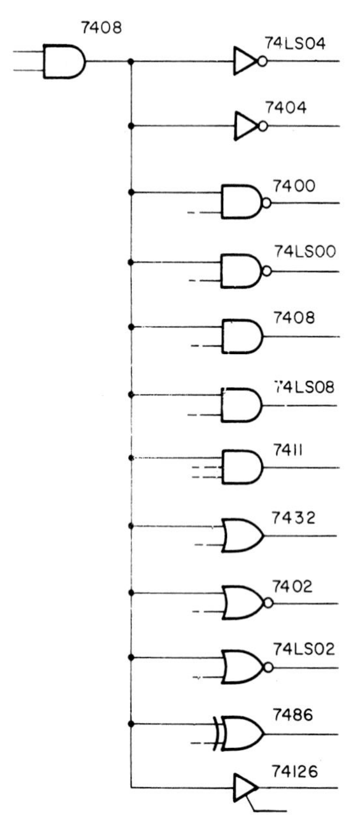

Fig. 14-19. A 7408 has a fanout of 10, yet in this circuit it is driving 12 inputs. This is permissible, since some of the driven elements are low-power chips.

LOW-POWER/HIGH-POWER COMBINATION CIRCUITS

Buffer/drivers are often considered to be special devices and may not be readily available. As a result, the circuit in Fig. 14-20 can be used. The microprocessor drives the low-power 74LS04. It, in turn, drives the standard TTL 7404. The low fanout of the microprocessor is raised to 10.

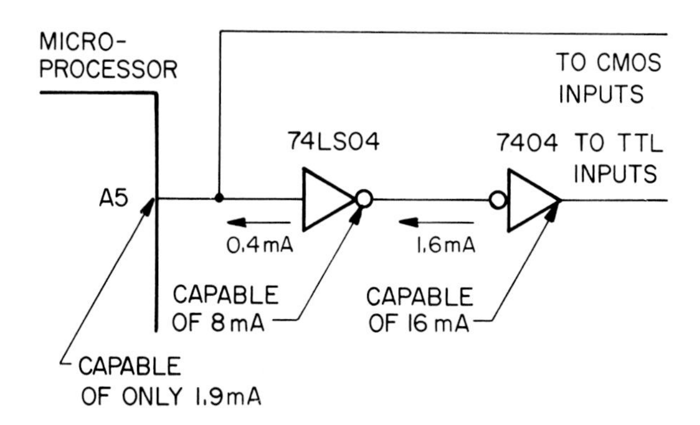

Fig. 14-20. A buffer can be constructed by placing a low-power NOT and a standard NOT in series with the low-power output.

DIVIDED BUS HELPS SOLVE FANOUT PROBLEMS

Fanout problems can be solved or the problem can be reduced in several ways. Fig. 14-21 shows one approach. The load has been divided between two outputs. Each output shares the load that is required. The cost of extra output elements is worth the improved performance.

BIDIRECTIONAL DRIVERS

Fig. 14-22 shows a microprocessor and two bus leads. The upper is an address lead (one of 16), and the lower is a data lead (one of 8). Note the difference in the buffer/drivers.

In most microprocessor applications, addresses originate within the microprocessor. As a result, addresses flow in only one direction, away from the microprocessor. A simple unidirectional buffer/driver can be used.

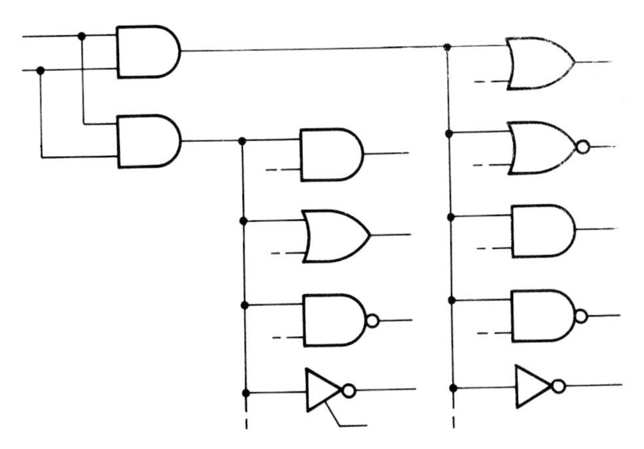

Fig. 14-21. When a load is greater than the fanout of a single element, the load can be divided between two outputs.

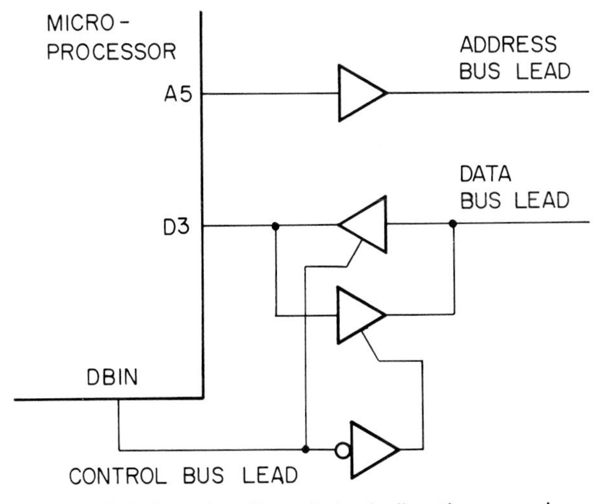

Fig. 14-22. When data flows in both directions on a bus, bidirectional buffer/drivers are used.

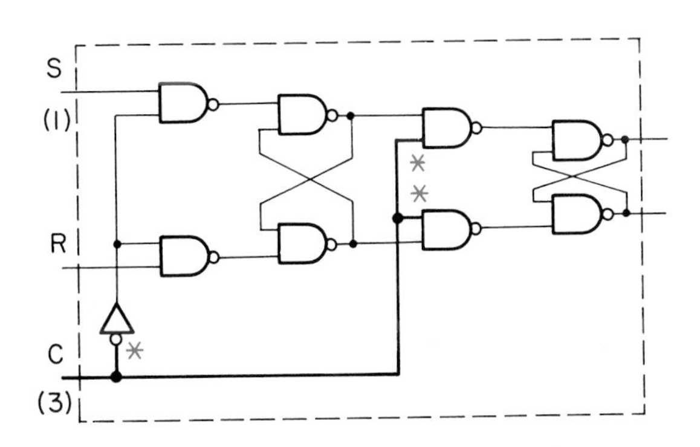

Fig. 14-23. The element that supplies the clock signal to this flip-flop must drive three inputs (see the three stars). Input C is said to have a *FANIN* of 3.

On the data bus, numbers flow to and from the microprocessor. This bus is bidirectional, so the more complex buffer/driver is needed.

Depending on the task being accomplished, DBIN (data bus in) will output a 1 or 0. When it is 1, data passes through the upper three-state element, while the lower element presents an open circuit to the bus. Data flows into the microprocessor.

When DBIN is 0, the opposite is true. The lower element is active, and data flows from the microprocessor. With a 0 applied to the enable of the upper element, it acts like an open circuit.

FANIN

Fanout describes the load-driving ability of outputs. The term *FANIN* describes the load an input places on an output. Input S on the flip-flop in Fig. 14-23 represents a single TTL input. Its fanin is 1. An inspection of lead C suggests a problem. This lead goes to three TTL inputs. Its driving current is three times that of S. The fanin of C is 3.

If three of these flip-flops were used to build the shift register in part "a" of Fig. 14-24, the fanin of the SHIFT ENABLE lead would be 9 (3 for each clock input). This control lead would take almost all the current sinking capacity of a TTL element.

To correct this problem, the noninverting buffer/driver in the lower circuit of Fig. 14-24 was added. Its fanin is 1. The two-bubble symbol was used to emphasize that SHIFT ENABLE is active-low. It is desirable to increase fanout and decrease fanin.

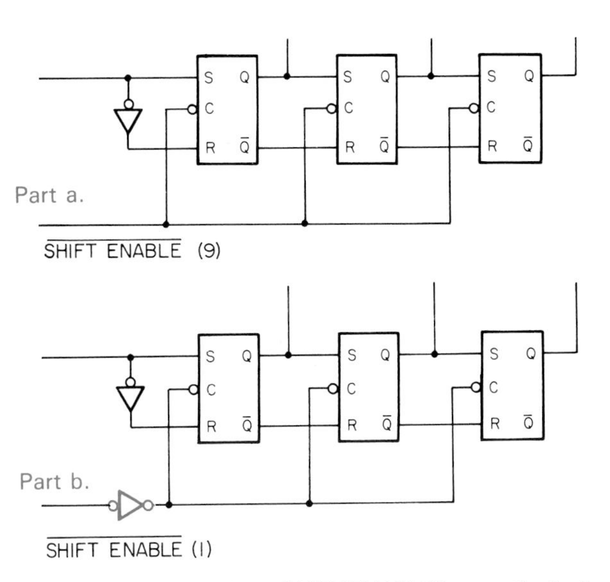

Fig. 14-24. The buffer in SHIFT ENABLE lowers the fanin of the lead shown from 9 to 1. Clocks are active-low and level-triggered.

SUMMARY

Buses are used to communicate between sections of computers. With many elements connected to each bus lead, problems result.

When more than one output is connected to a common lead, a wired AND is produced. Wired ANDs have several disadvantages, but they must be used when more than one driver is connected to a bus. To overcome these disadvantages, open-collector and three-state elements are used.

The fanout of an element indicates how many inputs it can drive. If the fanout of an element is exceeded, faulty operation results.

Fanin describes the load an input places on an output. Some elements have higher fanins than others. Buffers can decrease fanin and increase fanout.

IMPORTANT TERMS

Bidirectional driver, Buffer/driver, Current-sinking, Fanin, Fanout, Open collector, Pullup resistor, Stray capacitance, Three-state element, Wired AND.

TEST YOUR KNOWLEDGE

1. A number is sent from one portion of a computer to another on a __________ __________.
2. The job of informing the ALU that it is to take data from a bus originating at the memory is done by the __________ __________ (control bus, address bus).
3. Will a 1 or a 0 on a wired AND lead prevail?
4. The TTL family of logic elements is said to be a current- __________ family.
5. All conductors used to construct electronic devices have __________ capacitance.
6. With a weak output current, the capacitance of a circuit element takes a __________ (long, short) time to charge up to 3.5 V.
7. The performance of open-collector circuits is improved with the use of __________ resistors which are connected to a data bus.
8. A pullup resistor should have a resistance of __________ (50, 100, 1000, 5000, 50K) Ω to provide maximum performance.
9. Some buses use __________-state devices.
10. High __________ implies that little or no current can flow into or out of an element.
11. When outputting a 0, most TTL outputs can sink approximately __________ mA.
12. The fanout for the TTL family is __________.
13. The CMOS family has a fanout of __________.
14. The CMOS family is a __________ (low-power, high-power) family of logic elements.
15. To __________ means to isolate.
16. A __________ implies a high-power circuit.
17. A buffer/__________ is an element able to sink higher than normal currents.
18. Using a 7437 chip in place of a 7400 chip increases the fanout of the TTL family from 10 to ______.
19. Some low-power TTL elements are labeled with the code marking __________.
20. With a __________ (regulated, active-low, tertiary, divided, filtered) bus, the circuit load can be shared.
21. Data and addresses are handled in more than one direction with __________ drivers.
22. One TTL output can drive as many as ________ CMOS inputs.
23. __________ (One, Two, Three) NOTs in a row can be used as a buffer.
24. __________ describes the load an input places on an output.
25. It is desirable to increase ________ (fanin, fanout) and decrease ________ (fanin, fanout).

STUDY PROBLEMS

1. List the three buses found in most computers.
2. Copy the circuit in Fig. P14-1 onto a piece of paper. Using a series of arrows, indicate the current path from Vcc to ground. If you use electron flow, the path will be from ground to Vcc.

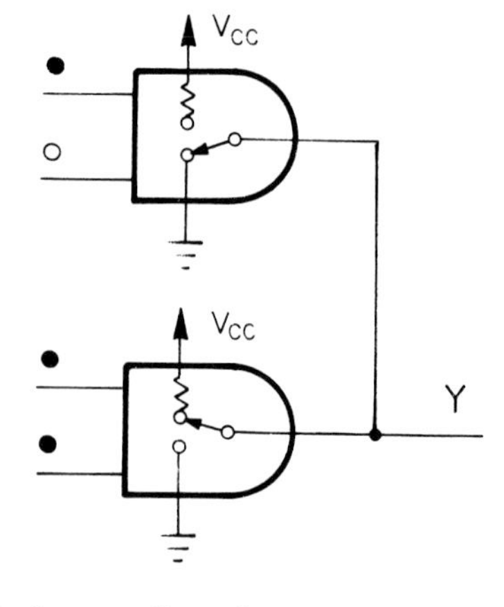

Fig. P14-1. Current flow directions can be determined for the wired AND shown. The current flow helps a technician see whether a 1 or a 0 will be on the output lead.

3. Based on the positions of the output switches in Fig. P14-1 from Problem 2, what signal will be

on the common lead (logic 1 or logic 0)?

4. For each set of switch positions in Fig. P14-2, indicate the signal (1 or 0) at Y.

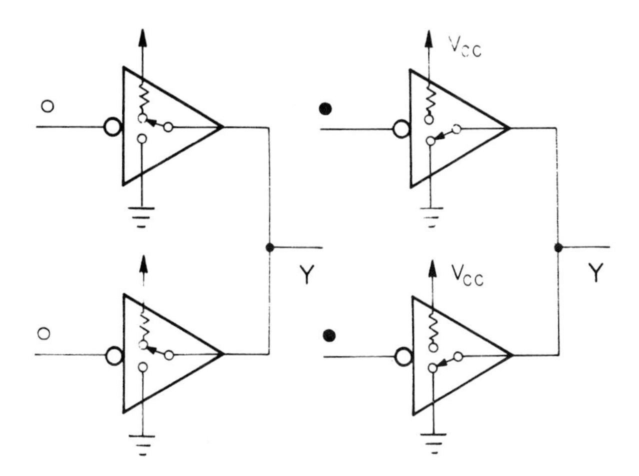

Fig. P14-2. Switch positions inside of NOT elements help in determining whether there is a 1 or 0 at Y.

5. Copy the circuit in Fig. P14-3 onto a piece of paper and add the symbol for a wired AND. Also write the Boolean expression for Y in terms of A, B, and C.

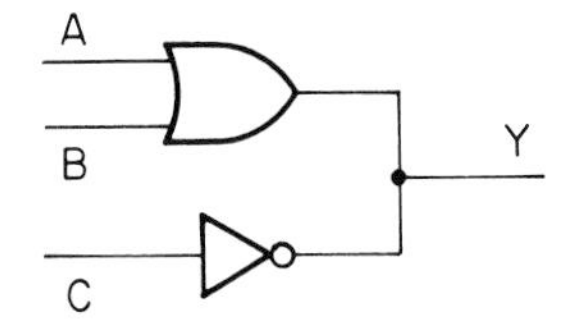

Fig. P14-3. Circuit containing OR element and NOT element is example of wired AND. Symbol for wired AND and Boolean expression for Y can be displayed.

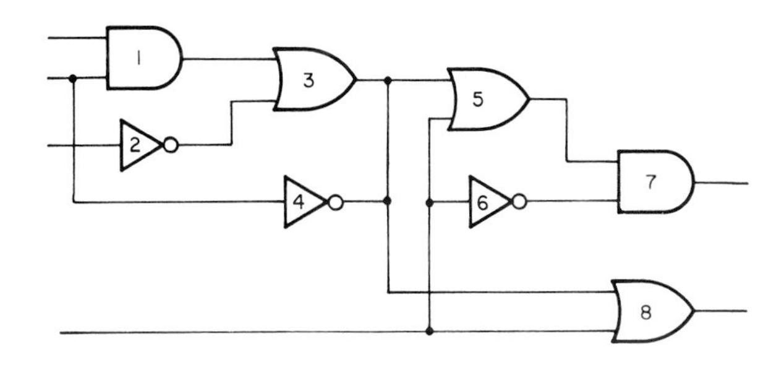

Fig. P14-4. Two logic elements involved in a wired AND can be found. A circuit with a pullup resistor can be drawn.

6. Which two elements from the circuit in Fig. P14-4 drive a wired AND?
7. Copy the circuit in Fig. P14-4 from Problem 6 onto a piece of paper. Assume the elements driving the wired AND are open-collector. Add a pullup resistor to your circuit. Label the end of the resistor that goes to the power supply Vcc.

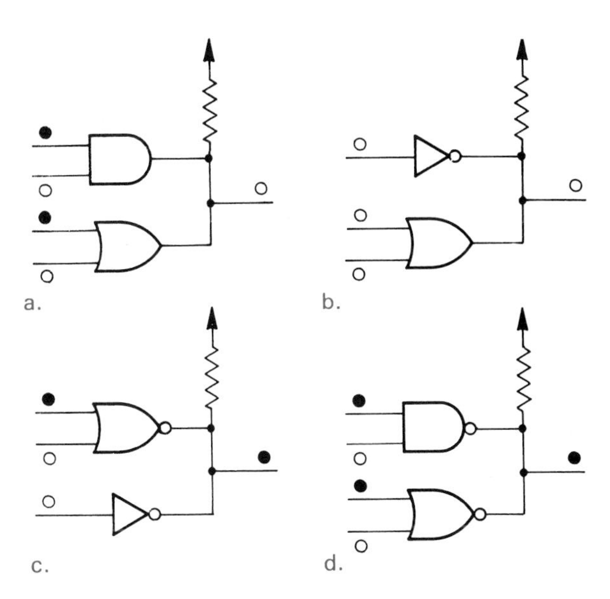

Fig. P14-5. Behavior of 4 wired ANDs can be checked.

8. What are the two primary functions of pullup resistors?
9. Measurements made on 4 wired ANDs are shown in Fig. P14-5. The elements are open-collector. Determine whether or not each output is correct, and list the circuits that are producing incorrect outputs.
10. The driving elements in both circuits in Fig. P14-6 are open-collector. Which circuit is most likely to have noise problems?

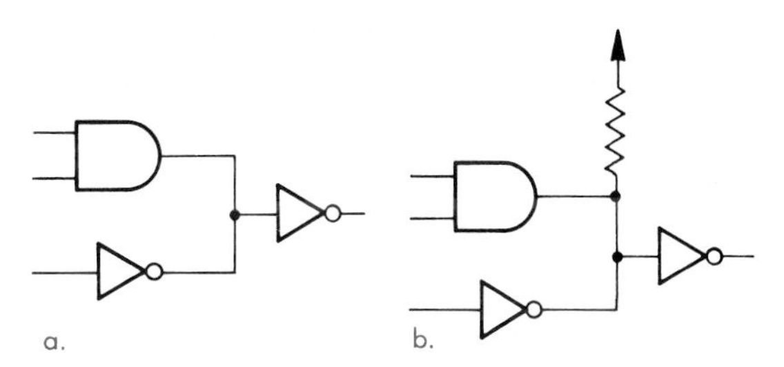

Fig. P14-6. For AND and NOT terminating in a wired AND, noise problems and charging times can be determined.

11. Refer to the circuits used in Fig. P14-6 from Problem 10. Which circuit is likely to take the longest time to go from a logic 0 to a logic 1?
12. Two wired ANDs are shown in Fig. P14-7. Based on the voltage measurements shown, in which circuit is the pullup resistor likely to be open?

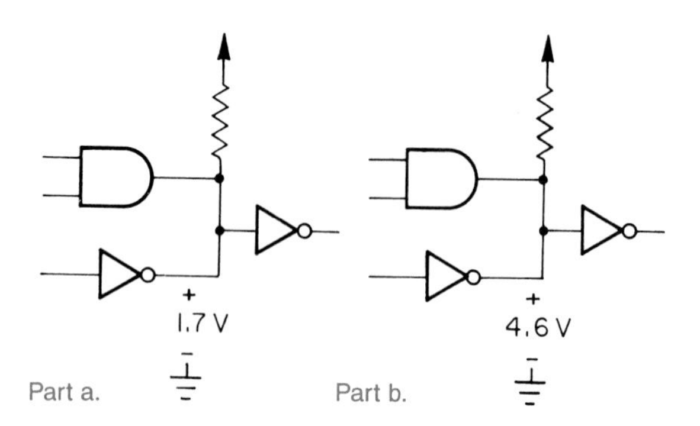

Fig. P14-7. For AND and NOT terminating in wired AND with pullup resistor, faulty behavior can be recognized.

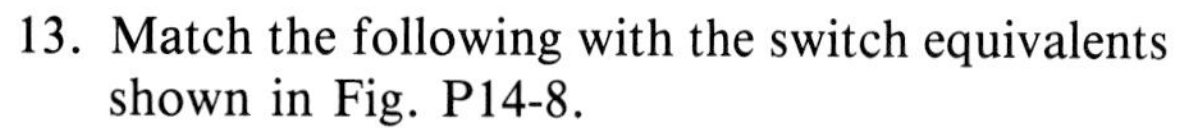

13. Match the following with the switch equivalents shown in Fig. P14-8.
    1. Open-collector.
    2. Three-state.

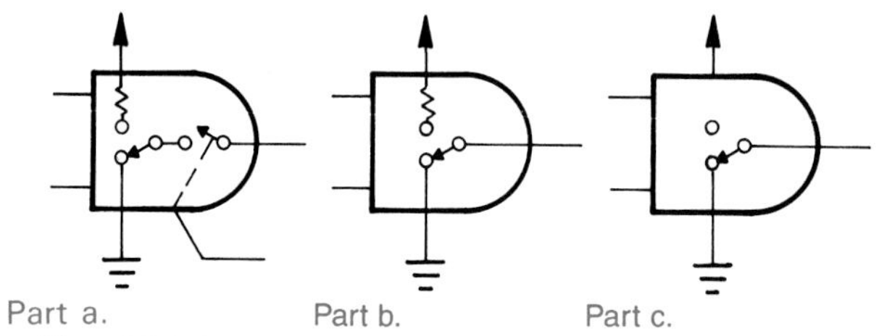

Fig. P14-8. Three types of circuit outputs can be recognized by their switch equivalents.

14. The diagram in Fig. P14-9 shows two sets of memory chips driving buses. Which set probably has three-state outputs? (The other set has open-collector outputs.)
15. Two leads marked A and B on the memory chip in Fig. P14-10 are attached to the data bus. Which is likely to be the chip's output?
16. Which element in Fig. P14-11 is most likely to have a three-state output?
17. What is the minimum fanout that the AND in Fig. P14-12 must have? What is the minimum fanout for AND number 1 in Fig. P14-11?

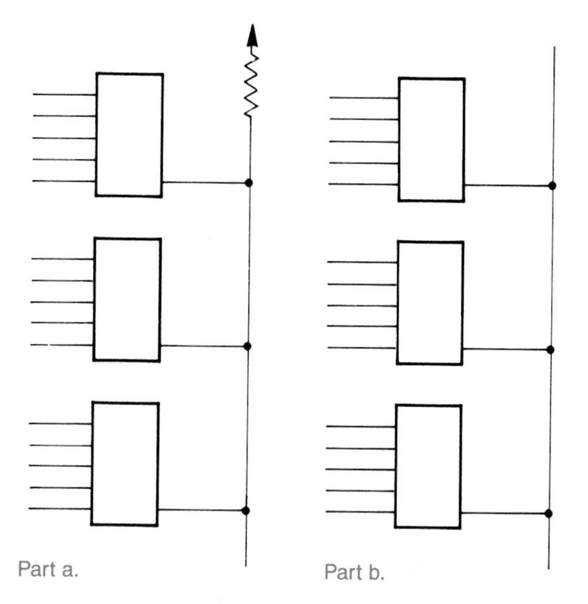

Fig. P14-9. Single bit is chosen from 1 of 3 memory locations. Left and right sets can be compared to decide which uses three-state outputs.

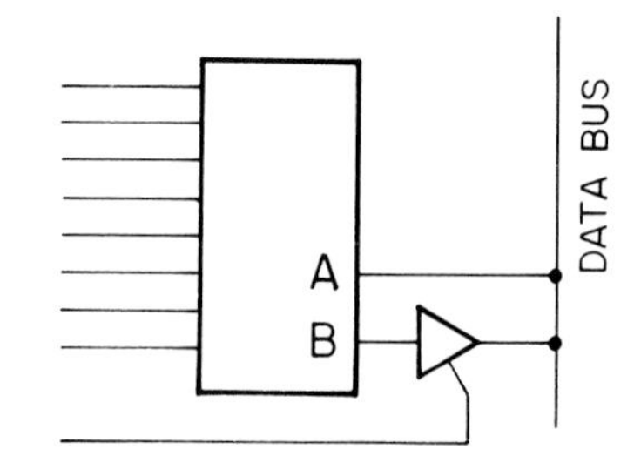

Fig. P14-10. Output of chip can be recognized.

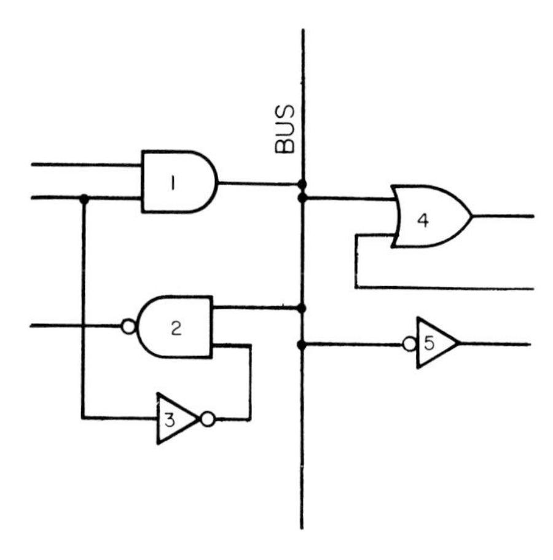

Fig. P14-11. Of 5 elements, one of them can be determined to have a three-state output.

18. Has the fanout of the 7408 in Fig. P14-12 been exceeded?

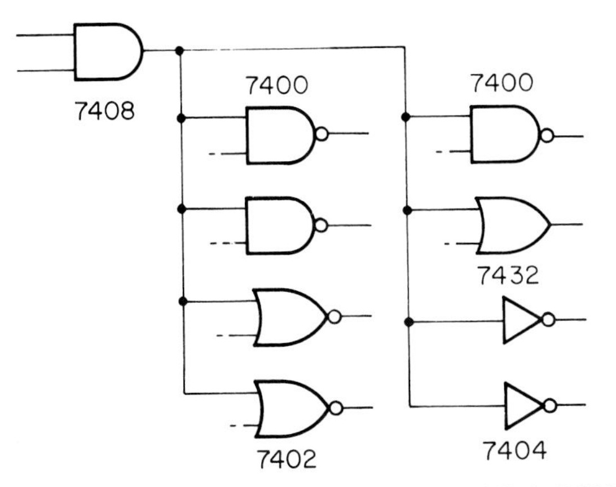

Fig. P14-12. The fanout for the upper left AND (a 7408) can be determined.

19. Has the fanout of the 7400 in Fig. P14-13 been exceeded?

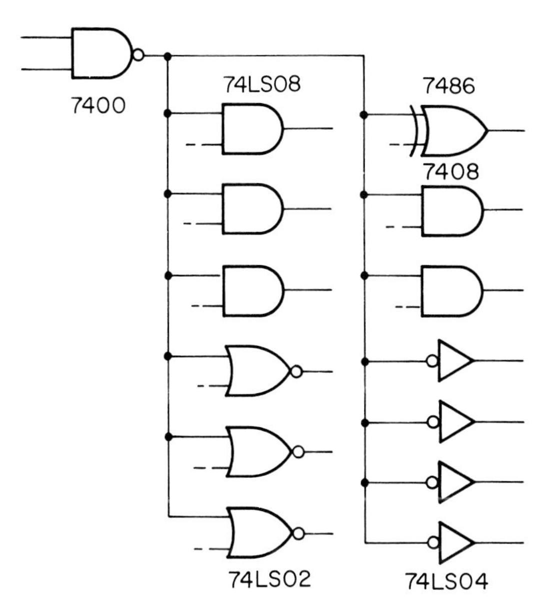

Fig. P14-13. The fanout for the upper left NAND (a 7400) can be determined.

20. Using Fig. P14-14, indicate the fanin of each of the following leads.
    a. DATA IN A.
    b. ADDRESS 1.
    c. ENABLE.

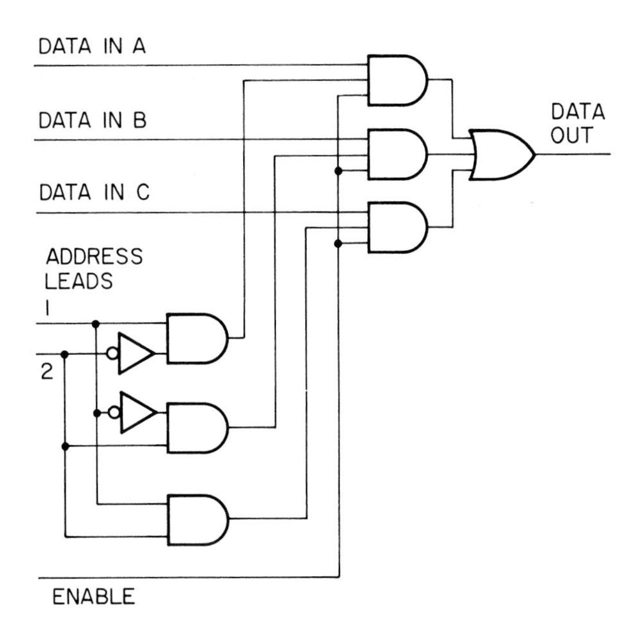

Fig. P14-14. Fanin of various leads can be determined.

21. For the control signals shown in Fig. P14-15 (DIRECTION = 0, ENABLE = 1), in which direction will data flow through the bidirectional buffer/driver?

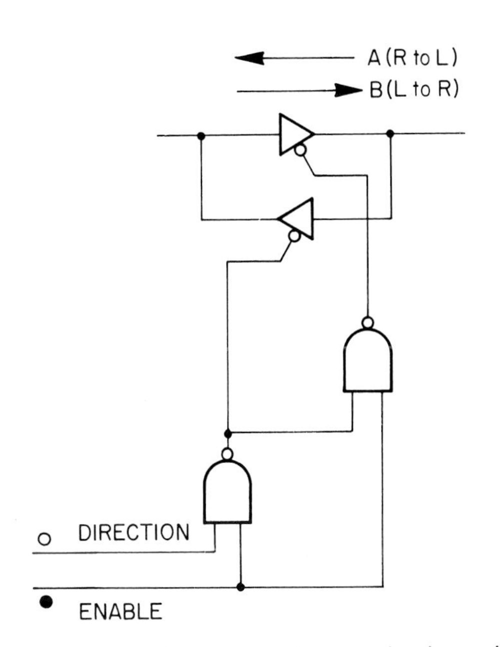

Fig. P14-15. Direction of data flow can be determined if the NANDs are examined one at a time. Use of active-low NAND outputs and active-low three-state inputs makes the job easy.

22. In Fig. P14-16 and Fig. P14-17, a NOT was added to reduce the fanin seen by lead E. In the original circuit, the upper data latch accepts new data at D1 when E goes from 0 to 1; the lower accepts data at D2, when E goes from 1 to 0. Which circuit change decreases the fanin and also retains the original relation between the two data latches?
 a. The change in Fig. P14-16.
 b. The change in Fig. P14-17.

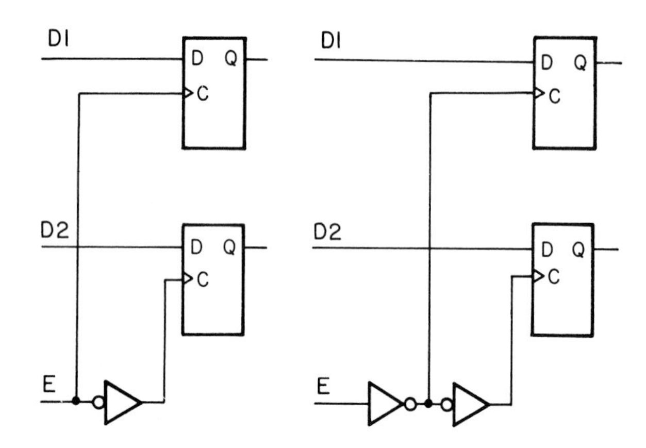

Fig. P14-16. Two NOTs in series catch one's attention. The purpose can be determined.

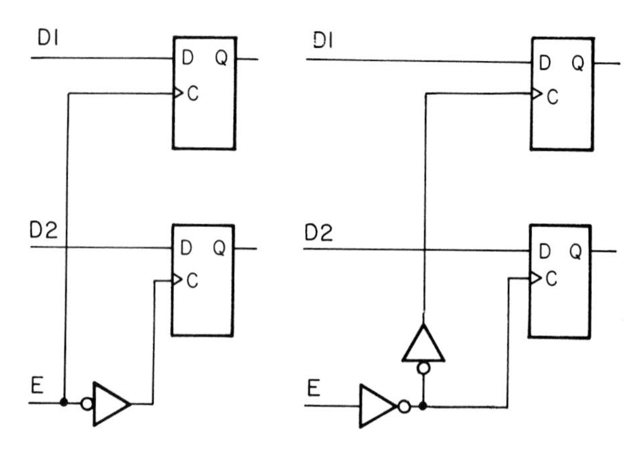

Fig. P14-17. Purpose for the use of two NOTs in series can be determined.

Chapter 15

Read-Only Memory

LEARNING OBJECTIVES

After studying this chapter and completing lab assignments and study problems, you will be able to:

- List two or more applications of ROMs.
- Define the terms *volatile* and *nonvolatile*.
- Estimate memory size based on the number of data and address leads.
- List the functions of the four circuit blocks usually found in ROMs.
- Describe the two most used memory array configurations.
- Describe the following read-only memories: mask-programmed ROM, PROM, EPROM, and EEPROM.
- Recognize and match the names with the symbols of the various electronic elements used in memory cells and state in which read-only memories they are used.

To accomplish even simple tasks, computers must be given step-by-step instructions. Lists of instructions are called programs.

Even the task of entering a program into a computer from a keyboard or tape must be under program control. Programs for entering instructions and data are called *BOOTSTRAP PROGRAMS.* These are usually resident within a computer. The word *RESIDENT* means that the bootstrap programs are built-in. This is usually accomplished through the use of READ-ONLY MEMORIES (ROMs). Such memories have the following characteristics:

1. Can be read by computers but cannot be altered. That is, stored numbers can be used by a computer, but those numbers cannot be changed. Programs in ROMs are protected from change by operators and programmers.
2. Are *NONVOLATILE.* The term means that numbers stored in ROMs are not lost when the power is turned off. This is not true of some memories. Numbers stored in volatile memories are retained only as long as power is applied. If power is lost, so is the stored information.

ROMs are to be read only and are nonvolatile. Based on these characteristics, ROMs are used for:

1. Resident programs that assist in the programming and operation of computers: Bootstrap programs are an example. Other programs that aid in the operation of computers (referred to as MONITORS, EXECUTIVES, and SUPERVISORS) are often stored in ROMs.
2. Programs in dedicated computers: A dedicated computer is one that does a limited group of tasks over and over. For example, the computer controlling an automobile engine is a dedicated computer. Its program is stored in ROM.
3. Tables: Mathematical tables are often in ROM. It is important that such tables be protected from accidental or intentional change.

ROM EXAMPLE AND ROM CLASSIFICATIONS

Fig. 15-1 shows the leads of a typical ROM. Its power and ground leads are at the top. Connections to the three computer buses are at the left. ROMs can be classified by some of the following properties:

- Memory size.
- Erasability.
- Diode, transistor, or shorted wire construction.
- Decoder arrangement.
- Output power.
- Input power.
- Mask programming.
- Speed.
- Cost.
- Flexibility as related to changeable programming.
- Matrix arrays, linear arrays, array diode uses.
- Control applications.

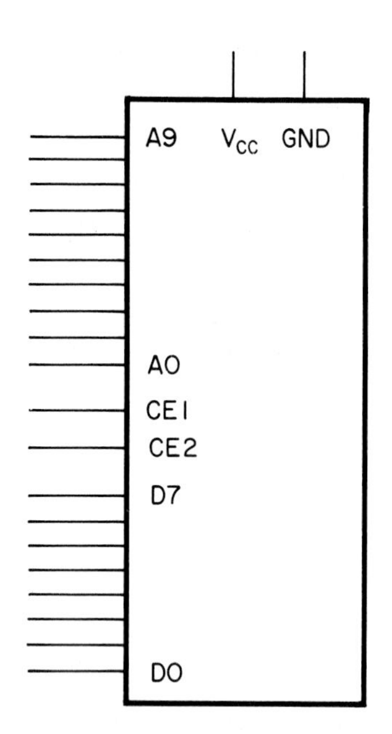

Fig. 15-1. In addition to power and ground leads, connections to the three computer buses are provided on ROMs. The three computer buses are: data bus, address bus, and control bus. The control leads are labeled CE1 and CE2 (control enable leads or chip enable leads).

MEMORY SIZE

The number of bits that can be stored in a given ROM is best determined from its data sheet. However, an estimate for the size of a memory can be obtained by counting leads. Three topics are involved with memory size. These are:

1. Word size.
2. Number of words.
3. Memory size description.

The first topic is important for designers and technicians and helps make further analysis possible, as in the following:

1. Word size: The memory in Fig. 15-1 has 8 data leads. The leads are numbered D0 through D7. The presence of 8 leads suggests that the memory is capable of storing 8-bit words. Design steps are well known. Beyond word length, the analysis becomes more complex.
2. Number of words: The memory in Fig. 15-1 has 10 address leads. This suggests that words can be stored at memory locations numbered as follows:
   00,0000,0000 through 11,1111,1111 (base 2) or
   000 through 3FF (base 16)

Because 00,0000,0000 is a valid address, the number of available memory locations is:

11,1111,1111 + 1 or 100,0000,0000 (base 2)

Note the lone 1 in position D10. This number can be converted to base 10 by raising 2 to the tenth power. This can be seen as follows:

100,0000,0000 (base 2) = 1024 (base 10)

A memory with 10 address leads can address 1024 words. This is usually referred to as a 1K memory (where K implies 1,000). The rounding of 1024 to 1000 simplifies the discussion of memory sizes.

A memory with eight address leads can store 256 words (2 raised to the eighth power). One with 16 address leads can store 65,536 words (2 raised to the sixteenth power). This is usually referred to as a 64K memory, since 64 times 1024 (the size of a 1K memory) equals 65,536 words.

3. Memory size description: In terms of the number of words and word size, the description of the chip in Fig. 15-1 would be 1024 × 8. This is usually rounded to 1K × 8.

In some older literature and when an attempt is made to make a memory sound larger than it is, the total number of bits might be listed. For example, a 1K × 8 memory might be described as an 8192 bit memory by this method.

INCREASING MEMORY SIZE

The obvious way to increase memory size is to use chips that can store more words. However, this is not always possible. Larger chips may not be available or may be too expensive. Also, it may be difficult to change chips in existing computers to expand memory. As a result, memory size is often increased by placing chips in parallel.

PARALLELING OF BIT-SIZE MEMORIES

Fig. 15-2 shows a 1K × 4 memory. Each chip is 1K × 1, so one bit of the word is stored in each chip. Chips are connected in parallel across the address and control buses. When a word is read from memory, all chips are activated at the same time.

PARALLELING OF WORD-SIZE MEMORIES

Fig. 15-3 shows a 2K × 8 memory. Each chip contains 1024 8-bit words.

The chips are again connected in parallel across the address bus. Chip 1 stores the first 1024 words; chip 2 stores the second 1024 words. The CHIP ENABLE leads determine which chip is active.

BLOCK DIAGRAM

Circuits within memory chips can usually be divided into four blocks or sections. The blocks are:

1. Memory array.

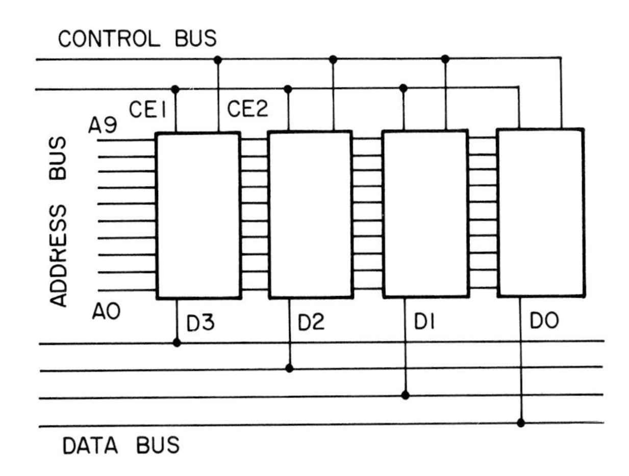

Fig. 15-2. In the circuit shown, four 1K × 1 ROM chips have been paralleled to provide a 1K × 4 memory.

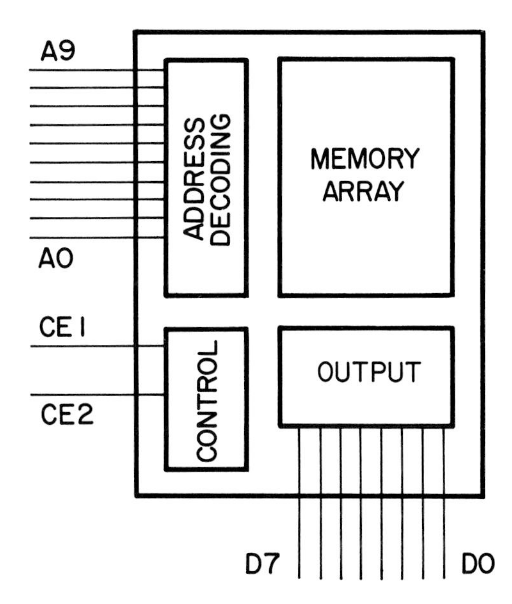

Fig. 15-4. Circuits within ROMs can usually be divided into four blocks or sections. The four blocks are: memory array, address decoder, output, and control.

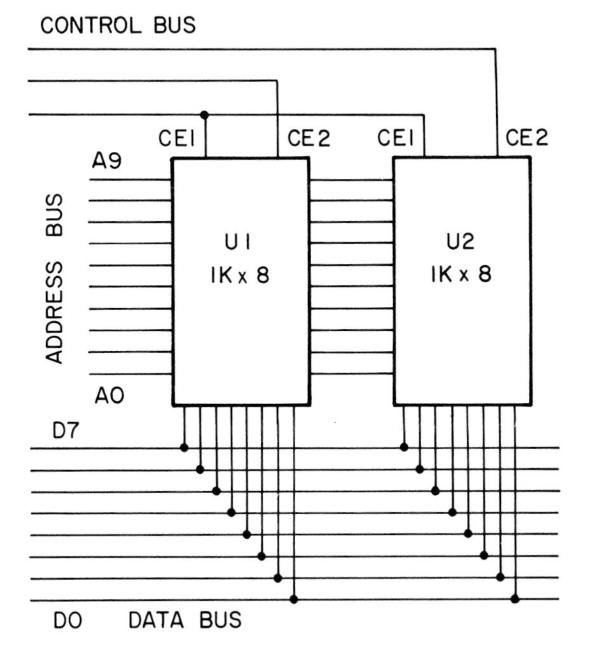

Fig. 15-3. Two 1K × 8 chips have been paralleled to produce a 2K × 8 memory.

2. Address decoder.
3. Output.
4. Control.

See Fig. 15-4. The function for the input part of an INPUT/OUTPUT section is not discussed because the memories are read-only memories. The functions of these blocks will be described in the following paragraphs. The simplest way to analyze a memory is to break it into understandable sections. Some sections require complicated discussion. Others require only a few simple diagrams to explain their operation.

MEMORY ARRAY

Individual 1s and 0s that make up the words stored in a memory are retained in units called *MEMORY ELEMENTS* or *MEMORY CELLS.* In their most complex form, each cell is a complete flip-flop. However, simpler elements are usually used. The nature of cells used in ROMs will be described later in this chapter.

There must be one memory cell per stored bit. For example, a 1K × 8 memory contains 8192 (1024 × 8) cells. An orderly arrangement of cells is called an *ARRAY.* Two types of arrays are in wide use. These types of arrays are:

- Linear array.
- Matrix array.

LINEAR ARRAY

Fig. 15-5 shows a linear array for a 4 × 3 memory. That is, it contains four 3-bit words. Although the memory is small, it demonstrates the general form of linear arrays.

Each horizontal row in this memory represents a word. As a result, such memories may be described as WORD ORGANIZED.

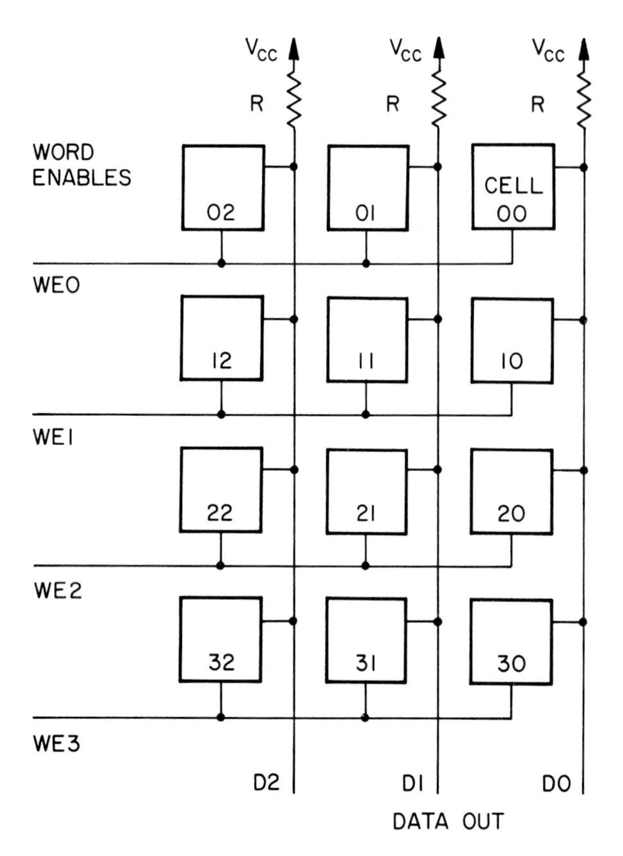

Fig. 15-5. In the linear array shown, memory cells have been arranged into words.

To output a given word, the appropriate WORD ENABLE is activated. Depending on the nature of the cells, this may be an active-high lead or an active-low lead. Only one enable lead is active at a given time for the memory.

All cells in a given column are connected to a single data lead. As a result, cell outputs must be either three-state or open-collector. In the circuit shown in Fig. 15-5, pullup resistors suggest the use of open-collector outputs.

In linear arrays, each word must have its own WORD ENABLE. In large memories, address decoding becomes a problem. As a result, linear arrays tend to be used only in small memories.

MATRIX ARRAY

Large memories are usually organized in matrix form, Fig. 15-6. Again, a small memory will be used. Also, data leads have been omitted for simplicity.

Each cell has two enables. Both enables must be active if a cell is to output a stored bit. For example, to enable cell 21, leads X1 and Y2 must be active.

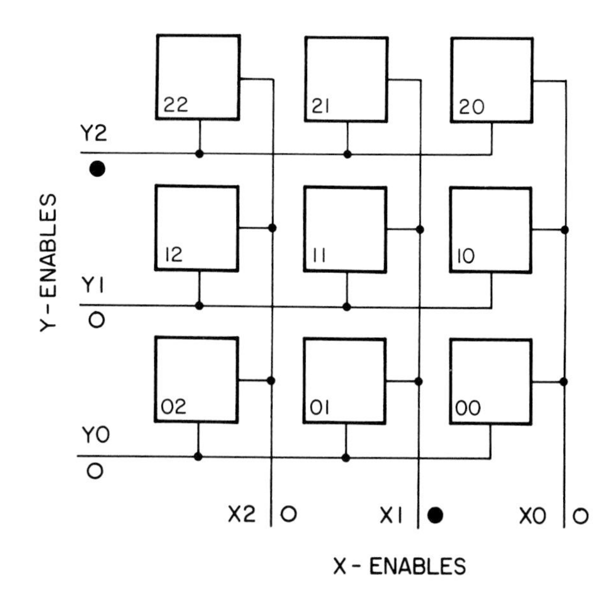

Fig. 15-6. The matrix of memory cells shown produces a bit-oriented array. Only the cell at the intersection of two active enable leads can output a stored bit.

Assuming that the leads are active-high, the numbers on the enable leads must be:

X2X1X0 = 010
Y2Y1Y0 = 100

Note that the 1 on X1 is also applied to cells 11 and 01. However, only cell 21 in this column will be activated. It is the only cell with 1s on both enables. A similar situation exists at cells 20 and 22.

In Fig. 15-7, the data-out or SENSE lead has been added. It has been drawn across the face of the array to emphasize that only one sense lead is used. All the cells in this array output on this one lead. Again, the cells must have three-state or open-collector outputs.

The memory in Fig. 15-7 is a 9×1 memory. Because the word length is 1, matrix arrays are often referred to as BIT ORGANIZED memories. To produce multibit words, matrix arrays are connected in parallel. See Fig. 15-8. The four planes are connected in parallel across the X and Y enable leads. For a given address, corresponding cells are enabled in all four planes. As a result, the memory in Fig. 15-8 is a 9×4 memory. Fig. 15-9 shows this memory drawn as a two-dimensional chip.

ADDRESS DECODER

Fig. 15-10 emphasizes the address decoder block. This circuit acts as an interface between the address bus and the memory array.

Addresses appear on the address bus in binary

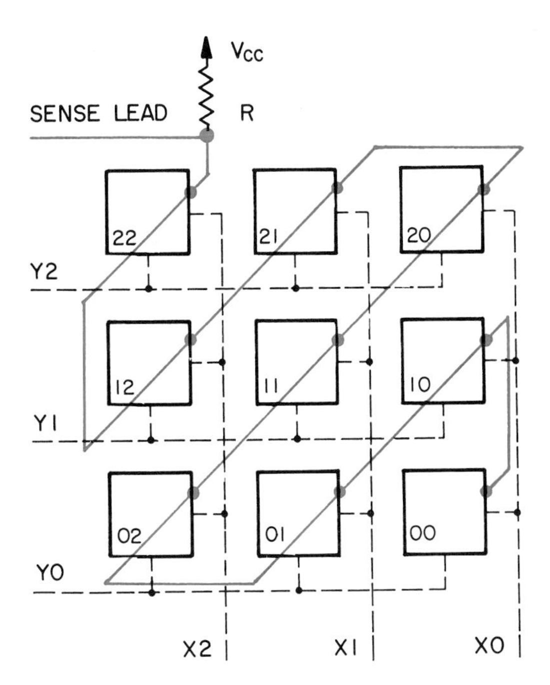

Fig. 15-7. Because only one cell can output at a given time, only one sense lead is used in this bit-oriented memory.

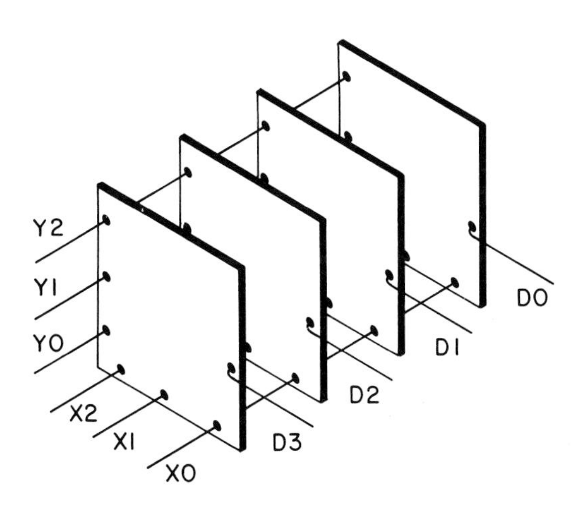

Fig. 15-8. In the circuit shown, four 9 × 1 arrays have been connected to produce a 9 × 4 memory.

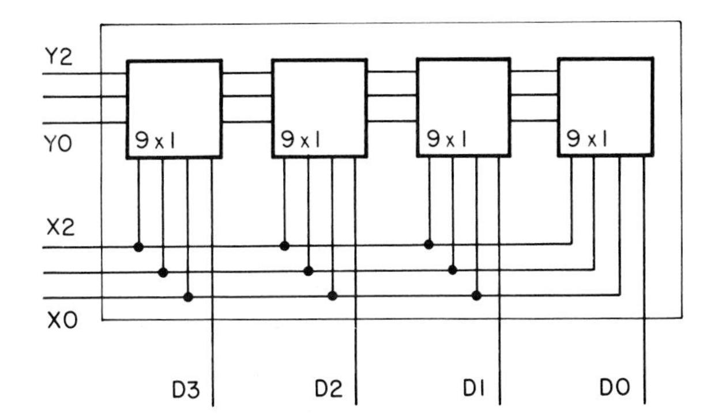

Fig. 15-9. The 9 × 4 memory from Fig. 15-8 can be represented in a single plane.

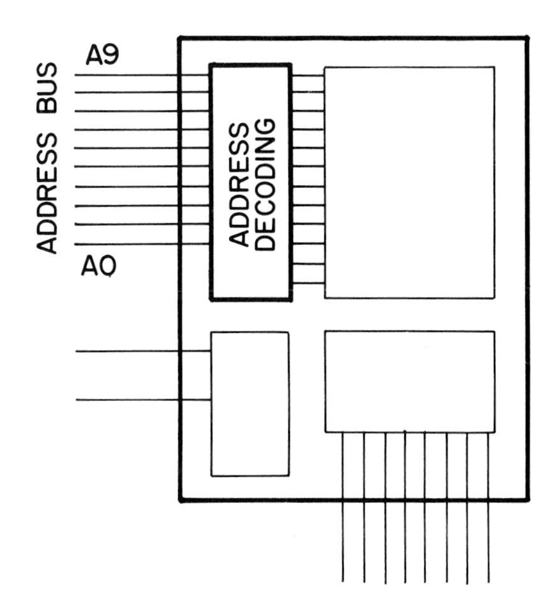

Fig. 15-10. In a ROM, the address decoder is an interface between the memory array and the address bus.

form. This is an efficient code, since a wide range of addresses can be transmitted on a relatively small number of leads.

However, addresses in binary cannot directly enable specific words or memory cells in an array. Binary addresses must be decoded before application to the array lead wires. A system that selects signal patterns is required. It can be easily designed.

ADDRESS DECODER FOR LINEAR ARRAY

Fig. 15-11 shows a 2-lead, combinational-logic decoder for a linear array. Leads A1 and A0 are the inputs of the decoder; the four enable leads are the outputs of the decoder. For each of the four possible addresses, a specific enable lead is activated.

Fig. 15-12 shows a decoder for a 4-bit address bus. It can enable a 16-word memory. Note its complexity. Imagine a decoder for a 1K memory.

ADDRESS DECODER FOR MATRIX ARRAY

Fig. 15-13 shows the decoder for a 16-bit matrix.

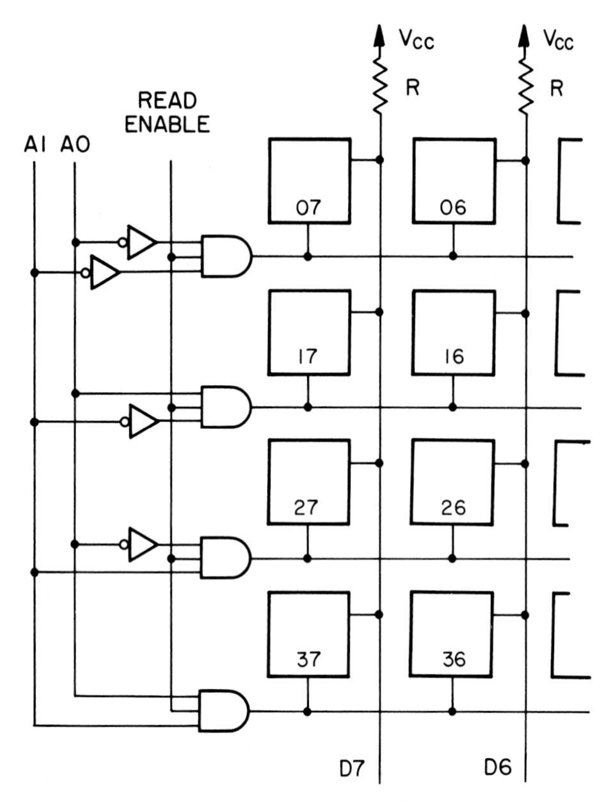

Fig. 15-11. ANDs and NOTs are used in the address decoder of the linear array shown. A 2-bit address is required for this 4-word memory.

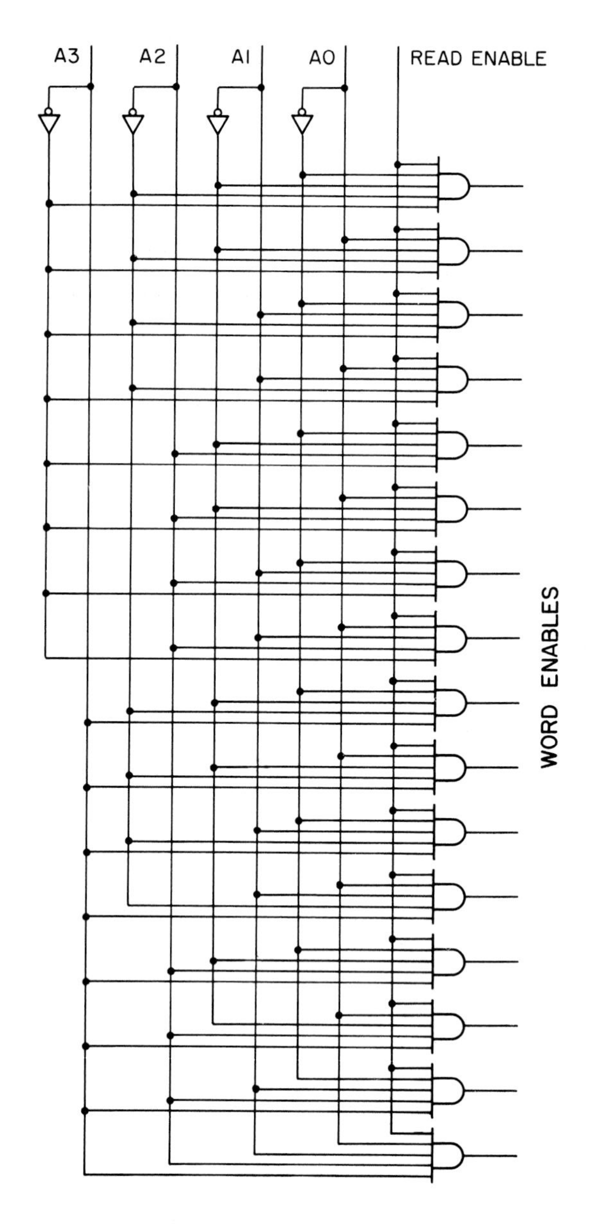

Fig. 15-12. Although only a 4-bit address is being decoded, the required circuit is complex.

Bits A1 and A0 are decoded to supply column enables; bits A3 and A2 supply the row enables. Only 8 ANDs are needed.

OUTPUT BLOCK OF A MEMORY

Fig. 15-14 shows the output block of a memory. Its purpose is to connect the memory array to the data bus at the proper time.

Fig. 15-15 shows a typical output circuit for a memory. Three-state buffers have been used. These are necessary to disconnect the memory from the data bus when the memory is inactive. The fanout of these elements must be high enough to drive the bus and drive its numerous loads.

CONTROL

Control functions in ROMs are simple. For example, the memory shown in Fig. 15-16 has only three control leads. All of the control leads are chip enables (sometimes called chip selects). Since all three must be active to activate the ROM, the control block on this chip might consist of a 3-input AND.

Chip enable leads have several functions. In multichip memories, this lead may be used to select the individual chip to be read. A second CE may be used to time the connection to the data bus. A third CE might be used in the decoding of the address. Fig. 15-17 illustrates these uses.

The memory shown in Fig. 15-17 contains two 1K chips. They might be referred to as PAGES or BLOCKS of memory. The computer has a 16-bit address bus, so it can easily address 2K of memory.

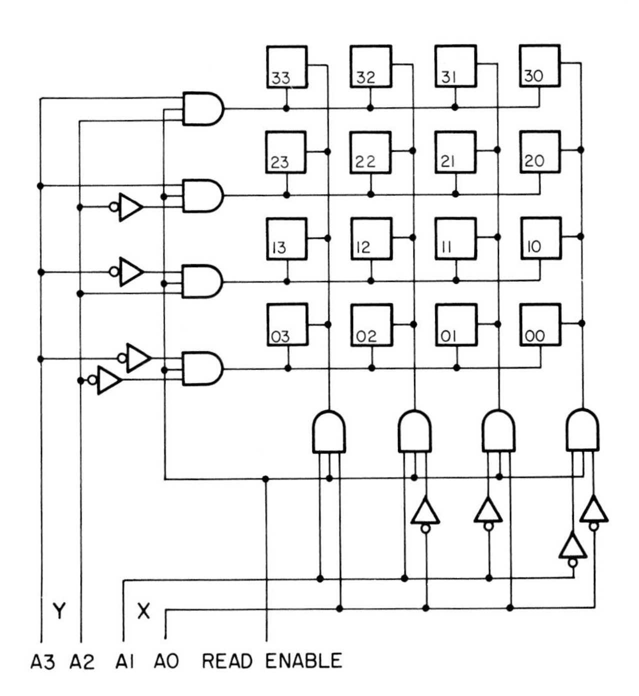

Fig. 15-13. The decoder for a 16-bit matrix requires only 8 ANDs. The sense lead has again been omitted for simplicity.

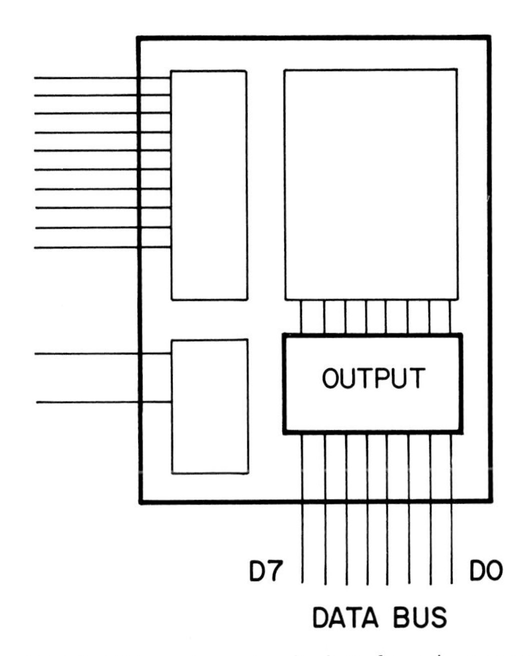

Fig. 15-14. ROM output circuits interface the memory array and data bus.

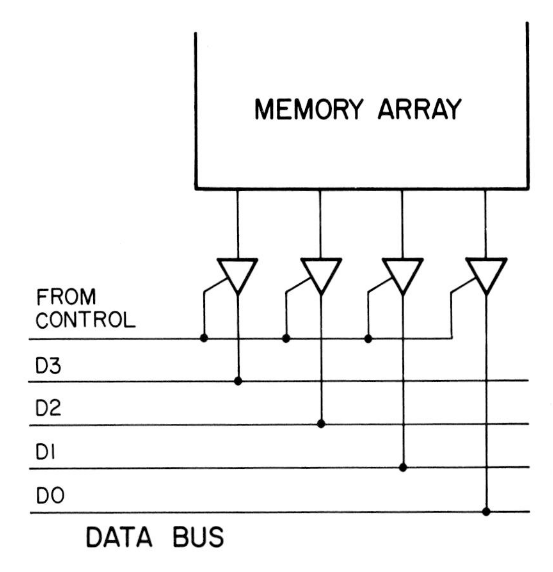

Fig. 15-15. The simple output circuit shown uses three-state buffers to connect the memory array to the data bus at the proper time.

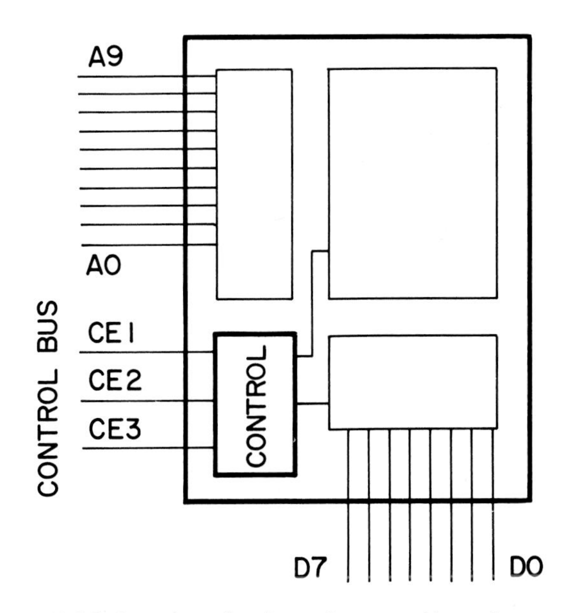

Fig. 15-16. Based on signals on the control bus, the control block determines when data is to be transferred to the data bus.

The ten address leads on the chips shown in Fig. 15-17 have been connected in parallel across leads A0 through A9. The portion of the address on these leads will be decoded simultaneously within the two chips. However, only the chip with three active CEs can output a stored number.

In the circuit in Fig. 15-17, A10 determines which chip will be activated. In terms of addresses, this

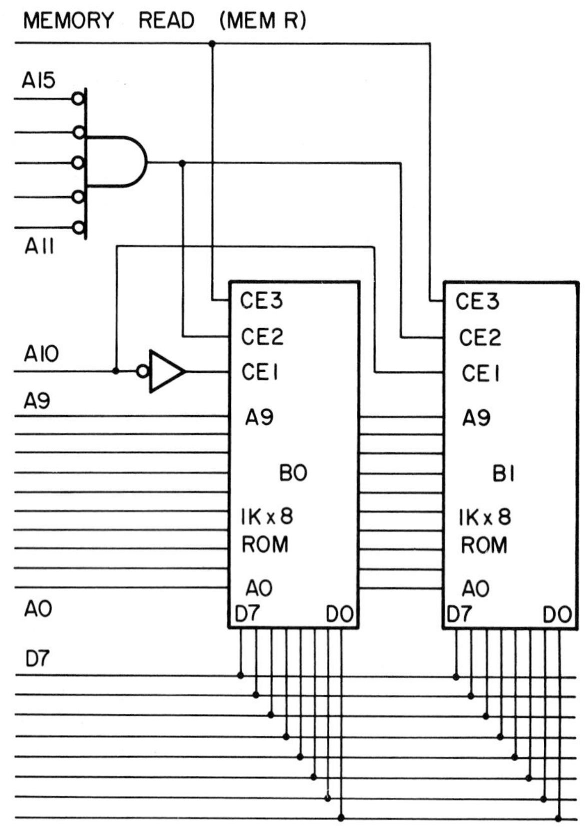

Fig. 15-17. Chip enable leads control the operation of the ROMs shown. When all three CE leads of a given ROM are active, that chip outputs stored data.

means the following:

Block B0 addresses (in base 2) are:

000,0000,0000 through 011,1111,1111

Block B1 addresses are:

100,0000,0000 through 111,1111,1111

The above addressing system assumes that 0s appear on the remaining address leads (A11 through A15 in Fig. 15-17). To insure that the chips are not activated by addresses above those indicated, the NOR was added. Only when 0s appear on address leads A11 through A15 will CE2 on each chip be enabled. For example, neither chip will respond to the address 0100,0100,1100,0101. A14 is a 1, so the output of the NOR will be 0.

With a 16-bit address, up to 64 blocks of 1K memory can be addressed. The upper 6 address leads (A10 through A15) would be decoded to select individual blocks. The lower 10 leads would point to a specific location within the selected chip.

The last chip enable (CE3) is used to time the operation of the ROM. When MEM R is high, the selected data is output to the bus.

MASK-PROGRAMMED ROM

The rest of this chapter will describe various read-only memories. Control, address decoding, and output circuits will be similar. Differences are found in the memory cells used.

A basic form of ROM is described as a *MASK-PROGRAMMED ROM.* The term *MASK* refers to a portion of the fabrication process. Mask-programmed ROMs are programmed during the manufacturing process. When purchased, they already contain permanently stored bits.

Because each ROM application requires specific stored data, mask ROMs must be custom fabricated. Users supply chip manufacturers with lists of the numbers to be stored. Manufacturers then fabricate chips to meet the needs of users.

MASK ROM PROGRAMMING

To reduce costs, mask ROMs are standardized as much as possible. Only during the last few steps in the manufacturing process are specialized tools needed. The programming of mask ROMs often takes place during the placing of the last set of conductors. See Fig. 15-18. In the chip shown in Fig. 15-18, all

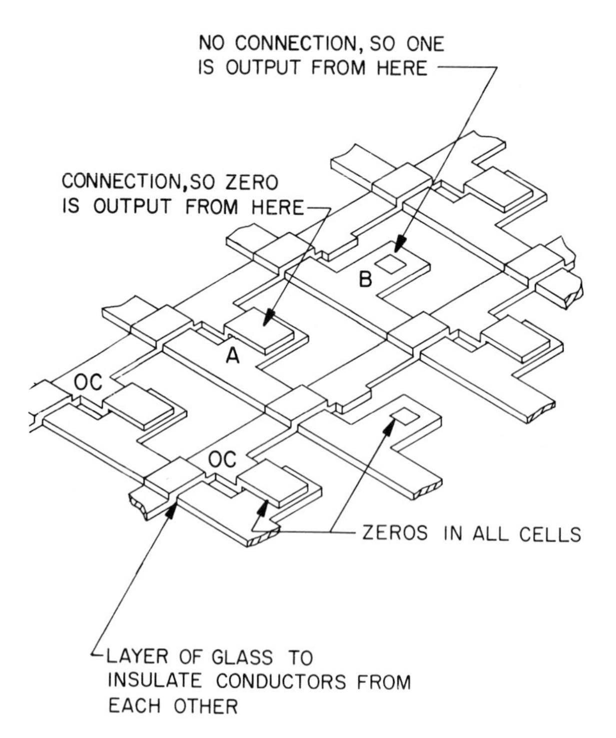

Fig. 15-18. In the mask-programmed ROM shown, all cells contain 0s. To store a 0, a connection is made between the cell and an output conductor (OC). See cell A. To store a 1, no connection is made. See cell B.

cells represent stored 0s. When a 0 is to be stored in a given cell, that cell is connected to the output conductor (OC). See cell A. If a 1 is to be stored, the cell is allowed to float. See cell B. The pattern of the final set of conductors is the only manufacturing step that must be altered to change the stored data.

ROM programming begins with the preparation of a large drawing of the required conductor pattern. This is photographically reduced to the size of the chip. The resulting photographic negative is called a *MASK* negative.

To transfer the conductor pattern from the mask to the chip, a thin metal film is deposited on the surface of the chip. By photographic methods, the conductor pattern is placed on the metal film. Acid is then used to etch away unwanted portions of the metal. What is left is the desired conductor pattern.

The diagram for a simple mask ROM is shown in Fig. 15-19. The heavy lines are the leads put down during the programming process. Some cells have been connected to the data-out leads; some have not.

To read the number stored in a given word, its WORD ENABLE is grounded. In Fig. 15-19, word 1 is being read.

The action of the memory can be described by treating each cell as a short circuit. With WORD ENABLE 1 grounded, leads D3 and D0 are shorted to ground. These leads will output 0s. During the masking process, cells 12 and 11 were not connected to their data leads, so D2 and D1 are not shorted to ground. These leads will output 1s. The number stored in word 1 is 0110.

Numbers stored in the other words are:

WORD	NUMBER
0	0011
2	0001
3	1101

DIODE ARRAY

In the above description, you were asked to picture the cells as short circuits. It turns out that this assumption is an oversimplification. Short circuits will not work as memory elements.

In Fig. 15-20, cells have been replaced by short circuits and word 1 is being read. The number output should be 0110. The measured number is 0010. The source of this error can be found by following the current path described by the series of arrows.

Fig. 15-19. In the mask ROM shown, cell outputs are connected to data leads to store 0s. To store 1s, they are left disconnected. The pullup resistor is 1000 Ω.

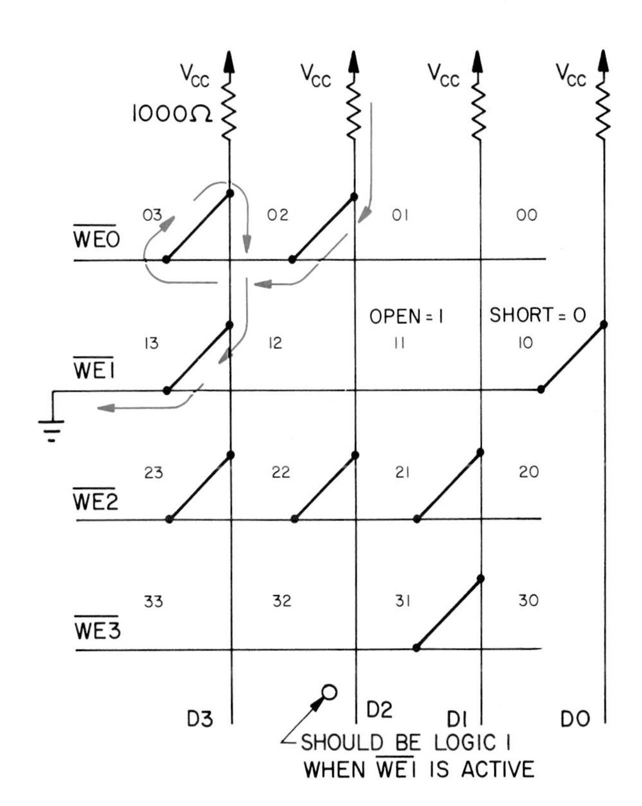

Fig. 15-20. Because unwanted current paths result, short circuits cannot be used as memory cells.

There should be no current flowing through the pullup resistor in output lead D2. However, a path exists through cells 02, 03, and 13 to the active word enable. $\overline{\text{WE0}}$ has been grounded by the erroneous current path that is present.

$\overline{\text{WE2}}$ has also been grounded. Cell 23 and cell 21 provide an erroneous current path. D1 should be 1, but it is 0.

$\overline{\text{WE3}}$ is at the proper level. No erroneous current paths exist. However, the next choice of word enable for reading may ruin the level of $\overline{\text{WE3}}$.

These problems can be corrected by preventing current from flowing in the wrong direction through cell 03 and cell 23. This can be accomplished by using diodes in place of short circuits.

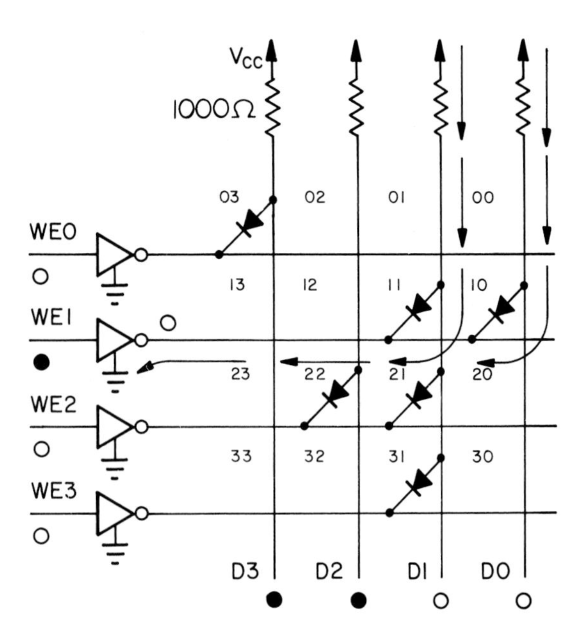

Fig. 15-22. Substituting diodes for short circuits solves the problem of unwanted current paths in simple ROMs.

REVIEW OF DIODE ACTION

Part "a" of Fig. 15-21 shows the symbol for a diode. Note the arrowhead. This device will conduct in the direction of this arrow (assuming conventional current flow). It will not conduct in the other direction.

In part "b" of Fig. 15-21, a CONDUCTING or *FORWARD BIASED* diode is shown. Note the polarity of the voltage across the diode. Its *ANODE* (the arrow portion of the symbol) is positive with respect to its *CATHODE* (the bar portion of the symbol). When a voltage with this polarity exceeds about 0.6 volts, the diode will conduct as shown.

In part "c" of Fig. 15-21, the diode is NONCONDUCTING or *REVERSE BIASED.* The diode's anode is negative with respect to its cathode. No current flows under these conditions.

OUTPUT ACTION OF DIODE ROM

Fig. 15-22 shows a 4 × 4 diode ROM. Each diode represents a stored 0. Omitted diodes represent stored 1s. To read a word, a 1 is applied to the proper WORD ENABLE. The output of the corresponding NOT goes to 0, and the cathodes of the diodes attached to that word line are grounded. The diodes conduct (their anodes are attached to the positive voltage at Vcc), and 0s appear on the proper output leads. Note that current cannot flow from lead D2 through cells 22, 21, and 11, since it cannot flow from the cathode to the anode of diode 21. Diode 21 has protected WE2.

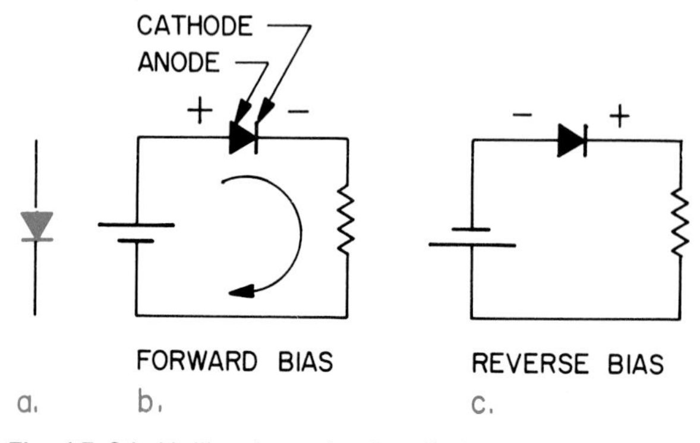

Fig. 15-21. Unlike short circuits, diodes conduct in only one direction—from anode to cathode.

As a self test, determine the numbers stored in the ROM in Fig. 15-22. Check your results against the following.

WORD	STORED NUMBER
0	0111
1	1100
2	1001
3	1101

Fig. 15-23 shows a mask ROM. All 16 diodes are in place. In the programming process, however, only those cells where 0s are stored have been connected.

DISADVANTAGES OF DIODE ROM

Between 0.3 and 0.7 V appears across a conducting diode. Allowing 0.2 V across the output of the driving NOT and 0.6 V across the diode, about 0.8 V will appear on the output lead when a 0 is being output. See Fig. 15-24. This is larger than permitted for reliable TTL operation. That is, problems may result when diode matrix memories are interfaced with TTL

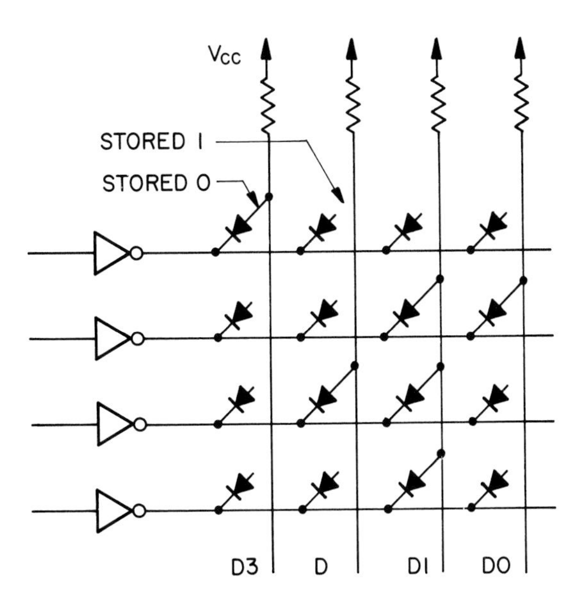

Fig. 15-23. In the mask ROM shown, diodes are placed at all intersections of the matrix. However, only those where 0s are stored have been connected.

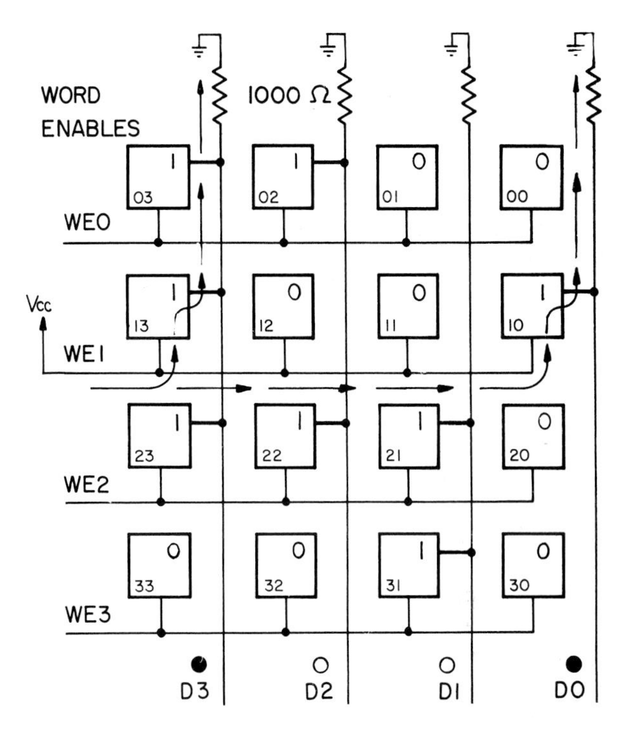

Fig. 15-25. Memory using short circuits and positive logic. The output of reading word 1 is 1001. Note how problems need to be solved with erroneous current paths, suggesting the need for diodes in a better design.

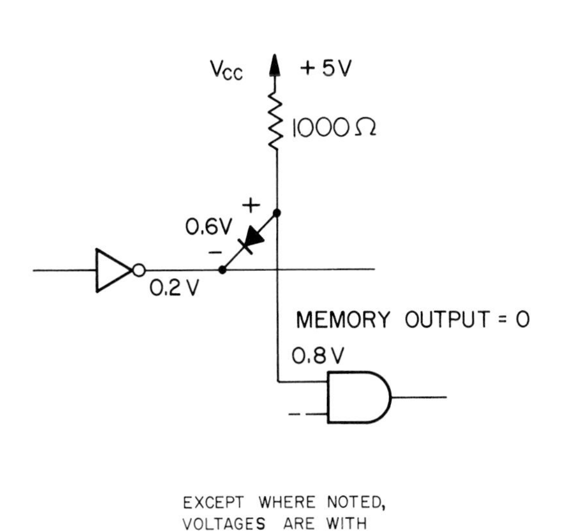

Fig. 15-24. Diode ROMs have a problem. Their output signals are not compatible with TTL Logic.

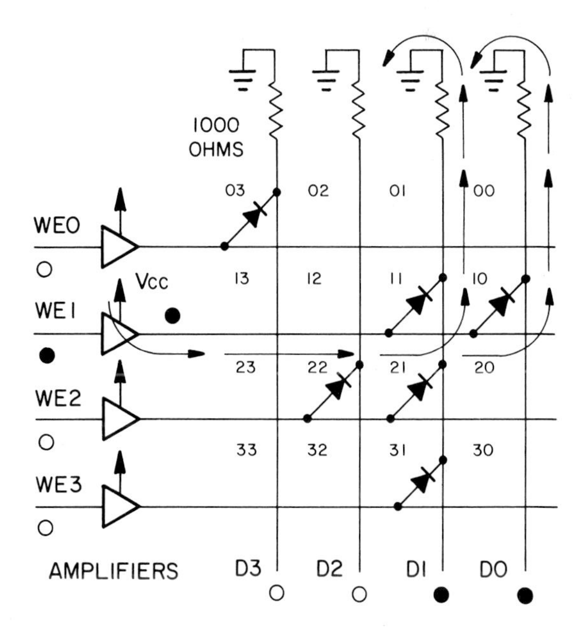

Fig. 15-26. Diodes solve problems with memory using positive logic. Note that the voltage drop of 0.8 V across the diodes does not affect the compatibility with TTL elements. A memory larger than 4 x 4 would need more amplifiers (one for each 4 x 4 block).

logic. To overcome this and other diode problems, transistors are often substituted for diodes. A possible solution using positive logic, short circuits or diodes, and power from Vcc is displayed in Fig. 15-25 and Fig. 15-26. With power from Vcc, a 1 is stored at a short circuit or a diode. A 0 is stored at the cells that are left disconnected.

As a self test, determine the numbers stored in the

ROM in Fig. 15-26. Check your results against the following.

WORD	STORED NUMBER
0	1000
1	0011
2	0110
3	0010

The stored numbers from Fig. 15-26 are the complement of the numbers from Fig. 15-22. The same relative positions for diodes were used in both Fig. 15-26 and Fig. 15-22.

BIPOLAR TRANSISTOR ROM

Two transistor types will be considered. These are:
1. Bipolar transistors.
2. Field-effect transistors.

Both transistor types may be thought of as electrically operated switches. As a result, you do not need to fully understand transistor action to follow descriptions of basic transistor ROMs.

BIPOLAR TRANSISTOR ACTION

A circuit component called a *BIPOLAR TRANSISTOR* may be thought of as containing two diodes or junctions. See part "a" of Fig. 15-27. The leads of the bipolar transistor are labeled *BASE, EMITTER,* and *COLLECTOR*. Its symbol is shown at b. The arrow distinguishes the emitter from the collector. Note that this arrow corresponds to the diode between the base and emitter on the drawing at a.

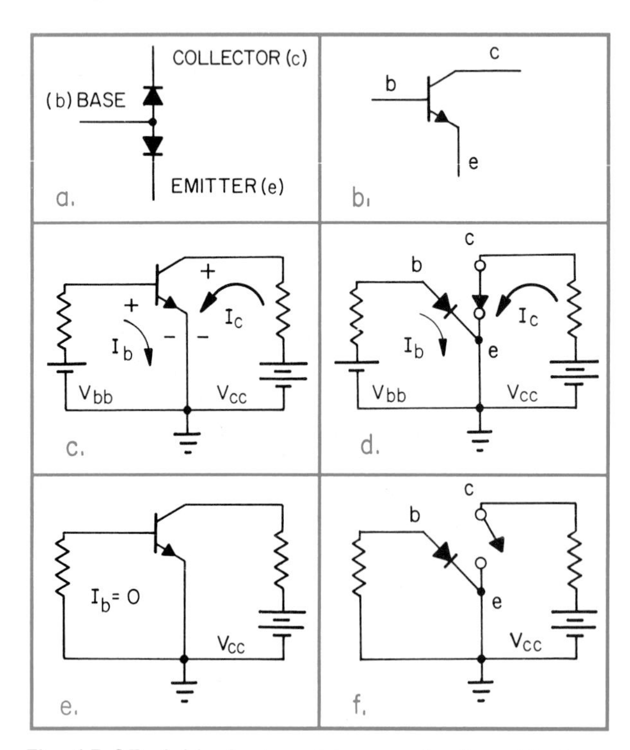

Fig. 15-27. A bipolar transistor may be thought of as a switch and diode. When current flows through the diode, the switch closes.

The base is usually the device's input; the collector is usually its output. In part "c" of Fig. 15-27, voltages have been applied to the transistor. The base-to-emitter diode has been forward biased, so it conducts, and a small current flows around the input loop. See the arrow labeled Ib. The base-to-emitter current has an internal effect on the collector conductivity. Due to transistor action, the flow of base current permits a large current to flow in the reverse direction through the diode in the collector. See current Ic.

The collector-to-emitter voltage of a conducting transistor is usually quite small (less than 0.2 V). This permits the output of a conducting transistor to be viewed as a closed switch, part "d" of Fig. 15-27. The 0.2 V is compatible with TTL elements.

In part "e" of Fig. 15-27, the voltage in the base circuit has been removed. As a result, there is no base current. With no base current there can be no transistor action, and the collector current stops flowing. The collector acts like an open switch, as shown in part "f" of Fig. 15-27.

MASK BIPOLAR-TRANSISTOR ROM

Fig. 15-28 shows a mask ROM with cells composed of bipolar transistors. All of the collectors are connected to output leads. However, during programming, some base connections were omitted.

To read the content of a given word, a positive voltage is applied to the proper WORD ENABLE. The number stored in that word appears on the output leads. In Fig. 15-28, WORD ENABLE 1 is energized. Transistors 12 and 10 are turned on and they short D2 and D0 to ground (outputting 0s). The bases of transistors 13 and 11 have not been connected so there can be no base current. In turn, these transistors do not turn on, so 1s appear on D3 and D1. In this whole array, only transistors 10 and 12 are turned on. As a result, there are no erroneous current paths to cause problems.

DISADVANTAGES OF BIPOLAR TRANSISTORS

Bipolar transistors are very fast, so they are used in some high-speed memories. However, they have several disadvantages. First, they require considerable space on IC chips. Second, they consume a great deal of power. As a result, they are seldom used in large memories. Small memories use bipolar transistors. The performance is good.

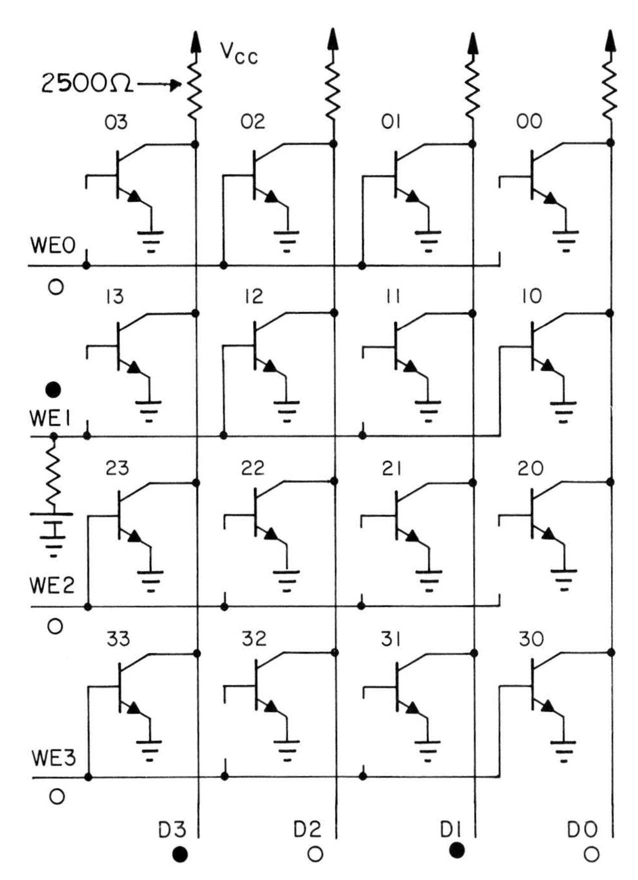

Fig. 15-28. Each cell in the mask bipolar transistor ROM shown is a bipolar transistor. If a transistor's base is connected, a 0 has been stored. If it is not connected, a 1 has been stored.

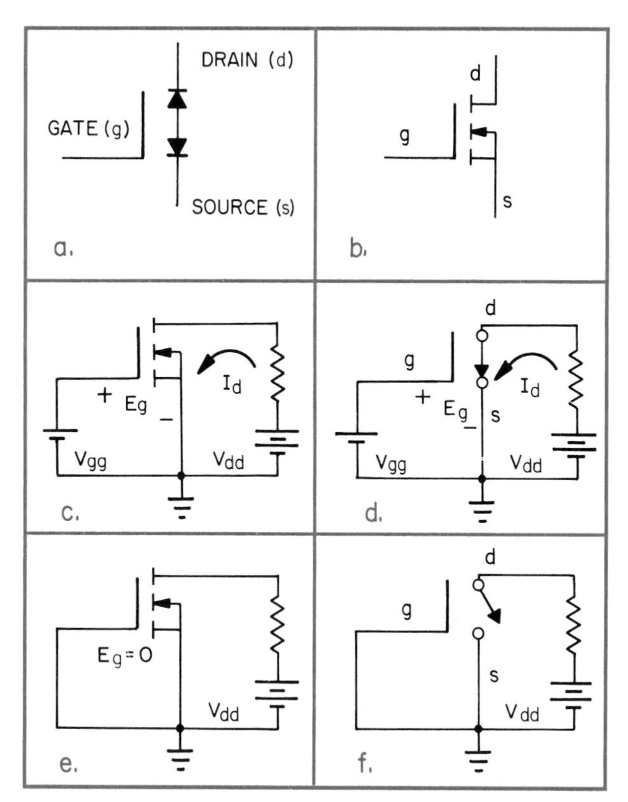

Fig. 15-29. Like the output of a bipolar transistor, the output of a field-effect transistor may be thought of as a switch. However, the switch is controlled by a gate that is operated by voltage (rather than by current). Voltage Eg, between the gate and source, must be positive to turn this transistor on. Zero or negative voltage will turn it off.

FIELD-EFFECT TRANSISTOR ROM

Bipolar transistors are current operated. Electronic devices called *FIELD-EFFECT TRANSISTORS* (FETs) are voltage operated. Because they take less space on chips and require less power, FETs are used to construct large memories.

FET ACTION

Part "a" of Fig. 15-29 shows the diode equivalent of an FET. Its leads are labeled *GATE, SOURCE,* and *DRAIN.* These correspond to the base, emitter, and collector of bipolar transistors. See part "b" of Fig. 15-29.

The gate of an FET is normally its input. Note that it does not make direct contact with the body of the device. When a positive voltage is applied between the gate and source of this FET, an electric field is produced between the gate and body. This electric field permits current to flow in the reverse direction through the diode at the drain. As long as the gate is sufficiently positive, a current labeled Id flows in the drain circuit. See part "c" and part "d" of Fig. 15-29 for a loop of current.

Removing the gate voltage stops the transistor action. See part "e" and part "f" of Fig. 15-29. Like bipolar transistors, FETs can be thought of as electrically operated switches. When the switch is closed, its resistance is low. When a transistor is off, it acts like an open switch. Its resistance is very high, and little or no current flows.

MASK FET ROM

Fig. 15-30 shows a mask-programmed FET ROM. Its operation is similar to that of a bipolar transistor circuit. The current levels are less than for bipolar transistors. Less power is consumed. The fanin levels are less and the fanout levels are greater. However, sensitivity to noise, cosmic ionizing radiation, and

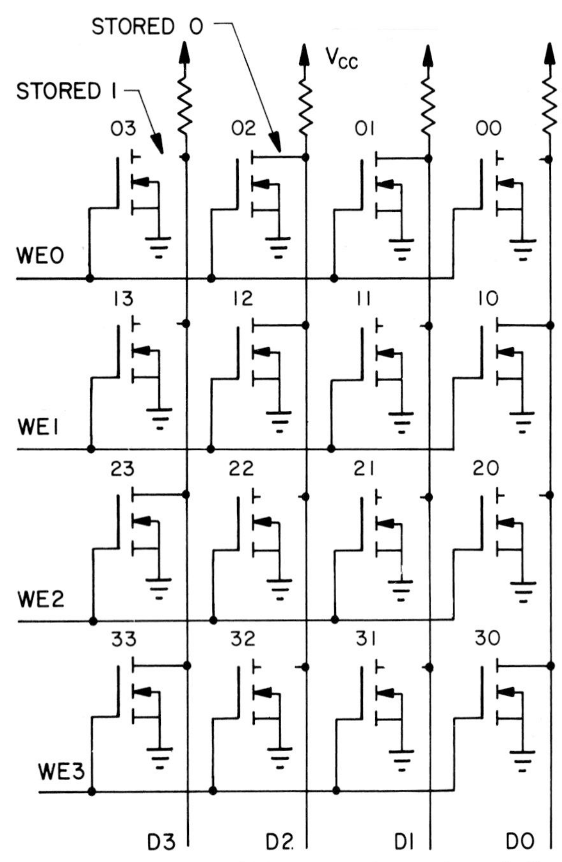

Fig. 15-30. FET mask ROMs are much like those built from bipolar transistors. In Fig. 15-28 and Fig. 15-31, the bases and emitters of disconnected transistors are open. This was done to reduce the load on word enable leads. In FET ROMS, this is not necessary. FET gates are voltage operated.

static electricity is greater. FETs can be damaged by sparks. Both bipolar transistors and FETs are damaged by nuclear radiation, such as would be encountered by instrumentation used near nuclear reactors. Bipolar transistors, which consist of bulk "N" and "P" junctions, have a slow decline in conductivity and a rapid increase in noise signals (known as "white noise" or "hiss") due to crystal destruction by nuclear radiation. FETs can have a total loss of signal control due to stray charges induced on the gate electrode from ionization of the electrode material and ionization of the SiO_2 (glass) insulator.

DIODE, BIPOLAR TRANSISTOR, AND FET MASK ROM DISADVANTAGES

Because mask ROMs must be programmed by the manufacturer, problems result. Some of the disadvantages of mask ROMs are:

1. The time between the ordering of a programmed chip and its delivery is long. About six weeks may be required. For production, this length of time is not unreasonable, but during design and development, this could slow a project.
2. When even small changes are made in stored numbers, a new chip must be manufactured. As a result, the turn-around time equals that for the design and development of a new chip.
3. In small quantities, mask ROMs are expensive. Masks are costly and must be written off against the chips being produced. If quantities are small, each chip must bear a larger portion of that cost.

PROGRAMMABLE ROM

To overcome the disadvantages of mask ROMS, *PROGRAMMABLE READ-ONLY MEMORIES* (PROMs) were developed. Such chips are programmed by users.

Fig. 15-31 shows a fusible-link PROM. When sold,

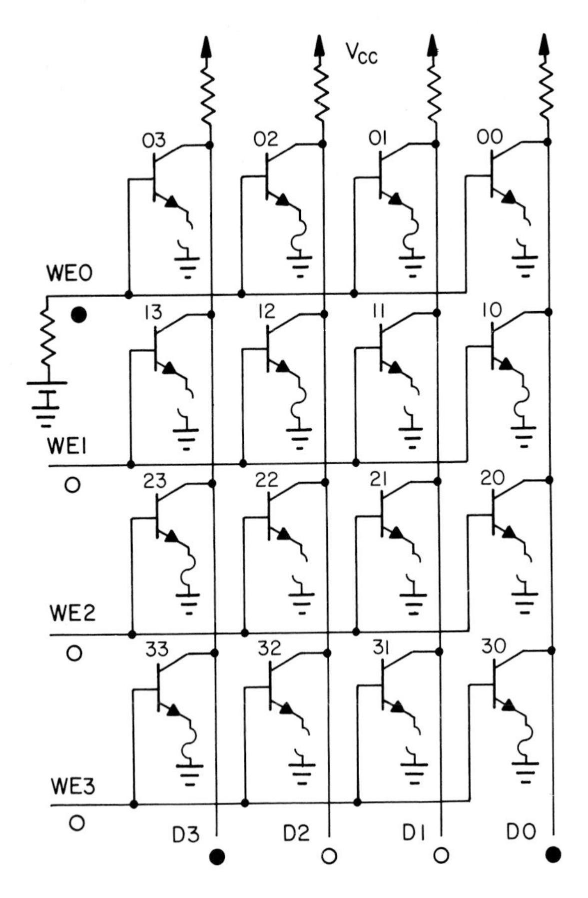

Fig. 15-31. The PROM shown uses fusible links. Zeros are stored in cells with intact fuses. Ones are stored in those with blown fuses.

all transistors have an intact fuse in their emitter circuits. These fuses are weak areas in the conductor pattern on the substrate.

PROGRAMMING A FUSIBLE-LINK PROM

When normal voltages are applied to a PROM (Vcc = 5 V), emitter currents are below the level needed to blow fuses. To program this memory, higher than normal voltages are applied. This results in emitter currents that are high enough to selectively blow fuses.

With a fuse in place, a cell contains a 0. To store a 1 in a given cell, its fuse is blown. For example, the number stored in word 0 in Fig. 15-31 is 1001.

Fig. 15-32 shows the storing of the number 1100 in word 1. A higher than normal voltage (+12.5 V) has been applied to leads D3 and D2. When WORD ENABLE 1 is activated, the four transistors in word 1 will turn on. Because the transistors attached to D1 and D0 are connected to only +5 V (through the pullup resistors), their emitter currents will be below the melting points of their fuses. This is not true of the transistors tied to D3 and D2. Their emitter currents will blow their fuses.

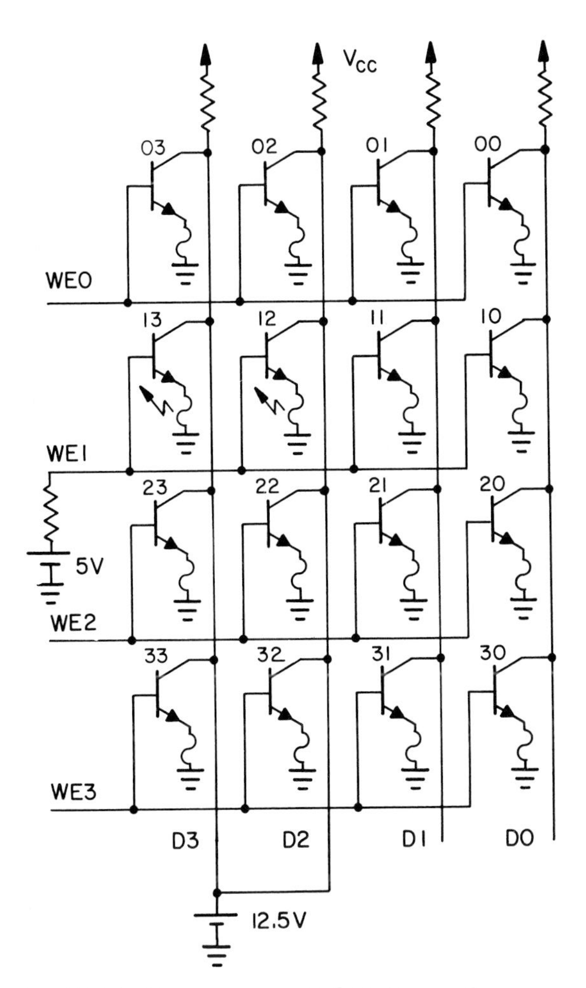

Fig. 15-32. To program a PROM, higher than normal voltages are applied. In this case, the fuses at 13 and 12 will be blown to store the number 1100 in word 1.

Fig. 15-33. PROM programmers can be either manual or computer controlled. This is a manual programmer. (Pro-Log Corp.)

Instruments for programming PROMs are available. Manual programmers usually have three sets of switches. One is used to input addresses. A second set inputs data to be stored. The third controls the storing process. With so many switches to be set, operators often make mistakes. When a mistake is made, the chip must be discarded and the process must be repeated from the beginning.

Fewer mistakes are made when computer-controlled PROM programmers are used. See Fig. 15-33. Data to be stored is placed in the computer's memory and checked for errors. Then the computer controls the programming of the PROM. If disks or tapes are used to store the data, changes can be made and new PROMs can be programmed without reentering all of the data previously checked for errors.

Programming fusible-link PROMs is a slow process. As each fuse blows, heat is produced. If programmed too rapidly, chips tend to overheat.

DISADVANTAGES OF PROMS

PROMs can be programmed only once. This is an obvious disadvantage. If an error is made during programming or if stored information is to be changed, a programmed PROM must be discarded. It cannot be reprogrammed.

A second disadvantage is less obvious. PROM manufacturers cannot completely test their product before delivery. They can test to see that 0s are present in all cells. However, they cannot be sure all locations are capable of storing 1s without destroying the chip. A fuse may be too strong to blow when it is needed to blow it. When an error appears after a user has programmed a chip, did the user make a mistake or was it a faulty chip? There is no way of knowing.

ERASABLE PROM

To overcome the disadvantages of PROMs, *ERASABLE PROGRAMMABLE READ-ONLY MEMORIES* (EPROMs) were developed. EPROMs are also referred to as electrically-programmable read-only memories. EPROMs are programmed in much the same way as PROMs. However, they can be erased and reprogrammed.

FLOATING-GATE FET

The cells in most EPROMs consist of FETs with two gates, part "a" of Fig. 15-34. The dashed line represents the second gate. Note that this gate is not brought out. That is, it is not connected to the outside world. It is completely surrounded by a layer of glass and electrically floats. See part "b" of Fig. 15-34.

With no charge on the floating gate, this FET functions like an ordinary FET. The select gate (the gate that is brought out) controls the flow of current in the drain circuit.

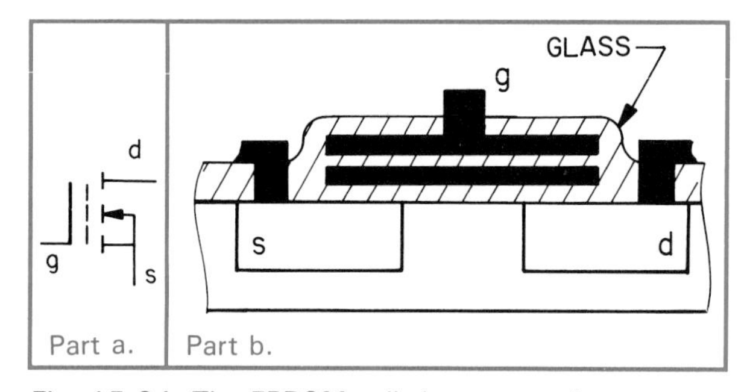

Fig. 15-34. The EPROM cell shown contains an extra gate. Because it is surrounded by glass, this extra gate is insulated from the outside world.

If, however, a negative charge is placed on the floating gate, the transistor is turned off. The negative charge shields the select gate, and it no longer controls the flow of current. In effect, the negative charge removes the transistor from the circuit.

Fig. 15-35 shows a highly simplified EPROM memory array. Negative signs indicate charged floating gates. When a word is enabled, only the uncharged transistors conduct and place 0s on their data lines at the bottom.

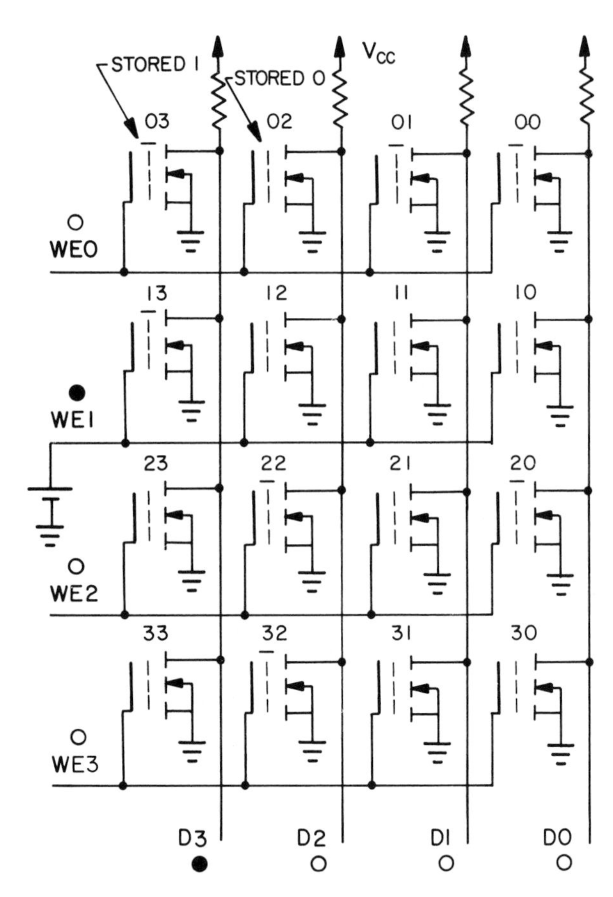

Fig. 15-35. In the EPROM shown, a charged floating gate represents a stored 1. Uncharged floating gates represent 0s.

In Fig. 15-35, WORD SELECT 1 has been activated. The transistor attached to D3 (transistor 13) cannot turn on, since its floating gate is charged. A 1 will appear on D3. The other three transistors will turn on, so lines D2, D1, and D0 will go to 0. The number stored in word 1 is 1000.

EPROM PROGRAMMING

As with PROMs, programming of EPROMs involves higher than normal voltages. To place a negative charge on the floating gate, electrons must be forced through the glass surrounding this element. This is accomplished by placing about 25 V between the gate and source and about 20 V between drain and source. Under these conditions, some electrons gain sufficient energy to cross the glass barrier. These are sometimes referred to as hot electrons. As electrons collect on the floating gate, the required negative voltage develops.

Because the floating gate is surrounded by glass, the deposited charge cannot escape. Until intentionally erased, this transistor is held in an off condition.

ERASING AN EPROM

The three devices with circular windows in Fig. 15-36 are typical EPROMs. The transparent windows are used in the erasing process.

Fig. 15-36. The three devices with circular windows contain EPROMs. (Gamry Instruments, Inc.)

To remove electrons trapped on the floating gate, the chip is exposed to a strong ultraviolet light. Such light contains great energy, and some of that energy is imparted to the stored electrons. When the electrons have enough energy, they will again pass through the glass barrier and dissipate the charge. This returns the control of the transistor to the select gate.

Because the entire chip is exposed to the ultraviolet light, all information is lost. Once erased, the whole chip must be reprogrammed.

EPROMs are not easily erased. Even with strong ultraviolet light, exposures of over 15 min. may be required.

Ordinary room light has little effect. About 70 percent of the original charge will still be on the floating gates at the end of ten years. However, it is standard practice to place opaque paper seals over EPROM windows to guard against accidental erasure.

EPROMS AS SUBSTITUTES FOR PROMS

EPROMs are often used in applications where they will never be erased. That is, they are often used in place of PROMs and ROMs. The erasable feature of EPROMs is helpful during the development of new products. When the circuit is placed in production, designers and engineers often prefer to stay with chips used during the design phases of a project. Thus, EPROMs are often used in production.

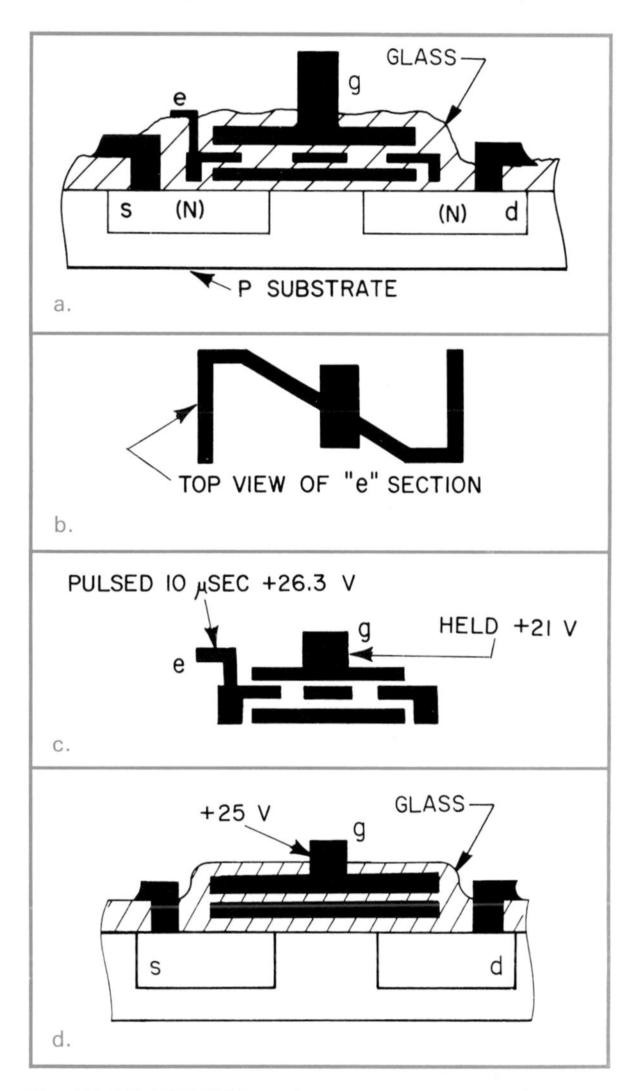

Fig. 15-37. EEPROM erasing systems are complex. a—Electrode at "e" is used for erasing. b—Top view of "e" section. c—Voltage required on "e" section is + 26.3 V. Voltage at g is held at +21 V. The "e" section is normally left disconnected to provide better isolation of the floating charge than the "g" lead can provide. d—Alternative EEPROM structure uses gate lead for erasing. To erase, +25 V is applied to "g" gate.

ELECTRICALLY ERASABLE PROM

When EPROMs are erased, all data is lost. *ELECTRICALLY ERASABLE PROGRAMMABLE READ-ONLY MEMORIES* (EEPROMs) permit selective erasure of words. Under computer control, individual words can be erased and new data can be stored. As in EPROMs, floating-gate FETs are used. A high positive voltage (about 25 V) on the control gate erases an EEPROM cell. See Fig. 15-37.

Erasure and writing require special circuits and the processes are time consuming. As a result, EEPROMs do not replace read/write memories. EEPROMs have been described as "READ-MOSTLY" memories.

SUMMARY

Bootstrap programs, programs in dedicated computers, and tables are some of the applications of read-only memories. ROMs can be read, but information stored in them is difficult or impossible to change. ROMs are also nonvolatile. Stored numbers are not lost when power is interrupted.

ROMs contain four subcircuits. They are: memory-cell array, address decoder, output, and control.

Stored numbers are placed in mask-programmed ROMs by the manufacturer. PROMs and EPROMs are programmed by users. Data stored in PROMs cannot be changed, but EPROMs can be erased and reused. EEPROMS can be erased selectively, save time, and are easy to use or reuse.

TEST YOUR KNOWLEDGE

1. Read-only-memories can be read by computers, but cannot be _________ (organized, checked for errors, duplicated, altered, improved, addressed).
2. In a(an) _________ (simple, nonvolatile, addressable, optical, active-low) memory, numbers are not lost when the power is turned off.
3. Resident programs for entering instructions and data are called _________ (dual, branched, synchronized, bootstrap, fixed, relative) programs.
4. A _________ computer is one that does a limited group of tasks over and over.
5. A memory with 8 data leads can store _______ (4-bit, 6-bit, 8-bit, 16-bit) words.
6. A memory with 10 address leads can address _________ (256, 512, 1024, 2048) words.
7. A memory which has a size of 1024 × 8 is said to be a _________ × 8 memory.
8. What are the 4 blocks within memory chips?
9. Individual 1s and 0s are stored in memory elements or _________ _________.
10. Two types of arrays are a linear array and a(an) _________ array.
11. In linear arrays, each word must have its own _________ _________ lead wire.
12. _________ (Large, Small) memories are usually organized in matrix form.
13. Each of the cells in a matrix array has _______ (1, 2, 3, 4) enable leads.
14. A memory can be made of 2 or more chips that are connected in _________ (series, parallel).
15. Memory output circuit elements must be of the 3-_________ type.
16. A photographic process and etching process are used to make _________-programmed ROMs.
17. Diodes in memory arrays are used to provide short _________, yet avoid _________ current paths in a device.
18. For compatability with TTL elements, diodes in memory arrays are often replaced with _______.
19. Transistors used in ROMs are of the bipolar type or of the _________ _________ type.
20. The collector-to-emitter voltage of a conducting bipolar transistor is about _________ V.
21. In a mask bipolar-transistor ROM, a 1 is indicated by a base that is _________ (connected, left disconnected).
22. In a mask FET ROM, a 1 is indicated by a _________ lead that is left disconnected.
23. About _________ weeks may be required to produce and deliver a mask ROM chip.
24. Can a fusible-link PROM be programmed more than once?
25. A floating-gate FET is a type of _________ (erasable, nonerasable) FET element.
26. A _________ (negative, positive) charge on a floating FET gate turns the FET off.
27. Electrons in a floating-gate FET that cross over the glass insulator are called _______ electrons.
28. To remove the electrons that are trapped on a floating gate, the chip is exposed to strong _________ light.
29. To erase an EPROM, exposures to ultraviolet light of as much as _________ min. are required.
30. If an EPROM is exposed to room light for 10 years, about _________ percent of the original charge will remain.
31. What is placed over EPROM windows to prevent accidental erasure?
32. EPROMS are often used in place of PROMs when a project moves from the design phase to the _________ phase.
33. An EEPROM is an _________ (energetically, edge-triggered, electrically, essentially) erasable programmable read-only memory.

STUDY PROBLEMS

1. For each of the following applications, indicate whether a read/write (R/W) memory or a read-only memory (ROM) would most likely be used.
   1. Storing the program that places identifying information on the screen when a computer is first turned on.
   2. Storing a program entered by a user of a personal computer.
   3. Storing the program in a computer that controls a microwave oven which is sold to the general public.
   4. Storing the desired settings in the computer for a microwave oven.
   5. Storing the program for doing multiplication in calculators.
2. What term is used to describe a memory that retains stored numbers when the power is removed?
3. Name the four subcircuits or blocks found in the ROM diagrammed in Fig. P15-1.

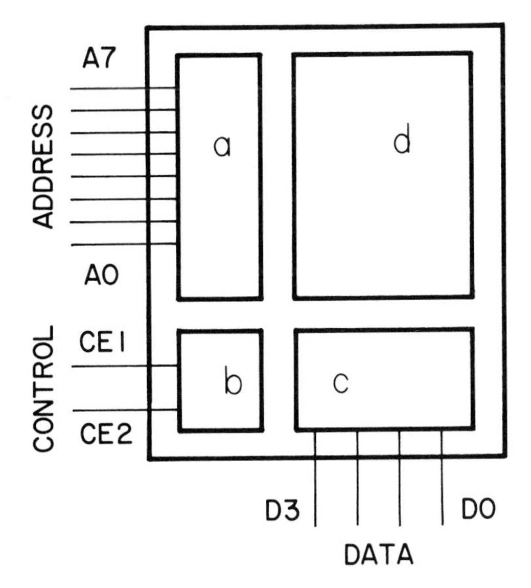

Fig. P15-1. The 4 subcircuits in the ROM shown can be named and the size of the memory can be determined.

4. Based on the number of leads in the diagram from Problem 3 (Fig. P15-1), estimate the size of the memory which is diagrammed.
5. Repeat Problem 4 for the memory in Fig. P15-2.
6. Repeat Problem 4 for the memory in Fig. P15-3.
7. Repeat Problem 4 for the memory in Fig. P15-4.
8. WORD ENABLE 0 of the memory array in Fig. P15-5 has been energized. What number will

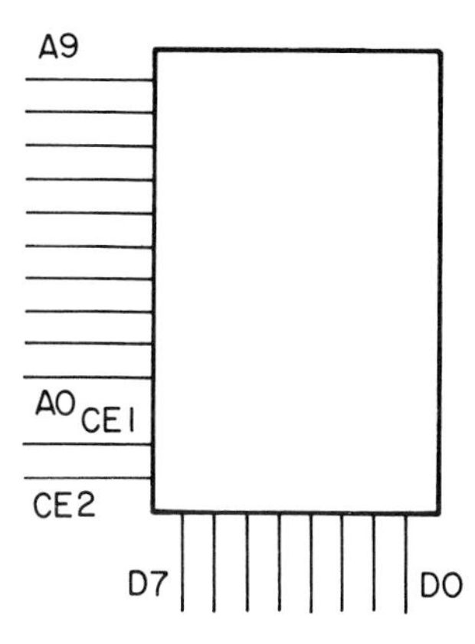

Fig. P15-2. The size of the memory shown can be determined.

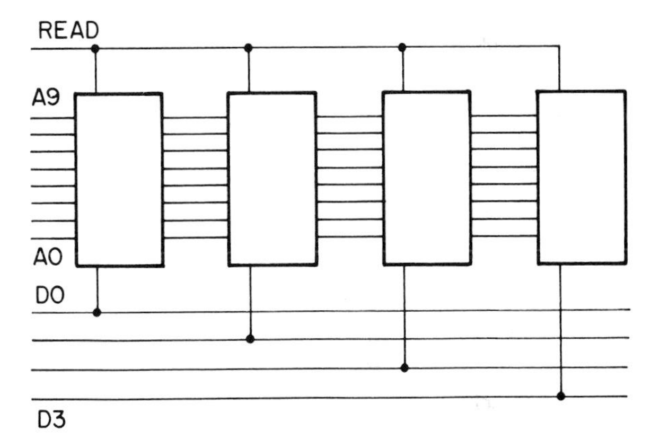

Fig. P15-3. The size of the memory shown can be determined.

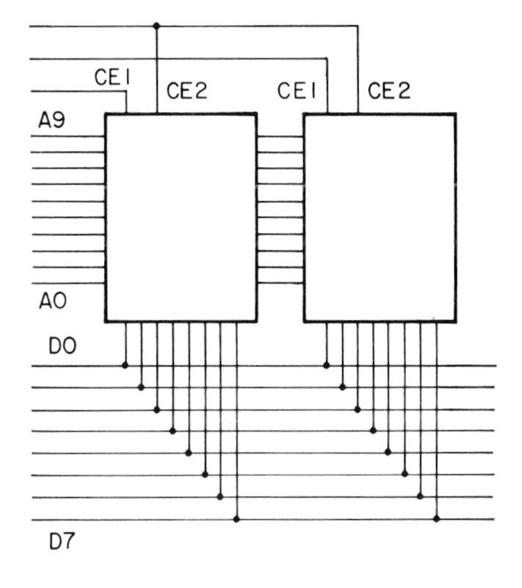

Fig. P15-4. The size of the memory shown can be determined.

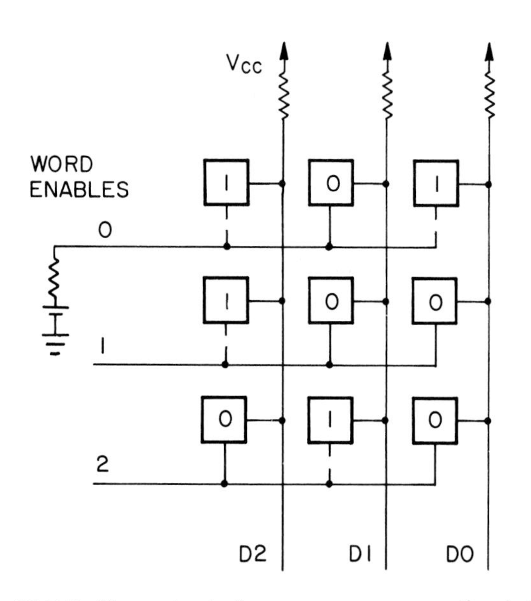

Fig. P15-5. The output of a memory array on the data leads can be determined.

appear at output D2D1D0?

9. Repeat Problem 8 for the memory array which is diagrammed in Fig. P15-6.

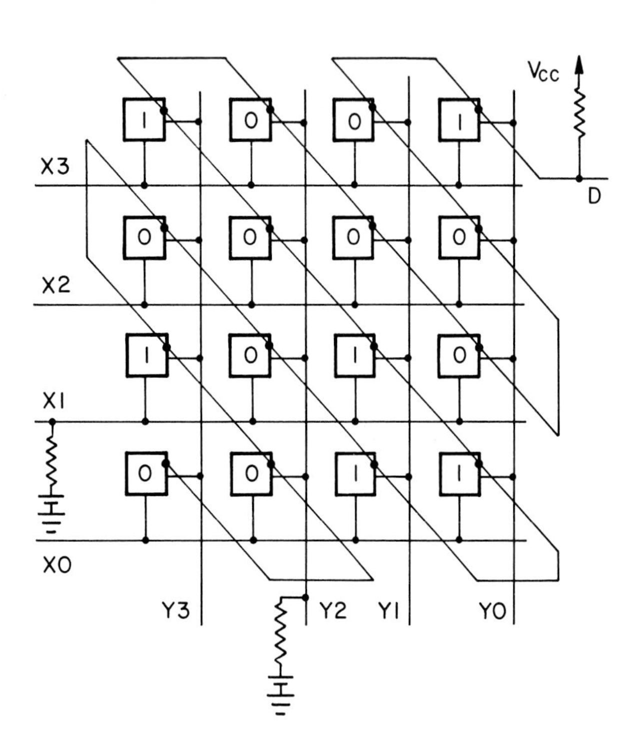

Fig. P15-6. The output of a matrix array memory can be determined.

10. Based on the number of leads shown, estimate the size of the memory in Fig. P15-7. The X and Y leads are not address leads. They are the leads to the 4 × 4 matrices.

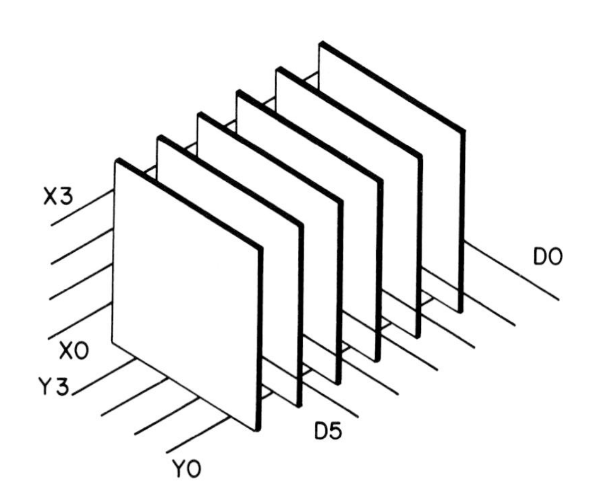

Fig. P15-7. The size of a parallel matrix array can be estimated.

11. For the signals on address leads A1 and A0 in Fig. P15-8, what number will appear on the output leads which are labeled D2D1D0?

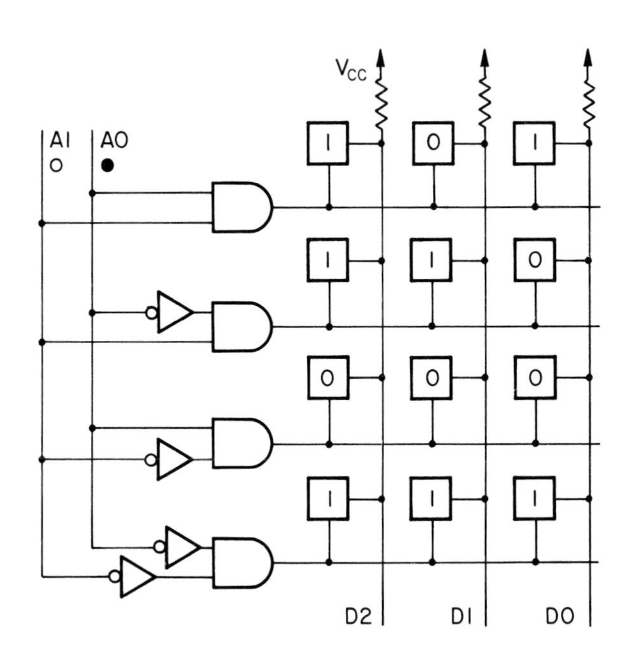

Fig. P15-8. The output of a memory with an address decoder can be determined.

12. For the signals on address leads A1 and A0 in Fig. P15-9, what number or signal will appear on the sense wire circuit?

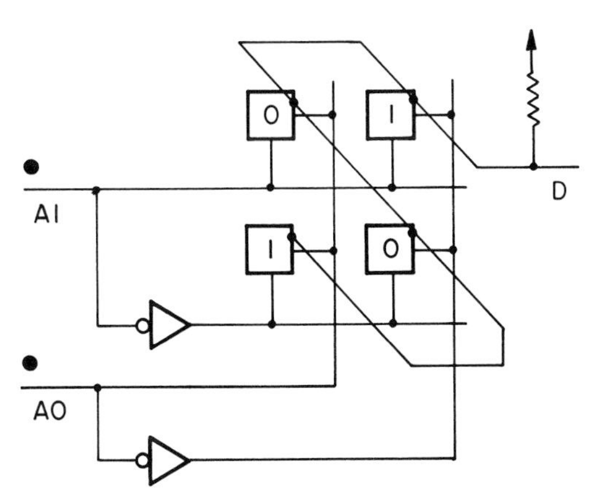

Fig. P15-9. With given address signals, the number on the sense wire of a memory can be determined.

13. For the address shown, what number will appear on output D2D1D0 of the ROM in Fig. P15-10?

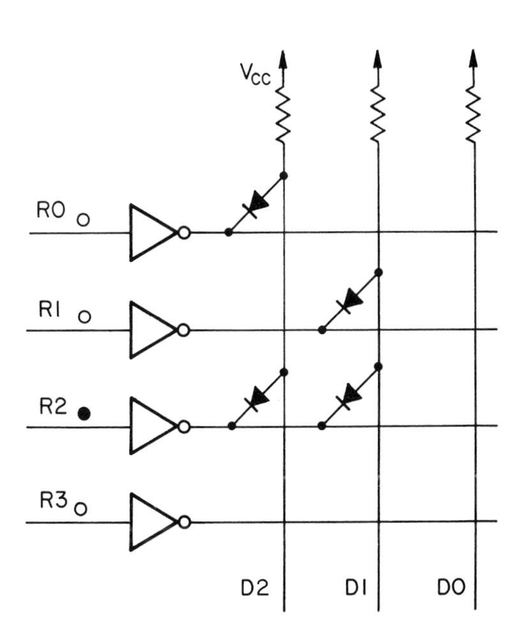

Fig. P15-10. With a given address signal, the output of a diode ROM can be determined. Diodes can be added to store the number D2D1D0 = 100.

14. Copy the last row of the ROM in Problem 13 (Fig. P15-10) onto a sheet of paper and add diodes to store the number D2D1D0 = 100.

15. Fig. P15-11 shows a mask-programmed ROM. For the signals on its word enables, what number will appear on the output leads D2D1D0?

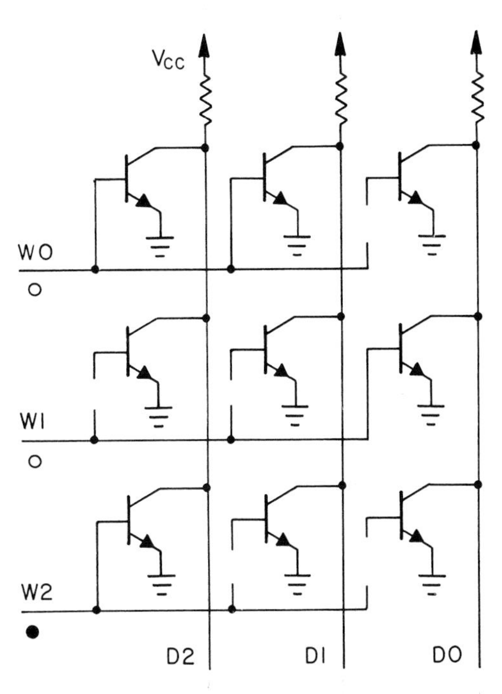

Fig. P15-11. With some transistor bases left disconnected during the mask and etch process, the output of the memory can be determined.

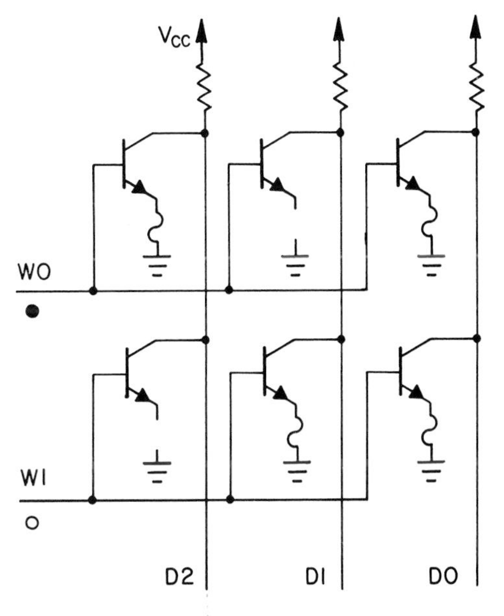

Fig. P15-12. With some emitter lead fuses blown, the memory can be read.

16. Fig. P15-12 shows a PROM. For the signals on its word enable leads, what number will appear at its output leads D2D1D0?
17. Match the following memories with the cells shown in Fig. P15-13.
    1. Mask-programmed diode ROM.
    2. Mask-programmed bipolar-transistor ROM.
    3. PROM.
    4. EPROM.

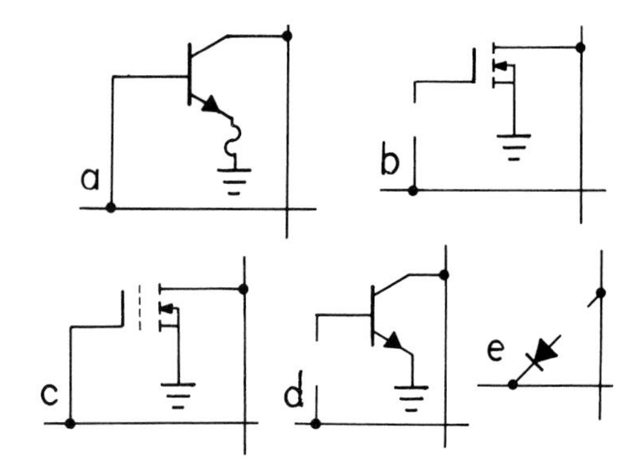

Fig. P15-13. Four types of memories can be identified.

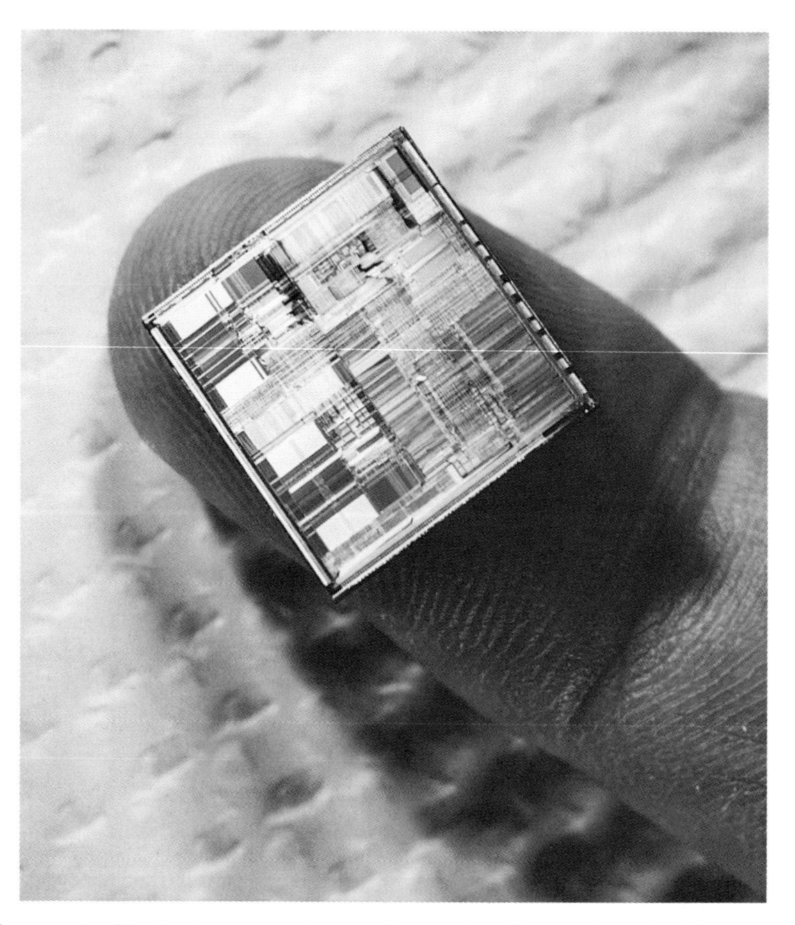

This chip contains thousands of logic elements and is small enough to fit on a fingertip. It is, essentially, a computer-on-a-chip containing RAMs, ROMs, an ALU, registers, timing and control circuits, counters, shift registers, and decoders. (Intel Corporation)

Random Access Memory

Chapter 16

LEARNING OBJECTIVES

After studying this chapter and completing lab assignments and study problems, you will be able to:

- Describe the difference between ROMs and RAMs.
- Describe the action of bistable multivibrators.
- List three concepts used in a description for the action of a TTL memory cell.
- Recognize circuits and name parts of circuits for MOS FET and CMOS FET memory cells.
- List components that are important in a description for the action of a dynamic memory cell and explain the need for refresh.
- Describe the relationship between memory size and memory speed.

The term RANDOM ACCESS MEMORY (RAM) is not an accurate description of the memory to be considered next. READ/WRITE MEMORY would be more meaningful. However, early read/write memories were called RAMs, and the term is still in use.

In general, RAMs are used for temporary storage. Programs, data, and results of computations can be stored in a RAM.

RAM storage is temporary because most read/write memories are volatile. When power is lost, so is stored data. Some modern RAMs require so little power that battery backup can be used. That is, batteries provide power to memory chips when the system power fails. A relay can switch to battery power.

RAM CHIP

Fig. 16-1 shows a typical RAM chip. Externally, it is much like a ROM. A difference is found in the presence of a read/write control lead.

Internally, RAMs and ROMs can both be divided into four blocks. See Fig. 16-2. In these two memories, address decoding and memory arrays are much the same. Differences are found in the control and output circuits and in the memory cells.

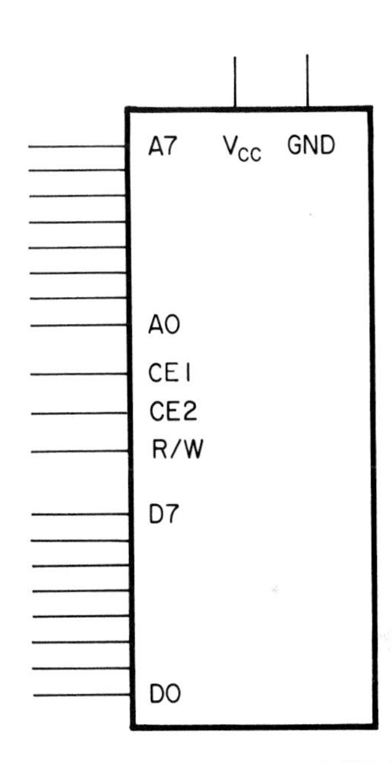

Fig. 16-1. Except for the control lead R/W, RAM leads are similar to those on ROMs.

INPUT/OUTPUT

Data must flow both to and from a RAM. As a result, input/output circuits in RAMs are more complex than those in ROMs. Fig. 16-3 shows a possible input/output circuit. Three-state output devices for the read mode simplify the circuit. AND elements isolate the read and write modes. What appears as a feedback circuit to the inputs of the ANDs causes no problem. The WRITE ENABLE lead keeps the ANDs off during a read process.

CONTROL

Fig. 16-4 shows a simple control circuit for a RAM. It permits three modes of operation: READ, WRITE,

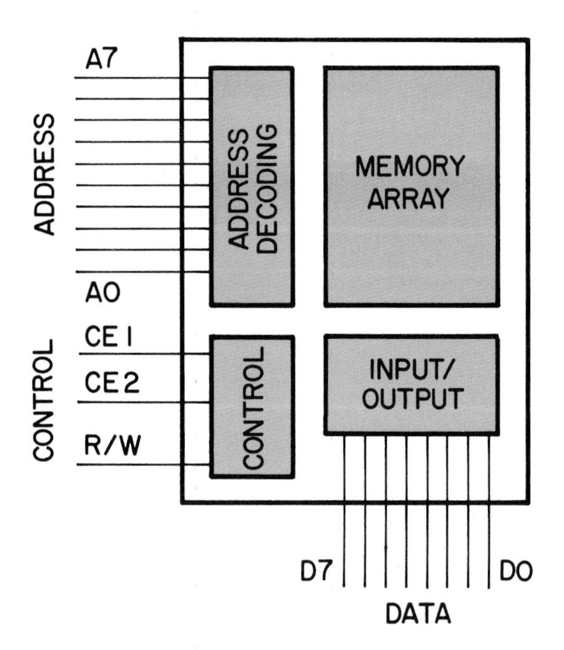

Fig. 16-2. Like ROMs, circuits within RAMs can usually be divided into four blocks.

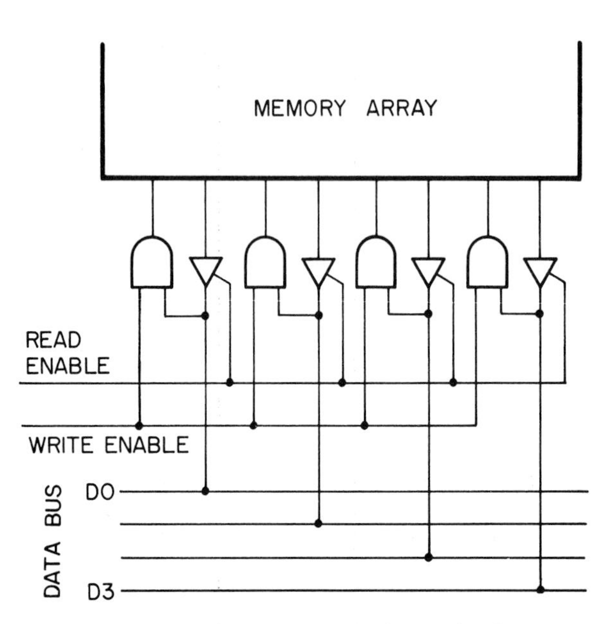

Fig. 16-3. RAM input/output circuits tend to be more complex than the output circuits of ROMs.

and STANDBY. CE1 and CE2 are chip enable leads. As indicated by the label "R/W," a READ process requires a 1 on the R/W lead and a WRITE process requires a 0 on the R/W lead.

Modes are accessed (activated) one operation or process at a time. Most units are fast.

READ MODE

In the read mode, stored numbers are output to the data bus. To accomplish this, enabling signals must be sent to the memory array to indicate that data is to be output. Also, the three-state buffers in the memory's output circuit must be enabled.

For READ ENABLE to be active, a 1 must appear on R/W. Also, both CE1 and CE2 must be active.

WRITE MODE

In the write mode, the number on the computer's data bus is moved into memory and stored. Signals must be sent to the memory array indicating that a store operation is to be accomplished. Also, gates at the circuit's input must be activated.

As before, both CE1 and CE2 must be active. A 0 must appear on the R/W lead to place the memory chip in its write mode.

STANDBY MODE

When one or both chip enable leads are inactive, the control circuit in Fig. 16-4 places the memory in its standby mode. The signal on R/W is meaningless, since neither READ ENABLE nor WRITE ENABLE can be active.

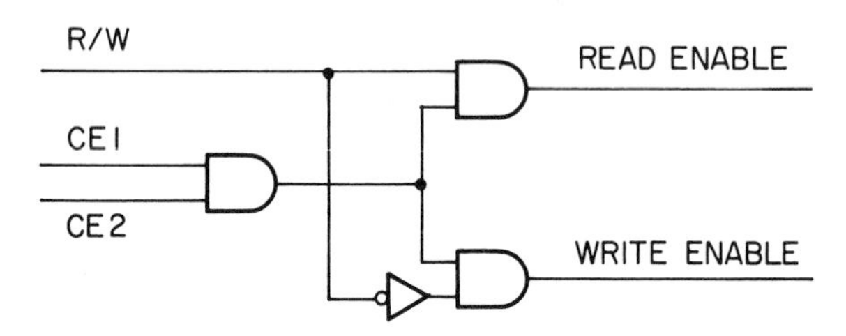

Fig. 16-4. The control circuit shown places the memory in one of three modes: read, write, and standby.

FLIP-FLOP MEMORY CELL

NAND-based flip-flops are used as memory cells in some RAMs. Such circuits have been described in Chapter 8, Fig. 8-18. As an aside, note that the memory in Fig. 8-18 was analyzed in terms of the four blocks described in Fig. 16-2.

Memories based on flip-flops are very fast. That is, numbers can be quickly stored and retrieved. However, each flip-flop may contain 16 or more transistors and 6 or more resistors. The result is so complex that only small memories can be constructed from flip-flops.

Memories based on data latches are often used as SCRATCHPAD MEMORIES. That is, they are used in ALUs to temporarily hold subtotals, etc. In such applications, high speed is more important than size.

BISTABLE MULTIVIBRATOR CELL

The circuit in Fig. 16-5 is the basis for a number of memory cells. It is called a *BISTABLE MULTIVIBRATOR*. The term ECCLES-JORDAN circuit (after its inventors) is also used. It may be called a flip-flop.

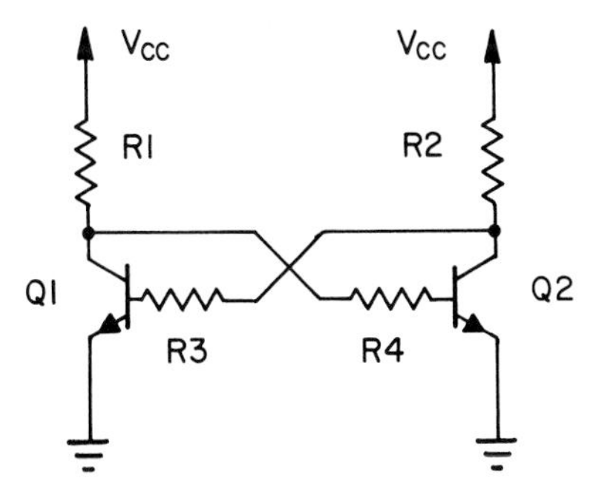

Fig. 16-5. The bistable multivibrator circuit shown is the basis for a variety of memory cells.

The term *BISTABLE* implies two stable states. These states are defined in terms of which transistor is conducting. If Q1 is on, Q2 will be off. If Q2 is on, Q1 will be off.

Like a NAND-based flip-flop, the bistable multivibrator circuit will remain in whichever state it is in. It will change only when an outside signal is applied or when power is lost.

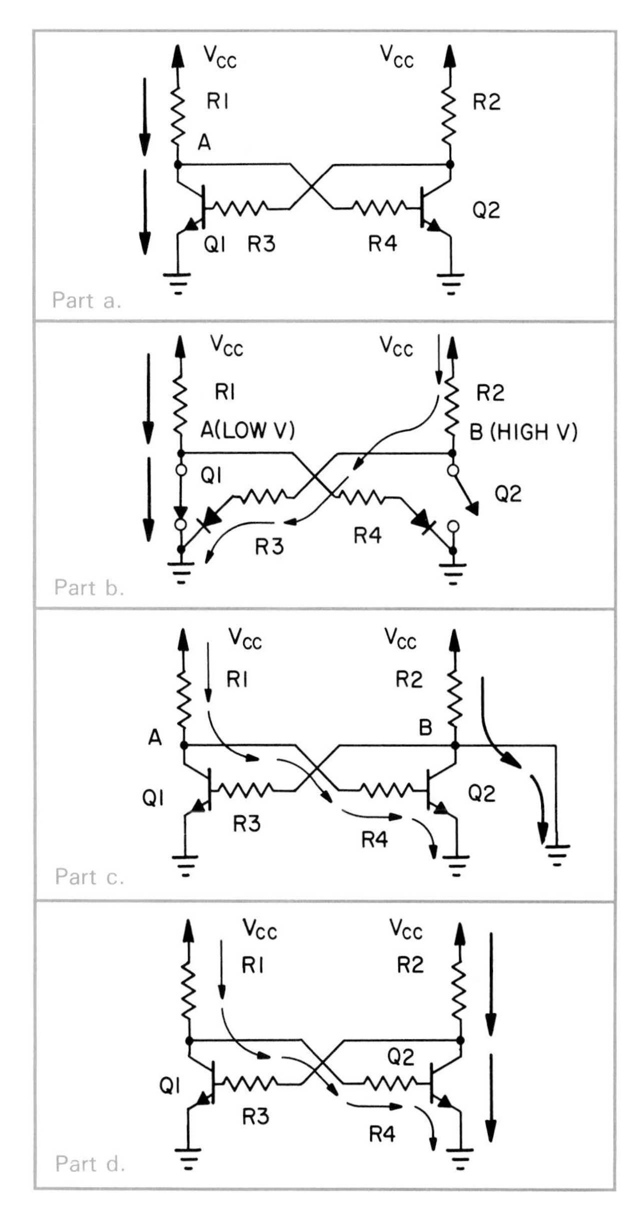

Fig. 16-6. When a given transistor in a bistable multivibrator circuit has been turned on, it remains on until an outside signal is applied to turn it off.

BISTABLE MULTIVIBRATOR ACTION

Part "a" of Fig. 16-6 shows a bistable multivibrator in one of its stable states. Q1 is conducting. The heavy arrows trace the flow of current from Vcc, through R1, and through Q1 to ground.

Because Q1 is turned on, the voltage across this transistor is small. Most of Vcc is dropped across R1, and the voltage at point A is near 0.

The voltage at A is applied to the base of Q2 through the cross-coupled leads. As a result, Q2 is turned off. See part "b" of Fig. 16-6.

With Q2 off, its collector current will be small, so there will be little voltage drop across R2. The voltage at B will be relatively high and positive. This voltage is applied to the base of the first transistor through the cross-coupled leads. Because it is positive, it will hold this transistor in its conducting state, and the circuit will be stable. The small current required by the base of Q1 flows through R2 as indicated.

Part "c" of Fig. 16-6 shows the storing of a new number. Point B has been grounded, so its voltage approaches 0. With no voltage applied to the base of

Q1, it turns off. With little current through R1, the voltage at A increases and is applied to the base of Q2. This turns on Q2. When the short to ground is removed, the new number will be available in the cell, part "d" of Fig. 16-6.

While this simple cell is capable of storing bits, it is difficult to connect it into an array. The following modification makes connection into an array possible.

MULTIPLE-EMITTER MEMORY CELL

Part "a" of Fig. 16-7 shows a multivibrator memory cell that uses transistors with leads called *MULTIPLE EMITTERS.* Any lead of a multiple emitter group can control the transistor current. Such transistors are found in TTL logic elements, so manufacturing methods are available for producing multiple-emitter cells. The extra emitters are used to connect such cells to memory array circuits. Three modes of operation need to be analyzed. These are: the standby mode, the write mode, and the read mode.

STANDBY MODE

Part "a" of Fig. 16-7 shows a cell in its standby mode. The memory cell contains a 0.

To provide a path to ground for the emitter current of Q1, transistor Q3 is conducting. This is accomplished by placing a 1 on CELL SELECT (CELL SELECT is active-low). In this mode the stored bit is retained.

WRITE MODE

Part "b" of Fig. 16-7 shows the storing of a 1. WRITE AMPLIFIERS have been added to the circuit. When Q4 or Q5 is turned on, the corresponding READ-WRITE lead is grounded.

To store a 1, READ-WRITE 1 is grounded (Q5 is turned on), and CELL ENABLE is active (Q3 is turned off). The only ground path for an emitter is one shown, so Q1 is turned off, and Q2 turns on. "READ-WRITE" is not same as "READ/WRITE."

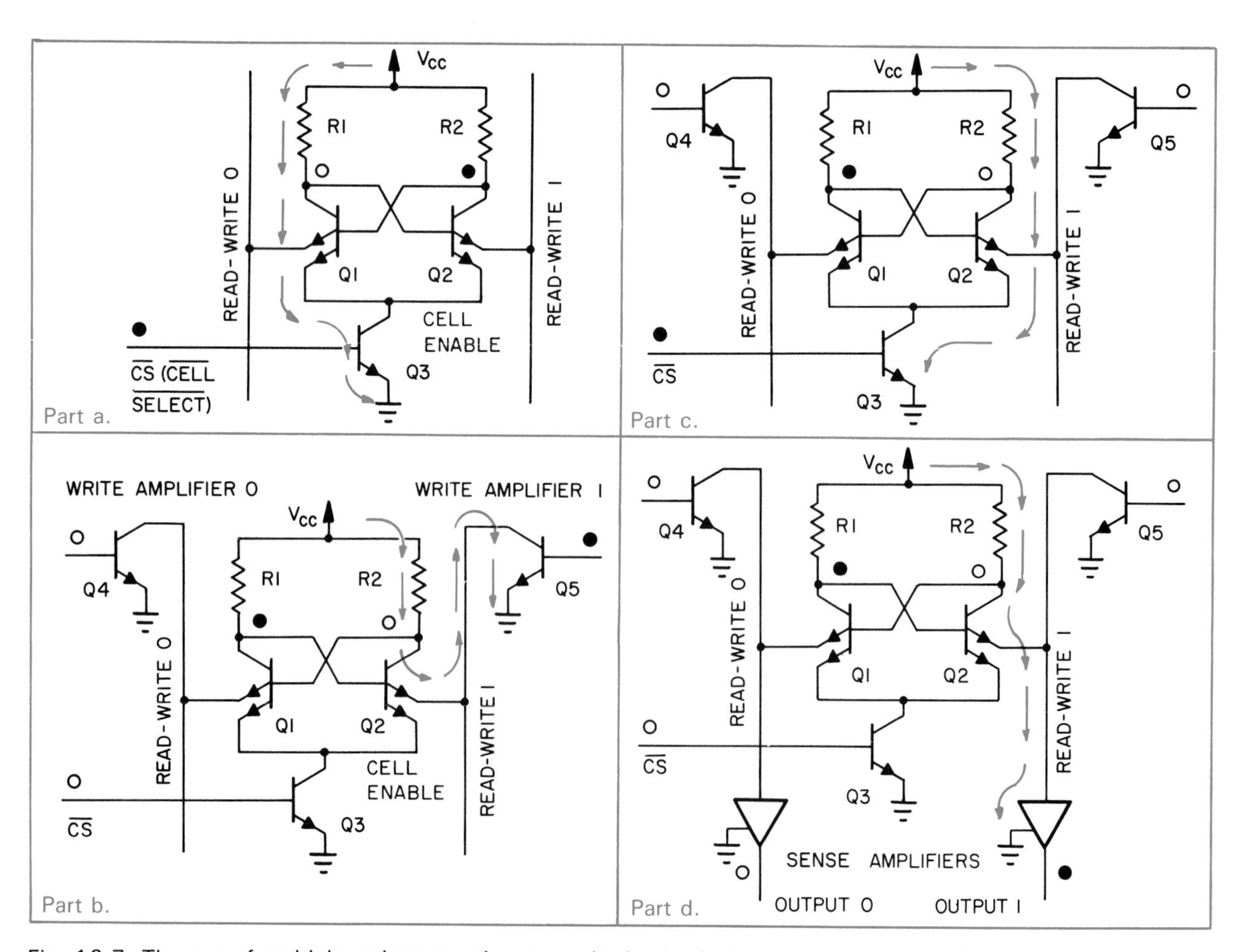

Fig. 16-7. The use of multiple-emitter transistors results in simple input and output circuits.

To complete the process, Q3 is turned back on (CELL SELECT returns to 1), and the grounding of READ-WRITE 1 is removed (Q5 is turned off). See part "c" of Fig. 16-7.

READ MODE

In part "d" of Fig. 16-7, SENSE AMPLIFIERS have been added. Such amplifiers have relatively low input resistances. When current flows into them, they output logic 1s.

To read a cell, CELL ENABLE is made active (a 0 is applied to the base of Q3). With Q3 turned off, the emitter current of Q2 must find a new path to ground. As shown in part "d" of Fig. 16-7, this path is through SENSE AMPLIFIER 1.

When the stored number has been output, CELL SELECT is returned to its inactive state (1), and Q3 again turns on. Because the resistance of Q3 when it conducts is less than that at the input of the sense amplifier, the emitter current of Q2 returns to Q3. The cell is back in its standby mode. Reading such a cell does not alter the stored information.

MULTIPLE-EMITTER, MULTIPLE-TRANSISTOR ARRAY

Many transistors of the multiple-emitter type may be used together in arrays. The group is called a multiple-emitter, multiple-transistor array. The memory cells described with the help of Fig. 16-7 may be used in linear (word organized) or matrix (bit organized) arrays. Their advantages and disadvantages depend on the size of the memory.

LINEAR ARRAY

Fig. 16-8 shows a simplified array using multiple-emitter cells similar to those in Fig. 16-7. They differ in that all the cells in a given word share a single enable transistor (Q3). Q3 has become $\overline{WS0}$. In the standby mode, emitter currents from all the cells in a given word flow through a common transistor. See word 0.

To read or write to or from a given word, its WORD SELECT goes to 0. This turns the WS transistor off and forces the emitter currents to flow in the read or write circuits. See word 1.

MATRIX ARRAY

Fig. 16-9 shows a typical multiple-emitter cell from a matrix memory array. Note the use of three emitters. This permits the use of two cell enables (X and Y). If one or both cell enables is inactive (high), the cell will be in its standby mode. If either Q3 or Q4

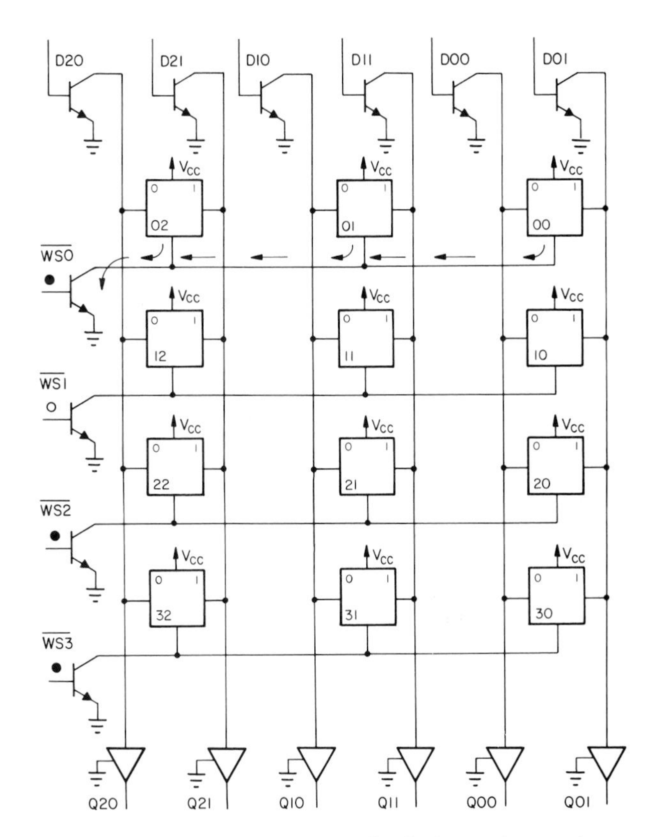

Fig. 16-8. Multiple-emitter cells find use in word-organized memories. The memory shown is a 4 x 3 array.

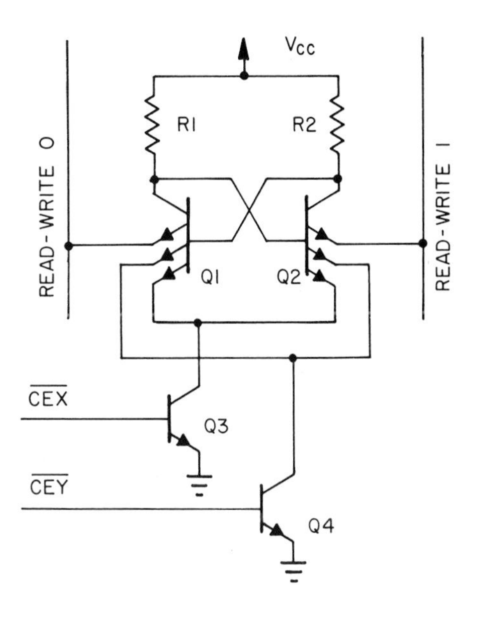

Fig. 16-9. The matrix cell shown has two CELL ENABLES. Both must be active (low) for the cell to be read or written into.

(or both) is conducting, no current is left to go through the read-write leads (R-W 0 and R-W 1). Only when both select leads are active (low) can current flow in the READ-WRITE leads.

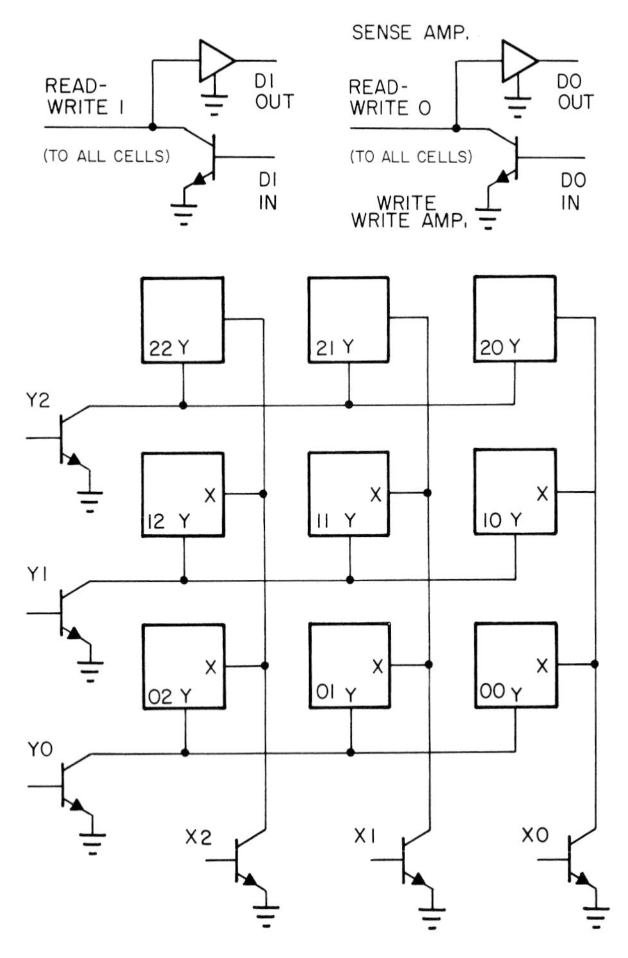

Fig. 16-10. Memory cells with two cell-select leads have been used in the bit-organized array shown. The memory is a 9 x 1 array.

Fig. 16-10 shows a bit-organized memory array using multiple-emitter cells. Leads of the multiple emitters are at X and Y in each cell. If either transistor Y2 or X2 (or both) conducts, cell 22 will be in standby. With both Y2 and X2 off, a read or a write occurs at cell 22. The two READ-WRITE leads in Fig. 16-10 run to all cells. The read-write leads have been omitted for simplicity.

FET MEMORY CELL

Because FETs are small and require less power than bipolar transistors, FETs are usually used in large

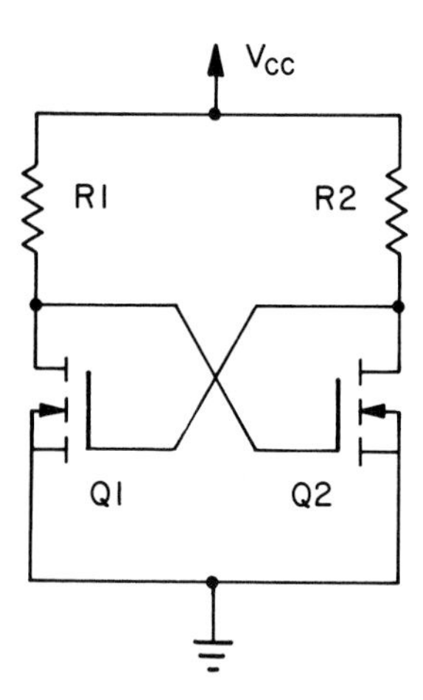

Fig. 16-11. Field-effect transistors have been used in the bistable multivibrator shown.

memories. Fig. 16-11 shows a FET bistable multivibrator. Its action is similar to that of bipolar transistor circuits.

PRACTICAL FET CELL

On IC chips, FETs require less room than do resistors. As a result, space can be saved if transistors are substituted for resistors. See part "a" of Fig. 16-12. As resistors, Q3 and Q4 can drop voltage much as resistors R1 and R2 do. The internal resistances of Q3 and Q4 have been substituted for the collector resistors. Because these transistors must be conducting at all times, their gates have been tied to the positive voltage, Vdd.

FETs are voltage-operated devices, so the enable circuits for FET memory cells differ from those of current-operated bipolar transistors. See part "b" of Fig. 16-12. To accomplish a read or write, 1s are placed on both X-ENABLE and Y-ENABLE. This turns on Q5, Q6, Q7, and Q8, and attaches the cell to the READ-WRITE leads. If either X-ENABLE or Y-ENABLE (or both) is 0, the memory cell will be in its standby mode.

In the circuit in part "a" and part "b" of Fig. 16-12, Q3 and Q4 act as passive devices. The term *PASSIVE* means that Q3 and Q4 function as resistors and are not switched on and off. As a result, Q3 and Q4 consume power but contribute little to the action of the circuit. These transistors can be made to take an active part in the circuit's operation if CMOS technology is used for 2 or more circuit elements.

CMOS FET CELL

The field-effect transistors described up to this

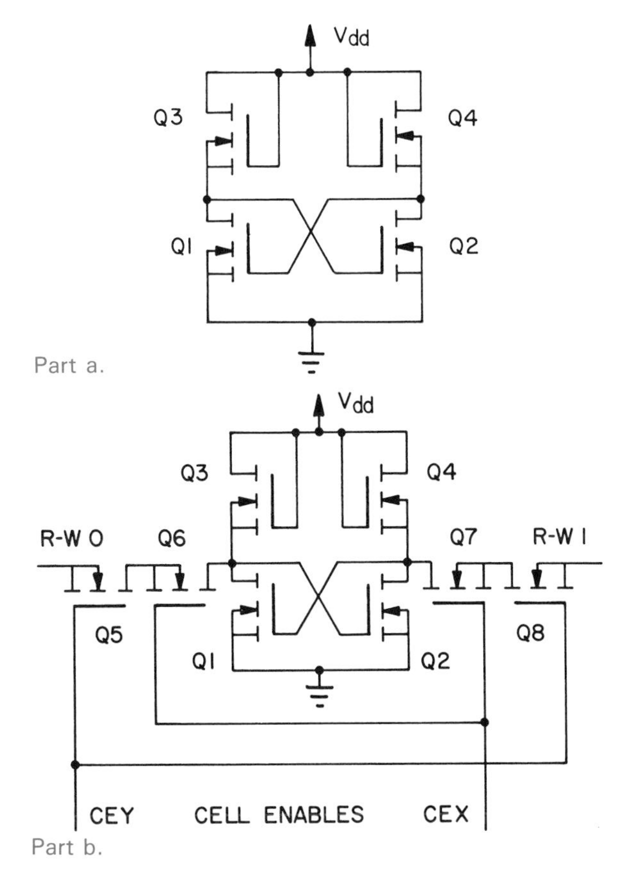

Fig. 16-12. In the circuit shown, Q3 and Q4 act as resistors.

point have been MOS FETs. MOS stands for *METAL-OXIDE SEMICONDUCTOR*. The metal oxide can be silicon dioxide (glass). SiO_2 insulates the gate from the transistor. Other examples of metal oxide insulators and/or substrates are aluminum oxide, lanthanum oxide, yttrium oxide/garnet, copper oxide, tin oxide, and bismuth oxide. Copper oxide is used in inexpensive semiconductors.

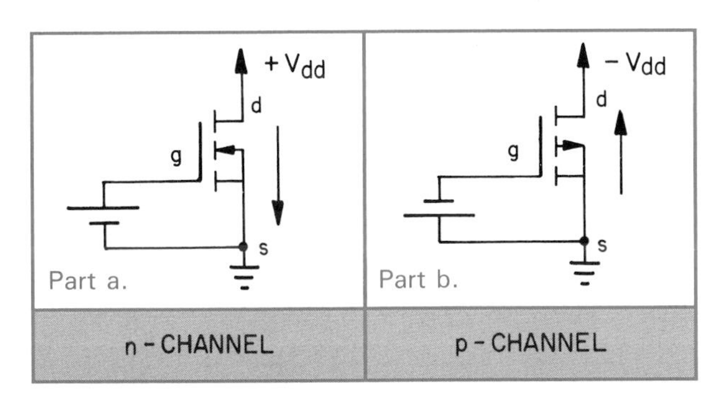

Fig. 16-13. The arrows in the displayed FET symbols indicate n-channel and p-channel units.

COMPLEMENTARY TRANSISTOR

The C in CMOS stands for complementary. The transistor in part "a" of Fig. 16-13 is turned on when its gate is positive with respect to its source. It is called an n-channel device. The complement of this transistor is shown at b. It is a p-channel MOS FET, and it is turned on when its gate is negative with respect to its source. Note the direction of conventional current flow through the p-channel device.

COMPLEMENTARY SYMMETRY

When n-channel and p-channel MOS FETs are used as shown in part "a" of Fig. 16-14, complementary symmetry results. Conventional current flows from drain to source in n-channel FETs and from source to drain in p-channel FETs. Therefore, the upper transistor is inverted in this circuit.

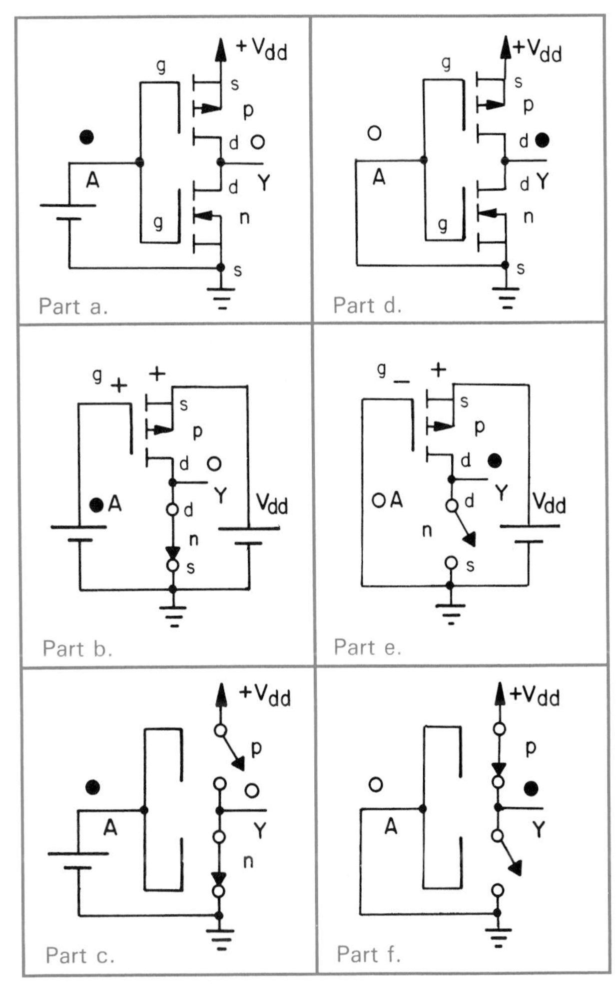

Fig. 16-14. In the complementary FET inverter shown, only one transistor is on at a given time.

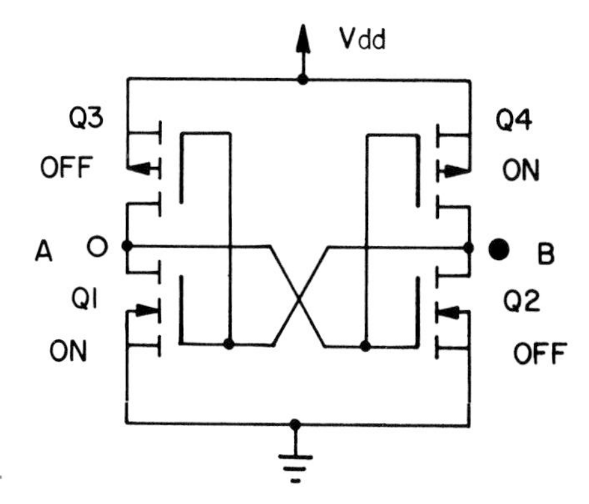

Part a.

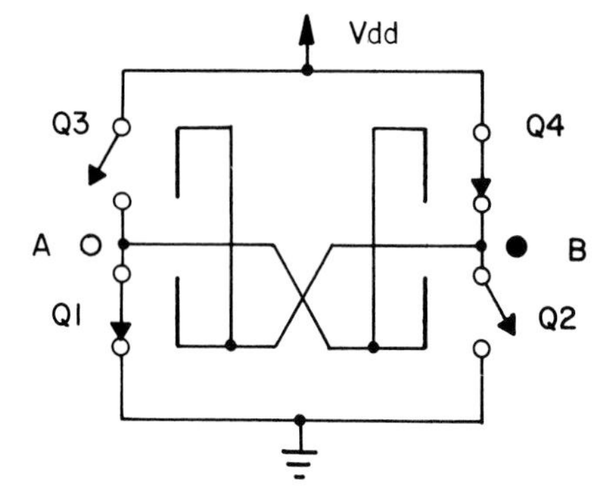

Part b.

Fig. 16-15. The bistable multivibrator shown uses CMOS technology.

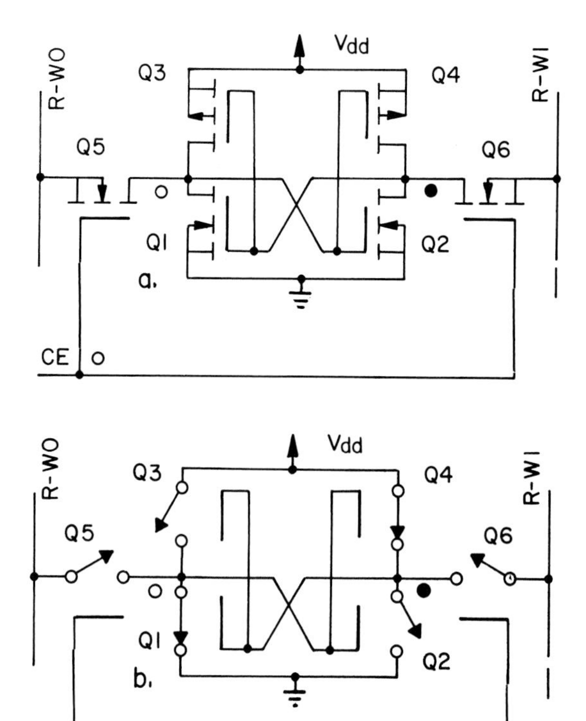

Fig. 16-16. When CELL ENABLE is inactive (0), this CMOS memory cell is isolated from other circuits.

In the circuits at the left in Fig. 16-14, A = 1 (a positive voltage). This turns on the lower FET (n-channel). Note the closed switch in part "b." For the upper (p-channel) FET to turn on, its gate must be sufficiently negative with respect to its source. Due to A = 1, this is not true, so the upper FET is off. Note the open switch in part "c" in Fig. 16-14. Y = 0, so this circuit is a NOT.

At the right in the figure, A = 0. The lower FET is turned off, and it appears as an open switch in parts "e" and "f" of Fig. 16-14. The 0 at A connects the gate of the upper FET to ground. Because ground is negative with respect to Vdd, the upper transistor turns on. See parts "e" and "f" in the figure. For A = 0, Y = 1. Note that when the n-channel is on, the p-channel is off. When the p-channel is on, the n-channel is off.

CMOS FET CELL CONSTRUCTION

Part "a" of Fig. 16-15 shows a CMOS FET memory cell. Q1 and Q2 are n-channel devices; Q3 and Q4 are p-channel devices.

When Q1 is on, point A is grounded. This low voltage is applied to the gates of Q2 and Q4, so Q2 is off and Q4 is on. See part "b" of Fig. 16-15. This means that point B will be connected to Vdd through the upper right transistor labeled Q4.

The high positive voltage at B is fed through the cross-coupled leads to the gates of Q1 and Q3. Q1 turns on and Q3 turns off. The high voltage from B holds Q1 and Q3 in the positions shown, so the circuit is in a stable state. It is in its standby mode and holds a stored bit.

When in the standby mode, the cell in Fig. 16-15 draws little current. In each path from Vdd to ground, there is an open transistor. Only when switching from one state to the other will sizable currents flow. As a result, CMOS FET cells use little power. For this reason, CMOS FET cells are used in larger memories.

When in standby, this cell is isolated from the outside world. See Fig. 16-16. With Q5 and Q6 off, the cell is disconnected from adjacent circuits.

DYNAMIC RANDOM-ACCESS MEMORY

The RAMs described to this point are referred to

as *STATIC RAMS.* The term static means unchanging. Static RAM circuits retain stored numbers as long as power is applied.

Another group of memories is called *DYNAMIC.* Even though power is not interrupted, such memories tend to forget. They must be *REFRESHED* at regular intervals or stored numbers will be lost.

While the refreshing of dynamic memories requires computer time and additional hardware, dynamic memory cells require essentially no power in the standby mode. As a result, this technology has been used in very large memories.

Fig. 16-17 shows both static and dynamic cells. In the static cell, the transistors are connected to Vdd through R1 and R2. To retain information, the voltage Vdd must be present.

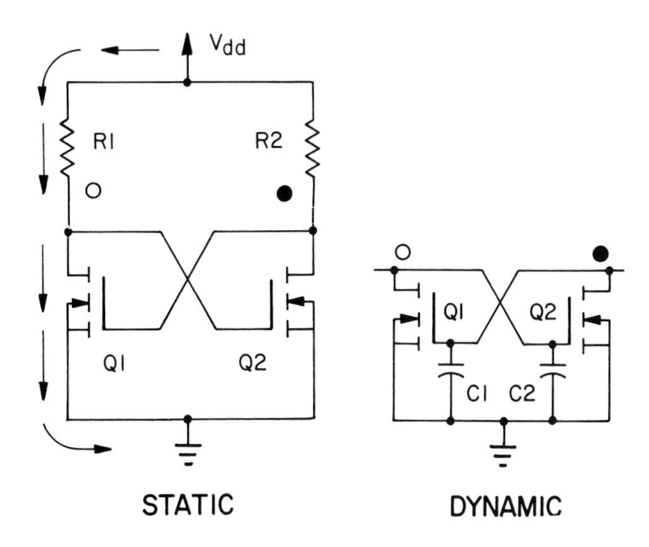

Fig. 16-17. In the dynamic memory cell, information is retained by charges stored in capacitors. The dynamic memory cell has no direct connection to Vcc.

In the dynamic cell, no direct connection is made to Vcc. Two components have, however, been added. C1 and C2 are capacitors, and the cell depends on the electric charges on these devices to retain information.

REVIEW OF CAPACITOR ACTION

A capacitor consists of two conducting surfaces separated by an insulator, part "a" of Fig. 16-18. When voltage is applied across these plates, charge flows onto one plate and off of the other.

As charge is delivered, a voltage appears across the capacitor. When this voltage equals that of the power source, the current stops. The capacitor is said to be fully charged, part "b" of Fig. 16-18.

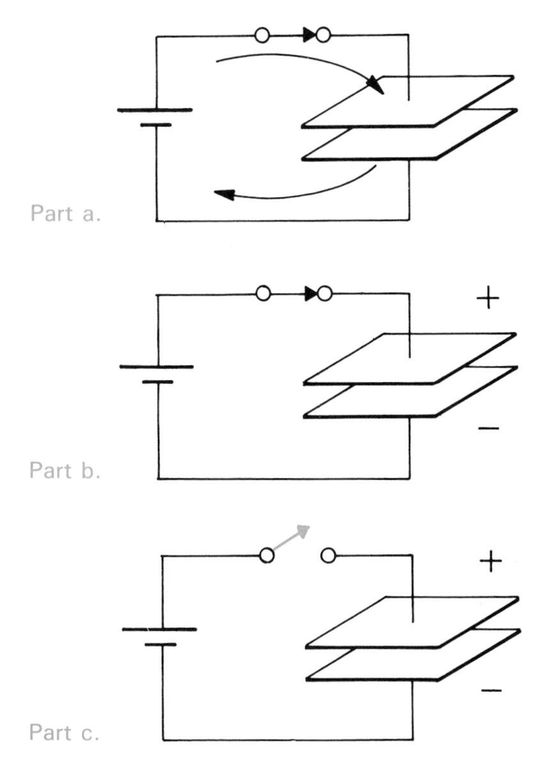

Fig. 16-18. Capacitors can hold an electric charge after the applied voltage has been removed.

At c the power source has been disconnected. However, the charge is retained, at least temporarily, on the capacitor.

The word temporarily is important. Capacitor plates cannot be perfectly insulated. There is always some leakage. The term *LEAKAGE* means a current as small as 1 μA (microampere) can flow. Small currents permit the charge to flow from one plate to the other, and the charge is slowly lost. As the charge decreases, so does the voltage.

There is a second problem. Capacitors require as much space on IC chips as 16 flip-flops. Capacitors take up so much room that they are avoided by chip designers.

DYNAMIC MEMORY CAPACITORS

The problem of capacitors on IC chips is solved by the nature of MOS FETs. Refer to Fig. 16-19. The gate (a conducting surface) of a MOS FET is separated from the body of the device (another conducting surface) by a thin layer of glass (an insulator). That is, each MOS FET contains a built-in capacitor.

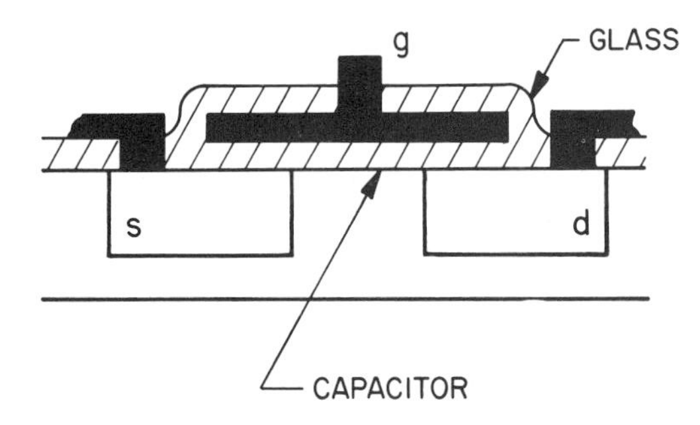

Fig. 16-19. The natural capacitance between the gate and body of a MOS FET is used in dynamic memory cells.

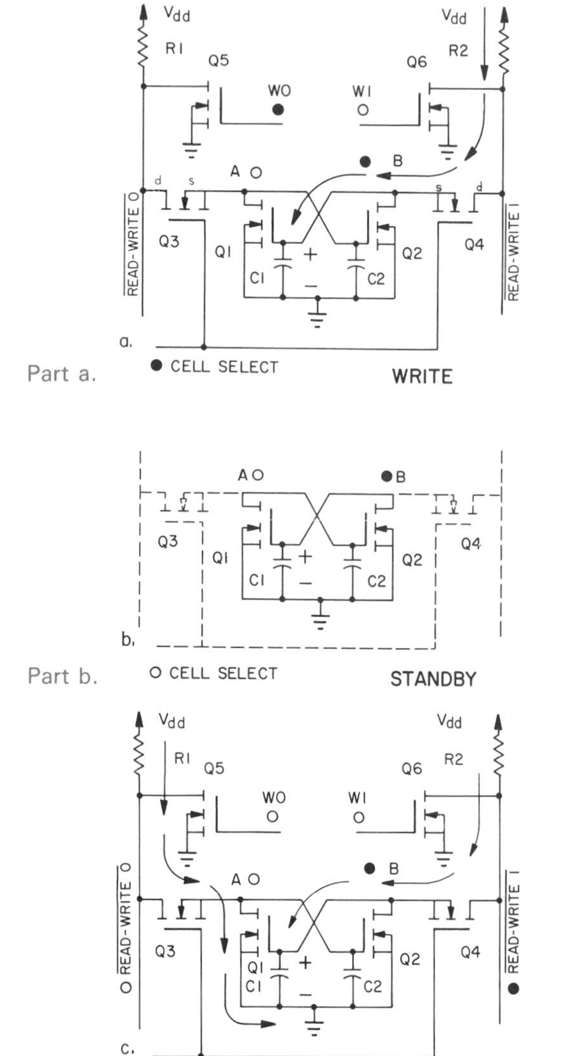

Fig. 16-20. For the dynamic memory cell shown, read and refresh are identical actions.

Capacitors do not need to be added to the integrated circuit for a MOS FET.

DYNAMIC MEMORY CELL ACTION

Part "a" of Fig. 16-20 shows a cell from a dynamic memory. Transistors Q1 and Q2 (and their gate capacitors C1 and C2) form the cell itself. Two lines called READ-WRITE 0 and READ-WRITE 1 have been added. A lead called CELL SELECT is used. The operation of the circuit can be analyzed after it is broken into subcircuits. Three topics are important. These are:

- Write mode.
- Standby mode.
- Read mode.

WRITE MODE

In the first circuit of Fig. 16-20, a 0 is being written into the cell. To do this, a 1 is applied to CELL SELECT. This turns on Q3 and Q4 and connects the cell to the READ-WRITE leads. READ-WRITE 0 is also made active (low). The following is done to make READ-WRITE 0 active-low: a 1 is applied to the gate of Q5 (W0 in part "a" of Fig. 16-20), so READ-WRITE 0 is grounded. W1 is off, so Q6 does not conduct.

With point A grounded, Q2 will not conduct, and the positive voltage on READ-WRITE 1 will appear at point B. Capacitor C1 is connected to point B, so current will flow and charge it to about Vdd. With a positive voltage at its gate, Q1 will turn on, but this will not alter the situation. Point A is already grounded through lead READ-WRITE 0.

STANDBY MODE

To complete the storage process and place the cell in its standby mode, CELL SELECT is returned to 0. This turns off Q3 and Q4 and isolates the cell. See part "b" of Fig. 16-20. Q1 conducts and maintains the grounded condition on lead A.

Due to the charge on C1, Q1 will continue to conduct and Q1 will short C2 and the gate of Q2 to ground. With no voltage on its gate, Q2 will not conduct, so it will not drain charge from C1. The circuit is stable.

READ MODE

In part "c" of Fig. 16-20, the cell is being read. Again, a 1 is placed on CELL SELECT to turn on Q3 and Q4. Q1 is still conducting, so READ-WRITE 0 is shorted to ground. The resulting 0 on the READ-

WRITE 0 lead indicates that a 0 had been stored. If a 1 had been stored, READ-WRITE 1 would have gone to 0.

REFRESH ACTION

During standby, part "b" of Fig. 16-20, the charged capacitor loses charge. It must be refreshed (recharged) from time to time. That is, if the voltage across C1 in part "b" of Fig. 16-20 becomes too low, Q1 will turn off. When an attempt is made to read the cell, neither Q1 nor Q2 will be conducting. A fault results.

The read mode shown in part "c" of Fig. 16-20 offers a method of refreshing the cell. Both W0 and W1 are off, so READ-WRITE 0 and READ-WRITE 1 are both inactive (high). During the read, the charged capacitor, C1, is connected to READ-WRITE 1. If its voltage is below that on READ-WRITE 1 (about equal to Vdd), current will flow and recharge C1. See the series of small arrows. That is, the act of reading the cell refreshes the stored information.

Capacitors in dynamic memories are very small. They are often only about 0.05 picofarads (0.000,000,000,000,05 Farads, where a Farad is the unit used to measure capacitance). Such small capacitors drain down quickly, so dynamic memories must be refreshed about every 2 milliseconds (0.002 seconds).

It is unlikely that every cell in a memory will be read each and every 2 ms, so a refresh cycle must be provided. That is, every 2 ms, every cell in a dynamic RAM must be read. The resulting numbers are, of course, not output. That is not the purpose of refresh. Efficiency is lower, but data is not lost.

To speed the refresh process, cells are refreshed in groups. An entire memory can often be refreshed in about 40 microseconds (0.04 ms or 0.000,04 seconds).

Refresh can be done under computer control. At intervals, special instructions are inserted to accomplish refresh. This is obviously a disadvantage, since the computer must stop whatever it is doing to refresh the memory.

Special chips are available for controlling refresh. Also, some dynamic memories have built-in refresh circuits. Both of these solutions add to the initial cost of a computer, but they relieve the computer of the refresh burden.

Although refresh seems to be a time consuming task, it is easily accomplished in modern computers. Dynamic memory cells require little space on IC chips and dissipate little power. As a result, very large memories can be built when this technology is used, but the production yield is often low.

SIZE VERSUS SPEED FOR FET AND BIPOLAR TRANSISTOR MEMORIES

Size and speed are important memory characteristics. Size implies the number of words that can be stored. If memory is small, program complexity is limited. Also, available space for data may be insufficient.

Speed relates to how quickly numbers can be moved in and out of memory. Slow memories reduce overall computer performance.

Memory size and memory speed tend to be inversely related. Large memories tend to be slow. In small memories (where cell size and power consumption are not critical), bipolar transistors are often used. Bipolar memories are fast.

MOS FETs tend to be slower. However, they take less space on ICs and consume little power. This is especially true of the CMOS FETs and dynamic cells used in large memories. As a result, the rule that large memories tend to be slow is supported.

SUMMARY

Random-access memories are used to temporarily store programs, data, and results of computations. Individual memory cells must have read/write capabilities. As a result, such cells take the form of flip-flops and bistable multivibrators.

Memories based on bipolar transistors are very fast, but these transistors require considerable space on IC chips and dissipate large amounts of power. As a result, they tend to be used only in smaller memories.

Cells built using FETs are small and use less power. Additional improvement is possible when CMOS FET technology is used.

RAMs are classified as static and dynamic. Static memories retain data as long as power is applied. Data stored in dynamic memories will be lost if cells are not refreshed at specific intervals. However, the size and power requirements of dynamic cells are so small that they are extensively used in very large memories.

IMPORTANT TERMS

Bistable, Complementary MOS, Complementary symmetry, Dynamic memory, Leakage, Refresh, Static memory.

TEST YOUR KNOWLEDGE

1. RAMs are used for __________ storage.
2. Some RAMs require so little power that ________ backup can be used.
3. A control circuit for a RAM provides three modes

of operation: read, write, and ________.
4. Are flip-flops useful in large memories?
5. The term ________ implies two stable states.
6. Any lead of a ________ emitter group can control the transistor current.
7. A(n) ________ (diode, capacitor, FET, wheatstone bridge, battery, inverter, switch) can be a substitute for a resistor.
8. What does the abbreviation CMOS stand for?
9. When n-channel and p-channel MOS FETs are used in the same circuit, complementary ________ can result.
10. A p-channel FET is turned on only when its gate is ________ (positive, negative) with respect to its source.
11. A zero voltage is ________ (more, less) negative than a positive voltage.
12. CMOS FET cells use little ________.
13. A dynamic RAM uses ________ (resistors, capacitors, diodes) to store numbers.
14. A dynamic RAM should have its cells ________ every 2 ms.
15. Dynamic RAMs use little power in the ________ (write, read, standby) mode.
16. In a dynamic memory cell, no direct ________ is made to Vcc.
17. Capacitor electrodes in a MOS FET are ________ by glass (SiO_2).
18. To speed the refresh process, dynamic memory cells are refreshed in ________.
19. Are special chips available for controlling and accomplishing refresh?
20. Memory size and memory speed tend to be ________ (inversely, directly, "un-") related.

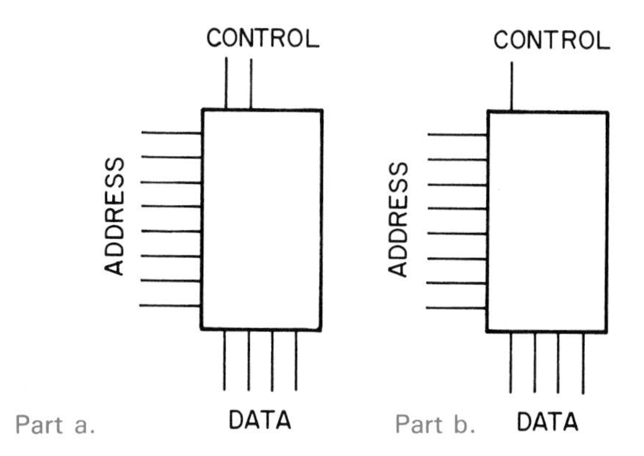

Fig. P16-1. A RAM and a ROM can be distinguished. Which memory has fewer leads?

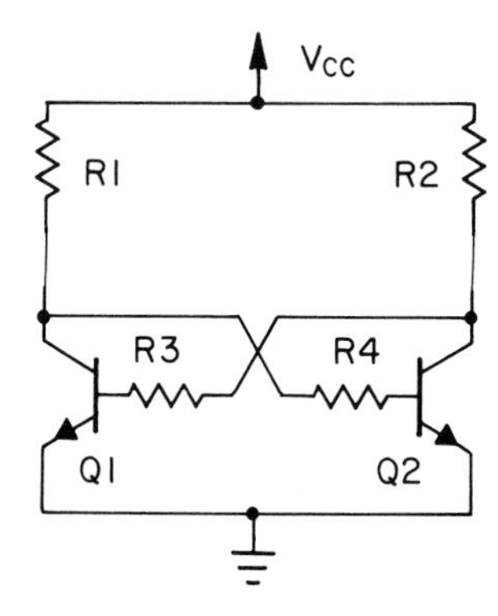

Fig. P16-2. Steps in analysis include finding current flow paths and voltages.

STUDY PROBLEMS

1. Fig. P16-1 shows a RAM and a ROM. Which is most likely the RAM?
2. Draw a schematic for a bipolar-transistor multivibrator. Label the components Q1, Q2, R1, R2, R3, and R4. Show the ground for the circuit and show Vcc.
3. Although they could be used as memory cells, data latches are seldom found in large memories. Why is this so? A person needs to choose from the following:
   1. They are far too slow.
   2. Data latches cannot have open-collector or three-state outputs.
   3. They take up too much room on a chip and dissipate too much power.
   4. They are subject to noise.
4. Copy the diagram in Fig. P16-2 onto a sheet of paper. Assume that Q2 is conducting. Use a series of heavy arrows to show the flow of current from Vcc, through the collector of the transistor, to ground. Also use a series of light arrows to show the current flow from Vcc, through the base of the transistor, to ground.
5. Add dots to the diagram of Problem 4 (Fig. P16-2) to show the signals at the two collectors. Use an open dot to show a signal near 0 volts; use a solid dot to show a positive voltage that might be used as a logic 1.
6. In the circuit in Fig. P16-3, a positive voltage has been applied to the base of Q1 (it is conducting). The resulting current is shown. When the switch is opened, which transistor will be conducting, Q2, Q3, or Q4? Note that this is a transistorized version of the loading method used in Fig. 16-6. The transistor, rather than a piece of wire, supplies the short to ground. Be aware that this circuit has active-low inputs.
7. Assume that Q1 in Fig. P16-4 is conducting (a

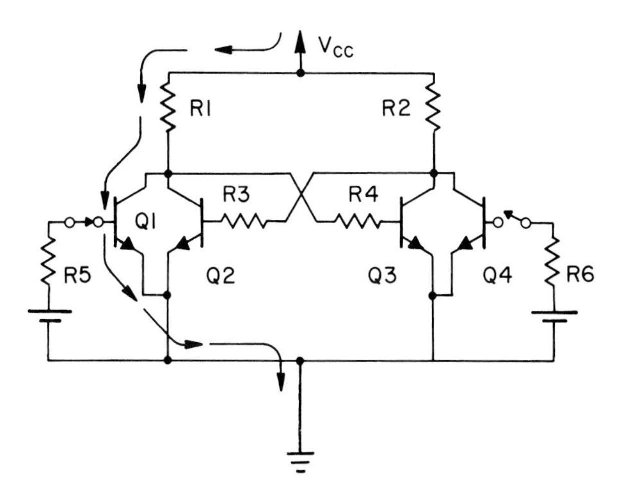

Fig. P16-3. A step in analysis is to find low and high voltage levels. A low voltage on the right can keep a transistor on the left from conducting, and vice versa.

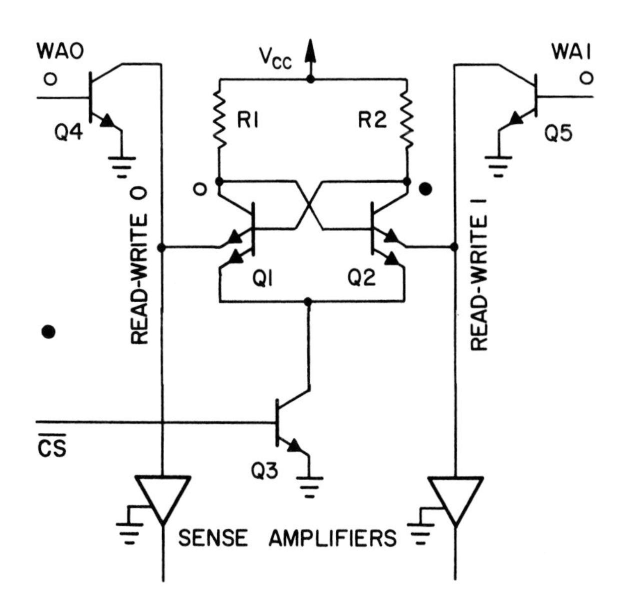

Fig. P16-4. A person can determine which of the 5 transistors in the memory cell is (are) conducting. A person can determine which mode the memory cell is in.

0 has been stored). For the applied signals, which of the following transistors are conducting: Q2, Q3, Q4, Q5? Also, what mode (standby, read, or write) is the displayed cell in?

8. Match the following with the cells shown in Fig. P16-5. A given cell may be used more than once. If more than one answer is possible, give only one.
   1. CMOS technology.
   2. Bipolar technology.
   3. Probably has the lowest standby power.
   4. Probably the fastest.
   5. Uses transistors as resistors.
9. Refer to circuit d from Problem 8 (Fig. P16-5). Which of the following best describes the physical location of the capacitors in dynamic memory cells?
   1. The capacitors are part of the FETs.
   2. The capacitors are fabricated on the chip next to each cell.
   3. The capacitors are mounted on the circuit board beside the chip.
10. Which of the following memories requires refresh?
   1. ROM.
   2. Static RAM.
   3. Dynamic RAM.

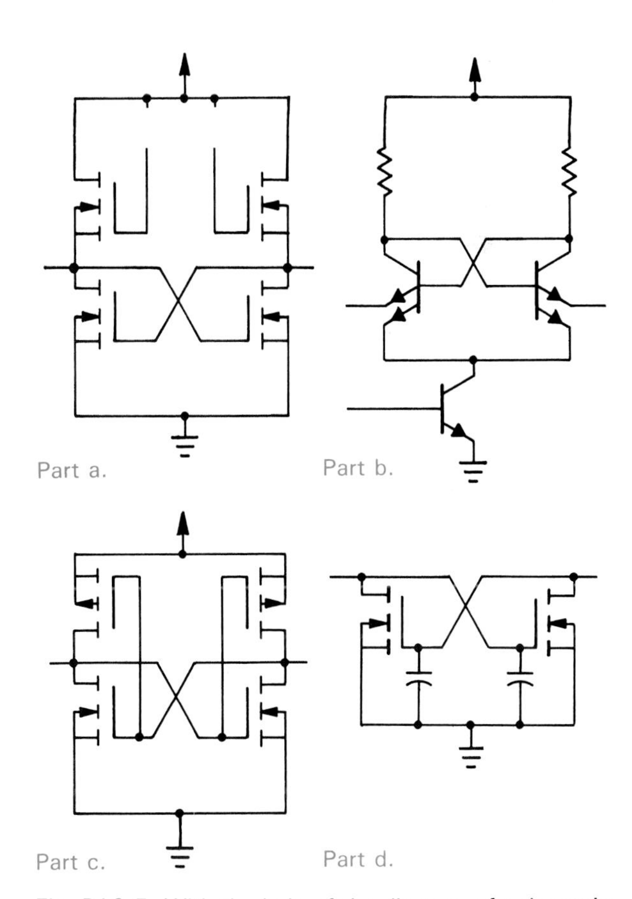

Fig. P16-5. With the help of the diagram of a dynamic memory cell, a person can recall the structure and/or locations of the capacitors.

This hard-disk drive is intended for use in mainframe computers. It has 14 disks, each 5.25″ in diameter. As a result, it can store 23 gigabytes. A gigabyte is a billion bytes. (Seagate)

Disk Storage

Chapter 17

LEARNING OBJECTIVES

After studying this chapter and completing lab assignments and study problems, you will be able to:

- List the factors that contribute to the strength of magnetic fields.
- Relate the terms *saturation, hysteresis,* and *residue magnetism to the magnetic curve of iron.*
- Describe three methods of obtaining relative motion between a wire and magnetic field to produce induced voltage.
- Describe the write and read actions of disk memories.
- List the do's and don'ts of disk care.
- Compare the storing of data on magnetic disks, compact-disk ROMs, writable compact disks, rewritable compact disks and tapes.
- Describe the advantages and uses of each storage method.

In RAMs, electric current and voltage are used to represent stored bits. Due to the properties of current and voltage, the characteristics of RAMs are tied to the nature of electricity. For example, RAMs are usually volatile. When the electrical power is removed, all of the stored data is lost.

When magnetism is used to store bits, memories with unique characteristics result. It is easy to produce nonvolatile magnetic memories. Magnetic read/write memories will be described in this chapter.

MAGNETISM

Electricity and magnetism are interrelated. Electric currents produce magnetic fields; magnetic fields can be used to produce electricity. To study magnetic memories, the relationships between electricity and magnetism must be understood. Some experiments help clarify principles of magnetism and electricity. Results can be easily duplicated and classified and new uses can be created.

MAGNETISM FROM ELECTRIC CURRENT

Every current-carrying conductor is surrounded by a magnetic field. The wire becomes an electromagnet. In Fig. 17-1, the presence of this field is indicated by the compass. Just as the compass aligns itself with the earth's magnetic field, this compass points in the direction of the field around this wire.

A number of factors determine the strength of a magnetic field. These include the following:

- Current.
- Coil size.
- Magnetic circuit.
- Magnetization curve.
- Hysteresis.

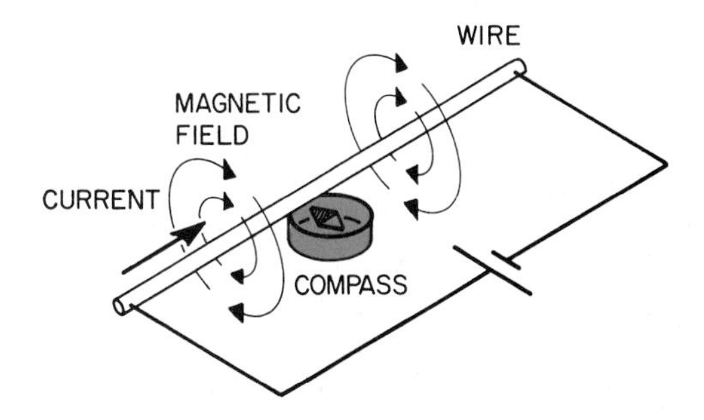

Fig. 17-1. A magnetic field surrounds a current-carrying wire.

RELATIONSHIP OF MAGNETIC FIELD STRENGTH TO CURRENT

The magnetic field strength near a wire is proportional to the current flow. The larger the current, the stronger the magnetic field. However, designers try to perform magnetic tasks with as little current as possible, for integrated circuits tend to be low-current devices. The next two sections describe methods for increasing the strength of magnetic fields without increasing current.

The direction of a magnetic field depends on the direction of the current that produced it. If the current is reversed, the magnetic field direction will be reversed.

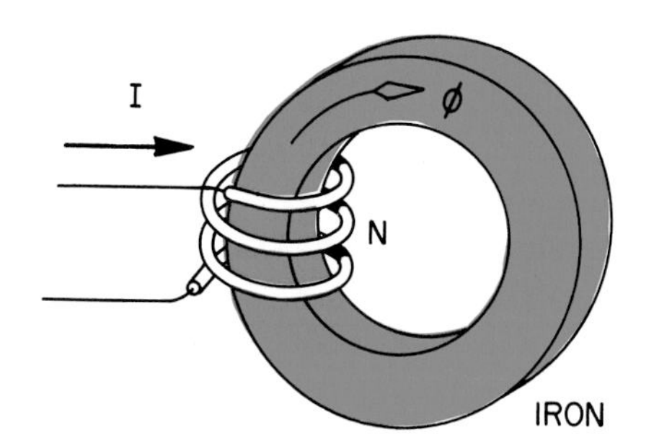

Fig. 17-3. Adding iron to the path of a magnetic field increases its strength.

COIL SIZE

The magnetic field strength can be increased by winding a current-carrying wire into a COIL or *SOLENOID.* A solenoid is a neatly wound coil of insulated wire. The wire must be insulated. The unit is an electromagnet. See Fig. 17-2. The magnetizing forces of the individual turns of wire are concentrated at the center of the coil. In addition to the current flow, the strength of the field depends on the number of turns in the coil. The force producing the field equals the product of N and I. N is the number of turns in the coil, and I is the current in amperes. NI is measured in ampere-turns.

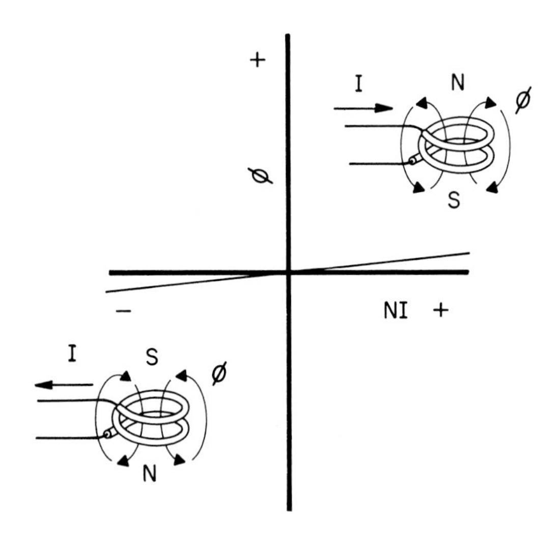

Fig. 17-4. Air has a straight, almost horizontal magnetization curve.

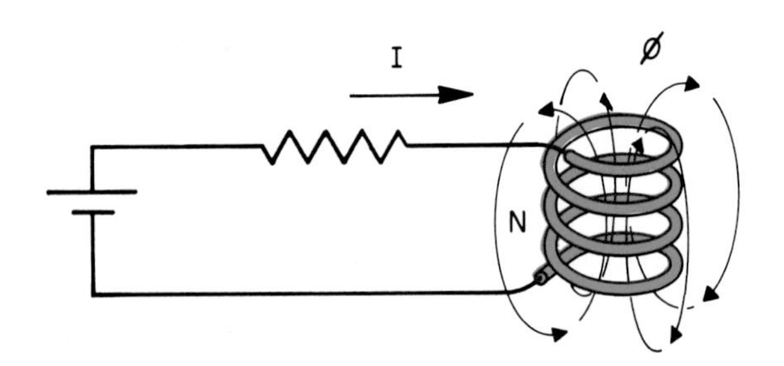

Fig. 17-2. Winding a wire into a coil increases the resulting magnetic field. Such a coil is called a solenoid.

MAGNETIC CIRCUIT

A stronger magnetic field can be produced by placing iron in the magnetic circuit. See Fig. 17-3. For a given number of ampere-turns, the magnetic field in iron can be many times that in air.

MAGNETIZATION CURVE

Fig. 17-4 shows a simplified magnetization curve for an air-core coil. The magnetizing force NI is plotted horizontally. The resulting magnetic field is plotted vertically.

The curve for air is a straight line. It is almost horizontal. That is, even when many ampere-turns of magnetizing force are applied, only a weak magnetic field results from the current.

Fig. 17-5 shows the curve for iron. Several features should be noted. These are:

1. Strength of field.
2. Saturation.
3. Hysteresis.
4. Residual magnetism.

The first of the features is important to save current and save space, as in the following:

1. Strength of field: For a given number of ampere-turns, the magnetic field in iron will be much stronger than it is in air. Compare the solid line with the dashed line in Fig. 17-5. Iron is used to produce smaller magnets than with air-core coils. Fields can be confined to a small space.
2. Saturation: For low values of NI, the field in iron increases rapidly. Beyond point B in Fig. 17-5, however, there is little increase. The iron is said

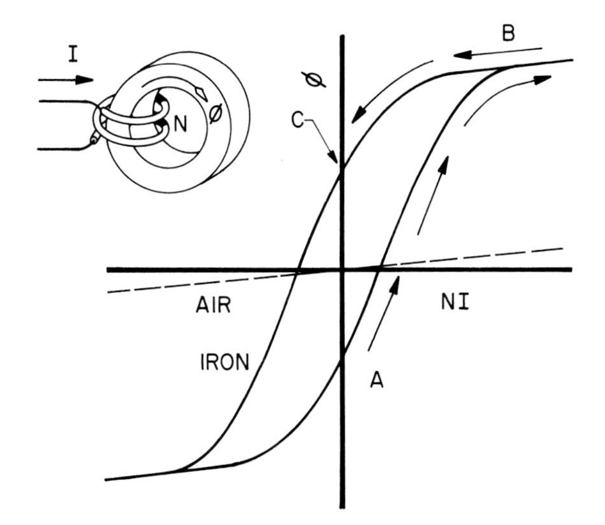

Fig. 17-5. The magnetization curve for iron displays hysteresis.

to be *SATURATED.* A further increase in magnetizing force produces little additional magnetism.
3. Hysteresis: The magnetizing curve for iron is not a single line. Rather, it follows one path when NI is increasing, and another path when NI is decreasing. This is called *HYSTERESIS.*
4. Residual magnetism: The series of arrows in Fig. 17-5 indicates the path followed when the current in the coil increases and then returns to zero. Note point C. When the magnetizing force was removed (NI returned to 0), magnetism remained in the iron. This is called *RESIDUAL MAGNETISM.* The iron acts like a permanent magnet.

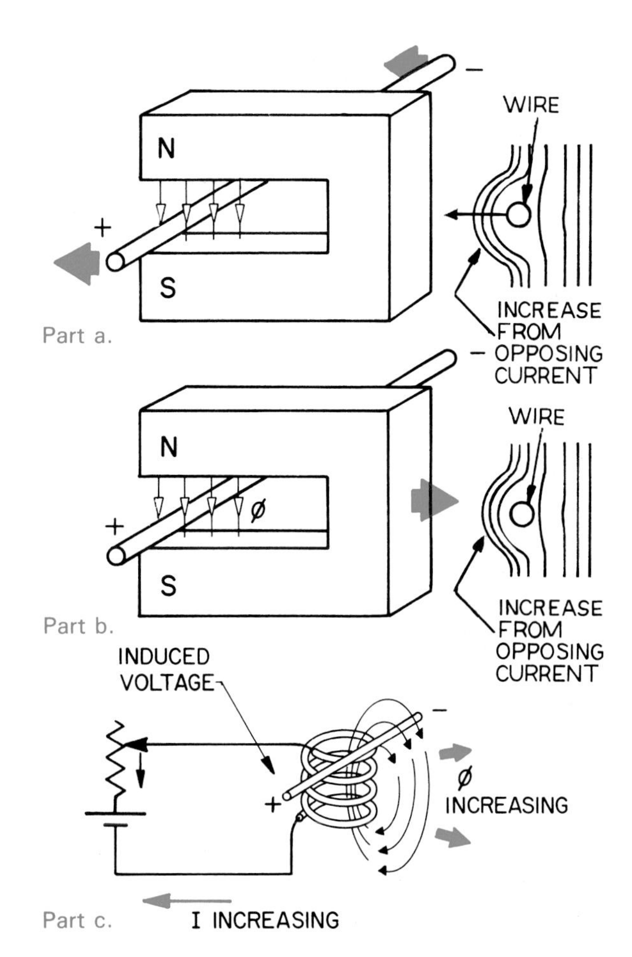

Fig. 17-6. For a voltage to be induced in a wire, there must be relative motion between it and the surrounding magnetic field. Three methods of producing that motion are shown.

INDUCED VOLTAGE

Just as electric currents produce magnetic fields, magnetism can be used to produce electricity. To do this, there must be relative motion between a conductor and a magnetic field. The effects of relative motion can be achieved in three ways:
1. Wire in motion.
2. Magnet in motion.
3. Changing field.

The first of the three topics suggests an experiment as follows:
1. Wire in motion: Part "a" of Fig. 17-6 shows a wire moving through a magnetic field. As it moves, it is said to cut magnetic lines. This induces voltage in the wire. INDUCE means to produce. If a voltmeter is across the ends of wire, voltage is detected. The stronger the field and the faster the wire moves, the greater the induced voltage.
2. Magnet in motion: In part "b" of Fig. 17-6 the wire is standing still, and the magnet is moving. Again, magnetic lines move past the wire, and a voltage is induced. Voltage is highest when the strongest part of the field is moving past the wire. This is usually near the poles. As an aside, this method is used to read magnetically stored data from tapes and disks. Small magnets on the tape or disk are moved past wires in the read head.
3. Changing field: Part "c" in Fig. 17-6 shows a third way of inducing voltage. Both the source of the magnetism (the coil) and the wire are fixed. Motion is obtained by increasing and decreasing the strength of the magnetic field. As the field grows or decays, its lines move past the wire and induce a voltage in the straight wire. Transformers use this method of inducing voltage. Their primaries produce changing fields that induce voltages in their

secondaries. In computers, core memories use this method of inducing voltage.

MAGNETISM AND DATA STORAGE

The nature of magnetism makes it highly useful in storing digital data. First, magnets are bipolar. A north pole might be assigned the value 1. A south pole could then represent a 0. Second, magnetic materials display residual magnetism. Data is stored as small permanent magnets. If power fails or a storage device it taken out of a computer, these magnets remain. Information is not lost, and data and programs can be stored magnetically for years. Third, electricity and magnetism interact. An electric current produces magnetism, and a changing magnetic field will induce a voltage in a wire. As a result, interfacing between electricity-based logic circuits and magnetic-based storage is relatively easy.

As with audio or video tapes, when new information is written in the same location as old information, the old information is lost.

Fig. 17-7. Access to the disk in this diskette is through rectangular holes in the upper and lower halves of the jacket. (3M Corporation)

FLOPPY-DISK STORAGE

Magnetic tape used to record sound and video signals has a coating of powdered iron oxide on its surface. A similar oxide coat is used on the disks used to store computer programs and data. Whereas a linear motion is used on tapes, a rotary motion is used on disks.

Magnetic disks come in many forms and sizes. However, the type called a diskette or 3.5 floppy is the most common. Adding "-ette" to the word disk suggests a small disk. The 3.5 refers to the outside dimensions of its square jacket. Billions are in use around the world.

The disk itself is thinner than a piece of paperhence the term floppy. It is so thin that it must be supported by a plastic jacket. See Fig. 17-7. Low-friction liners, or wiping fabrics, are placed on each side of the disk. These prevent scratches and stop the buildup of static electricity.

Disks are coated on both sides with magnetic oxide. Reading and writing on the disk is done through the rectangular windows in the top and bottom covers. The U-shaped metal part is a cover that protects the disk when the diskette is not in use.

WRITING ON A FLOPPY DISK

Fig. 17-8 shows a WRITE HEAD positioned on a floppy disk. The arrow indicates the motion of the floppy disk past the write head.

The head consists of a coil of wire (an electromagnet) and a magnetic circuit. When current flows into terminal A, a magnetic field, with the indicated polarity, results. Reversing the current direction reverses this magnetic field.

Note the air gap in the magnetic circuit in Fig. 17-8. Most of the magnetic field crosses the gap directly. However, some fringes (extends outside the gap) and enters the oxide coat. As the disk passes under the head, the fringing field drives the oxide to saturation. As each segment of the disk leaves the influence of the head, some magnetism is retained. A series of small magnets are laid end to end on the disk. The polarity of each depends on the direction of current flow during the time that it was under the head. See part "c" of Fig. 17-8. Magnets with one polarity represent stored 1s; those with the opposite polarity represent 0s.

READING A FLOPPY DISK

A read head looks much like a write head. In some machines, one head is used for both read and write. Part "a" of Fig. 17-9 shows a stored bit in the oxide coat passing under a read head. Part of the magnetic field is bypassed through the head and its coil.

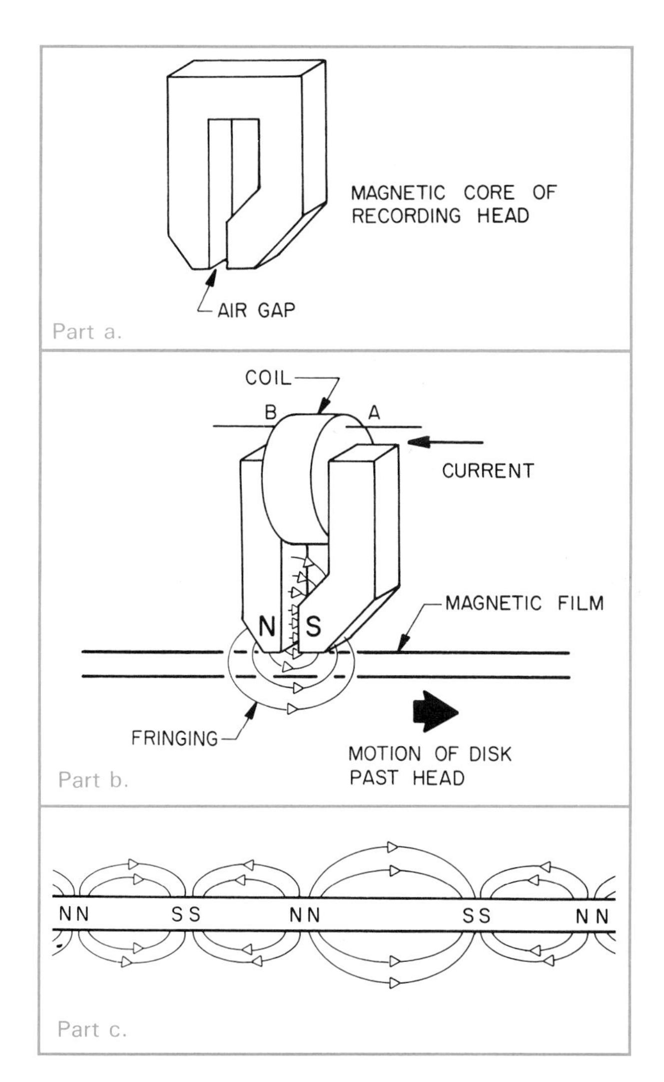

Fig. 17-8. The fringing magnetic field from the gap in a recording head places small magnets in the film on floppy disks. These stored magnets represent 1s and 0s.

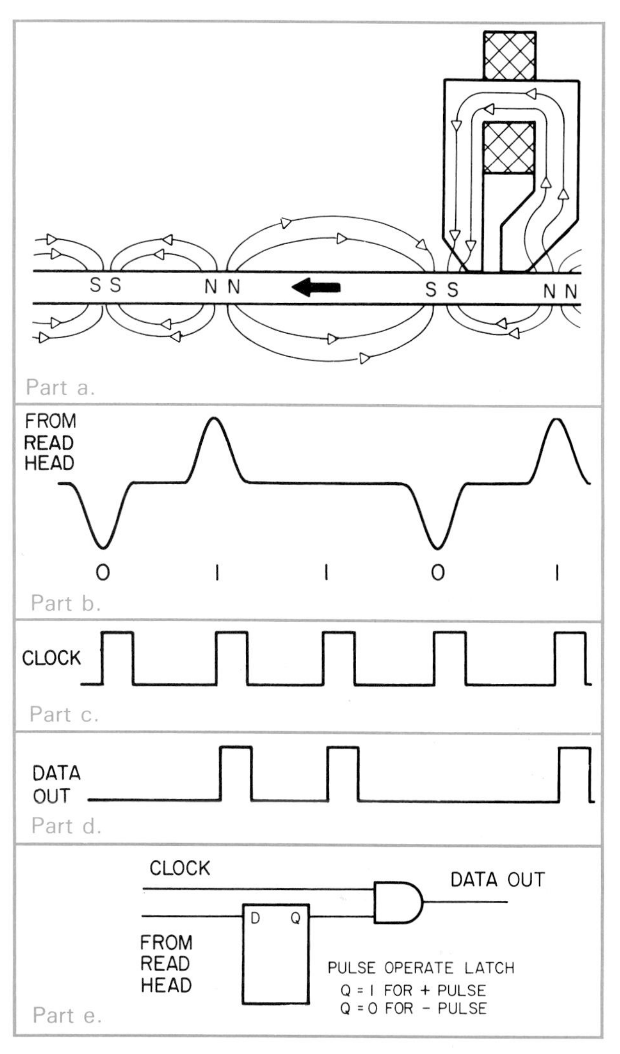

Fig. 17-9. When a pole of a stored magnet passes under a read head, a pulse is output. A positive pulse represents a stored 1; a negative pulse represents a stored 0. A region with no pulse (a horizontal line) represents a retention of the previous value.

Remember, only changing magnetic fields induce voltage. As a result, voltage is induced in the read coil only at the ends of the stored magnets. See graph b. At the end of each magnet, the direction of the magnetic field reverses.

The induced-voltage pulses are compared with a clock signal in circuits like the one at e. The read operation results in a series of 1s and 0s in binary such as those shown at d.

Stored numbers are not damaged by the read process. A disk can be read again and again. Nor does loss of power or the removal of a disk from a machine alter stored data. That is, disk are nondestructively read and nonvolatile. Disks are sometimes said to be "nondestructive read."

FLOPPY DISK DRIVES

The primary advantages of floppy disks are their simplicity and portability. Because they are uncomplicated, they are inexpensive and readily available. They are described as portable because they can be removed from drives. This portability dictates the design of these disks and their drives. Diskettes can be stored away from the computer and be used to transfer information from one computer to another.

Floppy disk drives are much like record players. They have motors for turning the disk and arms for

positioning the read/write heads. See Fig. 17-10. Parts of floppy disk drives include:
- Drive motor and spindle.
- Heads and head positioning mechanism.
- Index detector.
- Write-protect mechanism.

TURNING THE DISK

Diskette drives are often placed within computers—even laptop computers. As a result, size is important—the smaller, the better.

One of the largest parts in a drive is the motor that spins the disk. Fig. 17-10 shows one design for a disk drive assembly. While the outline of a diskette is shown, the actual disk is not depicted. The disk in the drawing is the motor. This motor is less than a quarter of an inch thick. In addition, it might be described as inverted. In most motors, the stator (the portion that does not turn), is on the outside and the rotor, that part that turns, is on the inside. The rotor on this motor is on the outside. The stator and the coils that produce the magnetic fields that cause the motor to turn are hidden underneath the disk-shaped rotor.

Each disk has a thin metal hub. Torque from the motor is delivered through two pins (B and C in Fig. 17-10) to that hub. Because there are so many concentric tracks on a disk, it must be accurately positioned.

HEAD POSITIONING

The heads that read, write, and erase the data stored magnetically on the disk are shown at A in Fig. 17-10. They have access to the disk through rectangular openings in the top and bottom of the cartridge. When a diskette is inserted into a drive, the protective cover is automatically moved to one side to expose the disk.

Because data is stored magnetically, tracks are not visible. However, concentric rings may be seen on the surfaces of some older disks. These rings are scratches, not data tracks.

There are 135 tracks per inch on a conventional diskette and 2490 tracks per inch on some advanced disks. As a result, heads must be accurately positioned. A STEPPER MOTOR is used to move the heads from track to track. When pulsed, such motors turn a fixed number of degrees. The angular motion

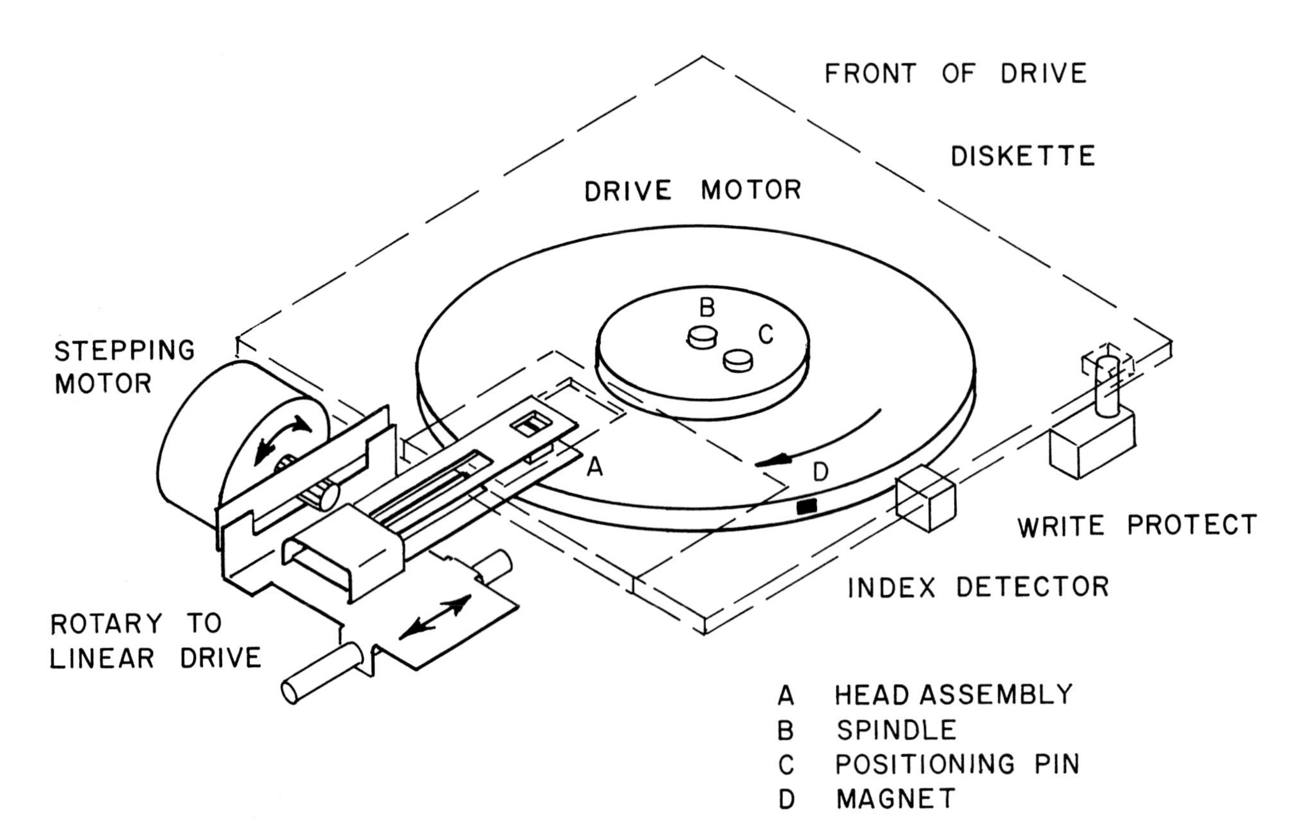

Fig. 17-10. In addition to the mechanical assemblies shown, disk drives contain complex electronic circuits. These control the storage and retrieval of data.

of the motor is converted to linear motion by the rotary-to-linear drive mechanism.

INDEX DETECTOR

On each track, data is organized into SECTORS. A sector is the smallest block of data that can be read or written. To alter a single word, the entire sector containing that word must be read from the disk. That block of data is then sent to the computer where the specific word is changed. The block is then returned to the disk when the SAVE function is used.

To read a given block of data, its track and sector must be specified. Positioning the head over the proper track is not difficult. Locating the proper sector on a spinning track is a problem. It is solved by using an INDEX DETECTOR. In the drive shown in Fig. 17-10, there is a magnet on the side of the drive motor at D. Nearby is the index detector. It contains a small coil of wire. Each time the magnet passes, a small voltage is induced in that coil. The resulting pulse indicates when the first sector will be at the heads.

Identifying numbers must be recorded on the disk at the beginning and end of individual sectors. This is called FORMATTING. Pre-formatted diskettes can be purchased. When unformatted disks are used, this process can be accomplished by the computer.

WRITE PROTECT

Like ROMs, some floppy disks are meant to be read only. To prevent the writing of data on such disks, diskettes have a WRITE-PROTECT FEATURE. A rectangular hole and sliding tab are provided in the upper right of the cartridge for this purpose. If the slide blocks the hole, both reading and writing are permitted. If the hole is open, only the reading of the disk is possible. Diskette drives have either a mechanical switch or a photo detector to determine if the hole is open or closed.

ADVANTAGES OF DISKETTES

If any storage methods can be described as standard, it is the 3.5 floppy disk. Almost all computers are equipped to read and write on this disk. As a result, diskettes are widely used to exchange information between computers.

Although other forms of storage offer far greater capacity than diskettes, a single disk can store a reasonable amount of data. A standard DS, HD (double-sided, high-density) diskette can store 1.44 megabytes (1.44 million bytes) of data. Fig. 17-11 shows a diskette called the LS-120. The use of ad-

Fig. 17-11. The capacity of this new disk is 80 times that of a DS, HD disk. (3M Corporation)

vanced technology allows it to store 120 megabytes. While a drive designed to use this new disk is required, such drives are backward-compatible with standard diskettes.

Diskettes are easy to store. Due to their protective jackets, they require little special handling. They can be used in offices, in schools, or on factory floors.

CARE OF FLOPPY DISKS

If treated with care, floppy disks are highly reliable. Error rates are remarkably low, and a disk can be used for years. However, faults do occur, so BACKUP is a must. That is, programs and data stored on a given disk should be copied on separated disks and stored in a safe place. At home, the loss of stored information can be inconvenient. In industry it can be costly and frustrating, for it may be difficult or impossible to replace lost data.

The thinness of the oxide coat and the density of data contribute to disk care problems. Thousands of tiny magnets are stored along each inch of track. As a result, a scratch, piece of lint, or a line from a finger print can hide many bits. Therefore, the disk must be protected from dust, dirt, smoke, and fingerprints. The metal slide that covers the access hole to the disk provides considerable protection. The temptation to move the metal slide aside and touch the disk should always be avoided.

While low temperatures are not as damaging as high temperatures, disks should be kept in the range of 50°F to 140°F. Stored information can also be lost if a disk is exposed to stray magnetic fields.

DISKETTES AS BACKUP

Random-access memory is VOLATILE. When power fails or is inadvertently turned off, all stored information is lost. Disk storage is NONVOLATILE. Stored information is not lost when power fails. The magnetic material on disks is selected to have a high residual magnetism. That is, information is stored as permanent magnets in the oxide coat. Unless a disk is mistreated or accidentally rewritten or erased, stored information is secure for the life of the disk.

In an industrial situation, it is essential that backup be provided for data and programs stored in RAM. One of the easiest ways to do this is to save such information on diskettes. Other methods of long-term storage will be considered later in this chapter, but for hour-to-hour backup, the diskette is very effective. It is better to develop the habit of using the SAVE function every fifteen minutes or so *before* you lose data and learn the hard way.

Some industrial computers used for control purposes, provide battery backup for RAM memory. See Fig. 17-12. Such batteries are not capable of operating the whole circuit, but they can power RAM until power is restored.

Fig. 17-12. The Lithium battery at the lower right of this board supplies battery backup to the RAM in this programmable controller. (Sylvia Control Systems)

HARD-DISK DRIVES

Floppy disks are thin and flexible; hard disks are thick and rigid. Because the hard disk in Fig. 17-13 does not need a jacket for support, it is visible in this photograph of a hard-disk drive. Most hard disks cannot be removed from their drives. They are a permanent part of the drive and cannot be transferred to another drive.

The drive shown is intended for use in a very small computer. Its disk is 2.5″ in diameter, and the whole drive is only 1/2″ high. Yet, it can store 256 megabytes. Because a hard disk is rigid and fixed to its drive spindle, it can spin faster than a floppy disk. Speeds of 7,200 rpm are not uncommon. Because the disk turns rapidly, a given point on a track comes under the read/write head quickly. This shortens ACCESS TIME. That is, it takes less time to find a given byte.

But the greatest advantage of a hard disk is the amount of data that can be stored. Capacities near ten gigabytes are common. A gigabyte is a billion bytes. One of the ways of increasing storage capacity involves placing the concentric tracks closer together. The rigidity and permanent attachment of a hard disk to its spindle contributes to the accuracy with which tracks can be laid down.

With closely spaced tracks, heads must be positioned with great accuracy. One approach to this involves simplifying the head-positioning mechanism. Refer to Fig. 17-13. Here, a pivoted arm (rather than straight-line motions used on floppy disk drives) is used.

Fig. 17-13. Because the hard disk in this drive does not need a jacket, the disk is visible. (Quantum Corporation)

The drive in Fig. 17-13 has only one disk. To increase storage capacity, 2, 3, or more disks can be used. For example, the drive in Fig. 17-14 has three, 3.5″ diameter disks. Each has two read/write heads (one for each surface). It is intended for use in personal computers, so small size was important, yet it can store 3 gigabytes. Very large capacities are possible with multiple-disk drives.

Some very large hard drives are designed so the stack of disks can be removed for storage. Being able to replace one stack of disks with another results in essentially unlimited storage capacity.

While hard disks are very reliable, it is still important to backup stored information. The high capacity of a hard disk can present a problem when diskettes are used to backup information. Many diskettes are required to backup all of the data on one hard disk because of their relatively low capacity. Other backup methods will be described later in this chapter.

Because a hard disk is fixed in its drive, it cannot be touched in normal use. Its temperature is the same as the equipment in which it is installed. Other than preventing the overheating of the equipment, the care for a hard disk is minimal. However, when doing maintenance on a hard-disk drive, care must be taken to keep the disks dust free. The disks must not be touched.

COMPACT DISKS

Early magnetic tapes were developed to record audio signals. Similar methods were then used to store digital information on tapes and disks. In a like manner, compact disks (CDs) were originally audio devices but are now used for digital storage as well. Fig. 17-15 shows such a disk ready to be inserted into a drive. This is a CD-ROM. The ROM portion of the description implies that it is a read-only device.

While CD-ROM technology is very similar to that of audio CDs, there is a major difference in the error rate. An occasional error in the reading of an audio CD would go unnoticed. A similar error in the reading of a stored computer program could cause the system to fail. As a result, elaborate error detection is built into CD-ROM systems.

Unlike floppy- or hard-disk storage, compact disks are not magnetic. Rather, digital information is placed on them mechanically. Super small PITS are pressed into the surface of a disk. Unlike the concentric tracks of floppy and hard disks, pits on a CD-ROM are arranged in a long, thin spiral up to seven miles long. That spiral can contain over 20,000 coils on a 120 mm disk (about 4 3/4″ in diameter). Spaces between the pits along the spiral are called LANDS. See Fig. 17-16. It is the pits and lands that store 1s and 0s. Disks are transparent, but a reflective coat of aluminum is applied to the surface that contains the pits.

To read a disk, it is placed in a CD-ROM drive and spun at high speeds. A beam of light is focused through the disk onto the bottom side of the mirror-like coating. This beam must be remarkably fine. See Fig. 17-17. One micron is 0.000.001 meters. One micron is to one inch as one inch is to 0.4 of a mile.

A low-power, semiconductor laser is used. This laser looks much like a light-emitting diode. Although its power may be only 5 milliwatts, it still presents a danger to the eyes. Appropriate precautions must be taken when the cover is removed from a CD-ROM drive.

Fig. 17-14. This hard-disk drive contains three disks and can store up to 3 gigabytes.

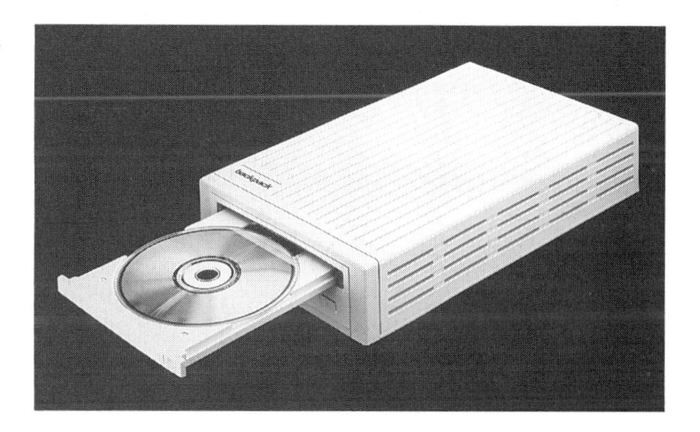

Fig. 17-15. This CD-ROM drive has a front-loading motorized tray. The CD is placed directly in the tray and no cartridge is needed. (MicroSolutions)

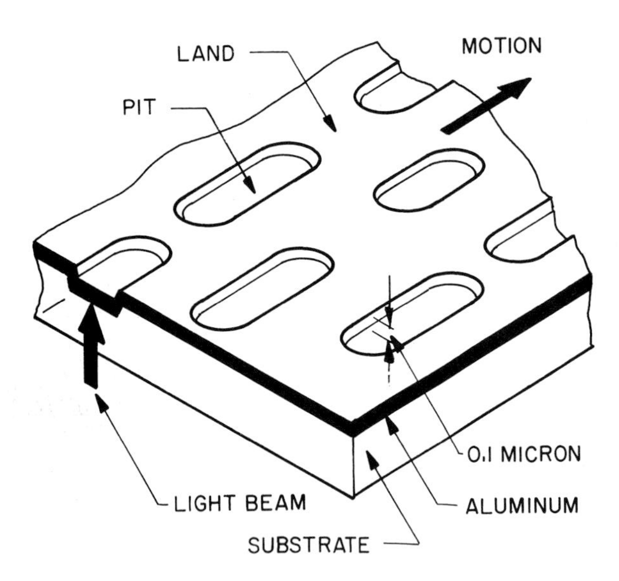

Fig. 17-16. The pits on this CD-ROM are only about 0.1 microns deep. This is less than the wavelength of the light that is used to read the disk.

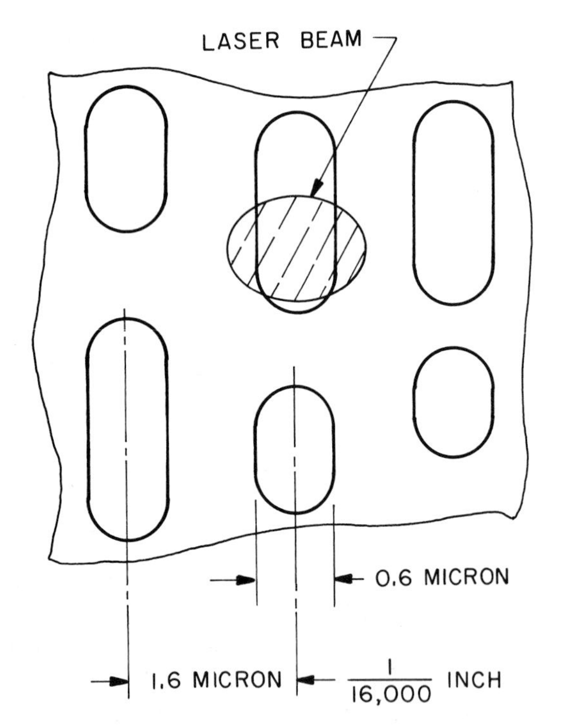

Fig. 17-17. The distance between tracks on a CD-ROM is so small that over 14,000 tracks can be placed in one radial inch.

A photodetector observes differences in the light reflected from the lands and pits. These differences are decoded to determine the 1s and 0s that make up the stored information.

A CD-ROM can house over 580 megabytes of digital data. This is hundreds of times more than the capacity of a standard floppy disk and is near to that of a very large hard disk. This large capacity allows audio and visual material to be stored on CD-ROMs along with programs and data. However, a moving picture contains huge amounts of data, so even this large capacity is not sufficient to store extensive video signals. There are, however, ways to compress signals to allow at least some video to be placed on CD-ROMs.

Information placed on a CD-ROM when it is manufactured cannot be altered. If the information is virus-free when placed on the disk, it cannot be tampered with.

CD-ROMs are random access, so computers do not have to search through the whole spiral to find a given file. Access time is reduced in present-day drives by turning disks at higher speeds. The top speed of a quad-speed drive is 2,120 rpm. Such a drive is often called an X4 drive. Devices designated as X8 are likely to be the standard in the near future.

Light is used to read CDs, so there is no physical contact with the disk and no wear. Because they are read-only, CD-ROMs cannot be accidentally erased or tampered with. As a result, they have a useful life of many years.

When it comes to dirt and scratches, audio compact disks are much more forgiving than traditional phonograph records. As a result, audio CDs gained the reputation of being very rugged. This is not true however, as the presence of dirt and scratches can lower playback quality. But CD-ROMs are even more sensitive to damage. The loss of a few bits in an audio signal might not be noticed. The loss of those same bits might make a major difference in the results of a computer run.

Like floppy disks, CD-ROMs are sensitive to high temperatures. Unlike floppy disks, CD-ROMs are not permanently mounted in jackets, so they must be protected from dirt, dust, and fingerprints. They should be kept in their original boxes and handled only by their edges and center holes. Commercial cleaning methods are available.

WRITABLE COMPACT DISKS

While CD-ROMs are read-only, disks that might be described as CD-PROMs are available. Often called CD-R (CD-Recordable) disks, these allow the transfer of digital data directly from a computer to permanent storage. A CD-R looks much like a compact disk. See Fig. 17-18. CD-R disks can be read on

Fig. 17-18. CD-ROMs and CD-Rs are very similar in appearance to audio disks. (3M Corporation)

standard CD-ROM drives. CD-R technology allows the recording of text, audio, and limited video.

The note at the bottom of the disk in Fig. 17-18 reads, "Use a felt tip permanent pen on printed surface only." This emphasizes the point that the laser light used to read this disk shines on the other side. Marks on that side would interfere with the reading of the disk.

One form of CD-R disk is made of a polymer plastic containing a dye that boils when struck by a laser beam. The boiling produces data pits in the reflective substrate. Once recorded, data on a CD-R cannot be altered.

A special drive is needed to record on CD-R disks. They can, however, be read on most standard CD-ROM drives.

REWRITABLE COMPACT DISKS

A CD-ROM is basically a read-only memory. The CD-R is similar to programmable read-only memories. CDs that function like a read-write memory (like a RAM) complete the set. Not only can data on such disks be stored and read, previously stored data can be changed.

These disks are often described as rewritable optical disks. One approach to such a device is described as PD. Depending on the manufacture, PD might stand for Phase-Change Dual, Phase-Change Device, or Phase-Write Dual.

In its normal state, the recording layer of a PD disk is said to be amorphous. That is, it has no specific pattern; it lacks a crystalline structure. To record 1s and 0s, a laser heats tiny spots to the transition temperature and small crystalline areas form in the material.

A lower-power laser, as in CD-ROM drives, is used to read stored information. With the aid of a photodetector, differences in reflectivity between the amorphous material and the crystalline spots is detected.

To prepare the disk for the storing of new data over old, a third level of laser power is used. Here the power is sufficient to locally melt the recording material. Upon cooling, it returns to the amorphous state, and the recording process can be repeated.

The drives used to store data on PDs can read both PDs and CD-ROMs. The disk in Fig. 17-19 is described as a PD 650. This means that it can store 650 megabytes—more than many hard drives. Unlike hard disks, PD disks are not fixed to drive spindles and can be removed. In effect, this results in almost unlimited storage capacity since additional disks can be inserted when a disk is filled.

Fig. 17-19. The rewritable disks can store 650 megabytes. This is more than the storage capacity of many hard-disk drives. (3M Corporation)

Often PD disks cannot be read on standard CD-ROM drives. However, a newly developed form of rewritable disk may solve this problem. It is called Compact Disk Erasable (CD-E). It uses the Phase-Change technology of PD, but it is intended to be readable on CD-ROM drives. The storage capacity of CD-E may be so high that it will be practical to record video programming on such disks.

TAPE STORAGE

Everyone is familiar with audio and video tapes. Digital data can be recorded on tape in a similar manner. The main advantage of tape storage is its large capacity. Fig. 17-20 shows a minicartridge that can store 120 megabytes. While this tape cartridge is only a few inches long, it can store the data from 80 standard diskettes. Slightly larger tape cartridges can store 525 megabytes, and capacities in the gigabyte range are common. However, these large storage capacities are being challenged by CD-R and PD disks.

The major disadvantage of tape storage is that it is slow. Data is put on tape sequentially and blocks of data are stored one after the other. To retrieve a given block, the tape must be forwarded or rewound until the desired location is reached. This is very time consuming. To reduce the search time, tapes have many parallel tracks and are capable of switching from one track to another, yet they are still slower than other magnetic memories.

Based on their high capacity but slow retrieval rates, tapes are used primarily for the long-term storage of large amounts of data. That is, data that is not presently in use but may be important for future reference.This type of storage is often referred to as archival storage.

SUMMARY

Access times for tapes, magnetic disks, and compact disks are much longer than that of RAM memory. As a result, data and programming are usually transferred out of storage and placed in RAM before a program is run. If this were not done, the execution of the program would have to stop to access information stored on disks or tape. This would slow the operation of the computer.

However, tapes, magnetic disks, and compact disks have three advantages:

- They can store vast amounts of data.
- They are nonvolatile.
- With the exception of hard disks, they can be stored outside the computer.

Fig. 17-20. Tapes are often used for backup and archival storage. This mini-cartridge can store 120 megabytes. (3M Corporation)

IMPORTANT TERMS

Electromagnetic, Magnetization curve, Saturation, Hysteresis, Residual magnetism, Induced voltage, Diskette, Write protect, Index detector, Stepper motor, Hard disk, CD-ROM, Writable CD, Phase-Change, Rewritable CD.

TEST YOUR KNOWLEDGE

1. Every electric current produces a __________ field.
2. When an electric current flows in a straight wire, the resulting magnetic field __________ (surrounds, is parallal to) that wire.
3. If the current is reversed, the magnetic field near the wire will __________ (reverse, remain in the same) direction.
4. If current-carrying wire is wound in the form of a coil, the strength of the resulting magnetic field will __________ (increase, decrease, remain the same).
5. Magnetizing force, NI, is measured in __________ __________.
6. If iron is placed within a current-carrying coil, the strength of the magnetic field will __________ (increase, decrease, remain the same).
7. The graph of the strength of a magnetic field is plotted against the magnetizing force, NI is called a __________ __________.
8. When more and more ampere-turns are applied to a piece of iron, the strength of the magnetic field will reach the point where it stops increasing. This is called __________.
9. The magnetization curve for iron is a loop rather than a single line. Such curves are said to display __________.
10. Even when the magnetizing force is removed from a piece of iron, some magnetism remains. This is called __________ __________.
11. When a wire moves through a magnetic field, an electric __________ (voltage, current) is induced in the wire.
12. What is the other name for a 3.5″ floppy disk?
13. The coating used on disks to allow them to store data magnetically is iron __________.
14. The small, permanent magnets that store 1s and 0s on a floppy disk are __________ (parallel, perpendicular) to the surface of the disk.
15. Liners within the jackets of floppy disks prevent scratches and stop the build up of __________ __________.
16. The motor that moves the read/write head to the proper track is called a __________ motor.
17. The write head in a floppy-disk drive is a(n) __________ (photocell, laser, electromagnet, stylus).
18. Concentric rings may appear on the surfaces of old floppy disks. These are __________, not data tracks.
19. The write-protect feature prevents __________ a floppy disk.
20. In general, hard disks __________ (can, cannot) be removed from their drives.
21. Of the two, a __________ (hard, floppy) disk generally has the greater storage capacity.
22. Data on CD-ROMs is read __________ (optically, magnetically).
23. A CD-R might be described as a __________ (ROM, PROM, RAM) memory.
24. Of the two, a __________ (floppy disk, PD) has greater storage capacity.
25. Of the two, a __________ (floppy disk, tape cartridge) is most likely used for archival storage.

STUDY PROBLEMS

1. At which point in Fig. P17-1 is the magnetic field likely to be the strongest?

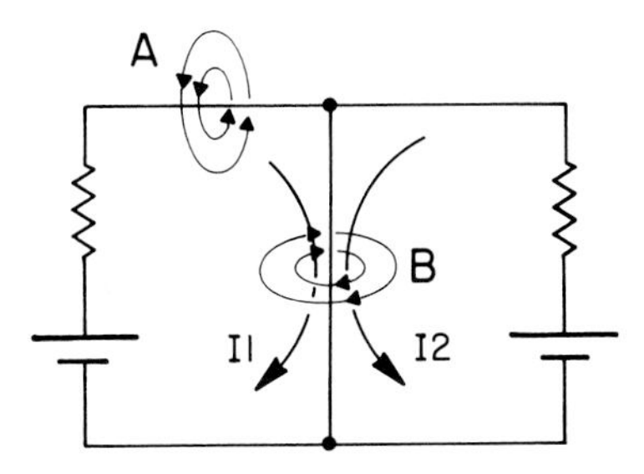

Fig. P17-1. Strengths of magnetic fields can be determined and compared.

2. Repeat Problem 1 for the 3 labeled parts of the circuit shown in Fig. P17-2.

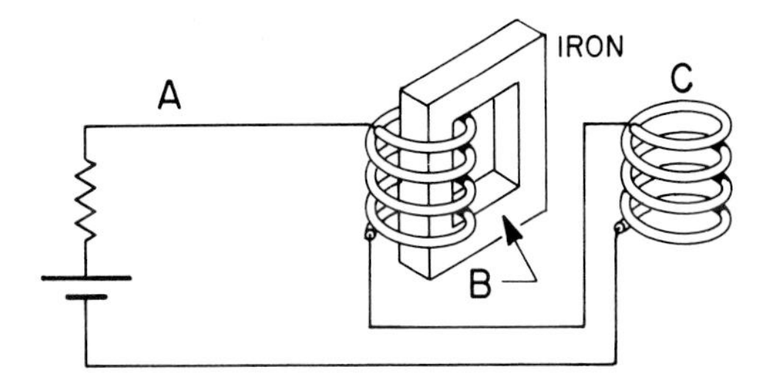

Fig. P17-2. The magnetic field strength can be determined and compared for three points.

3. Trace the magnetization curve in Fig. P17-3 onto a piece of paper and label its axes ϕ – resulting (total) magnetism and NI – ampere-turns.
4. On the curve traced for Problem 3 (Fig. P17-3), place the letters A and B to indicate the meaning of the following terms:
   a. Saturation
   b. Residual Magnetism
5. Draw a dashed line on the curve from Problem 3 (Fig. P17-3) to represent the magnetization curve of air.

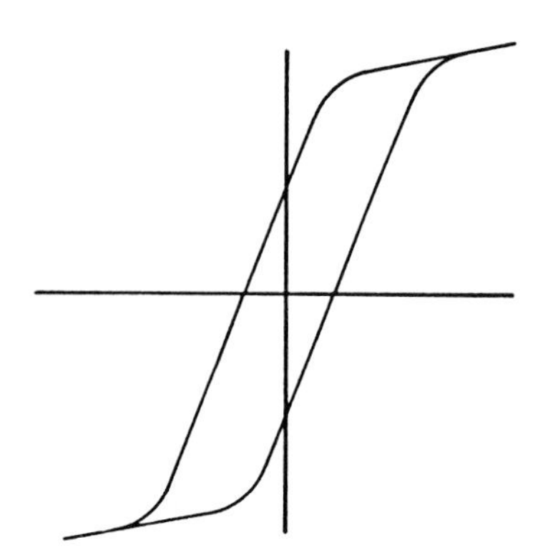

Fig. P17-3. Properties of a magnetization curve can be identified and labeled.

6. For NI = 2, which curve in Fig. P17-4 represents the greater magnetic field?
7. Which material represented by the curves for Problem 6 (Fig. P17-4) has the higher residual magnetism?
8. Describe three ways of obtaining relative motion between a wire and a magnetic field to induce a voltage into the wire.

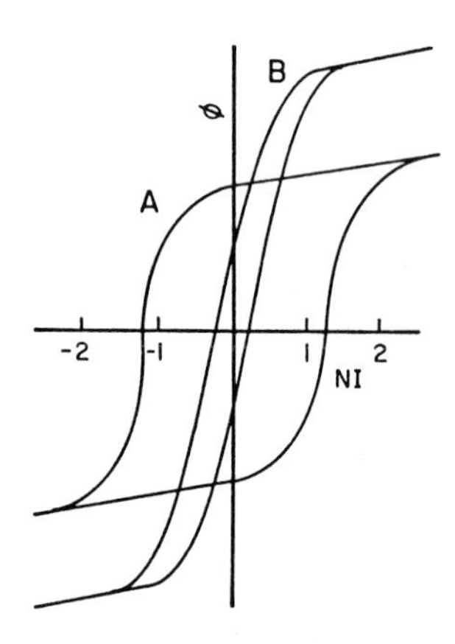

Fig. P17-4. The magnetic field strength at several values of NI can be determined from the curve which has hysteresis.

9. Wires A and B in Fig. P17-5 are within the magnetic fields produced by the currents in the adjacent wires. Which wire is likely to have higher induced voltage?

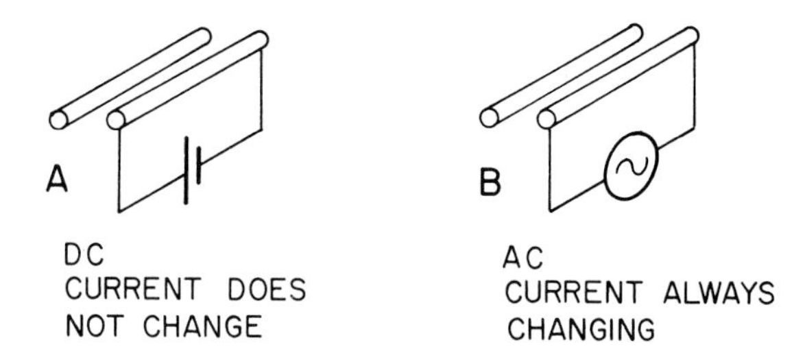

Fig. P17-5. The wire with the greater induced voltage can be determined.

10. Which write head in Fig. P17-6 represents the standard method of writing on disks?

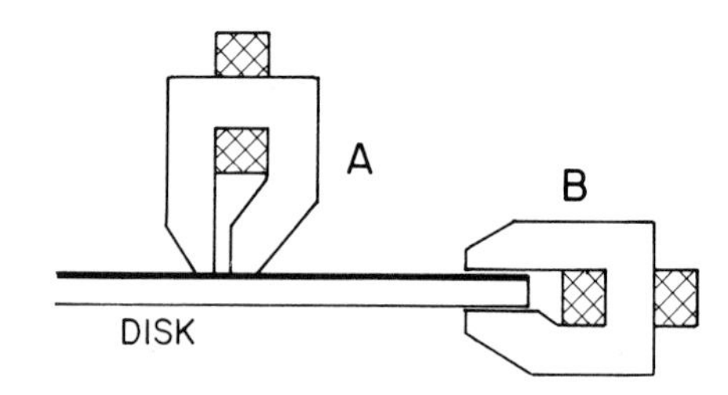

Fig. P17-6. The proper orientation of a write head for a disk can be determined.

11. At the instant shown in Fig. P17-7, which read head is likely to have the higher induced voltage?

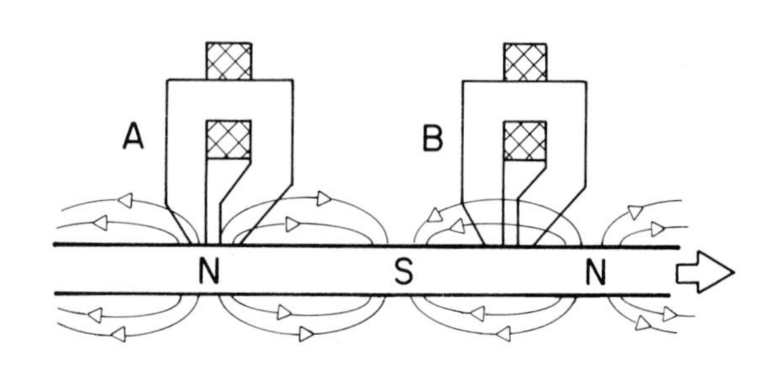

Fig. P17-7. The size of induced voltages in a read head can be determined from the positions.

12. On a floppy disk, which is the most likely material used for storing 1s and 0s?
    a. Silicon
    b. Polycarbonate
    c. Iron oxide
13. Which has the shorter access time?
    a. RAM
    b. Floppy disk
14. Disk storage is volatile. (True or False?)
15. It is okay to touch the surface of a floppy disk if you wipe it with a soft cloth afterwards. (True or False?)
16. In a floppy-disk drive, an index detector is used to determine ________.
    a. the rotational position of the disk
    b. the track over which the heads are located
    c. whether a disk is in the drive
17. Which is likely to have the higher storage capacity?
    a. Floppy disk
    b. Hard disk
18. Which is likely to have the larger number of tracks per radial inch of surface?
    a. Floppy disk
    b. Hard disk
19. Arms that carry read/write heads are shown in Fig. P17-8. Which is most likely from a hard-disk drive?

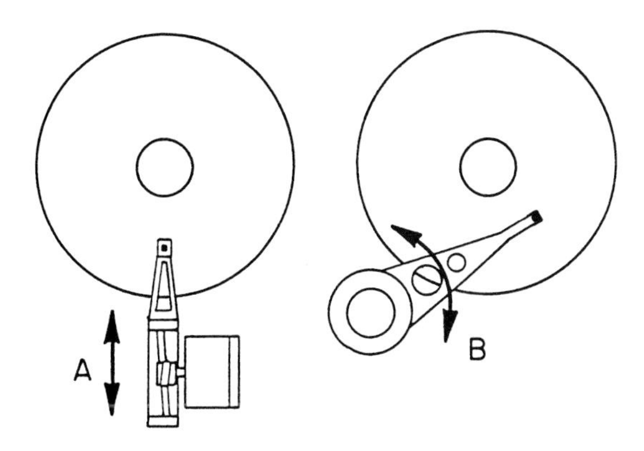

Fig. P17-8. Two arm styles are shown.

20. A hard-disk drive contains two disks. On how many surfaces can data be stored?
21. Assume you have purchased a new program for your computer. Of the two types of disks listed below, which is more likely to be used to send the program to you?
    a. Floppy disk   b. Hard disk
22. The primary use of CD-ROMs is to ________.
    a. transfer programs and information from manufacturers of CD-ROMs to users
    b. backup data generated by CD-ROM users
    c. replace hard-disks
23. Data is stored on a CD-ROM through the use of ________.
    a. small magnets
    b. electrical capacitors
    c. dents in a surface
24. The primary purpose of the aluminum surface on a CD-ROM is to ________.
    a. protect the pits and lands
    b. provide a surface on which identifying information can be printed or written
    c. function as a mirror
25. In which direction does the laser light shine on the CD-ROM in Fig. P17-9?

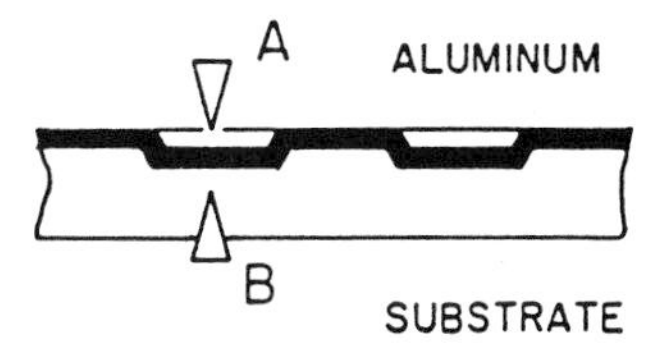

Fig. P17-9. This is a cross section of a CD-ROM.

26. Which of the following is likely to have large amounts of stored data?
    a. CD-ROM   b. Floppy disk
27. Which of the following is likely to have shorter access time?
    a. CD-ROM   b. RAM memory
28. Which of the following functions most like a PROM?
    a. CD-ROM   c. PD
    b. CD-R
29. Which of the following is most like a floppy disk with the write-protect activated?
    a. CD-R   b. PD
30. The P in PD stands for ________.
    a. polycarbonate
    b. protected-write
    c. phase-change
31. The main advantage of tape storage over floppy disk storage is the ________.
    a. low cost of tapes
    b. ease with which tapes can be stored
    c. large storage capacity of tapes
32. The primary use of tape storage is ________.
    a. backup of other forms of storage
    b. transfer of new programs to users

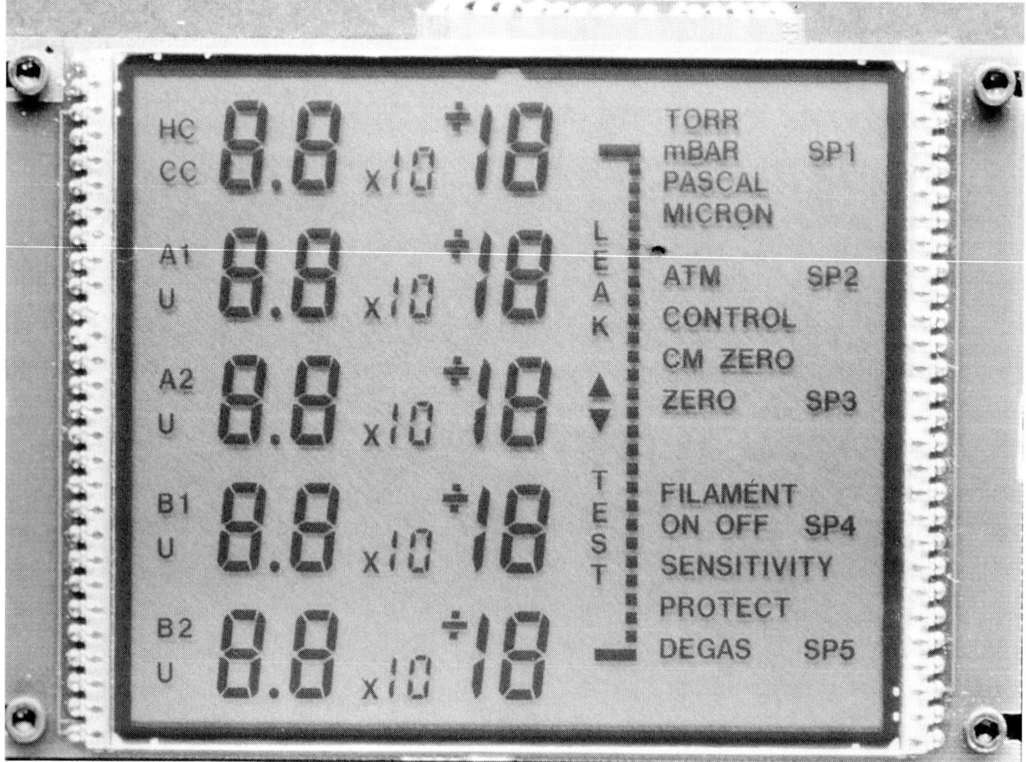

Top: Cathode-ray tubes do not lend themselves to miniaturization. As computers have become smaller, more and more liquid-crystal displays are being used. (Intelligent Peripheral Devices) Lower: This is a liquid-crystal display. The seven segments that make up the number 8 are used to create the numbers 0 through 8. This photograph emphasizes that LCDs can also exhibit letters and words. (Standish LCD)

Input/Output for Small Computers

Chapter 18

LEARNING OBJECTIVES

After studying this chapter and completing lab assignments and study problems, you will be able to:

- Describe the action of a port-oriented, input/output circuit.
- State the effect of the difference between a port-oriented and a memory-mapped input/output circuit.
- Explain the need for current-limiting resistors in LED (light emitting diode) circuits.
- List parts of an LED circuit, draw a circuit for driving an LED, and select a resistor that will result in a specified lamp current.
- Determine whether the 7-segment display in a computer circuit is of the common anode type or of the common cathode type.
- Compare the advantages and disadvantages of liquid crystal and LED displays.
- List parts of a multiplexed multi-digit display and describe how to outline steps that are used in an explanation of the action for a multiplexed multi-digit display.
- List parts of a multiplexed keypad or keyboard and describe how to outline steps that are used in an explanation of the action for a multiplexed keypad or keyboard.

To do useful work, computers and digital circuits must communicate with the outside world. In this chapter, input/output methods used on small microprocessor-based computers will be reviewed. Also, the displays and keypads used on many microprocessor trainers and prototyping boards will be described.

PORT-ORIENTED INPUT/OUTPUT

Computers must control the flow of data and instructions on their data buses, so external equipment cannot be directly connected to data buses. There must be an interface controlled by the computer. The next few sections describe port-oriented input-output.

Many topics are involved with port-oriented input/output. The following topics will be considered:

- Port definition.
- Port input/output instructions.
- Port accumulator input/outut.
- Output port.
- Output multiport circuit.
- Input port.
- Input and output ports compared.

PORT DEFINITION

A *PORT* is an inlet or outlet. It often suggests an interfacing of two media. For example, a seaport provides equipment (docks, cranes, etc.) at the interface between ocean and land transportation. An airport serves the same function for air and land transportation. A computer port interfaces computer circuits with those of the outside world.

PORT INPUT/OUTPUT INSTRUCTIONS

Computer instructions for inputting and outputting data through a port might take the forms IN 01 and OUT 01. IN and OUT tell the computer the task to be accomplished. The numbers indicate the addresses of the ports which are to be used.

Programmers must know what external equipment is attached to each port. A mixup in connections results in data flowing to and from the wrong equipment.

PORT ACCUMULATOR INPUT/OUTPUT

In some microprocessors, data to be output by means of the instruction OUT must be in the machine's accumulator. Also, data input by means of the instruction IN is delivered to the accumulator. For these reasons, port-oriented input/output is often referred to as accumulator input/output. Port-oriented input/output can make interfacing easy and fast.

OUTPUT PORT

Fig. 18-1 shows a simple output port. When the instruction OUT 03 is executed, the digital computer system does three things:

1. The number in the computer's accumulator is placed on the data bus (D0 through D7).
2. The port number (sometimes called the device number) appears on the lower eight bits of the address bus (A0 through A7).
3. The control lead OUT goes active.

The result is the storing of the number on the data bus in the 8 data latches. The number is then available on output leads D030 through D037.

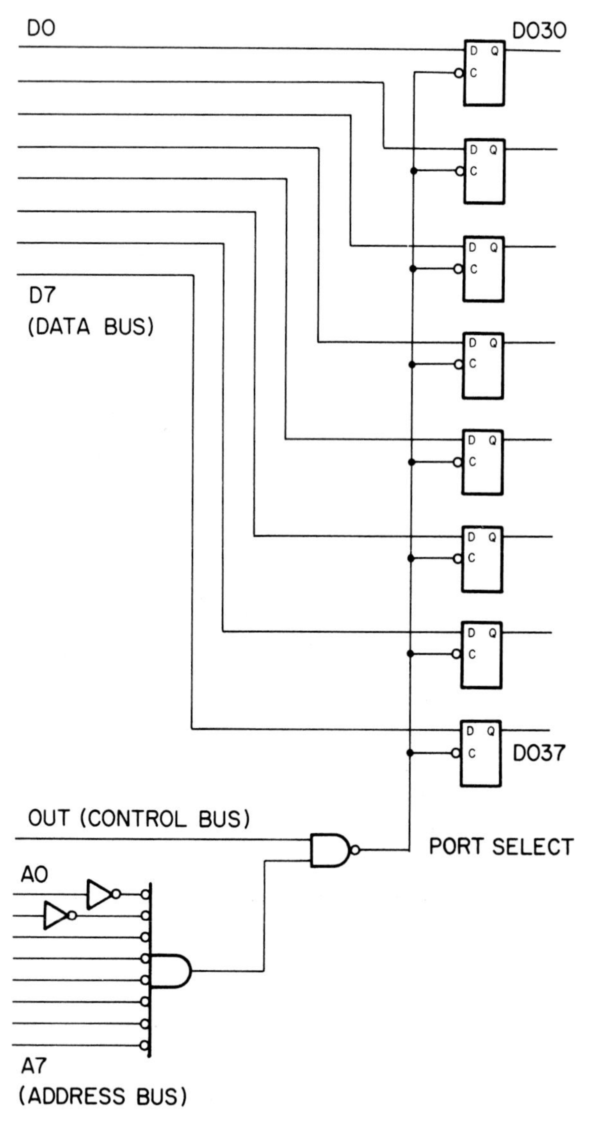

Fig. 18-1. In the circuit shown, data and control signals from a computer enter from the left. That data is available to the outside world at the right.

OUTPUT PORT DATA LATCHES

In most computers, data to be output is on the data bus for a microsecond (0.000,001 second) or less. It is difficult to design external equipment that will accept the data at exactly the right time. To overcome this problem, data latches are usually provided.

As a general rule, outputs are latched. Data stored in these flip-flops is available to external equipment even though the computer has gone on to execute other instructions. From the computer side, the interface looks much like a memory location.

CONTROL LEAD OUT

Control lead OUT has two functions:

1. It indicates that an output is to be accomplished. When a computer is in operation, many numbers appear on its address and data buses. It is only when OUT goes active that the circuit in Fig. 18-1 accepts the numbers on the bus leads.
2. OUT times the acceptance of the data on the data bus. OUT is active only during the time that the data and the port number are available.

PORT NUMBER DECODING

The NOR at the bottom of Fig. 18-1 (with the active-high leads A0 and A1 and the active-low leads A2 through A7) decodes the port number placed on the address bus. Based on the placement of the two NOTs, the elements in Fig. 18-1 make up Port 03.

OUTPUT MULTIPORT CIRCUIT

Fig. 18-2 shows a 3-port circuit. Multiport circuits can be analyzed by simple actions. Three topics are considered:

1. Multiport output data latches.
2. Control lead OUT.
3. Port number decoding.

MULTIPORT OUTPUT DATA LATCHES

Of the integrated circuits U0, U1, and U2 in Fig. 18-2, each contains 8 data latches. When a given clock input is activated, the number on the data bus is stored in the corresponding set of latches.

CONTROL LEAD OUT

Control lead OUT in Fig. 18-2 is active-high. When a 1 appears on OUT, the three AND gates can be opened. When it is low, no combination of inputs can activate a port select lead.

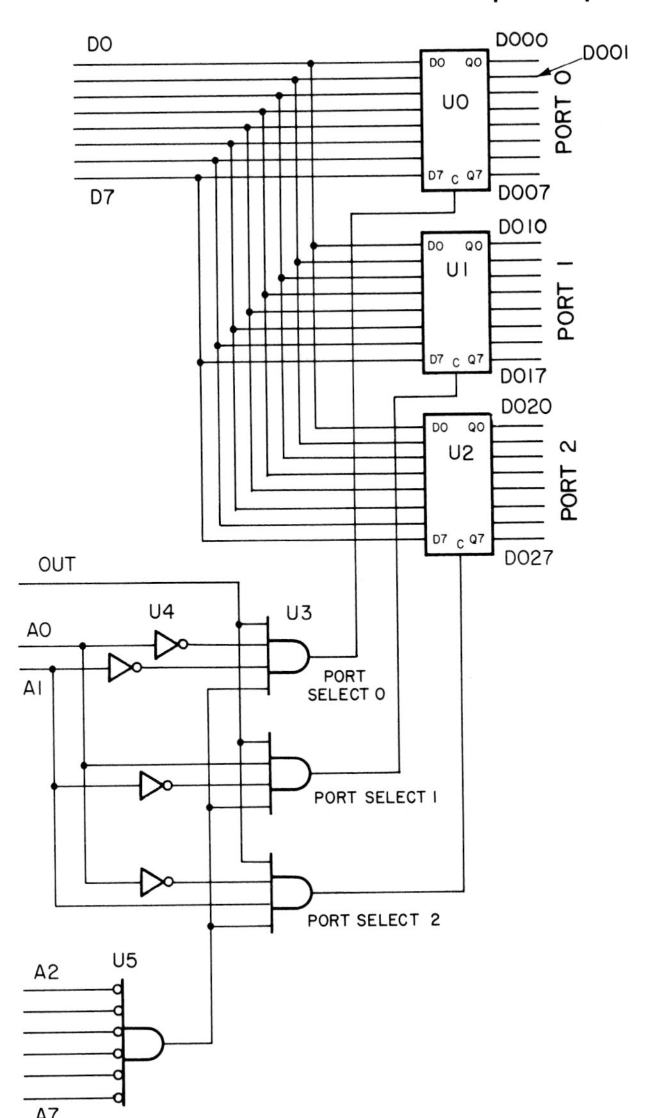

Fig. 18-2. The circuit shown contains three output ports. Of the elements U0, U1, and U2, each contains 8 data latches.

PORT NUMBER DECODING

As before, port numbers appear in binary form on address leads A0 through A7 in Fig. 18-2. ICs U3, U4, and U5 decode these numbers.

Because leads A2 through A7 will always be 0 when a valid port number appears on the address bus, these leads are decoded by the NOR at the bottom of Fig. 18-2. The ANDs and NOTs decode the lower order bits of the port address.

Assume the instruction OUT 02 is executed. The number in the computer's accumulator would appear on the data bus. The port number (0000,0010) would appear on the lower 8 bits of the address bus. Control lead OUT would go to 1. At the output of the decoder, a 1 would appear on port select 2, and the number on the data bus would be stored in the latches of port 2.

INPUT PORT

Fig. 18-3 shows an input port. The use of 3-state buffers is the difference between an input and an output port. Also, control lead IN is used in place of OUT. The port number decoding circuit is identical to that of an output port.

From the computer side, the input port interface looks much like a read-only memory location. Some of the interfaces may involve input devices like a mouse, a light pen, or another terminal.

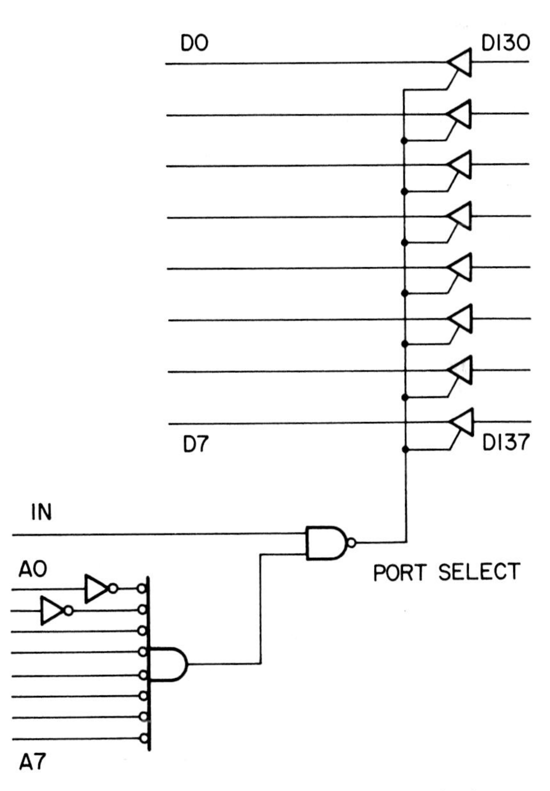

Fig. 18-3. Three-state buffers are used in the input port circuit shown.

INPUT AND OUTPUT PORTS COMPARED

Fig. 18-4 shows 3 output ports and 4 input ports. Outputs are treated differently than inputs. The difference is stated in the following:

1. Outputs are latched.

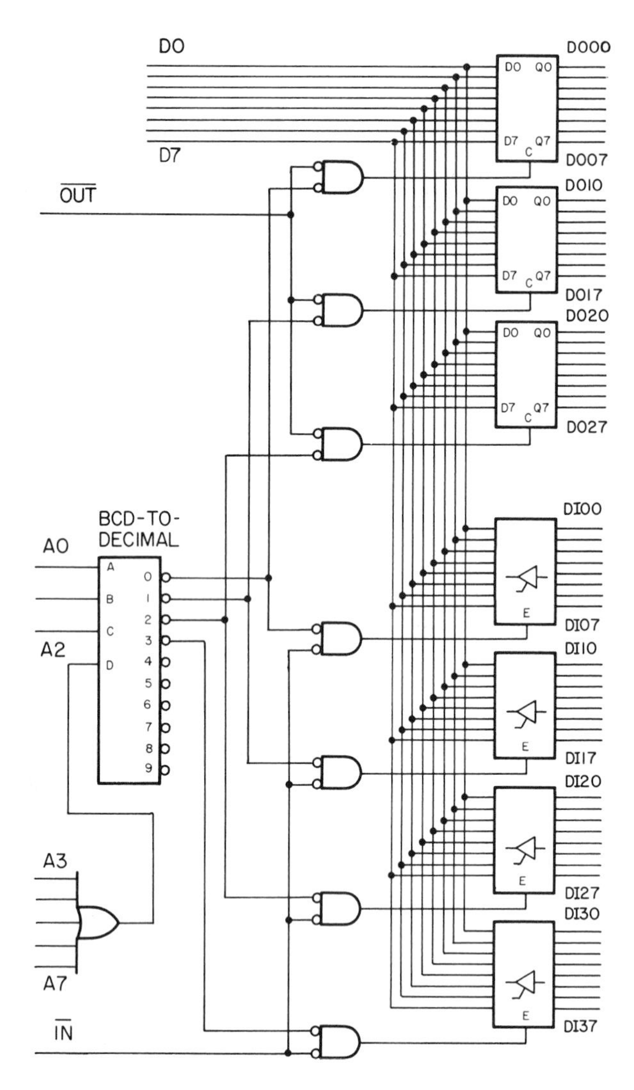

Fig. 18-4. The 3 output ports and 4 input ports of the circuit shown share a decoder.

2. Inputs are buffered.

Control leads IN and OUT are active-low in the circuit in Fig. 18-4. Only one decoding circuit is provided. Note the use of a BCD-to-decimal decoder as part of the circuit in Fig. 18-4.

Part "a" of Fig. 18-5 shows signals applied to the address and control leads of the circuit in Fig. 18-4. Control lead OUT is 0 (active) so an output is to be accomplished. Address leads A0 through A7 indicate that port 2 is to be used. The number on the data bus will appear at the output of Port 02.

Consider the signals at b. A valid port number appears on the address leads. However, control leads OUT and IN are both 1 (inactive). The computer is not executing an input/output instruction. The outputs will not change, nor will a number be input.

Part "c" of Fig. 18-5 represents a fault. It may be a programming mistake or it may be in the circuit. A 0 appears on OUT, so the number on the data bus should be output. However, the port number is faulty. Note the 1s on leads A4 and A5. Such a port is not available for the circuit in Fig. 18-4.

For the signals at d in Fig. 18-5, determine the resulting action. If you found that an input will occur through Port 03, you are correct.

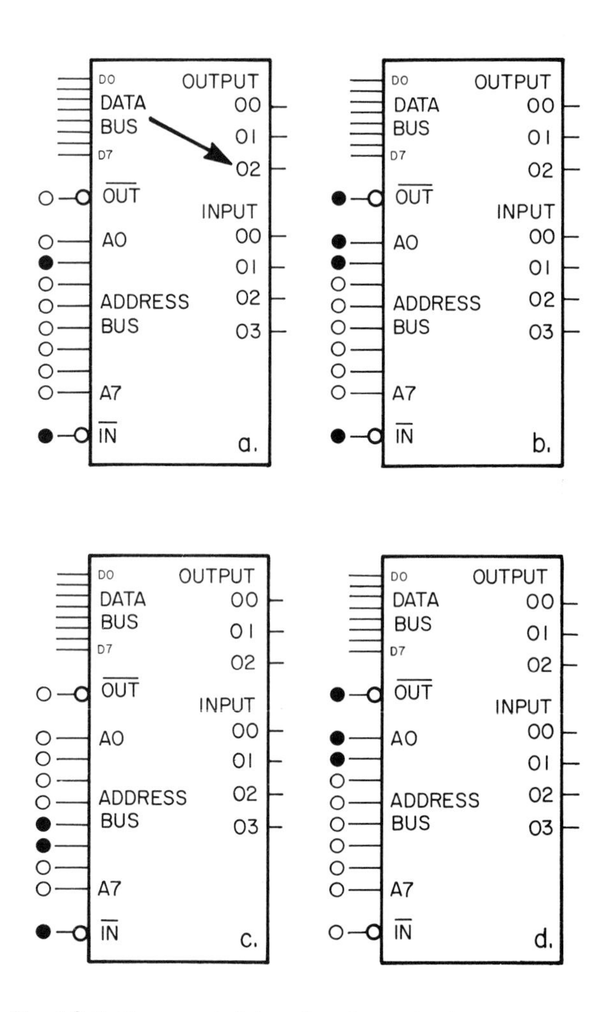

Fig. 18-5. To accomplish an input or output, proper combinations of control and address signals must be present.

MEMORY-MAPPED INPUT/OUTPUT

It has been suggested that input and output ports can be thought of as special memory locations. Memory-mapped inputs and outputs take this concept one step farther. Memory-mapped inputs and outputs are treated as ordinary memory locations.

They are given addresses, and read/write control leads are used to control the input/output processes in the memory-mapped system, Fig. 18-6.

Memory mapped inputs and outputs differ from ports in two ways. First, all address leads are decoded. Note the use of the 16 leads marked A0 through A15. Second, MEMORY WRITE (rather than OUT) and MEMORY READ (rather than IN) control the process.

From the computer's standpoint, an output circuit is a write-only memory location. An input is a read-only location.

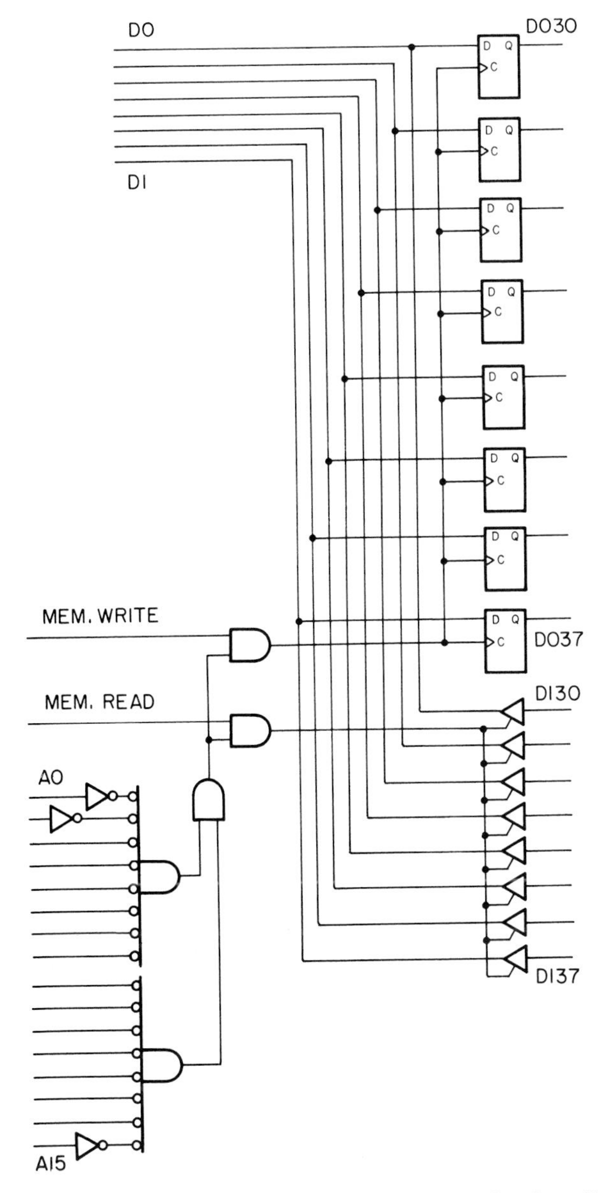

Fig. 18-6. In memory-mapped input/output circuits, all 16 address leads are used. Also, memory write and memory read are used in place of OUT and IN.

ADVANTAGES OF MEMORY-MAPPED INPUT/OUTPUT

Memory-mapped input/output is not accumulator oriented. Numbers to be output need not be in the accumulator (although they can be). Numbers that have been input do not have to go through the accumulator (although they can). Any memory-oriented instruction can be used to do an input or output. For example, the instruction for storing register B in memory location 8003 might be STB 8003. If the memory-mapped circuit in Fig. 18-6 were available, this instruction would output the content of B on leads D030 through D037. Because numbers do not need to pass through the accumulator, memory-mapped input/output is faster and more convenient.

Memory mapped input/output usually permits more input and output circuits than does port-oriented input/output. If the lower 8-bit section of the address bus is used for port numbers, 256 inputs and 256 outputs are available. In the same computer, about 65,000 memory-mapped inputs and 65,000 outputs could be supported. This results from the use of the whole address bus in the memory-mapped approach. In some applications, the additional input/output capability is important.

DISADVANTAGES OF MEMORY-MAPPED INPUT/OUTPUT

When memory-mapped input/output is used, all address leads are usually decoded. The resulting circuits are more complex than those for port-oriented input/output. In applications where cost is critical, this may limit the use of memory-mapped input/output.

Memory-mapped input/output displaces regular memory locations. In machines with small memories, this does not create a problem, since there will be many unused addresses. If, however, most or all addresses are used for RAM and ROM, a problem results. Memory must be removed to make room for memory-mapped input/output.

DIRECT MEMORY ACCESS

A third input/output system is called DIRECT MEMORY ACCESS (DMA). Because it is an advanced form of input/output, DMA will be described only briefly.

When DMA is used, the computer is temporarily turned off, and its memory is given over to a controller or another computer. The controller or second computer then writes and reads directly to and from the memory. When control is returned to the first computer, it finds new data in its memory.

DMA is used primarily when large amounts of data must be transferred. For example, programs for a number of small computers in a plant may be stored in a large, central computer. When a new program is needed in a small machine, it is downloaded from the large machine. The term *DOWNLOADED* means to transfer in large amounts. A complete program is loaded. DMA speeds the transfer of the large amount of data involved in such programs.

During the time that the large computer has the memory, the first computer is shut down. Because it cannot obtain instructions for its memory, it cannot operate. As a result, DMA cannot be used in some control applications.

RS-232 PORT

A type of port used for cable runs of less than 50 ft. is the RS-232C port. Data is sent and received in serial form. A 1 is represented by a +3 V signal and a 0 is represented by a −3 V signal. The data sequence of the voltage signal is shaped like pulses of varying width and with varying spaces between pulses.

LIGHT-EMITTING DIODES

Everyone is familiar with displays on calculators and digital clocks and watches. When glowing segments are used to produce numbers, LIGHT-EMITTING DIODES (LEDs) have been utilized. If dark segments appear against a reflective background, a LIQUID-CRYSTAL DISPLAY (LCD) has been used in the display.

LEDs and LCDs find wide use in digital circuits because their voltage requirements and power levels are similar to those of widely used logic families. Also, they are inexpensive.

LED

When current flows through a diode, light is produced. In most diodes, the amount of light is small, and the packages used do not permit direct viewing of the light.

The diode in part "a" of Fig. 18-7 is designed to be used as an indicator. It is made of gallium arsenide (a material that produces considerable light) and is housed in a transparent package. The gallium arsenide is allowed to cool into a single crystal structure. When the diode is forward biased, it glows. See part "b" of Fig. 18-7 for a typical LED and its circuit.

Although larger and smaller units are available, a typical LED is about 1/4 in. tall (exclusive of the leads). A flat on the rim of the LED may be used to indicate the cathode.

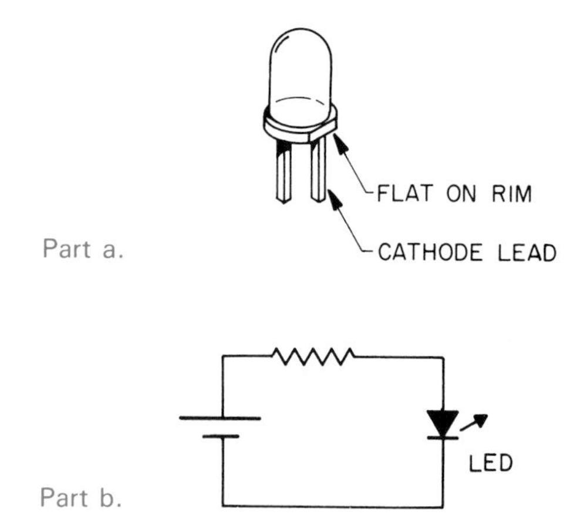

Fig. 18-7. Light-emitting diodes produce light when they are forward biased. That is, their anodes must be positive with respect to their cathodes.

LED CHARACTERISTIC CURVE

Fig. 18-8 is the characteristic curve of a typical LED. The term *CHARACTERISTIC CURVE* means the graph representing the properties of the device being analyzed. The characteristic curve shows the current flow for each value of applied voltage. Up to about 1.6 V, no current flows, and no light is produced. After 1.6 V, the current increases rapidly. The light produced is proportional to the current.

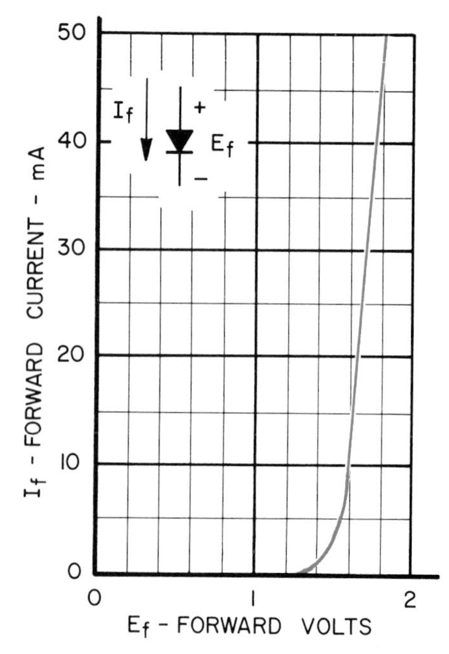

Fig. 18-8. The displayed curve is typical of the characteristics of LEDs.

LED CURRENT-LIMITING RESISTOR

The steep increase in current in an LED after 1.6 V has been reached is a source of difficulty. Small changes in applied voltage result in large current changes. As a result, voltage is seldom applied directly to LEDs. Rather, a current-limiting resistor is usually placed in series with LEDs, Fig. 18-9. The current-limiting resistor stabilizes diode operation. It minimizes the effects of changes in applied voltage, temperature, and diode characteristics.

LED currents are usually between 10 mA and 25 mA. The following equation can be used to determine the approximate current-limiting resistor:

$$R = \frac{Vcc - 1.8}{Id}$$

Vcc is the applied voltage (usually 5 V). Id is the desired current. The 1.8 represents the 1.6 V across the LED and allows 0.2 V across the logic element that might drive the diode.

For Vcc = 5 V and a current between 10 and 25 mA, the resistance range is 330 to 120 ohms. Some rounding was necessary to permit the use of standard resistors.

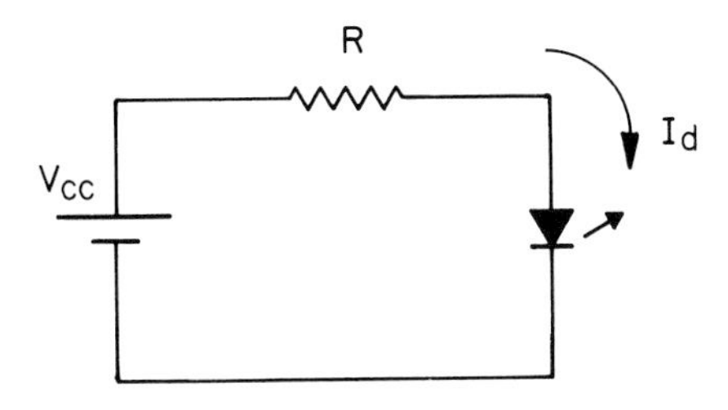

Fig. 18-9. Current-limiting resistors are usually used to stabilize the operation of LEDs.

LED DRIVERS

Part "a" of Fig. 18-10 shows a TTL element driving an LED. With a 1 at the NOT's input, the cathode of the LED is grounded. This completes the circuit from Vcc through the lamp to ground, and the LED device lights up properly.

Based on the use of a 220 Ohm resistor, the current will be near 15 mA. This is close to the current-sinking ability of the TTL NOT. As a result, the NOT can drive the lamp and only one additonal TTL input.

In part "b" of Fig. 18-10, a 0 has been applied to the NOT. Its output will be 1, so both ends of the lamp are connected to Vcc. No current flows, and the lamp will not light.

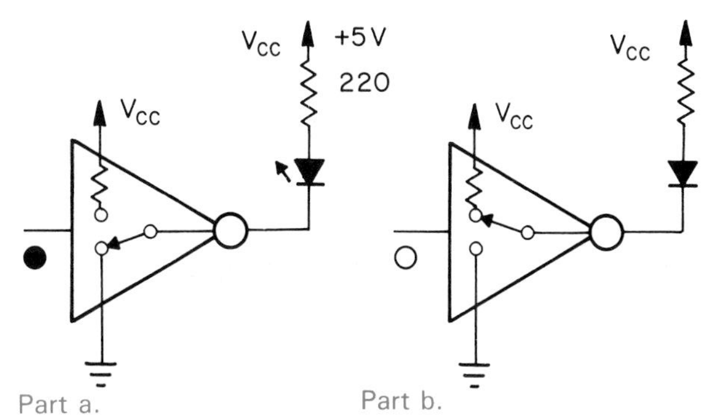

Fig. 18-10. The preferred method for using logic elements to drive LEDs is shown.

The circuit in part "a" of Fig. 18-11 can be used, but it tends to overload the driving element. Current will flow when the element outputs a 1. Note the 130 Ohm resistor within the TTL element. When lamp current flows, this resistor tends to overheat. As a result, this circuit should be avoided. It can, however, be used in some test circuits.

The circuit in part "b" of Fig. 18-11 overcomes the problem of the first circuit. An open-collector driving element is used. Because the current-limiting resistor is external, it can dissipate more power. When the element outputs a 0, the LED is shorted to ground, and it goes out. The short-circuit current through the element is limited by the 330 Ohm resistor. If a 1 is input, the buffer's output opens and current is no longer bypassed. The LED lights.

To permit even higher lamp currents, an external transistor can be used. See part "c" of Fig. 18-11.

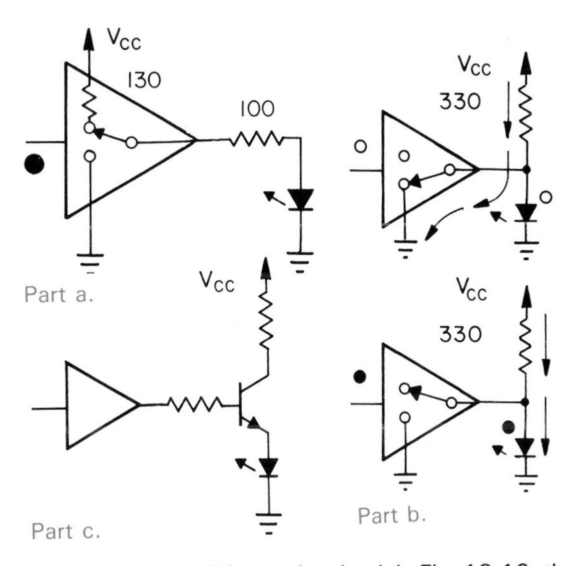

Fig. 18-11. In addition to the circuit in Fig. 18-10, these circuits are used to drive LEDs.

LED REVERSE BREAKDOWN

The diode in part "a" of Fig. 18-12 is reverse biased. That is, its cathode is positive with respect to its anode. If too much voltage is applied in this direction, it will break down. The voltage at which this occurs is called the *REVERSE BREAKDOWN VOLTAGE* or peak inverse voltage.

Ordinary diodes have very high reverse breakdown voltages (often in the hundreds of volts). This is not true of LEDs. Some may fail with as little as 3 V of reverse bias.

Because the diode in part "a" of Fig. 18-12 is reverse biased, no current flows. With no current, there is no voltage drop across the resistor, so all of Vcc is applied to the diode. This diode could fail.

As a general rule, reverse voltage should not be applied to LEDs. If it must be, the circuit in part "b" of Fig. 18-12 can be used to protect the LED. D2 is an ordinary diode.

The circuit in part "c" of Fig. 18-12 can also be used. No matter which polarity is applied, one diode will conduct. This limits the voltage to about 1.6 V, so the nonconducting LED is protected.

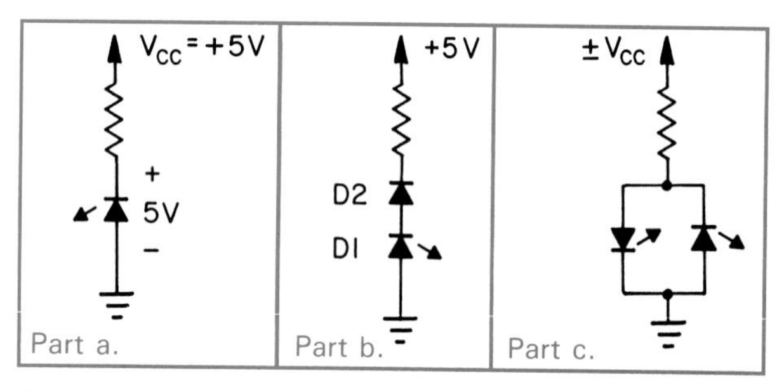

Fig. 18-12. Inverse voltages (cathode positive with respect to anode) as low as 3 V can damage LEDs.

LED OUTPUT INDICATOR

When troubleshooting computers, it is necessary to know what numbers are actually being output. Fig. 18-13 shows two methods for connecting LEDs to the outputs of a small microprocessor-based computer.

The circuit at the left is the obvious approach to indicating the numbers stored in the data latches. When a Q equals 1, the corresponding LED lights.

But there is a problem. Driving an LED from an active high tends to overload that output. The output can drive only the LED and one additional TTL input. The solution to this problem is shown at the right of Fig. 18-13.

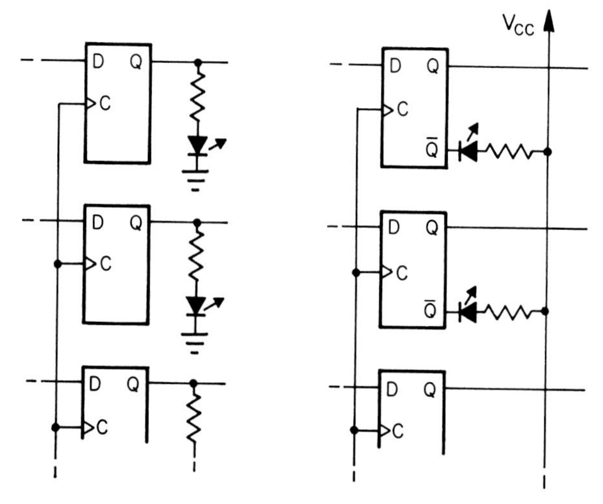

Fig. 18-13. Two methods of connecting LEDs to the output port are shown. The method at the right is preferred.

When Q = 1, $\bar{Q}$ = 0. When a given Q goes to 1, the corresponding LED is grounded through $\bar{Q}$. Although the lamp is not connected to Q, it lights and indicates the outputting of a 1.

7-SEGMENT DISPLAYS

Fig. 18-14 shows a 7-segment display. By activating appropriate segments, numbers between 0 and F can be produced. Note the potential confusion between the symbols 6 and b.

Each segment is either an LED or a liquid crystal. A decimal point is usually provided.

7-SEGMENT LED DISPLAY

7-segment LED displays are available in two forms—common anode and common cathode. See Fig. 18-15 for circuit drawings of both types.

To light a segment of a common-anode display, the appropriate lead must be grounded. When a common-cathode unit is used, the appropriate anode must be connected to Vcc. In both cases, current-limiting resistors must be provided. These may be external or part of the integrated circuits driving the display.

7-SEGMENT LED DECODER/DRIVER

7-segment displays cannot be driven directly by binary or BCD signals. There must be an interface—a device called a *DECODER/DRIVER*. See part "a" of Fig. 18-16 for block diagrams of decoders and lamps.

The decoder portion of the interface accepts the

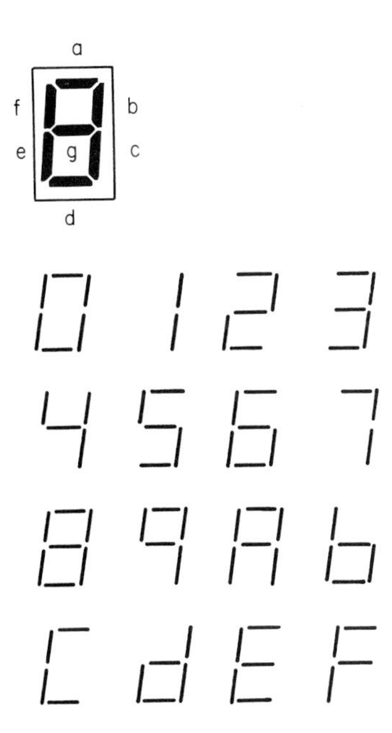

Fig. 18-14. By activating the proper segments of a 7-segment display, decimal and hexadecimal numbers can be formed.

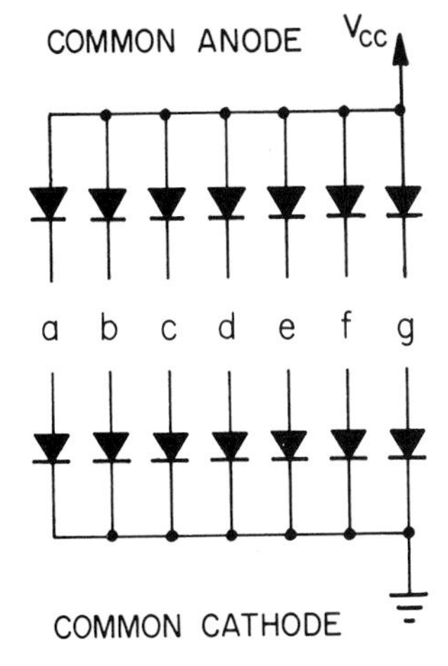

Fig. 18-15. 7-segment LED displays are available in both common-anode and common-cathode forms.

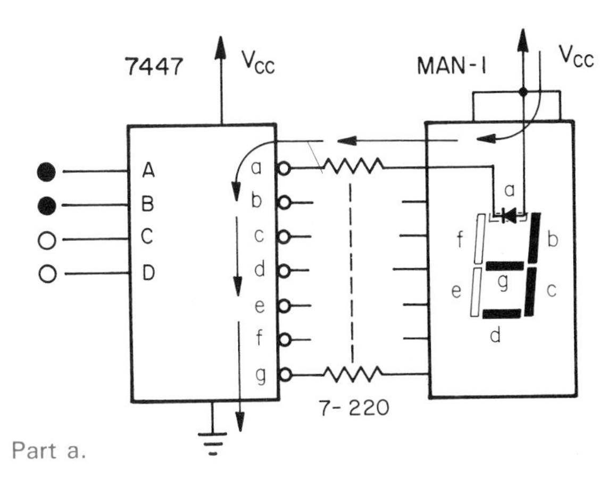

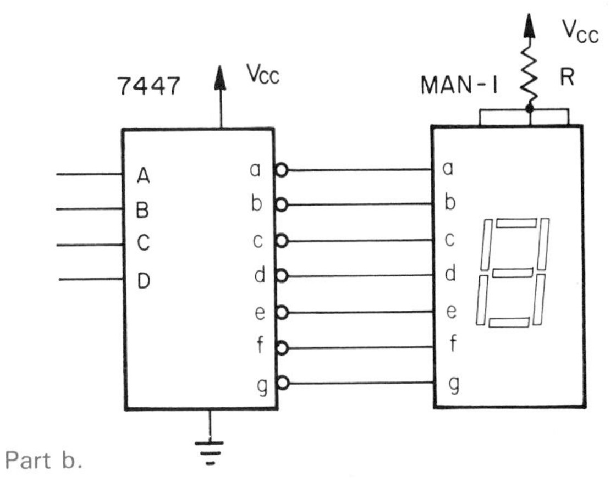

Fig. 18-16. The 7447 is a decoder/driver. It is intended for use with common-anode 7-segment displays. Two current-limiting resistor arrangements are shown.

binary or BCD number and outputs signals that will light the corresponding segments. For example, if DCBA = 0011, outputs a, b, c, d, and g become active. In this case, the outputs of the decoder/driver are active-low, since the display is common anode. The series of arrows shows the current path when segment a is on.

The term driver in the description of this interface suggests that the chip can sink the 20 mA needed to light the individual LEDs. When all segments are on, about 140 mA flows into the outputs of the decoder/driver.

Note the current-limiting resistors in series with the LEDs. 7 resistors are used (one for each segment). To simplify the drawing, only two are shown.

In part "b" of Fig. 18-16, the 7 resistors have been replaced by a single resistor in the common lead. This represents a major saving in cost and complexity, but there is a problem. Because two segments are lit when a 1 is displayed, only about 40 mA flows in this common resistor. When an 8 is displayed, this current increases to 140 mA. As a result, the voltage drop across the lone resistor varies with the number displayed. In turn, the brightness of the display changes as the numbers change. While a single resistor might be used in the laboratory (where workers can compensate for changes in brightness) the 7-resistor circuit is standard in industry.

BCD-TO-7-SEGMENT LED DECODER/DRIVER

When a hexadecimal-to-7-segment decoder/driver is used, all input combinations between 0000 and 1111 are decoded. In turn, all numbers between 0 and F can be displayed.

When a BCD-to-7-segment decoder/driver is used, it is assumed that the input's range will be 0000 to 1001 (decimal 0 to 9). If binary numbers larger than 1001 are applied, the displayed patterns appear meaningless. See Fig. 18-17. However, these patterns can be helpful when troubleshooting.

If the U-shaped pattern at 12 (1100) appears when input DCBA = 1000, the fault is probably in the decoder/driver or display (an 8 should appear). If, however, this U-shaped pattern appears with DCBA = 1100, the fault is not in the decoder/driver or display. A faulty BCD number is being supplied to the circuit.

7-SEGMENT LED ON-BOARD DECODER

Displays with built-in decoders, drivers, and data latches are available, Fig. 18-18. Such units tend to be expensive, but they reduce assembly time and require less space on printed circuits. They also tend to be more reliable than multichip circuits.

The latches permit the continued display of numbers after the originating signals have disappeared. Once latched, a number will be displayed until intentionally changed.

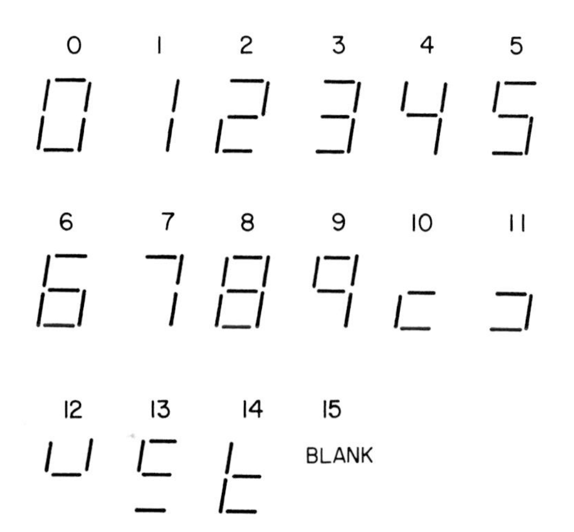

Fig. 18-17. When binary numbers larger than 1001 are applied to BCD-to-7 segment decoders, odd symbols are output. These can aid in troubleshooting.

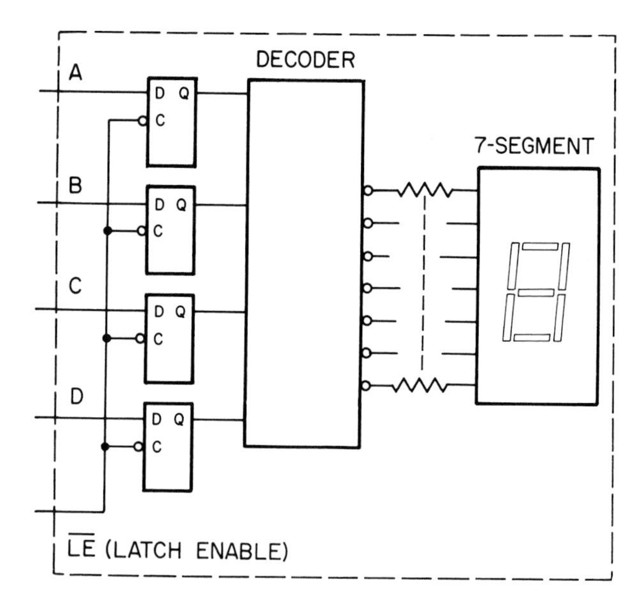

Fig. 18-18. Some displays have built-in decoders and latches. While such units make assembly easier, they cost more than displays without these circuits.

LIQUID-CRYSTAL DISPLAYS

Light-emitting diode displays have long been used in industry and student laboratories. Another form of display, LIQUID-CRYSTAL DISPLAYS (LCDs), is widely used on digital watches and hand-held calculators. LCDs are increasingly being used on industrial equipment and laptop computers.

LEDs are light producing and present numbers as glowing figures against a dark background. LCDs provide black numbers and letters against a bright background.

On the basis of power used, LCDs have an advantage because LEDs produce light and therefore consume a small amount of power. In contrast, LCDs use light already in the environment. As a result, they are all but passive devices. They draw only microwatts. It is this characteristic that dictates their use in battery-operated devices such as watches and calculators.

A second advantage has to do with the ease with which letters, words, and symbols can be provided. Both LCDs and LEDs use seven-segment displays to produce numbers. However, it is relatively easy to manufacture LCDs that contain words and even drawings. For example, calculators that allow the use of both degrees and radians usually display the abbreviation DEG (degrees) and RAD (radians). This would be difficult to do with LEDs. Present-day machines and processes are very complex so large amounts of information must be supplied to operators. LCDs can do this better than LEDs.

When the molecules in a solid are arranged in an orderly fashion that repeats itself in three dimensions, that material is described as a crystal. When a crystal is heated until it melts, this ordered structure is lost. It is no longer a crystal. However, there are some the liquid state. These are called LIQUID CRYSTALS.

In addition, the crystalline structures of certain liquid crystals will change when an electric field is applied. This is shown in Fig. 18-19. Without an applied electric field, the rod-like molecules at the left arrange themselves in twisted columns. At the right, a voltage has been applied, and the relationships between the molecules has changed. This response to an electric field is basic to the operation of LCDs.

LCDs use light-polarizing filters to produce dark images on a light background. Fig. 18-20 shows what happens when light shines through two polarizing filters. The filters on the right at A and B are positioned such that only vertically polarized light is present in the space between the two filters. At A, both filters are adjusted for vertical polarization. As a result, light that passes through the first filter will pass through the second. At B, the second filter has been turned 90°. With only vertically polarized light presented to the second filter, no light gets through.

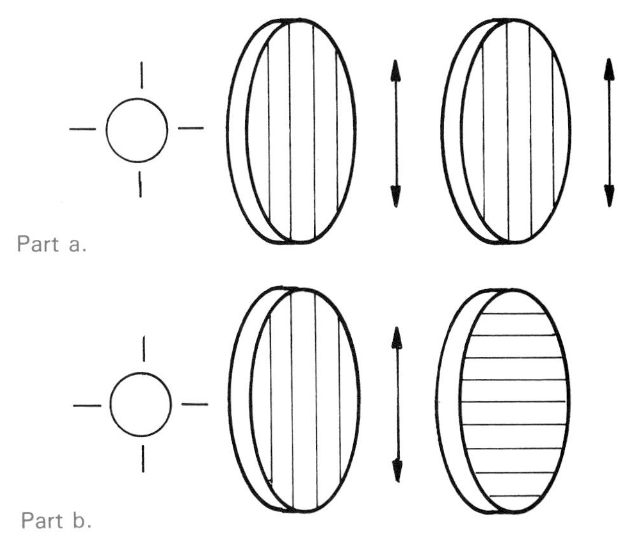

Fig. 18-20. When the planes of polarization of two light-polarizing filters are at 90°, light cannot pass through.

Fig. 18-21 is a cross section of a liquid-crystal display. Note the polarizers at the top and bottom. These are positioned so their faces are at 90° to each

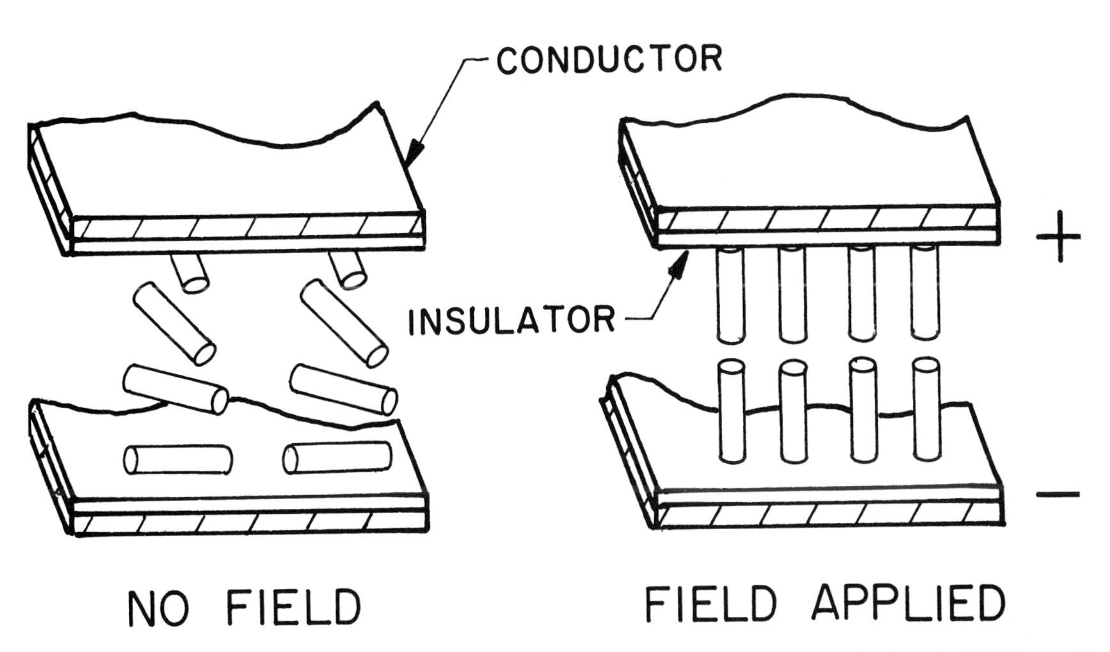

Fig. 18-19. The ordering of molecules within some liquid crystals change when an electric field is applied.

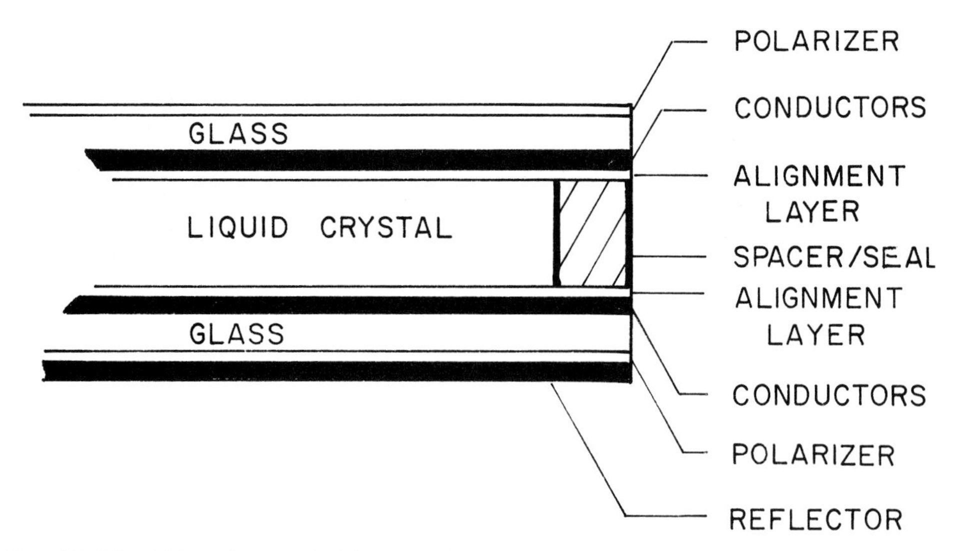

Fig. 18-21. Although remarkably thin, the internal structure of an LCD is complex.

other. Light coming from above passes through the display but is blocked by the second polarizer. (They are at 90° to each other.) That is, no light gets through to the reflector at the bottom of the display. If it were not for the liquid crystal, the whole display would be dark.

However, when a LCD is turned off, it is not dark. It has a slightly shiny appearance. Light does pass through the cell and out again.

This is where the liquid crystal comes in. Return to Fig. 18-19. With no applied voltage, the molecules arrange themselves as shown at the left of that drawing. Light that passes through the upper filter is indeed polarized, but the twisted columns of the liquid crystal shifts the polarization of that light 90°. It can now pass through the lower filter and be reflected by the mirror-like surface. On the return trip through the display, the liquid crystal again twists the light so it can pass back through the upper polarizer.

Note the alignment layers above and below the liquid crystal in Fig. 18-21. These surfaces ensure that the crystal has the proper 90° twist.

To produce a dark area, an electric field is applied across the LCD. Again refer to Fig. 18-19. The molecular arrangement at the right cannot produce the necessary twist to the light. Light that enters through the first polarizer is blocked by the second polarizer. No light gets to the reflector in the areas where the field exists, so they appear dark.

If an electric field is to be applied, conductors must be placed above and below the liquid crystal. If these were made of copper for instance, they would hide the action of the display. The solution is to use transparent conductors. Areas that are to represent numbers, letters, and leads, are made from indium tin oxide. This is a conductor that allows light to pass through. Only a thin coating is necessary because LCDs draw almost no current. In Fig. 18-21, the transparent conductors are in the layers marked conductors.

Signal voltages used to drive LCDs hold a bit of a surprise. While these are compatible with voltages used in logic circuits, LCDs are alternating-current devices. If direct current is applied, they deteriorate. Square waves are usually used to drive LCDs. These can be generated by standard logic circuits and do not contain a dc component.

KEYPAD AND KEYBOARD DIRECT CODING

Keypads and keyboards are used as direct inputs to computers and controls. Those with sixteen or fewer keys are usually directly encoded. When a key is pressed in Fig. 18-22, the proper 4-bit binary number appears on the data bus.

Because 16-to-4 encoders are not readily available, the circuit in Fig. 18-22 uses two 8-to-3 encoders. When an input (at the right) is grounded, these encoders output (at the left) a corresponding binary number. For example, if key 3 is pressed, the upper encoder will output CBA = 011.

Each encoder has two additional outputs. EO (enable out) goes to 1 when any key is pressed. KS (key sense) goes to 0 when any key is pressed.

To tell the computer that a key has been pressed, the KS outputs of these encoders are NANDed. When a 1 appears on KEY PRESSED, the computer knows

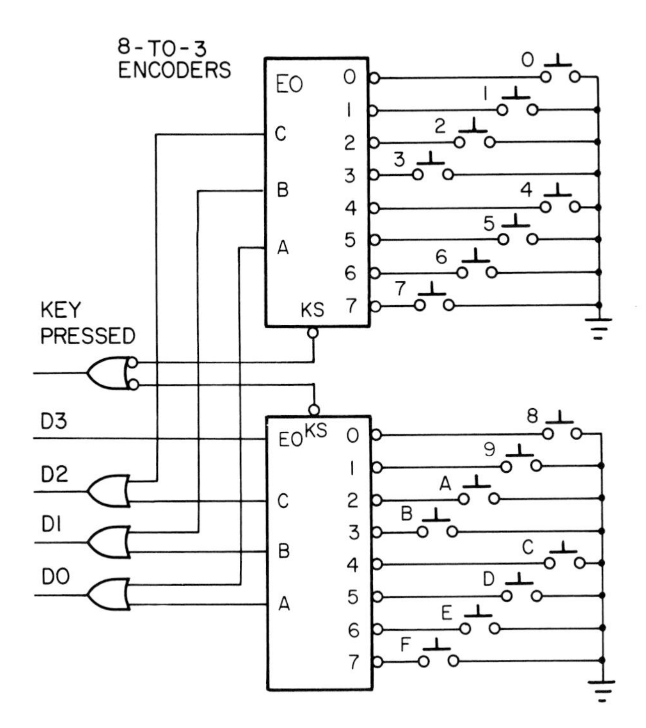

Fig. 18-22. In the circuit shown, combinational-logic circuits are used in the encoder. As a result, output signals are directly related to the pressing of a key.

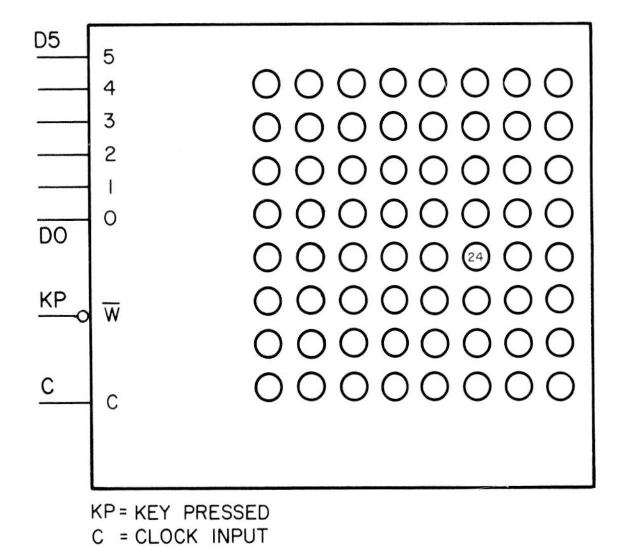

Fig. 18-23. Although the keyboard shown has 64 keys, it requires only eight signal and data leads.

that an attempt is being made to communicate a number to the microprocessor unit.

EO on the lower encoder is used for D3. When a key from 0 to 7 is pressed, the left side of the circuit in Fig. 18-21 outputs a number from 0000 through 0111. When a key from 8 to 15 is pressed, it outputs a number from 1000 through 1111. In either case, KEY PRESSED goes active.

Direct coding is not practical when large numbers of keys are involved. The need for one lead per key causes the number of leads to be excessive. For example, the 8-by-8 keyboard in Fig. 18-23 would require at least 64 leads.

KEYPAD AND KEYBOARD SCANNING

To reduce the number of leads, large keyboards are often *SCANNED*. That is, keys are tested, one after the other, to see if one is being pressed. Scan rates are high, so whole keyboards can be tested in less time than it takes to press and release a key.

SCANNING INPUT/OUTPUT SIGNALS

In the system shown in Fig. 18-23, one key is tested each clock cycle. A binary number corresponding to the key being tested at a given instant is output on data leads D0 through D5. These leads count from 000 000 through 111 111 again and again.

When a key is pressed, lead W goes to 0 (active) when the number of the pressed key appears on the data leads. Lead W will output a 1 (inactive) for each key that is not pressed. For example, assume key 24 (octal) is pressed. As the scan approaches and passes that key, the binary output will look like this:

Clock Cycles (in octal)	D5	D4	D3	D2	D1	D0	W
20	0	1	0	0	0	0	1
21	0	1	0	0	0	1	1
22	0	1	0	0	1	0	1
23	0	1	0	0	1	1	1
24	0	1	0	1	0	0	0
25	0	1	0	1	0	1	1

Note the 0 in Column W. It indicates that Key 010 100 is being pressed. This 0 can be used to activate data latches (not shown on this drawing) to capture and hold the number of the key being pressed.

As long as Key 010 100 is held down, the 0 will be output on W each time this key is tested. When it is released, W will return to 1 (inactive).

SCANNING CIRCUIT

Fig. 18-24 shows a circuit that might be used to scan an 8-by-8 keyboard. Only one switch is shown, but a NO (normally-open), pushbutton switch is at each intersection in the matrix.

U1 is a 0-to-64 ripple counter. As the clock runs, it counts from 000 000 to 111 111 again and again. Its count is directly output on data leads D0 through D5. The counter leads are also divided into two groups. The low-order three bits (0, 1, and 2) go to U3; the three high-order bits go to U2.

SCANNING THE MATRIX

During the first 7 clock cycles in Fig. 18-24 (000 000 through 000 111), the 3 high-order bits are 0. As a result, the decoder, U2, outputs a 0 on lead 0. This means that column 0 is to be scanned.

The 3 low-order bits are applied to the multiplexer. As the count proceeds from 000 000 to 000 111, the multiplexer, U3, connects each of the 8 row leads (one after the other) to lead W. If a key in Column 0 is pressed, a 0 will appear on W at the same time that the number of the key is output. If no key in this column is pressed, W remains inactive (at 1).

On the count following 000 111, the counter's output becomes 001 000. The 1 in the 3 high-order bits causes the decoder to output a 0 on lead 1 (001 represents 1 in the usual binary number system). As the count goes from 001 000 to 001 111, the second column is scanned by the multiplexer.

On the count after 001 111, the counter outputs 010 000, and the scan moves to Column 2 (010 represents 2 in the usual binary number system). Assume Key 24 (octal) is pressed. The 0 at output 2 of the decoder will be applied to input lead 4 of the multiplexer. When the count reaches 010 100, this 0 will be output on W. This 0 tells external circuits that the number on the data leads represents the pressed switch.

Note that only eight external data and control leads are used. Compared to the 64 needed for a direct decoding circuit, this is a small number. Note that the binary-to-decimal decoder at the top right in Fig. 18-24 is not a "BCD" decoder.

SUMMARY

For computers and other digital circuits to communicate with the outside world, interfacing must be provided. Ports, memory-mapped input/output, and

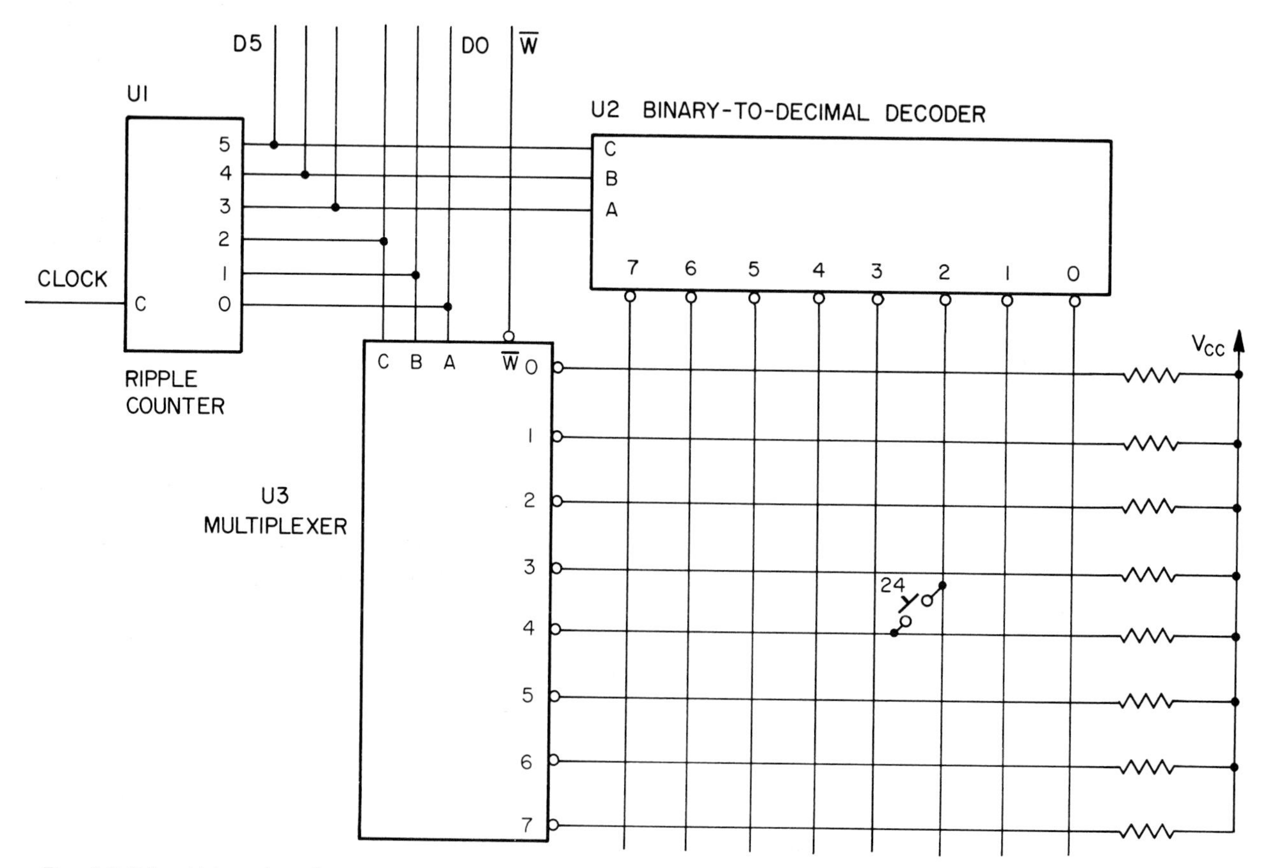

Fig. 18-24. Although only one pushbutton switch is shown, there is one at each intersection of the matrix.

direct memory access are used to accomplish input/output operations.

Light-emitting diodes and liquid-crystal displays are widely used to communicate at the outputs of small computers and other digital circuits. The voltage and power levels of these displays are compatible with those of widely used logic elements. Open-collector LED drivers are one use of high-state control.

Keypads and keyboards are often used as inputs to small computers and trainers. An interface must be placed between these devices and computers. These may take the form of simple encoders (for small keypads) or multiplexed systems (for larger keyboards).

IMPORTANT TERMS

Characteristic curve, Decoder/driver, Direct memory access, Download, LED, Liquid cryster, Memory-maped, Multplexed, Port, Reverse breakdown voltage, Scanning, Single crystal.

TEST YOUR KNOWLEDGE

1. A(n) __________ is an interface between a computer and the outside world.
2. Ports are identified __________ (by flags, randomly, by numbers, by parity, on a first-come-first-served basis).
3. Complete the following: Port __________ are latched, port inputs are __________.
4. Port-oriented input/output is often referred to as __________ (logic, memory-mapped, accumulator, asynchronous, coded) input/output.
5. Numbers from a data bus to an output port are stored in 8 or more __________ (data latches, matrices, diodes).
6. In most computers, data to be output is on the data bus for a __________ (second, millisecond, microsecond) or less.
7. Identifying a port number is called port number __________ (shifting, decoding, rectifying, diagramming).
8. In __________-mapped input/output systems, inputs and outputs are treated as ordinary memory locations.
9. Because numbers in memory-mapped input/output circuits do not pass through the accumulator, memory-mapped input/output is __________ (slower, faster) than port accumulator input/output.
10. When memory-mapped input/output is used, __________ (all 16, some of the 16) address leads are decoded.
11. To use direct memory access with a central controlling computer and a smaller receiving computer, the receiving computer is __________ (turned off, left on).
12. The term __________ (keyed, downloaded, accessed) means to transfer data in large amounts from a large computer to a smaller one.
13. What does LED mean?
14. What does LCD mean?
15. An LED can be made from __________ (gallium, indium, selenium) arsenide.
16. Gallium arsenide is allowed to cool into a(an) __________ (double, convex, single, amorphous) crystal structure.
17. A flat on the rim of an LED indicates the __________ (anode, cathode).
18. A __________-limiting resistor stabilizes LED operation.
19. LEDs are subject to damage by excess __________ (forward, reverse) breakdown voltage.
20. A 7-segment display can show numbers between 0 and __________ (D, E, F, G, H).
21. 7-segment LED displays are available in two forms—the common __________ type and the common __________ type.
22. When a given multidigit number is being displayed on the 7-segment LED display in Fig. 18-20, is each number held on continuously?
23. LEDs are current operated devices; LCDs are __________ operated.
24. LCDs require __________ (more, less) power than LEDs do.
25. Each __________ of a keypad or keyboard input is a SPST (single-pole, single-throw) pushbutton switch component.
26. Keys need some way to prevent errors due to switch __________.
27. Computer programs can be written that correct for key bounce. (True or False?)
28. A keyboard is scanned in __________ (more, less) time than it takes to press and release a key.
29. With the scanning type of system, one key is tested each __________ cycle.
30. In a keyboard scanning circuit, a NO (normally-open) pushbutton switch is at each __________ in the matrix.
31. Scanning of a keyboard uses __________ (more, fewer) leads than direct decoding does.
32. Pads with 16 or fewer keys are usually __________ (scanned, directly encoded).
33. Compared to LEDs, ordinary diodes have a __________ (low, very high) reverse breakdown voltage specification.
34. Do all diodes emit light?

STUDY PROBLEMS

1. The circuit in Fig. P18-1 contains an input port and an output port. Which lead, A or B, is control lead OUT?
2. In the circuit in Problem 1 (Fig. P18-1), is OUT active-high or active-low?
3. What is the port number for the input port in the circuit from Problem 1 (Fig. P18-1)? Give your answer in hexadecimal. What is the port number for the output port?
4. For the control and address signals shown in the illustration for Problem 1 (Fig. P18-1), which of the following actions will occur?
   a. Input.
   b. Output.
   c. Neither.
5. For the control and address signals shown in Fig. P18-2, which of the following actions will occur?
   a. Input.
   b. Output.
   c. Neither.

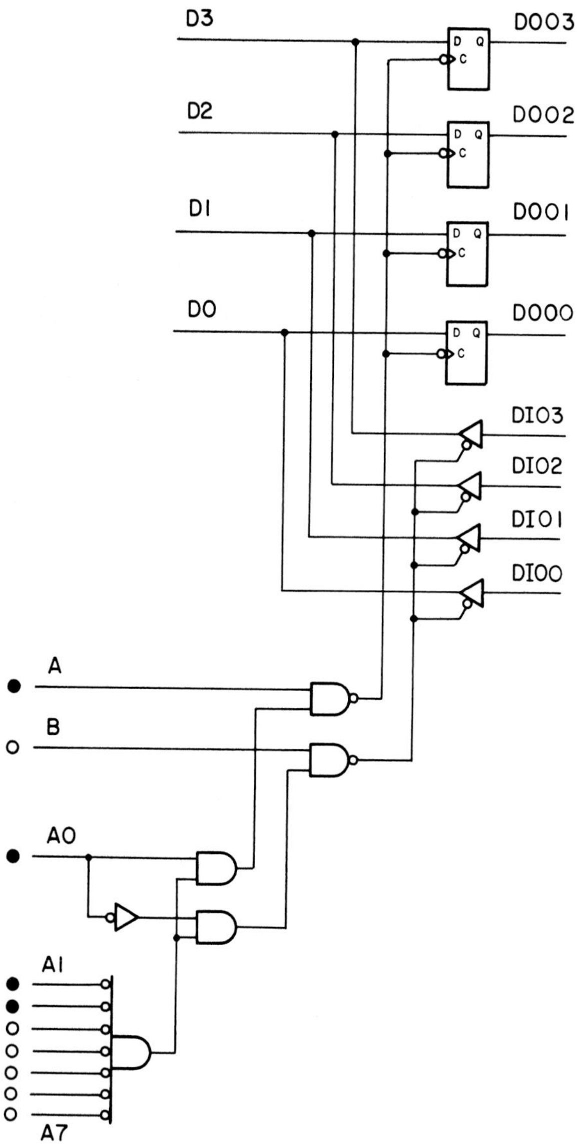

Fig. P18-1. Leads controlling the selection of an input port and an output port can be identified. Note that A0 belongs with A1 through A7.

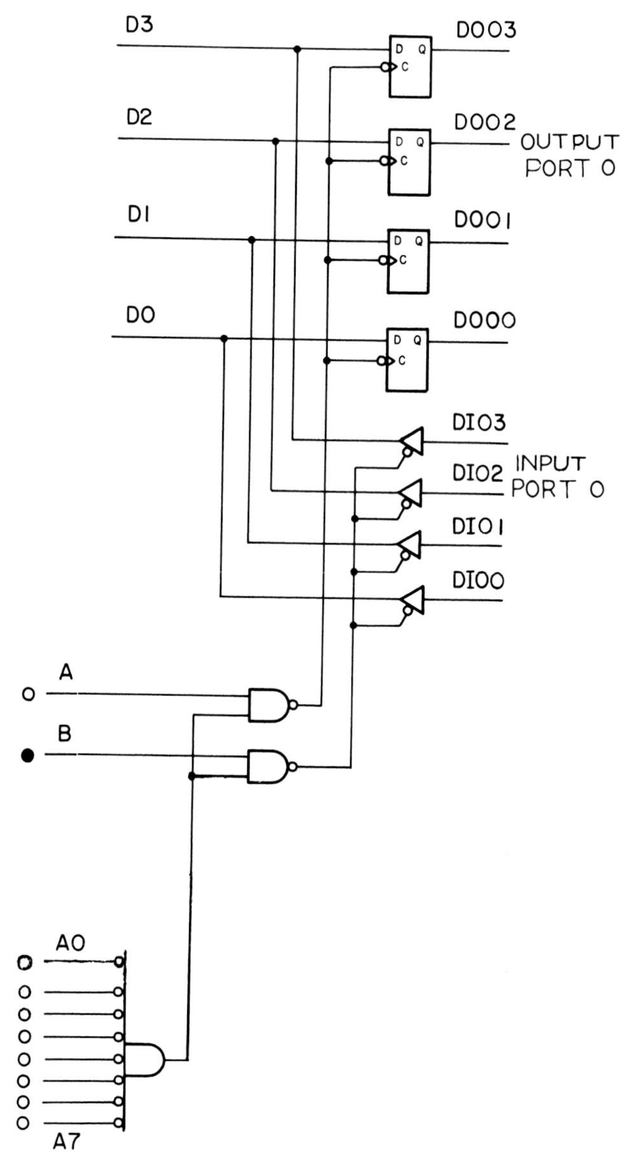

Fig. P18-2. For a circuit with an input port and output port, the action that occurs can be described.

6. Which of the following best states the general rule for input/output ports?
   a. Outputs are buffered; inputs are latched.
   b. Outputs are latched; inputs are buffered.
   c. Inputs and outputs must both be latched.
7. Fig. P18-3 shows a memory-mapped input/output circuit with input levels given. To which of the following computer control leads would the lead labeled B most likely be attached?
   a. OUT.
   b. IN.
   c. MEMORY WRITE.
   d. MEMORY READ.

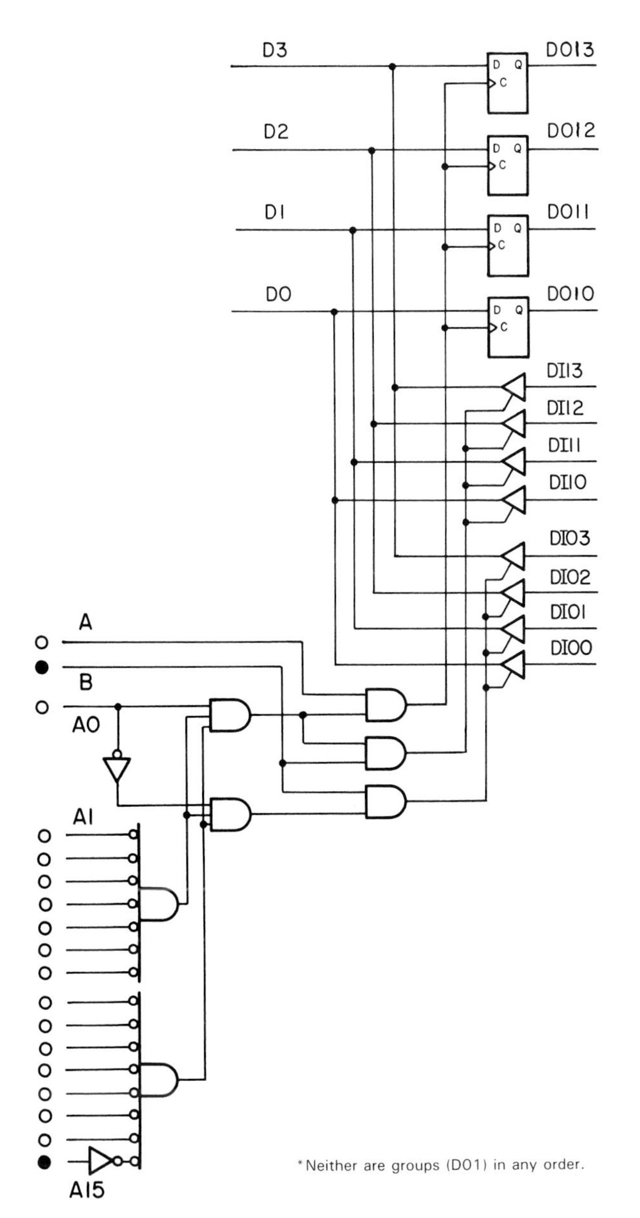

Fig. P18-3. Addresses and action for input port and output port can be determined.

8. Using lower left of Fig. P18-3, list in hexadecimal the addresses of the three ports. Do not look to subgroup (D013, DI13, DI03), for subgroup does not give group (D01, DI1, DI0) address.*
   a. Address of output grouping labeled D01.
   b. Address of input grouping labeled DI0.
   c. Address of input grouping labeled DI1.
9. Which instruction could be used to output data through DO1 of the circuit from Problem 7 (Fig. P18-3)? Ignore 3,2,1 from "D013," etc.
   a. OUT (output data through a port).
   b. STA (store the accumulator in memory).
   c. INC M (increment the number in a given memory location).
10. For the control and address signals shown in the illustration for Problem 7 (Fig. P18-3), which of the following actions will occur?
    a. Output through DO1.
    b. Input through DI1.
    c. Input through DI0.
    d. None of these.
11. Mark each of the following P to indicate an advantage of port-oriented input/output or M for an advantage of memory-mapped input/output.
    a. Many instructions available for accomplishing input/output.
    b. Can support larger number of input/output circuits.
    c. Two-byte instruction (other needs 3).
    d. Fewer address leads to decode.
12. Which form of input/output circuit, (a) port-oriented or (b) memory-mapped, is often referred to as accumulator input/output?
13. When is DMA (direct memory access) used for input or output?
    a. When an inexpensive form of input/output is needed.
    b. When small amounts of data must be handled rapidly.
    c. Two-byte instruction (other needs 3).
    d. Fewer address leads to decode.
14. Trace the circuit in Fig. P18-4 onto a piece of paper and add an LED symbol.

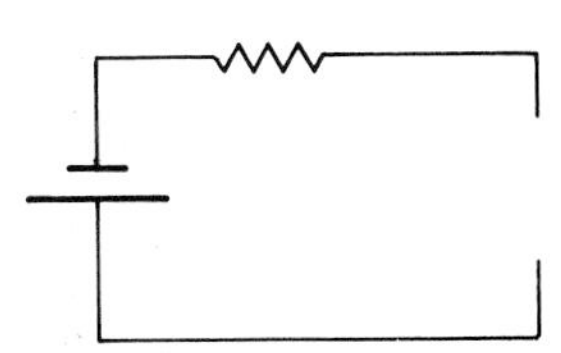

Fig. P18-4. Proper scale of a drawing and current direction for an LED can be created.

15. Trace the axes in Fig. P18-5 onto a sheet of paper and label them If—forward current in mA and Ef—forward voltage. Then sketch the approximate characteristic curve of a typical LED on your graph.

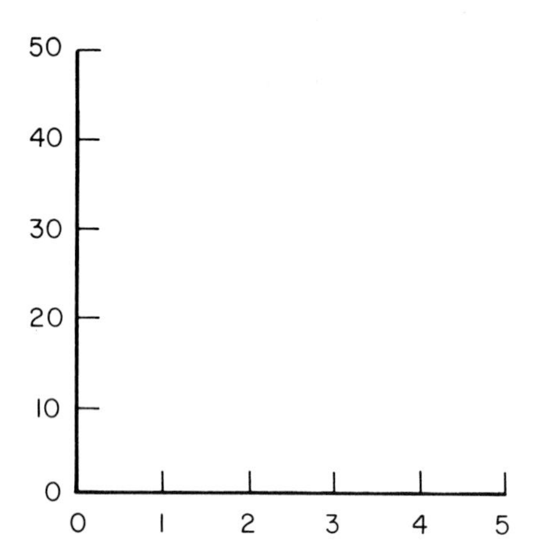

Fig. P18-5. Characteristic curve for an LED can be drawn.

16. Explain the need for a resistor in series with LED components.
17. To light the lamp in the circuit in Fig. P18-6, what signal, 1 or 0, must be applied at input A?
18. Using the equation in the text, determine R in the circuit in Problem 17 (Fig. P18-6) for each of the following diode currents:
    a. 10 mA (0.01 A).
    b. 18 mA (0.018 A).
    c. 30 mA (0.03 A).

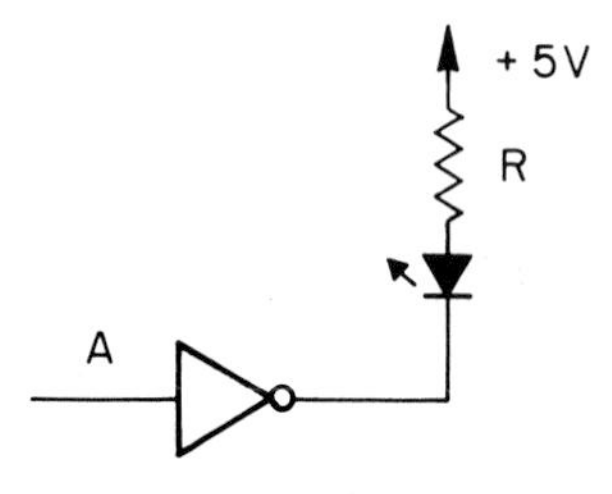

Fig. P18-6. Signal levels for an LED and values for a series resistor with an LED can be determined.

19. Which circuit in Fig. P18-7 contains a common-cathode 7-segment display?

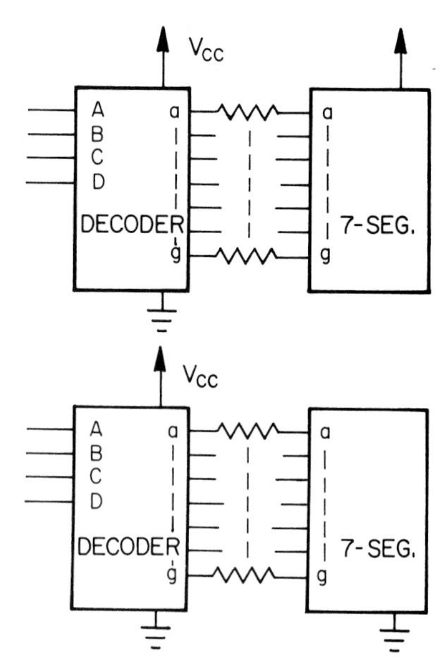

Fig. P18-7. Two types of common leads on 7-segment displays can be identified.

20. Describe the two functions of a decoder/driver in a 7-segment display circuit.
21. Referring to Fig. P18-8, the advantage of the single-resistor circuit for a 7-segment display is easy to see and describe. What is its disadvantage?

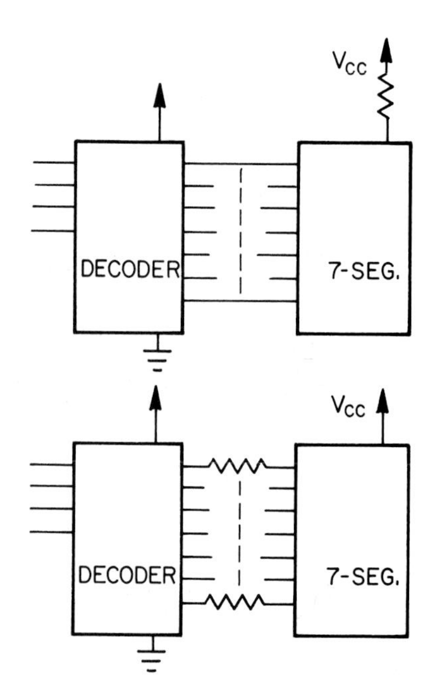

Fig. P18-8. Typical 7-segment displays with and without a common resistor to Vcc can be compared.

22. The diodes in Fig. P18-9 are reverse biased. Which is likely to be damaged?

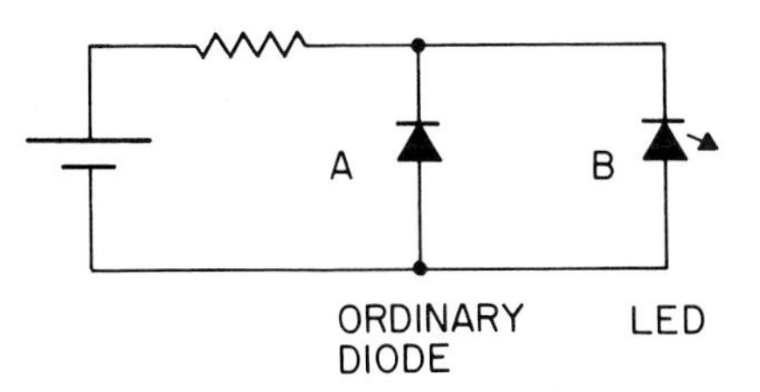

Fig. P18-9. A situation that might cause trouble or damage should be identified. Two types of diode can be compared.

23. Mark each of the following LED or LCD:
    a. Voltage operated.
    b. Requires the greater power.
    c. Self illuminating.
24. In multiplexed displays, each digit is on for only a portion of the time. However, each digit appears to be on all the time. Describe how this is possible.
25. Multiplexing is normally used on (a) larger or (b) small keypads and keyboards.

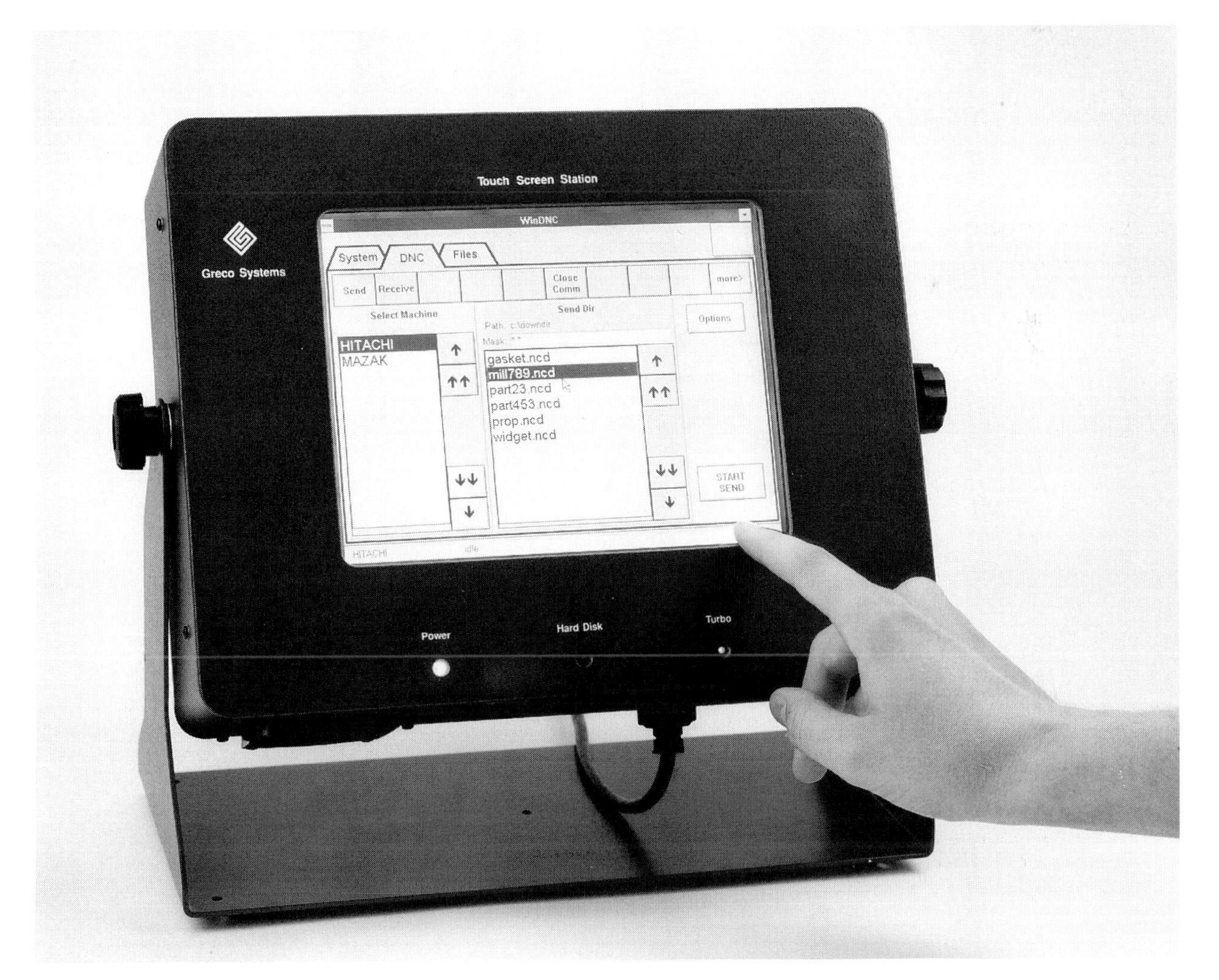

This compact industrial computer uses its screen for both output and input. It is a touch-screen device so the use of a keyboard is not required. (Greco Systems)

CD-ROM drives hold disks similar to music CDs. Many CD-ROM drives are now capable of playing music as well as CD-ROM material.

Chapter 19

Computer Organization and Microprocessors

LEARNING OBJECTIVES

After studying this chapter and completing lab assignments and study problems, you will be able to:

- Explain the difference between machine languages and high-level languages.
- Explain the difference between machine languages and high-level languages.
- Explain why all programs must be in machine language before they can be executed.
- List the steps in the approach needed to follow the action of a simple machine language program.
- Match the names of the basic sections of a microprocessor with the blocks of a simple block diagram.
- Based on a simplified block diagram, list procedures used to follow the action of a microprocessor as a program is executed.

This chapter introduces computer organization. It describes how the circuits studied in this book contribute to the operation of computers. The execution of a simple program will be used to show the interactions of these circuits.

Even small computers contain thousands of logic gates and flip-flops. If all elements were shown on computer diagrams, the resulting drawings would be too complex to be useful. As a result, block diagrams are used. By omitting details, interactions of major circuits are emphasized. You should remember, however, that you have studied the circuits that make up the individual blocks.

HIGH LEVEL LANGUAGES

A computer language consists of a set of symbols which represents tasks that can be performed by computers. Each language has a detailed set of rules that must be followed when writing programs.

You have probably heard of the computer languages called BASIC and FORTRAN. You may have written programs using these or similar languages. Such languages are described as HIGH LEVEL languages. They are designed to be convenient for humans rather than convenient for computers.

HIGH LEVEL LANGUAGE PROGRAMS

Fig. 19-1 shows a short program written in a high-level language. Even without a knowledge of programming, it is relatively easy to follow an explanation of what it accomplishes.

```
10 CLS
20 PRINT "WHAT IS YOUR FIRST NAME?";
30 INPUT A$
40 PRINT "WHAT IS YOUR LAST NAME?";
50 INPUT C$
60 PRINT "HELLO, "A$"!"
70 GO TO 100
   COMMENT--READ C$; ALPHABETIZE
80 END
```

Fig. 19-1. The symbols and rules used in high level computer languages are similar to those of ordinary English.

Execution begins at the top and proceeds step by step. Line 10 clears the screen. Line 20 places the words WHAT IS YOUR FIRST NAME? on the screen. Then line 30 causes the computer to wait until the user types his or her name (assume JANE of JANE DOE). It stores this information for future use. Whenever A$ appears in this program, the computer substitutes the name typed in at this point. Line 40 and line 50 put the person's last name (DOE) into the storage area C$. Line 60 places the message HELLO, JANE! on the screen. Line 70 starts a program dealing with the last name, such as alphabetizing all of the last names people have typed in that day. The computer returns to line 80. Finally, line 80 tells the computer that it has completed the program.

INTERPRETERS AND COMPILERS

While programs written in high level languages are understood by humans, they are not acceptable to computers. The English-like symbols cannot be directly used to control the action of a computer.

Before they can be executed, high level programs must be translated to a machine-compatible language. This is accomplished by special computer programs called INTERPRETERS and COMPILERS. These programs translate the words and numbers of the high level programs into machine language programs made up of the 1s and 0s that can be stored and manipulated within computers. See Fig. 19-2.

Many programmers work exclusively in high level languages. However, hardware personnel are responsible for what happens within computers. As a result, they should be familiar with machine language programming.

MACHINE LANGUAGE

Fig. 19-3 shows a program stored in the memory of a microprocessor-based computer. It is in machine language. That is, it is a series of binary numbers that can directly control the operation of the computer.

MEMORY LOCATION (HEX)	STORED NUMBER (BINARY)
00 00	0000 1100
00 01	0000 1000
00 02	0000 0100
04 08	1010 0011
04 09	0000 0101
04 0A	1000 0010
04 0B	0000 1000
04 0C	0110 0110
04 0D	0000 1000
04 0E	0000 0100
04 0F	0000 1100
04 10	0010 0000
04 11	0000 0100
04 20	BEGINNING OF REST OF PROGRAM

Fig. 19-3. In a machine language program, a given number may represent an op code, an address, data, or a port number.

Note the following: All programs must be in machine language before they can be executed. The designer must learn to think in machine language.

MEANING OF NUMBERS IN A MACHINE LANGUAGE PROGRAM

The binary number in a given memory location (other than the first) may represent an op code (operation code), an address, a byte of data, or a port number. Therefore, it is not possible to determine the meaning of a given number without reference to all the numbers that come before it in the program. At

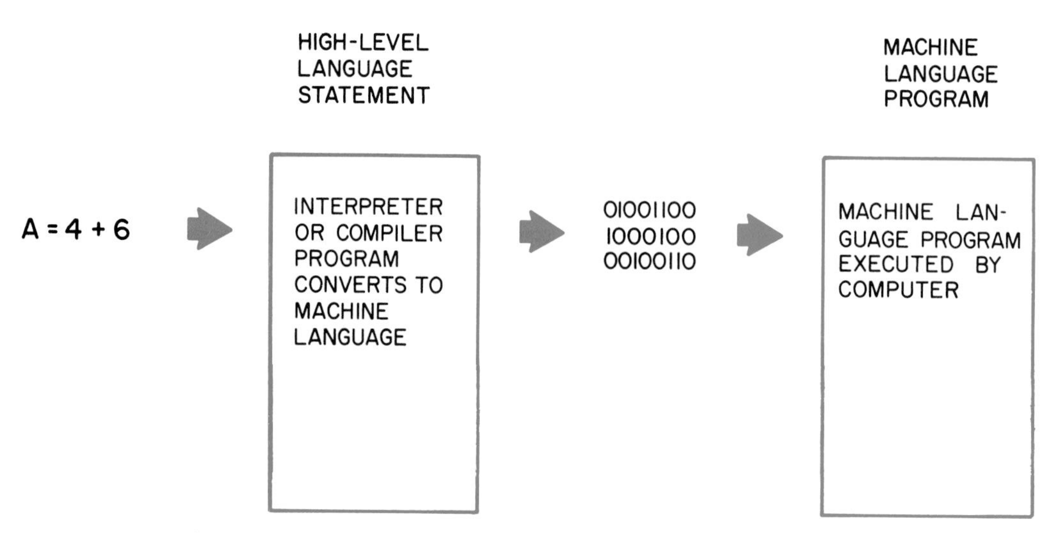

Fig. 19-2. High level language statements cannot be directly executed by computers. They must first be converted to machine language.

one point in a program, the number 0001 0101 might be an op code. At another, that same number 0001 0101, might be part of an address.

ANALYSIS OF A MACHINE LANGUAGE PROGRAM

To analyze a machine language program, a process similar to that used by the computer when executing the program can be applied. To describe this process, the program in Fig. 19-3 will be discussed. Assume that it is part of a larger program used to control a metalworking machine. The program is stored in ROM in the microprocessor-based computer shown in Fig. 19-4.

The purpose of the small program in Fig. 19-3 is to sense the pressing of START (a pushbutton switch that turns on the machine). When the computer is first turned on, the program samples the condition of START again and again. If it finds that it has not been pressed, it does nothing. When START is pressed, however, it signals the rest of the program to start and control the operation of the machine.

Fig. 19-5 shows START connected to lead DI3 of Port 05. For safety reasons, START is active-low. That is, lead DI3 is active-low. A 0 is delivered when START is pressed.

The binary numbers in the program are typical of those used in present-day microprocessors. However, they were not taken from the instruction set of a specific microprocessor. For this reason, there is no way to know their meaning beforehand. The concepts of machine language programming, not the actual numbers used, are important here.

A specification sheet for each unit is needed. It will save time, money, and "rework."

INSTRUCTION AT 00 00

When the signal is given to run the program in Fig. 19-3, the microprocessor automatically goes to memory location 00 00 for its first instruction. The programmer must design the program so that the number stored at 00 00 is the op code of the first instruction in the program.

Part "a" of Fig. 19-6 shows the signals that appear on the three buses during the fetch of this op code. The MPU (microprocessor unit) places the address 00 00 on the address bus. It also causes the memory-read lead in the control bus to go active. The ROM responds by placing the stored number on the data bus within a microsecond.

When the stored number is delivered to the MPU, it is decoded, and the computer learns what it is to do first. In this case, the op code 0000 1100 is a JUMP. That is, the computer is to look to another part of memory for its next instruction.

Other programs use a jump instruction for a simple program design. Many programs begin with a jump instruction. When the run signal is given, many microprocessors go to location 00 00 for the first instruction. Often, however, programs are stored elsewhere in memory. JUMP sends the computer to the proper place in memory.

The address of the next instruction requires 16 bits, so two bytes of memory are required. These two bytes are stored in the two memory locations following the op code for JUMP. See Fig. 19-7.

When 0000 1100 (the op code for JUMP) is decoded, the computer knows that the numbers in the next two memory locations must be brought in. Parts "b" and "c" of Fig. 19-6 show these address fetch operations. Based on the numbers 0000 1000 (low-order

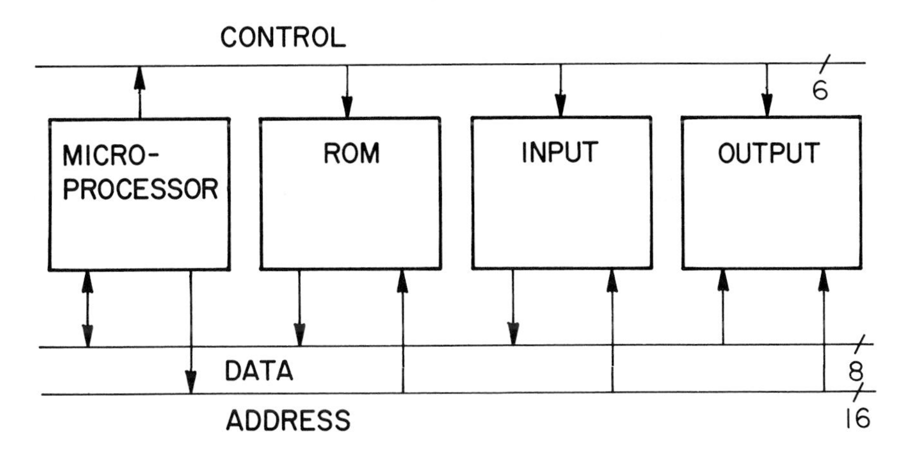

Fig. 19-4. A microprocessor-based computer such as this might be used to control a metalworking or assembly machine.

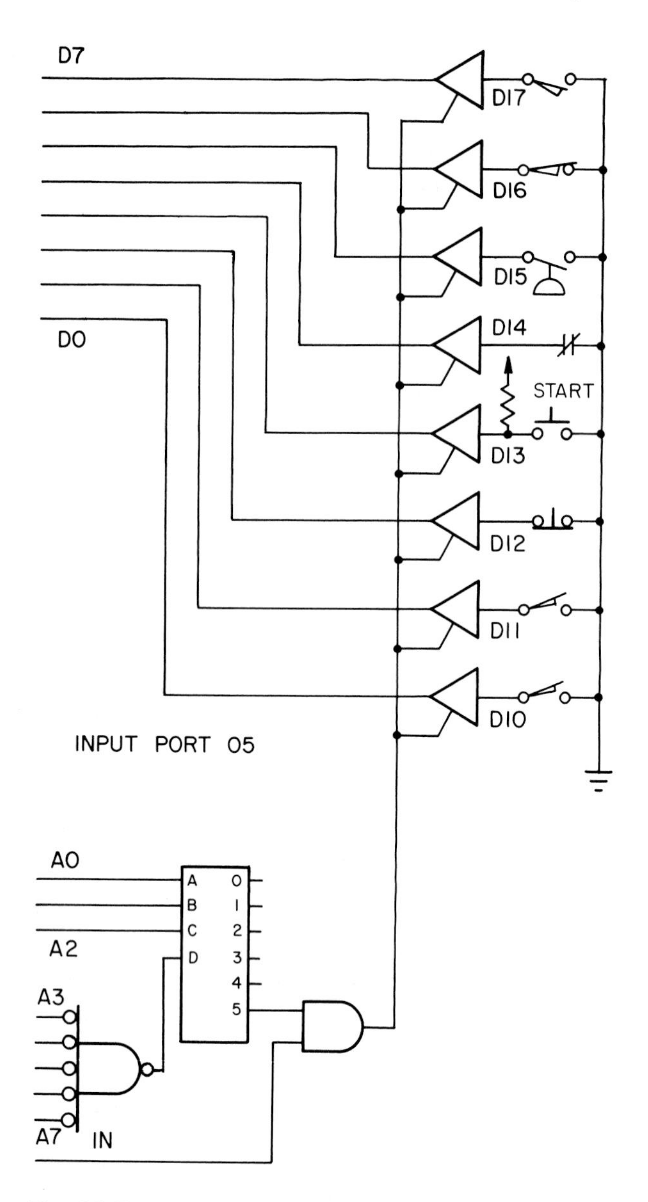

Fig. 19-5. The state of pushbutton START is input on data lead D13 through Port 05.

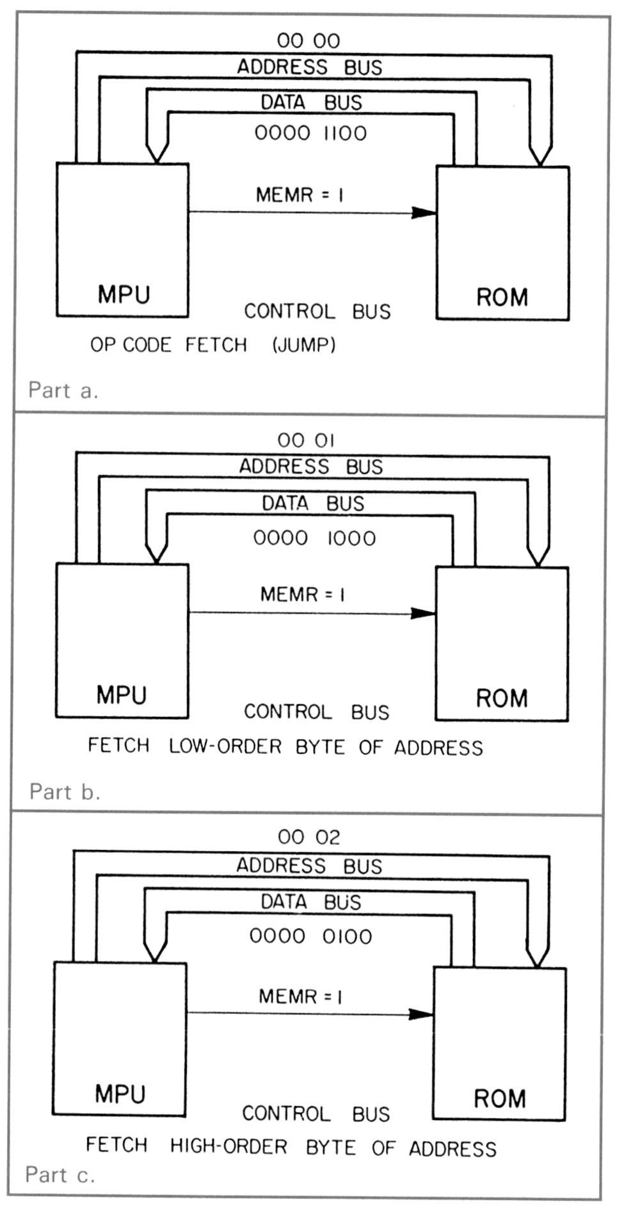

Fig. 19-6. When the RUN signal is given, the MPU goes to memory location 00 00 and brings in the first instruction.

byte) and 0000 0100 (high-order byte), the MPU now knows that its next op code is stored in memory location 04 08 (hexadecimal).

INSTRUCTION AT 04 08

The computer is ready for another instruction. Based on the JUMP, it knows that the instruction is located in memory location 04 08. Part "a" of Fig. 19-8 shows the fetch of the new op code.

When the number stored at 04 08 has been delivered to the MPU and decoded, the computer learns that

MEMORY LOCATION (HEX)	STORED NUMBER (BINARY)	
00 00	0000 1100	Op code, JUMP
00 01	0000 1000	Low-order byte of address
00 02	0000 0100	High-order byte of address

Fig. 19-7. The displayed JUMP instruction requires three bytes of memory—one for its op code and two for the address of its destination.

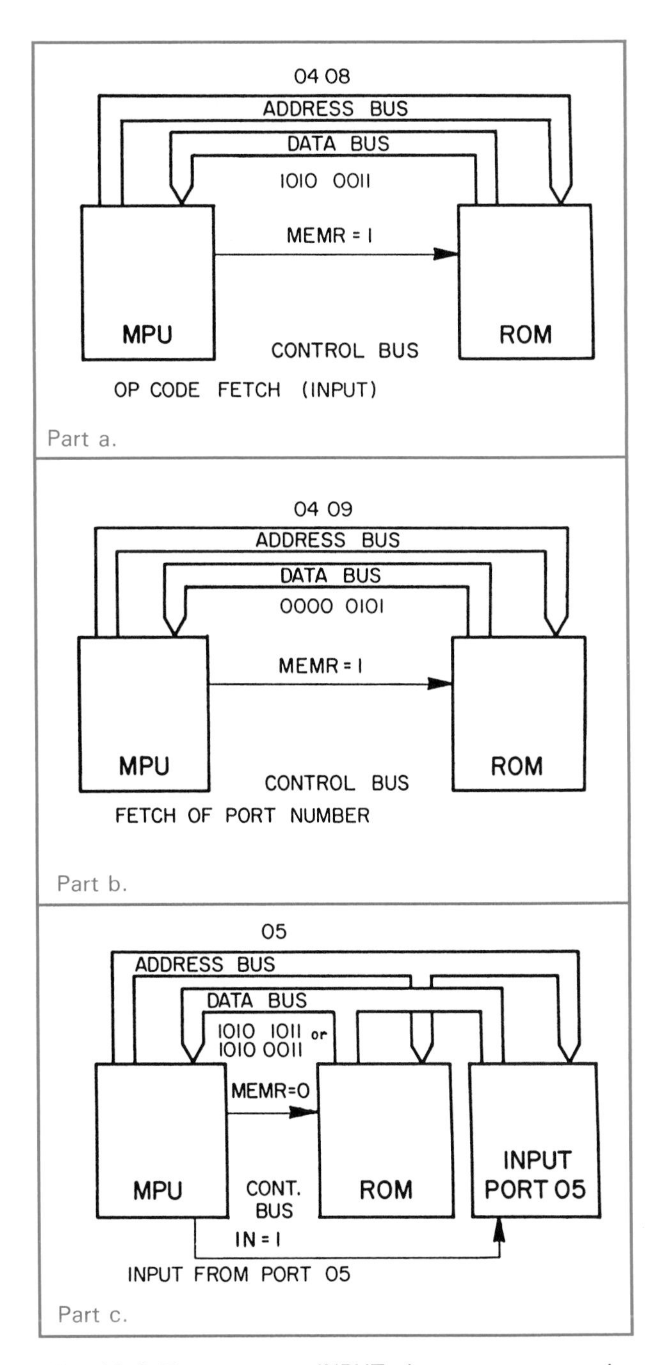

Fig. 19-8. To execute an INPUT, the computer uses the bus structure three times—twice to fetch the instruction and once to do the actual input.

MEMORY LOCATION (HEX)	STORED NUMBER (BINARY)	
04 08	1010 0011	Op code, INPUT
04 09	0000 0101	Port number for input (Port 05)

Fig. 19-9. The displayed INPUT instruction requires two bytes of memory. One contains the op code. The other stores the port number.

it is to do an input. This op code also tells it that the required port number is stored in the next memory location. Refer to memory location 04 09 in Fig. 19-9. Port 05 (0000 0101) will be used. Part "b" of Fig. 19-8 shows the fetch of the port number.

Part "c" of Fig. 19-8 shows the execution of the INPUT instruction. The MPU places the port number on the low-order byte of the address bus and causes control lead IN to go active. The circuit for Port 05 (see Fig. 19-5) responds by placing the number at its input on the data bus. Note that memory read is 0. The ROM is not used.

Based on the positions of the switches in Fig. 19-5, the number on the data bus will be 1010 1011 (assuming START has not been pressed). If START is pressed, the number will be 1010 0011. Note the change in bit D3.

This number is placed in the MPU's accumulator for later use. This completes the execution.

Before going on, note that JUMP requires three bytes of memory; INPUT requires only two. Circuits within the MPU take this situation into account. When the op code 0000 1100 (JUMP) has been decoded, the circuits know that the address of the jump is stored in the next two memory locations. They take the necessary action to fetch them. If the address is not there or is in the wrong order (the low-order byte must be first), an error will result.

When the op code 1010 0011 (INPUT) is decoded, the MPU is wired to fetch the port number from the next memory location. When programs are written for the computer just described, a port number must always follow an INPUT op code. That is, the circuits in the MPU dictate the way in which programs are written.

INSTRUCTION AT 04 0A

Upon completing the INPUT, the MPU looks to the next memory location (04 0A) for the next op code. See Fig. 19-10 and Fig. 19-11. This op code tells the computer to AND the number in the next memory location with the number in its accumulator.

Remember, the purpose of this small program is to detect the pressing of START. The number in the accumulator contains some information about the

MEMORY LOCATION (HEX)	STORED NUMBER (BINARY)	
04 0A	1000 0010	Op code, AND
04 0B	0000 1000	Number (MASK) to be ANDed with accumulator

Fig. 19-10. The displayed AND instruction requires two bytes of memory—one for its op code and one for the number to be ANDed with the number in the accumulator.

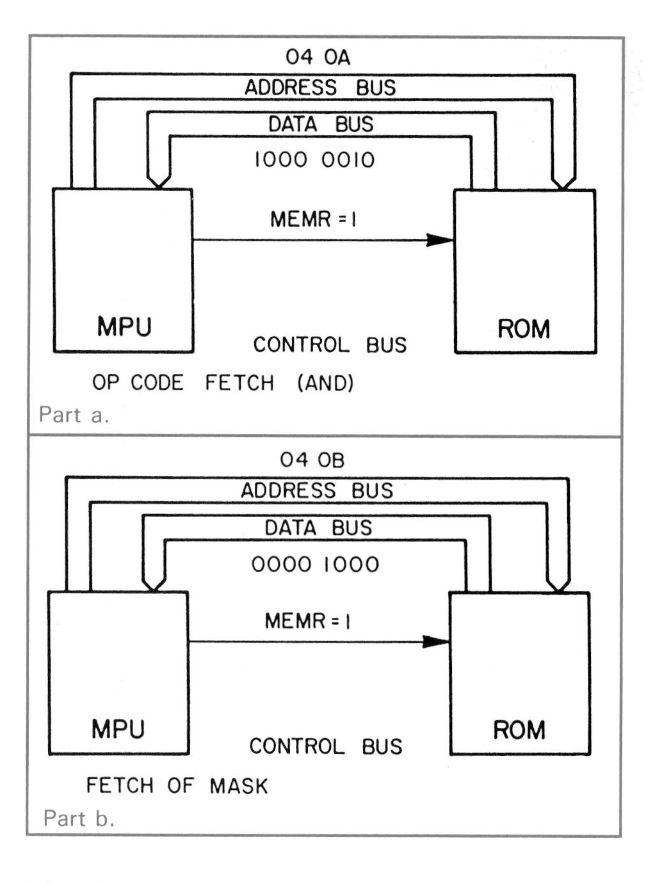

Fig. 19-11. To execute the AND instruction displayed in Fig. 19-10, the computer used the bus structure only twice. The ANDing takes place within the MPU.

condition of START. However, that number also contains unwanted information about the 7 other switches connected to Port 05.

To get rid of the unwanted information, a process called *MASKING* is used. That is, all bits of the number in the accumulator, except D3, are set to 0. This is done by ANDing the number in the accumulator with a MASK. In this case, the MASK is the number 0000 1000. Its use is shown below:

START not pressed:

Accumulator before ANDing	1010 1011
MASK to be ANDed with Acc.	0000 1000
Accumulator after ANDing	0000 1000

START pressed:

Accumulator before ANDing	1010 0011
MASK to be ANDed with Acc.	0000 1000
Accumulator after ANDing	0000 0000

If the accumulator is zero after the masking operation, START is being pressed. If it is nonzero (anything other than 0000 0000—in this case 0000 1000), START is NOT being pressed.

Because ANDing is a logic operation, flags will be set. Here, the ZERO FLAG (Z) is of interest. If START is not pressed, the accumulator equals 0000 1000 after the masking operation. As a result, $Z = 0$ implies START has not been pressed. If START is pressed, the result of the masking is 0000 0000 and $Z = 1$. That is:

$Z = 0$ implies START NOT pressed

$Z = 1$ implies START is being pressed

With the completion of this instruction, the MPU looks to the next memory location for another op code. The ROM must be read again.

INSTRUCTION AT 04 0C

At this point, the computer must make a decision. Was START pressed during its sweep through the program? In machine language programs, such decisions are made by testing flags. Flags are usually tested by using CONDITIONAL JUMP INSTRUCTIONS. That is, a jump is made if a certain flag condition is present. If it is not, the jump does not occur.

The op code at 04 0C calls for a conditional jump. See Fig. 19-12. If the ZERO FLAG is not set (if $Z = 0$), the computer makes the indicated jump. The address of the jump is stored in the next two memory locations. If $Z = 1$ (if this flag is set), the jump will not take place. Rather, the MPU will ignore the

MEMORY LOCATION (HEX)	STORED NUMBER (BINARY)	
04 0C	0110 0110	Op code, JUMP if Z = 0
04 0D	0000 1000	Low-order byte of address
04 0E	0000 0100	High-order byte of address

Fig. 19-12. The displayed instruction is a conditional jump. It takes place only if the ZERO FLAG is cleared ($Z = 0$).

JUMP instruction and go to the next op code (the one stored at 04 0F).

To accomplish the conditional jump instruction, the MPU brings in the op code and two bytes of address. See Fig. 19-13. It then looks at the ZERO FLAG. If Z = 0, it jumps; if Z = 1, it does not jump.

Assume that START has NOT been pressed (the machine should not start). When the conditional jump is executed, the computer finds Z = 0. It makes the jump. Note the address of that jump—04 08. This is the beginning of the small program. The computer is said to be in a *LOOP*. As long as START is not pressed, it will run through the program up to the conditional jump. Then it will start over. It will sample the condition of START again and again.

When START is pressed, the situation changes. The result of the masking will be 0000 0000, so Z will equal 1. The conditional jump will not be made, and the MPU will look to location 04 0F for its next op code. The computer has left the loop.

INSTRUCTION AT 04 0F

Fig. 19-14 and Fig. 19-15 show the next instruction and its transfer to the MPU. Note that it has the same op code as the instruction at 00 00. That is, it is an UNCONDITIONAL JUMP. Every time this instruction is encountered, the indicated jump will be made. Note how this differs from the conditional jump. A conditional jump is most often used.

The destination of the jump, 04 20, is the beginning of the program that starts and controls the machine. The little program has accomplished its task. Until START was pressed, the computer was in a loop, sampling the condition of START again and again. When the switch was pressed, the little program directed the computer to the beginning of the main program for more instructions. Remember, the little program made the decision to move to the main program by testing flags.

ACTION WITHIN THE MICROPROCESSOR

So far, little attention has been given to the action within the microprocessor. This will be the purpose of the next 7 sections. These sections are as follows:

1. Microprocessor block diagram.
2. Start of execution in the microprocessor.
3. Execution of jump instruction in microprocessor.
4. Execution of input instruction in microprocessor.
5. Execution of AND instruction in microprocessor.
6. Execution of conditional jump instruction in microprocessor.
7. Execution of unconditional jump instruction in microprocessor.

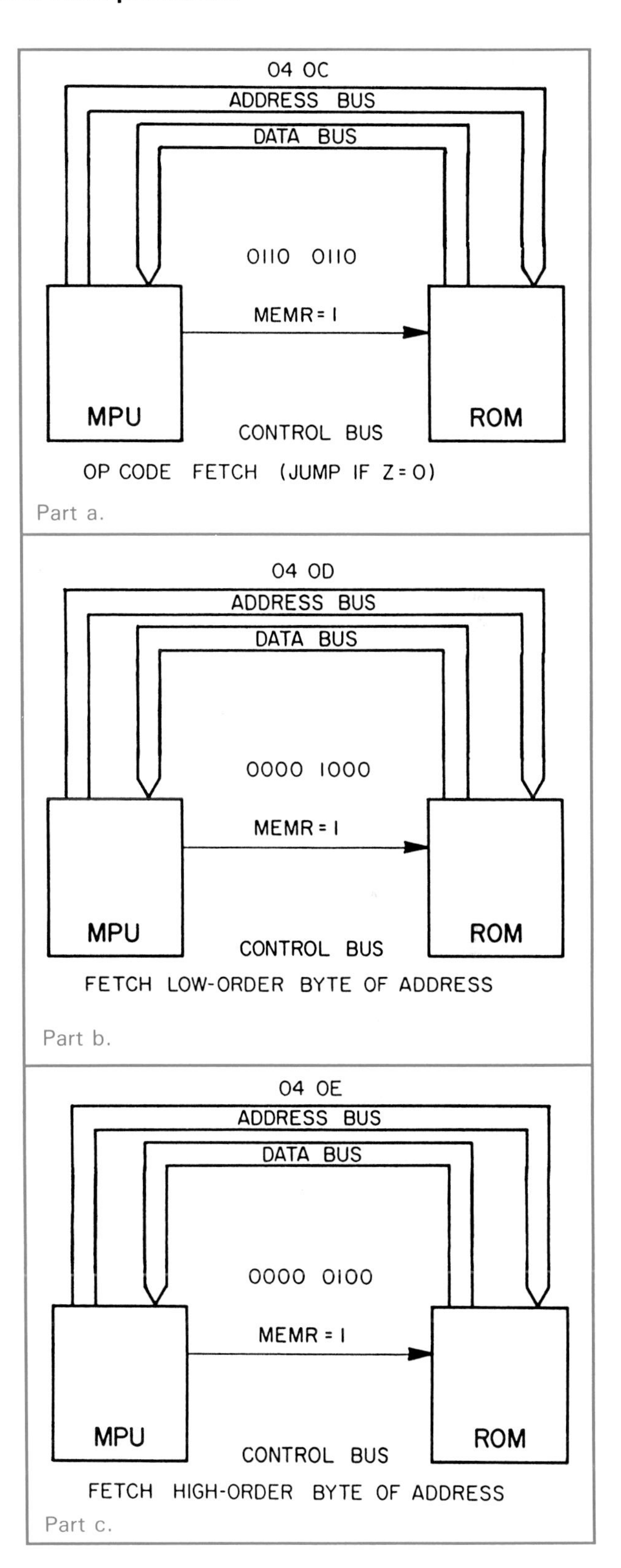

Fig. 19-13. To execute the conditional jump instruction displayed in Fig. 19-12, the op code and two bytes of address must be brought in from memory.

MEMORY LOCATION (HEX)	STORED NUMBER (BINARY)	
04 0F	0000 1100	Op code, JUMP
04 10	0010 0000	Low-order byte of address
04 11	0000 0100	High-order byte of address

Fig. 19-14. Except for the address of the destination, the displayed unconditional jump instruction is identical to that stored at 00 00.

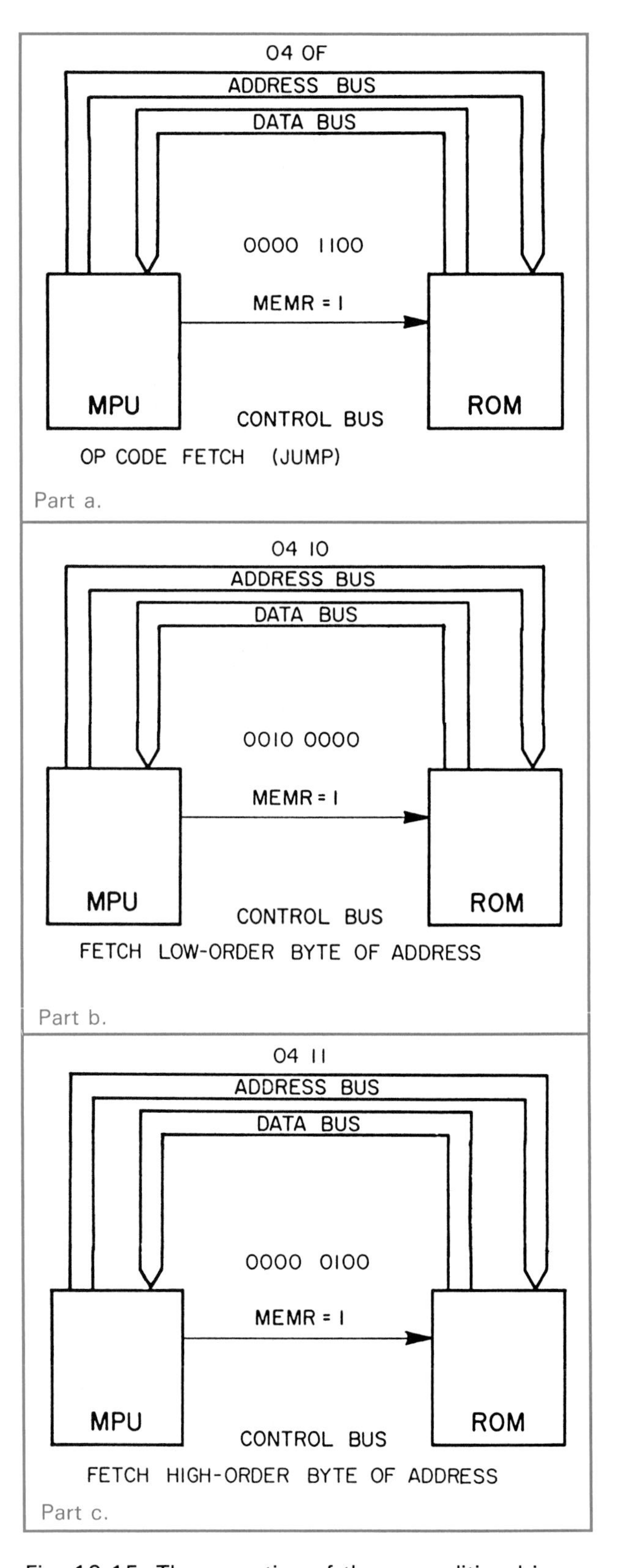

Fig. 19-15. The execution of the unconditional jump instruction displayed in Fig. 19-14 is identical to that of the unconditional jump at 00 00 in Fig. 19-7.

MICROPROCESSOR BLOCK DIAGRAM

Fig. 19-16 is a block diagram of a microprocessor. Although it is simplified, enough detail is given to permit a discussion of microprocessor operations.

The diagram may appear complex, but you are familiar with the circuits used in most blocks. For example, you know that the ALU (arithmetic/logic unit) contains circuits for doing addition, subtraction, ANDing, ORing, incrementing, decrementing, and shifting. You also know that these operations set flags in the FLAG REGISTER. These two blocks (the ALU and FLAG REGISTER) are not difficult to understand.

You also know that numbers to be added, ANDed, etc. must be stored (at least temporarily) in registers next to the ALU. This, then, is the probable function of the ACCUMULATOR and TEMPORARY REGISTER 3. You also know that the results of most arithmetic/logic operations are returned to the ACCUMULATOR.

You know that an op code is brought into the microprocessor at the beginning of each instruction. Note the block at the right marked INSTRUCTION REGISTER. You can probably guess that the op code is stored here while a given instruction is being executed.

Just below the INSTRUCTION REGISTER is the INSTRUCTION DECODER. This circuit determines the meaning of each stored op code. That is, when a number such as 0000 1100 (from the program just considered) appears in the INSTRUCTION REGISTER, this decoder determines that an unconditional jump is to be made.

Based on the output of the INSTRUCTION DECODER, CONTROL AND TIMING controls the execution of the instruction. It sends signals to circuits within the MPU and uses the control bus to tell the memory and the input/output circuits what to do.

In the center block above the program counter in Fig. 19-16, 6 registers are shown. The first two are called temporary. They are reserved for use by the

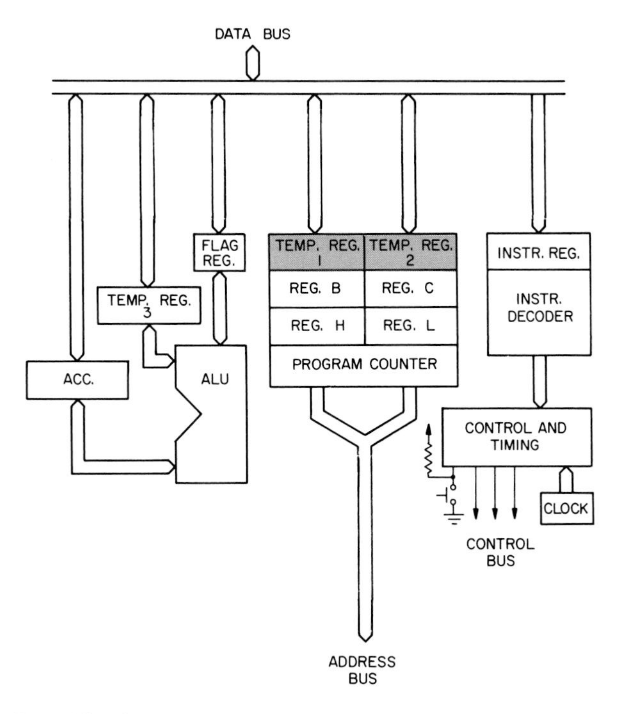

Fig. 19-16. The block diagram shown is typical of present-day microprocessors.

computer. The other 4 registers can be used by programmers.

The last register, the PROGRAM COUNTER, is special. Its function will be described in the following example.

The program counter is more important than other registers. One can easily examine its use.

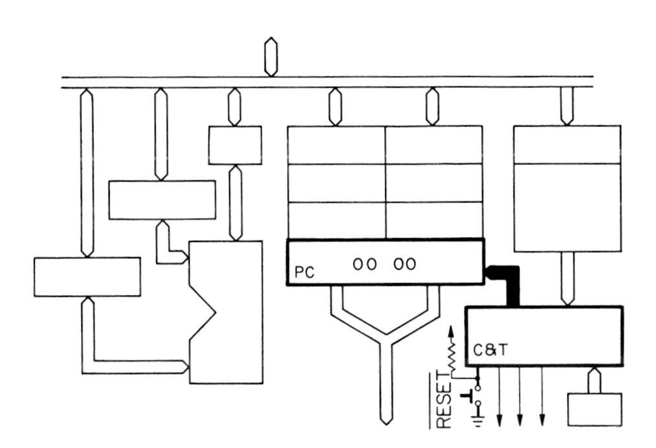

Fig. 19-17. When RESET is pressed, the control and timing section places the number 00 00 (hexadecimal) in the program counter.

START OF EXECUTION IN THE MICROPROCESSOR

Remember, the program in Fig. 19-3 is stored in ROM. All that is necessary to start its execution is to press RESET. See the lower right corner of Fig. 19-17 for a typical reset circuit.

When RESET goes active (in this case, low), a circuit in CONTROL AND TIMING automatically sends 0000 0000 0000 0000 to the 16-bit PROGRAM COUNTER (PC). That is, the hexadecimal number 00 00 is stored there. The machine is ready to fetch its first op code.

EXECUTION OF JUMP INSTRUCTION IN MICROPROCESSOR

To review, the first instruction in the program is:

Memory Location	Stored Number	
00 00	0000 1100	Op code for JUMP
00 01	0000 1000	Low-order byte of address
00 02	0000 0100	High-order byte of address

A step-by-step analysis shows the operation of the microprocessor.

FETCH OF OP CODE

The op code stored at 00 00 is brought in first. To do this, CONTROL AND TIMING sends a number of control signals. See Fig. 19-18. The dashed line represents control signals sent within the MPU. The heavy arrow indicates the activation of MEMR in the control bus.

Note the action of CONTROL AND TIMING. First, the number in PC is placed on the address bus by CONTROL AND TIMING. CONTROL AND TIMING also causes MEMR to go active. Memory responds by placing the number stored at 00 00 on the data bus. Then CONTROL AND TIMING signals INSTRUCTION REGISTER to latch the number on the address bus. The op code is now ready to be decoded.

DECODING OF OP CODE

The op code in INSTRUCTION REGISTER is delivered to INSTRUCTION DECODER. When decoding is complete, the MPU knows that it is to do an unconditional jump and the address of the destination is in the next two memory locations. No further instruction is necessary. The next two memory locations are given priority and treated as one address.

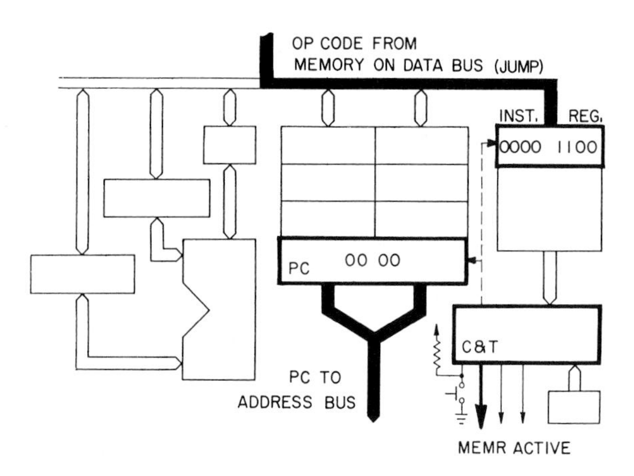

Fig. 19-18. To fetch the first op code, the number in PC is placed on the address bus and control lead MEMR is made active.

INCREMENTING PC

Before going on, CONTROL AND TIMING increments the PROGRAM COUNTER. See Fig. 19-19. The PC now contains the address of the next memory location. The next memory location contains the low-order byte of the destination of the jump.

FETCH OF LOW-ORDER BYTE OF ADDRESS

To fetch the low-order byte of the address of the jump, CONTROL AND TIMING sends the signals shown in Fig. 19-19. The number in PC is put on address bus, and MEMR is made active. The number from memory goes in TEMPORARY REGISTER 2.

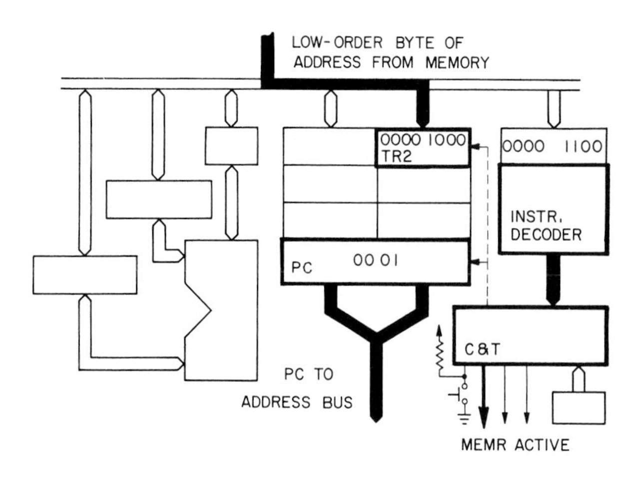

Fig. 19-19. Because the PC was incremented after the fetch of the op code, it contains the address of the low-order byte of the destination of the jump.

Again the PC is incremented. It now contains the address of the high-order byte of the destination of the jump instruction.

FETCH OF HIGH-ORDER BYTE OF ADDRESS

Fig. 19-20 shows the fetch of the high-order byte of the address. It is similar to the fetch of the low-order byte. Note that the number from memory is placed in TEMPORARY REGISTER 1.

The PC is again incremented. The incremented step (00 03) will not be used.

COMPLETION OF EXECUTION

Fig. 19-21 shows the completion of JUMP. The numbers in TEMPORARY REGISTERS 1 and 2 are

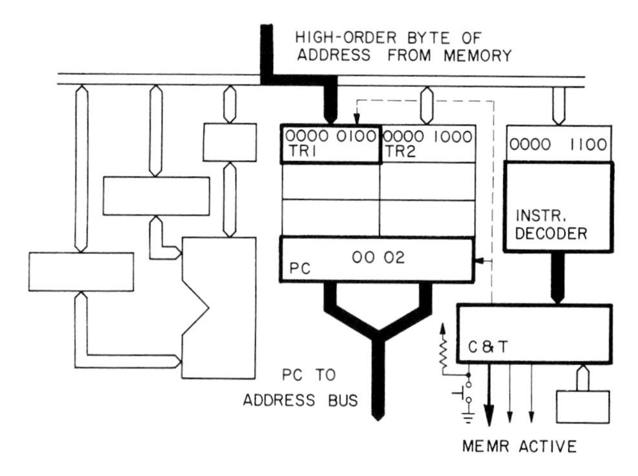

Fig. 19-20. After the fetch of the low-order byte of address, the PC is again incremented. As a result, it contains the location of the high-order byte.

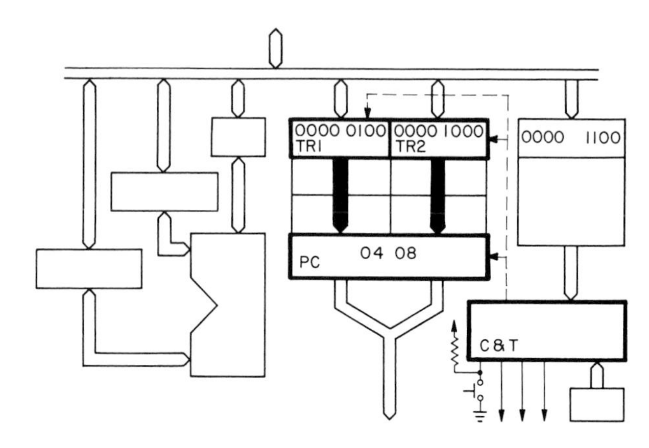

Fig. 19-21. To complete the jump, the two bytes of address are transferred to the PC.

transferred to the PC. This completes the instruction cycle, since the address of the next op code is available in the program counter.

EXECUTION OF INPUT INSTRUCTION IN MICROPROCESSOR

The second instruction is:

Memory Location	Stored Number	
04 08	1010 0011	Op code for INPUT
04 09	0000 0101	Port number for INPUT

FETCH AND DECODE OF OP CODE

The fetch and decoding of the op code after the jump is identical to that for the first instruction. See Fig. 19-22. Because 04 08 is in PC, the number at this location in memory is brought in and placed in the INSTRUCTION REGISTER. When decoded, the computer is told that an input is to be accomplished. It is also told that the port number is in the next memory location.

Before going on, CONTROL AND TIMING increments the PC. The memory address 04 09 is in the PC, ready to give the port number.

FETCH OF PORT NUMBER

The fetch of the port number is easily accomplished, since its address is in PC. See Fig. 19-23. CONTROL AND TIMING signals TEMPORARY REGISTER 2 to latch this number.

PC is again incremented. The memory address 04 0A is in PC.

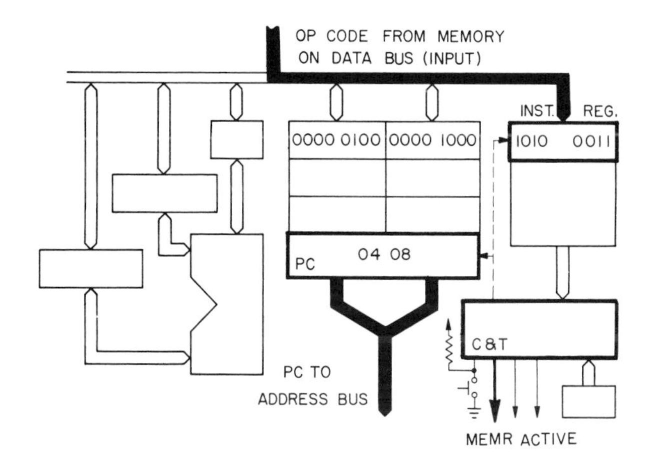

Fig. 19-22. Using the number in PC as an address, the fetch of the next op code is accomplished.

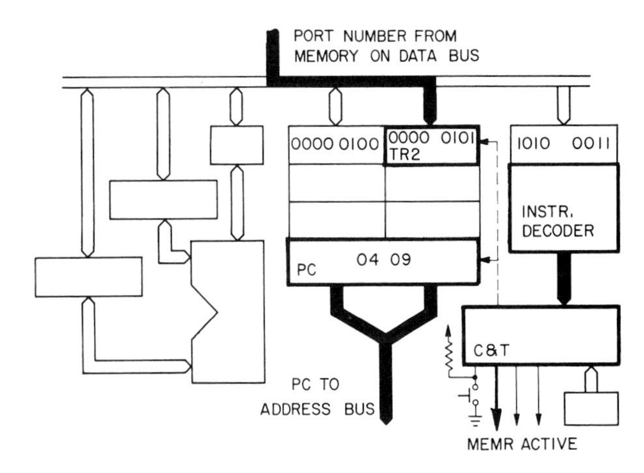

Fig. 19-23. The port number is brought in and placed in TEMPORARY REGISTER 2.

COMPLETION OF EXECUTION

Fig. 19-24 shows the execution of the input instruction. CONTROL AND TIMING sends a number of control signals. First, the port number in TEMPORARY REGISTER 2 is placed on the lower byte of the address bus. Second, control lead IN is made active. Finally, the number from the port is latched by ACCUMULATOR.

As you may remember, the number from the port will be 1010 1011 if START is not pressed. It will be 1010 0011 if it is pressed.

EXECUTION OF AND INSTRUCTION IN MICROPROCESSOR

The next instruction is:

Memory Location	Stored Number	
04 0A	1000 0010	Op code for AND
04 0B	0000 1000	Mask for bit 3

FETCH AND DECODE OF OP CODE

The fetch and decoding of the AND op code is the same as for earlier instructions. After decoding, the MPU knows that it is to AND the number in its ACCUMULATOR with the number stored in the next memory location.

The PC is incremented after the AND op code fetch. 04 0B is in PC.

FETCH OF MASK

With 04 0B in PC, CONTROL AND TIMING sends the necessary signals to fetch the mask. It is

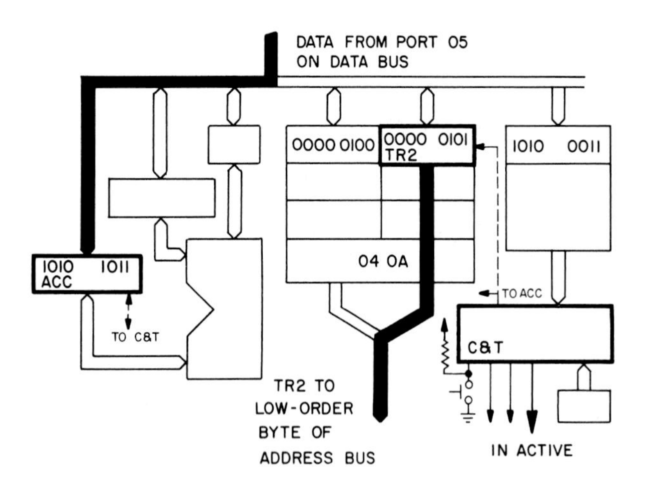

Fig. 19-24. To activate the input port, the port number in TEMPORARY REGISTER 2 is placed on the low-order byte of the address bus, and control lead IN is made active.

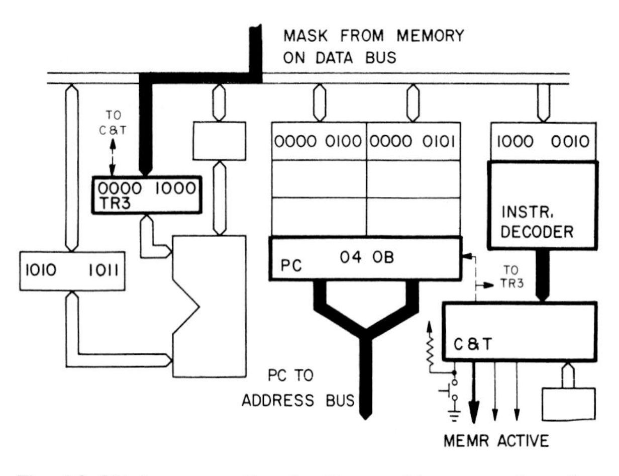

Fig. 19-25. In preparation for the masking operation, the mask is placed in TEMPORARY REGISTER 3.

placed in TEMPORARY REGISTER 3. Numbers to be ANDed are now in ACC and TR3, Fig. 19-25.

PC is again incremented. 04 0C is in PC.

COMPLETION OF EXECUTION

Based on signals sent by CONTROL AND TIMING, the ALU ANDs the numbers in the ACCUMULATOR and TEMPORARY REGISTER 3. See Fig. 19-26. The result is returned to the ACCUMULATOR, and appropriate flags are set.

If the result of the ANDing is 0000 0000, ZERO FLAG is set (Z = 1). If the result is 0000 1000, ZERO FLAG is cleared (Z = 0).

EXECUTION OF CONDITIONAL JUMP INSTRUCTION IN MICROPROCESSOR

This instruction is stored in the form:

Memory Location	Stored Number	
04 0C	0110 0110	Op code for COND. JUMP
04 0D	0000 1000	Low-order byte of address
04 0E	0000 0100	High-order byte of address

FETCH AND DECODE OF OP CODE

The fetch and decode actions are identical to those used to fetch and decode earlier op codes. At this point the computer knows that it is to do a conditional jump if Z = 0. It also knows that the destination of that jump is stored in the next two memory locations.

The PC is incremented. 04 0D is in PC.

FETCH OF LOW- AND HIGH-ORDER BYTES OF ADDRESS

In conditional and unconditional jumps, the address fetch is identical. The two bytes are brought in and placed in TEMPORARY REGISTERS 1 and 2.

After each fetch, PC is incremented. As a result, the number in PC after the fetch of the op code and the two bytes of address is 04 0F.

COMPLETION OF EXECUTION OF CONDITIONAL JUMP

At this point, CONTROL AND TIMING must make a decision. Should the jump be made? The op

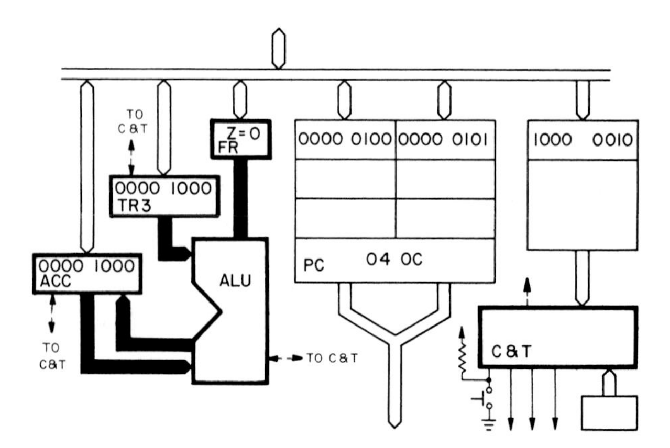

Fig. 19-26. When the ANDing is complete, the result is placed in the ACCUMULATOR and flags are set.

code indicates the basis for this decision. If Z = 0, the jump is to be made.

In part "a" of Fig. 19-27, Z = 0. The jump is to be made. To do this, CONTROL AND TIMING transfers the number in the temporary registers to PC. The jump is completed when the computer uses this number as the address of the next op code.

Part "b" of Fig. 19-27 shows Z = 1. Because the condition for the jump (Z = 0) has not been met, CONTROL AND TIMING takes no action. That is, the address of the next memory location, 04 0F, is left in PC. (See the description before the heading "COMPLETION OF EXECUTION OF CONDITIONAL JUMP.") 04 0F leads to an unconditional jump to 04 20. The program thus breaks out of the loop when Z = 1.

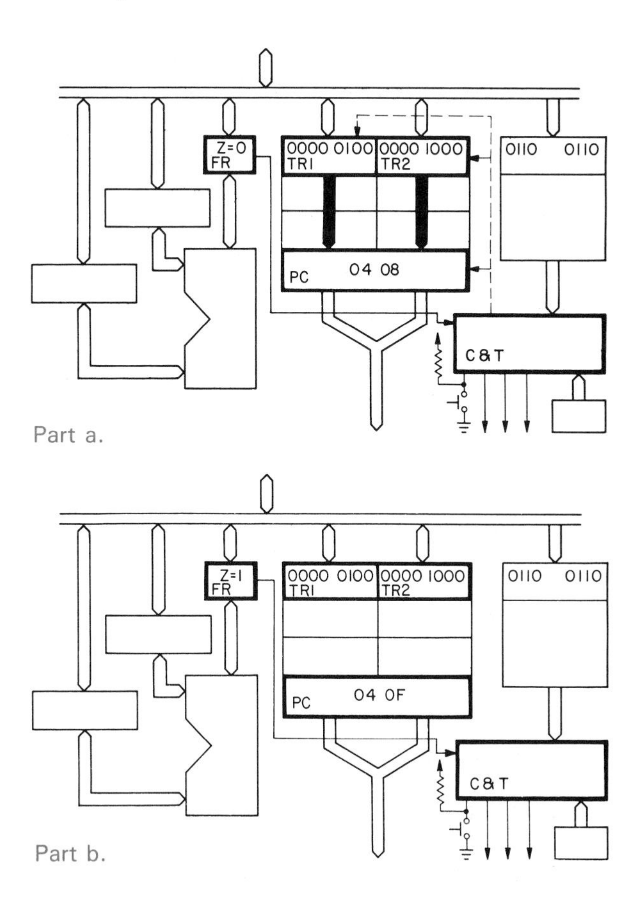

Fig. 19-27. If Z = 0, the jump is made. If Z = 1, it is not made.

EXECUTION OF UNCONDITIONAL JUMP INSTRUCTION IN MICROPROCESSOR

The last instruction in this portion of the computer machine language program is:

Memory Location	Stored Number	
04 0F	0000 1100	Op code for UNCON. JUMP
04 10	0010 0000	Low-order byte of address
04 11	0000 0100	High-order byte of address

The execution of this instruction is identical to that of the earlier unconditional jump. When completed, the address in PC will be 04 20. This is the beginning of the portion of the program that starts and controls the machine.

STRUCTURE OF INSTRUCTIONS

There is a common structure for instructions. Each can be divided into two parts. These parts are:
1. Operator.
2. Operand.

OPERATOR

An OPERATOR tells the computer what task is to be performed. In machine language programs, this takes the form of an op code. In high-level languages, a phrase or a word in English may be the operator. An operator gives directions and has a strictly defined form related to its use.

OPERAND

An OPERAND is the data upon which the operation is performed. Several methods are used to indicate the location of operands. These methods are:
1. Implied operand location.
2. Immediate operand location.
3. Indirect operand location.

IMPLIED OPERAND LOCATION

In some instructions, the location of the operand is indicated by the op code. For example, MOVE A,B moves the data in the ACCUMULATOR to REGISTER B. Because this takes place within the MPU, no additional data is needed. Move A, B is a one-byte instruction and has no formal operand. That is, the operator IMPLIES the location of the operand for the move instruction.

IMMEDIATE OPERAND LOCATION

In some instructions, the operand is placed in the memory location following the op code. The AND that is used to mask the number in the program just completed is an example. Such instructions usually require two bytes of memory.

INDIRECT OPERAND LOCATION

In some instructions, the address of the operand is supplied. The computer must go to that address to obtain the operand. LOAD ACCUMULATOR is such an instruction. The op code indicates that a number in memory is to be moved to the ACCUMULATOR. It also indicates that the next two bytes in memory contain the necessary address. Such instructions require three bytes of memory.

The forms taken by operands are called ADDRESSING MODES. Although there are many other addressing modes, the three described here are probably the most used.

NEED TO FOLLOW PRESCRIBED STRUCTURE FOR INSTRUCTIONS

In machine-language programs, instructions must follow the exact structure prescribed for each op code. If an op code requires an address in the next two bytes, that address must be there in the program. If a byte of data must be in the memory location following an op code, the program will not run if it is not there. For example, if the mask 0000 1000 were omitted in the following program, a fault would result:

CORRECT PROGRAM

04 0A	1000 0010	Op code for AND
04 0B	0000 1000	Mask to be ANDed
04 0C	0110 0110	Op code for COND. JUMP
04 0D	0000 1000	Low-order byte of address
04 0E	0000 0100	High-order byte of address

PROGRAM WITH MASK OMITTED

04 0A	1000 0010	Op code for AND
04 0B	0110 0110	Mask to be ANDed
04 0C	0000 1000	Next op code
04 0D	0000 0100	Etc.

Due to the omission, the numbers at 04 0B differ, but their function remains unchanged. The number following the op code 1000 0010 (AND) will be ANDed by the computer with the number in its ACCUMULATOR. The computer has no way of knowing that the stored number is incorrect.

The next instruction cycle results in an error when the second program is run. Upon completing the AND instruction, the PC will contain the address 04 0C. The number stored there, whatever it is, will be treated as an op code. The computer will decode it and attempt to execute the implied task. The result will be chaos. The machine will attempt to run the program which is, in effect, made up of random numbers.

In general, machine language programs for one type of microprocessor will not run on another. Each has its own set of op codes. Exceptions are sometimes found in MPUs available from one manufacturer. To promote sales, new products may be designed for a property called "upward compatibility." That is, programs written for older MPUs can be run on newer machines. The opposite may not be possible. Programs written for advanced MPUs may contain op codes not available on older units.

INSTRUCTION SET

The group of instructions that a given microprocessor can execute is called its INSTRUCTION SET. Certain basic instructions are available on almost all machines. However, differences exist, so in a given application, one type of MPU may do a better job than another.

In most microprocessors, the instruction set is hard wired. That is, the instructions that can be executed are built in by the manufacturer. In some computers, however, instruction sets can be modified by programmers. These computers are said to be MICROPROGRAMMABLE.

NUMBER OF INSTRUCTIONS IN INSTRUCTION SET

A microprocessor's instruction set is likely to contain over 200 instructions. Many of these, however, are nearly duplicates. For example, if MOVE A,B is in a set, its inverse, MOVE B,A is likely to be there also. Instructions for moving data from one register to another (MOVE C,D, MOVE B,D, etc.) can represent a large portion of the 200 instructions. For this reason, the number of unique instruction types that are available or can be devised is small.

NATURE OF INSTRUCTIONS IN INSTRUCTION SET

Machine language instructions must be compatible with the hardware of the computer. That is, only instructions that can be executed by the circuits in the MPU's block diagram can be in the set.

For the MPU in Fig. 19-16, the instruction MOVE A,B is likely to be available. It is unlikely that a machine language instruction for SINE (a function from the algebra of trigonometry) is in its instruction set. Such an instruction is too complex for the machine. If sine is needed, a program for generating this function would have to be written and stored in the memory (ROM or sometimes RAM).

TYPES OF INSTRUCTIONS IN INSTRUCTION SET

There are at least four types of machine language instructions in an instruction set. Machine language instructions can usually be classified under the following headings:

1. Data transfer.
2. Arithmetic/logic.
3. Branching.
4. Miscellaneous.

A discussion of each type of instruction helps to compare and contrast them. Some instructions, such as "JUMP 04 60," are commonly used to direct the operation to special locations.

DATA TRANSFER INSTRUCTION IN INSTRUCTION SET

Data can be moved within an MPU, between an MPU and memory, and between an MPU and the outside world. The instruction MOVE A,B is an example of a transfer within an MPU. Memory-oriented instructions such as STORE A or LOAD A transfer data between memory and an MPU. Instructions such as IN and OUT move data through input and output ports to points where data is needed.

ARITHMETIC/LOGIC INSTRUCTION IN INSTRUCTION SET

Most data manipulation takes place within an ALU. ADD and SUBTRACT are available on most machines. Newer microprocessors also have MULTIPLY and DIVIDE. These are not available on older machines, so programs are required to do these tasks.

Machine language instructions for ROTATE and SHIFT are usually included, as are INCREMENT and DECREMENT. AND, OR, EXCLUSIVE-OR, and COMPLEMENT are usually among the logic operations available.

BRANCHING INSTRUCTION IN INSTRUCTION SET

Upon completing an instruction, computers normally look to the next memory location for a new op code. Branching instructions permit op codes to be located elsewhere in memory.

As indicated earlier, JUMP is one form of branching. CALL SUBROUTINE is another. Fig. 19-28 suggests the difference.

When JUMP 04 60 is encountered in this program, the number 04 60 is placed in the computer's PC. This number is then used as the address of the next op code. Note the upper arrow in Fig. 19-28.

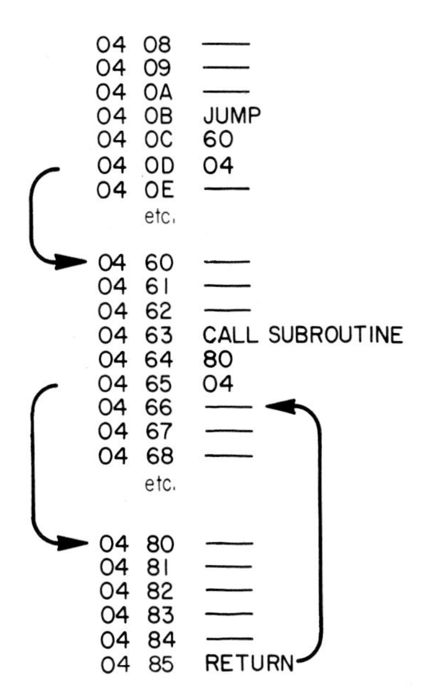

Fig. 19-28. The difference between JUMP and CALL SUBROUTINE is indicated above. The instruction RETURN can be used to return from a subroutine.

When JUMP is used, no provision is made for directly returning to the original section of the program. For example, to return to the op code stored at 04 0E, another jump instruction (JUMP 04 0E) would be necessary.

Instructions such as CALL SUBROUTINE 04 80 permit branching. They direct the computer to the next op code in the same way as JUMP. That is, the new address (in this case 04 80) is placed in PC. However, there is an important difference. Just before the new address is placed in PC, CALL SUBROUTINE instructions cause the number already there (in this case, 04 66) to be stored. (The PC was incremented to 04 66 during the two fetch operations.) The address 04 66 is thus available for future use.

The portion of the program at location 04 80 is called a *SUBROUTINE.* When it is completed, return to the main program can be accomplished by using the instruction RETURN. See line 04 85. This instruction takes the original address out of storage and returns it to PC. As a result, the computer looks to location 04 66 for its next instruction. It has returned from the subroutine and is again running the main program with its list of commands.

Conditional branches, both JUMP and CALL SUBROUTINE, are available. These are used to make decisions. Depending on the op code used, conditional branches can be based on any of the available flags.

It is interesting to note that the decision-making ability of a computer may be based on as few as four flags. The most common are:

Sign flag
Zero flag
Carry flag
Parity flag

MISCELLANEOUS INSTRUCTION IN INSTRUCTION SET

Instructions in this group are often used to control microprocessors. HALT is such an instruction. It is usually placed at the ends of programs to tell machines not to execute the random number in the next memory location. NO OPERATION is another such instruction. There can be no blank spaces in programs. If an instruction is removed from the center of a program, something must be put in its place. NO OP is often used. This instruction increments PC but takes no other action.

SUMMARY

Digital computers are machines called STORED PROGRAM machines. That is, instructions for doing a task are stored in a machine's memory and executed one after the other.

To be executed by a computer, instructions must be in machine language. Machine language programs consist of lists of binary numbers that can directly control computer action.

A few programs (mostly for small, dedicated machines) are written directly in machine language. Most are written in high-level languages. These must be converted to machine language before being executed. The translation is accomplished by programs called interpreters and compilers.

The numbers that make up machine language programs must be written and stored in precise order. The meaning of a given number depends on its position in a program. The first byte of an instruction (the operator) is always an op code. Numbers that follow (operands) can be data, addresses, or port numbers.

A register called the program counter keeps track of where a computer is in a program that is being run. After each fetch operation, PC is incremented. Therefore, the program counter points to the program entry that will be used next.

Circuits within microprocessors determine the instructions that can be executed by a given MPU. A list of these instructions is called an instruction set. Microprocessor instructions can be grouped under the headings: data transfer, arithmetic/logic, branching, and miscellaneous. Conditional branching instructions (both JUMP and CALL SUBROUTINE) are extensively used in computer decision-making.

IMPORTANT TERMS

Compiler, Conditional jump, Fetch, Increment PC, Initialize, Jump, Language, Loop, Machine language, Masking, Operand, Operator, Program counter, Reset, Return, Subroutine, Unconditional jump.

TEST YOUR KNOWLEDGE

1. A ________ (core, block, step) diagram simplifies a circuit.
2. FORTRAN is a ________ (low-level, high-level) language.
3. All programs must be in ________ language before they can be executed.
4. A ________ or interpreter converts a high-level language to machine language.
5. To get an op code, low-order byte of address, or a port number from a memory location means to ________ (call it up, fetch it, retrieve it, label it, decipher it, decode it).
6. When the run signal is given, many microprocessors go to location ________ for the first instruction.
7. When the op code for JUMP is received, the computer knows that the next ________ (one, two, three, four) memory location(s) contain(s) the address of the destination for the jump.
8. The instruction JUMP requires 3 bytes of memory; the instruction INPUT requires ________ byte(s) of memory.
9. An AND operation using a ________ (hole, complement, inverse, mask, window) isolates 1 bit of a multi-bit binary number.
10. The decision for a conditional jump is based on the number in which register (PC, instruction register, flag register)?
11. If a program tells a computer to keep cycling through the same set of instructions, the computer is said to be in a(n) ________.
12. The results of most arithmetic/logic operations are returned to the ________ (ALU, accumulator).
13. CONTROL AND ________ controls the execution of an instruction.
14. What does PC stand for?
15. After every fetch operation, the program counter is ________ (decremented, incremented).
16. There is a common structure for instructions. Each can be divided into the following two parts: operator and ________.

17. In this book, how many methods are used to indicate the location of operands?
18. An operand is the _________ upon which the operation is performed.
19. The instruction MOVE A,B (move the data in the accumulator to REGISTER B) has an _______ (immediate, implied) operand location.
20. "Upward compatibility" means that programs written for advanced MPUs can be run on older machines. (True or False?)
21. If an instruction set can be modified by programmers, the computer is said to be _________ (flexibly programmed, microprogrammable, macroprogrammable, conditionally programmed).
22. A microprocessor's instruction set is likely to contain over _________ (48, 56, 200, 1000, 4800, 5600) instructions.
23. JUMP is one form of branching, CALL _________ is another.
24. What command brings the operation of the computer back to the main program when a subroutine is finished?
25. There can be no _________ (fractional, linear, double, blank, temporary, conditional) spaces in a machine language program.

STUDY PROBLEMS

1. Explain the difference between machine languages and high-level languages.
2. Why must all programs be in machine language before they can be executed?
3. The program in Fig. P19-1 is written in a high-level language. The words in the brackets (GO TO) would not be part of the program but were included to suggest the meaning of the instruction. Analyze the program and determine what will appear on the computer's screen under each of the following conditions:
   a. When the number 22 is typed.
   b. When the number 112 is typed.

```
10  CLS
20  PRINT "WHAT IS THE SUM OF 10 AND 12?"
30  INPUT A
40  IF A=22 THEN (GO TO) 70
50  PRINT A;"IS INCORRECT. THE ANSWER IS 22."
60  GO TO 80
70  PRINT "THAT IS CORRECT."
80  END
```

Fig. P19-1. Typical program written in a high-level language. Uses for the program can be identified.

4. The program from Problem 3 (Fig. P19-1) is in a form that can be run directly by a computer. (True or False?)
5. A machine-language program is shown in Fig. P19-2. Spaces have been added to emphasize individual instructions. Assume that it will be run on the same computer used in this chapter. Analyze the program and predict the binary number that will be output.

MEMORY LOCATION (HEX)	STORED NUMBER (BINARY)	
00 00	1001 0001	Op code, LOAD ACC. from memory
00 01	0010 0000	Low-order byte of address
00 02	0000 0000	High-order byte of address
00 03	1011 0011	Op code, ADD next byte to ACC.
00 04	0101 0001	Number to be added to ACC.
00 05	0101 1101	Op code, OUTPUT ACC.
00 06	0000 0100	Port number for output
00 07	1111 1111	Op code, HALT
00 20	0001 0111	

Fig. P19-2. Sample program in machine language. The action of the program can be followed one step at a time.

6. In the program from Problem 5 (Fig. P19-2), through which port will the results of the computation be output?
7. In the program from Problem 5 (Fig. P19-2), which of the following best describes the number in memory location 00 20?
   a. Op code.
   b. Address.
   c. Data.
   d. Port number.
8. How many instructions are there in the program from Problem 5 (Fig. P19-2)?
9. How many fetches (from memory) must be made to execute (carry out) the program that is taken from Problem 5 (Fig. P19-2)?
10. Refer to the number in location 00 03 of the program from Problem 5 (Fig. P19-2). During the fetch of this op code, certain actions will be taken.

Answer the following:
a. What number will appear on the computer's address bus (in hexadecimal)?
b. What number is on the data bus (in binary)?
c. Which lead in the computer's control bus will be in an active state?

11. When the output instruction (stored at 00 05 and 00 06) in the program from Problem 5 (Fig. P19-2) is being executed (a number is being output) certain actions are taken. Answer the following:
a. What number will appear on the low-order byte of the address bus (in hexadecimal)?
b. What number is on the data bus (in binary)?
c. Which lead in the control bus will be active?

12. A simplified block diagram of a microprocessor is shown in Fig. P19-3. Match the following buses with the letters (A, B, and C) on the drawing. Example: Answers will look like "g.M."
a. Address bus.
b. Data bus.
c. Control bus.

13. Using numbers, match the following functions with the blocks in the diagram from Problem 12 (Fig. P19-3). This is the same diagram used in the text. Answers will look like "y.21."
a. CONTROL AND TIMING.
b. PROGRAM COUNTER.
c. ACCUMULATOR.
d. TEMPORARY REGISTER 3.
e. TEMPORARY REGISTERS 1 and 2 (two numbers to be given).
f. ALU.
g. FLAG REGISTER.
h. CLOCK.
i. INSTRUCTION REGISTER.
j. INSTRUCTION DECODER.

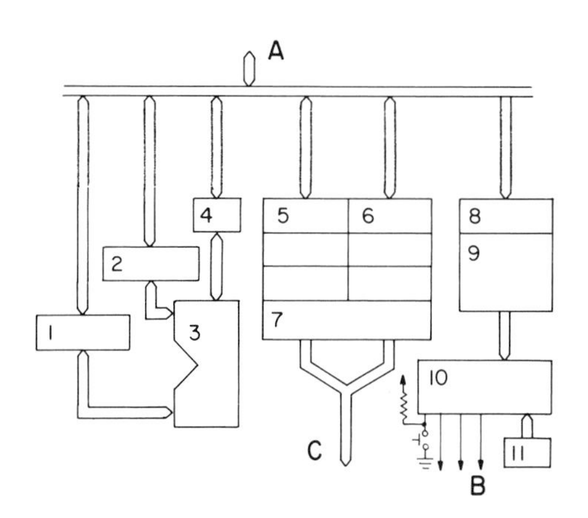

Fig. P19-3. Typical block diagram of a microprocessor. Components can be identified.

14. Refer to the program from Problem 5 (Fig. P19-2) and the block diagram from Problem 12 (Fig. P19-3). After completing the LOAD instruction, what numbers will be in the following registers?
a. ACCUMULATOR (in binary).
b. PROGRAM COUNTER (in hexadecimal).

15. Repeat Problem 14 for the following registers and the completion of the ADD instruction.
a. ACCUMULATOR (in binary).
b. TEMPORARY REGISTER 3 (in binary).
c. PROGRAM COUNTER (in hexadecimal).
d. The following flags in the FLAG REGISTER:
   1. ZERO FLAG.
   2. SIGN FLAG.
   3. CARRY FLAG.
   4. PARITY FLAG.

16. Refer to the program from Problem 5 (Fig. P19-2). Using the letter A for implied, B for immediate, and C for indirect, indicate the type of operand used in the following instructions:
a. LOAD instruction stored in the memory locations 00 00 through 00 02.
b. ADD instruction stored at 00 03 and 00 04.

17. If the numbers stored in memory locations 00 05 and 00 06 in the program from Problem 5 (Fig. P19-2) were interchanged (0000 0100 in 00 05 and 0101 1101 in 00 06), would the program run properly?

18. In this chapter, machine language instructions were classified under the headings:
A. Data transfer.
B. Arithmetic/logic.
C. Branching.
D. Miscellaneous.

Classify the instructions in the program from Problem 5 (Fig. P19-2) according to the above headings.
a. LOAD stored at 00 00, 00 01, and 00 02.
b. ADD stored at 00 03 and 00 04.
c. OUTPUT stored at 00 05 and 00 06.
d. HALT stored at 00 07.

19. Which of the following branch instructions permits returning to the main program merely by using the instruction RETURN?
a. JUMP.
b. CALL SUBROUTINE.

20. Which of the following best describes the way decisions were made in the machine language programs in this chapter?
a. Flags were tested using conditional branch instructions.
b. The decision-making circuits in the ALU were used for the operation.
c. The instruction DECISION was used.

TTL Logic Element Specifications and Operating Requirements

Chapter 20

LEARNING OBJECTIVES

After studying this chapter and completing lab assignments and study problems, you will be able to:

- Draw the schematic of the circuit of a TTL NAND and describe its action.
- Explain the term *current sinking.*
- Describe two methods for dealing with unused OR inputs and four methods for dealing with unused AND inputs.
- List typical values for the following TTL specifications:
  - Supply voltage, its tolerance and absolute maximum.
  - Signal voltages and noise margin.
  - Propagation delay and maximum clock frequency.

Most design projects and troubleshooting procedures can be accomplished without reference to the details for the types of electronics hardware on logic chips. However, hardware personnel should be familiar with the characteristics of widely used logic families.

This chapter describes the transistor circuits and characteristics of the standard TTL family. A later chapter will introduce other families.

REASONS FOR STUDYING TTL

In the past, standard TTL was the most used logic family. Today it is seldom designed into new equipment. Yet, much of modern digital technology is based on the advances made during the time that TTL was widely used. Many present-day families were designed to interface with standard TTL. As a result, familiarity with standard TTL logic can serve as an introduction to many present-day families.

To effectively use a given chip, an extensive knowledge of the transistor circuits within a logic element is not required. However, a certain basic understanding is desirable. The circuit within a TTL NAND is relatively easy to understand, so it is a good starting point for the study of such circuits.

STANDARD TTL CIRCUIT

TTL stands for TRANSISTOR-TRANSISTOR LOGIC. The transistors used in the standard TTL family are bipolar. The conductivity is high for bipolar transistor elements. This reduces heat. Bipolar transistors make circuits simple. There are seven topics that are important. These are:

- Review of transistor action.
- TTL output circuit.
- Phase splitter.
- Decision circuit.
- Clamping diodes on TTL inputs.
- Review of TTL NAND circuit action.
- Other TTL logic elements.

REVIEW OF TRANSISTOR ACTION

A brief review of the action of bipolar transistors is appropriate. Several conclusions about input signals can be made.

When a positive voltage greater than about 0.6 V is applied between the base and emitter of an npn bipolar transistor, a base current will flow. See part "a" of Fig. 20-1. When base current flows, the collector of the device conducts. That is, the collector-to-emitter portion of the bipolar transistor acts very much like a closed switch.

If less than 0.6 V or a negative voltage appears between the base and emitter, no base current flows. See part "b" of Fig. 20-1. With no base current, the collector acts like an open switch.

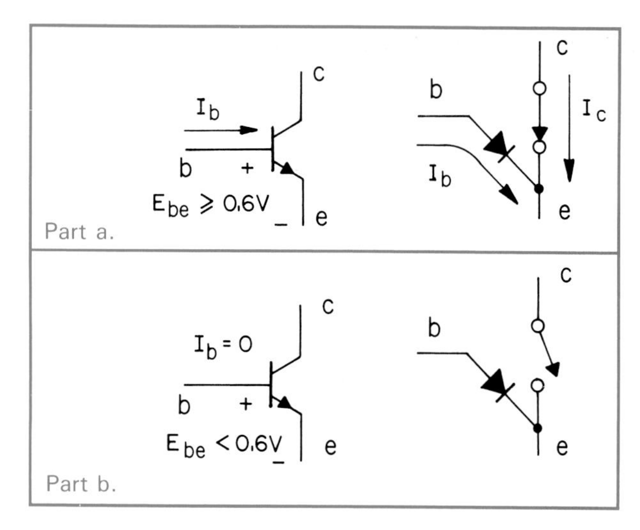

Fig. 20-1. Only when current flows into the base of a bipolar transistor will its collector conduct.

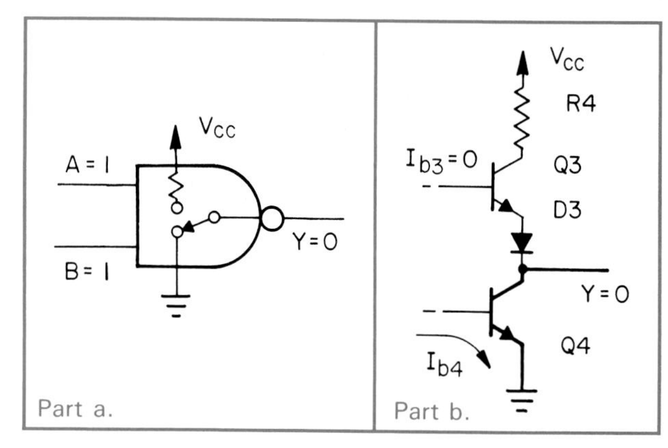

Fig. 20-3. When current flows into the base of Q4, its collector shorts output Y to ground.

TTL OUTPUT CIRCUIT

Fig. 20-2 shows the output circuit of a TTL gate. It consists of two transistors, a resistor, and a diode. Because the components are arranged vertically, the term *TOTEM POLE* is often used to describe this circuit. Cases can be examined for outputting a 0 and outputting a 1.

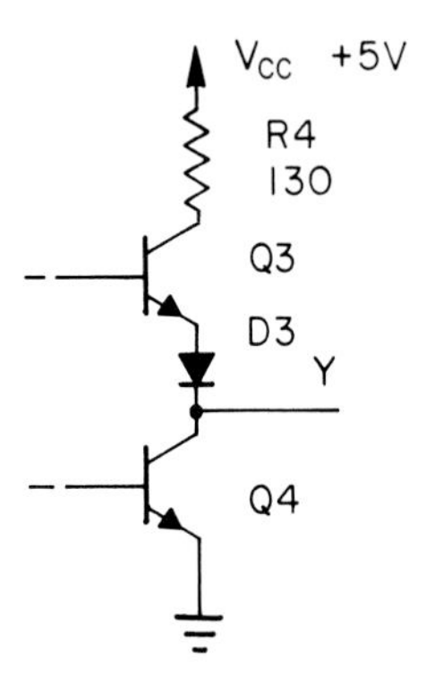

Fig. 20-2. The term totem pole is used to describe the output circuit of TTL elements.

OUTPUTTING 0

In part "a" of Fig. 20-3, a switch equivalent of a TTL NAND is shown outputting a 0. That is, lead Y is connected to ground.

Part "b" of Fig. 20-3 shows how the grounding of lead Y is accomplished by the output of an actual TTL element. When Ib4 (the base current of Q4) flows, transistor Q4 turns on. The voltage across the output of this transistor is only about 0.2 V, so Y is grounded. To emphasize that Q4 is conducting, its symbol has been drawn with heavy lines.

OUTPUTTING 1

In part "a" of Fig. 20-4, the switch equivalent of a NAND is shown outputting a 1. Lead Y is connected to Vcc through an output resistance Ro.

Part "b" of Fig. 20-4 shows how the production of a high state on lead Y is accomplished by the actual circuit. Current Ib4 is 0, so transistor Q4 does not conduct. However, the current Ib3 (the base current of Q3) causes the upper transistor to conduct. This connects Y to Vcc through D3, Q3, and R4. The voltage between Y and ground is about 3.4 V, well within the voltages accepted as a logic 1.

In this output circuit, Q3 and Q4 act as switches.

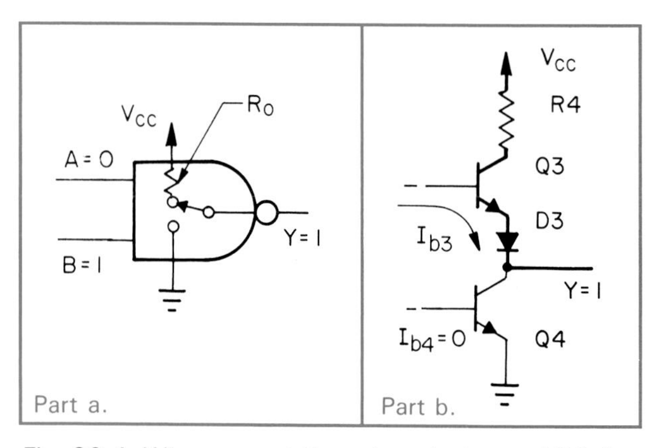

Fig. 20-4. When current flows into the base of Q3, its collector connects output Y to Vcc.

They connect lead Y to Vcc or to ground.

Resistor R4 limits the output current when Y = 1. Without this resistor, excessive current might flow under certain conditions.

Diode D3 is part of the biasing circuit of Q3. Its function will be described later. Note, however, the direction of the arrows on the diode and transistor symbols. Current (assuming conventional flow) can flow from Vcc, through both Q3 and D3 to output Y.

PHASE SPLITTER

Only one transistor in the totem pole may conduct at a given time. The circuit that turns on the proper transistor is called a *PHASE SPLITTER*. This circuit has been added in Fig. 20-5.

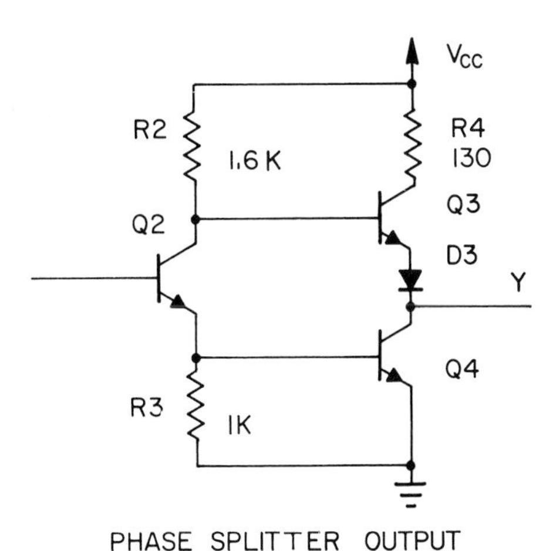

Fig. 20-5. The phase splitter determines which output transistor, Q3 or Q4, will conduct.

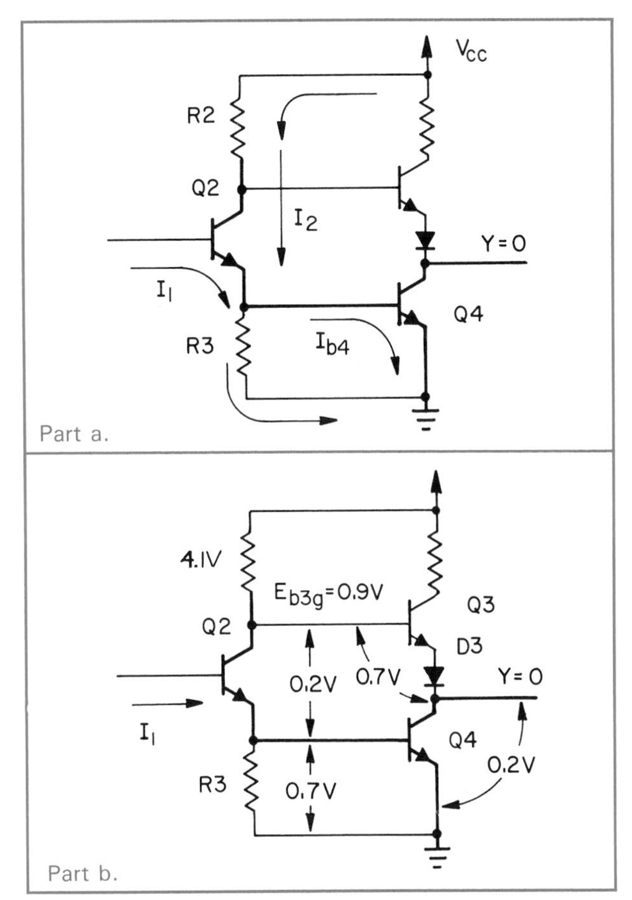

Fig. 20-6. The base-to-emitter voltage of Q4 is 0.7 V, enough to turn it on. The base-to-emitter voltage of Q3 is 0.35 V (half of 0.7 V). This is not enough to turn transistor Q3 on.

OUTPUTTING 0

Part "a" of Fig. 20-6 shows a current I1 flowing into the base of Q2. This transistor is turned on by that current, so its collector looks like a closed switch. Therefore, a current I2 flows from Vcc, through R2, Q2, and R3 to ground.

As I2 flows through R3, it produces a voltage that is high enough to cause Ib4 to flow into the base of Q4. This transistor is turned on, and the logic element circuit outputs a 0.

Turning on Q4 is only half the problem. Q3 must be turned off. As indicated in part "b" of Fig. 20-6, about 0.7 V appears across R3 (and the base-to-emitter junction of Q4). Another 0.2 V appears across Q2. This means that the base of Q3 is 0.9 V above ground. (The 0.2 V plus 0.7 V on the right equals the 0.7 V plus 0.2 V on the left.) However, the emitter of Q3 is connected to lead Y through D3. The voltage at Y is about 0.2 V, so the voltage between the base of Q3 and lead Y is about 0.7 V (0.9 − 0.2 = 0.7).

If this 0.7 V were applied directly between Q3's base and emitter, this transistor would turn on. However, this voltage is divided between the base-to-emitter junction of Q3 and diode D3. The resulting 0.35 V across the base-to-emitter junction of Q3 is too low to turn it on. Without D3 in the circuit, Q3 might turn on at the same time Q4 is on.

OUTPUTTING 1

In part "a" of Fig. 20-7, I1 is not flowing. This turns off Q2, so the current I2 cannot flow. With I2 = 0, there is no current through R3, so no voltage is applied to the base of Q4. It is also turned off.

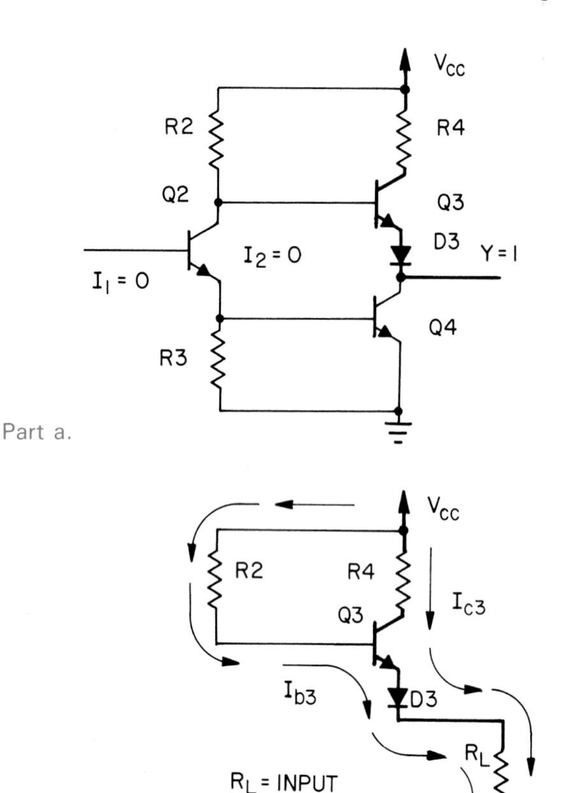

Fig. 20-7. When I1 does not flow into the base of Q2, output transistor Q3 is turned on.

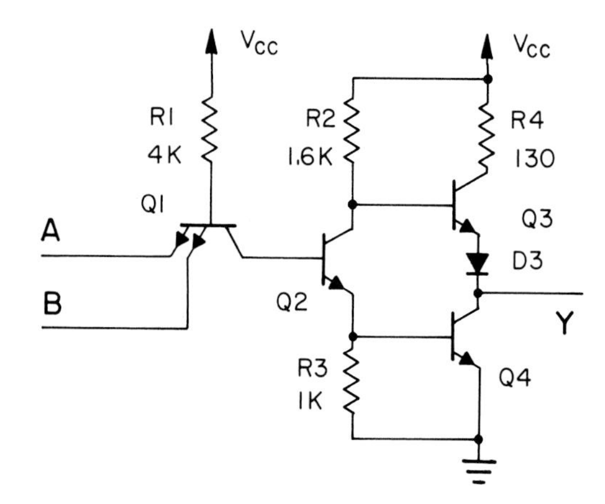

Fig. 20-8. Multiple-emitter transistors are used in the decision-making circuits of most TTL elements.

Because Q2 and Q4 are not conducting, they can be considered as open circuits and can be omitted. See part "b" of Fig. 20-7. With little current flowing in R2 (I2 is 0), little voltage is dropped across this resistor. As a result, the voltage at the base of Q3 is high enough to forward bias its base-to-emitter junction. The current Ib3 flows, and Q3 turns on. The circuit outputs a 1.

To review, when I1 flows into the base of Q2, transistor Q4 is turned on and Q3 is turned off. The circuit outputs a 0. When I2 stops flowing, Q4 is turned off, and Q3 is turned on. The circuit outputs a 1.

DECISION CIRCUIT

In Fig. 20-8, the decision-making portion of the circuit has been added. Note that the transistor at the far left has two emitters. It is called a MULTIPLE-EMITTER TRANSISTOR. Such transistors are found in most TTL elements. It is here that the circuit takes on the characteristics of a NAND.

Transistor Q1 can be thought of as containing three diodes. See part "a" of Fig. 20-9. Diodes Da and Db are emitters; the lone diode at the right represents the collector.

OUTPUTTING 0

In Fig. 20-9, batteries have been used to represent 1s at inputs A and B. With positive voltage applied to their cathodes, Da and Db are REVERSE BIASED. They will not conduct. Therefore, current I1 must flow into the base of Q2 as shown. Remember, when I1 flows, Q2 is turned on. This turns on Q4, and the circuit outputs a 0. Refer to the first row of the NAND truth table.

OUTPUTTING 1

In part "b" of Fig. 20-9, input A has been grounded (A = 0). It is easier for I1 to reach ground through diode Da (a base-to-emitter junction of Q1) than through the collector of this transistor. Therefore, diode Da steals I1 from the collector. Because A = 0, Q1 is turned on. The collector of Q1 is 0.2 V above ground. The collector of Q1 grounds the base of Q2.

With no base current, Q2 turns off. This in turn turns off Q4. However, Q3 turns on, and the circuit outputs a 1. See the second row of the truth table that is shown in Fig. 20-9.

When A or B or both are grounded, they steal I1. The result is the outputting of a 1.

The action of Q1 is a bit more complex than described above. When I1 flows from the base to emitter of Q1, this transistor is turned on, and its collector is shorted to its emitter. See part "c" of Fig. 20-9. Not only is I1 stolen, but the base of Q2 is grounded. This ensures that Q2 is turned off.

More than one view of the action of Q1 can be provided. While the action of Q1 involves shorting the base of Q2 to ground, the idea that the inputs steal I1 can be used when describing TTL action.

CLAMPING DIODES ON TTL INPUTS

In part "a" of Fig. 20-10, diodes have been added to the circuit's inputs. These are called *CLAMPS.* They stop the input voltage from going negative.

Under normal conditions, A and B do not go negative (logic 1 = +3.4 V; logic 0 = +0.2 V). Therefore, these diodes are usually reverse biased. They act like open circuits and do not enter into the action of the circuit.

In high speed circuits, however, the shift from 1 to 0 is not always clean. That is, instead of stopping when zero voltage is reached, ringing occurs. The term *RINGING* means a type of oscillation. See part "b" of Fig. 20-10. This is undesirable, since it slows circuit action.

At the negative peaks of the ringing, the clamping diode conducts. This clips the peak (see the dashed line) and reduces the ringing. That is, clamping diodes improve the high speed action of logic elements.

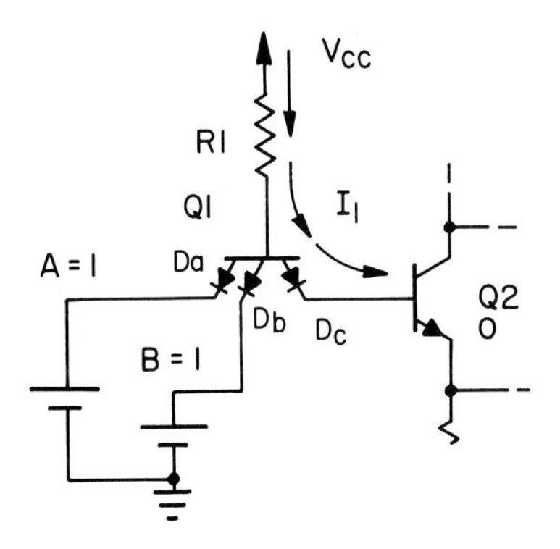

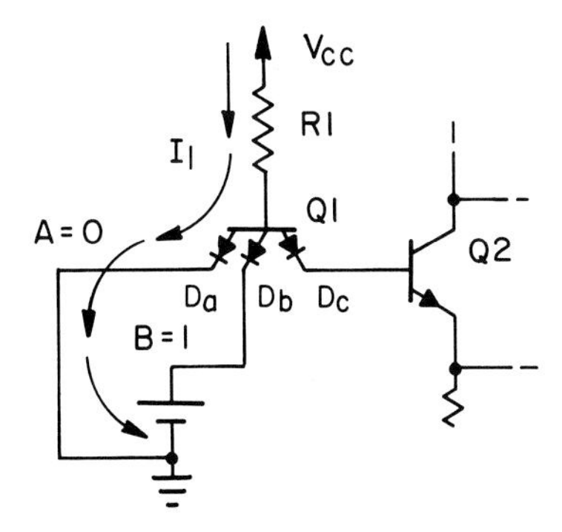

INPUTS A	INPUTS B	OUTPUT Y
1	1	0
0	1	1
1	0	1
0	0	1

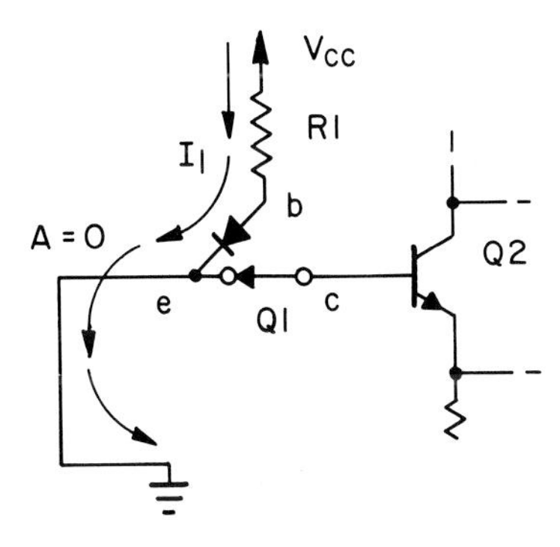

Fig. 20-9. If either or both inputs are grounded, current I1 is stolen from the base of Q2.

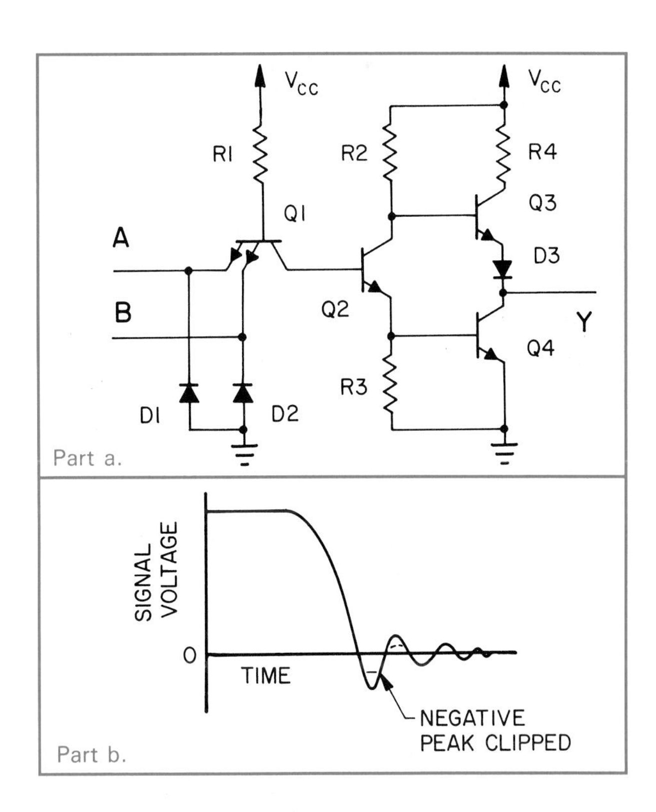

Fig. 20-10. Clamping diodes are used at TTL inputs to reduce ringing.

REVIEW OF TTL NAND CIRCUIT ACTION

When 1s are applied to all inputs of a TTL NAND, the emitters of the multiple-emitter transistor are reverse biased. Because they cannot conduct, current I1 must flow through the collector of Q1, part "a" of Fig. 20-11. This current turns on Q2 and allows I2 to flow.

Current I2 flows through R3 and produces enough voltage to turn on Q4. At the same time, so much of Vcc is dropped across R2 that the voltage at the base of Q3 is not enough to turn that transistor on. The circuit outputs a 0.

If a 0 is input at either or both inputs, I1 is stolen from the collector of Q1, part "b" of Fig. 20-11. With no current flowing into its base, Q2 is turned off. This stops the current I2, so Q4 is also turned off. Without I2, the voltage drop across R2 is reduced, and the voltage at the base of Q3 is large enough to turn this transistor on. The circuit outputs a 1.

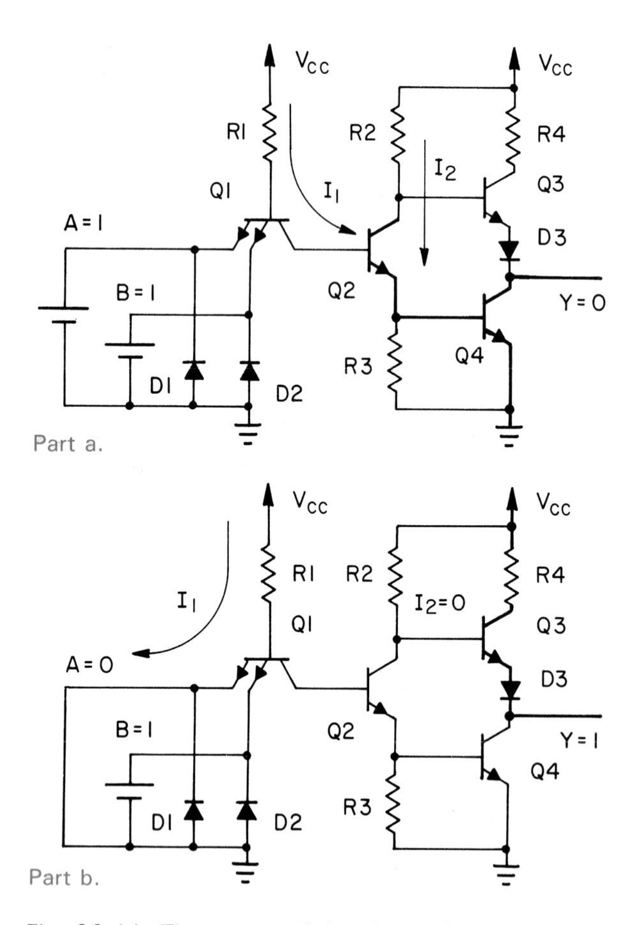

Fig. 20-11. The output of the circuit shown is determined by the path of I1. If it flows to Q2, Y = 0. If it flows to ground through either or both emitters of Q1, Y = 1.

OTHER TTL LOGIC ELEMENTS

Circuits for other TTL elements (ANDs, ORs, etc.) are similar to the NAND. The circuit in Fig. 20-12 is an example. An inverter (Q2) has been added between the decision circuit (Q1) and the phase splitter (Q3) to change the basic NAND circuit into an AND.

Open-collector TTL elements are formed by omitting D3, Q4, and R6. See Fig. 20-13.

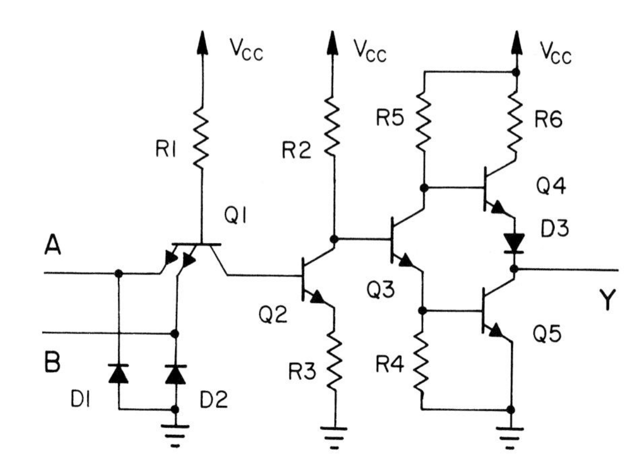

Fig. 20-12. The addition of an inverter (Q2) changes a NAND circuit into an AND.

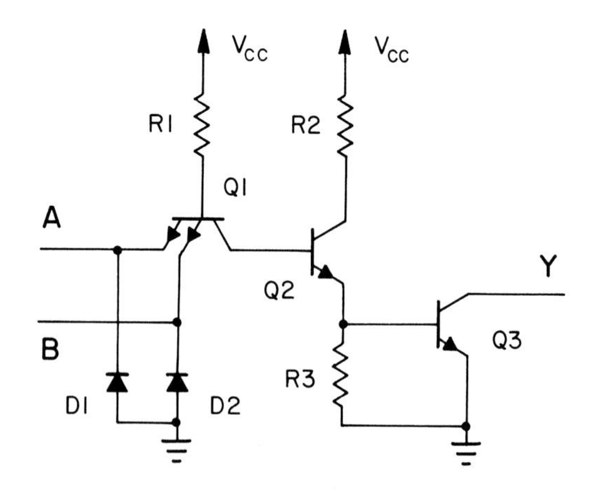

Fig. 20-13. If R6, Q4, and D3 are omitted, an open-collector circuit results.

CURRENT SINK AND CURRENT SOURCE

Chapter 14 discussed the terms sink and source. The meaning of current sinking can now be described in terms of the action of TTL circuits. Conclusions can be drawn. Safe load levels can be determined.

CURRENT SINK

Part "a" of Fig. 20-14 shows a TTL output driving a TTL input. To input a 0, current I1 must be stolen from the collector of Q1. To do this, the driving element must provide a path to ground. Current I1 must flow into the output of the driving element on its way to ground. That output is said to be a current sink.

CURRENT THROUGH Q4

TTL current ratings have been described earlier, but it is appropriate to review those ratings in terms of the transistors involved. The fanout can be checked.

I1 in part "a" of Fig. 20-14 is normally about 1 mA, but it can be as high as 1.6 mA. This means that Q4 must be capable of sinking 16 mA if a fanout of 10 is to be available.

VOLTAGE ACROSS Q4

When turned on, Q4 is a good conductor. However, there is a small voltage across it. Normally it is about 0.2 V, but it can be as high as 0.4 V and still be within specifications.

CURRENT SOURCE

When a 1 is to be input, the driving element changes from a current sink to a current source. See part "b" of Fig. 20-14.

Because emitter A is tied to a positive voltage (to Vcc through D3, Q3, and R4), it is turned off. I1 flows in the collector of Q1.

Although emitter A is turned off, a small reverse current still flows. Refer to the series of small arrows. This current is supplied by Q3, so the output functions as a current source.

CURRENT THROUGH Q3

The reverse current flowing into emitter A in part "b" of Fig. 20-14 is less than 40 microamperes (0.000,04 Amperes). To support a fanout of 10, Q3 must be capable of acting as a source for 400 microamperes (0.000,4 Amperes).

VOLTAGE AT OUTPUT

When a TTL output is operating properly, the voltage representing a 1 will be about 3.4 V. It can be as low as 2.4 V and still be within specifications. The difference between the output voltage and the 5 V at Vcc is dropped across D3, Q3, and R4.

SHORT-CIRCUIT CURRENT

Part "c" of Fig. 20-14 shows a short circuit across the output of a TTL element. If a 1 is being output, the element will act as a source and supply current to the short. Based on Vcc = 5 V, and R4 being 130 Ohms, this current will be about 34 mA. It can be as high as 55 mA. The components that carry this current (D3, Q3, and R4) will not be destroyed by this current. However, considerable heat is produced. As a result, only one gate in a multi-gate chip can be shorted at one time.

FLOATING LEADS

The action of a TTL circuit can be used to explain why floating leads act like logic 1s. For a 0 to be input,

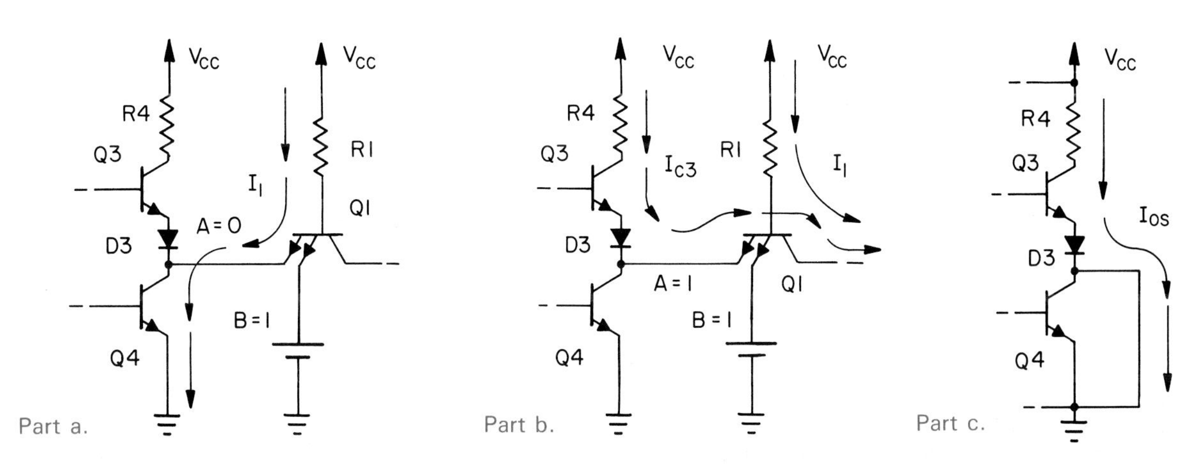

Fig. 20-14. TTL outputs function as both current sinks and current sources.

a path to ground must be provided for I1. Floating leads do not provide such a path, so they act as if a 1 is being input. See Fig. 20-15.

As indicated earlier, the voltage between a floating lead and ground is about 1.7 V. Although floating leads are treated as logic 1s, this voltage is in the undefined region between logic 1s and logic 0s. Because of the undefined values, floating leads are to be avoided in production equipment.

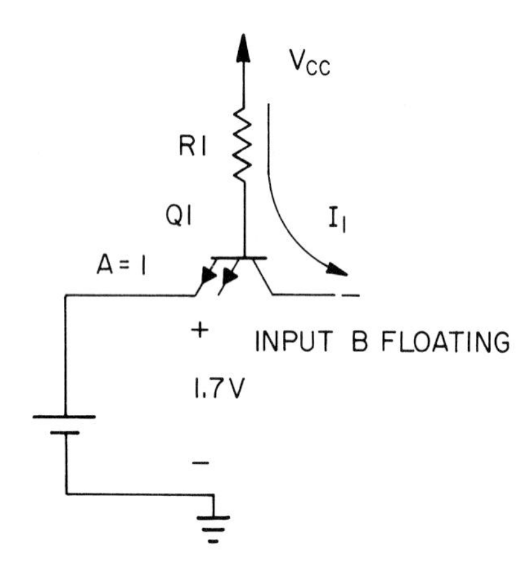

Fig. 20-15. A floating TTL input cannot steal I1, so it acts like a logic 1.

UNUSED LEADS IN LABORATORY CIRCUITS

Assume that the circuit in Fig. 20-16 is built and tested in a laboratory. Note the handling of unused inputs. Three topics must be examined. These are:

1. Unused leads on OR and NOR elements.
2. Unused leads on AND and NAND elements.
3. Unused leads on FLIP-FLOPS.

The first topic requires a note of caution, as in the following:

1. Unused leads on OR and NOR elements: Because floating inputs act like 1s, unused inputs on ORs must not float. If they do, the element's output will be s-a-1. In this case, the standard solution (tying unused leads to active leads) was used.
2. Unused leads on AND and NAND elements: In the circuit in Fig. 20-16, the unused AND lead was allowed to float. The state of the lead represents a permanent 1 and does not change the action of the element. In a laboratory, where electrical noise is seldom a problem, this is an acceptable solution.
3. Unused leads on FLIP-FLOPS: SET and CLEAR on the flip-flop in Fig. 20-16 have not been used. Because they are active-low, allowing them to float is often acceptable. In electrically noisy locations,

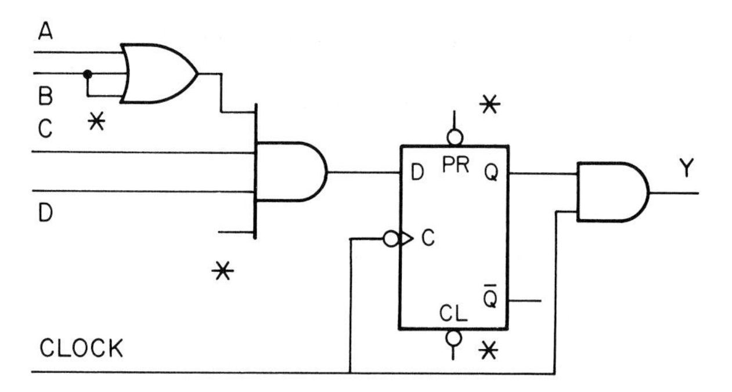

Fig. 20-16. In laboratory circuits, some unused TTL inputs may be allowed to float.

in circuits with poorly regulated power supplies, and when low quality chips are used, floating SET and CLEAR leads may cause faulty operation. If this is found to be true, active-low inputs should be connected to Vcc (a logic 1).

UNUSED LEADS IN PRODUCTION CIRCUITS

Floating leads should be avoided in equipment used outside the laboratory. Hardware personnel assume that problems caused by floating leads have been solved before equipment is sold. Faults caused by floating leads are difficult to locate and correct in the field. Two topics are important. These are:

1. Unused leads on OR and NOR elements.
2. Unused leads on AND elements, NAND elements, and active-low inputs.

The first topic looks at the fanin, as in the following:

1. Unused leads on OR and NOR elements: Fig. 20-17 shows two ways of dealing with unused OR and NOR leads. On printed circuits, it is usually easier to connect adjacent pins, so the method at the left is often used. It does, however, increase input current slightly when logic 1s are applied. If fanin is a problem, the method at the right should be used (the unused lead is grounded).

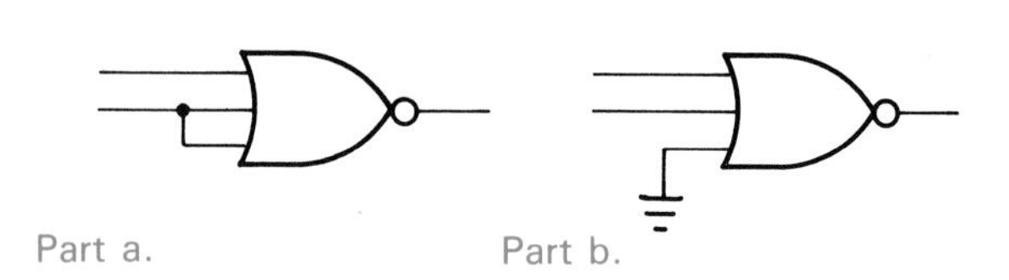

Fig. 20-17. In production circuits, unused OR and NOR inputs should be connected as shown.

2. Unused leads on AND elements, NAND elements, and active-low inputs: Fig. 20-18 shows four methods of dealing with unused AND, NAND, and active-low inputs. They are:
   a. Unused leads are connected to active leads. As before, this results in a small increase in fanin.
   b. Unused leads are connected to Vcc. If this method is used, Vcc must not exceed 5.5 V (the absolute maximum signal voltage).
   c. Unused leads are connected to Vcc through a 1K resistor. This protects elements from excess voltage (over 5.5 volts) at Vcc. Up to 25 inputs can be connected to one resistor.
   d. Unused inputs are connected to the output of an element with fixed logic 1. In some circuits, this would be cheaper than adding a resistor. Based on the fanout of the family, up to 10 inputs could be connected to one output.

TTL CHARACTERISTICS

Detailed descriptions of TTL elements are available in data books. Only selected specifications will be described in this section. There are 6 topics that need to be discussed. These are:

1. Supply voltage.
2. Signal voltages.
3. Power dissipation.
4. Propagation delay.
5. Half-voltage points.
6. Clock frequency.

SUPPLY VOLTAGE

For TTL elements, Vcc should be +5 V, plus or minus 5 percent. See Fig. 20-19. This voltage should be measured with the equipment connected to the power supply and operating. If Vcc is outside this range, TTL circuits will still operate, but the operation will not be optimum.

If Vcc exceeds 7 V, TTL chips are likely to be damaged. This is said to be the family's absolute maximum power supply rating.

SIGNAL VOLTAGES

TTL logic levels have been described in Chapter 3. Fig. 20-20 is shown here for review purposes. There are three topics that are important. These are:

1. Logic 1/logic 0.
2. Noise margin.
3. Absolute maximum signal.

LOGIC 1/LOGIC 0

When outputting a 1, the voltage at the output of a properly operating TTL element will never be less than 2.4 V. However, a TTL input will accept any voltage above 2 V as a 1.

A similar situation exists for logic 0. The voltage at the output of a properly operating TTL element will never be higher than 0.4 V when outputting a 0. TTL inputs accept any voltage below 0.8 V as a logic 0 signal level.

NOISE MARGIN

The difference between acceptable voltage levels at TTL inputs and outputs is said to be the family's

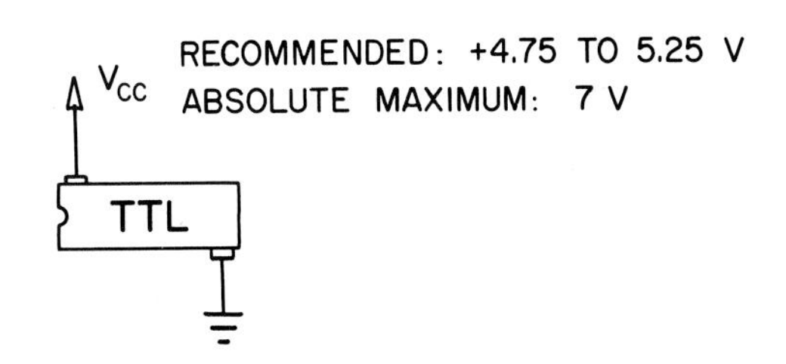

Fig. 20-19. The recommended range of the supply voltage for standard TTL chips is +4.75 to 5.25 V.

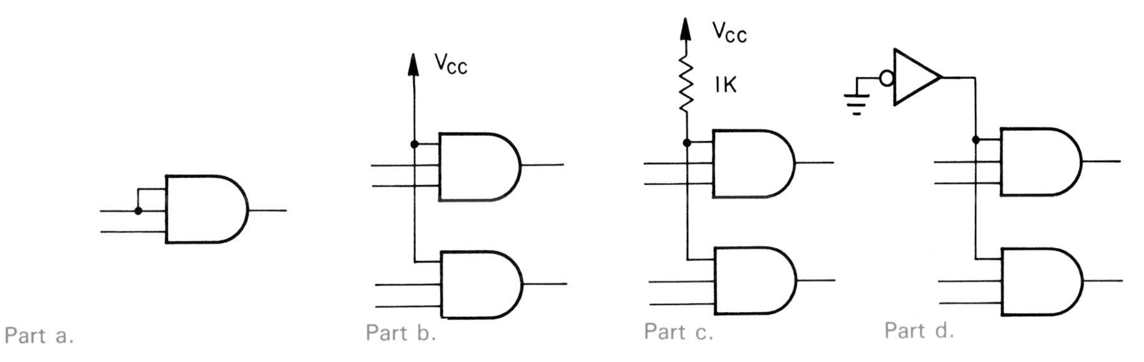

Fig. 20-18. Four methods of connecting unused inputs for ANDs and NANDs are shown.

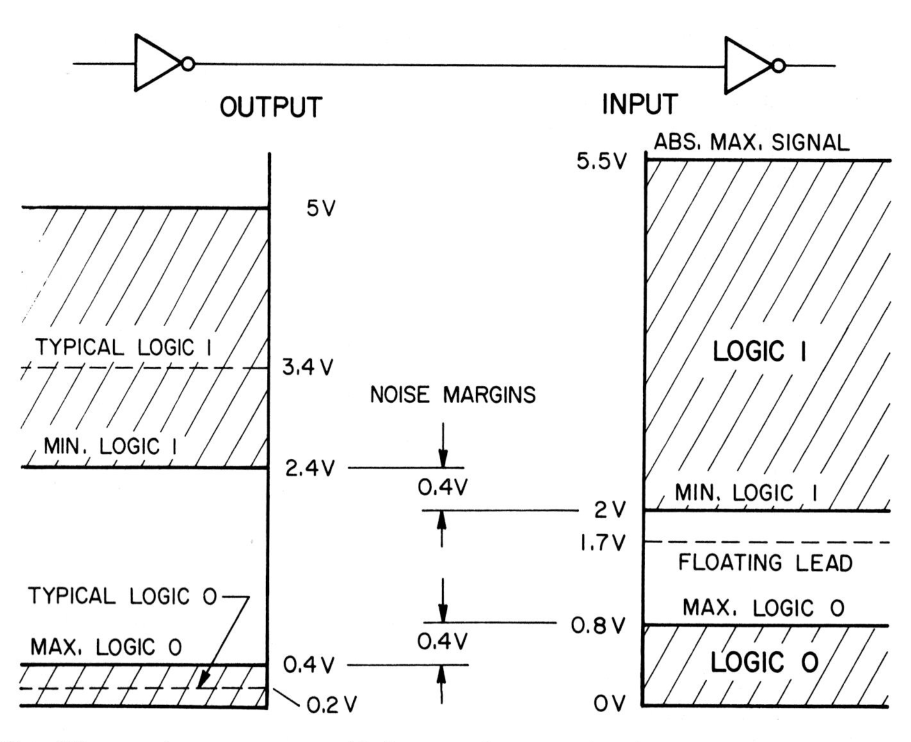

Fig. 20-20. The difference between acceptable input and output signal voltages is called the noise margin.

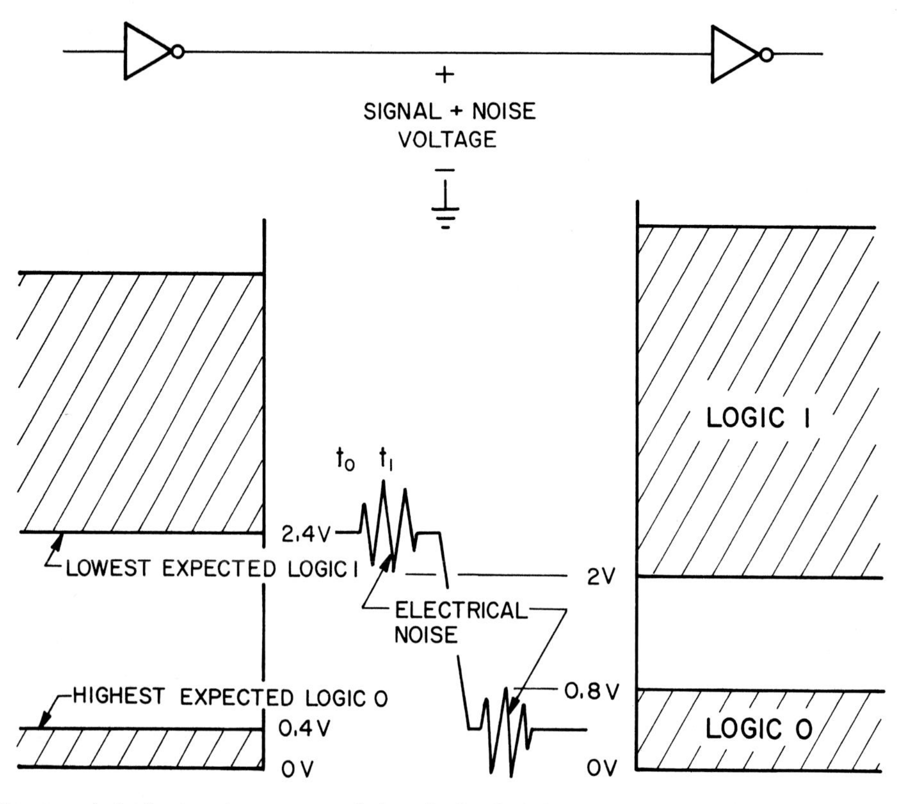

Fig. 20-21. Noise margin indicates the amount of electrical noise that can appear without producing unacceptable signals.

NOISE MARGIN. Fig. 20-21 suggests the importance of this concept.

At t0, the graph shows a low (but acceptable) logic 1 being output by a TTL element. At time t1, a 0.4 V noise pulse appears on the lead. Based on the TTL noise margin, the signal at t1 is still an acceptable logic 1, since the voltage did not go below 2 V.

A similar situation exists at logic 0. The voltage with the noise signal did not go above 0.8 V.

ABSOLUTE MAXIMUM SIGNAL

The highest signal voltage that can be safely applied to a TTL input is 5.5 V. Signals larger than this are likely to damage the logic elements.

POWER DISSIPATION

Electric power is measured in watts. The power delivered to a circuit or device is the product of current (in amperes) and voltage.

The current, and therefore the power taken by a gate, depends on whether it is outputting a 1 or 0. See Fig. 20-22. To estimate the current required by a group of gates, it is assumed that half will be outputting 1s while the rest are outputting 0s. Based on this average, a TTL NAND will draw about 2 mA of dc current.

The average power per gate equals this average current times Vcc. For a TTL NAND, this power is 2 mA x 5 V or 10 mW (0.01 W).

Individual logic gates use little current and power. However, a single chip or circuit may contain many gates, so the total power may be considerable.

Supplying large amounts of dc power is expensive in terms of both the power used and in terms of the cost of the power supply itself. In addition, equipment that consumes large amounts of power must be cooled, and this adds to the cost and the equipment cabinet size required.

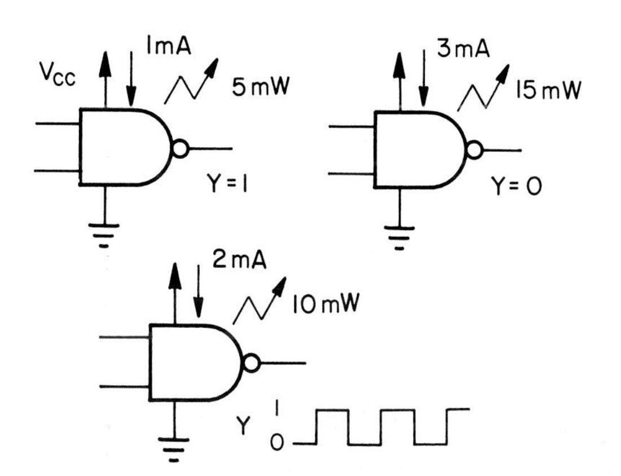

Fig. 20-22. The power dissipated by a TTL chip depends on its output signal.

PROPAGATION DELAY

When a change in signal is applied to the input of an element, its output does not change immediately. There is a slight delay. It is called a *PROPAGATION DELAY*. To propagate means to move.

Propagation delays are so short they must be measured in nanoseconds. Nano stands for 0.000,000,001. This is about the time it takes for light to travel 1 ft.

HALF-VOLTAGE POINTS

When the time scale of an oscilloscope is expanded enough to measure propagation delay, a problem becomes evident. The seemingly vertical edges of signal changes prove to be rounded. Due to the shape of the waves near zero, it is difficult to determine where a wave begins. To overcome this problem, propagation delay is measured between half-voltage points, Fig. 20-23.

Typical propagation delays for standard TTL elements are shown in Fig. 20-24. The average delay is about 10 ns. Although this is very short, signals in modern computers must pass through many elements.

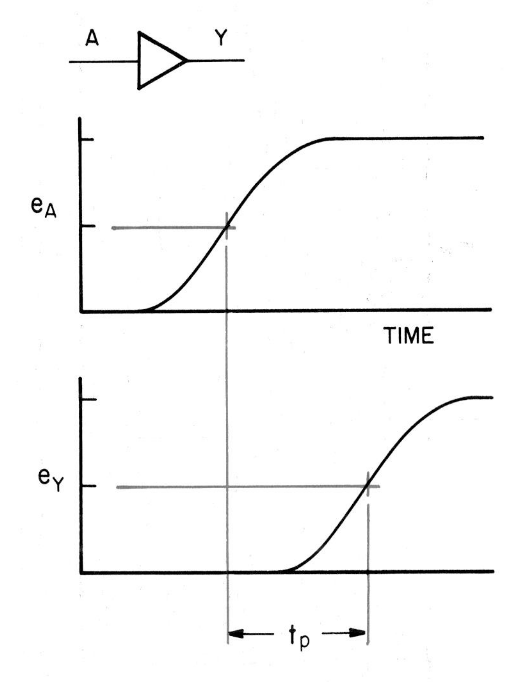

Fig. 20-23. Propagation delay is measured between input and output half-voltage points.

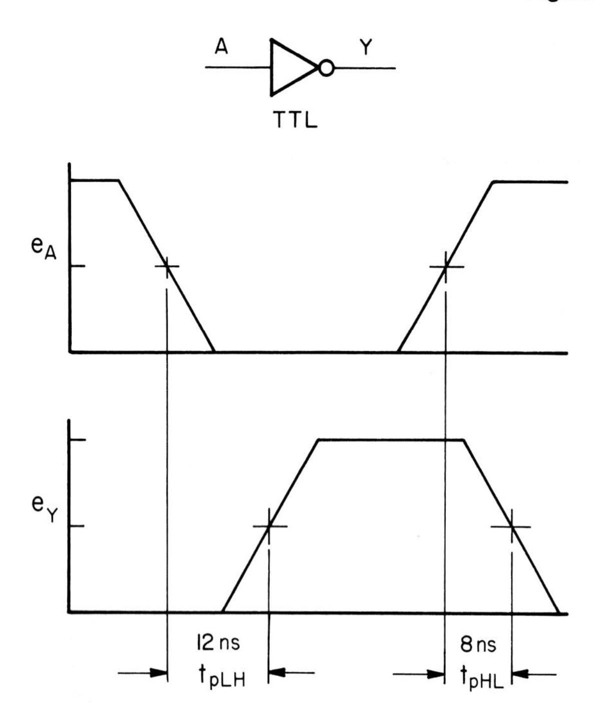

Fig. 20-24. The average propagation delay of TTL elements is about 10 ns. "LH" means a low-to-high transition. "HL" means high-to-low.

This delay can be a factor in determining the speed of digital and computer circuits.

CLOCK FREQUENCY

A measure of speed finds how many numbers per second are accepted by flip-flops of a family. One finds how many clock cycles can be applied per second without error. The less the propagation delay of a family, the higher its clock frequency.

TTL flip-flops can operate at about 35 MHz (35,000,000 cycles per second). That is, the number in a register composed of TTL flip-flops can be changed 35,000,000 times a second. It takes only 0.000,000,029 seconds to change the stored number.

SPECIFICATION SHEETS

Manufacturers of chips provide specification or data sheets to those who use their products. These contain very large amounts of data and, as a result, specification sheets appear to be difficult to use. With experience, however, you will learn to easily find needed information.

The next few pages contain parts of typical data sheets. The discussion will help tie what you already know about TTL logic to the information supplied by chip manufacturers.

PIN NUMBERS

Diagrams showing pin connections are probably the most used portions of specification sheets. Fig. 20-25 shows a number of such diagrams. The figure at the lower left is the logic symbol for a quad NAND. The symbol is based on a graphics standard established by the ANSI and the IEEE. Because this symbol tends to be confusing, it was not used in this book.

CIRCUIT DIAGRAMS

Part of the information supplied for a TTL quad NAND is shown in Fig. 20-26. The schematics depict circuits used in a number of families. The one of interest here is at the top. This circuit is the same as the one shown in Fig. 20-10. Note the clamping diodes across its inputs and the totem-pole output.

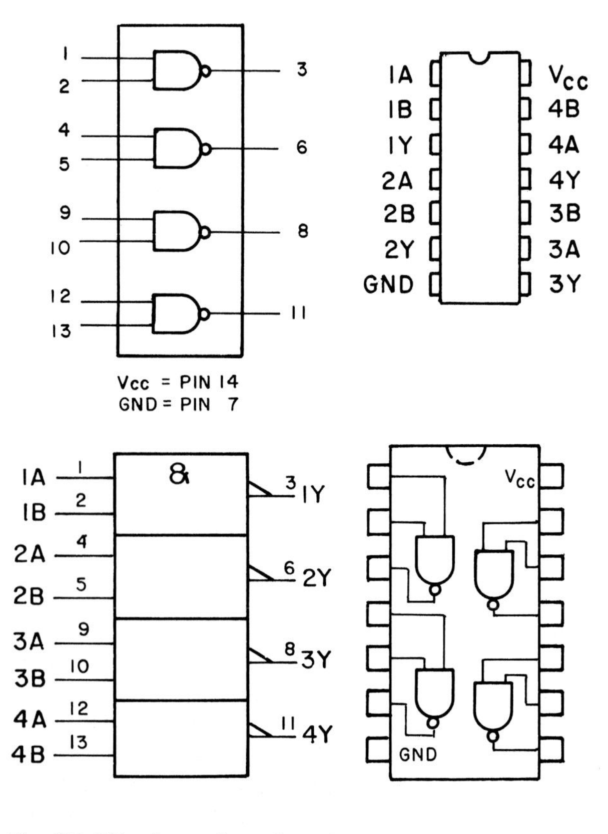

Fig. 20-25. A number of methods are used on sheets to indicate pin connections.

TTL Logic Element Specifications and Operating Requirements

SN5400, SN54LS00, SN54S00,
SN7400, SN74LS00, SN74S00
QUADRUPLE 2-INPUT POSITIVE-NAND GATES

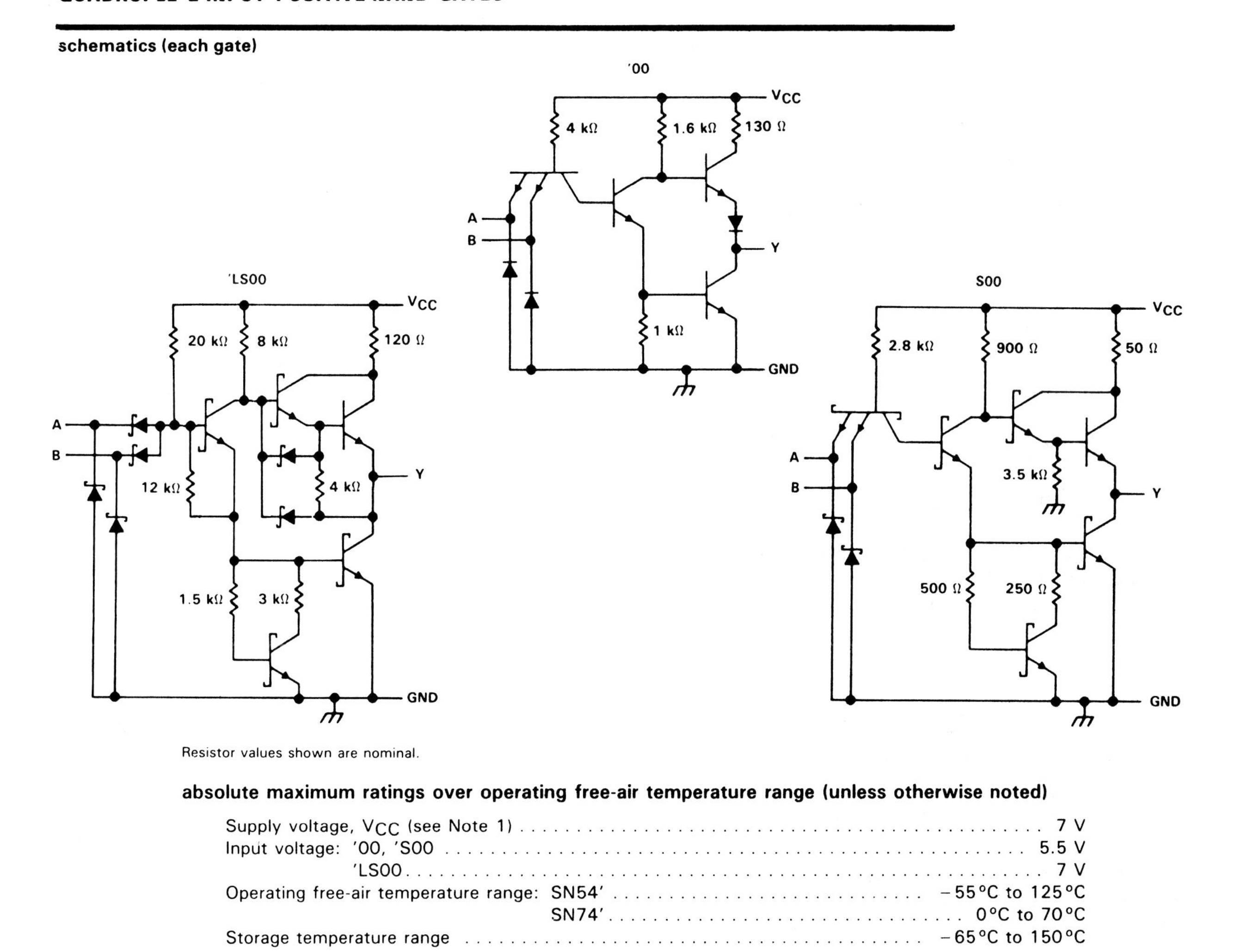

absolute maximum ratings over operating free-air temperature range (unless otherwise noted)

Supply voltage, V_{CC} (see Note 1)	7 V
Input voltage: '00, 'S00	5.5 V
'LS00	7 V
Operating free-air temperature range: SN54'	−55°C to 125°C
SN74'	0°C to 70°C
Storage temperature range	−65°C to 150°C

NOTE 1: Voltage values are with respect to network ground terminal.

Fig. 20-26. The upper schematic shows the circuit used in each NAND of 5400 and 7400 chips. (Texas Instruments, Inc.)

CHIP TYPE

In addition to describing the standard TTL NAND chips, SN5400 and SN7400, the sheet in Fig. 20-26 also describes the low-power Schottky and Schottky families. (These will be covered in the next chapter.) The '0 above the upper schematic implies that this circuit is used in the SN5400 and SN7400. The 'LS00 and 'S00 above the other schematics indicate that they are for the SN54LS00/SN74LS00 and SN54S00/SN74S00.

ABSOLUTE RATINGS

Below the schematics in Fig. 20-26 is a heading that reads, "Absolute maximum ratings over operating free-air temperature range..." The voltages and temperatures above those in this table are likely to damage the chip. The table indicates that V_{CC}, the supply voltage, should never be higher than 7 volts (V_{CC} is normally kept below 5.5 volts). The next line states that no more than 5.5 volts should be applied to the inputs of a 5400 or 7400.

The line indicating operating temperature suggests a difference between a 5400 and a 7400. The allowable temperature range of the 5400 is greater than that of the 7400. This is true for most of the characteristics of the 54' series. It is far more rugged than the 74' series. The 5400 is used in harsh environments. As would be expected, it is more costly. In most applications, 5400s can be substituted for 7400s, but the opposite is not necessarily true.

The next line shows the storage temperature. Like most electronic parts, these chips can tolerate a wide range of temperatures when power is not applied.

OPERATING CONDITIONS

The upper box in Fig. 20-27 shows the recommended voltages, currents, and temperature under which the 5400 and 7400 should be used. Note the use of the abbreviation NOM in the headings at the right. It stands for nominal. This is the preferred voltage. It is the one that is used in the design of circuits. No voltage is perfectly accurate, so a tolerance must be specified. The MIN and MAX suggest the allowable variation from the nominal voltage. Notice that the 5400 is again the more rugged chip. It can operate

SN5400, SN7400
QUADRUPLE 2-INPUT POSITIVE-NAND GATES

recommended operating conditions

		SN5400			SN7400			UNIT
		MIN	NOM	MAX	MIN	NOM	MAX	
V_{CC}	Supply voltage	4.5	5	5.5	4.75	5	5.25	V
V_{IH}	High-level input voltage	2			2			V
V_{IL}	Low-level input voltage			0.8			0.8	V
I_{OH}	High-level output current			−0.4			−0.4	mA
I_{OL}	Low-level output current			16			16	mA
T_A	Operating free-air temperature	−55		125	0		70	°C

electrical characteristics over recommended operating free-air temperature range (unless otherwise noted)

PARAMETER	TEST CONDITIONS†	SN5400			SN7400			UNIT
		MIN	TYP‡	MAX	MIN	TYP‡	MAX	
V_{IK}	V_{CC} = MIN, I_I = −12 mA			−1.5			−1.5	V
V_{OH}	V_{CC} = MIN, V_{IL} = 0.8 V, I_{OH} = −0.4 mA	2.4	3.4		2.4	3.4		V
V_{OL}	V_{CC} = MIN, V_{IH} = 2 V, I_{OL} = 16 mA		0.2	0.4		0.2	0.4	V
I_I	V_{CC} = MAX, V_I = 5.5 V			1			1	mA
I_{IH}	V_{CC} = MAX, V_I = 2.4 V			40			40	μA
I_{IL}	V_{CC} = MAX, V_I = 0.4 V			−1.6			−1.6	mA
I_{OS}§	V_{CC} = MAX	−20		−55	−18		−55	mA
I_{CCH}	V_{CC} = MAX, V_I = 0 V		4	8		4	8	mA
I_{CCL}	V_{CC} = MAX, V_I = 4.5 V		12	22		12	22	mA

† For conditions shown as MIN or MAX, use the appropriate value specified under recommended operating conditions.
‡ All typical values are at V_{CC} = 5 V, T_A = 25°C.
§ Not more than one output should be shorted at a time.

switching characteristics, V_{CC} = 5 V, T_A = 25°C (see note 2)

PARAMETER	FROM (INPUT)	TO (OUTPUT)	TEST CONDITIONS	MIN	TYP	MAX	UNIT
t_{PLH}	A or B	Y	R_L = 400 Ω, C_L = 15 pF		11	22	ns
t_{PHL}					7	15	ns

NOTE 2: Load circuits and voltage waveforms are shown in Section 1.

Fig. 20-27. This specification sheet details the characteristics of the 5400 and 7400. (Texas Instruments, Inc.)

effectively under a larger range of supply voltages than can the 7400.

The next line is a bit more confusing. It is designated VIH. Based on the V and I, it is an input voltage. The H indicates that it is the voltage of a high signal. That is, VIH is the voltage of a logic 1 at the input to this TTL gate. Looking to the right, you can see that only the minimum (2 volts) is stated. Fig. 20-28 will help explain this situation. This is the same diagram used in Fig. 20-20. At the right, on the input side, the minimum voltage this NAND will accept as a logic 1 is 2 volts. Any voltage greater than 2 volts is acceptable. There is no need to indicate a nominal voltage. Any voltage between 2 volts and the absolute maximum of 5.5 volts will be treated by the circuit as a 1.

The next line, VIL, states the low-level input signal. Here, only the maximum voltage is shown. Any voltage below 0.8 volts will be treated as a logic 0. Again refer to the right side of Fig. 20-28.

The next two lines refer to the amount of current the output can supply. IOH is the current that can be drawn when a logic 1 is being output. The negative sign indicates that this current flows out of the output. IOL is the amount of current the output can sink when a logic 0 is present.

The last line of this block is the operating temperature, TA. It is the same as the absolute temperature range given on the first page of these specifications.

ELECTRICAL CHARACTERISTICS

For many engineers and technicians, the first line of the next section suggests a difficult situation. There are times when you must ignore portions of data sheets. The line VIK is an example. It has to do with the inverse voltage that can be applied to an input lead. In most applications, this is of little importance, and should be ignored.

The next two lines, VOH and VOL, are output voltages for logic 1 and logic 0. Note that both nominal and maximum/minimum values are indicated. In a circuit that is operating normally, these voltages will be near the nominal value. Refer to the left side of Fig. 20-28.

The next line, II, is another parameter that is not of general interest. The next two are, however, important. IIH is the input current when a logic 1 is applied to an input. Compared to the other numbers in the table, IIH = 40 appears to be a rather large number. However, the last column (the units column) shows this to be only 40 microamperes (0.04 mA).

The negative sign in the line describing IIL emphasizes that this is a current-sinking famiy. That is, when a logic 0 is present at an input, current flows

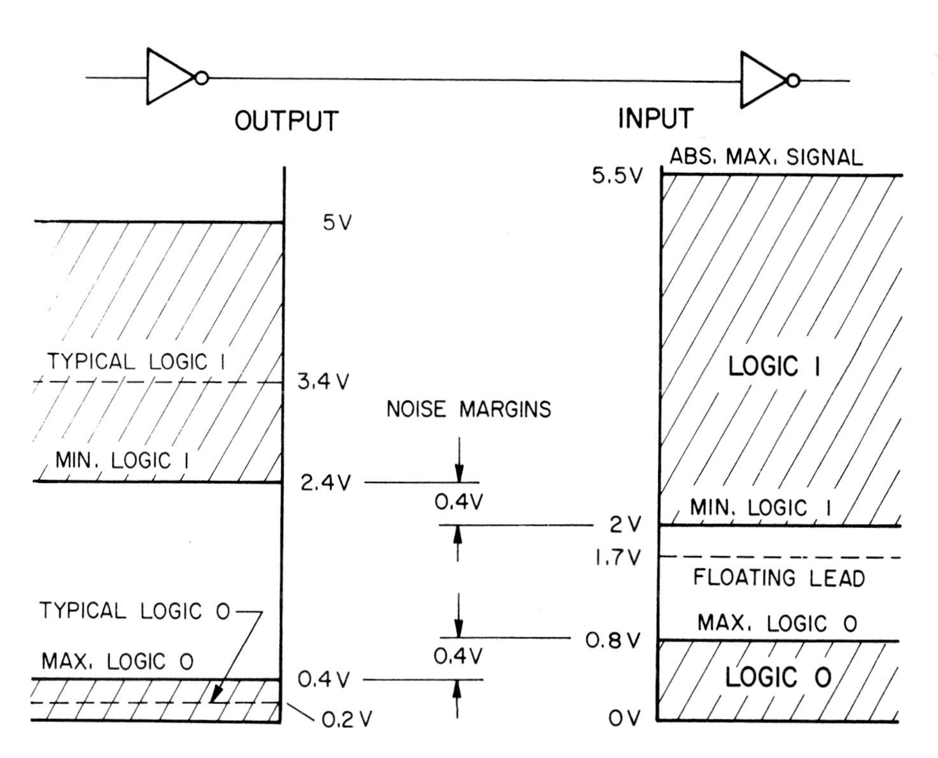

Fig. 20-28. This diagram emphasizes the relationships between input and output signals for standard TTL logic.

out of the input to ground through the output of the driving element.

The line for IOS shows the range of current flow that can be expected through a short to ground. Note the warning.

The next two lines have to do with the current the power supply must provide. For ICCH, all outputs are high. For ICCL, all outputs are low. These numbers are used in determining the required current rating of the power supply.

The last box specifies propagation delays that can be expected. Refer to Fig. 20-24 to review the meaning of the term propagation delay.

TIMING DIAGRAMS

Fig. 20-29 shows a timing diagram from a specification sheet. It is for a shift register. The dashed lines at the left of the diagram indicate that the outputs

SN54164, SN54LS164, SN74164, SN74LS164
8-BIT PARALLEL-OUT SERIAL SHIFT REGISTERS

typical clear, shift, and clear sequences

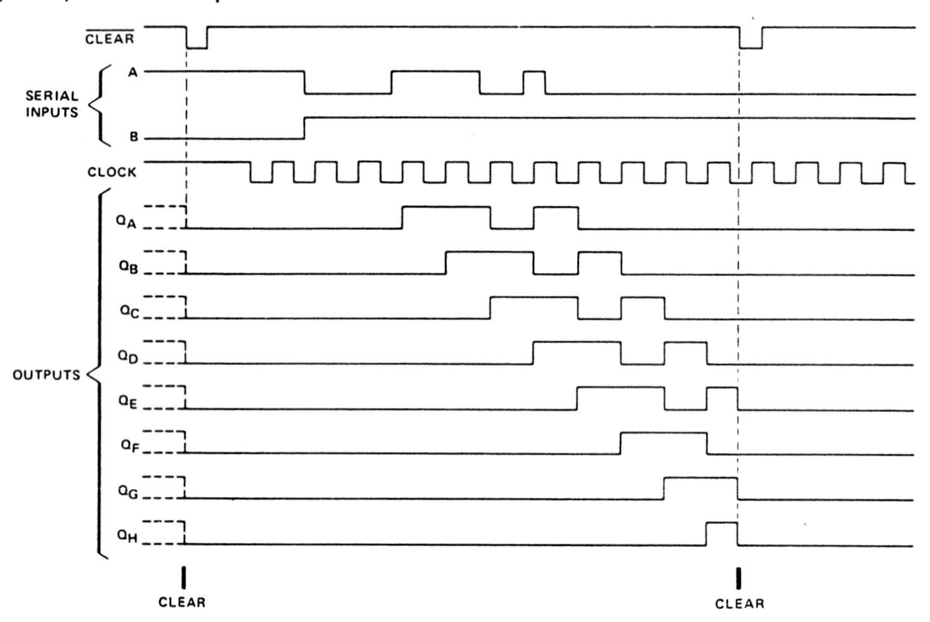

SN54164, SN54LS164, SN74164, SN74LS164
8-BIT PARALLEL-OUT SERIAL SHIFT REGISTERS

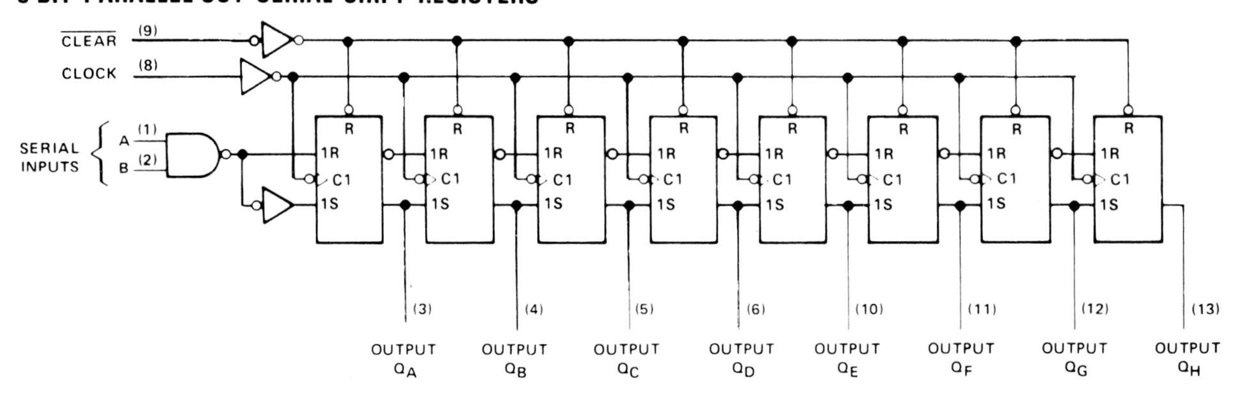

Pin numbers shown are for D, J, N, and W packages.

Fig. 20-29. Timing diagarams are often used on data sheets. (Texas Instruments, Inc.)

can be a mix of 1s and 0s when CLEAR is applied. After CLEAR has gone low, all outputs will be 0.

Refer to the heading on this data sheet. Note that this timing diagram applies to both the standard TTL and the LS TTL. Chips that have similar characteristics are often described on a single sheet.

The logic diagram below the timing diagram can be of considerable help in understanding the operation of a circuit. For example, drivers are provided in the CLEAR and CLOCK leads. Each must drive eight inputs, so these drivers were added to reduce fanin.

Also note that the driver on the CLEAR lead has bubbles at both its input and output. That is, it is a non-inverting amplifier. A low at its input will produce a low at its output. The draftsman added the bubbles to emphasize that the CLEAR is active low.

TRUTH TABLES

Fig. 20-30 shows a truth table from the data sheet for a positive-edge-triggered, D-type flip-flop. In this case, the truth table has been called a function table. To define the inputs and outputs, the logic diagram and logic symbol are shown.

The ANSI/IEEE standard was used for the logic symbol. It indicates that there are two data latches in the package. A triangle on a lead means that it is active low. The triangle on the inside of the symbol at lead 3 means that it is a positive-going edge-triggered input.

Note the seven Xs on the truth table. (The seventh is at the bottom of the D column.) These are don't cares. For the conditions specified in a given row, these inputs do not influence the outputs. It does not matter whether they are 1s or 0s.

The CLOCK on this flip-flop is triggered by a positive-going edge. This is indicated by the two arrows pointing upward in the CLOCK column.

The note under the table helps explain the marks beside the Hs in the output columns. When both PRESET and CLEAR are low at the same time, both outputs will go high. This is a temporary condition. When either PRESET or CLEAR goes high, one of the two outputs will go low.

SUMMARY

TTL stands for transistor-transistor logic and uses bipolar junction transistors. The TTL gate is often called a totem pole. It is capable of producing logic 0 and logic 1 outputs. Many TTL gates contain multiple-emitter transistors.

Only one transistor in the totem pole can conduct at a time. The circuit that controls the transistors is a phase splitter. The phase splitter turns on one of the transistors and turns off the other. Diodes are used as clamps in TTL circuits. The diodes stop the input voltage from going negative. The input voltage attaining a negative value is a risk in high speed circuits. The negative voltages come from a type of oscillation called ringing.

In laboratory circuits, there should be no leads left unused in OR and NOR TTL elements. Unused leads should be tied to active leads. Unused leads in AND and NAND circuits can be left floating. The SET and CLEAR leads on flip-flops can be allowed to float unless they are in an electrically noisy situation. In circuits used in equipment outside the laboratory, floating leads should be avoided. Unused leads should be connected to active leads.

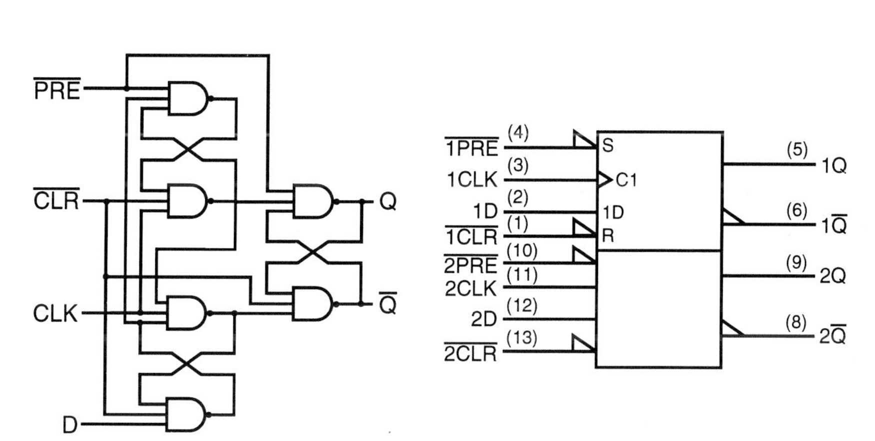

FUNCTION TABLE

INPUTS				OUTPUTS	
$\overline{PRE}$	$\overline{CLR}$	CLK	D	Q	$\overline{Q}$
L	H	X	X	H	L
H	L	X	X	L	H
L	L	X	X	H†	H†
H	H	↑	H	H	L
H	H	↑	L	L	H
H	H	L	X	Q_0	$\overline{Q}_0$

† The output levels in this configuration are not guaranteed to meet the minimum levels in V_{OH} if the lows at preset and clear are near V_{IL} maximum. Furthermore, this configuration is nonstable; that is it will not persist when either preset or clear returns to its inactive (high) level.

Fig. 20-30. Truth tables are often used to describe the actions of logic elements and circuits. On this sheet, the truth table is called a function table. (Texas Instruments, Inc.)

V_{CC} should be +5 volts in TTL circuits. Exceeding 7 volts is likely to damage TTL circuits. A TTL will except any voltage above 2 volts as a logic 1. The output of a properly working TTL will be above 2.4 volts for a logic 1. Any voltage below 0.8 volts is read as a logic 0. A properly working TTL will not go above 0.4 volts when outputting a logic 0. Input signals to TTLs should never rise above 5.5 volts.

Specification sheets contain a wide variety of useful information about logic chips. Diagrams showing the pin connections are probably the most used portions of the specification sheets. The specification sheets also include circuit diagrams, chip type, the absolute maximum ratings, recommended operating conditions, electrical characteristics, timing diagrams, and truth tables.

IMPORTANT TERMS

Bipolar, Half-voltage points, Multiple emitter, Noise margin, Phase splitter, Propagation delay, Totem pole.

TEST YOUR KNOWLEDGE

1. TTL stands for __________.
2. When a positive voltage greater than ________ V is applied between the base and emitter of an npn bipolar transistor, a base current will flow.
3. The term __________ __________ is used to describe an output circuit having two transistors and a diode arranged vertically.
4. The circuit that turns on the proper transistor in a totem pole arrangement of a TTL logic element circuit is called a(n) __________ __________.
5. Which of the following components in a totem pole stabilizes its behavior?
   a. Multiple emitter.
   b. Diode.
   c. Vcc.
   d. Charge-coupled device.
   e. Bleeder resistor.
6. In high-speed circuits, an oscillation called __________ can occur.
7. A diode can be used as a(n) __________ (fuse, input, clamp, divider) to help prevent ringing at an input.
8. Open-collector TTL outputs are formed by __________ (adding, omitting) components.
9. Can one ever say that current is "stolen" from an input lead?
10. The counterpart of a current source is usually called a current __________.
11. Unused leads on an OR logic element must not be allowed to __________.
12. Unused leads on an AND logic element in a laboratory circuit can be allowed to float. (True or False?)
13. If Vcc exceeds __________ V, TTL chips are likely to be damaged. This is the absolute maximum power supply rating.
14. Which of the following is closest to the noise margin of TTL chips?
    a. 0.2 V.
    b. 0.4 V.
    c. 0.8 V.
15. What is the average power dissipation for one TTL logic gate?
16. What is the average propagation delay for a TTL element?
17. TTL flip-flops can operate at a frequency of about __________ MHz.
18. Propagation delay is measured on a curve using the __________-voltage points.
19. The TTL logic family is seldom used in VLSI and even LSI because its logic elements take up too much room and because it:
    a. Dissipates too much power.
    b. Is not sufficiently reliable.
    c. Cannot be fabricated on silicon chips.
20. What does ns stand for?

STUDY PROBLEMS

1. What emitter-to-base voltage must be applied to an npn transistor to turn it on?
2. Draw the schematic for the totem pole circuit of a TTL NAND. Label its parts Q3, Q4, D3, and R4. Show and label the two input currents Ib3 and Ib4. Label the circuit's output Y. Include a ground and indicate Vcc.
3. Current is flowing into the base of the upper transistor in the TTL output circuit shown in Fig. P20-1. What number (1 or 0) will be output at Y?
4. Which of the following best describes the function of resistor R4 in the circuit that is taken from problem 3 (Fig. P20-1)?
   a. It acts as a load for Q3 when the circuit outputs a logic 1.
   b. It insures that Q3 will be off when Q4 is on.
   c. It limits the circuit's output current.
5. Repeat Problem 4 for diode D3 in Fig. P20-1.
6. Redraw the circuit from Problem 2. Then add the circuit for the phase splitter used in TTL NANDs. Label the components in the phase splitter Q2, R2, and R3. Indicate ground and Vcc.
7. Copy the circuit in Fig. P20-2 onto a sheet of paper or use the circuit you drew for Problem 6. Add arrows to show the paths of I1 and I2. Some

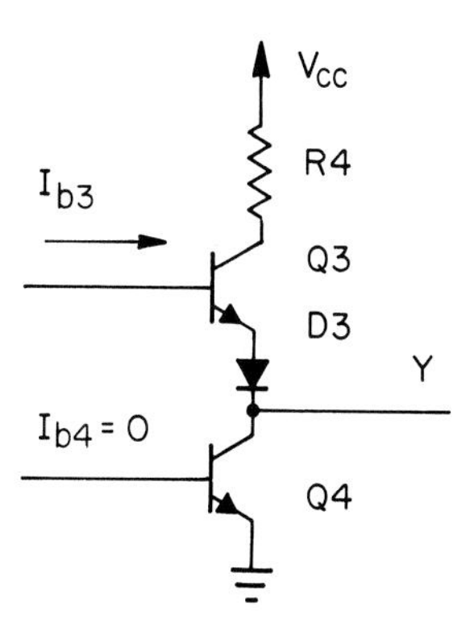

Fig. P20-1. The effects of current levels and the purposes of components for a totem pole circuit can be determined.

arrows can show current reaching the ground. Circle the transistors that will be turned on by the indicated currents.

8. If I1 in Problem 7 (from Problem 6 or from Fig. P20-2) were off, which transistor or transistors would conduct?

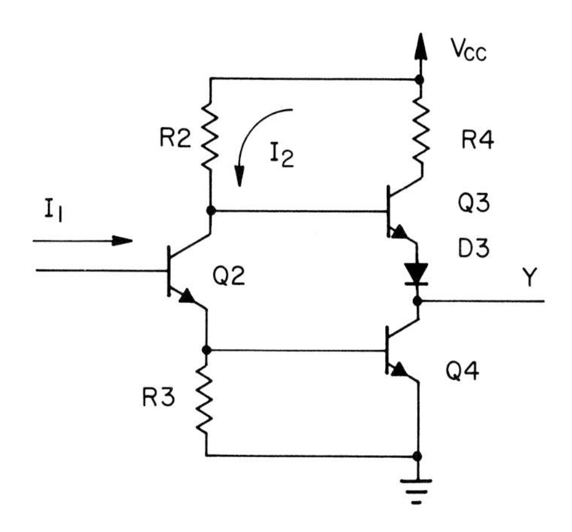

Fig. P20-2. The effects of different currents (I1 on or I1 off) on a phase splitter can be determined. Making a drawing can help bring some concepts from memory.

9. Draw the complete schematic of a TTL NAND from memory. Include clamping diodes, and label the signal leads A, B, and Y. Also label Vcc.
10. Copy the circuit in Fig. P20-3 onto a sheet of paper or use the circuit you drew for Problem 9. For the input signals shown, indicate the path of current I1 to ground. Also circle all transistors that will be turned on by the given signals.

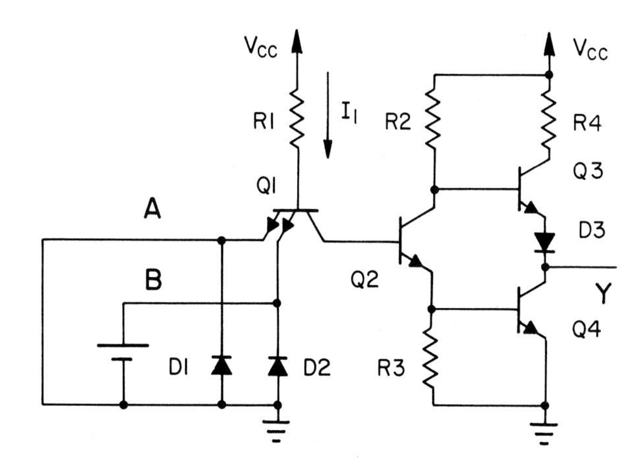

Fig. P20-3. A complete TTL NAND circuit without input levels can be drawn from memory and analyzed.

11. If both inputs to the circuit from Problem 10 (from Problem 9 or from Fig. P20-3) were floating, which transistors would be conducting? Also, what signal (1 or 0) would appear at Y?
12. When a TTL input is grounded as shown in Fig. P20-4, what is the maximum current that is likely to flow out of that input to ground? Give your answer in mA.

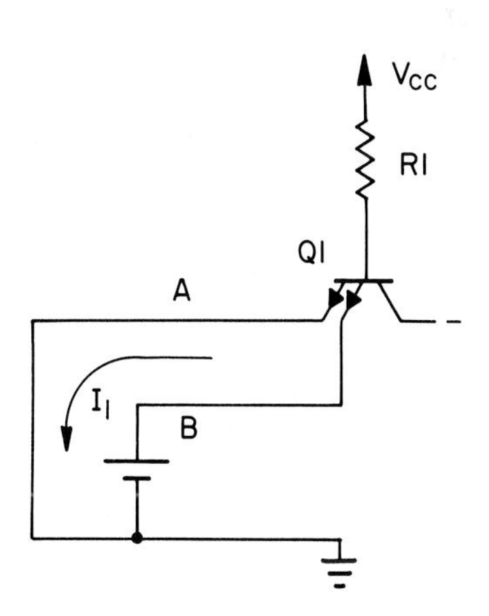

Fig. P20-4. The maximum amount of current flowing out of a TTL input to ground can be determined.

13. The circuit in Fig. P20-5 shows a standard TTL output driving two TTL inputs. Assume Q4 is conducting. What is the maximum current (in mA) that Q4 will have to sink?

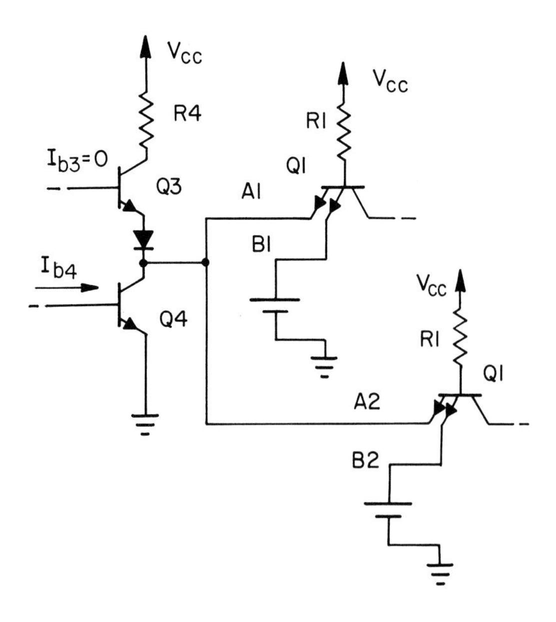

Fig. P20-5. For a TTL output (Q4) driving 2 TTL inputs with the output of Q4 at 0, the current that the output Q4 must sink can be determined.

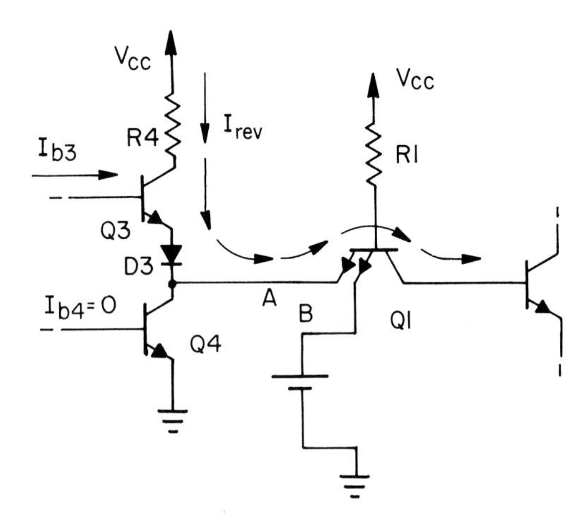

Fig. P20-6. The following two current levels involved with TTL inputs and outputs can be determined: 1. The reverse current into input A when the output driver is outputting a 1. 2. The short-circuit current to ground when the output driver outputs a 1.

14. What is the maximum current that Q4 in Problem 13 (Fig. P20-5) can sink and remain within its rating? Give your answer in mA.
15. The series of arrows in Fig. P20-6 indicates the flow of reverse current into input A when a 1 is applied to this lead. What is the maximum reverse current that can be expected to flow into a TTL input? Give your answer in microamperes.
16. If the output of the driver in Problem 15 (Fig. P20-6) were shorted to ground, which of the following best describes the short-circuit current that would flow through Q3?
    a. 1 mA.
    b. 50 mA.
    c. 500 mA (0.5 A).
    d. 1 A.
17. Draw two 3-input NOR symbols. Assume only inputs A and B will be used. Show two methods for dealing with the unused leads.
18. Draw four 3-input NAND symbols. Assume only inputs A and B will be used. Show four methods for dealing with the unused leads.
19. Indicate the values of the following parameters for a standard TTL NAND.
    a. Nominal (typical) power supply voltage.
    b. Absolute maximum supply voltage.
    c. Nominal output signal voltage for logic 1.
    d. Minimum output signal voltage for logic 1.
    e. Minimum input voltage accepted as logic 1.
    f. Absolute maximum positive signal that can be applied without damage.
    g. Nominal output signal voltage for logic 0.
    h. Maximum output signal voltage for logic 0.
    i. Maximum input voltage accepted as logic 0.
    j. Noise margin for standard TTL in volts.
    k. Average propagation delay in ns.
    l. Maximum clock frequency for standard TTL flip-flops in MHz.

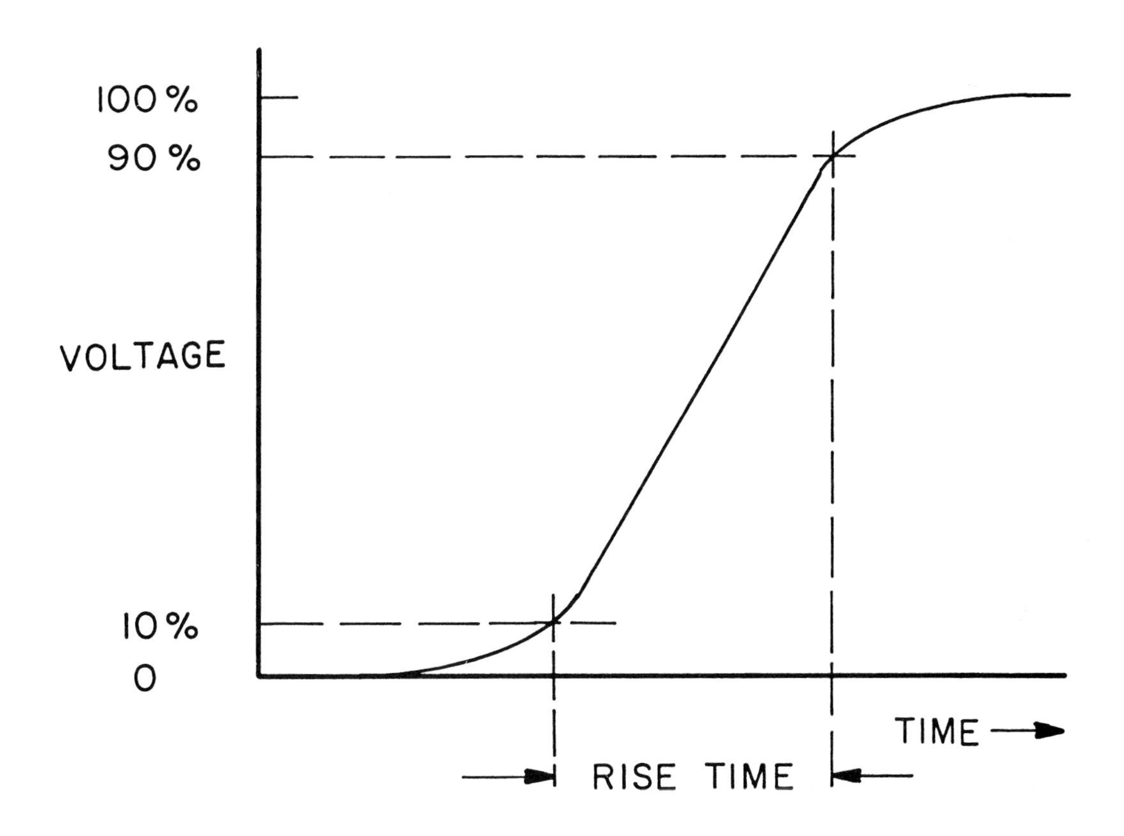

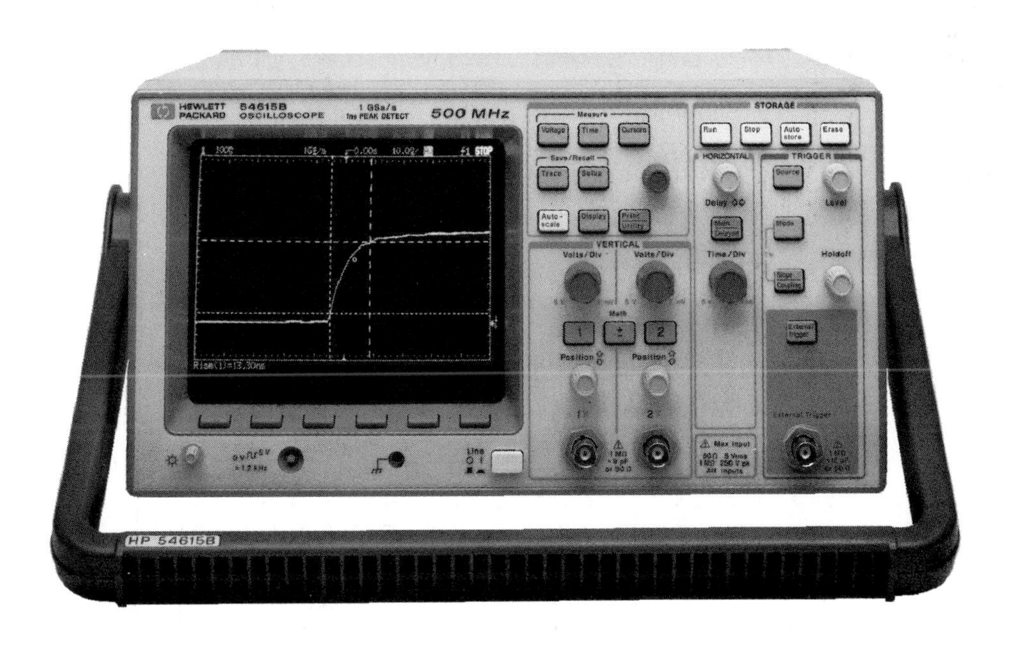

The time it takes for a voltage to go from one level to another is called rise time or fall time. Due to the roundness of the curve, it is difficult to determine when a voltage starts to change and when it reaches its destination. To solve this problem, rise and fall times are measured between the 10% and 90% points. See the upper figure.
The photograph shows an oscilloscope being used to measure rise time. (Hewlett-Packard Company)

Probes such as this are used to connect instruments to circuits under test. Note the use of the 74LS family. (Tektronix, Inc. 1996©)

Chapter 21

Logic Families Other Than Standard TTL Family

LEARNING OBJECTIVES

After studying this chapter and completing lab assignments and study problems, you will be able to:

- Describe the general relationship between the speed and power dissipation of a logic family.
- Compare the 74LXX series with standard TTL.
- Recognize the symbol of a Schottky transistor and describe how this transistor produces faster action.
- Compare the 74SXX series with standard TTL.
- Compare the 74LSXX series with standard TTL.
- Describe the action of CMOS NOT.
- State how many steps there are in a description for the action of a CMOS logic gate.
- List the precautions that should be taken when working with CMOS chips.
- Compare the CMOS family with standard TTL.
- Recognize the circuit of an ECL gate and list the parts in the circuit of an ECL gate.
- Compare the ECL family with standard TTL.
- Classify the following according to speed and power dissipation: 74XX, 74LXX, 74SXX, 74LSXX, CMOS, and ECL.

The characteristics of the standard TTL family can be described as moderate. These chips are fast enough for most applications; their power consumption is reasonable; and they are inexpensive. For these three reasons, this family is widely used.

When an application requires special characteristics, other families are used. Some of these are faster than TTL; others require less power. Several of the more common families will be introduced in this chapter.

In general, the faster a family, the more power it requires. In turn, low power families tend to be slow. This relationship between speed and power will be emphasized, and each family will be compared with standard TTL.

LOW POWER TTL

Identification on standard TTL chips takes the form 74XX. The XX represents numbers assigned to specific chip types. Chips identified as 74LXX are low power TTL devices. Such chips use about 1/10 the power of standard TTL devices.

LOW POWER CIRCUIT

For both standard and low power TTL, Vcc = +5 V. Because of the fixed supply voltage, reduced power can be obtained only by lowering the current. In the 74LXX series, this is done by using resistors with larger values, Fig. 21-1. 74LXX resistors are 5 to 10 times those in 74XX circuits (decision circuit has 40 K and thus, power goes down by 10 times). In Fig. 21-1, a 500 Ω resistor replaces 130 Ω. Current and power go down. E^2/R can go down by 3.8 times.

CHARACTERISTICS

The general rule relating power and speed applies to the low power TTL series. 74LXX chips use less power than do 74XX chips, but they are slower. Three topics are important. These are:

1. Power.
2. Propagation delay.
3. Fanout.

First topic is a power calculation as follows:

1. Power: Standard TTL NANDs draw about 2 mA per gate; low power TTL NANDs draw only 0.2 mA. See Fig. 21-2. The product of 5 V and 0.2 mA results in an average power per gate of only 1 mW.
2. Propagation delay: With larger resistors and reduced current, it takes longer to charge and discharge the capacitance within low power chips. Longer propagation delays result.

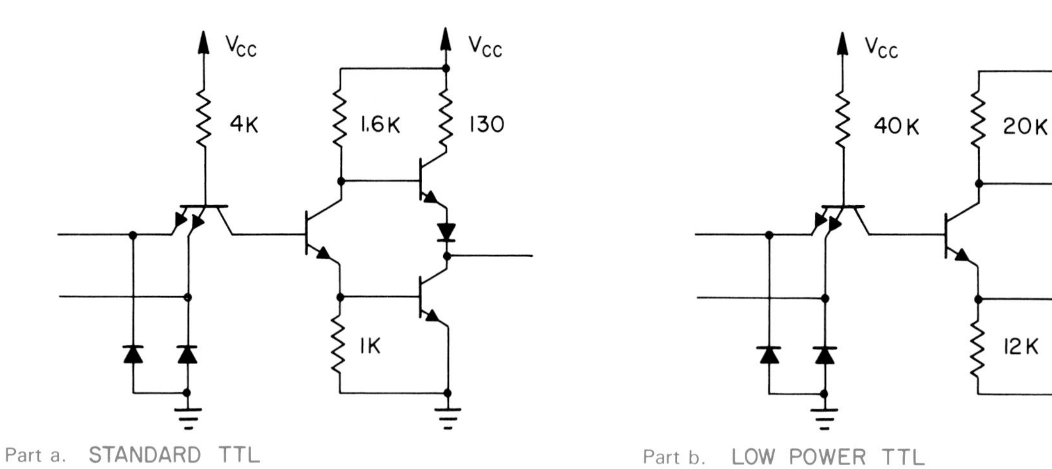

Fig. 21-1. Resistance values in 74LXX circuits are higher than those in 74XX circuits.

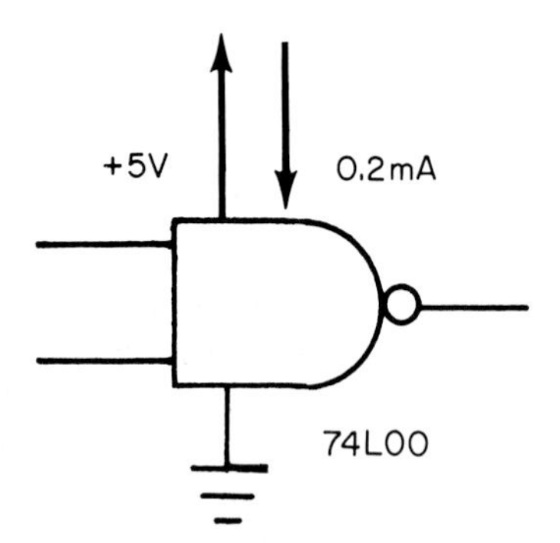

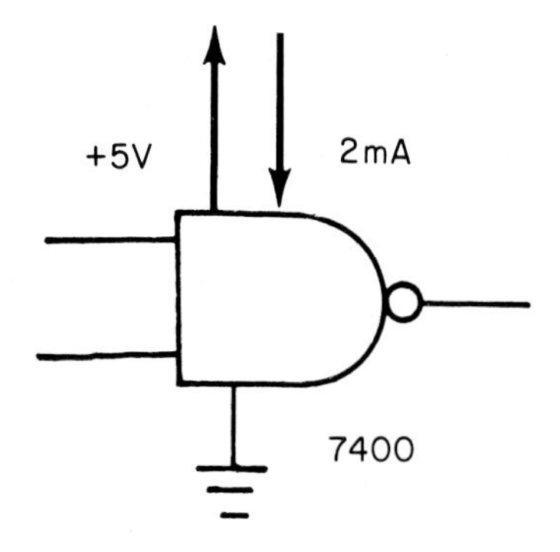

Fig. 21-2. The average supply current per gate (and therefore the power) for 74LXX elements is about 1/10 that for 74XX elements.

For a low power NAND, this delay averages 33 ns—3 times the delay of a standard TTL NAND. Refer to Fig. 21-3.

Due to their longer propagation delays, low power flip-flops are slower. Clock frequencies of only 3 MHz can be expected. This is less than 1/10 that of standard TTL flip-flops.

From the standpoint of speed, standard TTL chips can usually be substituted for low power units. The opposite is not always true.

3. Fanout: Fig. 21-4 shows input and output current ratings for standard and low power TTL elements. Although the current-sinking ability of low power outputs is less than that of standard TTL, the current coming from low power inputs is also lower. Because of the low values, the fanout of the low power TTL series is 20.

	STANDARD TTL 74XX	LOW POWER TTL 74LXX	
Supply Voltage	+5	+5	V
Propagation Delay	10	33	ns
Clock Frequency	35	3	MHz
Power Dissipation	10	1	mW
Fanout	10	20	

Fig. 21-3. Comparison of 74XX and 74LXX series logic elements.

INTERFACING WITH STANDARD TTL

Although signal levels differ slightly, low power and standard TTL chips can be intermixed in circuits. However, fanout and fanin must be considered. Due

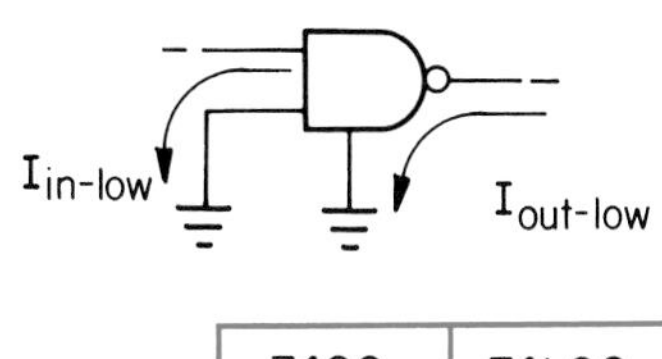

	7400	74L00
$I_{in\text{-}low}$	1.6mA	0.18mA
$I_{out\text{-}low}$	16mA	3.6mA
FANOUT	10	20

Fig. 21-4. Rated signal currents for the 74LXX series are much lower than those for the standard TTL series. The ratio of Iout-high to Iin-high for the 74L00 is also 20 to 1. That is, fanout is the same whether based on high or low signals.

to their high current-sinking ratings (16 mA) when the output is in the low state, standard TTL outputs can drive 88 low power inputs (at 0.18 mA per input). However, a 74LXX output (with a current-sinking rating of only 3.6 mA) can drive only two standard inputs (at 1.6 mA each).

In the laboratory, low power chips and standard TTL chips are often substituted for one another. In the field, care must be exercised when considering such substitutions. If a low power chip is used in place of a standard chip, its fanout may not be sufficient. If a standard chip is substituted for a low power chip, its fanin may overload its driver. It is better to replace faulty chips with chips from the same series.

USE OF 74LXX SERIES

As expected, the low power series finds use in equipment where power must be conserved. For example, 74LXX chips are found in battery-powered equipment.

Low power chips are usually more expensive than standard TTL devices. However, lower power permits smaller power supplies. The savings in power supply costs, size, and weight may offset much of this additional cost. Low power also means reduced heating. With less heat per chip, more chips can be mounted in a given space, and cooling requirements are reduced.

74LXX chips are used as an interface between low power devices and standard TTL logic, Fig. 21-5. The circuit at the left would have difficulty sinking the 1.6 mA from a standard TTL input. However, with the low power NOT as an interface, it needs to sink only 0.18 mA on the average.

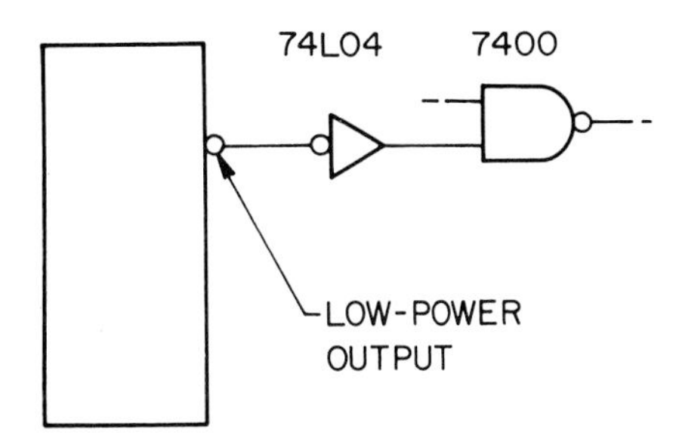

Fig. 21-5. Elements from the 74LXX series may be used to interface low power chips and standard chips.

SCHOTTKY TTL

Chips identified as 74SXX are from the Schottky series. These are high speed chips. As would be expected, they require more power than standard TTL chips. A Schottky device uses a junction of aluminum (n) and silicon (p). Look ahead to Fig. 21-8.

SATURATED/NONSATURATED OPERATION

In most logic families, transistors are operated SATURATED. That is, base currents higher than necessary to cause collectors to conduct are applied. High base currents result in reliable closing of the collector switch, part "a" of Fig. 21-6. However, such operation results in problems related to speed.

SATURATED OPERATION STORAGE TIME

When excess base current flows, charge carriers are temporarily stored in the transistor's base. When an attempt is made to turn the transistor off, the stored charge causes a problem. Although the signal at the base of the transistor has gone to 0 (the base current has stopped flowing), the stored charge will permit the collector current to flow, part "b" of Fig. 21-6.

The time required to dissipate the stored charge and permit the opening of the collector switch is called *STORAGE TIME.* It contributes to propagation delay. See part "c" of Fig. 21-6. To increase speed, the storage time must be reduced.

NONSATURATED OPERATION

If transistors are operated nonsaturated, the storage of charge carriers can be avoided and propagation delay can be reduced, part "e" of Fig. 21-6. With nonsaturated operation, just enough base current is applied to turn the transistor on, so saturation and the storing of charge carriers is avoided.

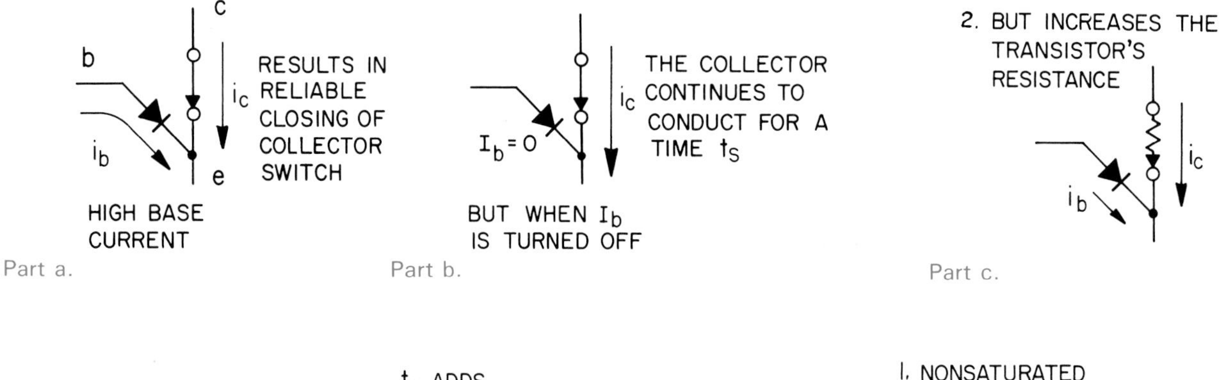

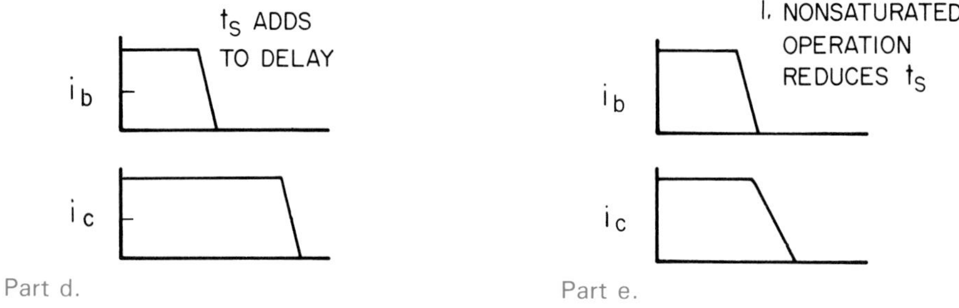

Fig. 21-6. Saturated transistors display storage time that increases propagation delay. This situation can be improved by using nonsaturated transistors. However, power dissipation increases.

However, nonsaturated operation has problems. Without excess base current, the collector-to-base switch does not close tightly. See part "d" of Fig. 21-6. The resistance of the collector-to-emitter portion of the transistor increases, and it dissipates more power than does a saturated transistor. To gain speed, power dissipation is increased.

Remember, propagation delay is measured at the half-voltage points. See Fig. 21-7. Also note that the delay when the output goes from 0 to 1 (tPLH) is likely to differ from the delay when the signal goes from 1 to 0 (tPHL).

SCHOTTKY TRANSISTOR

It is difficult to obtain reliable nonsaturated operation. One approach involves the use of Schottky transistors. The subject of Schottky transistors can be broken into two parts. These are:

- Schottky diode.
- Schottky diode and transistor.

The first part compares two types of diodes, as in the following:

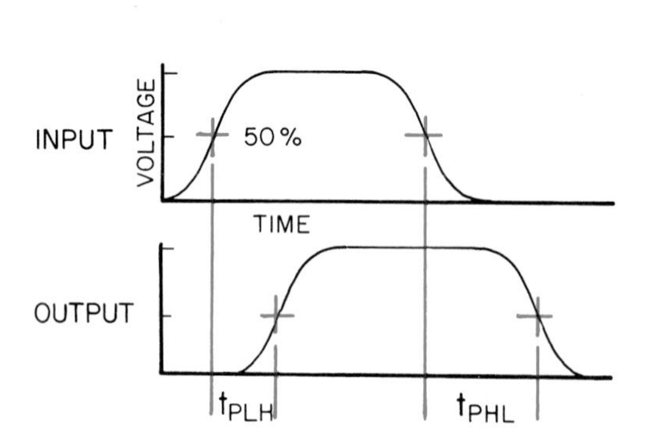

Fig. 21-7. Propagation delay is measured between half-voltage points.

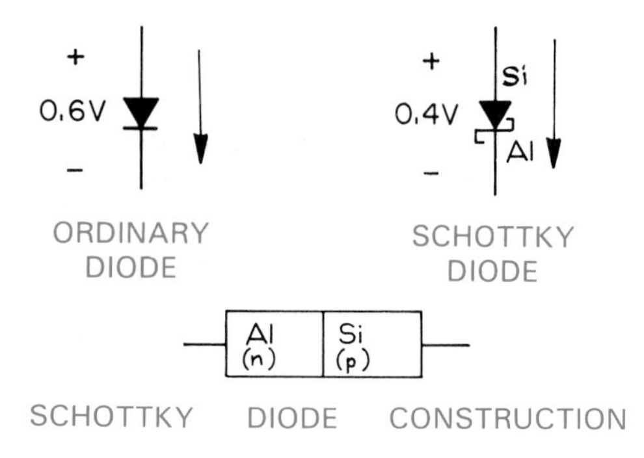

Fig. 21-8. The forward voltage of a Schottky diode is less than that of an ordinary diode. A Schottky diode consists of a junction between aluminum (n) and silicon (p).

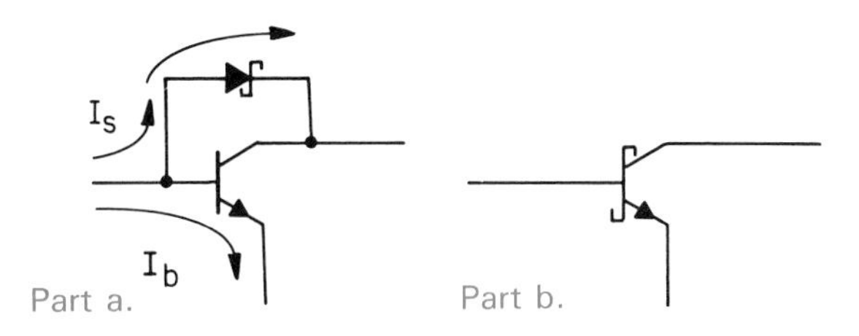

Fig. 21-9. A Schottky transistor consists of an ordinary transistor with a Schottky diode between its base and collector.

1. Schottky diode: Ordinary diodes conduct when a forward voltage of about 0.6 V is applied. The forward drop across a Schottky diode is only about 0.4 V. See Fig. 21-8. Some fast diodes are called hot carrier diodes. A hot carrier diode only drops a voltage of about 0.35 V.
2. Schottky diode and transistor: When a Schottky diode is placed between the base and collector of a transistor, excess base current is shunted as shown in part "a" of Fig. 21-9. Only the current necessary to turn on the transistor flows into its base. The rest of the current (the current that would cause saturation) flows through the diode. The result is a transistor that turns off faster than one without the diode.

The symbol for a diode and transistor is shown in part "b" of Fig. 21-9. The diode between the base and the collector is represented by the line which is bent twice on the transistor symbol.

SCHOTTKY CIRCUIT

Fig. 21-10 shows a Schottky NAND. There are three features that promote high speed. These are:

1. Schottky transistor use.
2. Darlington configuration.
3. Low resistance.

The first feature compares speed and power with TTL devices as follows:

1. Schottky transistor use: Because the Schottky transistors operate nonsaturated, they are faster than the bipolar transistors in standard TTL circuits. Remember, however, nonsaturated transistors take more power than do those that are saturated.
2. Darlington configuration: Note in Fig. 21-10 that the collectors of Q3 and Q4 are connected. Also note that the emitter of Q3 drives the base of Q4. This pair of transistors is said to be *DARLINGTON CONNECTED.* Before the current from the phase splitter reaches Q4, it is amplified by Q3. This drives Q4 harder and turns it on and off

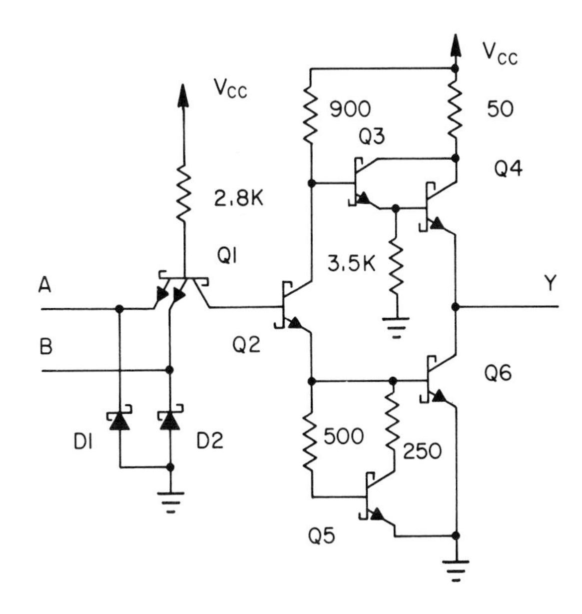

Fig. 21-10. The NAND shown uses Schottky transistors to obtain high speed operation.

faster. While this increases speed, the extra transistors use more power.

3. Low resistance: In general, resistance values in Schottky circuits are lower than those in standard TTL. For example, the resistor at the collector of Q4 is only 50 Ohms. In standard TTL circuits, it is 130 Ohms.

Small resistors imply higher currents; high current results in increased power dissipation. This is the price that must often be paid for increased speed. The use of low resistance, Schottky transistors, and the Darlington configuration results in a fast logic family.

CHARACTERISTICS

Fig. 21-11 compares the 74SXX series with standard TTL. Note that the Schottky series is faster but requires more power. There are three topics that are important. These are:

1. Propagation delay.
2. Power.
3. Fanout.

The first topic shows advantages of 74SXX devices compared to standard TTL devices, as in the following:

1. Propagation delay: Propagation delays of 3 ns are typical of the 74SXX series. This is 1/3 the delay of standard TTL. Flip-flops in the 74SXX series operate at 125 MHz. This is 3.57 times the

	STANDARD TTL 74XX	LOW POWER TTL 74LXX	SCHOTTKY TTL 74SXX	
Supply Voltage	+5	+5	+5	V
Propagation Delay	10	33	3	ns
Clock Frequency	35	3	125	MHz
Power Dissipation	10	1	19	mW
Fanout	10	20	10	

Fig. 21-11. Comparison of 74XX and 74SXX series logic elements.

operating frequency of standard TTL components. Some designers prefer to use 74SXX devices in place of standard TTL devices.

2. Power: Small resistors, nonsaturated transistors, and extra stages cause the 74SXX series to dissipate more power than standard TTL. The average current to a Schottky AND is 3.8 mA, and the power per gate is 19 mW. This is almost twice that of a standard TTL NAND.
3. Fanout: Fig. 21-12 shows signal currents for a 74S00. Based on the ratio of these currents, the 74SXX family has a fanout of 10.

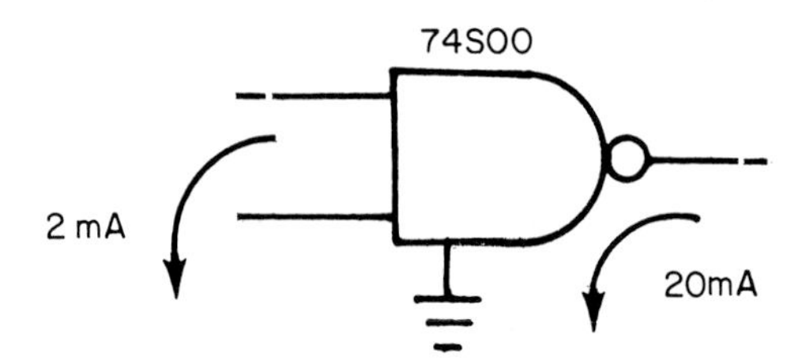

Fig. 21-12. Based on rated signal currents, the fanout of the 74SXX series is 10.

INTERFACING WITH STANDARD TTL

Signal levels in the 74SXX series differ slightly from those of standard TTL. However, the two are close enough that 74XX and 74SXX chips can be used in the same circuit.

Due to their high current ratings, 74SXX outputs can drive 12 standard TTL inputs. However, 74XX outputs can drive only eight 74SXX inputs.

74SXX chips can be substituted for those of the standard series. However, care is necessary when standard chips are substituted for the 74SXX series. The S-series is used only when its high speed characteristics are needed. Substituting a slower standard chip may degrade circuit performance.

USE OF 74SXX SERIES

A common use of the 74SXX series is in high speed circuits. Components in the 74SXX series should not be used unless their speed is required.

Signals in high speed circuits tend to have steep edges. The signals tend to ring, Fig. 21-13. Such signals can cause problems in digital circuits.

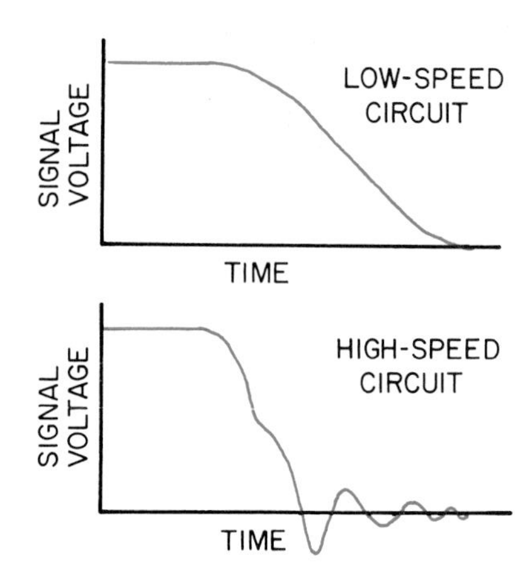

Fig. 21-13. Signals in high speed circuits tend to ring.

In high speed circuits, power supply leads and ground leads can act as unwanted signal paths. To prevent signals from traveling on power supply leads, capacitors are often placed as shown in Fig. 21-14. These capacitors bypass unwanted signals to ground.

CARE OF CAPACITORS

The capacitors used to bypass power supply leads often stand above other components on printed-circuit boards. See C7, C8, etc. in Fig. 21-15. Due to their height, capacitors are easily bent to one side. If

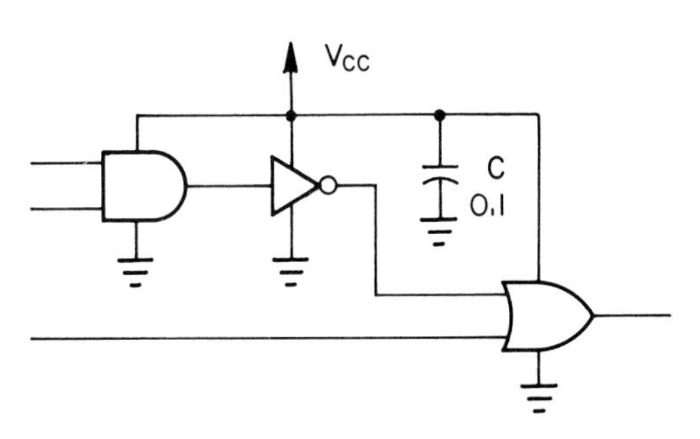

Fig. 21-14. Capacitors are often used to stop unwanted signals from traveling on power supply leads.

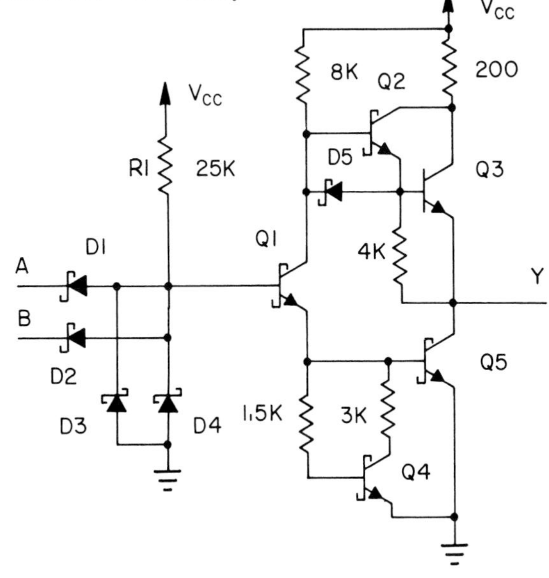

Fig. 21-16. Except for decision-making portions, 74LSXX circuits are similar to those for 74SXX circuits.

Fig. 21-15. Bypass capacitors (see C3 at the upper edge of this board) often stand above other components. When they are bent to one side, they should not be straightened. (Jameco Electronic Components)

straightened a number of times, their leads will break. Problems resulting from the loss of these capacitors are difficult to troubleshoot. A rule of good maintenance is, do not straighten components that have been bent to one side or the other. Leave them in whatever position you find them.

LOW POWER SCHOTTKY TTL TECHNOLOGY

In the 74LSXX series, Schottky transistors are used to increase the speed of a low power circuit. The result is a family with speeds near those of standard TTL but power ratings near those of the low power series.

LOW POWER SCHOTTKY CIRCUIT

The output portion of the circuit in Fig. 21-16 is similar to that of the 74SXX series of logic elements. However, the input portion differs. There are two topics that are important. These are:

1. Diode logic.
2. Diode-transistor logic.

The first topic calls for a simple circuit design, as in the following:

1. Diode logic: Diode logic is used in the decision-making section of the low power Schottky circuit instead of a multiple-emitter transistor. Diodes D1 and D2 and resistor R1 in Fig. 21-16 function as an AND. The action of a diode-logic AND is fairly easy to understand.

Part "a" of Fig. 21-17 shows a simple diode AND. With 1s (positive voltages) applied to the cathodes of both diodes, neither diode will conduct. With no current through R1, there is no voltage drop across this resistor. This means that the voltage at output Y will be near Vcc. This action is described in the first line of the truth table in Fig. 21-17.

In part "b" of Fig. 21-17, A = 0. That is, input A is grounded. Note the current flow. This current produces a voltage drop across R1, and the voltage at Y approaches 0. The truth table shows that if either or both inputs are grounded, output Y goes to 0.

2. Diode-transistor logic: When Q1 in Fig. 21-16 is added to the diode circuit, an efficient circuit is created. The name *DIODE-TRANSISTOR LOGIC* (DTL) can be used to describe the circuit. The 74LSXX series combines the characteristics of the DTL and TTL families. Diode-transistor logic saves power. R1 in the decision-making section is larger than in TTL, so it saves heat.

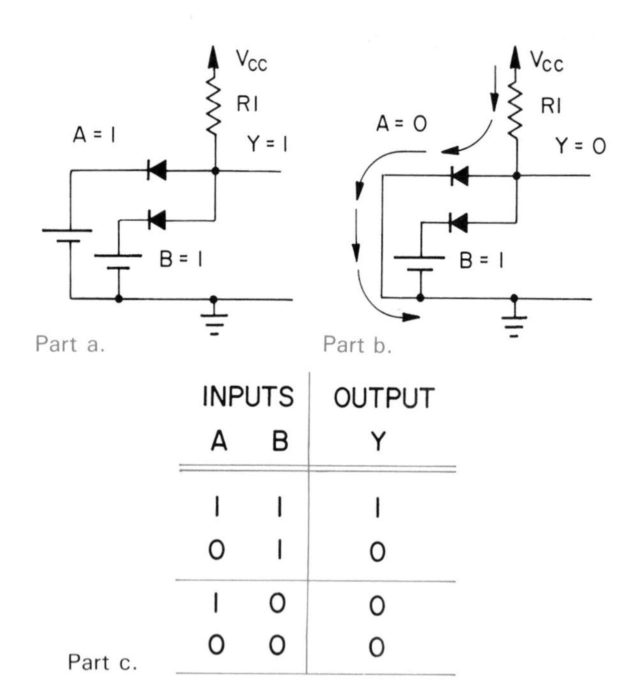

Fig. 21-17. Diodes can be connected to function as an AND. Therefore, AB = Y.

CHARACTERISTICS

Except for power, the characteristics of the 74LSXX series are similar to those of the standard TTL series. See Fig. 21-18. It has been suggested that this series may replace the 74XX series as the standard of the industry. There are three topics that are important. These are:

1. Power.
2. Propagation delay.
3. Fanout.

The first topic discusses a good balance of current and voltage behavior, as in the following:

1. Power: In the 74LSXX series, low power is obtained by using large resistors. High speed is obtained by using Schottky transistors.

The average power supply current is about 0.4 mA per 74LSXX gate. This represents a power of 2 mW. While this is twice the 1 mW dissipation of the low power series, it is much less than the 10 mW of the standard TTL series.

2. Propagation delay: The average propagation delay of a 74LS00 NAND is about 9 ns. This is similar to that of the 7400 device. Most of the flip-flops in the lower power Shottky series have clock input frequencies of about 45 MHz.
3. Fanout: Fig. 21-19 shows the signal currents at the input and output of a 74LS00 NAND. Based on these numbers, the series has a fanout of 20.

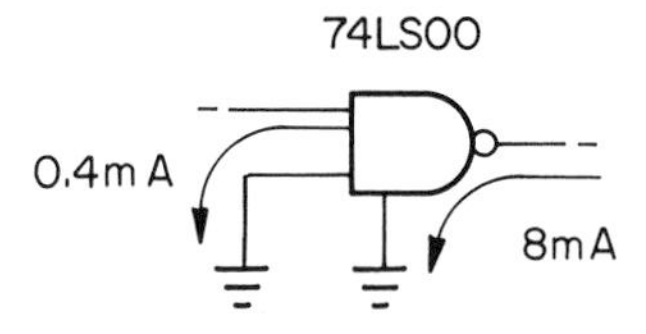

Fig. 21-19. Based on the rated signal currents, the fanout of the 74LSXX series of logic elements is 20.

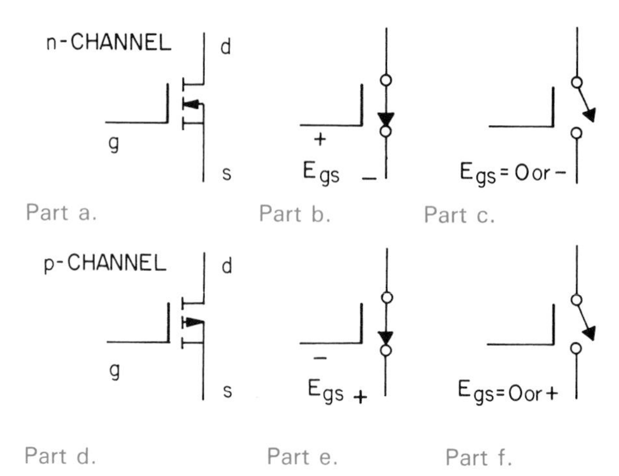

Fig. 21-20. A positive voltage between the gate and source of the n-channel MOS FET shown turns it on. A negative voltage must be applied to turn the displayed p-channel MOS FET on. The label "Egs" stands for E between gate and source.

	STANDARD TTL 74XX	LOW POWER TTL 74LXX	SCHOTTKY TTL 74SXX	LOW POWER SCHOTTKY 74LSXX	
Supply Voltage	+5	+5	+5	+5	V
Propagation Delay	10	33	3	9	ns
Clock Frequency	35	3	125	45	MHz
Power Dissipation	10	1	19	2	mW
Fanout	10	20	10	20	

Fig. 21-18. Comparison of 74XX and 74LSXX series logic elements.

INTERFACING WITH STANDARD TTL

The 74LSXX series of logic elements can be used with standard TTL elements. A standard TTL output can drive 40 low power Schottky inputs. However, a 74LSXX output can drive only 5 standard TTL logic element inputs.

USE OF LOW POWER SCHOTTKY LOGIC ELEMENTS

The low power Schottky series of logic elements is a low power replacement for standard TTL logic elements. As the price of the 74LSXX series has fallen, it has appeared in more and more circuits.

CMOS FAMILY

Metal-oxide semiconductor field-effect transistors are used in the CMOS family. The C indicates the use of complementary symmetry.

The CMOS family is a low power family. Under some conditions, a NAND gate may dissipate as little as 0.000,5 mW (1/2 μW). As would be expected, CMOS devices tend to be slow.

REVIEW OF MOS FET TECHNOLOGY

Part "a" of Fig. 21-20 displays the symbol for an n-channel MOS FET. The next two drawings show its action. When its gate is positive with respect to its source, the path from the drain to the source acts like a closed switch. When the gate-to-source voltage is negative or zero, this switch opens.

The symbol in part "d" of Fig. 21-20 represents a p-channel MOS FET. Its action is the complement of an n-channel device. With a negative voltage applied between its gate and source, it conducts. With positive or zero volts, it is turned off.

Remember, a MOS FET is a voltage operated device. Almost no current flows into the gate.

REVIEW OF COMPLEMENTARY SYMMETRY

Part "a" of Fig. 21-21 shows a circuit containing both an n-channel FET and a p-channel FET. This represents complementary symmetry. Refer to Chapter 16. There are two conditions to be analyzed. These are:

1. Outputting 0.
2. Outputting 1.

The first condition shows a well-matched circuit action, as in the following:

1. Outputting 0: In part "b" of Fig. 21-21, a 1 (positive voltage) has been applied at A. Q2 is n-channel, so it is turned on by the positive voltage at its gate. Output Y is grounded.

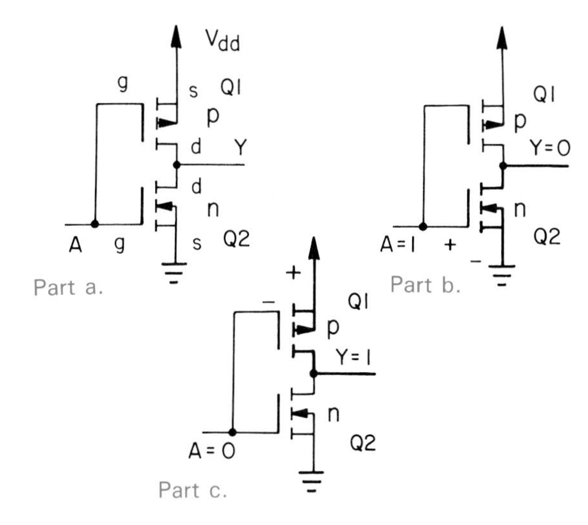

Fig. 21-21. When a 1 is applied to the CMOS NOT that is shown, Q2 turns on, and Y = 0. When a 0 is applied, Q1 is turned on, and Y = 1. Note the heavy lines drawn for the transistors that are on.

Q1 is p-channel, so the positive voltage at A turns it off. None of Vcc gets transferred to the output at Y. The output at Y is 0. With a 1 at its input, the circuit outputs a 0. It functions as a NOT.

2. Outputting 1: In the circuit at c in Fig. 21-21, A = 0 (grounded). With no voltage between its gate and source, Q2 is off. No grounding of the output at Y occurs.

The 0 at A grounds the gate of Q1. Note, however, that its source is connected to Vcc (a positive voltage). If this transistor's source is positive with respect to its gate, its gate must be negative with respect to its source. When its gate-to-source voltage is negative, the FET turns on. This connects output Y to Vcc, and a 1 is output.

CMOS CIRCUIT

Fig. 21-22 shows a CMOS NAND. Note that the n-channel transistors are connected in series. That is, output Y will be grounded only when both A and B are 1. See row 1 in the truth table.

The p-channel transistors are parallel connected. If either A or B is 0 or both are 0, one or both of these transistors will conduct and connect Y to Vcc. See the last three rows in the truth table.

SUPPLY AND SIGNAL VOLTAGES

CMOS chips operate with any supply voltage between +3 V and +18 V. Note that +5 V (the

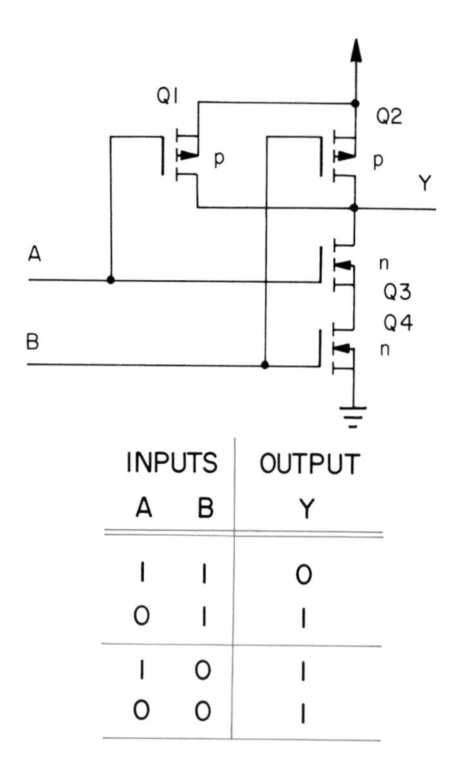

INPUTS A	INPUTS B	OUTPUT Y
1	1	0
0	1	1
1	0	1
0	0	1

Fig. 21-22. The CMOS NAND shown uses two n-channel transistors and two p-channel transistors.

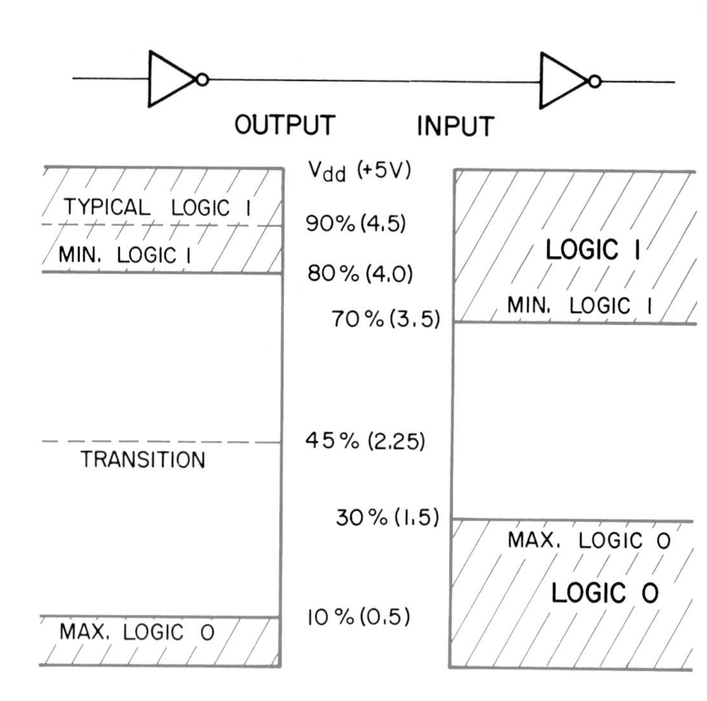

Fig. 21-23. CMOS logic levels are defined as percentages of Vdd.

supply voltage for TTL) is within this range. Also, 12.6 V, the voltage of automotive batteries, is included.

In CMOS circuits, the supply voltage is often indicated by Vdd. The letter d in the label dd stands for drain.

CMOS signal voltages are defined in terms of percentages of the supply voltage, Fig. 21-23. To aid in comparing the CMOS and TTL systems of components, signal voltages for Vdd = +5 V are shown.

CMOS CHARACTERISTICS

The CMOS family is a low power family. Due to the low power used in the CMOS family, the family tends to be slow. See Fig. 21-24. There are three topics that are important. These are:

1. Power.
2. Propagation delay.
3. Fanout.

The first topic compares static effects with dynamic effects, as in the following:

1. Power: CMOS elements dissipate less power than do those of the 74LXX series. Power depends upon how often signal levels change.

The term *STATIC* implies unchanging. CMOS circuits require little power when their signal levels are static. Fig. 21-25 shows a CMOS output driving a CMOS input. Note that there is no complete circuit between Vdd and ground. Except for very small leakage currents, the output does not supply any current. As long as the output remains a 1 or 0, the driving element may require as little as 0.000,5 mW (1/2 μW) on the average.

Change is suggested by the term *DYNAMIC*. When signals in CMOS circuits change from 1 to 0 or 0 to 1 (a dynamic situation), additional power is required.

Fig. 21-26 shows the inherent capacitance within a CMOS element. When the input signal goes from 0 to 1, this capacitance must be charged. That is, current must flow from Vdd through Q1. See time t1 on the current curve.

When the signal returns to 0, the capacitance must be discharged. See the pulse of current at t2. It flows to ground through Q2.

Although the current pulse at each edge is small, the average current can be high if enough signal changes are made per second. At 1 MHz, supply current is near 0.2 mA and represents a power dissipation of 1 mW. This is about the same as low-power TTL. As the rate with which signals change increases, so does dissipation. For example, at 3.5 MHz, dissipation will be about 1.8 mW.

2. Propagation delay: The power-speed relationship suggests that the CMOS family is slow, due to the low power dissipation. Propagation delays of 40 to 90 ns are typical. This is slower than the 74LXX series. However, recent developments suggest that faster CMOS chips will become available in the

Logic Families Other Than Standard TTL Family

	STANDARD TTL 74XX	LOW POWER TTL 74LXX	SCHOTTKY TTL 74SXX	LOW POWER SCHOTTKY 74LSXX	CMOS	
Supply Voltage	+5	+5	+5	+5	+3 to 18	V
Propagation Delay	10	33	3	9	65	ns
Clock Frequency	35	3	125	45	5	MHz
Power Dissipation	10	1	19	2	.0005 to 1	mW
Fanout	10	20	10	20	50 or more	

Fig. 21-24. Comparison of 74XX and CMOS families of logic elements.

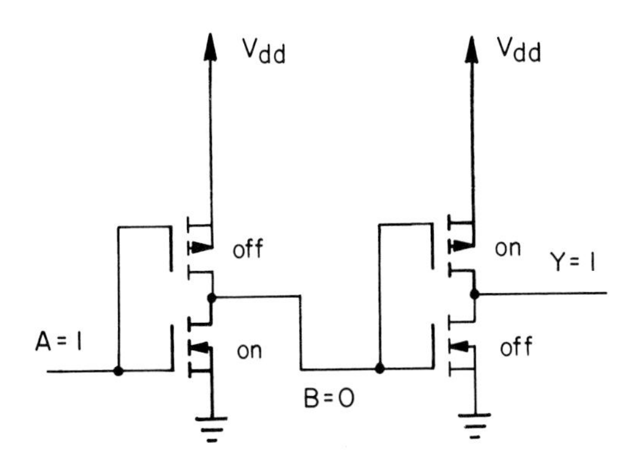

Fig. 21-25. When signals applied to CMOS circuits are static, little current flows, since there are no paths between Vdd and ground.

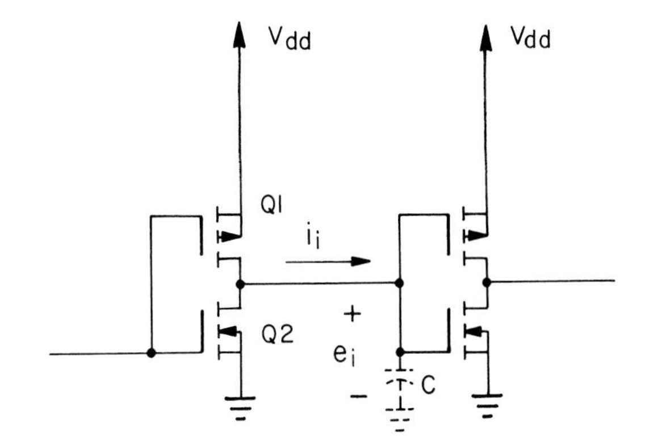

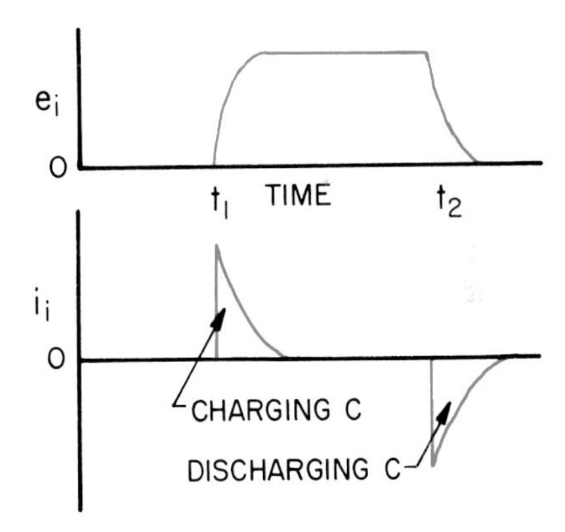

Fig. 21-26. Current flows in CMOS elements only during transfers from one logic level to another.

future. The average clock input frequency is 5 MHz. This is slightly faster than the 74LXX series.

3. Fanout: Because CMOS inputs require little current, the family has a high fanout. One CMOS output can drive 50 or more CMOS inputs. As inputs are paralleled, however, propagation delays increase.

INTERFACING CMOS ELEMENTS WITH STANDARD TTL ELEMENTS

TTL elements and CMOS elements interface easily. However, power and signal levels are not identical. There are two combinations to check:

1. TTL to CMOS.
2. CMOS to TTL.

The first combination involves extra circuit requirements, as in the following:

1. TTL to CMOS: At logic level 0, TTL and CMOS are compatible. There is, however, a problem at logic 1. CMOS inputs require 3.5 V or more for a valid logic 1. TTL outputs may be as low as 2.4 V. To overcome this problem, a pullup resistor can be used, part "a" of Fig. 21-27. The pullup resistor insures a voltage near +5 V for logic 1.

A TTL output can drive 75 CMOS inputs. Some effect on propagation delays is involved with the interface. Delays in the TTL unit go up as inputs are added. If all 75 inputs are used, a capacitive load on the TTL element can distort outputs by adding spikes, negative pulses, or sloping "ramps."

2. CMOS to TTL: Under ideal conditions, a CMOS output can drive one standard TTL input. Part "b" of Fig. 21-27 shows the use of a low power TTL NOT to protect the CMOS output. The 74L04 can drive two standard TTL inputs.

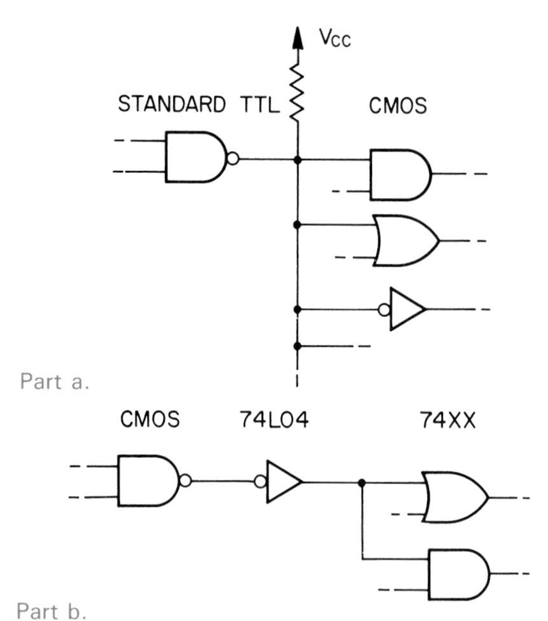

Fig. 21-27. a—One component is needed to interface a TTL output to a CMOS input. b—One component is needed for CMOS output to TTL input. A total of two types of interfacing components is needed for part "a" together with part "b."

USE OF CMOS COMPONENTS

Due to their low power dissipation, many CMOS elements can be placed on a single chip without producing excessive heating. Because of the low heat dissipation of CMOS elements, this family is widely used in large and very-large scale (VLS) integrated circuits. CMOS elements are widely used on memory chips because their static power drain is very low.

WORKING WITH CMOS COMPONENTS

Methods used to build and troubleshoot CMOS circuits differ little from those used with TTL. However, there are several precautions that should be noted.

The thin layer of glass between the gate and body of MOS FETs is easily punctured electrically. See part "a" of Fig. 21-28. Static electricity stored on the human body is more than enough to break down this thin barrier.

To protect input transistors, chip manufacturers often use protective diodes, part "b" of Fig. 21-28. Positive voltages larger than Vdd cause D1 to conduct. Negative voltages larger than about 0.6 V cause D2 to conduct. Because of the controlled conduction of the diodes, voltages applied to the gate are limited. In addition to the protection offered by these diodes, manufacturers recommend the following:

1. Care in handing from person to person.
2. Carė in grounding.
3. Care in storing.
4. Avoidance of static generating material.

The first of these suggestions from manufacturers helps keep circuit quality high, as in the following:

1. Care in handing from person to person: Differences in static charges on two people may be great enough to damage CMOS chips. Do not hand chips directly from one person to another. Rather, lay chips on a nonconducting table and have the second person pick them up. With a metal table, touch the hand to the table and then to the CMOS IC chip leads before putting the chip down. The person taking the CMOS IC chip should touch the hand to the metal table and then to the CMOS IC chip leads before picking up the chip.
2. Care in grounding: Workers, tools, and circuits should be connected (at least momentarily) to a common ground before CMOS chips are inserted into circuits.
3. Care in storing: Electrically conducting plastic foam is available. When CMOS IC chips are stored, CMOS leads should be inserted into such foam for protection.
4. Avoidance of static generating material: Nylon and some plastics (notably of the type used in drinking cups) generate static electricity. Such materials should be kept away from CMOS chips.

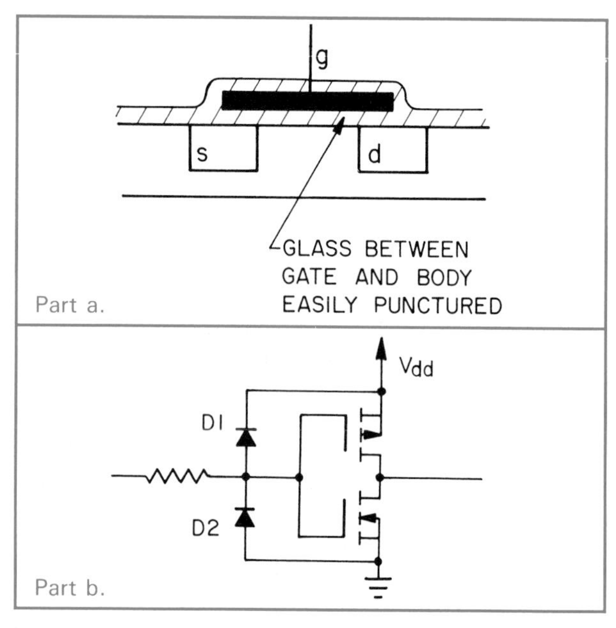

Fig. 21-28. Because CMOS inputs are easily damaged by static charges, protective diodes are often placed across input leads.

UNUSED LEADS ON CMOS ELEMENTS

The action of floating CMOS leads cannot be predicted. They may represent 1s or 0s. Because of the lack of signal level prediction, unused CMOS leads should be connected to Vdd or ground.

ECL FAMILY

ECL stands for *EMITTER-COUPLED LOGIC.* The ECL series of logic devices, with or without an integrated circuit structure, is a relatively new family. It is very fast and is used in high speed digital communication and data processing equipment. Some forms of instrumentation hardware use ECL circuit elements. Analog circuits, digital circuits, analog-to-digital circuits, and digital-to-analog circuits are available.

ECL CIRCUIT

A detailed study of the ECL circuit is not necessary at this time. Rather, only a few comments about its action are needed to provide an understanding of its characteristics.

The emitter coupling referred to in the name of the ECL family can be explained in terms of resistor R2 in Fig. 21-29. It is in the emitter circuits of both Q1 and Q2. It couples, or transfers, the signal from transistor Q1 to transistor Q2.

ECL CIRCUIT WHEN INPUTTING 0

When A = 0, the base of Q1 in Fig. 21-29 is grounded and this transistor turns off. With little current flowing through R3, the voltage at $\bar{Y}$ approaches Vcc. That is, a 1 appears at this output. For the signal path from input A to output $\bar{Y}$, this circuit acts like a NOT.

D1 is a special diode (called a Zener diode). It holds the voltage at the base of Q2 constant.

The base-to-emitter voltage of Q2 (Vbe2) equals the voltage across D1 (Vb2) minus the voltage across R2 (Ve). That is:

$$Vbe2 = Vb2 - Ve$$

With A = 0, the current through Q1, and thus through R2, decreases. This causes Ve (the voltage across R2) to decrease. Because Ve is subtracted from Vb2, the base-to-emitter voltage of Q2 (Vbe2) increases. Note the conditions needed for the following:

1. A decrease in current.
2. An increase in voltage.

When its base voltage increases, Q2 turns on harder, and the current through R4 increases. With a larger voltage drop across R4, the voltage at output Y goes down. This is the noninverting output, since when A goes to 0, Y decreases. Because the output of this circuit cannot even approach 0 V, this decrease is treated as a 0.

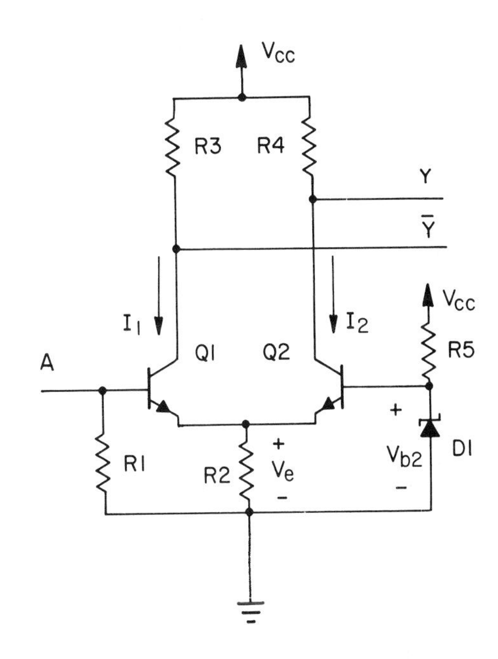

Fig. 21-29. Emitter-coupled circuits are used in ECL logic. Resistor R2 couples the signal from the emitter of Q1 to the emitter of Q2.

ECL CIRCUIT WHEN INPUTTING 1

When A = 1, the opposite action occurs. The positive voltage at the base of Q1 in Fig. 21-29 increases the current through Q1 and R2. The increased current through R3 results in a larger voltage drop, so the voltage at $\bar{Y}$ drops. When A = 1, $\bar{Y}$ = 0.

When I1 increases, the voltage across R2 (Ve) increases. This decreases the base-to-emitter voltage (Vbe2) applied to Q2. (Remember that Vbe2 = Vb2 − Ve. When Ve increases, Vbe2 must decrease.) This in turn tends to turn Q2 off. With less current through R4, the voltage drop decreases, and voltage at Y increases. That is, when A = 1, Y also equals 1. Note the conditions for the increase and decrease of voltages.

The circuit in Fig. 21-29 is used because its transistors operate in the mode which is called *NON-SATURATED.* Due to the nonsaturated operation, charge carriers are not stored in the bases of the transistors, and the circuit has a short propagation delay.

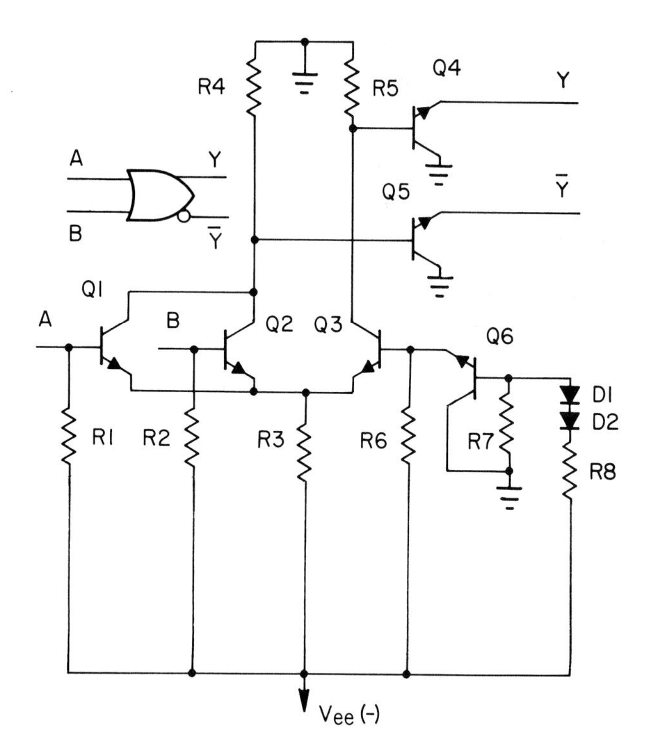

Fig. 21-30. The ECL circuit shown functions as both an OR and a NOR.

DIAGRAM OF COMPLETE ECL CIRCUIT

Fig. 21-30 shows an ECL OR/NOR gate. Transistors Q4 and Q5 are output drivers. Because only one transistor is used at each output, these are open-collector outputs.

ECL SUPPLY AND SIGNAL VOLTAGES

The collectors of all transistors in the ECL OR/NOR circuit in Fig. 21-30 go to ground or toward the ground lead (through resistors connected to ground). For noise rejection reasons, this circuit operates with a NEGATIVE POWER SUPPLY. Voltages between −3 V and −8 V can be used, but −5.2 V is usually specified.

The negative power supply results in negative signal voltages. For Vee = −5.2 V, signal voltages more negative then −1.75 V are treated as logic 0s. Those more positive than −0.9 V are logic 1s.

ECL CHARACTERISTICS

The ECL series of logic elements is very fast. As indicated in Fig. 21-31, the ECL family is faster than any of the families studied here. It also dissipates more power. There are two topics that are important. These important topics are:

1. Propagation delay.
2. Power.

The first topic discusses high speed ringing effects, as in the following two paragraphs:

1. Propagation delay: Propagation delays of only 1 ns are possible with the ECL family. ECL flip-flops can operate at clock frequencies as high as 400 MHz. Race conditions become important at frequencies above 400 MHz. 100 Ω resistors in series with inputs help reduce sensitivity to outside signal interference.

As noted earlier, high speed often causes ringing and other problems. Leads between ECL elements must be kept short, and terminating resistors are often necessary at the end of long leads. Leads must be kept shorter than 3.6 in. (9.1 cm). Care must be taken to shield an ECL circuit so that it does not radiate electromagnetic waves to the outside.

2. Power: To gain speed, transistors in ECL circuits are operated nonsaturated. Also, ECL circuits contain many resistors. Both result in high power dissipation.

INTERFACING ECL CIRCUITS WITH STANDARD TTL CIRCUITS

ECL and TTL signals are not compatible. They even have different polarities. The term FANOUT has little meaning with the ECL family in any kind of connection to TTL circuit systems. To interface the ECL family with the TTL family, special chips are used.

	STANDARD TTL 74XX	LOW POWER TTL 74LXX	SCHOTTKY TTL 74SXX	LOW POWER SCHOTTKY 74LSXX	CMOS	ECL	
Supply Voltage	+5	+5	+5	+5	+3 to 18	−5.2	V
Propagation Delay	10	33	3	9	65	1	ns
Clock Frequency	35	3	125	45	5	400	MHz
Power Dissipation	10	1	19	2	.0005 to 1	60	mW
Fanout	10	20	10	20	50 or more		

Fig. 21-31. Comparison of 74XX and ECL families of logic components and instrument hardware.

USE OF ECL

ECL is used only in very high frequency circuits. For example, test equipment must operate at speeds higher than those of the circuits under test. Because of the high speeds required, ECL circuits are used in such instruments.

An ECL circuit is difficult to troubleshoot. Attaching test equipment to high speed circuits often disturbs their operation. Due to the circuit sensitivity of ECL systems and ECL instruments, experience is needed to design and repair high speed equipment.

OTHER LOGIC ELEMENT FAMILIES

Families other than those described here are available. Some have been in use for many years and are presently found only in older equipment. Diode logic, DTL (Diode-Transistor Logic), and RTL (Resistor-Transistor Logic) are examples of older families.

Other families are new and are found only in special applications. IIL (Integrated-Injection Logic) is an example. This family has a high value for a property that is called *PACKING DENSITY*. Many IIL gates can be placed in a small space. (About ten IIL gates can be placed in the space occupied by one TTL gate.) As larger and larger numbers of elements are placed on chips, the IIL family may become common.

While each logic family has its special characteristics, basic logic concepts remain relatively unchanged. Workers in the digital electronics field who are versed in present-day logic families should have little trouble working with new families.

SUMMARY

In the past, standard TTL was the standard of industry. When new equipment was designed, there had to be a good reason for using any family other than TTL. Today, few new designs incorporate standard TTL. Among others, versions of CMOS and low power Schottky are being used, and new families are presently being developed.

If standard TTL is outdated, should it be used in student laboratories? Should it be studied? A primary function of a first course in digital circuits is to present the basic concepts of digital logic. The rugged, reliable operation of standard TTL is well suited to this objective. Concepts learned using TTL apply to all families, since most logic families are compatible with standard TTL. Personnel experienced in TTL circuits will have little difficulty working with other families.

With each new work assignment, there is a period of adjustment during which a worker gets used to the equipment and logic family required by the new task. How are unused leads handled? Will connecting test equipment to a circuit alter the circuit's action? An understanding of TTL is a good general foundation on which to build.

The specific logic family used in your student laboratory is not as important as what you learn about basic logic circuits. As a worker in the digital industry, you will use the things you learned (either directly or as a foundation for advanced learning) almost every day.

IMPORTANT TERMS

Darlington connection, DTL, ECL, Nonsaturated operation, RTL, Saturated operation, Schottky diode, Schottky transistor, Storage time.

TEST YOUR KNOWLEDGE

1. The faster a logic family is, the more ________ it requires.
2. TTL chips identified as 74LXX use about ________ (1/2, 1/10, 1/20, 1/6) the power of standard TTL devices.
3. The time it takes to charge and discharge the stray capacitance in a logic element is one cause of ________ delay.
4. From the standpoint of speed, can low power TTL chips be substituted for standard TTL chips?
5. Standard TTL chips can drive ________ (16, 88, 38) low power inputs.
6. 74LXX chips are used as a(an) ________ (gate, block, interface, net) between low power devices and standard TTL logic.
7. A Schottky logic element uses a junction of ________ (beryllium, tin, aluminum, gold) and silicon.
8. Which of the following terms means that high base currents in a well-conducting transistor are used?: Saturated, nonsaturated.
9. A transistor operating in the nonsaturated mode dissipates ________ (more, less) power than a saturated transistor.
10. When a Schottky diode is placed between the base and collector of a transistor, excess base current is ________ (dropped, shunted, opposed, stored).
11. Does a Darlington connection make a Schottky element faster?

12. Extra transistor stages in Schottky devices cause the device to use more _________.
13. Signals in high speed circuits tend to _________ (decay, draw too much current, ring, improve the fanout).
14. Capacitors on Vcc leads in high speed circuits can _________ (reduce power consumption, bypass unwanted signals to ground, "regenerate" a weak signal).
15. A helpful rule for handling circuits is "_________ (Do, Do not) straighten components on PC boards that have been bent to one side."
16. In the 74LSXX series, low power is obtained by using _________ (capacitors, resistors) with large values.
17. The low power Schottky series of logic elements is a low power _________ (replacement, branch) for standard TTL logic elements.
18. A standard TTL output can drive _________ (4, 40) low power Schottky inputs.
19. The "C" in "CMOS" stands for _________ (conductive, current-operated, complementary symmetry, collector-operated).
20. The CMOS family is a _________ (high, low) power family.
21. Because a CMOS element uses little power, it is _________ (fast, slow).
22. A MOSFET is a _________ (current, voltage)-operated device.
23. CMOS chips operate on any supply voltage between _________ (+6 V and +30 V, +1V and +18 V, +3 V and +30 V, +3 V and +18 V).
24. CMOS signal voltages are defined in terms of _________ (percentages, multiples, analogs) of the supply voltage.
25. Except for very small leakage currents, a CMOS output does not supply any current to a CMOS input. (True or False?)
26. A surge of current flows in CMOS elements when they _________ (change from one output to another, output a 1, output a 0).
27. How many standard TTL inputs can a CMOS output drive?
28. The thin layer of glass between the gate and body of MOS FETs is easily _________ (formed, punctured, charged) electrically.
29. To protect CMOS chip input transistors, chip manufacturers use protective _________.
30. Are differences in static charges on two people great enough to damage CMOS chips?
31. Why is grounding important for CMOS chips that are being put into a circuit?
    a. Lowers the cost.
    b. Drains static charges.
    c. Allows time for correct circuit orientation to be achieved.
    d. Adjusts the state of internal memory elements.
32. Plastic foam to contact CMOS chip leads is made conductive to _________ (heat, electricity, sound, light, gases).
33. The action of floating CMOS leads can be predicted. (True or False?)
34. ECL stands for _________________.
35. The ECL family of logic elements is _________ (fast, slow).
36. Instrumentation hardware uses ECL circuits because of their _________ (simplicity, speed, negative logic).
37. ECL transistors are operated _________ (saturated, nonsaturated).
38. ECL circuits use a _________ (negative, positive) power supply voltage.
39. ECL flip-flops can operate at frequencies as high as _________ (400 MHz, 10 MHz).
40. One must shield an ECL circuit so that it does not radiate _________ waves to the outside.
41. Does the term "fanout" have meaning for an ECL circuit?
42. Is an ECL circuit easy to troubleshoot?
43. Attaching test equipment to high speed circuits often _________ (improves, disturbs) their operation.
44. About _________ (2, 10) IIL gates can be placed in the space occupied by one TTL gate.

STUDY PROBLEMS

1. Based on the relationship between speed and power, which of the following is likely to dissipate the most power?
   a. Family A with a propagation delay of 10 ns.
   b. Family B with a propagation delay of 3 ns.
2. What is the relationship between a family's propagation delay and the maximum clock frequencies of its flip-flops?
   a. The longer the delay, the higher the frequency.
   b. The shorter the delay, the higher the frequency.
3. Fig. P21-1 shows two TTL output circuits. Which is most likely the output for the low power TTL series of logic elements?
4. Copy this table onto a sheet of paper and use the following letters to indicate the relative delay and power of the 74LXX series.

   L = considerably lower than standard TTL
   H = considerably higher than standard TTL
   E = about equal to standard TTL

	Propagation delay	Power per gate
74XX	10 ns	10 mW
74LXX	a._____	b._____

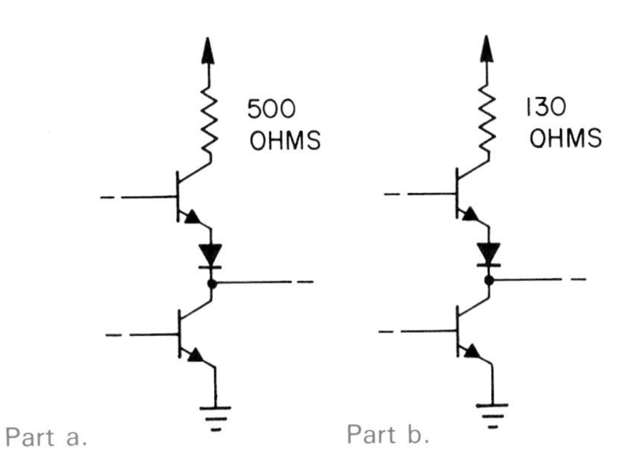

Fig. P21-1. Out of two similar circuits, a low power circuit can be identified.

5. Based on the signal currents shown in Fig. P21-2, how many standard TTL inputs can a NAND from a 74LXX series drive?

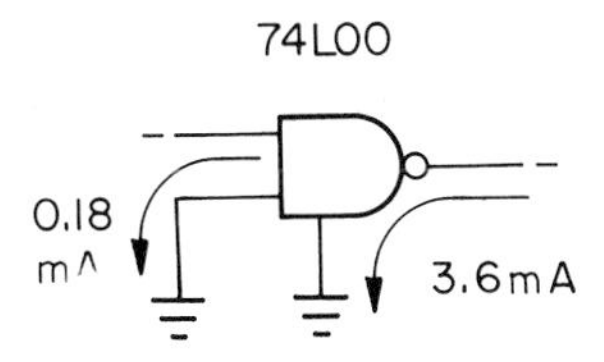

Fig. P21-2. Signal current levels help determine interfacing abilities of low power TTL elements with standard TTL elements.

6. Which symbol in Fig. P21-3 would most likely be referred to as a Schottky transistor?
7. What is the advantage of a Schottky transistor over the transistors used in standard TTL chips?
   a. Dissipates less power.
   b. Operates nonsaturated.
   c. Draws no base or gate current.
8. Which type of transistor operation is likely to produce the shorter storage time?
   a. Saturated.
   b. Nonsaturated.
9. Repeat Problem 4 for the 74SXX series.
10. For the input signals shown in Fig. P21-4, what number (1 or 0) will appear at the output of the diode logic circuit?
11. Repeat Problem 4 for the 74LSXX series.
12. In the CMOS circuit shown in Fig. P21-5, which transistor (Q1 or Q2) is conducting?
13. Based on the input and output signals shown in

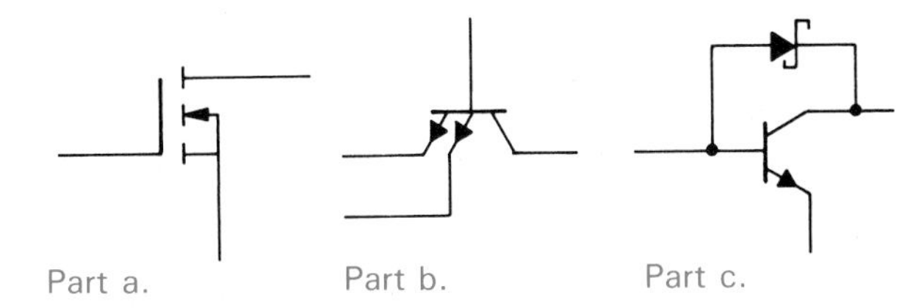

Fig. P21-3. A Schottky transistor can be identified.

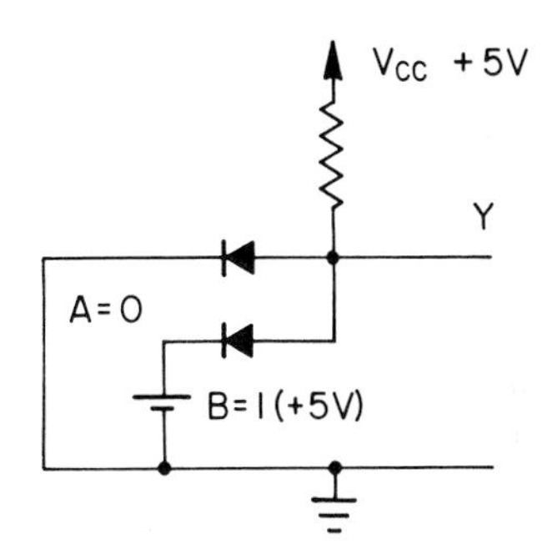

Fig. P21-4. Diode logic circuit can be analyzed.

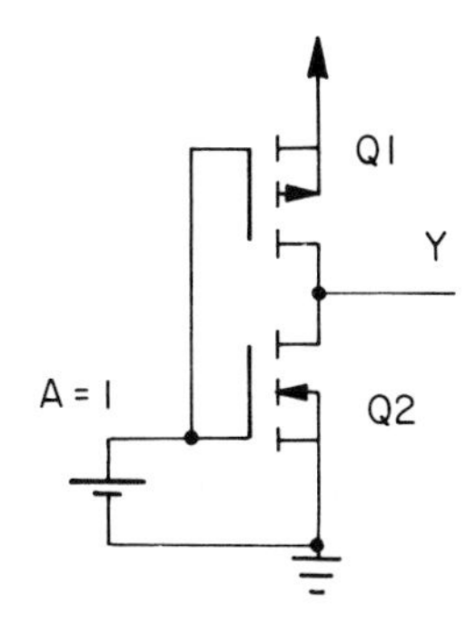

Fig. P21-5. The on and off states of transistors in a CMOS circuit can be determined.

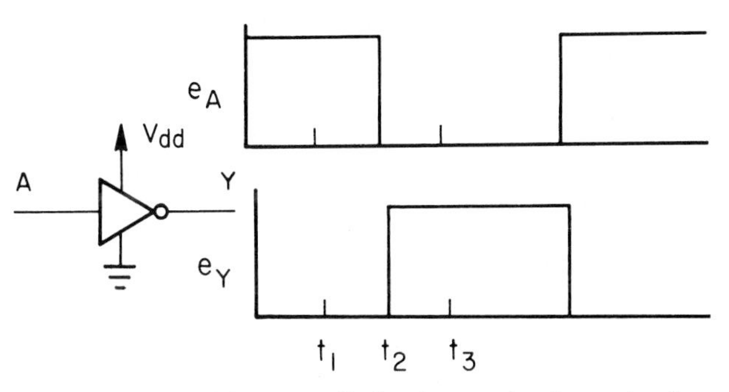

Fig. P21-6. CMOS power dissipation can be determined.

the graph in Fig. P21-6, at which time (t1, t2, or t3) will the CMOS logic element dissipate the most electrical power?

14. Repeat Problem 4 for the CMOS family of logic circuit elements.
15. Describe the four precautions listed in the text that relate to the handling of CMOS chips.
16. Which circuit in Fig. P21-7 is from the ECL family of logic elements?
17. Repeat Problem 4 for the ECL family.
18. The chip in Fig. P21-8 is from what family? What is the approximate clock speed?

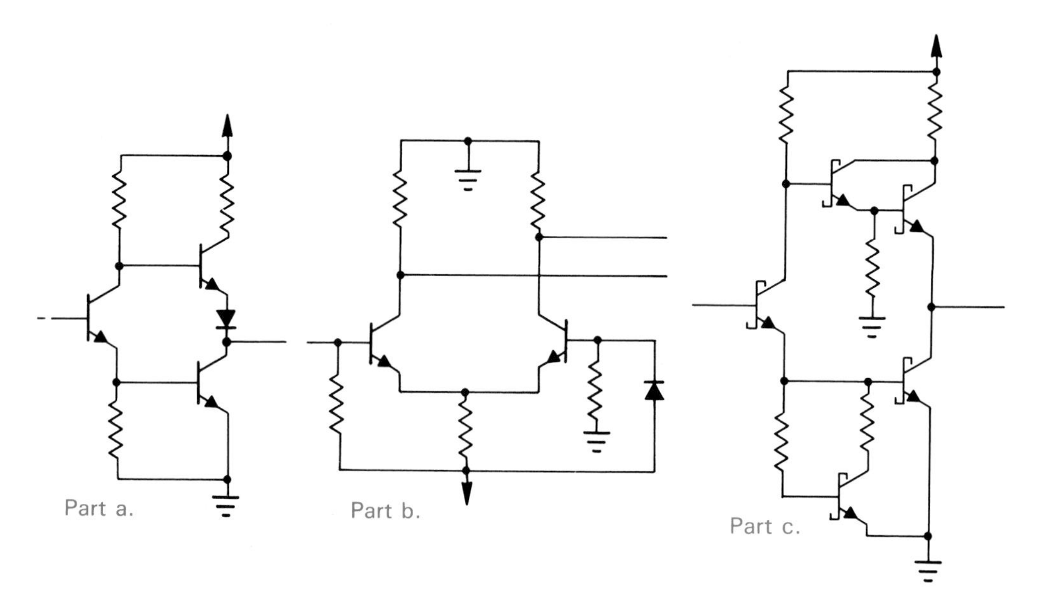

Fig. P21-7. Out of three circuit samples, an ECL circuit can be identified.

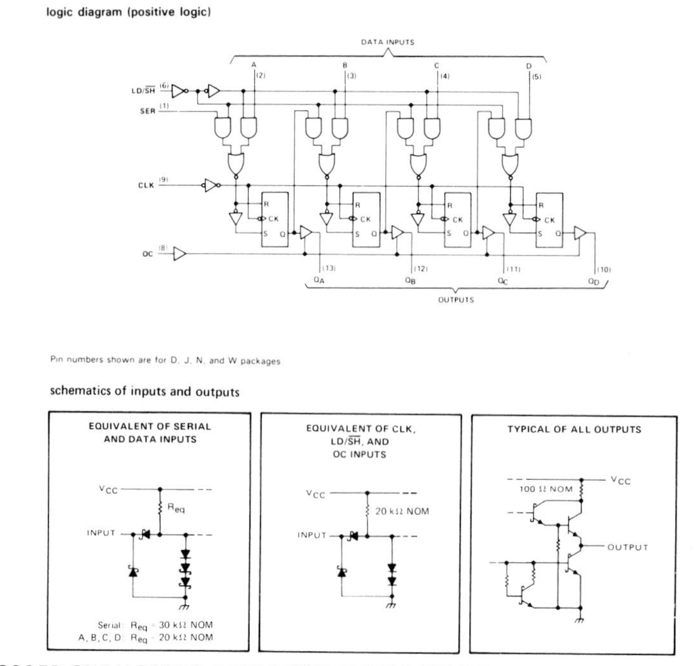

TYPES SN54LS295B SN74LS295B 4-BIT RIGHT-SHIFT LEFT-SHIFT REGISTERS WITH 3-STATE OUTPUTS

Fig. P21-8. Circuit specifications can be identified. (Texas Instruments, Inc.)

Relay Logic and Programmable Controllers

Chapter 22

LEARNING OBJECTIVES

After studying this chapter and completing lab assignments and study problems, you will be able to:

- Recognize and match the names with symbols for pushbutton switches, limit switches, disconnect switches, control relays, time-delay relays, overload relays, fuses, control transformers, solenoids, and three-phase induction motors.
- Analyze and list steps for the action of simple relay-based control systems based on their ladder diagrams.
- Explain why those working with programmable controllers should be familiar with relay logic and ladder diagrams.
- Draw a block diagram of a basic programmable controller.
- List at least four advantages of programmable controllers over hard-wired relay logic.

Digital logic has long been used to control machines and processes. Originally, relays were the basic logic element. In the 1960s, attempts were made to substitute transistorized logic elements, but relays continued to dominate the field. In the early 1970s, programmable controllers were developed. Today, few relay-based control systems are being built. Rather, programmable controllers are being used.

Programmable controllers are computers programmed to simulate relay logic. Control circuits are still designed as if relays were to be used. Standard relay-logic diagrams, called ladder diagrams or ladder logic, are drawn. These are then used in the programming of programmable controllers.

The use of ladder diagrams suggests that workers involved with digital circuits in manufacturing and process control must be competent in both relay logic and integrated circuit logic. Relay logic will be introduced in this chapter.

SEQUENCE CONTROL

In self-timing sequence control, the end of one portion of a machine cycle signals the beginning of the next. The machine does one operation after another. The beginning of each operation is indicated by the completion of the previous operation.

The machine in Fig. 22-1 will be used to describe a self-timing sequence control. It resistance welds

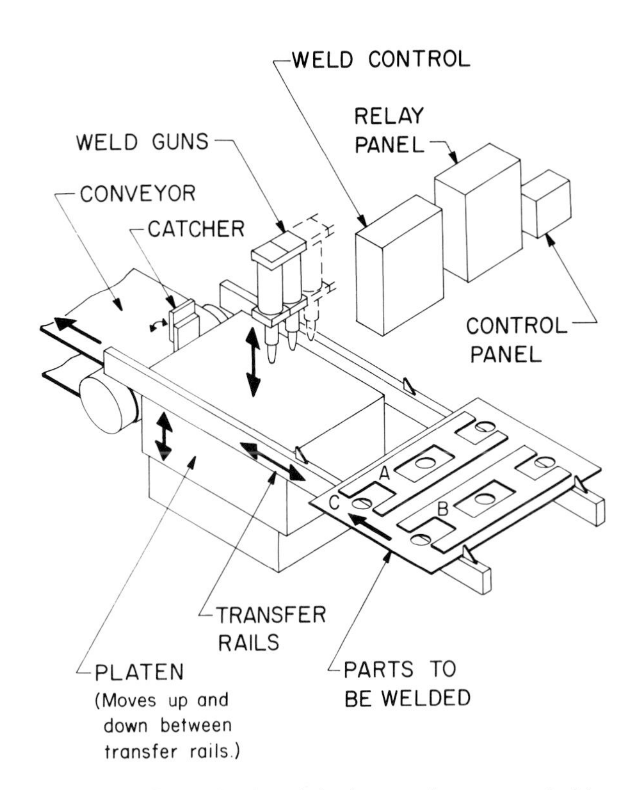

Fig. 22-1. Relay logic might be used to control this machine.

(often referred to as spot welding) two parts (A and B) to a larger part (C). Its machine cycle is:

1. Two operators load the three parts onto the transfer (the two horizontal bars extending out of the machine). They then move away from the machine and press two foot switches 1FS and 2FS.
2. The transfer moves forward, positioning the parts under the weld guns. When all the way forward, the transfer activates a limit switch. Fig. 22-2 shows such a switch. This signals the completion of this portion of the machine cycle.

Note the self-timing nature of this arrangement. If the transfer were to move slower than usual, this would not disturb the rest of the cycle. The next operation cannot begin until the transfer is fully forward.

3. The platen (the flat top of the table) then moves upward, lifting the parts from the transfer. When it reaches its uppermost position, another limit switch is activated. This limit switch signals the completion of step number 3.
4. Now, two things happen. The transfer is returned so the operators can load it in preparation for the next machine cycle. Also, the weld process begins.

The making of the welds is under the control of a separate controller. When the welds are completed, control is returned to the relay circuit.

5. The platen is lowered. This places the completed parts on the forward portion of the transfer, ready for delivery to the conveyor. Again, a limit switch senses the completion of the operation. It is activated by the platen in its down position. This completes the machine cycle.

When the next cycle begins, the completed parts will be on the forward portion of the transfer; the parts placed on the transfer by the operators will be on the rear portion. When the foot switches are pressed, the completed parts will be delivered to the conveyor; the new parts will be placed under the weld guns.

The system's self timing is important. The completion of one step signals the beginning of the next. Few external timers are needed in such control systems.

Fig. 22-2. This limit switch is activated by the motion of a part of a machine. Based on the symbols, it provides one set of normally-open and one set of normally-closed contacts. (Allen-Bradley Company, Inc.)

RELAY-BASED CONTROL

Fig. 22-3A and Fig. 22-3B show the control circuit for the welding machine previously discussed. To simplify the drawing, certain features have been omitted. For example, there is no emergency stop, and no reset circuit has been provided.

LADDER DIAGRAM

Drawings like those in Fig. 22-3A and Fig. 22-3B are called *LADDER DIAGRAMS*. Their similarity to ladders is apparent. Numbers at the left are line numbers. They refer readers to portions of the drawing. A ladder diagram is called ladder logic.

RELAY-BASED CONTROL POWER SECTION

Refer to lines 1, 2, and 3 in Fig. 22-3A. This is the power section of the circuit. In this case, power is obtained from a 460 V source.

DISCONNECT SWITCH

The first symbol on lines 1 and 2 represents a disconnect switch. Its purpose is to prevent application of power to the circuit during repair, setup, etc. It should not be used to start and stop the machine. When large motors are involved, the use of this switch to stop a machine could result in major damage and injury. The disconnect switch normally is not designed to be opened under load.

The dashed line indicates a mechanical connection. The blades of this switch open and close together.

FUSES

Protective fuses are shown next. Because there is no neutral (neither side of the 460 V service is grounded), both lines are fused.

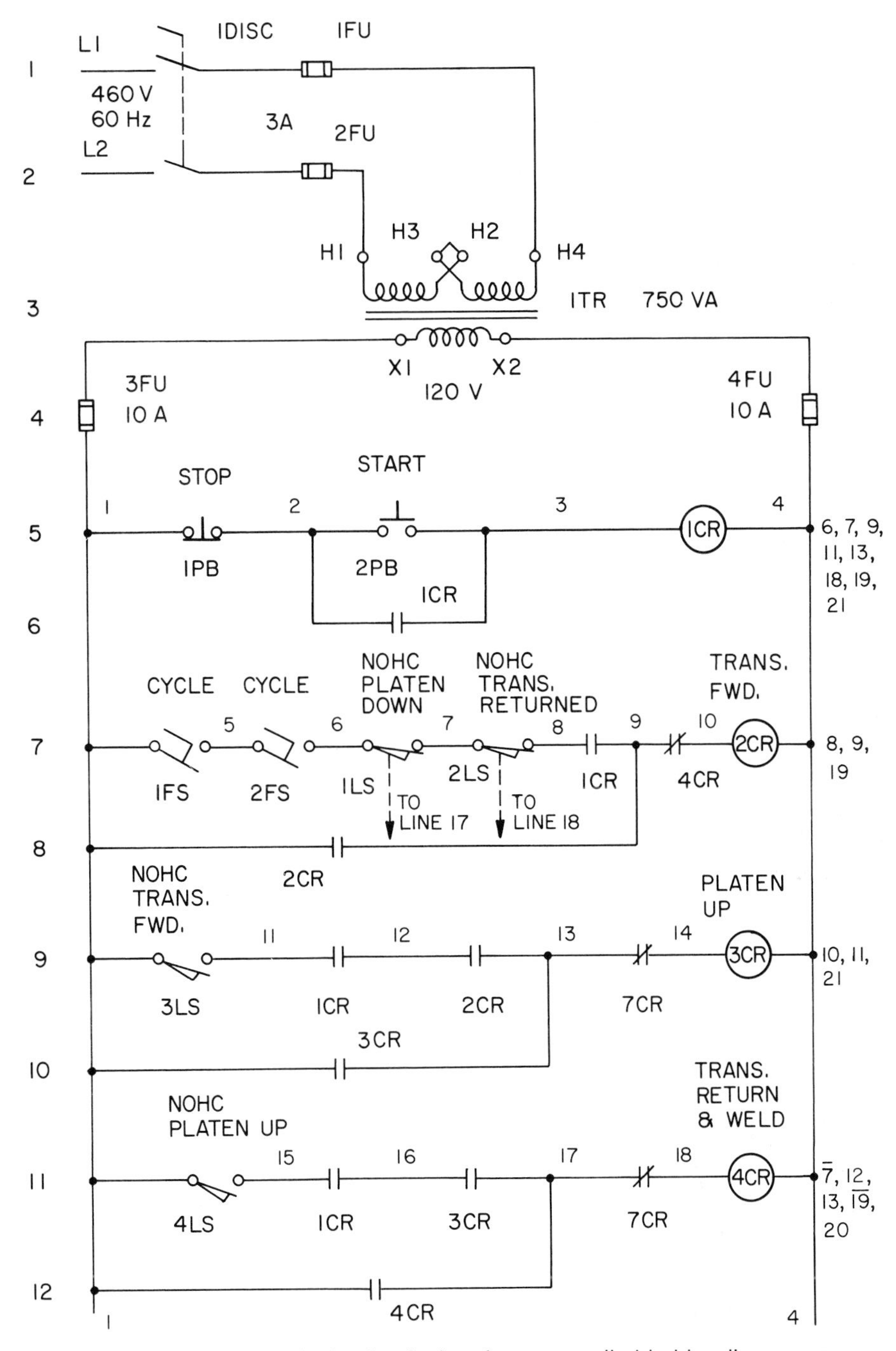

Fig. 22-3A. Relay-logic circuit drawings are called ladder diagrams.

CONTROL TRANSFORMER

The relays and solenoids used in this circuit operate at 120 V, 60 Hz. The transformer lowers the 460 V to the required 120 V.

The voltage between the vertical leads at each side of the control circuit will be 120 V. That is, the voltage across each rung of this ladder diagram will be 120 V.

CONTROL-CIRCUIT FUSES

The control portion of this circuit is also fuse protected, line 4. Note the 10 Ampere rating of each fuse.

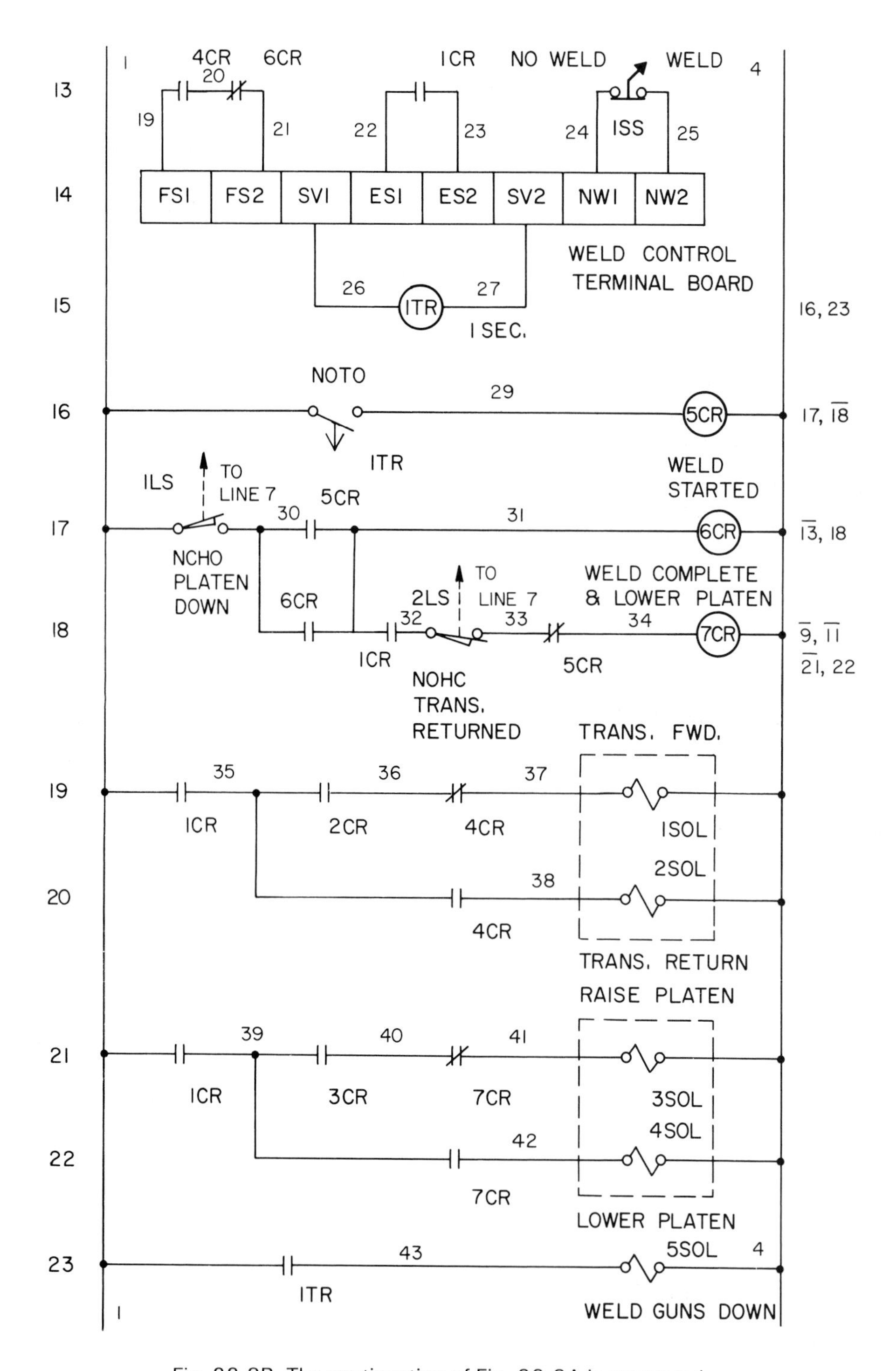

Fig. 22-3B. The continuation of Fig. 22-3A is presented.

This is considerably larger than the 3 Ampere rating of those in the primary circuit.

A transformer that steps voltage DOWN, steps current UP. The transformer has a voltage ratio of about 4/1. Therefore, it has a current ratio of 1/4. Its secondary current (the current through its output) will be about 4 times its primary current (current on the power side).

RELAY-BASED CONTROL START/STOP

Lines 5 and 6 in Fig. 22-3A and Fig. 22-3B contain the START/STOP circuit. Fig. 22-4 shows a relay that might be used in this circuit.

Fig. 22-4. The control relay shown might be used in the circuit in Fig. 22-3. (Allen-Bradley Company, Inc.)

REVIEW OF CIRCUIT ACTION

The START/STOP circuit in Fig. 22-3A and Fig. 22-3B acts like a flip-flop. It remembers which button, START or STOP, was pressed last.

Pressing 2PB completes the circuit and allows current to flow through the coil of the relay. With 1CR energized, the contacts on line 6 close. When 2PB again opens, the circuit remains energized, since current continues to flow through the closed contacts of 1CR. Remember, contacts used in this manner are often called *HOLDING* or *SEALING* contacts.

Once activated, current will continue to flow through 1CR until STOP is pressed. Switch 1PB is normally closed. When pressed, it opens and breaks the circuit. This deactivates the relay, and the contacts on line 6 open.

When STOP is released (and closed), 1CR remains de-energized. Although voltage is applied, both 2PB and 1CR are open.

UNDERVOLTAGE PROTECTION

The START/STOP circuit in Fig. 22-3A and Fig. 22-3B is widely used, because it offers undervoltage protection. When a machine stops due to a power outage, it is important that it not restart when power is restored. Control circuits that require the operator to restart the machine are said to have undervoltage protection.

This feature is of major importance to the safety of machines and personnel. When power is interrupted, control circuits usually forget where they are in the machine cycle. If a machine were to restart after an outage, control could be lost. Also, parts could be jammed in the machine, and operators and repair personnel might be working on it. Imagine the injury and damage that could result from a machine starting on its own.

Note how the circuit on lines 5 and 6 provides this protection. Loss of power has the same effect as pressing STOP.

NUMBERS AT RELAY

Numbers at the right of relay coil 1CR are line numbers. They indicate where the contacts on that relay are to be found. For example, one set is on line 7. Look for 1CR on that line.

RELAY-BASED CONTROL FOR TRANSFER FORWARD

With the pressing of START, the machine is ready to begin its cycle. The first step is to move the parts to be welded under the weld guns. This is done by moving the transfer forward. The energizing of relay 2CR signals this action. Refer to line 7.

To energize 2CR, all switches and contacts on this line must be closed. At this point in the cycle, relay 4CR is not energized, so its normally-closed contacts on line 7 will be closed.

CONTACTS 1CR

Contacts 1CR on line 7 may be thought of as enabling contacts. Only when the start circuit is energized can the cycle start. Assume START has been pressed and contacts 1CR are closed.

TRANSFER IS IN RETURNED POSITION

At the beginning of the cycle, the transfer should be in its returned position. Limit switch 2LS insures that this condition is met.

Note the symbol for this switch. Its blade is below

the contact at the right. If it were not held closed, it would fall open. This indicates that it is a normally open switch.

Also note the ramp. This tells you that it is a limit switch—a switch that is activated by the motion of machine parts. As a machine part moves against this ramp, it activates the switch.

The notation, NOHC TRANS RETURNED, indicates the condition for its closing. It is a normally-open switch (NO), held closed (HC) by the transfer in its return position.

Ladder diagrams are drawn with switches and contacts in the position they should be in at the beginning of the cycle. 2LS is closed, so the transfer must be returned at the beginning of the cycle. If it is not, the machine will not operate.

A person should learn the symbols for limit switches. Fig. 22-5 shows the four commonly used limit switch symbols.

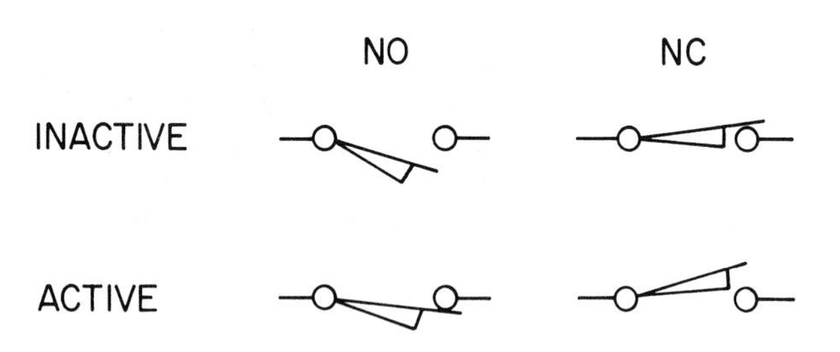

Fig. 22-5. The four possible positions for limit switches on ladder diagrams are shown.

PLATEN DOWN

The next switch, 1LS, indicates another condition for starting the cycle. The platen must be down. If it were not down and holding this switch closed, the circuit through the coil of 2CR could not be completed.

It is important that the platen be down when the transfer moves forward. If it were not, parts on the transfer would be jammed against the platen.

Note the arrows and numbers below 1LS and 2LS. These limit switches contain more than one set of blades, and the arrows and numbers indicate where these other blades are used in the circuit.

FOOT SWITCHES

To force the operators to move away from the machine before the transfer starts forward, two foot switches are used. They are placed away from the machine and must be pressed at the same time. They are labeled 1FS and 2FS.

Note how the symbol suggests the action of such switches. The switch is activated by placing the foot on the upper bar and tilting it forward.

HOLDING CONTACTS

There is a problem. When the transfer starts forward, 2LS will open. To retain an electrical record that the signal to move the transfer has been given, 2CR should remain energized. This is accomplished by contacts 2CR on line 8. Any of the contacts within the loop formed by the 2CR holding contacts (1FS through 1CR) can open without de-energizing 2CR.

DE-ENERGIZING CONTACTS

At some point in the cycle, the transfer will be returned, so provisions must be made to de-energize 2CR. This is the purpose of 4CR. These contacts are outside the influence of the holding contacts. When relay 4CR is energized, 4CR on line 7 will open and turn off 2CR.

SOLENOID VALVE

The numbers beside the coil symbol for 2CR indicate that one of its contacts is on line 19. Refer to that line. It is here that the hydraulic valve is shown. The angular symbol at the right (shaped like a sawtooth wave) is the solenoid of the valve. When activated, oil is supplied to the cylinder that moves the transfer forward.

Relay 4CR has not yet been activated, so its normally-closed contacts are still closed. It has been assumed that 1CR is on, so when 2CR is energized, current flows through the solenoid, and the transfer starts forward.

RELAY-BASED CONTROL FOR PLATEN UP

When the transfer arrives at its forward position, the platen is to start moving up. This is signaled by relay 3CR on line 9.

Note the required conditions. Enabling contacts 1CR and 2CR must be closed. Timing is provided by 3LS. This limit switch remains open until the transfer arrives at its forward position. Proper timing of the signal to raise the platen is assured.

Contacts 7CR will de-energize this relay at the proper time, and the contacts 3CR on line 10 are holding contacts.

Contacts 3CR on line 21 complete the circuit through the solenoid to raise the platen. Refer to this line in the diagram for Fig. 22-3B.

RELAY-BASED CONTROL FOR TRANSFER RETURN AND WELD

As the platen raises, it lifts the parts from the transfer and then supports them during the weld operation. The circuits on lines 11 and 12 determine when the platen is all the way up and then signal the return of the transfer for reloading. This circuit also signals the weld controller to start the weld. Limit switch 4LS senses the position of the platen.

DE-ENERGIZE 2CR

At this point, 2CR can be turned off. This is accomplished by the normally-closed contacts, 4CR, on line 7. When the coil of 4CR on line 11 turns on, 2CR is turned off.

RETURN TRANSFER

Contacts from relay 4CR can be found on lines 19 and 20. When this relay is energized, its contacts on line 19 open and those on line 20 close. Current is shifted from the forward solenoid to the return solenoid.

START WELD

Line 14 represents the terminal board on the weld control. This is a separate piece of equipment, and its internal circuits are not shown on the ladder diagram. Only the electrical signals to and from its terminal board are considered.

To start the weld process, terminals FS1 and FS2 are shorted. This happens when 4CR on line 13 closes.

The use of some symbols for welders must be explained. FS stands for foot switch. On early welding machines, foot switches were used to start the weld process, and the terminology is still used. ES stands for emergency stop. These terminals must be shorted for the control to operate. NW stands for no weld. During setup and troubleshooting, it may be desirable to cycle the machine without making welds. Opening 1SS (selector switch) permits this mode of operation. SV stands for solenoid valve. The weld control outputs 120 V between these terminals when the weld guns are to be lowered.

In this circuit, the weld control does not act directly on the solenoid valve. Rather, the voltage between SV1 and SV2 is applied to the coil of 1TR. See line 15.

Contacts on 1TR appear on line 23. When these contacts close, the solenoid is energized, and the hydraulic system lowers the guns.

This solenoid valve differs slightly from those used for the transfer and platen. When this solenoid is energized, the guns go down. When it is de-energized, they return. (The valves used for the transfer and platen must be energized in both directions. See lines 21 and 22. To raise the platen, the solenoid on line 21 is energized. To lower the platen, the solenoid on line 22 must be energized.)

In addition to lowering the weld guns, the 1TR contacts on line 16 energize relay 5CR. These are special contacts, and their action will be described later.

When 5CR is energized, the circuit through the coil of 6CR on line 17 is complete. Limit switch 1LS is closed, because the platen is up. (It is a normally closed switch, held open only when the platen is down.) Relay 6CR prepares the weld control for the next cycle by opening the short circuit between FS1 and FS2. See line 13.

WELD CYCLE

The weld control times the weld cycle. It waits for the weld electrodes to descend to the parts to be welded and then provides additional squeeze time. That is, it lets the electrodes press the parts tightly together. It then applies current to the electrodes. This flows during the weld time. Finally it allows for a hold time (time for the welds to cool). It signals that the welds are complete by removing the voltage between SV1 and SV2.

RELAY-BASED CONTROL FOR RAISING GUNS AND LOWERING PLATEN

When the welds are complete, the coil of 1TR is de-energized. Several tasks must be performed.

RAISE GUNS

When 1TR is de-energized, its contacts on line 23 open. With no current in its solenoid, the valve causes the hydraulic system to raise the guns.

LOWER PLATEN

Lowering the platen presents a problem. The downward force of eight weld guns is considerable. If the platen were released before this force were removed, the platen might be damaged. Therefore, the release of the platen must be delayed. This is the purpose of the special contacts on line 16.

This set of contacts closes instantaneously when the coil of 1TR is energized. However, when 1TR is de-energized, these contacts remain closed for a predetermined time. In this case, the delay would be set at a little less than a second.

In general, time delays apply in only one direction.

The arrow on the contacts on line 16 indicate that the delay is on its opening. Fig. 22-6 shows other combinations of instantaneous and delayed action.

Time delay relays also have instantaneous contacts. Those on line 23 are an example. They close and open immediately when power is applied and removed from the relay's coil.

Due to the time delay, 5CR remains energized for a short period after the guns start up. When it does de-energize, its contacts on line 17 open. However, these are bypassed by the holding contacts from 6CR on line 18. In addition, the contacts from 5CR on line 18 close. (They are normally-closed contacts that were held open while 5CR was energized.) When the transfer gets all the way back, it will close 2LS. Therefore, 7CR is energized.

Relay 7CR accomplishes several tasks. First, it lowers the platen. See lines 21 and 22. Second, it turns off 3CR and 4CR. See the normally-closed contacts from this relay on lines 9 and 11.

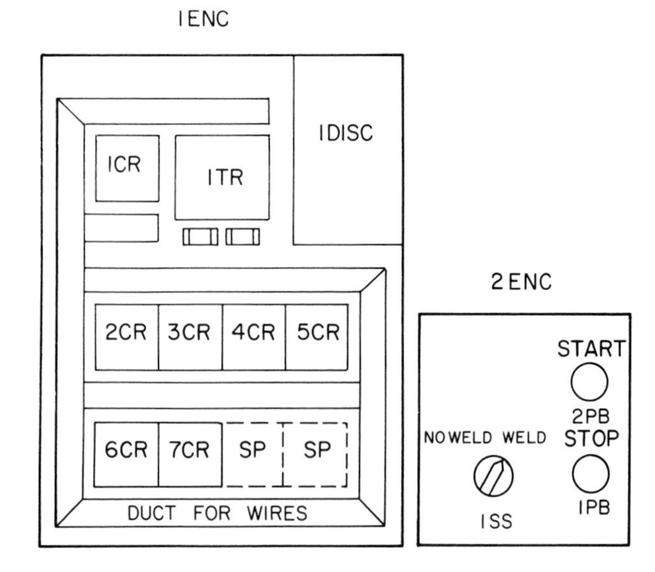

Fig. 22-7. Panel drawings show locations of relays, switches, and other components in the relay panel (1ENC) and control panel (2ENC).

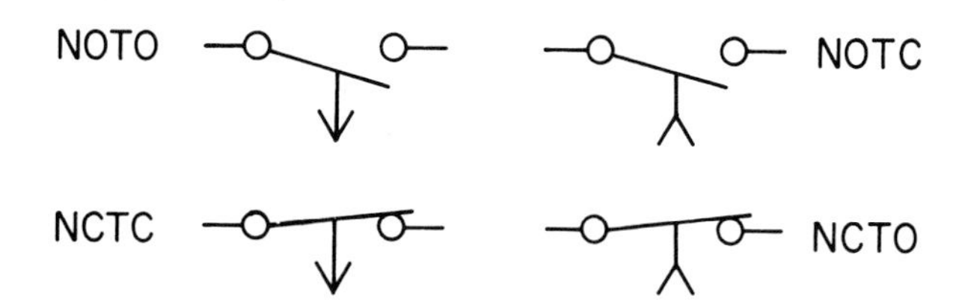

Fig. 22-6. The time delay of a TR relay can be on the opening or closing of the delayed contacts.

COMPLETING THE RELAY-BASED CONTROL CYCLE

When the platen is all the way down, it will activate 1LS. Its contacts on line 7 will close, but its contacts on line 17 will open. Power to the coils of 6CR and 7 CR will be removed, and the cycle will be complete. All relays, except 1CR, are off, and the platen, transfer, and weld guns are in their home positions. The machine is ready for another cycle.

WIRING OF RELAY CIRCUITS

Relay panels are usually wired directly from ladder diagrams. Often, the only additional information supplied is plans for the layouts of the enclosures. See Fig. 22-7.

Because detailed wiring diagrams are not used, panels representing the same circuit may differ. However, a method has been developed that permits easy circuit tracing even in circuits that are not identical.

Refer to line 11 of the ladder diagram in Fig. 22-3A and Fig. 22-3B. Note the leads between contacts 3CR, 4CR, and 7CR. The number 17 is a wire number. This number will appear at the ends of each wire used to connect this portion of the circuit. Small plastic sleeves with numbers on them may be slipped on each wire. Adhesive strips may be used. With another method, numbers may be permanently formed in the insulation of the wire.

When wiring this portion of the circuit, any of the wire paths in Fig. 22-8 might be used. They are electrically identical.

Wire numbers also simplify the identification of relay contacts. When attempting to locate a specific set of contacts, the worker merely finds the relay and then looks for contacts with appropriate wire numbers at its terminals. For example, contacts 3CR on line 11 will be on relay 3CR and have wire numbers 16 and 17 at its terminals.

The use of wire numbers also aids circuit modification. In Fig. 22-9 a switch is to be added. It can be connected to any terminal where wire number 17 appears.

MOTOR-STARTING CIRCUIT

Relay-type circuits are used in many motor starters. See Fig. 22-10. This circuit starts, stops, and reverses

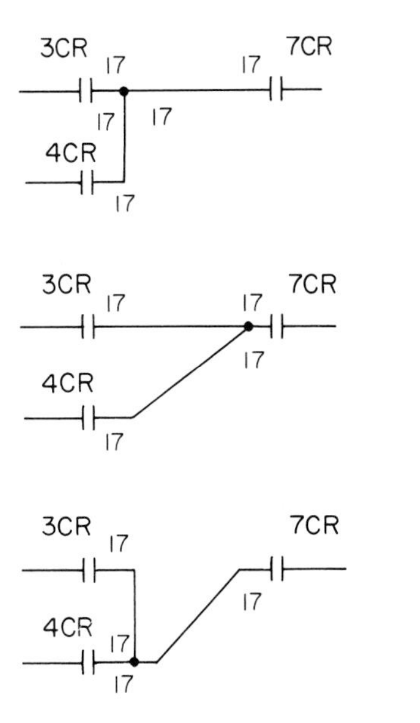

Fig. 22-8. The displayed wire paths result in the same electrical connections.

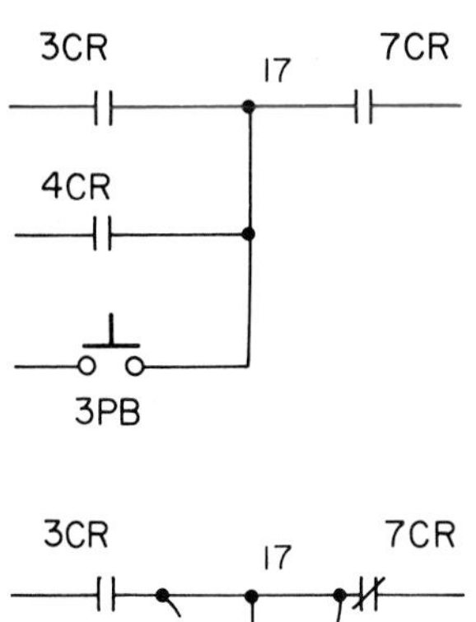

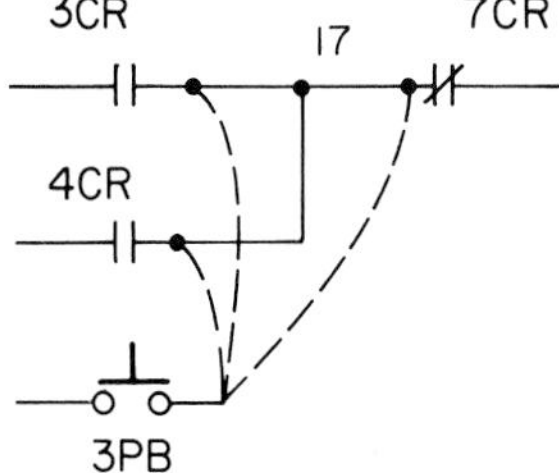

Fig. 22-9. When 3PB is added, one of its leads can be connected to any terminal where a number 17 wire appears.

the motor. It also provides overload and undervoltage protection.

THREE-PHASE MOTOR

The symbol at the upper right in Fig. 22-10 indicates that a three-phase motor is to be controlled. Such

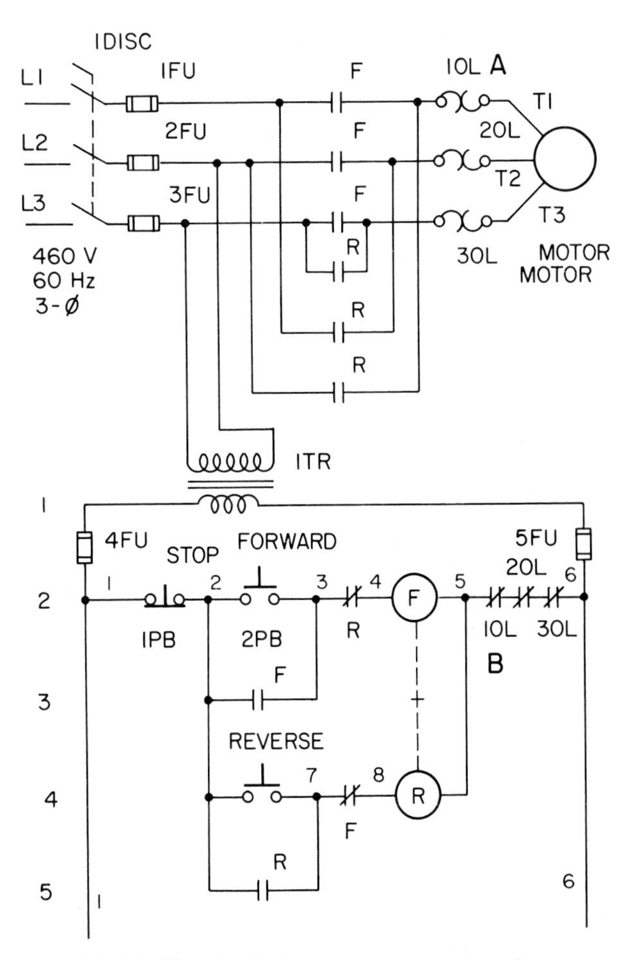

Fig. 22-10. The circuit shown starts, stops, and reverses a three-phase motor.

motors are widely used in industry, since they have good starting characteristics, require little maintenance, and are relatively inexpensive.

Three-phase power is available in most industrial and commercial locations. The relationships between the line-to-line voltages in such a system are shown in Fig. 22-11. Homes and industries do not use the same form of electrical power. Three-phase power is seldom found in homes. Although three wires may be brought in, such services are almost always single-phase power.

REVERSING CIRCUIT

If the simplest connections are made (L1 to T1, L2 to T2, L3 to T3 in Fig. 22-10) this motor will turn in a known direction (say clockwise). If any pair of leads are reversed (for example, L1 to T2 and L2 to T1), the motor will reverse (turn counterclockwise).

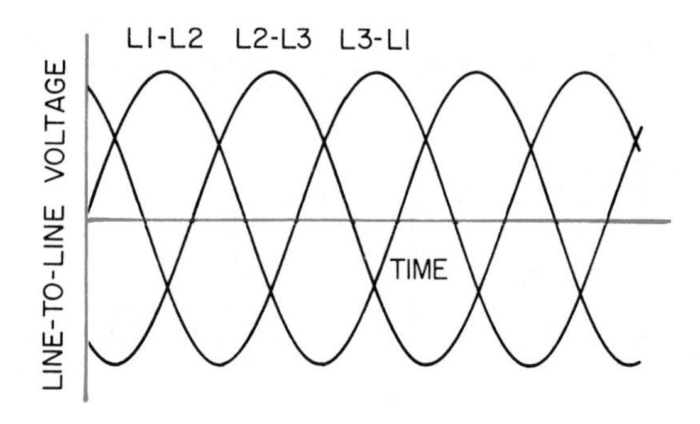

Fig. 22-11. Voltages between lines of a three-phase system vary in a systematic way.

Note the connections in the power portion of the circuit in Fig. 22-10. If contacts F are closed, the motor runs clockwise; if contacts R are closed, the motor runs counterclockwise.

CONTROL CIRCUIT

The control circuit consists of two START/STOP circuits. They share a common STOP button. Once FORWARD or REVERSE has been pressed, it can be released, since holding contacts have been provided.

CONTACTORS

Although the symbols for F and R are identical to those of control relays, these devices are called contactors. Their actions are the same as those of relays, but they are able to control the high currents and voltages in the power portion of the circuit. Due to the high current and voltage requirements for contactor relays, they are much larger than control relays. Although not indicated in the drawing, the contacts in the power portion of the circuit are large and capable of controlling high currents. Contacts in the control circuit are much like those on control relays.

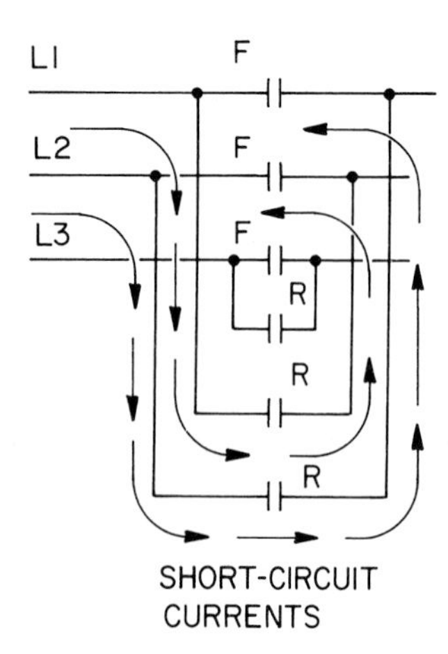

Fig. 22-12. If contacts F and R are closed at the same time, lines L1 and L2 are shorted.

PROTECTIVE FEATURES

The circuit in Fig. 22-10 contains five protective features. These are:

1. Interlocks.
2. Disconnect switch.
3. Fuses.
4. Overload.
5. Undervoltage.

The first protective feature provides safe power usage, as in the following:

1. Interlocks: A major safety problem exists in this circuit. If contact sets F and R in the power circuit were to close at the same time, lines L1 and L2 would be shorted. See Fig. 22-12. To prevent this, both electrical and mechanical interlocks are provided.

Note the normally-closed contacts on lines 2 and 4. When F is energized, its contacts on line 4 open, and pressing REVERSE can have no effect. The circuit through the coil of R cannot be completed. A similar situation exists when the REVERSE pushbutton is pressed first.

The dashed line between the coils of the contactors is a mechanical interlock. When one device is energized, this link blocks the motion of the other contactor.

When running in one direction, pressing the opposite button will not reverse the motor. STOP must be pressed before the direction of rotation can be changed. This helps reduce wear on the motor and the machine due to attempts to reverse the motor from a full speed forward condition.

2. Disconnect switch: Again, the disconnect switch must not be used to start or stop this motor. While closed, it can handle the motor current, but it is not designed to be opened and closed under load. It is to be activated only when both F and R are de-energized.
3. Fuses: Fuses in the power circuit protect the power service. If a major fault occurs in the power portion of this circuit, these fuses disconnect the circuit from the system. These fuses have high ratings, since they carry the motor current.

Normally, a short circuit in the control portion of this circuit will not draw enough current to blow the fuses in the power circuit. To provide protection at low current levels, fuses with appropriate ratings have been placed in the control circuit.

4. Overload: Overload protection has been provided for the motor. Each overload relay (OL) has two sections. A heater is placed in series with the motor leads. As current flows through 1OL at A, heat is produced. If the relay becomes hot enough, the contacts in the control circuit open. See 1OL at B. This de-energizes the active contactor and stops the motor before damage can occur.

A small overload that lasts a long time can do as much damage as a large overload that lasts a short time. If the overload relays are matched to the characteristics of the motor, protection from both types of overload can be obtained.

5. Undervoltage: Voltage for the control circuit was intentionally taken from the power portion of the circuit. If power is lost, the active contactor is de-energized, and its holding contact opens. When power is restored, a start button must be pressed to restart the motor.

As a safety note, control circuits must not be connected to separate, single-phase sources. Such sources may not fail at the same time as the three-phase source, so undervoltage protection would be lost.

PROGRAMMABLE CONTROLLERS

Computers that can be programmed to simulate relay logic are called *PROGRAMMABLE CONTROLLERS.* See Fig. 22-13. They have replaced relays in many sequence control applications.

Fig. 22-14 shows a block diagram for a typical programmable controller installation. It consists of four basic blocks. These are:

- Input.
- Output.
- Programmable controller section.
- Programming console (usually can be detached and used only when needed).

INPUT

Signal levels inside and outside programmable controllers differ greatly. Because of the different signal levels within the controller and at the input and output devices, an interface is needed. Interfaces are used for both INPUT and OUTPUT.

INPUT DEVICES

Symbols for typical input devices are shown at the left in Fig. 22-14. These sense the condition of the machine or process being controlled and deliver this information to the controller.

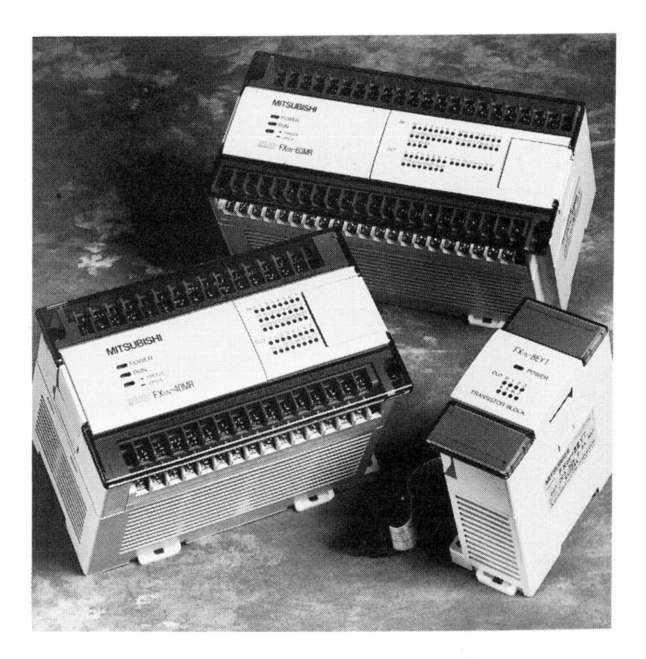

Fig. 22-13. Programmable controllers are widely used in industry to control machines and processes. In addition to simulating relay-logic circuits, many programmable controllers provide analog control capabilities. (Industrial Automation Division, Mitsubishi Electronics America, Inc.)

SIGNAL LEVELS

The computers used in programmable controllers operate at about the same signal levels as other computers. Outside the protected confines of the computer, however, such low signal levels are not practical. Electrical noise from welders, the opening and closing of contactors, etc. can induce false signals far greater than the 5 to 20 V used in programmable controllers. To overcome this problem, higher signal voltages (such as 120 V ac) are usually used. Safety becomes more important at 120 V.

To interface the external signals to those within the programmable controller, input modules are provided. See Fig. 22-15.

OUTPUT

Signals within programmable controllers are usually too low to drive external devices. Output modules capable of switching the voltages and current used to operate solenoids, contactors, and lamps are provided.

Units called *SOLID-STATE RELAYS* are used. They have no moving parts.

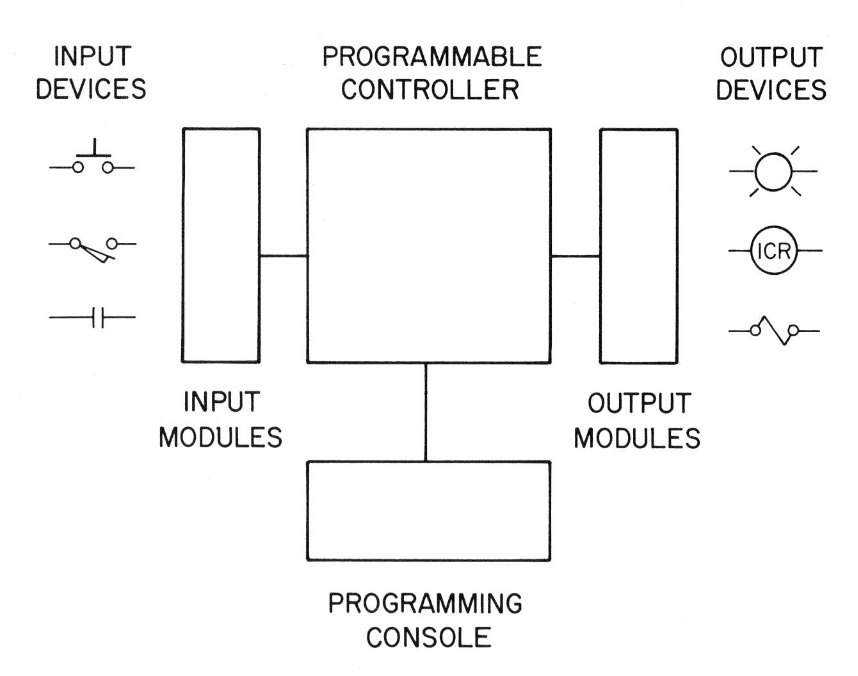

Fig. 22-14. Block diagrams of programmable controllers usually contain four main sections.

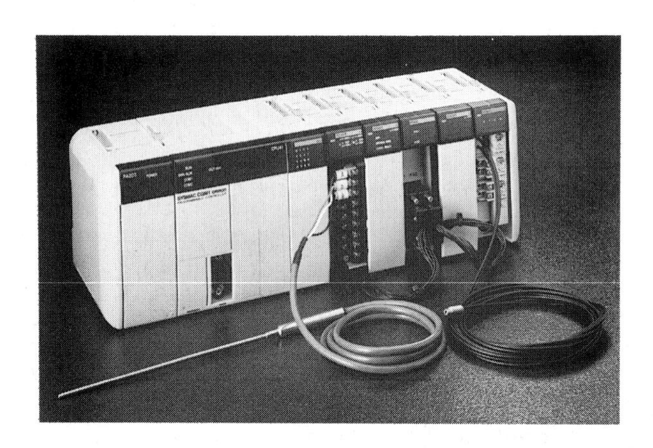

Fig. 22-15. The five I/O modules at the right interface between the programmable controller and the machine or process to be controlled. (Omron Electronics, Inc.)

PROGRAMMABLE CONTROLLER SECTION

The programmable controller is a complete computer. It has all the blocks, including input/output, found in other computers. It simulates the logic of relay circuits. Analog simulation can be done. A programmable controller using PID (proportional, integral, derivative) control can control a system going through smooth transitions. The system tests itself. A person can simulate process responses.

RELAY-LOGIC ACTION TO BE COMPARED TO PROGRAMMABLE CONTROLLER ACTION

Fig. 22-16 shows several lines of a ladder diagram. Note the logic associated with the decision to activate or deactivate solenoid 4SOL.

If the switches and relay contacts on line 23 are all closed, relay 3CR is energized. This closes contacts on lines 24 and 30. The holding contacts on line 24 merely permit 4PB, 1CR, and 1LS to open without de-energizing 3CR. Those on line 30 output a signal to energize the solenoid.

When relay 4CR is energized, its normally-closed contacts on line 23 open and de-energize 3CR. This removes the signal from the solenoid. Relay 3CR will remain inactive until the switches and contacts on line 23 are again closed at the same time.

PROGRAMMABLE CONTROLLER ACTION

The following description is highly simplified. It does, however, suggest how a computer can simulate the action of a relay circuit.

In the relay circuit in Fig. 22-16, the condition of a switch or relay is indicated by the position of its contacts. The direct simulation of open and closed contacts is not possible in a computer, so memory locations will be substituted. A memory location will be assigned to each device on the ladder diagram. For

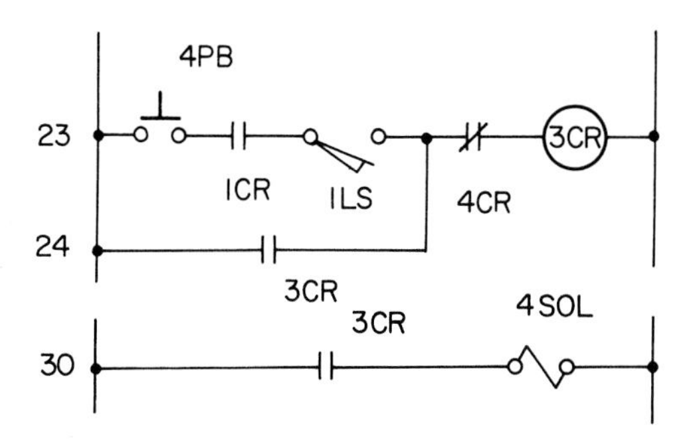

Fig. 22-16. The displayed relay circuit can be simulated by a program within a programmable controller.

the circuit being considered here, let the following notation be used:

Memory location	Switch or relay
A	4PB
B	1CR
C	1LS
D	4CR
E	3CR

If relay 3CR is energized, a 1 will be stored at memory location E; if it is de-energized, a 0 will appear at that location. When 4PB is pressed, a 1 will be input and appear at location A; when it is released, a 0 will appear at A.

A program, based on the complete ladder diagram, will be stored in the computer's memory. When the portion of the program dealing with lines 23 and 24 is reached, the computer will use the same logic as that of the relay circuit. Stored in the program will be the Boolean expression:

$$(ABC + E)\bar{D} = E$$

In the Boolean expression, A, B, C, D, and E represent numbers stored in corresponding memory locations. Note that E appears twice. The E on the left is the present value stored in location E. The one at the right is the value that will be written into this location after the expression has been evaluated.

On the first pass through the program, assume that A and E are both 0. Upon evaluating the expression, the computer would determine that E = 0. This number would be stored in location E. Because the number there was already 0, no change would occur.

On a later pass, assume that A, B, and C are all 1 and that E and D are 0. The computer would now find that the expression shows E to equal 1. This 1 would be written over the 0 already stored in location E in the memory.

On the next pass, A, B, and C do not need to be 1 to keep E equal to 1. With a 1 stored in location E, ABC will be ORed with a 1 at E, so the number inside the brackets will always be 1. As long as D = 0 (that is, $\bar{D}$ = 1), E will remain 1.

On some future pass through the program, D will be 1 (relay 4CR will be energized). When the expression is evaluated, E will be 0 (implying that 3CR is de-energized).

In the relay circuit, line 30 represents the output circuit. It delivers 120 V to the solenoid through the contacts of 3CR. Such a line might be included in the computer program, but it is not necessary. The output portion of the controller could function directly from memory location E. Whenever a 1 appears in this location, 120 V can be output to the solenoid; when E is 0, this voltage is removed.

PROGRAMMING CONSOLE AND SEQUENCING FOR PROGRAMMING EQUIPMENT

Several methods are used to load programs into programmable controllers. These are:

1. Portable console.
2. Tape.
3. Other computers.

The first method is flexible, as in the following:

1. Portable consoles are available. See Fig. 22-17. On some consoles, ladder diagrams are displayed on a CRT screen to aid in converting relay-based drawings into computer programs. See Fig. 22-17. Such consoles are also used for troubleshooting.
2. Programs can be placed on tape and loaded directly.
3. Programs can be stored in other computers and sent to programmable controllers by telephone line. This process is known as downloading. A *MODEM* (modulator/demodulator) is used at each end of the telephone line.

With only a few hours of instruction, technical personnel can learn to convert ladder diagrams to programs. A knowledge of both relay and integrated circuit logic is needed to effectively troubleshoot programmable controller circuits.

ADVANTAGES OF PROGRAMMABLE CONTROLLERS OVER RELAY-BASED CONTROL

Cost is basic to any discussion of advantages and disadvantages of one technology over another. There are a number of reasons why programmable controllers are less expensive than relay-based sequence controls. These involve the following:

1. Programming time.
2. Logic changes.
3. Salvage.

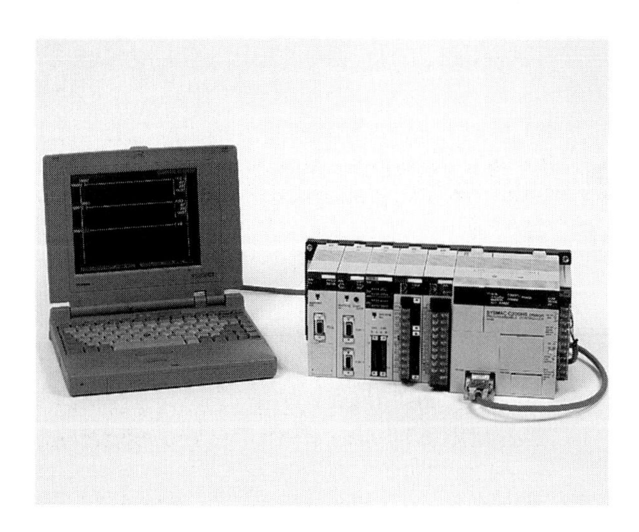

Fig. 22-17. Programming of programmable controllers is usually accomplished through a console such as this. Either ladder diagrams or mnemonics can be used. (Omron Electronics, Inc.)

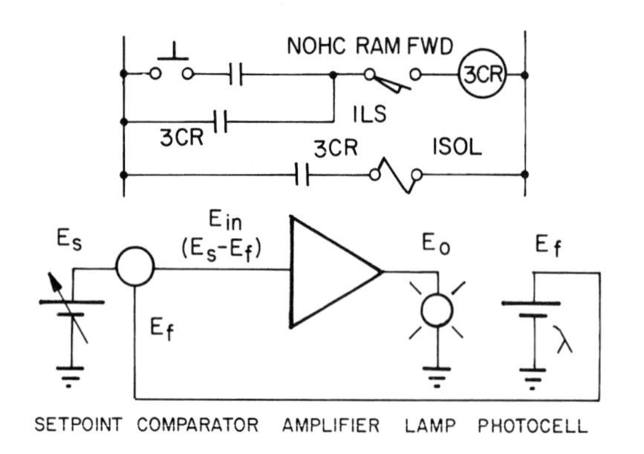

Fig. 22-18. Relay and analog control circuits are shown. Many programmable controllers can accomplish both forms of control.

4. Interfacing with computers.
5. Analog control capability.
6. Arithmetic operations.
7. Fault diagnosis.

PROGRAMMING TIME

Wiring a relay panel is a form of programming. It involves connecting relays and switches into a decision-making circuit. However, hard wiring of logic circuits is slow and costly.

Programming of programmable controllers is quickly and easily done. It can be accomplished on the factory floor or in an engineering office.

Relay panels are difficult to change. Wires and relays must be added or removed, and the chance of error is high. Machines and processes being controlled must usually be down during such changes. If a change does not result in improved operation, the process of returning the logic to its original form requires almost as much time as the change.

Programs in programmable controllers are easily changed. Machines and processes need to be down for only a few minutes, and it is easy to return to the original program.

SALVAGE

Relay-based controls are seldom worth salvaging. They are usually scrapped with the machines they control.

Controllers and their input/output modules adapt easily to new tasks. It is usually cost effective to move them from an old machine to a new machine.

INTERFACING WITH COMPUTERS

Programmable controllers communicate easily with other computers. Programs can be downloaded from computers to controllers. Information about the number of parts produced, downtime, and quality control can be stored in controllers and sent, upon request, to a central computer.

ANALOG CONTROL CAPABILITY

In addition to digital control, many programmable controllers are capable of analog control. Fig. 22-18 shows both forms. The relay circuit moves a ram forward. When 3CR is energized, the solenoid valve, 1SOL, starts the ram forward. When the ram reaches its forward position, it activates 1LS, so 1CR and 1SOL are deactivated. The relay signals the start of the operation, and 1LS indicates that it has been completed. This is a form of feedback. The circuit controls the beginning and end of the action, but it cannot control points in between. Its control is intermittent and stepwise. However, it is relatively inexpensive.

The analog circuit controls the brightness of the lamp. The amplifier amplifies the voltage Ein, and delivers a larger voltage, Eo, to light the lamp. The lamp shines on the photovoltaic cell which produces a voltage, Ef, proportional to the brightness of the lamp. Ef is fedback (hence the f) and is compared with Es. That is, Ein = Es − Ef. Es is called the setpoint and is used to indicate how much light is desired.

When the control is operating, Ef, from the photocell, will be just enough less than Es to produce Ein which, when amplified, will produce just enough Eo to make the lamp bright enough to produce Ef. Yes, we could go around and around this loop.

Now assume that the lamp has aged and is becoming dim. Ef will decrease, so the difference between Es and Ef will increase. As Ein increases, so will Eo. That is, as the lamp dims, the applied voltage will increase and compensate for the dimming. The lamp will remain almost at its original brightness. Where relay control is intermittent and stepwise, analog is continuous and smooth. It is also expensive.

ARITHMETIC OPERATIONS

Because they contain computers, programmable controllers can also do arithmetic operations. This ability can be used as part of the control process or for reducing data for recordkeeping purposes.

FAULT DIAGNOSIS

Programs that help with troubleshooting can be built into programmable controllers. These help repair personnel find faults in complex systems.

SUMMARY

Self-timing control and sequence control is used on many machines and processes. In such controls, the end of one operation signals the beginning of the next. This results in simple and reliable control systems.

Digital concepts are used to construct sequence control circuits. Until recently, relay-based circuits were used. Today, programmable controllers are replacing most relays.

Ladder diagrams depict relay-based control circuits. These diagrams are the basis for programming programmable controllers. As a result, those working with such controllers should be familiar with relay-based logic and integrated logic.

IMPORTANT TERMS

2CR, Analog control, Boolean logic, Downloading, Enabling contacts, Holding contacts, Interface, Interlock, Ladder diagram, Ladder logic, Limit switch, Modem, NCHO, NOHC, NOTO, Programming console, Relay-based logic, Safety system requirements, Sealing contacts, Self-timing control, Sequence control, Solid-state relay, Time-delay relays, Undervoltage protection, Wire numbers.

TEST YOUR KNOWLEDGE

1. In self-timing sequence control, the _________ (beginning, end) of one portion of a machine cycle signals the _________ (beginning, end) of the next.
2. A resistance weld can also be commonly known as a(an) _________ weld.
3. Relays are controlled by a system known as _________ (band, reinforced, ladder, repeatable) logic.
4. The voltage used for control relay logic is usually _________ (higher, lower) than the voltage used to power electric motors.
5. A transformer that steps voltage down steps current _________ (down, up).
6. Is a type of feedback required for the continuous current action of holding contacts and sealing contacts?
7. In undervoltage protection circuits, loss of electrical power has the same effect as pressing _________ (START, STOP).
8. What does NOHC mean?
9. Ladder diagrams are drawn with switches and contacts in the position they should be in at the _________ (beginning, end) of the cycle.
10. Can a relay have more than one set of contacts?
11. "Enabling" contacts are usually placed _________ (in series, in parallel) with the portion of the circuit to be enabled.
12. A normally-closed contact can be used to de-energize the circuit after the circuit has completed its action. (True or False?)
13. A time delay in a relay can be on its _________ or its closing.
14. _________ marked on wires help a technician trace them.
15. If any pair of leads on a three-phase motor are reversed, the rotation direction of the motor will _________ (reverse, stay the same).
16. Contactors handle _________ (more, less) voltage and current than control relay contacts.
17. A(an) _________ (holding contact, interlock, normally-closed contact, interface) is a protective feature.
18. For safety reasons, control circuits must not be connected to sources of single-phase power that are separate from the source supplying power to a motor or other machine. (True or False?)
19. Programmable controllers _________ (simulate, provide backup for, control, provide protective features for) relay logic control.
20. The four blocks that make up a programmable controller are: 1. _________. 2. Output. 3. Programmable controller section. 4. Programming

console and sequencing.

21. Signal levels inside and outside programmable controllers differ greatly. Because of the different signal levels, a(an) __________ (interlock, interface) is needed.
22. Which of the following is an example of the Boolean logic used in programmable controllers?
    a. 4CR
    b. $(ABC + E)\ \overline{D} = E$
    c. 1LS + 3CR = 1CR
    d. $\overline{NOHC} = NOTO$
23. The output portion of a programmable controller (usually an output module) can output 120 V (it can handle 120 V) based on the 0 or 1 state of a logic element such as one which is labeled "E." (True or False?)
24. Several methods are used to load programs into programmable controllers. Some of these are: 1. Portable console. 2. Tape. 3. __________. (Relays, Karnaugh maps, Other computers).
25. The process of loading a program from a computer into a programmable controller or another computer is called __________.
26. If a programming console displays ladder diagrams on its CRT screen, this can aid in converting relay-based drawings into computer programs. (True or False?)
27. Ladder diagrams on a CRT screen help a technician troubleshoot a system. (True or False?)
28. What does the term MODEM stand for?
29. Does programming of a programmable controller take more or less time than hard wiring a relay logic circuit?
30. Is it more costly or less costly to make logic changes in a programmable controller than in hard wired relay logic?
31. A(An) __________ (analog, digital) control, such as that used to control light bulb voltage, changes its state continuously and smoothly.
32. A(An) __________ (analog, digital) control, such as that used to simulate the on-off states of relay contacts, changes its state abruptly from 0 to 1 or from 1 to 0.
33. When a machine is to be replaced (because it is old or no longer used), which form of controls usually has the higher salvage value, relay-based controls or programmable controllers?
34. Which form of control, analog or digital, provides continuous feedback?
35. The following Boolean related expression is to be evaluated by a computer.

    $$(AB + E)\ (B + D) = E$$

    The variable E appears on both sides. One E is the variable's present value and is stored in memory. The other is its value after the computation. Which, the one on the left or the one on the right, is the value before the computation?

STUDY PROBLEMS

Problems 1-18 refer to conveyor switch in Fig. P22-1. Like a railroad switch, it sends objects to one of two chutes. Fig. P22-2 and Fig. P22-3 show control and power circuits used with this switch. If pushbutton RIGHT is pressed, motor drives the switch to chute at the right. If LEFT is pressed, it goes left. See pushbuttons 3PB and 4PB.

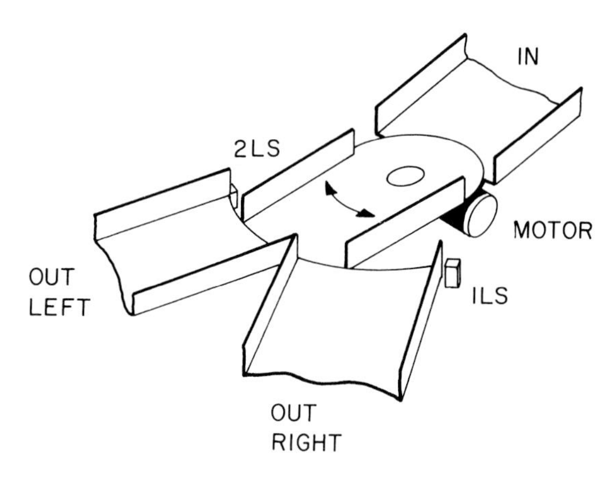

Fig. P22-1. Conveyor and chute system for Problem 1 through Problem 18. This conveyor switch and railroad switches serve similar functions.

Only the control portion of the circuit is shown. The power section will be considered later. R and L are contactors (large relays). Contacts on these devices are used in the power portion of the circuit to apply power to the motor. When R is energized, the motor turns in the direction necessary to move the conveyor to the right. L causes it to turn in the other direction.

1. What name is given to diagrams such as the one shown in Fig. P22-2?
    a. Switching diagram.
    b. Relay diagram.
    c. Logic diagram.
    d. Ladder diagram.
2. What is the most likely voltage between A and B?
    a. 5 V.
    b. 120 V.
    c. 440 V.
3. Before the conveyor can be moved, relay 1CR must be energized. Describe the action of the circuit on lines 2 and 3. Emphasize that push-

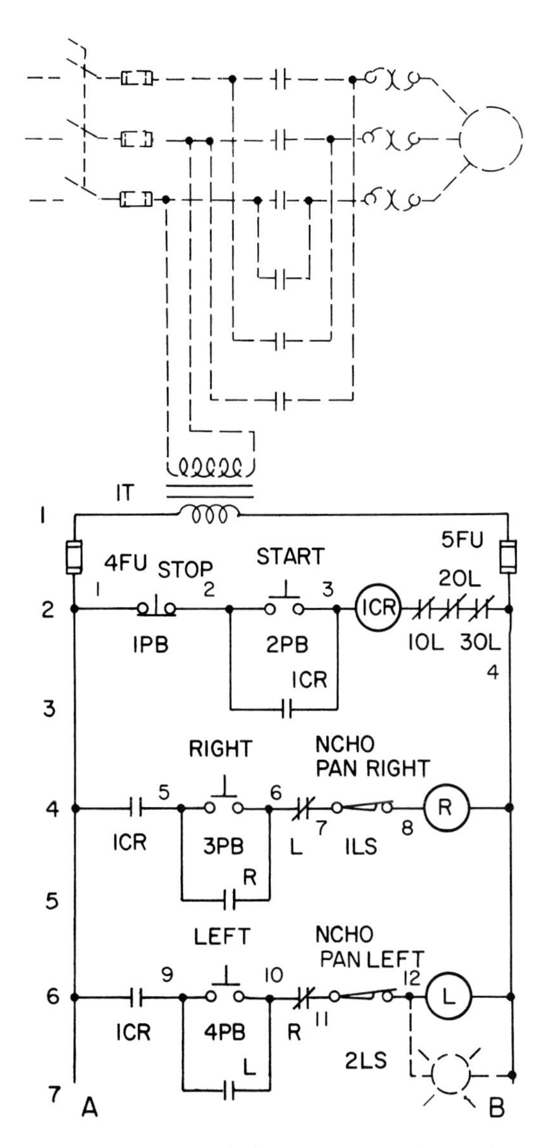

Fig. P22-2. Control circuit for conveyor and two chutes. Information on its operation can be derived.

buttons do not need to be held in their active positions to energize and de-energize 1CR.

4. Which contacts of 1CR are holding contacts?
   a. See line 3.
   b. See line 4.
   c. See line 6.
   d. See line 7.
   e. See line 1.
   f. See line 2.
   g. See line 5.
5. Indicate the condition (E = energized, N = not energized) of each switch and relay on line 4 necessary to energize the coil of contactor R. (Assume R is not energized, so line 5 contacts are open.) Do the labeling for the following:
   a. Relay 1CR.
   b. Pushbutton switch 3PB.
   c. Contactor L.
   d. Limit switch 1LS.
6. Based on line 4, contactor R cannot be energized if contactor L is energized. (True or False?)
7. Is limit switch 1LS normally-open or normally-closed?
8. Ladder diagrams are drawn with switches and contacts in the position they should be in at the beginning of the machine cycle. Based on the positions of 1LS and 2LS, what is the normal position of the conveyor?
   a. At the right.
   b. Somewhere in the middle.
   c. At the left.
9. Which best describes the way 3PB must be pressed to move the conveyor to the right?
   a. It can be pressed and released.
   b. It must be held down until the conveyor arrives at the right.
10. If the conveyor is at the right, what will happen when 3PB is pressed?
    a. Nothing. No current will flow through 1LS and the coil of R, so the motor will not start.
    b. R is energized, but motor will not start, because conveyor is against the fixture stop.
11. Assume R is energized and the motor is moving the conveyor to the right. Explain what would happen if 4PB were pressed.
12. If STOP were pressed while the conveyor was moving to the right, what would happen?
    a. Nothing. The conveyor would continue to move until it arrived at the right. Then it would stop. However, START must be pressed before it can move to the LEFT.
    b. The conveyor would stop at once. START must then be pressed before moving to the RIGHT or LEFT.
13. What wire numbers appear on the leads to 1LS?
14. If a lamp were added as shown in dashed lines, what wire numbers will be associated with the terminals to which it is attached?

In Fig. P22-3, the power circuit has been added to Fig. P22-2 from Problem 1 through Problem 14. Fig. P22-3 is to be used with Problem 15 through Problem 18 as a group.

15. Which best describes the function of switch 1SW?
    a. It starts and stops the motor.
    b. It stops the motor in emergency situations.
    c. It removes the line voltage during repair, etc.
16. If the motor were overheating due to a small overload that lasted for a long time, which group

of protective devices would probably activate?
a. 1FU 2FU 3FU
b. 4FU 5FU
c. 1OL 2OL 3OL

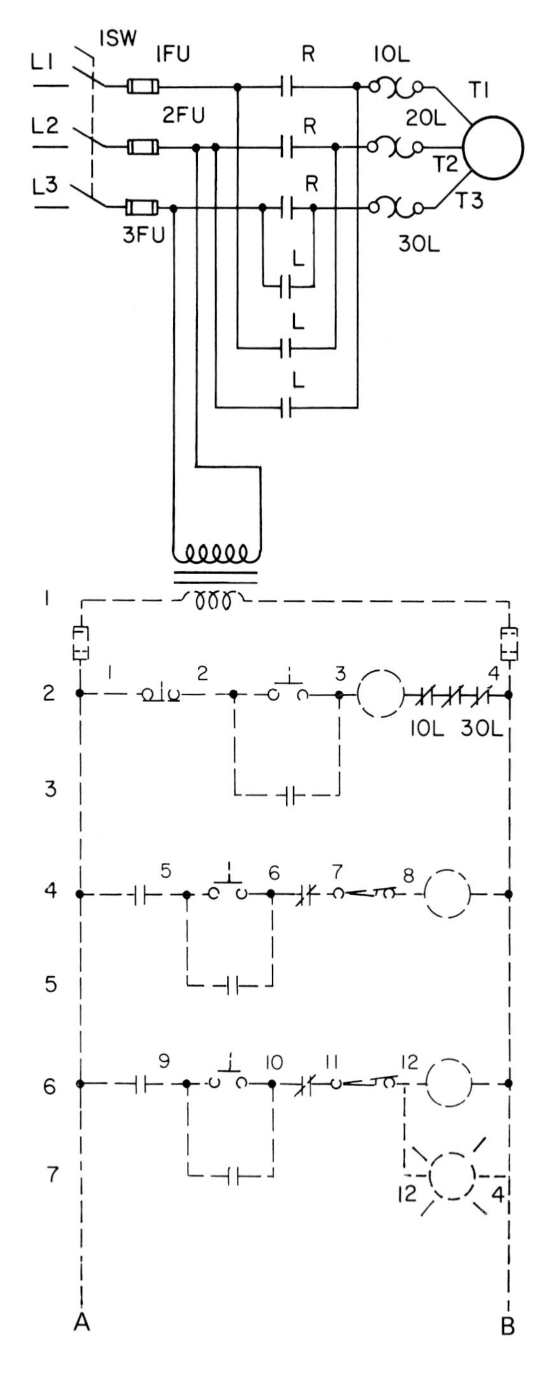

Fig. P22-3. Power circuit has been added to the control circuit in Fig. P22-2. The power portion is emphasized. Note that the power and control portions are linked through the contacts on contactors R and L and also through the contacts on the overload relays.

17. A three-phase induction motor (the type used in this circuit) will reverse if any two of its leads are reversed. Complete the following:
    a. Indicate which line (L1, L2, L3) will be connected to each motor terminal (T1, T2, T3) when contactor R is energized.
    b. Repeat part a for contactor L energized.
18. What name is given to the transformer in this diagram?
    a. Isolation transformer.
    b. Power transformer.
    c. Control transformer.
    d. Interstage transformer.
19. Draw the circuit (power and control) for a motor starting circuit. Use a three-phase induction motor. Include a disconnect switch, overload relays, and fuses. Label the contactor M, the STOP switch 1PB, and START 2PB. The circuit is to have undervoltage protection. The three-phase voltage is to be of the type which is listed as 460 V, 60 Hz.
20. Describe what would happen if both 1PB and 2PB in the circuit that is taken from Problem 19 were pressed at the same time.
21. Redraw the control portion of the circuit from Problem 19. Use two START pushbutton switches, 2PB and 3PB. Pressing either (or both) START buttons will energize M.
22. Which symbol in Fig. P22-4 represents a normally-closed limit switch in its inactive state?

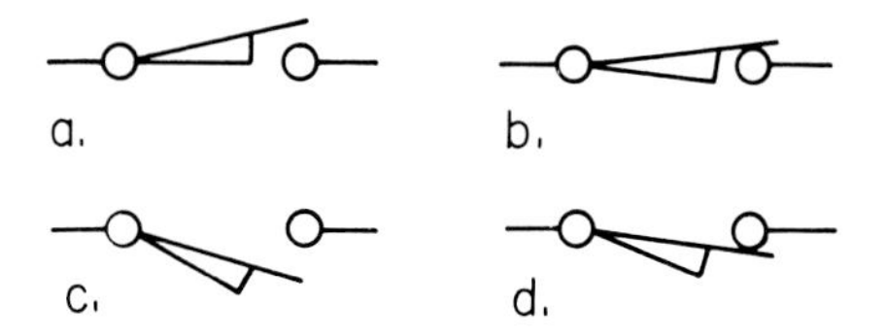

Fig. P22-4. Normally-closed and normally-open limit switches in active and inactive states can be identified.

23. Match the symbols in the drawing from Fig. P22-5 with the following names. A given symbol may be used more than once.
    a. Limit switch.
    b. Disconnect switch.
    c. Control relay.
    d. 3-phase induction motor.
    e. Fuse.
    f. Overload relay.
    g. Contactor.
    h. Pushbutton switch.
    i. Lamp.
    j. Control transformer.

24. Match the following with the blocks of the programmable controller system in Fig. P22-6.
    a. Programmable controller.
    b. Programming console.
    c. Output modules.
    d. Input modules.

25. List four advantages of programmable controllers over hard-wired relay logic. Discuss some reasons for your choices and responses. Be as specific as you can be about possible future developments.

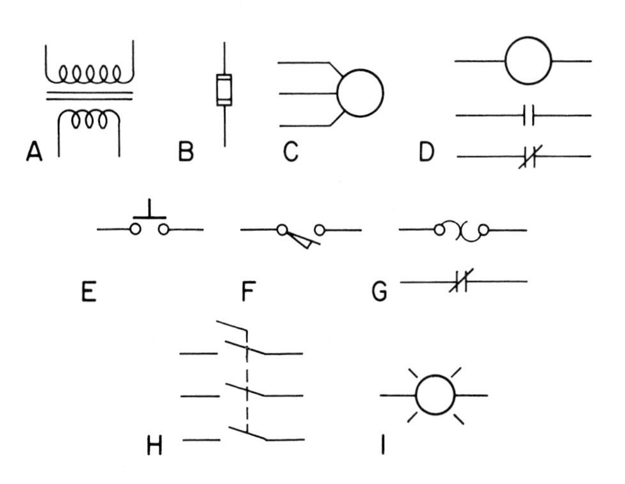

Fig. P22-5. Controller circuit components and power circuit components can be identified.

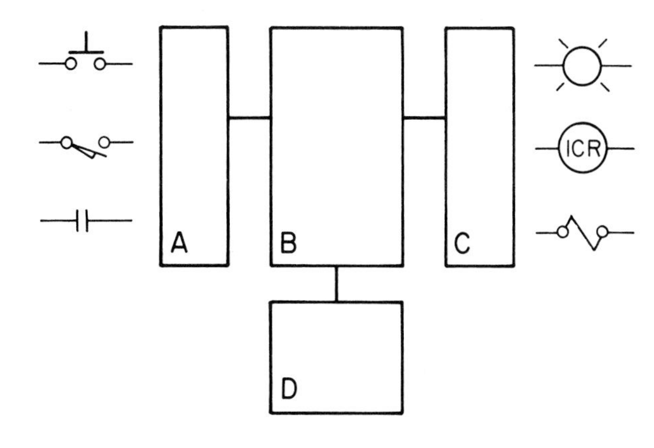

Fig. P22-6. Blocks in a programmable controller can be identified.

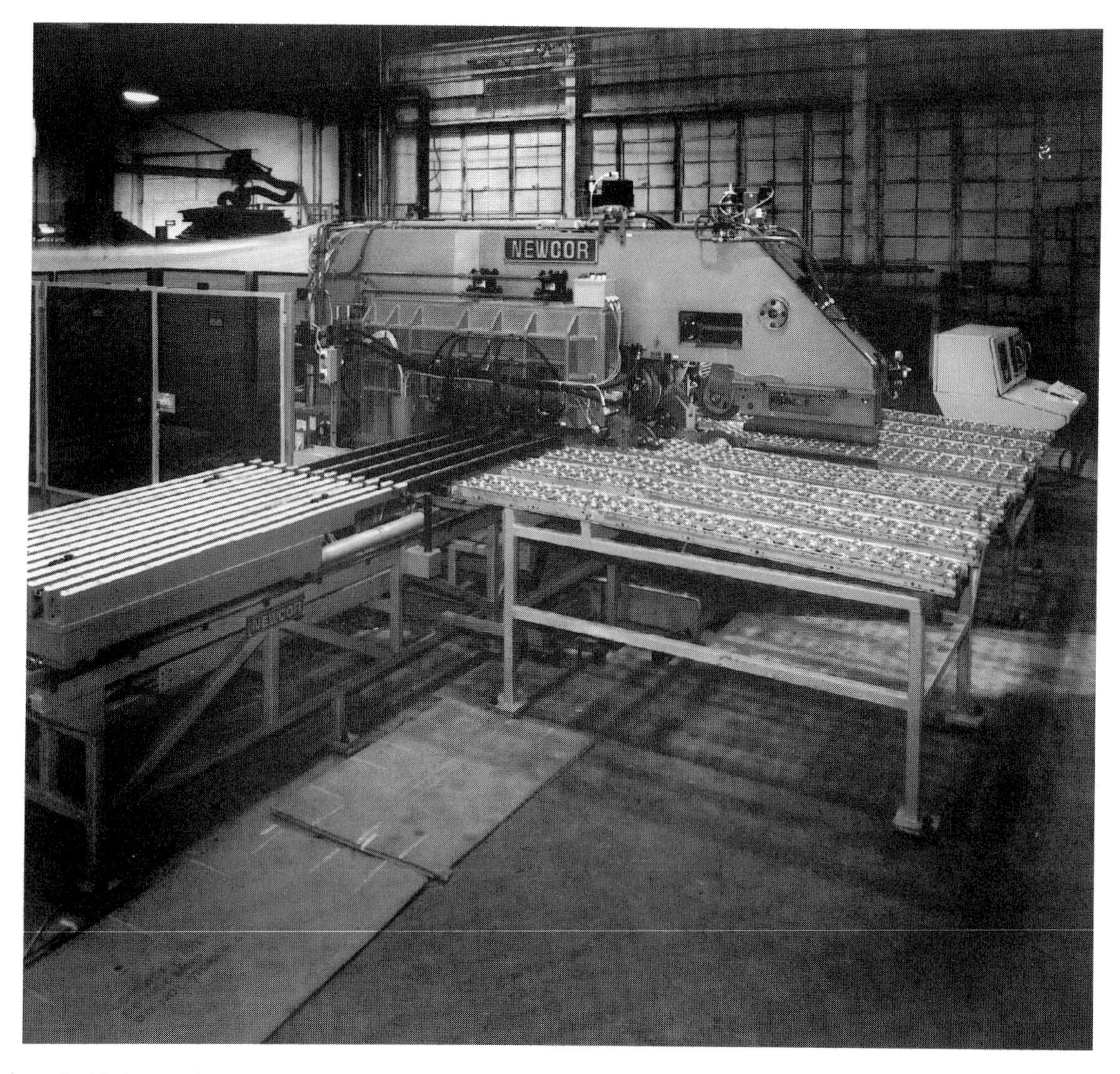

Although this is not the same machine, it is similar to the one described earlier in Chapter 22. Note the transfer and the housing for the welding guns. (Newcor Bay City)

Index

A

B

C

D

E

I

J

K

N

R

T

U

V

W

X